T0137118

Lecture Notes in Networks and Systems

Volume 80

The series "Lecture Notes in Networks and Systems" publishes the latest developments in Networks and Systems—quickly, informally and with high quality. Original research reported in proceedings and post-proceedings represents the core of LNNS.

Volumes published in LNNS embrace all aspects and subfields of, as well as new challenges in, Networks and Systems.

The series contains proceedings and edited volumes in systems and networks, spanning the areas of Cyber-Physical Systems, Autonomous Systems, Sensor Networks, Control Systems, Energy Systems, Automotive Systems, Biological Systems, Vehicular Networking and Connected Vehicles, Aerospace Systems, Automation, Manufacturing, Smart Grids, Nonlinear Systems, Power Systems, Robotics, Social Systems, Economic Systems and other. Of particular value to both the contributors and the readership are the short publication timeframe and the world-wide distribution and exposure which enable both a wide and rapid dissemination of research output.

The series covers the theory, applications, and perspectives on the state of the art and future developments relevant to systems and networks, decision making, control, complex processes and related areas, as embedded in the fields of interdisciplinary and applied sciences, engineering, computer science, physics, economics, social, and life sciences, as well as the paradigms and methodologies behind them.

**** Indexing: The books of this series are submitted to ISI Proceedings, SCOPUS, Google Scholar and Springerlink ****

More information about this series at http://www.springer.com/series/15179

Michael E. Auer · Kalyan Ram B.
Editors

Cyber-physical Systems and Digital Twins

Proceedings of the 16th International Conference on Remote Engineering and Virtual Instrumentation

Editors
Michael E. Auer
Carinthia University of Applied Sciences
Villach, Austria

Kalyan Ram B.
Electrono Solutions Pvt. Ltd.
Whitefield, Bengaluru, India

ISSN 2367-3370 ISSN 2367-3389 (electronic)
Lecture Notes in Networks and Systems
ISBN 978-3-030-23161-3 ISBN 978-3-030-23162-0 (eBook)
https://doi.org/10.1007/978-3-030-23162-0

This Springer imprint is published by the registered company Springer Nature Switzerland AG
The registered company address is: Gewerbestrasse 11, 6330 Cham, Switzerland

Preface

The REV conference is the annual conference of the International Association of Online Engineering (IAOE) and the Global Online Laboratory Consortium (GOLC).

REV2019 was the 16th in a series of annual events concerning the area of Remote Engineering and Virtual Instrumentation. The general objective of this conference is to contribute and discuss fundamentals, applications, and experiences in the field of Remote Engineering, Virtual Instrumentation, and related new technologies like Internet of Things, Industry 4.0, Cyber Security, M2M, and Smart Objects. Another objective of the conference is to discuss guidelines and new concepts for education at different levels for abovementioned topics including emerging technologies in learning, MOOCs & MOOLs, Open Resources, and STEM pre-university education.

REV2019 has been organized in cooperation with B.M.S. College of Engineering, Bengaluru, India and the Indo Universal Collaboration for Engineering Education (IUCEE) from February 04 to 06, 2019.

REV2019 offered again an exciting technical program as well as networking opportunities. Outstanding scientists and industry leaders accepted the invitation for keynote speeches:

- **Padma Shri. B. Dattaguru**, Chairman of the Aerospace Engineering Department, IISc, India
- **Klaus Hengsbach**, Phoenix Contact, Blomberg, Germany
- **S. Asokan**, Indian Institute of Science, Bangalore, India
- **Dominik May**, University of Georgia College of Engineering, Athens GA, USA
- **Saurabh Chandra**, Co-founder, Ati Motors, Bengaluru Area, India
- **Vasanthi Srinivasan**, IIMB, Bangalore, India
- **Thrasyvoulos Tsiatsos**, Aristotle University of Thessaloniki, Greece

It was in 2004 when we started this conference series in Villach, Austria together with some visionary colleagues and friends from around the world. When we started our REV endeavor, the Internet was just 10 years old! Since then, the situation regarding Online Engineering and Virtual Instrumentation has radically changed. Both are today typical working areas of most of the engineers and are inseparable connected with

- Internet of Things
- Cyber-physical Systems
- Collaborative Networks and Grids
- Cyber Cloud Technologies
- Service Architectures

to name only a few.

With our conference in 2004, we already tried to focus on the upcoming use of the Internet for engineering tasks and the problems around it—and very successfully, as we can see.

The REV 2019 conference takes up the following topics in its variety and discusses the state-of-the-art and future trends under the global theme "Smart Industry & Smart Education":

- Online Engineering
- Cyber-physical Systems
- Internet of Things & Industrial Internet of Things
- Industry 4.0
- Cyber Security
- M2M Concepts and Smart Objects
- Virtual and Remote Laboratories
- Remote Process Visualization and Virtual Instrumentation
- Remote Control and Measurement Technologies
- Networking, Grid, and Cloud Technologies
- Mixed-reality Environments
- Telerobotics and Telepresence
- Collaborative Work in Virtual Environments
- Smart City, Smart Energy, Smart Buildings, Smart Homes
- New Concepts for Engineering Education in Higher & Vocational Education
- Augmented Reality & Human Machine Interaction
- Biomedical Engineering
- Standards and Standardization Proposals
- Applications and Experiences

The following submission types have been accepted:

- Full Paper, Short Paper
- Work in Progress, Poster
- Special Sessions
- Workshops, Tutorials

All contributions were subject to a double-blind review. The review process was very competitive. We had to review near 230 submissions. A team of 180 reviewers did this terrific job. Our special thanks go to all of them.

Due to the time and conference schedule restrictions, we could finally accept only the best 110 submissions for presentation. The conference had again about 160 participants from 34 countries from all continents.

REV2020 will be held at University of Georgia, College of Engineering Athens GA, USA.

Michael E. Auer
REV General Chair
Kalyan Ram B.
REV2019 Chair

Committees

General Chair

Michael E. Auer	Founding President and CEO of the IAOE, CTI Frankfurt/Main New York, Vienna

REV2019 Chair

Kalyan Ram B.	IAOE President, Electrono Solutions Pvt. Ltd., India

REV 2019 Co-chair

Dr. B. V. Ravishankar	Principal in Charge, BMSCE, India

International Advisory Board

Abul Azad	President Global Online Laboratory Consortium, USA
Philip Bailey	MIT, Cambridge, MA, USA
Denis Gillet	EPFL Lausanne, Switzerland
Bert Hesselink	Stanford University, USA
Zorica Nedic	University of South Australia
Teresa Restivo	University of Porto, Portugal
Cornel Samoila	University of Brasov, Romania
Franz Schauer	Tomas Bata University, Czech Republic
Tarek Sobh	University of Bridgeport, USA
Vasant Honavar	Pennsylvania State University, USA
Krishna Vedula	IUCEE, India

Program Chairs

Doru Ursutiu	IAOE Past-President, Transilvania University of Brasov, Romania
Dr. Ravishankar Deekshit	Vice Principal, BMSCE, India

Technical Program Co-chairs

Sebastian Schreiter	IAOE, France
C. Lakshminarayana	BMSCE, India
P. Meena	BMSCE, India

IEEE Liaison

Manuel Castro	UNED, Madrid, Spain
Abhishek Appaji	IEEE, Bangalore Section

Workshop and Tutorial Chair

Andreas Pester	CUAS, Villach, Austria
Umadevi V.	BMSCE, India

Special Session Chair

Raviprakash Saligame	Aptiv Automotive, India
Anil Pandit	IUCEE, India

Publication Chair and Web Master

Sebastian Schreiter	IAOE, France
Sandeep Verma	BMSCE, India

Demonstration & Poster Chair and Start up Zone Chair

Suma H. N.	BMSCE, India
Indiramma M. S.	BMSCE, India

International Program Committee

Akram Abu-Aisheh	Hartford University, USA
Laiali Almazaydeh	Al-Hussein Bin Talal University, Jordan
Yacob Astatke	Morgan State University, USA
Gustavo Alves	ISEP Porto, Portugal
Nael Bakarad	Grand Valley State University, USA
David Boehringer	University of Stuttgart, Germany
Michael Callaghan	University of Ulster, Northern Ireland
Manuel Castro	MIT Madrid, Spain
Torsten Fransson	KTH Stockholm, Sweden
Javier Garcia-Zubia	University of Deusto, Spain
Denis Gillet	EPFL Lausanne, Switzerland
Olaf Graven	Buskerud University College, Norway
Ian Grout	University of Limerick, Ireland
Christian Guetl	Graz University of Technology, Austria
Alexander Kist	University of Southern Queensland, Australia
Vinod Kumar Lohani	Virginia Tech, VA, USA
Petros Lameras	Coventry University, UK
Sergio Cano Ortiz	Universidad de Oriente, Cuba
Carlos Alberto Reyes Garcia	INAOE Puebla, Mexico
Ananda Maiti	University of Southern Queensland, Australia
Dominik May	University of Georgia, Athens, USA
Zorica Nedic	University of South Australia, Australia
Ingmar Riedel-Kruse	Stanford University, USA
Franz Schauer	Tomas Bata University, Czech Republic
Juarez Silva	University of Santa Catarina, Brazil
Matthias Christoph Utesch	Technical University of Munich, Germany
Igor Verner	Technion Haifa, Israel
Dieter Wuttke	TU Ilmenau, Germany
Katarina Zakova	Slovak University of Technology, Slovakia
Stefan Marks	Auckland University of Technology, New Zealand
Prabhu Vinayak Ashok	Nanyang Polytechnic, Singapore
Vinay Kariwala	ABB, Bengaluru
Sudarsan S. D.	ABB, Bengaluru

Contents

Part I

Industry 4.0 & Cyber Physical Systems

Intelligent Online Interface to Digital Electronics Laboratory with Automatic Circuit Validation and Support

Ananda Maiti(✉), Andrew D. Maxwell, and Alexander A. Kist

Faculty of Health Engineering and Sciences, University of Southern Queensland, Toowoomba, Australia
anandamaiti@live.com, andrew.maxwell@usq.edu.au, kist@ieee.org

Abstract. Digital electronics laboratory activities involve setting several circuits using logic gates. The circuits are created according to a specific Boolean algebra model or computing components. Such activities include setting up basic gates such as AND OR, NOT etc. followed by complex circuits e.g. Adder, Flip-Flops, Registers. These activities involve substantial amount of wiring between specific ports or end-points of components. Students often consume significant amount of time trying to make a circuit rather than analyze the circuit and its operation. This work proposes an intelligent online interface supported by a new dedicated hardware architecture to map the circuits while they are being designed by the students. The hardware is based on low-cost microcontrollers and port expanders. Such a system is capable of matching the current circuit with a target circuit in a web-based environment as well as find the possible faults in the current circuit. The time taken for the overhead procedure of mapping circuits is shown to be within acceptable limits for a real-time learning experience. Once the circuit is mapped it is re-produced on-screen in a web-page so that students can determine the actual status of their circuits in real-time.

Keywords: e learning · Circuit design · Micro-controllers · Digital electronics · Cyber-physical systems

1 Introduction

Laboratory activities are a core part of engineering education. In an on-site laboratory, students often 'connect' different components to make an experimental setup. In case of electronics or similar labs the connections are typically purely in form of wired connectors between different end-points of components e.g. anode of a battery to one end of a resistor. A key part of the learning is successfully making the electronics circuit by sequentially building it from basic components to complex circuits [1]. This usually involves multiple steps and the students may fail to create the circuits in those intermediate steps often extending the learning time.

Digital electronics laboratories typically require students to make various circuits using logic gates and record their outputs. Such circuits often involve connecting

M. E. Auer and K. Ram B. (Eds.): REV2019 2019, LNNS 80, pp. 3–18, 2020.
https://doi.org/10.1007/978-3-030-23162-0_1

various types of logic gates such as AND, OR, NOT along with more complex ones such as XOR and Flip/Flops. Usually students are given pre-fabricated boards with the individual circuits pre-assembled on them, or they place individual gate ICs on a breadboard themselves. The next steps involve wiring the inputs and outputs of such gates to create more complicated circuits such as logic Adders, Subtracters, Flip-Flops, and Serial Counters for instance. One drawback of the using digital circuits is differentiating between the obtained output and the correct output, as the output is often determined by discrete states of LED(s). This means the students can create an incorrect or sub-optimal circuit and still obtain sufficiently convincing data that would seem to be correct for a given circuit. In many cases these errors are checked for by an instructor in the physical laboratory.

In this paper, to help with the learning process the hardware is made autonomous enough to detect the current circuit being built and provide feedback quickly. To do this, a special hardware board is designed that can detect the current circuit connections using a micro-controller. The current circuit is represented as a graph and sent to an online platform where it is matched with a desired circuit also in form of a graph which the students are trying to build. Checking the circuits using an online system also records each experiment event for each user. This information could be relayed to the lab instructor, providing heuristics regarding student needs, for instance if a student needs more attention. Overall the online system engages real time methods, so it can additionally prevent students from falling behind due to being inattentive or experiencing difficulties with understanding the experiment concepts. It can be used for collaboration as well if different groups are making separate circuits.

The rest of the paper is organized s follows: Sect. 2 discusses the nature of digital electronics laboratories, circuits and the visualization of circuits. Section 3 describes the basic components of the system architecture involving the internet-based components and the hardware boards used along with the assumptions. The design of the new hardware board is presented in Sect. 4 followed by the proposed algorithms and software to match the circuits, provide support and visualization in Sect. 5. The utility of such a system is discussed in Sect. 6.

2 Related Works

This section discusses the related work in the fields of digital electronics laboratories, circuit matching and visualization of circuit diagrams.

2.1 Digital Circuits Laboratory Activities

Digital electronics covers the courses where students are taught about circuits which are part of computing architectures. Such courses involve both classroom and laboratory work [1]. The major emphasis is often on the students to make circuits on their own. Then they construct and verify a truth table for a given set of inputs. This process can become complicated when many wires are involved and confuse the students if a point to point connection has been made or not. Any wrong circuit must be repaired or reconstruct increasing the learning time.

Basic circuits of digital electronics are nowadays constructed with virtualization [2–4]. Several software can provide gates or ICs that can be wires up in a virtual environment allowing the students to easily rectify any error they make. But it is still important to have real hands-on experience with real circuits to understand the flow of data/bits int he circuits with respect to external inputs. Figure 1 shows an *electronics board* used for teaching where several gates are drawn in the top and the inputs/outputs are connected to the actual ICs hidden from the students view.

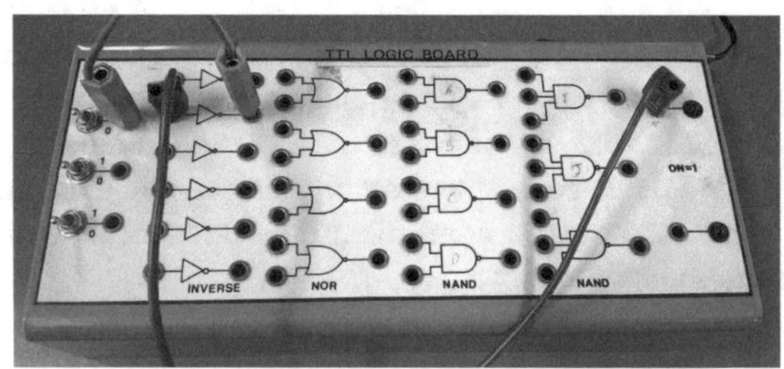

Fig. 1. Digital electronics board

The aim of this work is to make these boards intelligent. The boards have an additional layer consisting of a micro-controller which maps the current circuit in real time. Often with many wires used for a circuit, the board become difficult to analyze. With an online interface showing the current circuit, students can easily understand the current status in real time.

2.2 Circuits and Graphs

Electronics Circuit are easily represented as graphs as in graph theory [5] and consequently graph algorithms have many uses in Electronic Design Automation (EDA). These mainly relates to solve a circuit:

- analysis of its composition and determining the output of the circuit at certain points [6] and
- determine equivalences with *graph isomorphism* [7].

Graph isomorphism has been a key research are for electronic design automations [8]. Such approaches usually rely on matching netlists of circuits designed in a software. Special case of graph isomorphism has been shown to be fast and practically implemented in real time [9].

This work uses a simplified solution to the color-preserving graph isomorphism problem as the graph is generated automatically from an electro-mechanical subsystem. This means that there is no way to annotate the graph with additional information other than what is measured by the device itself i.e. no netlist is available. Each circuit is represented as a colored graph based on its properties. Another issue that differs from EDA is that students are not supposed to connect certain components *physically*, even though they are connected *logically* for matching purposes. The proposed system handles them in separate graphs and joins them when required.

2.3 Visualization of Circuits

Visualization is a key part of student learning. In this case students must be able to see the circuit they are making in real-time. Most EDA software allows students to draw wires between the components. Even remote laboratories such as VISIR provide a visually accurate depiction of the components placed on a bread board [10]. In EDAs, the wires are usually mapped on a grid to avoid cluttering. However, in all these cases, the circuit is drawn or constructed int he virtual environment by the students.

In the current context, the circuit is not drawn by the students. It is mapped by hardware which is then depicted on the screen. This causes challenges as often interconnected components show extra connections that are not part of the real circuit. These extra connects must be removed.

3 The System Description and Assumptions

This section discusses the basic components of the proposed intelligent online platform which works in conjunction with a Micro-Controller Unit (MCU) as a controller to monitor students' circuit construction and provide real-time feedback on their work. This intelligent software has the ability to monitor students' wiring and determine if the circuit is being built incorrectly. This software essentially provides real time feedback in place of reliance on a laboratory instructor. It also enables students to check individual connection lines which is important if there are many connecting wires. There exist several IC circuit verification systems based on commonly found MCUs, but they check ICs which have an individual input and output combination, not a circuit. In the proposed architecture, the students can use any component end-points in different combinations of physical layout of the circuit, which are all equivalent and acceptable. Following are the components of the system as shown in Fig. 2:

- A hardware board that is capable of being re-wired to create multiple circuits. It has several logic gates marked on them with open ports at their end-points.
- MCU based circuit mapper which creates a graph in form of a matrix of the current circuits
- A software comprising of online application which receives the graph for processing and a web-based User Interface to show the current circuit and its status.

The MCU connects to the online learning platform via WebSocket and sends the graph of the current circuit to the online systems which can then generate a suggestion based on these circumstances and display the resultant feedback on the web-based user interface (UI). As the students construct the circuit, the corresponding wires appear on the online UI screen. For simple matching, the MCU can check it on the board locally, but if the circuit is big with complicated connections, checking the graph could take time after creating it. This would pause the MCU for lengthy periods and potentially

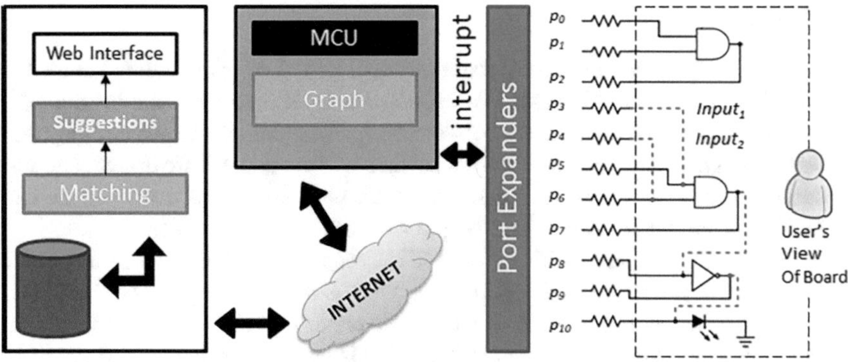

Fig. 2. The system architecture

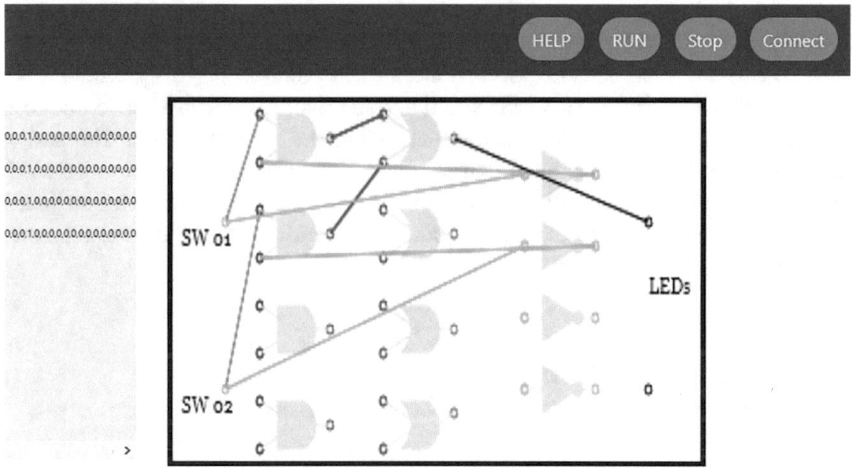

Fig. 3. The web user interface (UI)

hamper the real-time experience of the students. Hence most of the checking is done online.

A typical web UI is shown in Fig. 3. This interface has a canvas area, where the circuit is drawn, along with are some other components to connect to the server and a HELP button to request for help.

This paper will use the case of XOR gate circuit. In this experiment, the students use basic logic gates such as AND and NOT to ultimately create an XOR gate. As XOR gates can also be created with NAND gates the software can match either of these circuit possibilities.

3.1 The Device

The circuit matching boards have a fixed physical layout of components with a Wi-Fi enabled Micro-Controller Unit (MCU) connected to a set ports and contains digital ICs. The student adds the physical wires between the pins of the ICs. However, they do not see the ICs and instead interact with a flat surface where the gates are drawn, and inputs and outputs have connector pins. These connector pins are the end-points of the circuit e.g. INPUT of an AND Gate or and output of a OR gate.

The boards are designed with a NodeMCU based on ESP8266, which has multiple control ports. For most practical applications, there is not enough ports on the MCU, so it is essential to use port expander which can be either digital or analog. In the current context, a I2C digital port expander (MCP23017) is used which can add 16 GPIO ports individually. MCP23017 has 3-bit I2C addressing system which means that 8 such ICs can be used in a bus simultaneously, giving the MCU 128 ports. Other methodology may be used as well such as multiplexers or SPI based expanders, which can have a larger number of ports. Each port represents a vertex in the graph and it is determined whether each port is connected to any other port to obtain the graph.

The board has fixed pairs of port and circuit end-points Q = (P, V), where P is the set of ports and V is the set of vertices in the graph. Normally, P will have one-to-one mapping with V, but this could change depending on the application. The software knows the placement of each fixed components' end-points e.g. the output of a AND gate. By measuring the connectivity between two ports on the experimentation board, it can be determined if the corresponding components end-points are connected as well. as shown in Fig. 2. This is done as follows:

```
Procedure FindCircuit(Ports P)
  Disable LOGIC GATES ICs in Layer 2
  Enable Port Expanders in Layer 1

  For each port p ∈ P:
     set p as OUTPUT mode
     set p as logical LOW
     set all q ∈ (P - p) to INPUT mode
     set all q ∈ (P - p) with a PULL UP as logical HIGH

     if any q ∈ (P - p) is LOW, then p, q is connected
     else p, q is not connected

  Disable LOGIC GATES ICS

  Disable Port Expanders in Layer 1
  Enable LOGIC GATES ICs in Layer 2
```

The MCP23017 already has a PULL UP resistor in built, so it does not add to complexity of the hardware design. The output of this procedure is a matrix Mt(Q) with size of P × P measured at time t. For digital circuits the procedure is simple as outputs are discrete and there is low probability of noise in the circuit. For other applications, this process can be complicated due to the nature of wiring which could allow for noise to affect any encoded signal. However, the MCU cannot take any significant time to complete this step either.

A matrix $M_t(q)$ is generated for the Q_t at each interval *t* when a change is made to the circuit or after a certain interval of time. The change can be detected by an interrupt on the MCU that is controlling the experiment session. When the circuit is to be determined, the MCU initiates to create a new updated graph matrix which involves the following steps:

a. Shutting down power supply;
b. Run FindCircuit (P). Only a subset of the ports may be checked if required.
c. Re-enabling the power supply if the connections are deemed safe. For safety, the MCU determines if the current circuit would cause any serious electrical failures.

The experimentation board hardware also has a set of diodes present to protect the components. For instance, if a wire is connected incorrectly, damage to the board and components is avoided. This issue is not significantly prevalent in a digital circuit where the inputs/outputs are usually discrete. The students can use any of the gate end-points to wire up with the other circuits, but as the MCU already has the information on the end points connected to the ports, it can still produce the circuit graph irrespective of the actual end points used in the hardware For e.g. in the system diagram shown in Fig. 1, the students can use either of the AND gates in any combinations to create different physical layout of the same circuit on the board.

An example of a board setup with 4 AND gates, 4 OR gates and 4 NOT gates is shown in Fig. 4. This setup used a 7408, 7404 and 7432 ICs for the purpose. Also, 3 MCP23017 expanders were used. The mapping from the expander ports to the circuit components i.e. gates are fixed—p0 → AND1_INPUT1, p1 → AND2_INPUT1, … p4 → AND1_INPUT2 … p8 → AND1_OUTPUT and so on. A total of 36 ports are used—20 gate INPUTS, 12 gate OUTPUTS, 2 SWITCH INPUTS and 2 LED OUTPUTS. Note that the students only connect a single wire from and to the switches and LEDs which are grounded inside the board.

Also, 6 *relays* are used to connect or disconnect the power supply in Layer 2 and 1. When the circuit is mapped by the MCU, the gate ICs and SWITCHES/LEDs in Layer 2 are disconnected from the power supplies and ground in order to isolate and disable them. When the circuit is not being mapped, the MCP23017s in Layer 1 are disabled so that they don't interfere with the output of the experimental circuit. The user only sees the Layer 3 and connects various end points (solid black circles).

This example only shows a small setup board with 12 gates. The method used can be scaled up to include many more ICs and expanders. The board design could vary a

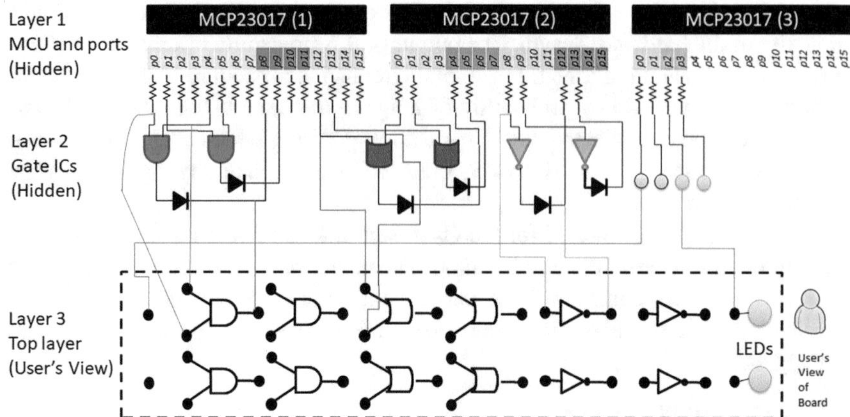

Fig. 4. The digital circuits experiment board design. Not all gates have been drawn here to maintain clarity

lot as well according to the physical properties of the components used and the desired learning outcomes. But it is assumed that hardware is able to generate a matrix M representing a graph where m1, m2 ∈ M = 1 if there is a connection (e.g. a wire) between the two end points in the actual physical setup. For the rest of the paper, this example board is used to illustrate the graph matching and visualization of circuits.

To test the feasibility of measurement time, a test was done to measure the time taken for mapping the circuit. A critical part of the data collection procedure is turning the port expanders on and measuring the ports individually. The MCU used is the Wi-Fi enabled NodeMCU 1.0 operating at 3.3 V. Figure 5 shows the time taken for mapping the circuit goes from 1 to 6 s with the number of end-points from 32 to 64. The time take is quadratic. The procedure can be fastened by reading the registers

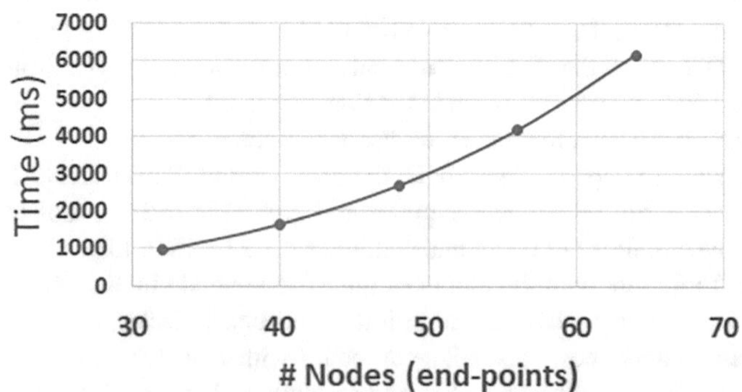

Fig. 5. Time taken for mapping the circuit

directly from port expanders. At 6 s with 64 end-points is still usable for periodic or 'on-demand' checking in a 2 h laboratory session. Number of MCP23017 expanders used were from 2 (for 32 end-points) to 4 (64 end-points).

3.2 The Software

The software is an online application which run in a server on the internet. It runs a WebSocket server periodically receiving the circuit graph from he MCUs. It then proceeds with the following three steps:

i. The graph matching to determine if the circuit that is currently built on the board equals to the desired circuit (or target circuit) that the student is trying to make.
ii. Generating a difference of the circuit under construction and the target circuit to support the user.
iii. Draw the circuit on the web interface.

4 Colored Graph Matching

This aspect of the software matches the current circuit with the *target circuit*. Any given circuit can be represented as a graph $G_c = (V_c, W_c, L)$ where V is the "end points" of each component and W are the connections (or wires) between those end points and L is a set of *colors* i.e. unique individual numbers assigned to vertices in V depending on their types. As current flows in both directions, this graph is undirected. An algorithm based on colored-graph isomorphism checks whether the current connections in Gc' are equivalent to the *target* circuit i.e. $Gc' \cong Gc$. If not, then the software performs a difference operation $(Gc - Gc')$ to determine the edges that are not required and identify corrective measures to optimize the circuit.

4.1 The Matching Algorithm

The matching algorithm needs some pre-processing of data. The target circuit graph must be prepared to be matched with the current circuit graph. The colors of the graph are determined by the type of vertices they are. For example, in the Fig. 2, there are 8 colors: AND_INPUT, AND_OUTOUT, OR_INPUT, OR_OUTPUT, NOT_INPUT, NOT_OUPUT, SWITCH and LED. They are all unique constant values.

Note that in order to match the correct circuits, some fixed additional values other than the wiring connections have to be added to $G_c\prime$ in run time. The values are stored in a separate matrix called the *adjustment* matrix/graph α. These values represent a logical connection between two nodes which are otherwise not connected by the students. This include a connection between the inputs and outputs of the gates. These is required to ensure the matching algorithm can verify the inputs from a preceding set of components are correctly connected to the succeeding set of components.

These fixed connections are shown in Fig. 6 which generate the adjustment matrix α for this board. The solid lines indicate the wires that are connected or expected to be connected with wires i.e. the *target circuit graph* of XOR gate. The dotted lines are never measured and not expected to be connected. They are part of α. They simply relate the inputs and output of the same gate. This matrix is board specific and always constant. Once $G_c\prime$ is obtained, $\Psi c'$ is calculated as

$$\Psi'_c = G'_c + \alpha$$

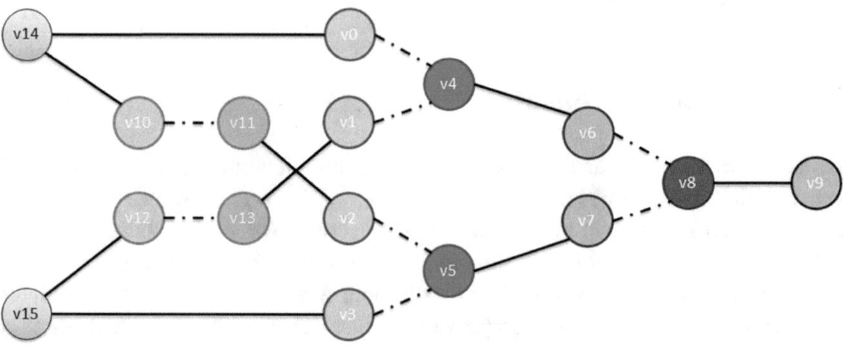

Fig. 6. The target circuit graph (G_c) for XOR gate using AND, OR and NOT gates (solid lines). The dotted lines are in α

These values of α are added to the matrix G'_c before matching in run time. Also, all self-loops are removed from the graphs.

The colors for each vertex measured for the graphs are always going to be the same as the end-points are fixed with the ports. The graph matching algorithm is as follows:

```
Procedure Matching (Gc, Gc')
    if the number of edges |W| or connections in Gc ≠
    Gc' or there exists w ∈ Gc' such that w ∈ α as well
    then return false

    else
      set Ψc = Gc + α
      set Ψc' = Gc' + α

      for each pair a, b ∈ {Ψc, Ψc'}
          set empty array visited_a and visited_b
          set visited_a[a] = true
          set visited_b[b] = true

          if for any (a,b), isEquiv(Ψc, Ψc', a, b,
          visited_a, visited_b) = true, return true

          else return false

Procedure isEquiv(Ψ, Ψ', a, b, visited_a, visited_b)
    if color of a and b are different, return false
    else
          if size of neighborhood of a and b in Ψ and
          Ψ' are different return false
          else
            for each pair of e,f ∈ (Ψ, Ψ')
              if e and f have the same number of
              adjacent vertices with the same color
              then
                set visited_a[a] = true
                set visited_b[b] = true

                if isEquiv(Ψ, Ψ', e, f, visited_a,
                visited_b) = true
                      push e,f into a stack

            if each pair e, f stored in the stack are
            unique and there is no e, f' or e', f in
            the stack return true

            else return false
```

The very first line of procedure Matching checks the similarity between the number of edges of the raw graph that is obtained by measurement G_c' with the unadjusted circuit graph G_c. This is done because it is possible that students might connect two end-points that are otherwise never supposed to be connected e.g. wiring the AND INPUT to its own OUTPUT. If there is any such wire, then obviously the students circuit is wrong.

The procedure used has a very high complexity. However, with a maximum of 60–70 nodes on a board and up to 20 colors the procedure can be done in real-time. Also, the graph is expected to be sparsely populated as the students are never going to connect all possible combinations of the end-points. If they do use a lot of wire, then checking of the number of edges at the very beginning will fail and the procedure return false immediately. For all the tests done for the XOR gate with 36 end-points, the time taken was between 1 and 2 s for returning a result of matching.

The graph matching procedure can be run asynchronously i.e. the graph G_c' may be relayed to the Web UI immediately, but the matching algorithm can run separately and return the result afterwards. The complexity of the algorithm is high which means it is best done online in the cloud where more computational resources can be dedicated if required. If the graph matches the UI notifies the students, otherwise an analysis is done on the difference between the graphs to provide support.

4.2 The Help Algorithm

The 'help' procedure is invoked when the users click the help button, or the system could ask if the users need some help. The primary task of this procedure is to suggest a connection that is not present in the circuit but required or vice versa. In this paper a simple algorithm is presented that checks the status of the circuit linearly every time the graph is updated. A matrix β is kept for each target circuit graph G_c. β_{ij} contains the total number of connections between colors i and j in target graph G_c. $\beta(G_c')$ is the matrix of total connections between each pair of colors in current graph G_c'.

```
Procedure HELP(Gc')
    if ∃i, j ∈ β(Gc') such that βij > βij(Gc')

          Suggest that a node from color i to color j
          can be connected

    else if ∃i, j ∈ β(Gc') such that βij < βij(Gc')

          Suggest that a node from color i to color j
          should be disconnected
```

The above procedure would suggest students to connect any missing wires such as

"*You can connect between AND OUTPUT and OR INPUT*"

without actually suggesting the specific end-points. Further works can concentrate on more advanced methods of detecting students' behavior and provide help.

The students can make changes to the circuit in real-time and may follow the suggested directives from the software. However, they are also free to ignore them as well further exploration opportunities. The software may not issue a warning/directive on every mistake that it detects, but instead allows for enough time to determine if the students require corrective assistance and provide support.

4.3 Drawing the Circuit

Drawing the circuit involves drawing components of the digital circuit and the wires between them. This requires a component array and the adjustment matrix. Note that the adjustment matrix contains additional values representing logical connections which are essential for matching the graphs but are not supposed to be shown on the circuit diagram on the web interface. Thus, α is subtracted from the Ψc received from the WebSockets. These connections can be shown optionally if requested by the students.

The component array contains 36 elements corresponding to the ports and component endpoints. Each element has 3 attributes each—X and Y coordinates, Color (real RGB color of the component) and identification name e.g. 'AND 1 INPUT 1'. The coordinates describe where the end-points are to be placed on the screen. The color is used to differentiate between types of components e.g. AND has blue color while OR is red color.

The web UI is created in Javascript and uses a canvas to draw the circuit. Whenever an updated graph is received from the server, the web UI clears the canvas does the following:

i. draws the end points with the coordinates and colors
ii. draw the components with the coordinates and colors
iii. draw wiring lines according to ($\Psi c - \alpha$) along with any other adjustments.

One drawback of the circuit mapping mechanism described in Sect. 3.1 is that if end-points/ports (p1, p2) and (p2, p3) are physically connected, then p1 and p3 is also connected even though the students may not have placed a physical wire between them. There is no way to determine if p1 and p3 are actually connected or not. This create loops as the circuit mapping will always determine that all of (p1, p2, p3) are connected to each other. This however has an advantage in term of learning, as the students can see connection that they may not have built otherwise. While this is an advantage, unnecessary connections could be confusing and consume space. To get around this, the input circuit matrix Ψc is processed as:

```
Procedure PurgeLoop (Gc')

     For each pair e, f ∈ Gc' with Gc'(e, f) = true i.e.
     e and f are connected

          if there exists g such that e, g and f, g are
          also connected, then remove e, f from Gc'
```

This procedure simply removes all instances of triangular loops. If there are 4 ports/endpoints (p1, p2, p3, p4) that are interconnected, then it basically contains multiple triangular loops which are removed one after another until there is no loops left. This is procedure can be further improved and adapted to the specific boards. Figures 6 and 7 shows the difference between before and after removing the loops. The main aim of such a procedure is make the circuit less "cluttered".

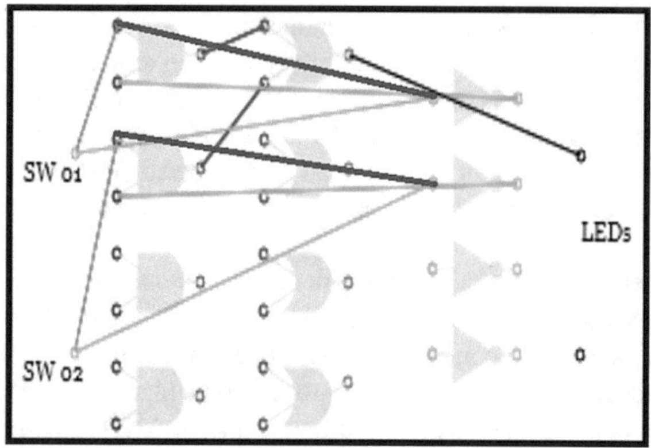

Fig. 7. A drawing of the circuit diagram with the loops present compared to Fig. 3

An example of this is shown in Fig. 7. Compared to Fig. 3, there are two extra lines between inputs if AND GATES and the NOT GATEs. This because even with two wires between the switch—SW 01 (or SW02) to corresponding AND and NOT, the connections are actually between all three points—switch, AND gate and NOT gate. The above algorithm removes the third line to de-clutter the picture as in Fig. 3.

5 Uses of the System

This section describes the possible usability of the proposed model in a digital electronics practical course.

5.1 Digital Electronics—Single User

The circuit checking mechanism is scalable and can be used for any electrical laboratory involving test circuits where the students must wire up different components. However, if the circuits have analogue outputs, the effectiveness of the checking system is reduced with respect to the output. Although it can determine the correctness of the circuit, it could be difficult to determine if the output is optimal compared to the intended learning outcomes of the activity due to slight imperfections of actual behavior of each individual component. In such cases, the analogue output from the circuit must also be transmitted to the online software, which needs to be checked to determine if the students are observing the correct output.

In terms of *learning*, the online system can provide immediately relevant feedback and if necessary inform the students what to do next. However, for learning purposes, these suggestions should not be immediately accessible even if the help button is pressed, so as to ensure the student develop diagnostic skills. There exists a possibility that a student following immediate feedback from the online system may not achieve the required learning for that activity. This can be resolved if the online platform:

- uses a fixed minimum timeout (perhaps minutes) before the next suggestion is given; and
- records a specific number of events (i.e. circuit changes) before the next suggestions.

5.2 Digital Electronics—Group Work

The boards can serve a different purpose in context of internet enabled group work. As the boards have specific input and output end-points i.e. the switches and the LEDs, they can pass around bits between multiple boards. In this scenario, each students group make a circuit on their boards. The inputs to this circuit may come in from another board through the network and be replicated by the MCU on behalf of the switch. The output of the circuit in the LEDs can then be passed on as the input to another board. This would enable a classroom wide network of boards to operate in parallel.

6 Conclusions

This paper proposed a cyber-physical system including a hardware-based methodology to determine student activities. This hardware may be regarded as a multi-point multimeter. Once the student activities i.e. the circuits that they are preparing are determined, they a validated again at a desired set of values i.e. the target circuit graph. The system has been shown considering a prototype board for digital electronics which can be scaled to implement bigger interfaces and more complex activities. This proposed architecture can be used in analog circuits as well as other control systems experiments with additional automations to determine if a connection has been made. This architecture can greatly improve the learning outcomes and reduce the learning time.

References

1. Ugarte, I., Fernández, V., Sánchez, P.: Motivation of students in the learning of digital electronics through the double integration: Remote/presential work and theoretical/laboratory classes. In: 2012 Technologies Applied to Electronics Teaching (TAEE), pp. 63–67 (2012)
2. Dolgov, A., Miao, B., Zane, R., Maksimovic, D.: GUI-based laboratory architecture for teaching and research in digital control of SMPS. In: 2006 IEEE Workshops on Computers in Power Electronics, pp. 236–239 (2006)
3. García-Zubía, J., Angulo, I., Rodríguez, L., Orduna, P., Dziabenko, O., Güenaga, M.: Integration of a remote lab in a software tool for digital electronics. In: 2013 2nd Experiment@ International Conference (exp.at'13), pp. 174–175 (2013)

4. Sapula, T.: Interactive web-based laboratories in digital electronics. In: International Conference on Education and e-Learning Innovations, pp. 1–5 (2012)
5. Marques-Silva, J.P., Sakallah, K.A.: Boolean satisfiability in electronic design automation. Presented at the Proceedings of the 37th Annual Design Automation Conference, Los Angeles, California, USA (2000)
6. Toscano, L., Stella, S., Milotti, E.: Using graph theory for automated electric circuit solving. Eur. J. Phys. **36**, 035015 (2015)
7. Luks, E.M.: Parallel algorithms for permutation groups and graph isomorphism. In: 27th Annual Symposium on Foundations of Computer Science (sfcs 1986), pp. 292–302 (1986)
8. McKay, B.D., Piperno, A.: Practical graph isomorphism, II. J. Symbolic Comput., vol. 60, pp. 94–112, 2014/01/01/ (2014)
9. Wolinski, C., Kuchcinski, K.: Identification of application specific instructions based on subgraph isomorphism constraints. In: 2007 IEEE International Conference on Application-specific Systems, Architectures and Processors (ASAP), pp. 328–333 (2007)
10. Fischer, T., Scheidinger, J.: VISIR—Microcontroller extensions. In: 2015 12th International Conference on Remote Engineering and Virtual Instrumentation (REV), pp. 177–179 (2015)

Cyber-Physical Control and Virtual Instrumentation

Sonal Varshney(✉), S. S. Rohit, Sunidhi, Anirudh Gururaj Jamkhandi, D. Thanya, and Meena Parathodiyil

B.M.S College of Engineering, Bangalore, India
{sonalvrshny, rohitshrothrium.ss1997, sunidhi.adiga97, anirudhgururaj, thanya97}@gmail.com, pmeena.eee@bmsce.ac.in

Abstract. Over the years, industries have been growing at an enormous rate and a large percentage of these industries are put up in remote and isolated areas. Such areas, as compared to urbanized locations, pose a number of issues. This paper presents a cloud-based remote access solution integrated with a virtual laboratory to control the processes and devices in isolated industries. With the use of LoRa as a communication link, it is possible to send the data acquired by end devices located at the field to cloud and retrieve that data on an HMI (Human Machine Interface) which can be controlled from a suitable place for humans. The major purpose of this technology is to implement a cloud based virtual machine that would use LoRa as the connectivity protocol to control on-site equipment without actually being present at locations which prove to be hazardous.

Keywords: IoT (Internet of things) · Lora (Long range) · AI (Artificial intelligence) · SCADA (Supervisory control and data acquisition)

1 Introduction

4 In the current scenario, there are several industries that are located in remote, precarious areas making it inconvenient for employees. The issues posed include connectivity to the location, deleterious environment, lack of communication network, etc. Therefore, it becomes necessary to find an efficient method to tackle these issues. With the advent of AI (Artificial Intelligence), Cloud Computation and IoT (Internet of Things), an affordable and robust solution is proposed in this paper. In the current technology, there exists a DCS (Distributed Control System) that is responsible for the complete control of a process industry. The industries have a control room where monitoring of the entire process is handled. Thus, there is a requirement of hardware engineers to be present on the field. But, due to lack of safety and effective communication, human presence in such areas is not advisable.

The solution offered in this paper is lucrative, dependable and inexpensive. To overcome the lack of connectivity, infrastructure expenses and locomotive

M. E. Auer and K. Ram B. (Eds.): REV2019 2019, LNNS 80, pp. 19–27, 2020.
https://doi.org/10.1007/978-3-030-23162-0_2

expenses, LoRa gives an ad-hoc solution. A virtual laboratory in a convenient location is used to control the various industrial processes making it easily accessible and safe for the employees.

The prototype consists of an end device which can be an actuator or a sensor that is tethered to the virtual lab via a LoRa-cloud infrastructure. Data acquired by the sensor can be viewed on the virtual lab through this infrastructure. Control signals from the virtual lab are sent to actuators located at the plant via the same framework. The technology proposed will enable industries to automate their process with the advantage of mobile access to control the industrial process. There is an added advantage of cost reduction due to the use of LoRa and reduced labour requirement.

2 Related Work

There exists a virtual lab which provides an HMI through which the controls engineer can monitor and control the system process. This VM is created on cloud using IaaS infrastructure. Virtualization is abundantly used in IaaS cloud in order to integrate physical entities in an ad-hoc manner [1]. Virtual machines are conventionally used for virtualizing real-world entities and hardware [2–4], which makes them a suitable option for HMI. The virtual lab is developed on the virtual machine. The emulation of the physical process gives the controls engineer a better perspective of the operations, making it convenient and easier to control the industrial process. With the use of this technology, the issues of safety and employee transportation are solved. The virtual lab is located in a place appropriate and convenient for the employees to work from. LabVIEW provides an optimal environment for custom-written software-hardware interfacing and GUI (Graphic User Interface), complying to our requirement of HMI [5]. The GUI presents a logical and visual overview of the essence of the entire process, makes program development and maintenance effortless, and reduces programming errors to a large extent. The peripherals of National Instruments myDAQ proffer input and output streams that are used to acquire data and reciprocate by sending suitable control signals. This data acquisition device also caters to our data logging requirements.

LoRa is a long-range transceiver module which can communicate over a range of 10–15 km tested in urbanized locations [6]. It has private network capabilities, hence enabling an ad-hoc network setup [7]. This provides a communication link between the equipment on site and the controls engineer operating from a safe and suitable location. LoRa provides a strong 128 bits AES Encryption unlike other low power long range technologies such as Sigfox [8]. Although there exist long range technologies such a NB-IoT with high LTE based encryption, they do not offer private network capability and hence, cannot be setup ad-hoc. Therefore, LoRa provides a compelling blend of low power, long range, highly secure and flexible setup as a communication link between the process industry and the virtual lab. LoRaWAN with repeated stations is deployed to increase the range of operation. It establishes the networking protocol for LoRa based devices [9]. LoRa satisfies the need for communication in Cyber-Physical Systems with added advantages.

Cyber-Physical System, even in its embryonic stage, has many applications [10,11]. Its potential to diminish other existing and developed technologies makes it the center of shifting trends in the field of industrial automation. The integration of computation and communication proficiency with monitoring and control of industrial processes makes it a niche architecture for automation based companies [12]. Computing techniques are integrated with real-time objects in physical world with the use of sensors and actuators which comprise the end device present on field [13]. This makes it possible for the virtual lab to emulate the industrial process giving the controls engineer a better idea so that an informed decision can be made regarding the control signals to be transmitted.

3 Existing Technologies

There are various technologies in place viz. SCADA (Supervisory Control and Data Acquisition), DCS and PLC which are generally used for industrial automation applications. However, these systems lack flexibility which hinders further expansion making reconfiguration difficult. SCADA uses complex hardware units which are expensive and difficult to maintain, thereby decreasing cost-effectiveness. Also, due to the mission critical nature of SCADA, if one distributed system fails, all the processes supervised by this system are affected and there is a considerable probability of loss of sensitive data [14]. Due to the centralized architecture of SCADA, interoperability is thwarted. Reconfiguration of the control system is a difficult task and upgrading process is time-consuming. Often, the customer is confined to a specific hardware manufacturer for legacy device migration to newer versions. Due to the lack of options, the purchase of newer equipment is made expensive by the few available manufacturers [15].

4 Framework

4.1 Cyber-Physical Systems

Cyber-Physical System incorporates embedded architecture which obtains data from the end devices. This data is amplified and fed back to the virtual lab via LoRa-Cloud infrastructure. The virtual lab responds to this signal by transmitting a suitable control signal via the same communication link.

4.2 Embedded Architecture

Embedded architecture primarily consists of two sides, the first being the field side and the other being the HMI. The field side architecture comprises of end devices that can be sensors or actuators. The HMI side has the data acquisition system which comprehends the data received from the field and reciprocates by sending a suitable control signal to maintain operation.

4.3 Communication Architecture

To establish communication between the field devices and HMI, there is a requirement for a robust and secure communication link. A low power, long range ad-hoc network is established for the aforementioned purpose. LoRa is an inexpensive and energy efficient substitute to the present communication techniques. This technology runs on Chirp Spread Spectrum modulation where a chirp signal controlled by Spread Factor values generates a spread spectrum that provides variable frequency [16]. The required data signal is modulated onto this chirp signal at a higher data rate. This technology provides a simple wireless control mechanism aiding the communication between controller and end devices (Fig. 1).

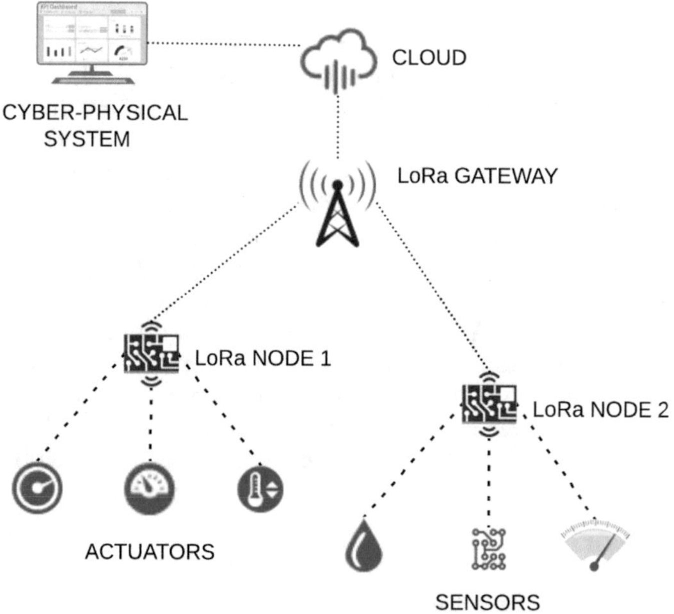

Fig. 1. Cyber-physical system implemented using LoRa

4.4 Cloud Architecture

A virtual machine is created using IaaS (Infrastructure as a Service) and data acquired from the field is pushed to cloud. Since IaaS is utilized to create the virtual machine, the cost-effectiveness increases [17]. The abstraction of coupling in HMI is achieved with the help of virtualization [18].

Another method of implementation is to utilize SaaS (Software as a Service) by suitable providers that provide virtual infrastructure along with built-in applications for data visualization and controls implementation. Data is retrieved from LoRa gateway and sent to the cloud provider where further processing of data occurs.

4.5 Virtual Lab

The virtual lab receives signals via LoRa-Cloud infrastructure from the end devices located on field. It reverts with suitable control signals which are sent over the same communication link.

4.6 Artificial Intelligence

AI driven approach is generally data intensive. The data that is acquired from the plant is used to make the plant smarter. Furthermore, to enhance user experience, chat-bots like GEs digital twin are used [19]. Analysis and monitoring of on-site equipment can be taken forward using computer vision techniques which also help in detecting faults and hazards. For a power generation plant, load-side forecasting can be implemented, thereby ensuring sturdy and reliable working of the plant [20].

5 Methodology

This prototype consists of two pumps installed at distinct geographical locations as shown in Fig. 2. Each end device is located at a distance of 11 and 14 km from the LoRa gateway respectively. Each pump is tethered to an embedded system along with multiple sensors collecting necessary data such as electro-mechanical parameters of the pump, flow and pressure in the pump valves and water level in the tank. The data acquired by the sensors is processed by the embedded system and via the SPI (Serial Peripheral Interface) this data is transferred to

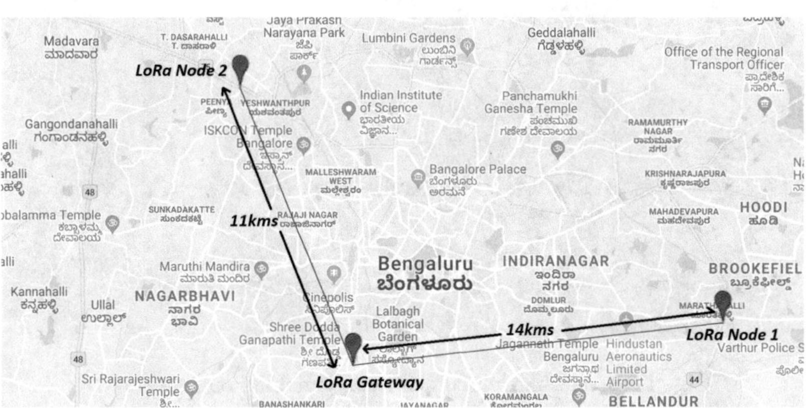

Fig. 2. Range of LoRa

the LoRa transceiver which further sends the data to a LoRa gateway. Since the setup is in a remote location, it becomes necessary to transmit data over several kilometers. But since LoRa has a maximum range of 15 km, a LoRaWAN can be setup with repeated stations in order to boost the signal along the path. The data acquired by the LoRa gateway is then pushed to a cloud infrastructure. The cloud platform hosts a virtual machine where various processes are carried out such as data visualization, data analytics, semantic learning and data logging operations.

The virtual lab is hosted on the virtual machine. It provides UI (User Interface) for a controls engineer to build relevant control logic. This logic is responsible for the complete control and synchronization of the field devices. The controls engineer can also analyze various plots on the virtual machine and gain complete insight to the current parameters of the field equipment. This virtual machine can be accessed by any person with the credentials for the cloud at any given time.

The feedback mechanism deployed is such that the virtual lab hosted on the virtual machine sends a signal to trigger the actuators to perform the directed action. This triggering signal is sent to the actuator via LoRa architecture. Figure 3 shows a virtual lab implemented on LabView through which the control and monitoring of the process can be performed real-time. Virtual lab is a software platform that synchronizes all the field processes.

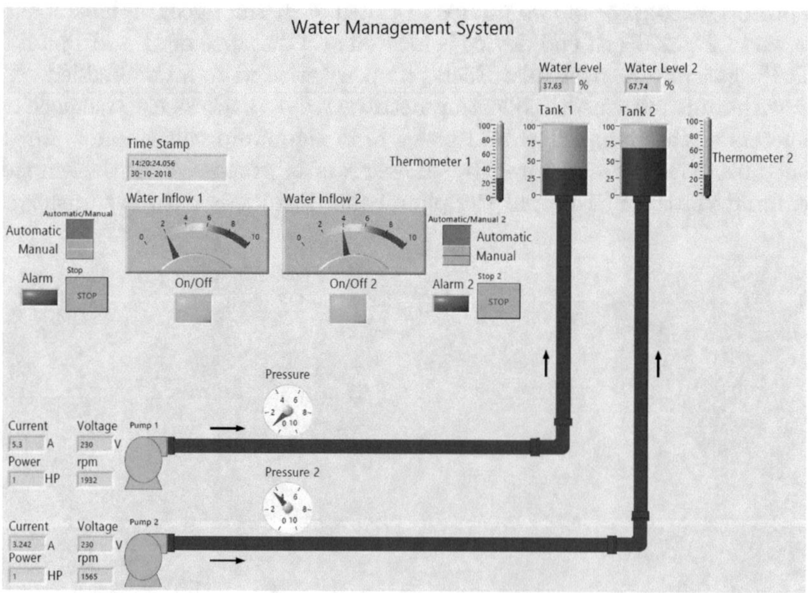

Fig. 3. Virtual instrumentation implemented on LabView

Any hardware upgrades can be implemented easily by just replacing the end device as opposed to the replacement of the entire system. Upgrades to the virtual machine are purely of software domain, making it versatile and allowing it to be hosted on any machine. It can be migrated from one platform to another with ease. Put together, it forms a reliable Cyber-Physical System.

6 Results and Conclusion

We tested the prototype proposed for a range of 13 km in remote locations as shown in Fig. 2. This setup proved to be cheaper than setting up an internet connection. Since the end devices function independently, upgrading them becomes easier and cost-effective. Cloud services also increase flexibility allowing users to use the resources as per their requirements. Usage is metered and customers pay only for what they use. The capital cost is reduced as the licenses for the hardware and software do not have to be purchased. As a Cyber-Physical System is used, interoperability is enhanced. A total of 120 end devices can be attached to LoRaWAN, increasing the scalability of the Cyber-Physical System.

This technology enables an easily accessible remote system which allows controls engineers to supervise a process on site from a location of their choice. As this replaces the existing expensive and complex mechanism in place, it proves to be a cost effective and simple solution. Furthermore, it cuts down on labour due to complete automation in the process industry.

Figures 4 and 5 show graphs with water flow, speed (rpm) and current plots against time with data being logged every 5 min. It can be inferred from the graph that when fault occurs, the current shoots up rapidly due to sudden increase in rpm. Meanwhile, the water flow in the valve abruptly decreases, thereby triggering reverse response. Here, the data logging time is shown from 14:15 and fault occurs at 14:20. This is depicted at 300 s on the graphs. When the fault occurs, the data from the previous 5 min is saved and the fault data begins to log. The virtual lab reciprocates by sending a suitable control signal to turn off the pump, hence preventing any damage to it. Also, Control engineers can set the time duration for logging data in order to optimize the usage of memory. If logging time duration before the fault occurs is more, memory consumed

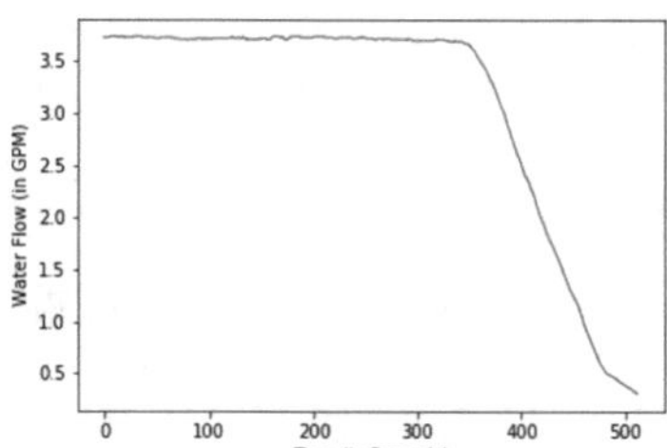

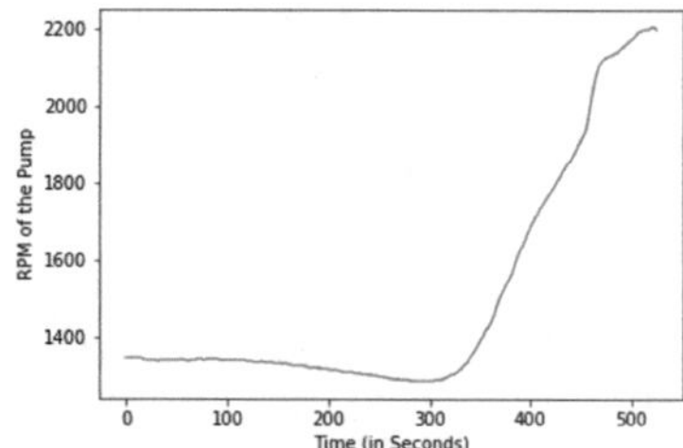

Fig. 4. (Left) Graph of water flow versus time, (Right) graph of RPM of the pump versus time

increases. Therefore, it becomes necessary to find a balance between the amount of data logged and memory consumption.

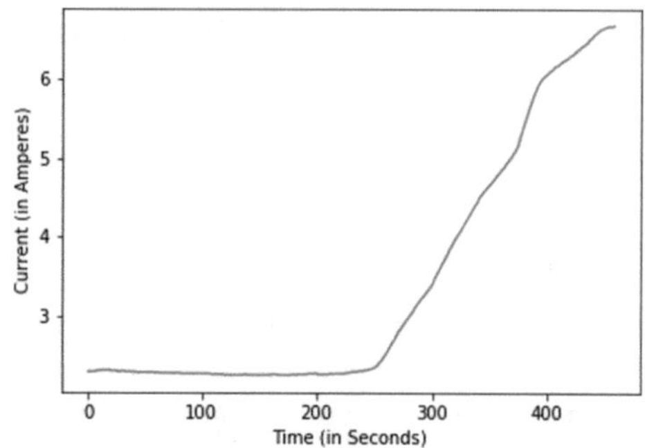

Fig. 5. Graph of current versus time

7 Future Work

Rapid technological advancement in Cloud Computing, Remote Sensing and Artificial Intelligence has greatly enhanced Cyber-Physical Systems and Virtual Instrumentation. Deep reinforcement learning can be modelled and deployed on the cloud to optimally control sequence of actions taking place on site by using the acquired data from the site and uploading it onto the cloud. These systems can be used to track and monitor industrial assets by retrieving real time data such as location and its ambient conditions. Such technologies can also be further extended to space research. For various exploration projects, such ad-hoc networks can be established in order to create a successful communication link between the exploration rover and the base station.

References

1. Dillon, T., Wu, C., Chang, E.: Cloud computing: issues and challenges. In: 2010 24th IEEE International Conference on Advanced Information Networking and Applications (AINA). Ieee (2010)
2. Levis, P., Culler, D.: Mat: A tiny virtual machine for sensor networks. ACM Sigplan Not. **37**(10). ACM (2002)
3. Dickman, L.I.: Small virtual machines: a survey. In: Proceedings of the Workshop on Virtual Computer Systems. ACM (1973)
4. Bugnion, E., et al.: Disco: running commodity operating systems on scalable multiprocessors. ACM Trans. Comput. Syst. (TOCS) **15.4**, 412–447 (1997)
5. Wang, C., Gao, R.X.: A virtual instrumentation system for integrated bearing condition monitoring. IEEE Trans. Instrum. Measur. **49**(2), 325–332 (2000)

6. Mahmoud, M.S., Mohamad, A.A.H.: A study of efficient power consumption wireless communication techniques/modules for internet of things (IoT) applications (2016)
7. Mekki, K., et al.: A comparative study of LPWAN technologies for large-scale IoT deployment. ICT Express (2018)
8. Xevgenis, M.G., et al.: The Virtual Lab (VLAB) Cloud Solution. Globecom Workshops (GC Wkshps), 2016 IEEE. IEEE (2016)
9. Aras, E., et al.: Exploring the security vulnerabilities of lora. In: 2017 3rd IEEE International Conference on Cybernetics (CYBCONF). IEEE (2017)
10. Lee, E.A.: Cyber-physical systems-are computing foundations adequate. Position Paper for NSF Workshop On Cyber-Physical Systems: Research Motivation, Techniques and Roadmap, vol. 2. (2006)
11. Lee, E.A.: Cyber physical systems: design challenges. In: 11th IEEE Symposium on Object Oriented Real-Time Distributed Computing (ISORC). IEEE (2008)
12. Cardenas, A.A., Amin, S., Sastry, S.: Secure control: towards survivable cyber-physical systems. In: 28th International Conference on Distributed Computing Systems Workshops, 2008. ICDCS'08. IEEE (2008)
13. Stankovic, J.A., et al.: Opportunities and obligations for physical computing systems. Computer **38**(11), 23–31 (2005)
14. Coates, G.M., et al.: A trust system architecture for SCADA network security. IEEE Trans. Power Deliv. **25**(1), 158–169 (2010)
15. Abbas, H.A.: Future SCADA challenges and the promising solution: the agent-based SCADA. Int. J. Crit. Infrastruct. **10**(3–4), 307–333 (2014)
16. Augustin, A., et al.: A study of LoRa: Long range & low power networks for the internet of things. Sensors **16**(9), 1466 (2016)
17. Bhardwaj, S., Jain, L., Jain, S.: Cloud computing: a study of infrastructure as a service (IAAS). Int. J. Eng. Inf. Technol. **2**(1), 60–63 (2010)
18. Rimal, B.P., Choi, E., Lumb, I.: A taxonomy and survey of cloud computing systems. In: NCM'09. Fifth International Joint Conference on INC, IMS and IDC, 2009. Ieee (2009)
19. Dale, Robert: The return of the chatbots. Nat. Lang. Eng. **22**(5), 811–817 (2016)
20. Warrior, K.P., Shrenik, M., Soni, N.: Short-term electrical load forecasting using predictive machine learning models. In: 2016 IEEE Annual India Conference (INDICON). IEEE (2016)

Application for Monitoring and Prediction of Energy and Water Consumption in Domestic Cyber-Physical Systems

Alberto Cardoso[1(✉)], Joaquim Leitão[1], Daniel Azevedo[1], Paulo Gil[1,2], and Bernardete Ribeiro[1]

[1] CISUC, Department of Informatics Engineering, University of Coimbra, Coimbra, Portugal
{alberto,jpleitao,bribeiro}@dei.uc.pt, daniel96.azevedo@gmail.com, psg@fct.unl.pt
[2] Department of Electrical Engineering, FCT, NOVA University of Lisbon, Lisbon, Portugal

Abstract. The growth of cyber-physical systems in the context of smart homes and the Internet of Things (IoT) offers an opportunity to specify and develop useful applications to different purposes, based on data acquired remotely and supported by data science algorithms. This article aims to present a contribution for the development of solutions for monitoring and prediction of energy and water consumption in smart homes, which can be used, for example, by households to visualize, monitor and optimize their consumption, as well as reduce costs.

Keywords: Cyber-Physical systems · Data science · Smart homes · Internet of things · Remote monitoring · Home energy and water management · Energy optimization

1 Introduction

Domestic consumption of energy and water is a matter of great importance to individual consumers and supplier companies for resource optimization and cost reduction purposes. With the advances in the development of cyber-physical systems in the context of the smart homes and the Internet of Things (IoT) [1], including in domestic environments, data acquired from local and remote measurement systems can be used to monitor the entire system and predict consumption based on the human behavior, the forecast of the environmental conditions and the cost of energy over time.

The home energy and water resources management is a topic of great relevance due to its economic, environmental and social impacts, and various research works have been published presenting different approaches to explore these resources in a more efficient way [2, 3].

In [4], an optimized management of water and energy resources in a domestic environment has been proposed, exploring the relationships between these resources

M. E. Auer and K. Ram B. (Eds.): REV2019 2019, LNNS 80, pp. 28–37, 2020.
https://doi.org/10.1007/978-3-030-23162-0_3

and scheduling the operation of certain appliances in the household so that energy costs are reduced, while assuring that the electrical and water needs are met over time.

In this context, it is important to design and develop applications to visualize and predict the energy and water consumption, which can be used, for example, by households to monitor and optimize their consumption, as well as reduce costs.

Considering a familiar house with a measurement system to acquire data of energy and water consumption and energy production (when a system for energy production is installed), a cyber-physical system can be developed to integrate all available data, analyze and process it to generate and provide information that can be very important to help families make decisions concerning the resources optimization and, consequently, reducing the energy bill.

Assuming that data is available over time with a given sampling period, the approach proposed in this article considers the development of an application to integrate all measurements, the weather forecast and the algorithms to predict consumption over, for example, a week, and to minimize the costs optimizing the resources according to patterns representing the family's expected behavior in terms of energy and water use.

Therefore, several optimization methodologies have been tested and evaluated [4] in order to minimize costs through an optimized schedule of the use of some electric machines, which are also water consumers (i.e. washing machine or dishwasher), satisfying the requirements and constraints pre-defined by the family for a certain period of time and taking into account the price of energy, combining external and internal sources.

This work aims to contribute with an application that can be considered to visualize and monitor the consumption of energy and water in domestic environments and to help families to manage their consumption, promoting an optimized use of the resources, namely the locally produced energy, and the reduction of costs, supported by an appropriate timeline of some of the major household tasks, based on the definition by the family of a set of temporal requirements and constraints.

The next sections will present details about the application, namely showing the interfaces developed to define the system, configure the typical family's behavior, and to visualize the results from different perspectives as, for example, the prediction of energy and water consumption for a given period of time.

2 Application Description

This work considers a model for individual houses, where a typical family (5 individuals, for example) lives, with water and energy requests that change over time and which must be fulfilled.

The Alliance for Water Efficiency (AWE) [5] identifies five home appliances responsible for most of the indoor water uses: washing machines, dishwashers, toilets, showers and faucets.

Therefore, the model assumes that the house may have several electrical machines (washing machine, dishwasher, drying machine, air conditioner, etc.) and a water tank, filled with groundwater by means of a pumping system, which is used to assure the occupants' demands.

The water level of the tank is controlled to guarantee an adequate supply pressure.

Additionally, it is also possible to integrate a photovoltaic system, considering weather forecast information to predict local energy production over the prediction horizon.

Based on this model, an optimization problem is formulated aiming to control the pump's operation and to schedule the main electrical machines, in a given time horizon, and minimizing an efficiency metric based on the overall energy cost. At the same time, the water level in the tank should be kept close to the desired target level. Figure 1 shows a typical scenario.

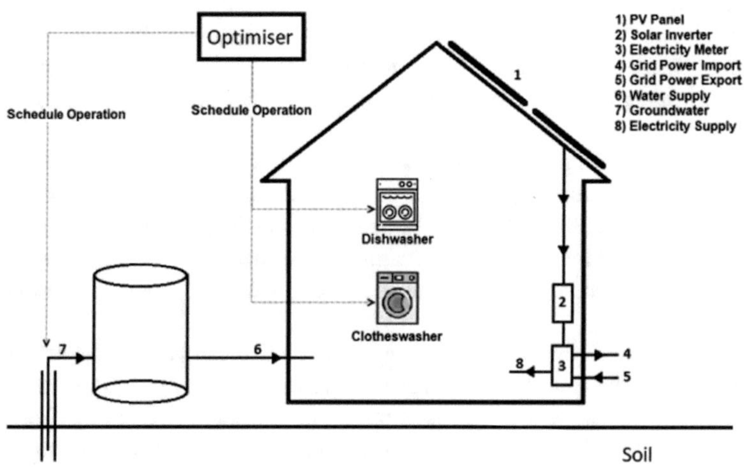

Fig. 1. Typical problem scenario [4].

To solve the optimization problem, the Particle Swarm Optimization (PSO) algorithm [6] was considered. Using different parameters, the PSO was able to compute valid solutions with minimum values of the objective function, determining the best scheduling for the appliances and exploring periods of time during which energy is cheaper.

In order to visualize, monitor and optimize the energy and water consumption, an application (app) with different functionalities was designed and developed. With this app it is possible to simulate and find optimal solutions for energy and water consumption in a given house.

Considering friendly interfaces, the app has several configurable fields, that can be changed by the user according to the household, with the objective of create different household scenarios and configure the model to obtain energy and water consumption values similar to the real ones.

The app offers also the possibility of visualize graphs that display the energy and water consumption at an hourly rate, indicating which of the tasks the energy/water has been spent on.

In terms of the app functionalities, the main interface (see Fig. 2) can be divided into 6 different parts: General fields, Machines, Household, Output, Export/Import and Real Data vs Optimized Data.

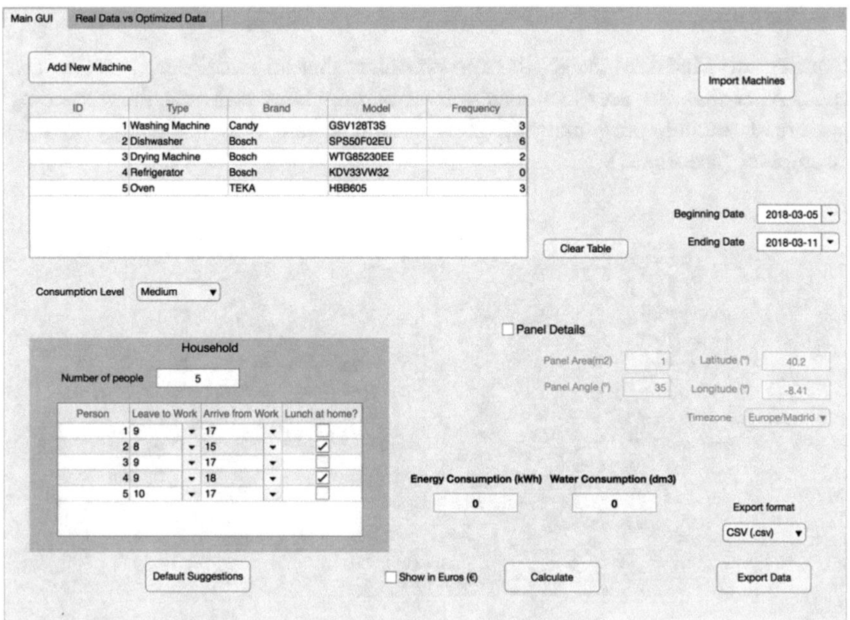

Fig. 2. Main application interface.

2.1 General Fields

The application has some general fields that can be changed, like:

- Beginning Date
- Ending Date.

In case where the house has Photovoltaic Panels, the Panel Details checkbox should be checked, and the following panels can be changed:

- Panel Area
- Panel Angle.

Regarding the house location and taking into account that the app uses an algorithm to predict the solar exposition, the user should indicate the:

- Latitude, Longitude, Time zone.

The user can also change the Consumption Level, using a dropdown button, to indicate the family's tendency to consume energy/water.

2.2 Machines

Regarding the Machines usage, the app considers that an initial configuration can be loaded. After that, the user can change information about each machine regarding its type, brand, model, programs and their durations, and also the energy and water consumption (see Fig. 3).

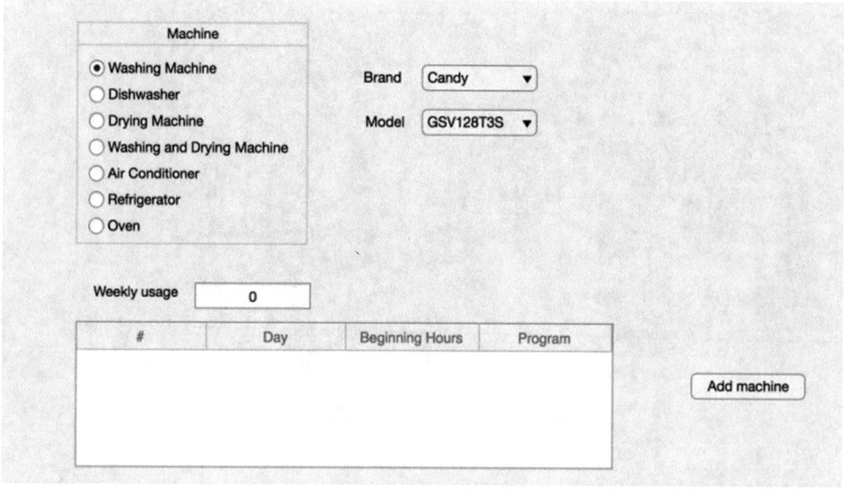

Fig. 3. The app interface for machines configuration.

The interface gives the possibility to add the following equipment:

- Washing Machine
- Dishwasher
- Drying Machine
- Washing and Drying Machine
- Air Conditioner
- Refrigerator
- Oven.

Trying to make the configuration as realistic as possible, some assumptions have been made:

- For machines of type “Air Conditioner” and “Oven” the configuration is standard and can’t be modified;
- The Refrigerators have the same consumption per hour and, since they are considered to be running continuously, the weekly usage table is not displayed;
- For Air Conditioners, their use is daily and not weekly, and daily consumption is applied to every day of the week.

When all the changes and configurations are completed, the user just needs to click on the **Add machine** button to add a new machine instance or change an existing one.

2.3 Household Information

One important definition for this work is about the elements of the family household and about their usual tasks during the period of time in consideration. Thus, the app allows to define the number of family members and some significant aspects of their behavior, represented by their main and usual activities (see Fig. 4).

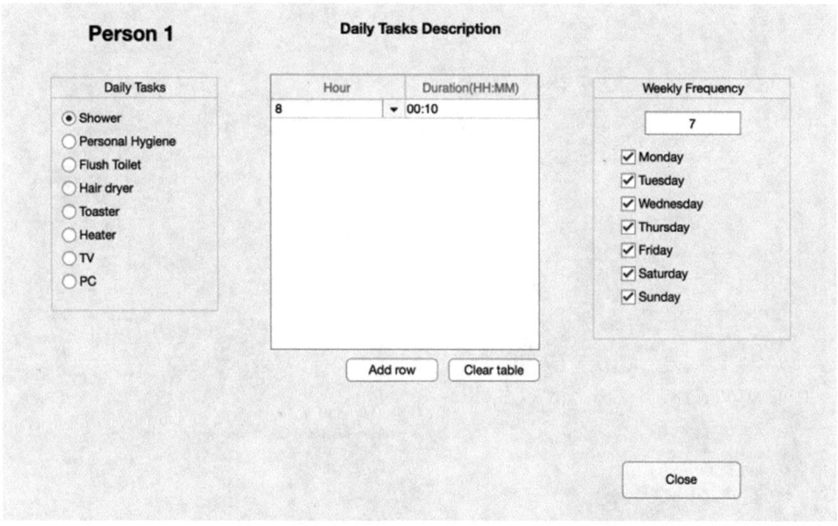

Fig. 4. The app interface for person information.

This interface considers the following daily tasks:

- Shower, Personal Hygiene, Hair dryer, Toaster, Heater, TV, PC.

For the Flush Toilet task, the hours of use were generated following a Poisson’s law to obtain more accurate results. For all the Daily tasks (depending on the task), a default energy or water consumption value was considered (for a specific amount of time).

The person's information includes the setting of work hours (the time each person leaves and arrives home) and also the indication whether the person has lunch at home or not.

For a more accurate description, it is possible to specify more advanced individual tasks for each person, using another specific interface.

2.4 Output

After the configuration of all parameters using the different app interfaces, the user can simulate the cyber-physical system and obtain information about, for example, the energy and water consumption during a given period of time (one day, one week, …).

For example, considering the scenario presented in Fig. 5, with a family of 5 persons, photovoltaic panels, a Washing Machine, a Dishwasher, a Drying Machine, a Refrigerator and an Oven, the energy consumption and the water consumption during one day can be observed in Figs. 6 and 7, respectively.

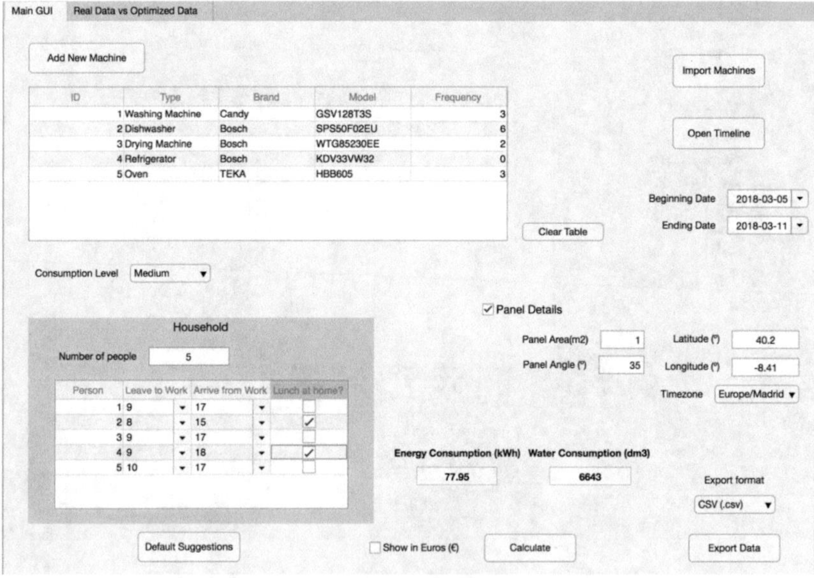

Fig. 5. Scenario example.

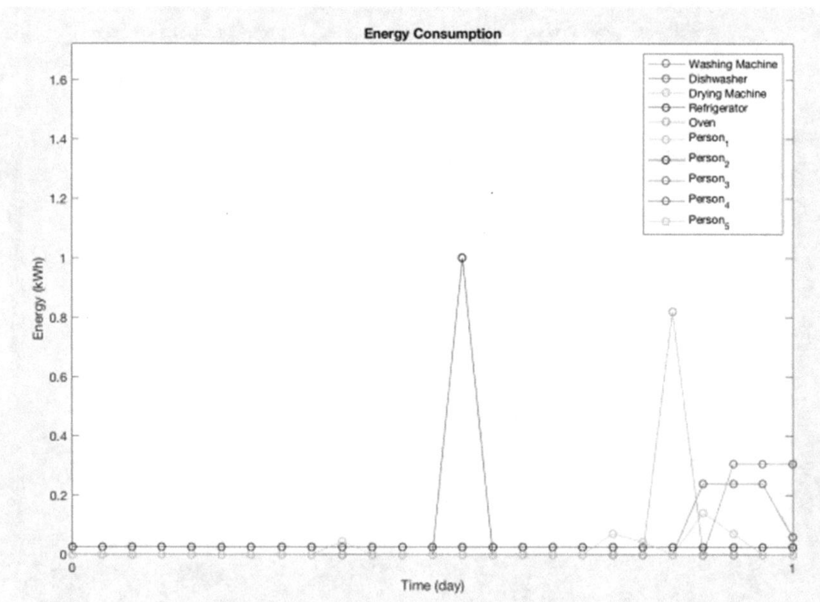

Fig. 6. The app output with the energy consumption during one day.

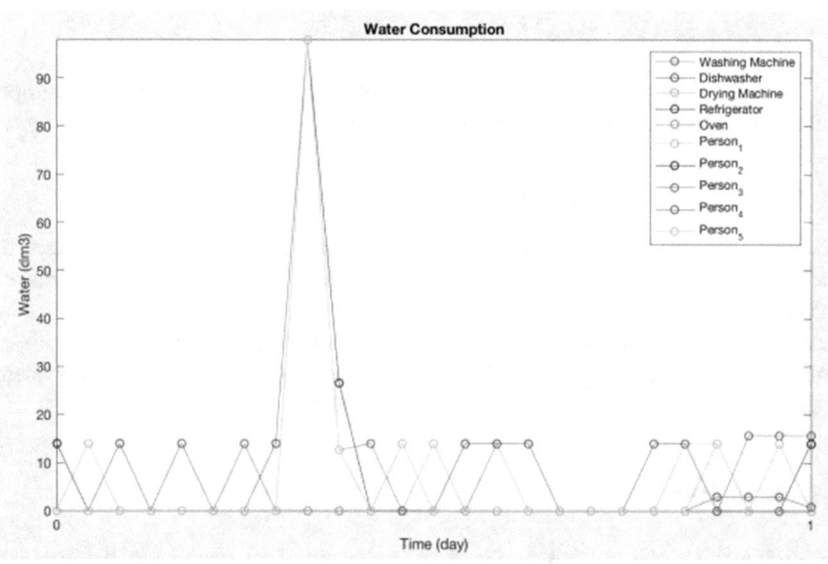

Fig. 7. The app output with the water consumption during one day.

If the house in consideration has photovoltaic panels, the app can also presents the graph with information about the energy production (see Fig. 8).

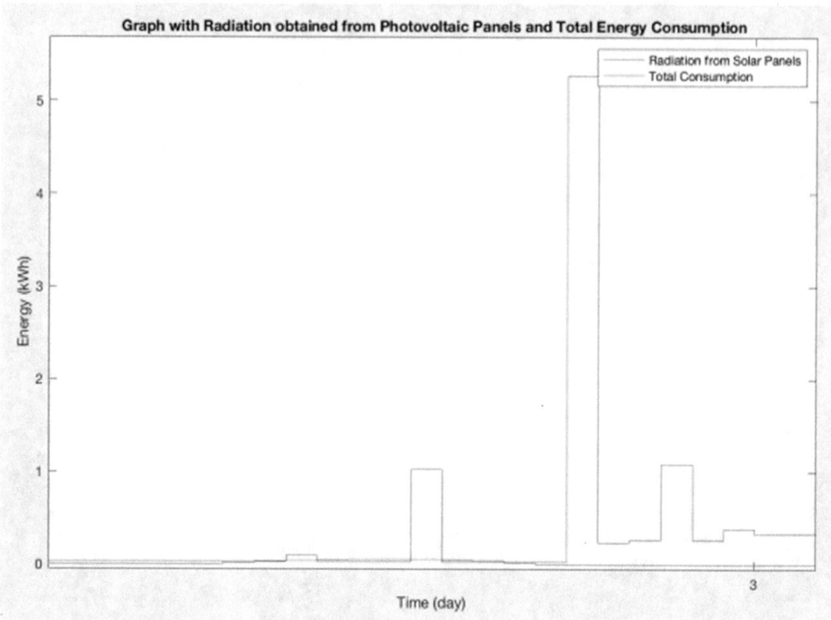

Fig. 8. The app output with the energy production using photovoltaic panels.

2.5 Export/Import

With the app, it is possible to export data with the energy and water consumption details and import the machine configuration details.

2.6 Real Data Versus Optimized Data

A relevant aspect of this application is the functionality of comparing data generated by the simulation based on the user configurations and data resulting from the optimization of machines uses.

As discussed in [4], some solutions were computed to solve the optimization problem, defined according to the formulation based on the configuration parameters. For example, the Particle Swarm Optimization (PSO) algorithm was implemented.

3 Conclusion

Considering the significant advances of the cyber-physical systems in the context of IoT and the importance of contributing to a sustainable world, the proposed application expects to contribute to a widespread use of technology to help families to manage and plan their domestic activities while minimizing the energy and water costs.

The application assumes that household appliances that consume water and energy, along with a pumping system to fill a tank that supplies the house with water, are controlled during a specific time period.

The problem of management and scheduling of electrical and water appliances in a home environment was formulated as an optimization problem and solved using the Particle Swarm Optimization (PSO) algorithm.

The results obtained with the optimization process provide significant expectations about the applicability and the interest of this application and approach.

The application was designed and developed for home systems but can be extended to be used by supplier companies to analyze, manage and plan the consumption of a home network in a given region.

Acknowledgements. This work was also partially supported by the Portuguese Foundation for Science and Technology (FCT) under the project UID/EEA/00066/2013 and the PhD scholarship SFRH/BD/122103/2016.

References

1. Weinreich, A.: The Future of The Smart Home: Smart Homes & IoT: A Century In The Making, Forbes Media LLC (2017)
2. Ocampo-Martinez, C., Puig, V., Cembrano, G., Quevedo, J.: Application of predictive control strategies to the management of complex networks in the urban water cycle. IEEE Control Syst. **33**(1), 15–41 (2013)
3. Pedrasa, M.A.A., Spooner, T.D., MacGill, I.F.: Coordinated scheduling of residential distributed energy resources to optimize smart home energy services. IEEE Trans. Smart Grid **1**(2), 134–143 (2010)
4. Leitão, J., Gil, P., Ribeiro, B., Cardoso, A.: Improving household's efficiency via scheduling of water and energy appliances. In: Proceedings of the 13th APCA International Conference on Automatic Control and Soft Computing (CONTROLO2018), pp. 253–258, IEEE (2018)
5. Alliance for Water Efficiency. http://www.allianceforwaterefficiency.org/ (2017). [Online; accessed 14 Dec 2018]
6. Kennedy, J., Eberhart, R.: Particle swarm optimization, In: Proceedings of the IEEE International Conference on Neural Networks, vol. 4, pp. 1942–1945 (1995)

The Need for a System to Benefit the Implementation of *Digital Twin*, by Helping Visualize the *Virtual Dynamics* Remotely

Srinivas K. Badkilaya[1,2(✉)] and Hari Prasad Bhat[2]

[1] Cymbeline Innovation Pvt Ltd, Bengaluru, India
badkilayasrinivas@gmail.com
[2] Vdesign, India
hariprasad.bhat@vdesignautomation.com

Abstract. **"The Future for Industrial services is DIGITAL TWIN"**, as the mentioned quote says, the next big revolution in the Industrial sector is going to be the Implementation of the Digital Twin Technology, which would make many processes simpler when it comes to the Industrial services, if it successfully achieved. Digital Twin is the virtual representation of a system's elements and dynamics, where the virtual representation of the physical system and the motion of the same is very much required. A platform with a simulating tool, a designing tool and a data acquisition system functioning simultaneously would facilitate to analyze the 3D's (Design, Data and Development) of any system in a single Interface. The 3D's mentioned, plays a vital role in the functioning of any system. This would help the user to properly analyze the Virtual Dynamics of the system and make the design modifications before getting the system fabricated, which is nothing but the development of the system. The system proposed is found to be more and more efficient, if all these processes are accessed remotely anywhere by anyone. If the virtual representation of the dynamics of a physical system is visualized and analyzed remotely, this would facilitate the Remote Engineering technology also by helping the user in some ways. The outcome of this study would be a Platform, with the following systems (a) A designing tool, (b) A simulating tool, (c) A data acquisition system functioning on a single Interface which would help to achieve the Digital Twin of a system in a pretty easier and simple way. The proposed system will be complex, since there are 3 different tools or the software functioning simultaneously and a lot of data exports and imports will be in process.

Keywords: Digital twin · Remote engineering · Virtual dynamics · System's dynamics · Design · Simulation · Data acquisition · Virtual representation

A Propose for a platform to ease the process of Digital Twin, which would facilitate the analysis and study of the System through its Virtual Representation rather than examining it physically.

M. E. Auer and K. Ram B. (Eds.): REV2019 2019, LNNS 80, pp. 38–50, 2020.
https://doi.org/10.1007/978-3-030-23162-0_4

1 Introduction

Digital Twin is undoubtedly the most booming technology at present, the Future Industrial services will be made more and more efficient with the help of these technologies. The usage and Implementation of these technologies would make the World smarter, simpler and efficient. The paper mainly talks about using the above technologies in a very practical way and to bring them into reality. In the present world, there is a higher need for the Modelling and Simulation of any physical system, component. The result of this modelling and simulation is expected to be a virtual representation of the same system, but this virtual representation would be more effective if the actual design of the system is also loaded. Therefore, the actual system is being modelled and it is simulated to work as in the real time conditions with the actual design inherited resulting in a Digital Twin, through which not only the end results but also the system's dynamics, mechanisms are also visualized in a way more proficient way. To inherit the design, a 3D representation or the 3- dimensional design of the system is very much needed.

Therefore, to reach the above-mentioned result, there are mainly three areas of concern, where a proper study must be made. They are: (1) Modelling and Simulation of the system, (2) Inheriting the visual representation of the system (the mechanical design), (3) Data handling and Analysis (Input/Output) to utilize them to make the System with the design inherited run digitally as a virtual representation. A prolonged research has been done to make the above three areas upend running, and to connect them together to work mutually, simultaneously with the common interface to make the digital simulation of the system with the actual representation of the physical system.

The paper also talks about the Remote Engineering Technology, the Digital twin or the virtual representation of a system, if it is visualized and is accessed remotely, the implemented technology would be more effective and efficient. This would add a greater value for the users while using it and benefits them in many ways. The user can have the access to the system remotely. If the remote engineering is successfully implemented with the previously explained result, the user gets to visualize, analyze, study the system remotely and make improvements, modifications both in designing and modelling aspects based on the data, the simulation emits.

As explained, the study focuses on developing a system or a product with the results anticipated as above. The paper speaks about many Tools and Software, that are used to develop a system beneficial to many Design Engineers and to help achieve the Digital Twin concept in a pretty simpler way. An explanation for all the tools' usage and the answer to why that tool is being used will be made as the paper progresses. The Remote Engineering technology take this system to a higher level, makes it versatile to many and many applications. This would also make the system feasible enough to many and many multi user applications where the remote access is needed in higher priority.

A simple representation for the Architecture of the system proposed is shown in the following figure (Fig. 1).

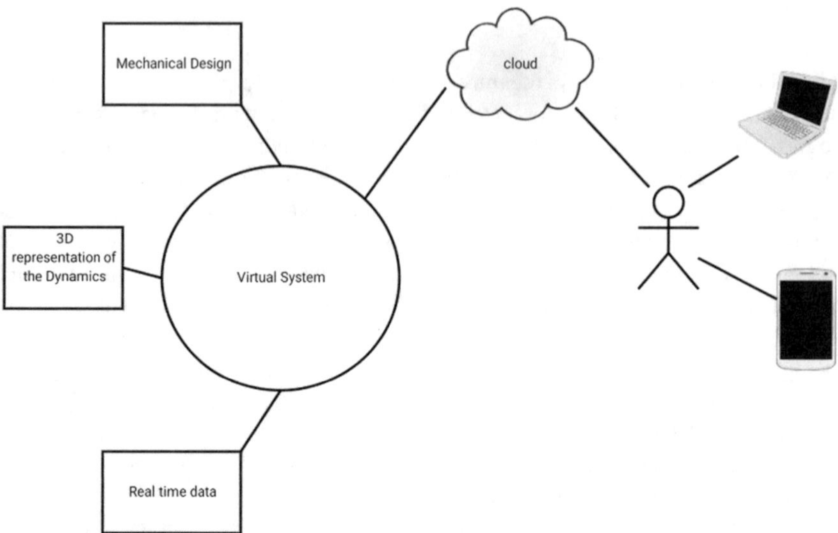

Fig. 1. Block diagram representation of the proposed system.

As shown in the figure, A system with 3D representation of the Dynamics masking with the actual Mechanical Design with sensors upon it putting out the real time Data of various parameters is visualized virtually and it is bridged to the cloud, through which the user connects and have the authority over it to develop the existing system is the anticipated outcome of the study. The block diagram itself explains the whole need or the concept to which this study particularly is aiming at.

2 Approach and Working

In this section, the path followed to get a Virtual System as explained in the previous sections is briefed. The algorithm of the approach pursued here is, shown in the Fig. 2. The first and the foremost step as shown, is to Develop the Mechanical Design and to achieve a 3D representation of a system. Any CAD tool or a software can be used to get the 3D design of a system, a machine, or a component. There are many CAD tools that are available such as CATIA, SOLIDWORKS, AUTOCAD, FUSION360, Creo and a lot more. The CAD tool that must be used to get the anticipated result and the reason why it is being used will be discussed further. This is one of the most important steps, since this creates a platform, where a user can have the view of the system's actual design digitally.

As the next step shown in the Flowchart, modelling and simulation of the system comes into picture. There are many barriers, that a designer must overcome while getting the modelling of a whole system done. These points will be explained in detail later in the paper. The modelling is a stage of higher priority, since it imports life to the bare Mechanical Design, which helps in getting the visualization of the actual dynamics in a simulated environment rather than in the real world. The most difficult

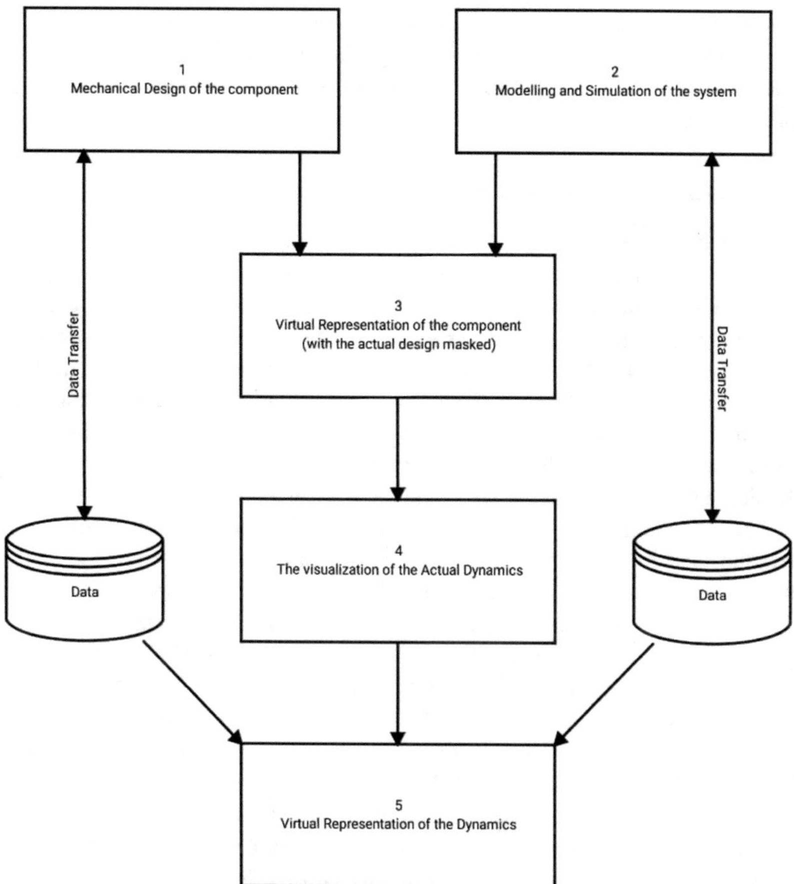

Fig. 2. The Algorithm/Flowchart for the approach followed.

factor here, is to achieve the collaboration of these Mechanical Design and Modelling together, so as to work mutually with a single interface, so a proper import/export will be done between them, enhancing the model of the system simulated to run with the mechanical 3D design displayed externally. A combination of a CAD tool and a simulation tool is used to determine the reach the above-mentioned result.

As explained before, fetching of the data is very much necessary to envisage the representation of the dynamics. For this to happen, two processes are needed to be take care of, one is to pull the data of various parameters with the integration of various sensors, hardware to the actual physical system, and the other is to import this data received to the modelled system, so as the system is made to run in the simulated environment with the real time data it is being imported. Therefore, even here, a pair of two systems is proposed, which is going to be discussed further.

These above sections give an overall general picture of the approach followed here, the upcoming sections provides a thorough description, where the reader gets the complete information of the process followed to achieve the Digital Twin explained in the previous sections.

2.1 Mechanical Design Aspects

As explained in the 'Approach' segment, the development of the Mechanical 3-Dimensional Design is the first and the foremost stage. This can be achieved with the help of any CAD tool. But by considering the future issues that a developer or a designer would face to pursue the Digital Twin, the CAD tool called 'SOLIDWORKS' by 'DASSAULT SYSTEMS' is preferred over other different tools. The fundamental reason to go with the Solidworks here is that, it has this advantage of speaking, basically sharing of the Data with a simulation tool, which would make the future integration of the Design and the Digital Model easier. The ability of the Solidworks tool, to export the Mechanical Design generated to the Simulation software makes it very distinguishing over other similar software. This is the reason why Solidworks would be rather used and it forms the basis for which the same software is being used here in our study, research.

But there are certain aspects that must be considered, and proper methods must be adapted while designing. Basically, the final design must be in a way such that, each component can be studied separately. This mean to say that the Design as a whole assembly, should be capable of making its each and every part ready to be studied or analyzed as a separate component, so that while the design integrated with the modelling is being simulated and made to run, the dynamics of each and every component can be visualized as a very separate entity and any modifications or the development to any particular component's design can be done based on the motion the simulation exhibits. The user can have this supreme power, he will be in that position to view the motion of the system in a simulated world, but with the actual physical representation the system has in its real world. A simple sketch numbered as 'Fig. 3' is shown below, the figure displays a shot captured while designing a simple assembly. A simple pendulum which oscillates in a axis is designed in the 'SOLIDWORKS' platform. The assembly designed can be easily exported to a very beautiful software, which is 'MATLAB and SIMULINK' by 'MATHWORKS', a tool which is exclusively used to model a system and simulate and analyze it in a very detailed way.

There are certain procedures that are supposed to be followed, to make the Data Transfer between the Solidworks and Matlab happen. They are listed as follows, (1) Download the Simscape Multibody package or the Add-on to the Matlab, if it is not installed. (2) Run the following commands on the command prompt in the Matlab window "install_addon ('the downloaded package file folder.zip')", "regmatlabserver", and *smlink_linksw". (3) In the Solidworks workbench, add the Simmechanics link Add-In from the Add-In's toolbar. 4)In the tool's menu on the Solidworks window, under Simmechanics link, click on extract, and then either Simmechanics First Generation or Simmechanics Second Generation. (4) The assembly designed should be saved in .xml format. (5) Finally, by running smimport('filename.xml') or mech_import('filename.xml'), the assembly designed is successfully imported to the Matlab and Simulink Workbench.

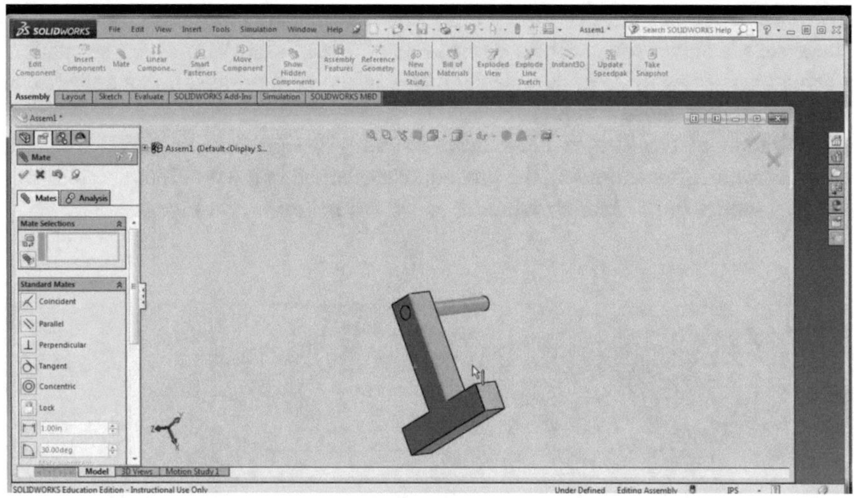

Fig. 3. A simple assembly designed in the Solidworks workbench.

The control to the designed Mechanical system can be designed in the Matlab workbench itself, with the prior knowledge of using the Simscape from Simulink can help to build the model and add control to it. A proper representation of the mechanical system can be made in the Matlab, by furnishing the system with different color gradients to beautify it. Certain measures have to be taken while designing the system in Solidworks, the design should be done in the front plane itself to ensure its better handling in future.

2.2 Modelling Aspects

Modelling and Simulation in simple terms can be considered as the substitute for a physical experimentation, in which the Computers are used to compute the results of some physical phenomenon. In this firstly, the mathematical modelling of the system must be made by a computer, which contains all the parameters of a physical model. The modelled system is then made to run digitally in a simulated environment, which is used to calculate many variables changing on the mathematical model. A software tool called "MATLAB and SIMULINK", is exclusively used here for this Modelling and Simulink purpose, as explained in the previous segment.

The Mathematical modelling of any system is a very complex process, since it involves the application of several differential equations defining the dynamics of each element. The physics of an element in each component in a system is represented using several complex higher order differential equations, then each component is connected to each other, resulting in the model of complete system. There are certain things that are followed while modelling a system. The system or the state variables of the system is expected to be in two states, that are Steady state (equilibrium state) and the transient state. Here in particular, the transient state of the system is very much concerned rather than just the steady state. The transient state analysis of the system would result in the study of Dynamic behavior of the system, where the steady state analysis would point at the system's statics only. Since, our goal is to achieve the Virtual representation of the Dynamics of any system, the transient state of the system must be modelled, which makes the process more complex.

The import of the Simple Pendulum Assembly designed in the CAD tool to the Matlab software, after following the procedure explained in the previous section, would result in a model in the Matlab window as shown below in the Fig. 4.

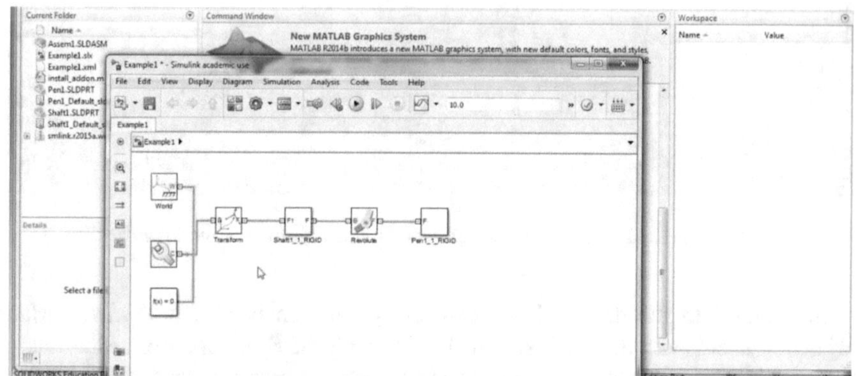

Fig. 4. The representation of the assembly in Matlab window.

As shown, the Mechanical Design when imported to 'Matlab and Simulink' tool, will be converted to a model with many simscape blocks. This is made to run in the Matlab and Simulink software tool itself, where any parameters can be configured by analyzing. If the model shown in the figure above is made to run on the Simulink, this would generate a new window called Mechanics explorer, where again the system with actual motion is running. This representation is shown in the following figure numbered Fig. 5.

Figure 4 represents the assembly in the form if Simulink blocks, but once this Simulink model is made to run for a specific simulation time, the model exhibits the dynamics with the actual mechanical geometry itself as shown in the Fig. 5. Any kind of changes, modifications can be made on the Simulink model like, restricting the oscillation, constraining the force for a direction, the geometry modifications, along with the mentioned changes, the color of the model can also be modified for the better visualization. These are some Modelling aspects that are achieved while during the study.

2.3 Data Aspects

Data is everything in the present world, the data analysis or the usage in each technology, takes it to higher level. The above modelled and simulated Pendulum Assembly, with the actual geometry synced or imported to it, if it connects to the physical Pendulum Assembly in motion running in the real world with several sensors integrated on it which fetches data of many parameters, then both the real and simulated systems will be dependent on each other. This results to a condition where the actual physical system gets controlled by the simulated model of the system digitally on a computer and the other way around. Therefore, the visual representation of the actual dynamics is achieved virtually. This condition is defined as the Digital Twin.

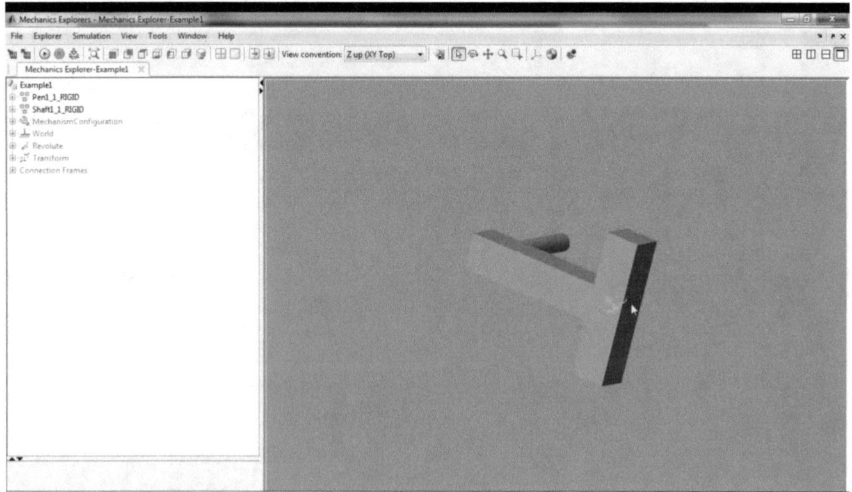

Fig. 5. The representation of the dynamics of the assembly in Matlab window.

Graphical User Interface (GUI) is one of the most advanced tools in the Matlab and Simulink software, which can be used for this situation where there is a need of sending inputs and getting the outputs as data between the real and simulated system. This would need Micro Controller boards or any Data Acquisition devices as an interface for the import, export of the data. GUI represents the front panel of a system, where it has this ability to call the Simulink model to run on the backend for a specific amount of time given by the user. The GUI also facilitates the user to pass the required input parameters necessary for the system to run. Basically, GUI acts as a Data Interface and bridges between the real and simulated world.

The proper import/export of the real time data among the Real and simulated world must be maintained properly to visualize the actual dynamics in a very accurate way, with high precision. The accurate data transfer ensures the simulated system to run as the actual system does.

As explained in the previous sections, the Matlab software's GUI tool and the combination of Matlab and Simulink is enough to take the Data Transfer part. So, the Solidworks and Matlab combination can derive the anticipated outcome, which is Digital Twin. Therefore, both Solidworks and Matlab functioning as a single system would benefit the Implementation of Digital Twin by helping visualize the Virtual Dynamics of any system in a very simpler way.

The same result can also be achieved by Creo (CAD tool) and Thingworx (An IoT platform) combination. They both are the products of PTC. PTC Creo is used as a substitute for Solidworks to get the mechanical design done. Similarly, Thingworx is a platform, with a beautiful architecture Interface. It is a compete, end to end technology platform designed for the Industrial Internet of Things. It is finely developed to deploy powerful Augmented Reality (AR) applications. Figure 6 is a snapshot used from a public source, it showcases the usage of both Creo and Thingworx to achieve the Digital Twin.

Fig. 6. The Creo-Thingworx platform, exhibiting Digital Twin.

Figure 6 provides a clear representation of the Digital Twin process. As shown, there is a system (Bicycle) which is in the real world, there are several sensors on it to fetch and transmit data of various parameters, this is fed to the virtual system running on the Creo platform, the platform exhibits the accurate motion, what the real system has through the data sensor gives. Thingworx has the advantage of displaying the sensor measured values too, therefore the motion and the measure of the motion can be clearly visualized in the Creo-Thingworx interface. The concept is clearly exhibited with the help of the Creo-Thingworx combination.

A looped process between the real and digital world would add more benefits to the Digital Twin Technology, that is, the change in the real world is recorded digitally and similarly if the real system is made to change its Dynamics through the digital input. This is about the Data aspects that must be considered.

3 Remote Engineering Incorporation

As we all know, Remote Engineering is a current trend in engineering and science, aiming to allow and organize a shared use of equipment and resources. Remote Engineering ensures that the technology or work subjected to it, is accessed by any number of users remotely. So here in our case, irrespective the place where the real system exists, if the virtual system is accessed remotely, this becomes very beneficial to the users in many users.

There are many methods to implement this Remote Engineering Tech to our system. One is sharing the access of the host PC with the other user and allow that user to work on the host's PC with the installation of any simple Remote Access Software like AnyDesk, Chrome Remote Access or any such. The other method is to send all the Data that is been transferred during the process to the cloud with a proper interface, headings. So, the user will be able to speak to the system by connecting to the cloud. The user now is capable to visualize the Virtual Dynamics remotely, rather than analyzing the real actual system itself. This would delete the Logistical or the Travelling issues which the user faces, this also saves usual system analysis duration. Figure 1 gives the clear indication of how the whole technology along with the Remote Engineering Technology works.

4 Other Benefits of Digital Twin

Along with the already explained benefits of the Digital Twin, there are many other things that would make the technology very important in the future. One of such most important benefits of the Digital twin is that, this would lead to the elimination of the Fabrication Process of the Mechanical System in the first stage. At present, what each automotive industry or any industry for that matter does is, they firstly design a component, a machine, they get the fabrication done without even evaluating the dynamics or the motion of the machine. However, if the machine is not built to meet the requirements, it needs to be scraped and the process must get repeated. But the Digital Twin Technology helps the person getting the Machine, to have a virtual representation of the Machine, even before getting it fabricated. By accurately designing the virtual machine, one can visualize the actual dynamics of the machine, through which it can be tested for whether it is meeting the requirements or not. Then the required design level modifications must be made till it reaches the requirement. And the Final Design gets fabricated. This saves a lot of capital, since there is no risk of getting the fabrication done repeatedly and scraping it several times. Along with the elimination of the repeated fabrication, the design level development of the component can also be achieved with the help of the view of the Dynamics. This is one of the most important functionalities of the Digital Twin technology.

5 Digital Twin and Cyber Physical Systems

The Cyber Physical Systems can be defined as the mechanism that is controlled by the computer-based programs. This needs a very deep intertwining of the Physical and software components. This is nothing but the Digital Twin improving its performance based on the algorithm coded to it. Cyber Physical System needs Digital Twin as its basis, as its first step. The development of the digital Twin leads to the Cyber Physical System. The Cyber Physical system can be referred to a physical system twinned with its Digital System, which is capable of monitoring and controlling itself based on the requirement and performance of the Machine coded digitally. A simple representation of the Cyber Physical System is shown in the figure named Fig. 7. The Cyber Physical System involves multidisciplinary approaches, it includes cybernetics, mechatronics, design and process science. They are typically designed as a network of interacting elements with physical input and output instead of as standalone device. Ongoing advances in science and engineering improve the link between computational and physical elements by means of intelligent mechanisms increasing the adaptability, autonomy, efficiency, functionality, reliability, safety, and usability of cyber-physical systems. This will broaden the potential of cyber-physical systems in several directions, including intervention, precision, operation in dangerous or inaccessible environments, coordination, efficiency and augmentation of human capabilities.

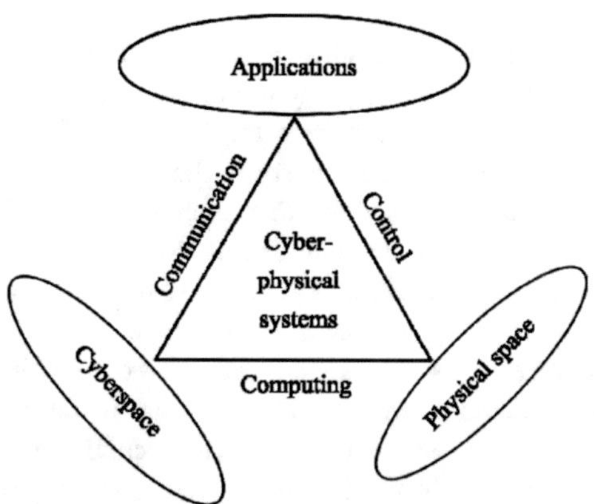

Fig. 7. Cyber physical system.

Cyber-physical models for future manufacturing- with the motivation of cyber-physical system, a "coupled-model" approach was developed. The coupled model is a digital twin of the real machine that operates in the cloud platform and simulates the health condition with an integrated knowledge form both Data driven analytical algorithms as well as other available physical knowledge. The coupled model first constructs a digital image from the early design stage. System information and physical knowledge are logged during product design, based on which a simulation model is built as a reference for future analysis. Initial parameters may be statistically generalized, and they can be tined using data from testing or the manufacturing process using parameter estimation. The simulation model can be considered as a mirrored image of the real machine, which is able to continuously record and track machine condition during the later utilization stage. Finally, with ubiquitous connectivity offered by cloud computing technology, the coupled model also provides better accessibility of machine condition for factory managers in cases where physical access to actual equipment or machine data is limited. These features pave the way toward implementing cyber manufacturing.

6 Outcomes

As explained previously, the keen outcome from this study would be the development of Digital Twin, to obtain the mirror image of the real physical system digitally. This would in turn help to achieve the Cyber physical system, which is going to be the future in the Manufacturing Industries. This would help the user in several ways. Since, Industry 4.0 is the current standard expected in all the manufacturing industries, the deliverables of the Digital Twin, Cyber Physical System will make the implementation of the required standard much easier. This results in a very higher standard process with very accurate results. Even though lot of research and development must be made in this field, to get solutions to very complex problems, this study can be served as the basis for all such developments.

7 Conclusions

The paper started with the explanation of Digital twin in depth, the paper also speaks about the huge need of bringing this technology to reality. A simple way which would help building this technology has been explained in a very structured way. The paper also gives explanation on a very important concept called cyber physical system, which can be considered as the byproduct of Digital Twin. The paper talks about achieving this cyber physical system also. The other benefits of the Digital Twin are explained under a separate column. The implementation of this technology needs the combined excellence of different streams like Mechanical Design, MBSE (Model Based System Engineering), Mathematical Modelling, Industrial IoT and such. With this full-fledged technology, the goal, Digital Twin can be reached. As mentioned earlier, the future of the Industrial services relies on Digital Twin, and henceforth this study would be the first step towards reaching the destiny. The reader would find two ways here, one with the combination of SOLIDWORKS, MATLAB and SIMULINK, and the other with the help of CREO and THINGWORX from PTC. More combinations of a Designing tool and a Modelling or an IoT platform can be used. The first combination out of the listed variants is highly proposed in the paper. A step by step procedure, right from the Installation of the add-ons, linking SOLIDWORKS to MATLAB is explained in detail. The Author believes, that the reader has basic knowledge of designing, mathematical modelling, and certain coding basics and henceforth this study would help build the reader a simple Digital Twin.

Acknowledgements. The Author wish to extend thanks to everyone, who served their support in building this study. He wishes to thank the software builders MATHWORKS, DASSAULT SYSTEMS and PTC, which helped the Author to make this study easier.

References

1. Numerical Methods for Engineers and Scientists using MATLAB by RAMIN S ESFANDIARI
2. Simulink—Developing S-FUNCTIONS by MATHWORKS
3. SIMSCAPE guides by MATHWORKS
4. Building GUI's with MATLAB by MATHWORKS
5. Advanced Engineering Mathematics -10th Edition by ERWIN KREYZIG
6. Modern Control Engineering—5th Edition by KATSUHIKO OGATA
7. Modern Based Systems Engineering: Fundamentals and Methods (Control, Systems and Industrial Engineering Series) by PATRICE MICOUIN
8. GORDON PARKER's webinars
9. TOM IRVINE's webinars
10. SOLIDWORKS guide by DASSAULT SYSTEMS

Machine Health Monitoring of Induction Motors

Panchaksharayya S. Hiremath[1], Kalyan Ram B.[1], Santoshgouda M. Patil[1,2(✉)], V. Sabarish[1], Preeti Biradar[1], and S. Arunkumar[1]

[1] Electrono Solutions Pvt LMT, Bengaluru, India
santosh@electronosolutions.com
[2] Cymbline Innovation Pvt LMT, Bengaluru, India

Abstract. The reason behind the induction motor health monitoring is Induction motor especially three phase induction motor plays vital role in the industry due to their advantages over other electrical motors which is lesser in cost. Therefore, there is a strong demand for their reliable and safe operation. If any fault and failures occur in the motor it can lead to excessive downtimes and generate great losses in terms of revenue and maintenance. Therefore, an early fault detection is needed for the protection of the motor. In the current scenario, the health monitoring of the induction motor are increasing due to its potential to reduce operating costs, enhance the reliability of operation and improve service to the customers. The health monitoring of induction motor is an emerging technology for online detection of incipient faults. The on-line health monitoring involves taking measurements on a machine while it is in operating conditions in order to detect faults with the aim of reducing both unexpected failure and maintenance costs. The best way to avoid machinery failures is to know they're coming. This is precisely what condition monitoring enables. Condition monitoring is the process of determining the condition of machinery while in operation. The three major steps in a condition monitoring system are data acquisition, data processing, and data assessment for maintenance decision-making and fault diagnostics and prediction. Successfully implementing a condition monitoring programme enables the repair of problem components prior to failure. This not only helps reduce the possibility of catastrophic failure, but also allows you to order parts in advance, schedule manpower, and plan other repairs during the downtime. In the present paper, a comprehensive survey of induction machine faults, diagnostic methods and future aspects in the health monitoring of induction motor has been discussed.

Keywords: Industrial 4.0 · Continuous monitoring · Prediction

1 Introduction

Some of them have the doubts regarding the "What's the reason behind Monitoring Health of the Machines", the reason to this is "TIME" and TIME is equals too "MONEY" so approximately half of all operating costs in most processing and manufacturing operations can be attributed to maintenance. This is ample motivation for studying any activity that can potentially lower these costs. Machine condition

M. E. Auer and K. Ram B. (Eds.): REV2019 2019, LNNS 80, pp. 51–56, 2020.
https://doi.org/10.1007/978-3-030-23162-0_5

monitoring and fault diagnostics is one of these activities. Machine condition monitoring and fault diagnostics can be defined as the field of technical activity. Once again segregating this machine into Servo motors and Induction motors in most of the machines the induction Motors are used due to its advantages compared to the other motors so monitoring of Induction motors is more important in present scenario, To monitor a Induction motors we need to know the physical parameters which can predict the conditions of the motors. the physical parameters from which can predict are Ambient temperature, Vibration, Mechanical, Electrical etc. from this parameters we can predict the health of the motor. This paper is regarding the electrical parameters through which the health of the machine can be predicted.

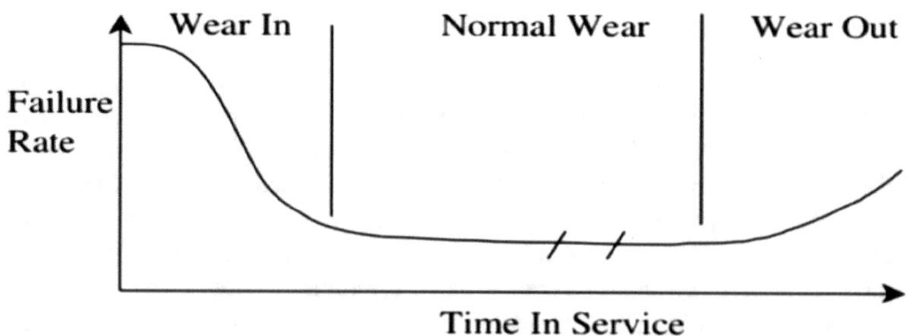

Typical bathtub curve.

This figure represents the frequency of failure of machine also called as the "Bathtub Curve" The beginning of a machine's useful life is usually characterized by a relatively high rate of failure. These failures are referred to as "wear-in" failures. They are typically due to such things as design errors, manufacturing defects, assembly mistakes, installation problems and commissioning errors. As the causes of these failures are found and corrected, the frequency of failure decreases. The machine then passes into a relatively long period of operation, during which the frequency of failures occurring is relatively low. The failures that do occur mainly happen on a random basis. This period of a machine's life is called the "normal wear" period and usually makes up most of the life of a machine. There should be a relatively low failure rate during the normal wear period when operating within design specifications. As a machine gradually reaches the end of its designed life, the frequency of failures again increases. These failures are called "wear out" failures. This gradually increasing failure rate at the expected end of a machine's useful life is primarily due to metal fatigue, wear mechanisms between moving parts, corrosion, and obsolescence. The slope of the wear out part of the bathtub curve is machine-dependent. The rate at which the frequency of failures increases is largely dependent on the design of the machine and its operational history. If the machine design is such that the operational life ends abruptly, the machine is under designed to meet the load expected, or the machine has endured a severe operational life (experienced numerous overloads), the slope of the curve in the wear out section will increase sharply with time. If the machinery is overdesigned or experiences a relatively light loading history, the slope of this part of the bathtub curve will increase only gradually with time.

1.1 Need and the Existing Machine Learning and Industrial 4.0

Previously if any catastrophic accidents occurs in the machine the maintenance staff couldn't able to find the problem inside the machine which use to take a lots of time due to which the production has to be stopped and by Machine health monitoring the maintenance staff of the machine can get to known problem before the machine stops from Energy data collected from machine every second depending on the energy consumed by machine for every single cycle and the continuous data 24/7 getting from the machine from this sitting at any part of world operator can check the machine health by comparing the energy consumed per cycle for the reference one cycle it is equal to manufacturing of one component.

2 Architecture

2.1 Block Diagram

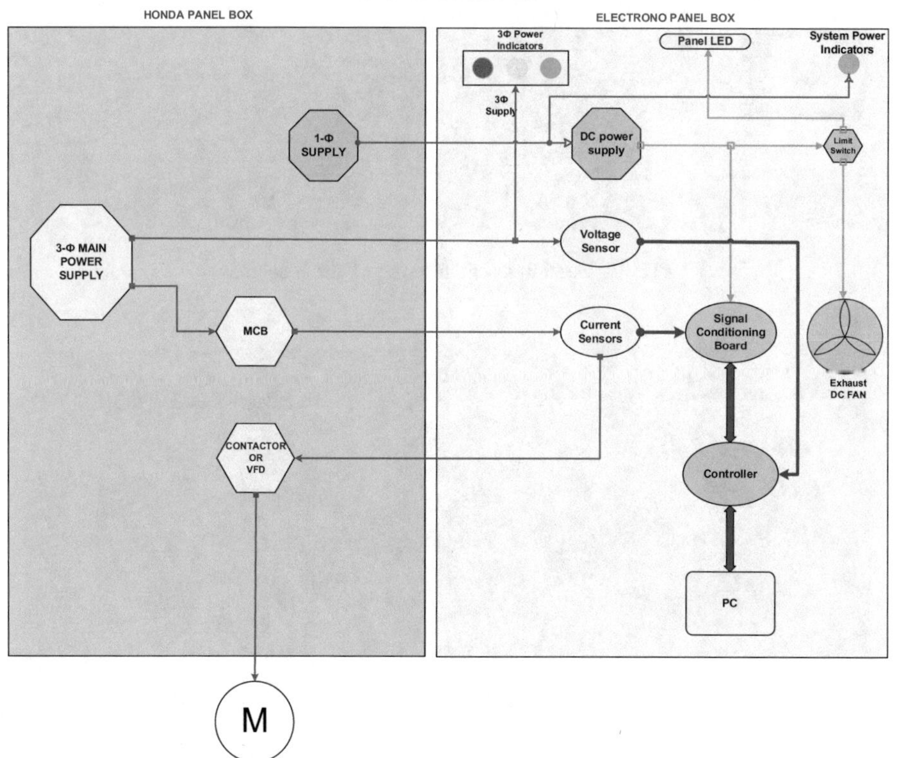

2.2 Features

- No of induction motors Monitored-13
- Energy Calculation
- Leakage current
- Total Harmonic Distortion
- Data logging and Report generation.

We are logging the 3 phase voltage and 3 phase current of each motor and computing the energy of each cycle, Each cycle represents the manufacturing of one components (Figs. 1, 2 and 3).

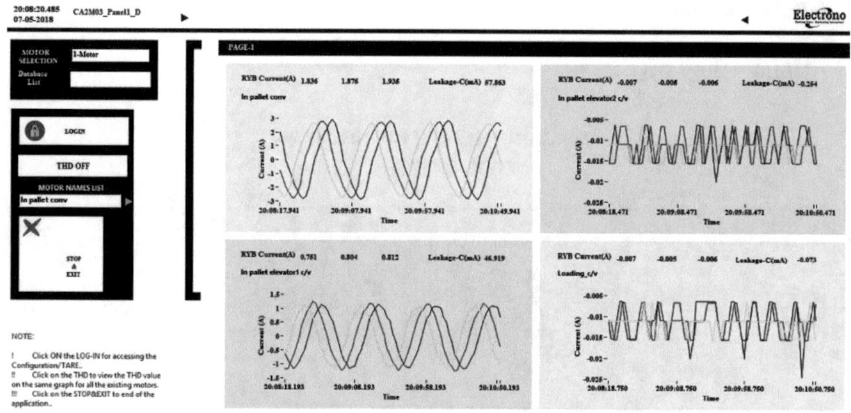

Fig. 1. Continuous monitoring of the machines

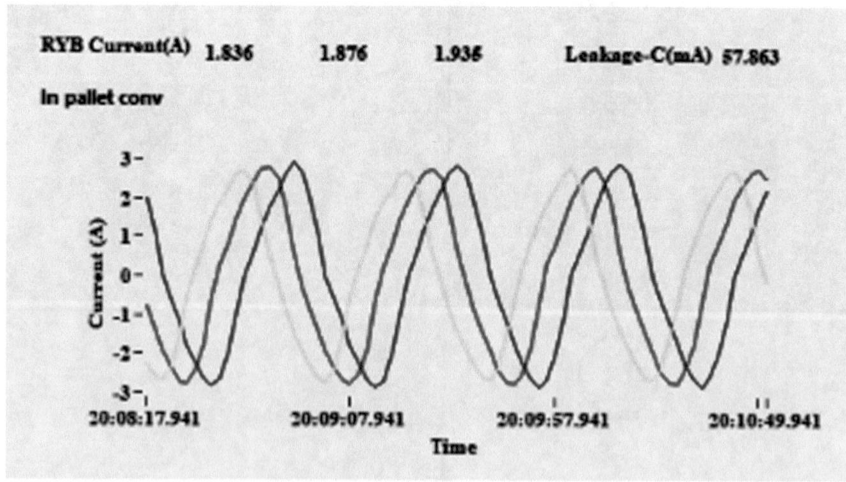

Fig. 2. Current drawn by the motor

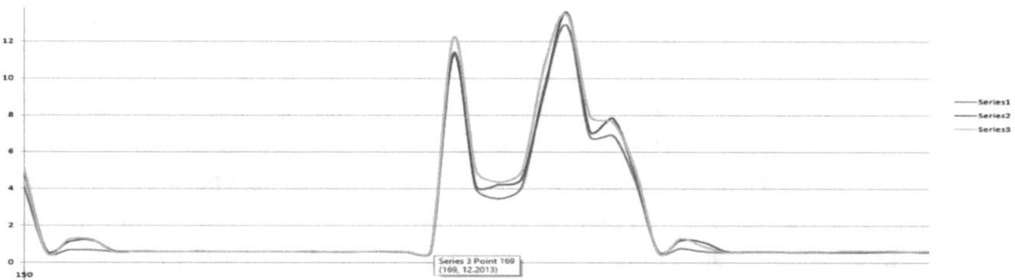

Fig. 3. To verify the data comparing with the timing diagram of the machine

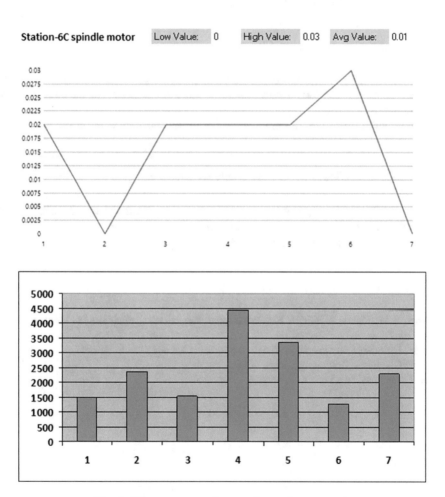

Fig. 4. Energy consumed per cycle per day per week

This represents one cycle so whenever the change in the energy consumption of motor for each cycles it doesn't vary continuously but day by day the energy consumption increases so we need to compare it with the previous data and after that we can predict whether motor is properly working properly or not (Fig. 4).

This is the observation of the spindle motor for a week the variations can be seen but whenever the variation goes above the threshold value we can predict the machine is not working properly and if any catastrophic accidents occurs we can check the graph and predict the actual point in the machines where accidents has occurred. The main thing is it can be accessible remotely by the maintenance staff sitting somewhere in corner of the world they can check the health of machine which is implemented in Honda Plant present in Narsapur Industrial area Karnataka.

3 Actual or Anticipated Outcomes

- To reduce the breakdown time when the machine occurs failure.
- Helps in continuous monitoring through continuous acquiring of the data from the machine.
- Intimation is given before the machine failure so that production doesn't stops.

4 Conclusions/Recommendations/Summary

Machine health monitoring is intent to predict machine failure through the energy data collected from machine every second depending on the energy consumed by machine for every single cycle from this faulty can be detected.

References

Siddiqui, K.M., Sahay, K., Giri, V.K.: Health monitoring and fault diagnosis in induction motor-a review

Machine Learning Documentation Initiative-Kenneth Chu and Claude Poirier

Machine Condition Monitoring and Fault Diagnostics

Chris, K.: Mechefske Queen's University

Remote Analysis of Induction Motor

A. Venkatesh Prashanth(✉), Sai Prithvi Raju, Nikhil Janardhana, and Santoshgouda M. Patil

B.M.S Institute of Technology and Management, Bangalore, KA 560064, India
{venkateshprashanth6, saiprithviraju}@gmail.com,
nikhil@cymbelinein.com,
santosh@electronosolutions.com

Abstract. Since the Industrial Age Started, A lot of Machinery was introduced and **Induction Motors** Were a Major part in the industry as it was reducing the work load of many workers and it complete the work of 10 workers in a short period of time. A Time came where-in it was necessary to look after the machine and lot of failures occurred which made it hard for the industry to function and hence we have come up with a solution for the problem and make the induction motor a independent and self-monitoring machine through remote engineering and IOT.

Keywords: Remote engineering · IOT · Smart machine

1 Introduction

Extensive use of Technology to solve problems has led to many such innovative and creative solutions that can not only reduce the expenses of the Customer but also Can Make Sure the Device can Function at its Very Best. This is Basically What Engineer Are Meant to do. One Such Technology that is Fast Rising and In the Limelight is Remote Engineering. This Gives Remote Access to the User who is Sitting anywhere in the world. He Can Access the Device from any corner in this world and get the data to his device at an instant. This technology is used here extensively to provide solutions for induction motors. In Recent Time, Many Industries are Facing problems from Induction.

Motors Failures and Breakdowns and Due to this a lot of Companies are Facing Losses. The Solution Provided by Our Technology will Give a Solution to the Industries and Solve their problems. They will be having Timely Updates on the machines and will be having the knowledge of the health of the machines. They will be having the 3-D Representation of their machines and will be having the Knowledge of their Machine Functioning from the 3-D View (MT-Linki Remote Tool). This Can Save The Industries a Lot of Time and They Can Plan their Maintenance based on the data provided. We Can Also Provide the Company Valuable Information regarding the event of failures of the machines Based on the Data Given by the Machines and The Company Can Rectify the Issues and Can Function Smoothly. This Solution Definitely Introduces the Concept of Smart Machines As it Making the Induction Motor a self Sufficient and Independent machine which is communicating to a server telling its

M. E. Auer and K. Ram B. (Eds.): REV2019 2019, LNNS 80, pp. 57–64, 2020.
https://doi.org/10.1007/978-3-030-23162-0_6

status and its health updates on a regular basis. Hence making the induction motor a Person like Machine. This Surely will Provide Some Shift in the way machine work and will change some workings in the industries and the way they function will be different. The Basic Architecture of the Solution Our Technology is providing is given in the Block Diagram below (Fig. 1).

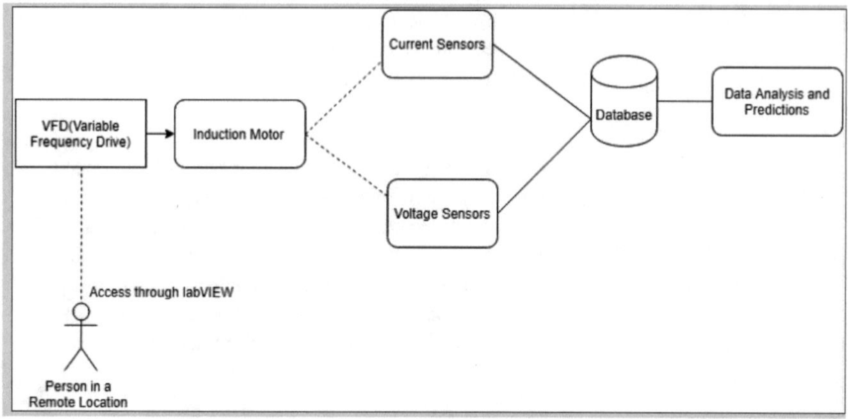

Fig. 1 Architecture of the induction motors

2 Approach

First of all, Let us Know the Comparison of the Induction Motors With and Without our Solution. To Get a Brief Understanding and a Total Comparison Using the Block Diagram (Fig. 2).

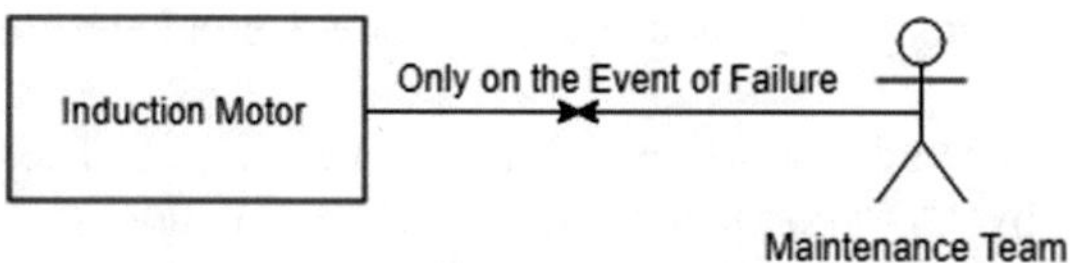

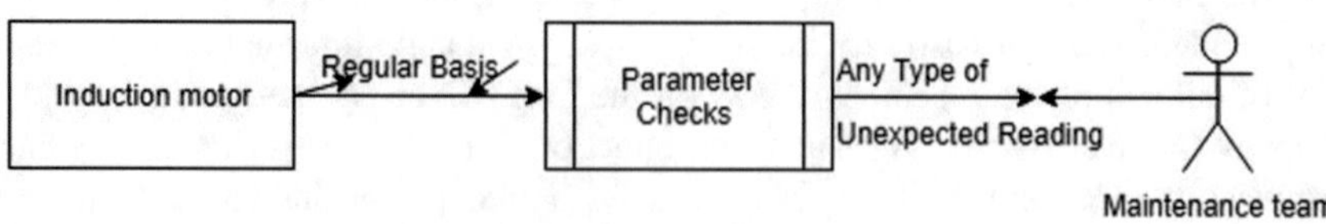

Fig. 2 Review of the machines

Now Without Our Approach. The Induction motor would be Performing at its normal Functionality. But Only During the Event of Breakdown or some Failure, The Maintenance Team Comes to the knowledge of the Part or Device that is Malfunctioning.

But with our Approach, We Can tap into the Device by Logging some Parameters Such as Current, Voltage etc. And Make Predictions Based on that on a daily Interval and Come to the Conclusion based on some Calculations the Event of Failure. This Not only can be done from the facility it is operating but it can done from any corner of the world as the Data parameters are logged onto a server which can be easily accessible and also we can create a system where-in we can send message and emails based on the predictions and we can Prevent the Failure before hand and also Save the Industry of some Hours of Crisis and Profits.

But With the Infrastructure in the Industry. The Team will have limited Access to the induction motors and will have to Check on a regular basis to get to know the faults and also then sometimes its not possible to prevent the possibility of failure.

In the Coming abstracts we will discuss more about the (1) The Hardware Used to Achieve this (2) The Hardware Setup (3) The Working.

2.1 The Hardware Used

First of all it was necessary for us to select the proper infrastructure so we needed the know the Induction motors to be controlled and By our Observations we came to the conclusion that spindle motors and Lubricant Motors were to be controlled as it was possible to control these Motors with a CNC and Hence we had the Chance to Implement our Solutions to these Motors.

Now Lets look at the Hardware used in the Following:

1. **Voltage Transformer**: Voltage Transformers Are Devices which Measure the Voltage of the Device Connected to it and also it used to step down the Voltage from A High Voltage to A Low Measurable Voltage. Most of the VTs Used are To Step down 230 V AC Down to 1 V or 3 V to Be able to Measure the Voltage Without any type of Damage or Danger. It is also Isolated from Each other(Primary windings to the Secondary Windings).It Also Indicates the Accurate phase Relationship.

2. **Current Transformers**: Current Transformers are used to Measure Alternating Current. It Produces A Equivalent Current in its Secondary Coil Proportional to its Primary Coil. This also is Insulated in Between The Two Coils Hence Preventing any Events of Shock.

3. **Variable Frequency Drive**: Variable Frequency Drive is a device used to Control the Speed of the Device By altering the Frequency in the events of Steps. It Also Controls the Voltage of the Device. Hence Increasing the Possibility of Remote Access to any motors you want to Control. This does not affect the torque, Magnetic Flux, Impedance and many other parameters (In Case of Induction motors).

4. **Multiplexers**: Multiplexers are The Devices that Are given Many Inputs but gives only one output depending on the select lines. Now Lets Consider 8 Machines, We Only have One Data Acquisition device and hence we need to Take all the Machine Data. Hence we Use a Multiplexer to Get the Readings of all the 8 Machines at an interval of 1 s.

This Provides us the Ability to Reduce the cost of the Data Acquisition Devices.

5. **DAQ 6001**: DAQ 6001 Is a Device From National Instruments that is Used For Data Acquisitions and Many More. This Device is used to Acquire Data from the Multiplexer And Display on a Computer for the Remote Access And Data Analytics and Many More.

2.2 Working

First Off All we Need to Understand the No Of Machines And the Number of Sensors to be Integrated in order to Tap into the Parameters of the Machines.

The Number of CTs and VTs to be integrated must be considered on the effect of cost and every aspect that could affect the Current Reading.

Now the Components were First Tested and then Integrated on to the sensor individually and then Were Integrated onto the Machines.

The First Step Was Integrating the Sensors onto the Machines. The Current Transformers were First Integrated. It Is A Loop Through type Current Sensors which is basically Built on the Principle of Hall Effect. And Then the Voltage Transformers were Integrated onto the Machines And then We are Integrating all the Three Phases (R, Y, B) onto Both the sensors.

Now the second Step was to Get the Data onto the Devices but the Problem which aroused was that the All the Machine Data could not be integrated onto the NUC parallelly as it would require that much amount of Data Acquisition Devices which would mean that the Cost would shoot up and would be a loss of money. So we Have Come up with a Multiplexer Board Which will give us the Take in all the Data of the Machines And will give the Data on to Data Acquisition Devices in time Intervals of 1 s. So we Can Use only one DAQ to Bring in the Data of the Device. So It was a Vital Step in this Matter.

The Next Stop was To Integrate the DAQ to Receive the Data With a Calibrated Value. Since the DAQ is from National Instruments. Their Software Called LabVIEW is Used to Get the Data into the Computer And Get the Value to the User in a Remote Accessible Way.

Now the Most Crucial stuff is Giving Access to Any User in the World. This Can be done By integrating VFDs and Many more Devices with the NUC and to the Machine. By Building a Suitable VI to the user And A Good Front Panel for the User to Interface. We Can Give Remote Access to any user in this world Using these Technologies.

Now When we Have Got the Data onto the Device. There is a Software Called MT-Linki which will gives the 3-D Representation of the Machines that will be Running on Field and will Indicate to the User the Live Data that will be coming from the machines and based on the Data it will be programmed in a way to Indicate the Normal Working, Emergency, Stopped, Failures. And will Give A clear View to the User and The Machine can be Replaced and also can be prevent before hand.

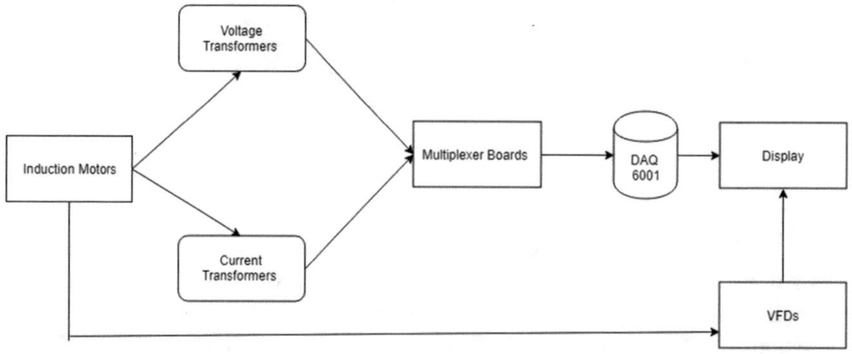

Architecture of working of induction motors

2.3 Outcome

When We Set Up this Tech in a Industry. It Started Tapping into the Device and got the required parameters to predict the analysis of the Data. Now When we Started working on the data produced we started recognizing the Parameters required to Tap in order to determine the **health of the Machine. We Also Could Find Out the Time when the machine was switched on** and switched off and at what time was the Machine started and stopped and the reason for the Unexpected delay and also some parameters can be derived on this Depending on the Requirement.

Since we Installed this Tech in a Certain Industry. The Obvious Outcome was that the Machines were talking to the Humans About their health and were telling all the States of their health and were made Independent of Human Help and Were Made Accessible to all the people that would like to know the Info about the Motors and The Machines were Now Smart enough to indicate to us the Health and Many more stuff and hence Making them a grid of Smart Machines.

2.4 Digital Twin by General Electric: A Vision for the Future

Digital Twin Created and Patented by the General Electric Put Forth a Foundation for the Future of Industry 4.0 And made it a Benchmark In the industry For IOT Machines and Smart Machines.

It Basically Creates a **Mathematical Model** and A **Real time Model.** Based on the Mathematical Models Ideal Readings and Conditions. A Central System Derives the Conditions for the Real time System to Perform in. Under the Same Conditions, A Central System Makes Predictions Based on the Readings from the Mathematical Model and Hence has the Ideal conditions for the Real time System.

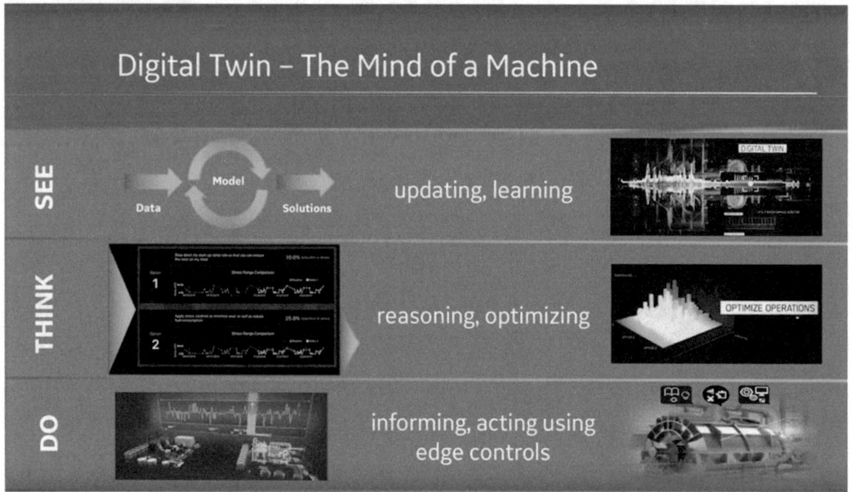

2.5 Conclusion

Remote Analysis of Induction Motors Has Brought The Industry the Concept of Industry 4.0 and Also Has Made it Easy for the Team to Monitor and Manage the Induction Motors and Provide more Features in this in order to Enhance the Possibility of New Technology which could Gives us more and more Solutions which is apt and more suitable for the Situation to be there. Moreover We Can Access More and More Of the Induction Motors Parameters to enhance the Working and More of the Health Monitoring possibilities.

References

1. Bonnet, A.H., Souku, G.C., Pilloni, A. et al.: Causes and analysis of stator and rotor of squirrel cage induction motors In: Fault detection in induction motors, AC electric motors control: Advanced Design Techniques and Applications, pp. 275–309 (2013)
2. Seera, M., et al.: Fault detection and diagnosis of induction motors using motor current signature analysis and a hybrid FMM-CART model. IEEE Trans. Neural Netw. Learn. Syst. **23**(1), 97–108 (2012)

Optimized Additive Manufacturing Technology Using Digital Twins and Cyber Physical Systems

Sreekanth Vasudev Nagar, Arjun C. Chandrashekar, and Manish Suvarna(✉)

Department of Mechanical Engineering, B.M.S. College of Engineering, Bull Temple Road, Bengaluru 560019, India
snv.mech@bmsce.ac.in, manishsuvarna2608@gmail.com

Abstract. The latest industrial revolution, Industry 4.0, has called upon for the needs of integrating the processes involved in the production systems with the advanced information technologies powered by artificial intelligence and data driven analytical solutions. In the past decade, we have witnessed how the simulation models developed by analyzing the Big Data generated by the manufacturing units have aided in boosting the productivity of the industry and give rise to the concept of smart manufacturing; especially in the manufacturing sectors lead by additive manufacturing technology. A 3D printer is a classic example of an additive manufacturing machine and hence has been considered as the framework of study in this research. Deployment & development of digital twin technology will engage the manufacturing systems in a heuristic cyber domain that will help the manufacturing industries to achieve larger productivity with reduced downtime. The proposed digital twin model of the 3D printer shows a framework for developing a machine learning module to reduce and replace the standard defects and reducing the data transmission and data overload in wireless networks.

Keywords: Cyber physical systems · Additive manufacturing · Digital twins · Industry 4.0

1 Introduction

Additive manufacturing is the next generation manufacturing domain which has shown promise in the field of manufacturing and development of new products at a rapid rate. There has been a plethora of opportunities in this field due to its nascent stage of development. On the other side, unfortunately, the traditional quality control in AM is largely limited to offline processing. Existing mitigation strategies, such as weighting test and redundancy, were intended to detect accidental manufacturing defects based on statistical analysis for large volume production, not suitable for those deliberately hidden flaws without any geometry/weight change and small-batch AM fabrications. The development of the additive manufacturing technology has been greatly attributed to the nature of the technology which demands digital data for manufacturing and product realization. The need for digital data has ensured the skills of the designers to

M. E. Auer and K. Ram B. (Eds.): REV2019 2019, LNNS 80, pp. 65–73, 2020.
https://doi.org/10.1007/978-3-030-23162-0_7

be improved from drafting on drawing boards to digital mockups of their designs after needed engineering analysis. The technology itself required minimum human intervention and this has led to expectations in the industry for automation in all the stages of the manufacturing environment [1].

Typically the additive manufacturing process occurs in six stages as follows:

Stage 1: Design of model by Computer Aided Modeling Software (CAD software)
Stage 2: Saving the CAD file in.STL format
Stage 3: Slicing of the.stl file into layers using an image processing software
Stage 4: Converting the sliced software into G-Code -input format for machines
Stage 5: Transferring the data to the machine
Stage 6: Setting the machine for printing
Stage 7: Printing of the model in the machine
Stage 8: Removal of the finished part and post processing.

During these operations most of them is computer controlled. However, the stage 7 of the manufacturing process is based on the hardware availability and the environment that is provided for the printing process. This stage of the manufacturing is concerned with the maintenance of the machine and the reliability of the machine for achieving a efficient output of the finished part. The maintenance of the machine is a very important aspect of the surface finish obtained in any machine [2].

The technology being new has opened up a huge market for investors, hobbyists and industries to explore various business opportunities by utilization of this technology. This in turn has made it very challenging to machine builders and distributors to develop reliable machines due to the lack of testing period of the machines. The machines are designed and manufactured at a very fast rate and distributed to customers release after release rather than working on the development of an organized service plan or maintenance schedule of the machines. Though industrial machines are equipped with operation manuals and time tested reliable spare parts and service manuals the low cost 3D printer market is slowly losing shine due to the complex operational procedures and varied technologies involved in printing (Fig. 1).

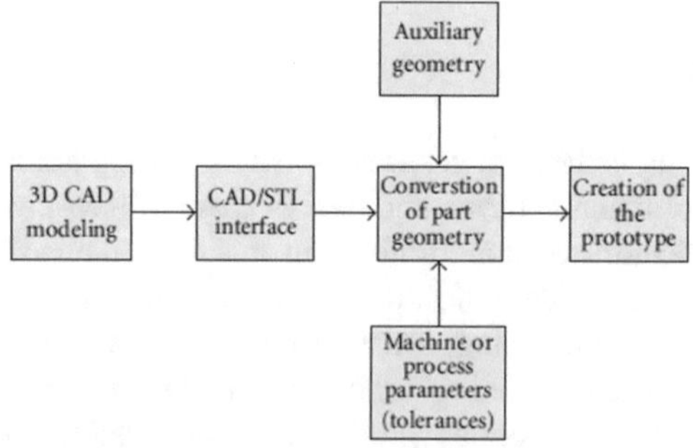

Fig. 1. Data flow in a 3D printer

1.1 Technologies in AM

To provide a glimpse of this the various available techniques in this industry are:

a. Extrusion based techniques
b. Digital light processing techniques
c. Powder bed Fusion technique
d. Binder jetting techniques
e. Material jetting techniques
f. Sheet lamination technique
g. Aerosol technique
h. Direct material deposition technique.

Extrusion based techniques involves the extrusion of materials in the form of filaments through a nozzle heated to a temperature which allows the materials to change its viscosity and become visco-elastic in nature. This material is then forced into a nozzle which is guided by actuators in three dimensions to develop the prototype layer by layer [1, 3]. These machines have been popularized by a revolution called Reprap revolution which involves manufacturing a replicator using a replicator which utilizes bare minimum accessories for development of 3D printers. The open source revolution has helped many entrepreneurs, designers and machine manufacturers to work on the development of these machines to serve designers as desktop machines. Unlike the large manufacturing enterprises that often use expensive industrial powder-based 3D printers (e.g., SLS), more affordable consumer-grade FDM (Fused Deposition Modeling) 3D printers are still the most widespread desktop technology and thus are favored by small businesses and end users. However, the maintenance of the machines have been neglected leading to low cost machines having very less reliability and product life leading to huge amount of non-operational 3D Printers in the market.

The concept of digital twin in the industry helps in development of data model from physical systems for achieving strategic insights, veritable simulations, closed loop product design, evaluate products and develop innovation. The digital twin concept becomes very complex as the number of variables increase [3]. These variables can be modeled using multi objective optimization to bring in a relationship between the digital and physical systems [4]. This concept is very well applied in the robotics for development of maximum interrelation between the systems [5].

Artificial Intelligence is related to providing smartness in systems for better decision making. Machine learning is considered to be one of the techniques that is applied in order to develop the intelligence in the systems using various methodologies and models that have proven to become more and more efficient in the recent years. This has opened up the application of machine learning into many fields of engineering and allied industries.

The integration of advanced analytics with communication systems in very close interactions of physical systems are called as Cyber Physical Systems [2, 5]. The various physical systems can be integrated with the computational models using cyber physical systems due to the advances in ICT and production systems leading to Industry 4.0. Driven by the Industry 4.0 vision and the development of big data analytics faster algorithms, increased computation power, and amount of available data enable the simulation with ability of real-time control and optimization of products and production lines, which is referred to as a Digital Twin, using a digital copy of the physical system to perform real-time optimization. The breakthrough is achieved from two aspects: (1) Calculation time has gone from hours to minutes, which has made it possible to explore the solution space searching for the global optimum; (2) Increased use of sensors and on-line measuring equipment are making it possible to reuse simulation models during both pre-production and production phases, now with real context data rather than estimated or historical data as input. This will allow for adjustment of machine settings for the work-in-process (WIP) in line based on simulations in the virtual world before the physical changeover, reducing machine setup times and increasing quality. The digital twin's ability to link enormous amounts of data to fast simulation by creating a closed control loops also makes it possible to perform real-time optimization of products and production processes. Particularly in design, realistic product and production process models are essential to allow the early and efficient assessment of the consequences, performance, quality of the design decisions on products and production line. The supervisory controls of the CPS systems have been explored in a human and robot virtual simulations on Java as an Eclipse RCP plug-in [6]. The advantages of cyber physical systems are many and the architecture to develop cyber physical systems essentially follows 5C structure Smart connection, data to info conversion, Cyber, Cognition and configuration tools. Most of the physical systems which can be designed for maintenance and support can follow a similar architecture. In the production environment there the concept of CPS has been defined by Lee as "Cyber-physical systems (CPS) are integrations of computation and physical processes. Embedded computers and networks monitor and control the physical processes, usually with feedback loops where physical processes affect computations and vice versa" [7]. With this vision of implementing digital twin in additive manufacturing machines like 3D printers, a feedbacking mechanism that communicates between the physical and cyber domain is proposed, in order to facilitate better maintenance and improved Product Lifecycle Management (PLM) of the 3D printer.

2 Digital Twin for Maintenance Activities

Additive manufacturing or 3D printers have gained a lot of traction in the recent years [8]. This necessitates the manufacturers to train the industry personnel to install and commission the machines at the site. The larger industrial grade machines will have large profit margins which will allow the manufacturers to absorb the installation and commissioning costs. However, the smaller companies which cater to the maximum number of customers in households, universities and labs and design schools face a challenging situation. A model is proposed to have an emulator to efficiently train the customer and help in virtual commissioning [8]. This poses a challenge on the customer side as it requires a minimum amount of knowhow of the technology to make the best use of the emulators. This will work in the reconditioning of the machines as discussed in the previous researches carried out in the field of maintenance of a core making machine using HMI interface and PLC. The additional cost of maintenance will also affect the profit margins of most of the low cost 3D printers that are making inroads to most of the households and desktops in the offices of many design houses. The effect of after sales maintenance for many companies has led to the increase in cost of the machines or making the companies run out of business for the inability to have service personnel cater to many customers. There is a gaping hole in this area which needs to be addressed for the entrepreneurs who would like to explore the opportunities of machine building in the field of Additive manufacturing.

3 Methodology

The application study presented in this research mainly deals with the use of a framework to place an array of sensors [8] in a low cost additive manufacturing 3D printer machine available locally in the Indian market and two printers with the similar technologies imported from the US with a lab made 3D printer using RepRap Marlin firmware. The Indian based machines were installed five years ago and the machine required a complete maintenance due to excessive wear and tear of the machine in the five years of service life. The other machines were procured more recently but all of the machines work on extrusion based technology. One machine was built in the lab using lab based components and few stepper and an Arduino ATMEGA kit. This wear and reliability characteristics was compared with the surface characteristics of the machines to have a complete physical modeled data. These parameters are studied to plan for the development of the digital twin as an emulator for maintenance related system modeling. The motivation behind this work comes from the improved Product Lifecycle Management technology (PLM) which has allowed the industries to work at low cost process, high product reliability, improved quality and faster product development. This technology when combined with additive manufacturing process, further catalyzed the benefits of additive manufacturing technology itself. However, for a successful deployment of PLM in a manufacturing unit, a highly skilled human intervention is a necessity. By introducing Digital Twins technology, this will be an attempt to make a 3D printer become self aware and manage its condition on its own and understand its KPIs in order to increase its reliability and productivity with the least degree of human intervention.

4 Case Study

The low cost 3D printers manufacturing and maintenance has been a challenging area which is recognized as an additional cost for the manufacturers. To avoid these additional costs, the new model as proposed in the Fig. 2 can be utilized to develop a seamless digital twin model including the various components as show in the Fig. 2. In this model the thrust is on the movement of data related to major component levels of the 3D Printer and to make the data lighter using machine learning approaches and developing the model for CPS. For collection of data from the physical systems an array of sensors will be first placed at the important points of the 3D printer as shown in Fig. 3 [7]. The following sensors are planned in this model to be placed on the machine.

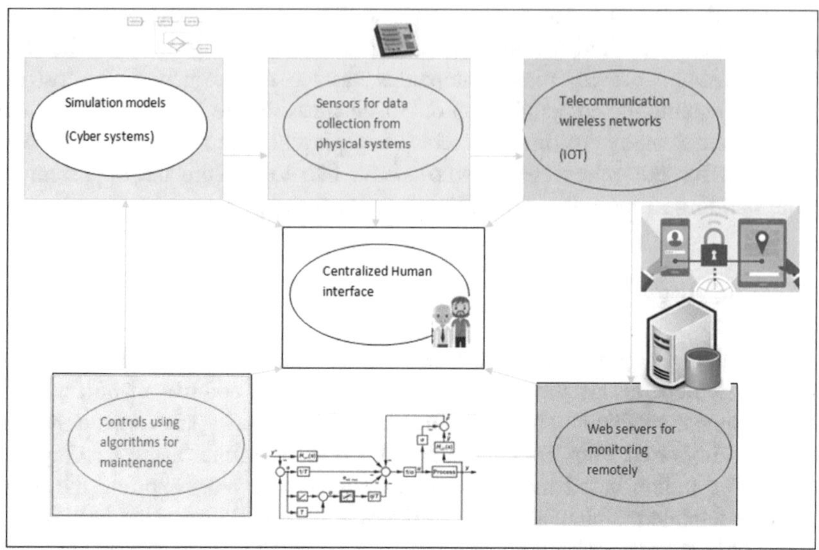

Fig. 2. Proposed model for maintenance using cyber physical systems

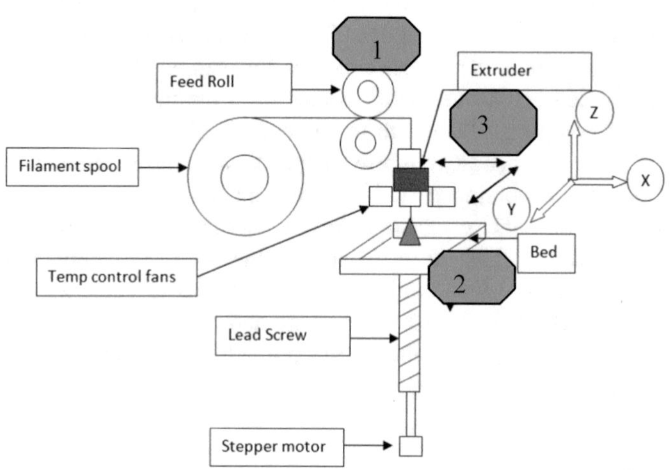

Fig. 3. Schematic of a low cost extrusion based 3D printer with sensor placement plan

1. A strain gauging sensor is planned at the entrance of the filament which comprises of the rollers that are used to feed the printer. Here, the pressure of the rollers has to be maintained and the variation of the pressure will lead to the reduction in the print quality or missing of print layers in the final product.
2. The second position of the sensor is planned at the belt used for driving mechanisms. The tensions between the belts are another important parameter which affects the print quality for which an indigenously developed sensor will be mounted on the printer.

3. The third sensor is planned to be mounted on the build plate or the bed of the printer. The bed level at each stage will lead to variation of the print output for which it is planned to incorporate an accelerometer on the bed to show the variation of the bed level at each layer steps.
4. Fourth sensor that is planned for acquiring the overall vibration of the machine, three accelerometers on the machine is placed at vital points of the printer.

As the data is collected from the sensor we propose a memory unit which stores the data and transmitted through a Zigbee XBee Module S2C 802.15.4 2mW with Wire Antenna XB24CZ7WIT- module to transmit the data into wireless network which will then move on to the web based servers. At this level it is proposed to get saved in the form of xml formats on the web. This allows for a comparison of the encapsulated data with a control algorithm developed for variations of the outputs measured through the sensors. Any change in the input will be sent as an input parameter to the model developed for the maintenance of the system. The instructions from the maintenance models developed using a digital twin will then be passed across as the input to the machine for minor adjustments. If the model behavior monitored is beyond the control limit then a human interface will be notified through a virtual Andon. This will open a ticket to the service agent and will book an appointment with the customer for repair. During the repair the service agent feeds in the service manual the changes and the same will be used as an input for the machine and stored into a common knowledge base. This knowledge base will be updated as variable in the model that is developed and will lead to online repair and maintenance in the next cycle of breakdown. To reduce the overloading of the data a machine learning module is planned to be introduced at the machine level which will be able to filter the repetitive data reduce the load on the wireless module during data transmission. Once the maintenance schedule is completed the KPIs are measured and further the model will be tuned for next cycle. For validation of the KPI and building a model 3D Experience software will be used which has the ability to conduct close to real world simulations (Table 1).

Table 1. Sensors list for selection

Model	Type of sensor	Size in cm	Sensor
GY-29	Accelerometer	2.4 × 1.90 × 0.30	ADXL345
GY-61	Accelerometer	1.57 × 2.03 × 0.12	ADXL335
GY-50	Gyro	2.30 × 2.30 × 0.33	L3G4200D
GY-85	Gyro+Acc+Mag.	2.12 × 1.68 × 0.30	ITG3200, ADXL345, HMC5883L

Specifications of Zigbee XBee Module which is planned to be used:

- 3.3 V @ 40 mA; @@@250kbps Max data rate; 2 mW output (3dBm).
- 400ft (120 m) range; Built-in antenna; Fully FCC certified.
- 6 10-bit ADC input pins; 8 digital IO pins.
- 128-bit encryption; AT or API command set.

5 KPI Identification

Some of the KPI that will be in focus in this study for the inputs to be compared with the outputs will be the surface finish of the finished part, the hardness and the bending strength in the parts that are used as structural parts. For the initial study and validation of the model the above parameters will be considered. The mounting of the sensors once validated on a single printer will be replicated to other printers in the lab to make a sample production environment for validating and generating more data as proposed in the model for further validation and improvement in the quality of the data.

6 Conclusion

In this technical report, we have presented how a Digital Twin of a manufacturing systems (with fused deposition modeling based additive manufacturing as a test case) can be modeled by taking various working parameters of a 3D printers as the input for the Digital Twin fed by the respective sensors. We then presented how dynamic data-driven application systems concepts can be used to re-rank, and re-train the Digital Twin model. This model update is based on measurement of KPIs varying from the set control limit, which measure the difference in predicted and real key performance indicators (for example nozzle temperature and tension in the belt). We showed how monitoring. This methodology is scalable to create Digital Twin for multiple key performance indicator prediction, and towards other manufacturing systems as well. To the best of our knowledge, this is the first work that demonstrates how dynamic data-driven application systems enabled feature re-ranking method can help in keeping the Digital Twin up-to-date. The use of CPS and Big data that is proposed in the given model on low cost 3D printers will allow startups and small concerns in the service and manufacturing industries to enhance their productivity by completely relying on data that is generated through our system. The data will help the industries to develop new models and scale it into a newly evolved maintenance through digital twin. Using the data and AI methodologies like machine learning will help the in achieving Industry 4.0 into a reality in low cost environments. This model can be utilized to develop algorithms for effective control in developing systematic maintenance with very less Human intervention and reducing the cost of machines further for manufacturers and in turn prices of machines for customers.

Acknowledgements. The authors would like to acknowledge BMS College of Engineering 3D printing Lab and Product Innovation lab funded by Dassault Systemes and 3D PLM Bangalore for providing the infrastructure and software for conducting the research.

References

1. Gibson, I., Rosen, D., Strucker, B.: Additive Manufacturing Technologies 3D Printing, rapid Prototyping and Direct Digital Manufacturing 2nd Edition (2015)
2. Qi, Q., Tao, F., Zuo, Y., Zhao, D.: Digital twin service towards smart manufacturing. Procedia CIRP **72**, 237–242 (2018)
3. Lee, J., Bagheri, B., Kao, H.A.: A cyber-physical systems architecture for industry 4.0-based manufacturing systems. Manuf. Lett. **3**, 18–23 (2015)
4. Lee, J., Ardakani, H.D., Yang, S., Bagheri, B.: Industrial big data analytics and cyber-physical systems for future maintenance & service innovation. Procedia CIRP **38**, 3–7 (2015)
5. Tao, F., Zhang, M.: Digital twin shop-floor: a new shop-floor paradigm towards smart manufacturing. IEEE Access **5**, 20418–20427 (2017)
6. Baumann, F., Roller, D.: Additive manufacturing, cloud-based 3D printing and associated services—overview. J. Manuf. Mater. Process (2017)
7. Baumann, F., Schön, M., Eichhoff, J., Roller, D.: Concept Development of a Sensor Array for 3D Printer. 3rd International Conference on Ramp-up Management (ICRM). Procedia CIRP **51**, 24–31 (2016)
8. Ayani, M., Ganeback, M., Ng, A.H.: Digital twin: applying emulation for machine reconditioning. Procedia CIRP **72**, 243–248 (2018)
9. Zhang, H., Liu, Q., Chen, X., Zhang, D., Leng, J.: A digital twin-based approach for designing and multi-objective optimization of hollow glass production line. Special section on key technologies for smart factory of industry 4.0. **5**, 26901–26911 (2017)
10. Krolczyk, G., Raos, P., Legutko, S.: Experimental analysis of surface roughness and surface texture of machined and fused deposition modelled parts. Tehnički vjesnik **21**(1), 217–221 (2014)
11. Nikolakisa, N., Sipsasa, K., Makris, S.: A cyber-physical context-aware system for coordinating human-robot collaboration. Procedia CIRP **72**, 27–33 (2018)
12. Lee, E.A.: Cyber physical systems: design challenges. In: Technical Report No. UCB/EECS-2008-8, 2008, Electrical Engineering and Computer Sciences, University of California at Berkeley
13. Cho, S., May, G., Tourkogiorgis, I., Perez, R., Lazaro, O., de la Maza, B., Kiritsis, D.: A Hybrid Machine Learning Approach for Predictive Maintenance in Smart Factories of the Future. APMS 2018, IFIP AICT 536, pp. 311–317 (2018)
14. Susto, G.A., Wan, J., Pampuri, S., Zanon, M., Johnston, A.B., O'Hara, P.G., McLoone, S.: An adaptive machine learning decision system for flexible predictive maintenance. In: 2014 IEEE International Conference on Automation Science and Engineering (CASE)
15. Wang, L, Wang, G.: Big data in cyber-physical systems, digital manufacturing and industry 4.0. I.J. Eng. Manuf. **4**, 1–8 (2016)

Process Parameter Monitoring and Control Using Digital Twin

Nihal Desai(✉), S. K. Ananya, Lalit Bajaj, Anupriya Periwal, and Santosh R. Desai

Department of EIE, BMSCE, Bengaluru, India
{nihaldesai29, ananya.karandikar, lalitbajaj88, anupriyaperiwal}@gmail.com

Abstract. With the growing deployments of Internet of Things (IoT) systems, the importance of the concept of a digital avatar of a physical thing has gathered significant interest in the recent years. Digital twin means a virtual copy of a system in operation to measure, monitor, and analyze the operational performance through continuous collection of real-time data. Evolution of sensor technology, investment in infrastructure to capture digital data of physical product, and innovation in analytical software platforms over the years is helping adoption of digital twin technology in the industry. Here we are using the open modelica software for simulation of the model. In this paper, a non-linear model-based approach is developed for controlling pressure of an actuator chamber. Through sliding mode control approach, the controller utilizes a on/off solenoid valve to implement pressure control task.

Keywords: OpenModelica · Digital twin · Process control · Automation · Pressure control

1 Introduction

The purpose of our study is to develop small sized pressure control system. In this study, we proposed and tested a small sized pressure control system using on/off valves. We investigated the optimal operation of on/off valves in the control system to maintain a constant pressure in the cylinder.

With the help of microcontroller, the pressure inside the cylinder can be maintained at the desired setpoint. As the pressure inside the cylinder varies above the setpoint, the solenoid valve can be actuated to control the pressure. The above mentioned system is also modelled on the Open Modelica software.

2 Procedure

2.1 Review Stage

The concept behind our project was taken under consideration when we encounter with such a problem in the small scale and big scale industries. And even the voidness of

M. E. Auer and K. Ram B. (Eds.): REV2019 2019, LNNS 80, pp. 74–80, 2020.
https://doi.org/10.1007/978-3-030-23162-0_8

many existing solutions for the problem. Pumps play a major part in many industries and monitoring and maintenance was an issue which was never considered and thereby affecting the economy of that particular industry or company. Cost effectiveness of the module to be created also played a major role in the consideration of the project idea. As it is being implemented with simple components and of minimal cost. At first, the digital twin model of the system is designed using the open modelica software. With various tools available in the software, we can develop a digital twin of the proposed system. After the simulation of the model, we construct the hardware.

In this stage the cylinder was operated manually to work as an on/off controller. When the pressure inside the cylinder increased beyond the desired setpoint, we manually operated the solenoid valve to release the excess pressure.

2.2 Final Stage

In this stage, the manual operation was made automatic with the help of a microcontroller. We used arduino uno as the microcontroller. The input given to the board was the input coming from the pressure sensor. Since the output of the pressure sensor was in terms of ampere, we used a I–V converter to convert the current to voltage. This voltage was supplied as input to the arduino. The output is the automatic control of the solenoid valve. The microcontroller is programmed such that it actuates the solenoid valve when the pressure inside the cylinder crosses a desired setpoint.

Figures

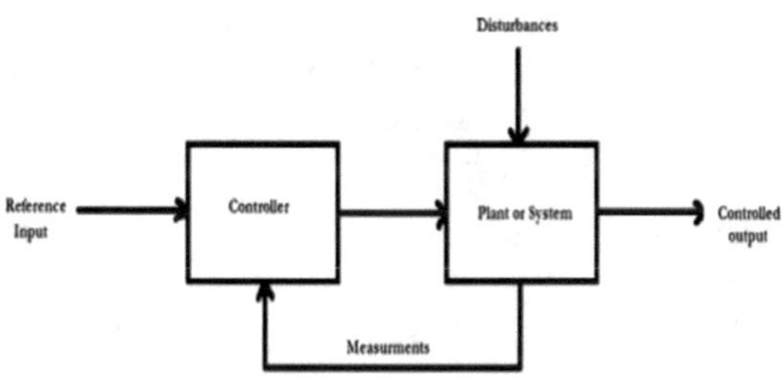

Basic feedback system

Cylinder: A 5 kg lpg gas cylinder is used in the physical model. It has been modified to accordingly to fit the pressure sensor, solenoid valve and non return valve.

Pressure sensor: A high-sensitivity differential pressure sensor is used here, with an analogue current output. The sensor is ideal for checking the performance of dust and air conditioning filters, and duct sensing applications. The pressure measurement inside the cylinder is done with the help of a pressure gauge. The pressure is measured in terms of psi.

Inlet: The input to the cylinder is coming from the manually operated air pump. The air is pumped into the cylinder, in order to prevent the loss of air, a non return valve is fitted at the inlet.

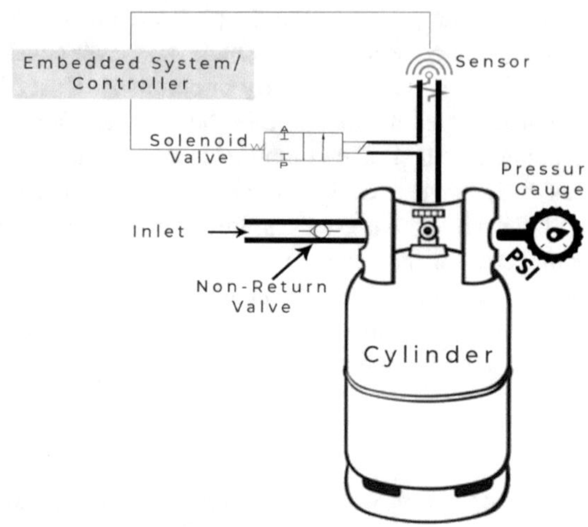

Non-Return Valve: A non-return valve allows a medium to flow in only one direction. A non-return valve is fitted to ensure that a medium flows through a pipe in the right direction, where pressure conditions may otherwise cause reversed flow.

Solenoid Valve: A one way, two position solenoid valve is used here. It is ideal for use with air, inert gases and potable water. They help in controlling the back flow and act as shut offs and release valves in many applications.

Arduino Uno: Arduino Uno is a microcontroller base on ATmega328P. It is being used here to be able to automate the whole process.

Specifications of the actual model:

Radius of the cylinder: 11.95 cm
Diameter: 23.88 cm
Circumference: 75 cm
Height: 22 cm

SI No.	Pressure (in cm^3)	Current (mA)
1	0.2	11.50
2	0.4	13.38
3	0.6	13.92
4	0.8	14.74
5	1.0	15.67
6	1.2	16.53
7	1.4	17.26
8	1.6	18.24
9	1.8	18.97
10	2.0	19.76

(*continued*)

(*continued*)

SI No.	Pressure (in cm^3)	Current (mA)
11	2.2	20.55
12	2.4	21.62
13	2.6	23.41
14	2.8	24.28
15	3.0	25.04

The above table shows the current variation with respect to the pressure change inside the cylinder.

3 I to V Converter

Here we have proposed the usage of current to voltage converter as the further processing and the automation of the data becomes easier as we have planned to use a microcontroller in the further stages. So a voltage input for the Arduino being used will be taken and the comparisons and everything will be done according to the required values by the user and the set point we assign.

An I to V converter or also called as a Transresistance Amplifier.

Here in the adjoining table where the Pressure versus Voltage graph has been plotted, a minimal value of Resistance has been assumed in order to simplify the calculation process in the conversion of current of voltage values.

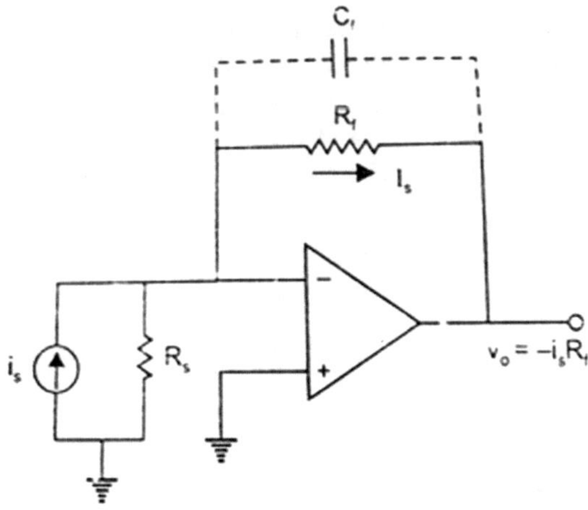

Here is a diagram of a simple I to V converter:

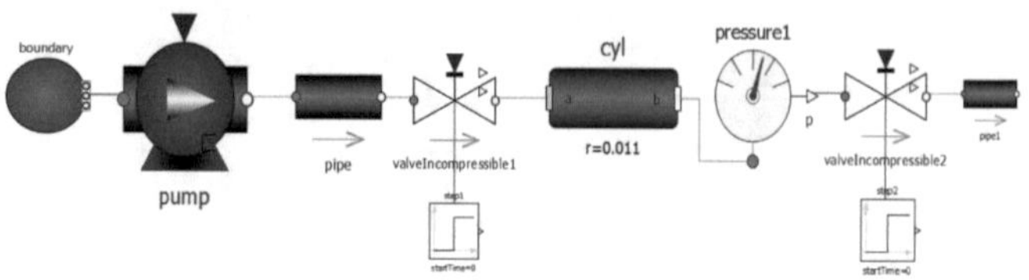

Block Diagram of the model simulated in Open Modelica Software:

Open Modelica is an open-source modelling software where such types of modelling can be done. Due to a few software and other constraints the equipments being used here do not have the exact specifications and have been altered and the best has been tried to match the actual model.

The cylinder being used here is a container whose dimensions and specifications were given accordingly. Pump being used here is a substitute to the manual pumping device being used in the actual model.

The pipe is further connected to the Non Return Valve to stop the reverse flow of the fluid being pumped in into the container and thereby avoiding unrequired wastage of energy.

Pressure gauge to monitor the amount of fluid in the container and a solenoid valve here acting as a simple on/off controller device.

Further processing which will be done to automate the process using the microcontroller to be able to easily automate the on/off process in order to maintain the pressure in the container.

Open Modelica being an OSS where the continuous monitoring of such a process can be done even remotely and thereby reducing the manual intervention in the process and making it easier in the respective field of application.

Here even for the further updation of the process, other components such as PID controllers can be applied for the smooth inflow and outflow of the fluid in order to

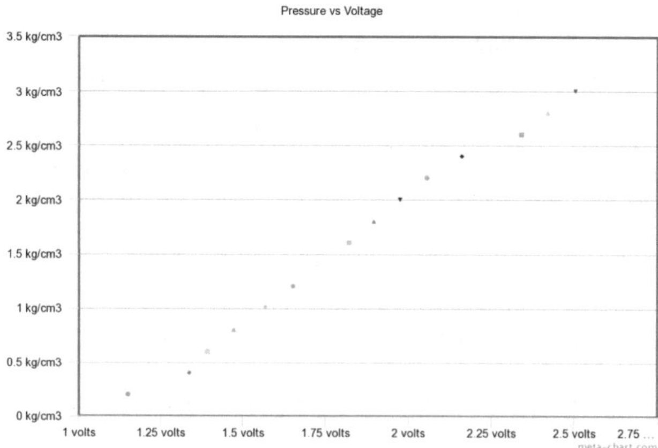

maintain the pressure. And thereby increasing the accuracy and precision of the whole process. Which will inturn increase the fields of application of the device.

The above figure represents the results of simulations

4 Conclusion

- The use of digital twin enables remote controlling and monitoring of the process in huge manufacturing units or industries where physical presence is not always viable.
- The digital twin helps us to understand the entire lifecycle of the system. This increases the reliability of the the equipments used.
- This model provides a warning about the emergency that may occur in the plant. We can run "what if" scenarios to optimize daily decisions or model new operating states. This is one of the major advantage.
- The digital twin can be used to provide training to plant supervisor and other employes so that they have complete knowledge about the plant.
- The process can be controlled from any remote locations with the help of IoT.
- At the finish of the final product, these sensors allow for the implementation of predictive maintenance. They can determine when maintenance is necessary based on use and wear and tear. This helps avoid expensive and unexpected repairs.
- This makes the device economically viable for various industrial applications.

Acknowledgements. We are very grateful to our Professor Dr. SANTOSH R DESAI who provided insight and expertise that greatly assisted the project.

We thank Mr. PAVAN for assistance with Open Modelica Software to make the digital twin of the physical model. We would also like to show our gratitude to the Mr. CHANDRABABU for sharing their pearls of wisdom with us during the course of this project.

Symmetric solutions for supporting us with the components and their support and maintenance.

We want to grate everyone, who has supported the creation of this work.

References

http://www.arresearchpublication.com/images/shortpdf/1459435178_1159B.pdf
Automatic Tyre Pressure Control in a Vehicle
https://www.grundfos.com/service-support/encyclopedia-search/non-return-valve.html
https://en.wikipedia.org/wiki/Solenoid_valve
http://www.ques10.com/p/5298/prove-that-op-amp-can-be-used-as-current-to-volt-1/

Digital Twin of the Robot Baxter for Learning Practice in Spatial Manipulation Tasks

Igor Verner(✉), Dan Cuperman, Sergei Gamer, and Alex Polishuk

Faculty of Education in Science and Technology, Technion—Israel Institute of Technology, 3200003 Haifa, Israel
{ttrigor, dancup, gamer}@technion.ac.il, alepole@ed.technion.ac.il

Abstract. In the study presented in this paper we developed a laboratory exercise in which first-year industrial engineering students operate the robot Baxter to perform spatial manipulations of objects. We implemented the exercise on a digital twin of Baxter which simulates robot manipulations in the Gazebo virtual environment. For this purpose, we developed a virtual robotic cell for practicing spatial manipulations with oriented blocks. The digital twin was calibrated to mimic the physical properties of the Baxter and correctly simulate its operations. The exercise was delivered to a class of 25 students as part of the robotics workshop in the Introduction to Industrial Engineering course. We administered post-workshop questionnaires with focus on the analysis of the learning outcomes and students' spatial difficulties. The students noted that the workshop and particularly the exercise effectively exposed them to industrial robotics and raised their spatial awareness in robot operation.

Keywords: Virtual twin · Baxter · Robot manipulations · Freshman engineering students

1 Introduction

An industrial robot was traditionally defined as "an automatically controlled, reprogrammable, multipurpose manipulator programmable in three or more axes, which can be either fixed in place or mobile for use in industrial automation applications" [1]. This definition undergoes a radical change in the conceptual framework of the fourth industrial revolution where robots are considered as essential components of cyber-physical systems (CPS) [2]. The two fundamentally new capabilities that extend the functionality of the traditional robots are: (1) internet communication for information exchange and collaboration with other devices of the CPS; (2) planning and taking actions, and interaction with humans based on intelligent technologies. Some of the basic functionalities that a modern autonomous robot should include are objects recognition, enhanced dexterity of spatial manipulations, and safe human-robot interaction.

To answer the need to prepare the new generation of engineers for Industry 4.0 in the current transition period, engineering education has a challenge to impart knowledge and competences in both conventional and state-of-the-art industrial technologies [3]. This requires developing educational strategies that meet the above-mentioned

M. E. Auer and K. Ram B. (Eds.): REV2019 2019, LNNS 80, pp. 81–92, 2020.
https://doi.org/10.1007/978-3-030-23162-0_9

challenge on the basis of fostering generic competencies required by engineers. Among these competencies is spatial awareness which is based on the abilities of spatial perception, mental operation and visualization.

Our ongoing research proposes and investigates an approach to the development of the spatial abilities through practice with robot-manipulators. The research focuses on educating novice engineering students and is conducted in the robotics laboratory of the Technion Faculty of Industrial Engineering and Management. We engage students in planning and operating robots to execute complex spatial manipulations and evaluate the outcomes of this learning practice.

In the previous stages of research, we explored the approach in physical, virtual and remote environments based on conventional robot manipulators, using Scorbot 5 DOF and SCARA 4 DOF robots [4]. At the current stage presented in this paper we explore the opportunity to engage novice students in practice with the new generation industrial robot Baxter.

Baxter has a pair of 7 DOF arms and is designed to work safely and intelligently perform various production tasks in close proximity to people. The robot can adapt to its environment by sensing and controlling force, position, and torque at every joint, and by computer vision. Baxter serves as a platform for robotics research and advanced engineering education. Recently, experiments have begun on its use for educating novice students, e.g. first attempts to utilize Baxter as a robot-teacher in an elementary school class [5].

2 Planning Robot Manipulations

Planning robot manipulations is considered a complicated problem because of the need to select among a large number of possibilities for grasping and moving an object [6–8]. In this section we will discuss the aspects of the planning and the problems that students encounter when controlling a robot to manipulate objects.

2.1 The Rotation Manipulation Task

The pick-and-place manipulation of an object is a basic task in robotics. It is discussed in literature, usually regarding one of the two main topics: grasp planning or path planning. While the former focuses on a stable grasping of the object, the later focuses on the robot's post-grasp trajectory. Nevertheless, grasping consideration should address not only stable grasping but also the possible need for object rotation during the manipulation [6, 7]. This rotation is determined by the change in the orientation of the gripper between picking and placing the object.

To address the robot pick-and-place subject in a laboratory exercise for students, we selected a robot manipulation task in which the location and orientation of an object are defined in its initial and final positions. In this exercise, the students need to find within the reachable workspace of the robot the right way to grasp and rotate a simple oriented object i.e. a cube with symbols on its faces. To enable the students to easily define cube grasping and rotation, we developed a code language for such manipulations. In our previous studies [4, 9] we developed and used a limited code language which was tailored for a 5 DOF robot. For the current study we developed a new generic language suitable for cube rotation with each and every robot. This Rotation Manipulation Language (RML) is based on the notion that the cube is picked, rotated and then placed with a different orientation in its initial location. Thus, RML defines the initial orientation (while grasping the cube), and the final orientation (while ungrasping the cube). The user needs to consider grasp planning and ungrasp planning while the robot plans the path between the two. In other words, the rotation of the cube is a product of its change in orientation. RML relates to the world Cartesian coordinate system (X, Y, Z) of the robot. The coordinate system is defined for any robot so that the XY plane is horizontal and its origin is located at the center of the robot base, as in Fig. 1a, b. The X axis is directed along the workspace symmetry line, towards the robot front workspace (see the top views). Manipulations are performed with the cubes placed in front of the robot and orthogonal to the coordinate axes.

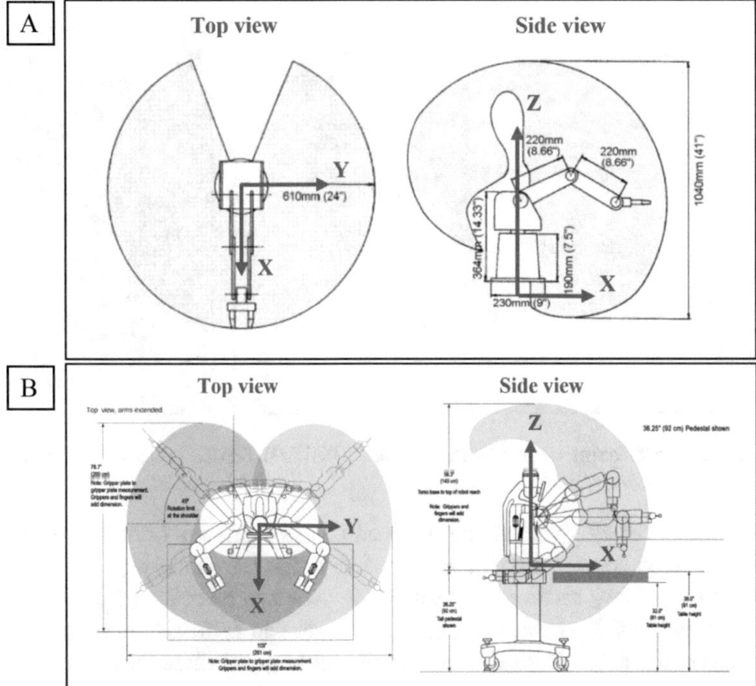

Fig. 1. Top and side views of the workspaces and world coordinate systems: **A** Conventional robot Scobot [10]; **B** New generation robot Baxter [11].

The RML code consists of four characters in the format “**S**α**F**β”. Here:

S—stands for the direction of the gripper axis when grasping a cube. It can get the values “X”, “Y”, or “Z”, which refer to the positive directions of X and Y axes, and the negative direction of Z axis.

A—determines the orientation of the gripper fingers when grasping the cube. It stands for the angle of rotation of the gripper around its axis. α and can get the values "2", "1", "0", or "–", which represent counter clock-wise rotations by 180°, 90°, 0°, and −90° correspondingly.
F—stands for the direction of the gripper axis when ungrasping the cube after the rotation, and can get the values {X, Y, Z} with the same meaning as S.
B—stands for the angle of rotation of the gripper around its axis, with the same meaning as α.

In the example illustrated in Fig. 2a, the word "X0Z1" represent a rotation that starts with grasping the cube while the gripper axis is in line with X axis, and the gripper is not rotated around its axis. The cube is placed while the gripper axis is in line with the negative direction of Z axis, and the gripper is rotated 90° around its axis. Figure 2b presents a robot Baxter arm in this position.

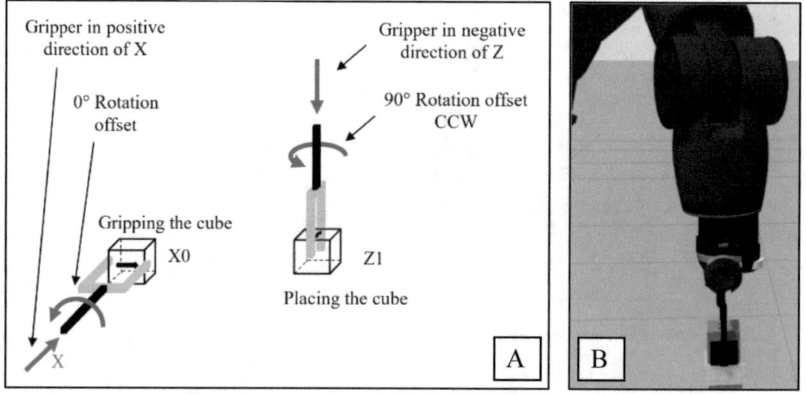

Fig. 2. **A** The meaning of the rotation code "X0Z1"; **B** Baxter in Z1 placing position.

Before picked or after placed, the cube can be in 24 different orientations, thus there are 576 orientation combinations. It can be shown that for all these combinations there are only 24 elementary rotations, by which the cube can be rotated from any initial to any final orientation. This can be done if the robot can approach and grasp the cube, which rests on the table, from all five possible orthogonal directions.

2.2 Rotation Manipulations with Robot Baxter

The arm of the robot Baxter is presented in Fig. 3. It consists of seven joints marked with the following notations: Shoulder (S0 Roll, S1-Pitch), Elbow (E0-Roll, E1-Pitch) and Wrist (W0-Roll, W1-Pitch, W2-Roll).

From experimentation with Baxter, we found that there are certain limitations in rotating a cube through a pick-and-place manipulation by a single arm. For instance, the cube, wherever placed, cannot be grasped from both positive and negative

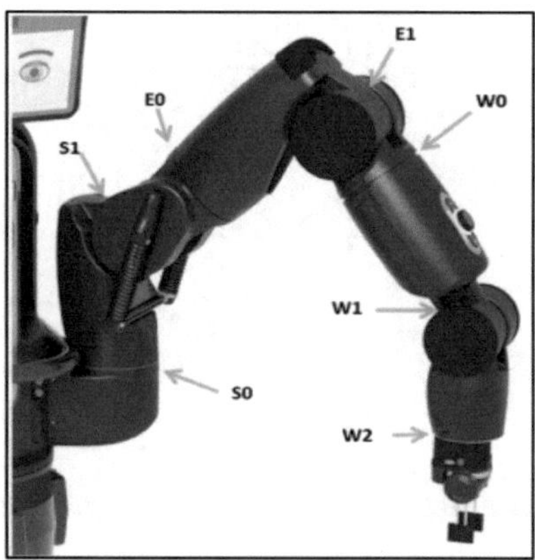

Fig. 3. The arm of the robot Baxter [12].

directions of the X axis. Also, choosing to approach the cube from both +Y and −Y directions significantly narrows the area where the cube can be placed. Therefore, in the exercise we excluded −X and −Y directions and programmed Baxter to handle the cube from three directions: +X, +Y, and −Z. So, not all 24 elementary rotations, described above can be achieved in the exercise with Baxter. Therefore, the cube cannot always be rotated from an initial to a final orientation in one pick-and-place manipulation. As Baxter is capable to grasp the cube from three directions, it can be shown that 19 out of 24 elementary rotations and 456 out of 576 manipulations of the cube are still possible.

3 The Developed Workcell Environment

3.1 Baxter and Its Workcell

The Baxter robot embodies core concepts of modern intelligent robotics. Thus, engaging novice engineering students in practice with Baxter could be of great value. However, Baxter is a complex engineering system that requires knowledge and skills far beyond that the novice students have. This fact raises the need for a simple user interface and an instructional strategy to make learning practice with Baxter accessible and effective for novice students. Baxter is operated and controlled via Robot Operating System (ROS) running in a Linux environment. Figure 4a presents a block diagram of the Baxter system in which the embedded computer uses ROS to receive feedback from the robot's sensors and to control limb actuators via motor controller.

The Baxter includes a software developers kit allowing users to develop custom software for the robot. The developed software can be run either on the physical robot or on its digital twin, within the Gazebo simulator, which models the robot and its environment. Figure 4b presents the diagram of the user workstation which can operate the physical and virtual robots. The Gazebo simulator can host the digital twin and models of objects in its environment.

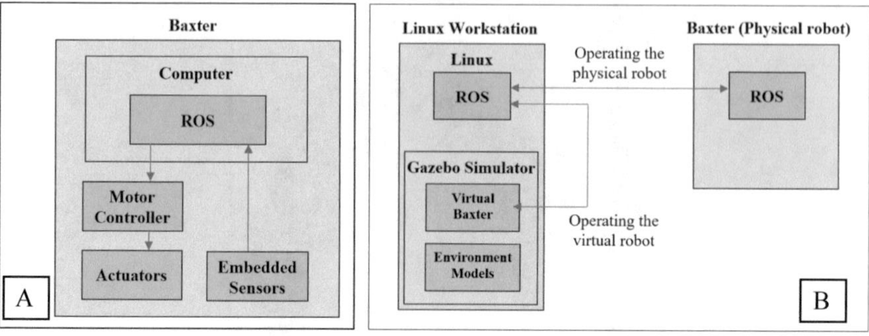

Fig. 4. **A** Baxter's block diagram; **B** User workstation system diagram.

For cube rotation exercise with the physical robot Baxter, we constructed two tables and placed them near the robot (Fig. 5a). A utility table on which we place the cubes needed for the exercise is located to the left of the robot. In front of the robot we located a second table to be used as a "buffer" on which a cube is placed for immediate rotation by the robot. The buffer was located in a spot in which the robot rotation dexterity is high.

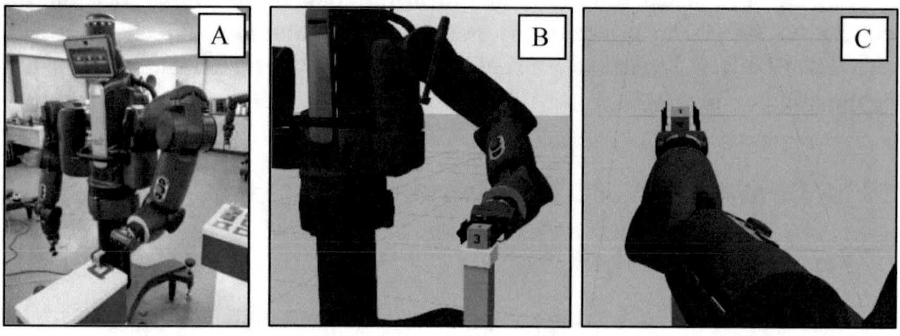

Fig. 5. **A** Baxter places a cube; **B** Virtual workcell; **C** Baxter's point of view.

For the exercise with the digital twin we developed a virtual workcell within the Gazebo simulator. We imported to Gazebo a virtual model of Baxter and added models of the cubes to be rotated and stands for placing the cubes (Fig. 5b). In this workcell, the user can choose the desired point of view by rotation the entire workcell and zooming in or out. Thus, the manipulated cube can be seen by default from a point of view of an observer in front of the Baxter (Fig. 5b) or, for example, from a point of view of an observer standing in Baxter's place (Fig. 5c).

To rotate the virtual cubes, the user opens two Linux terminals: one for running Gazebo, and the other for controlling the digital twin by means of GUI we developed. While in a physical environment we start an exercise after placing real cubes on the

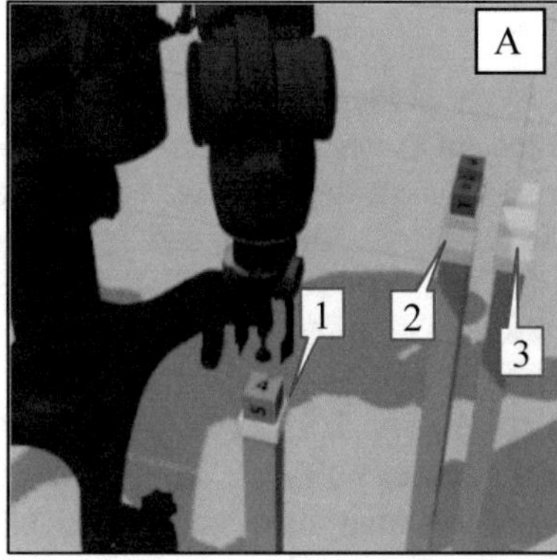

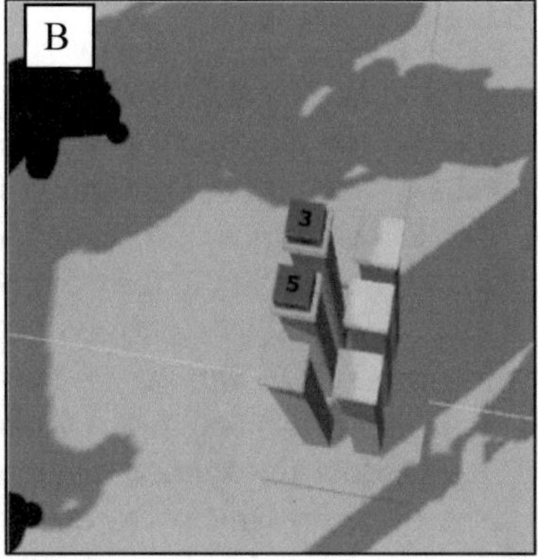

Fig. 6. **A** Virtual workcell: buffer (1), storage (2), destination (3); **B** Blocks in the storage.

utility table, in the virtual environment, the user needs to select the initial orientation of the virtual cubes. In the example presented in Fig. 6, the initial orientation of the 3 cubes was defined by the digits 453.

The user can set up the initial orientation of the cubes for the exercise by using the GUI presented in Fig. 7a. The orientations of the second and third cubes can be seen in Fig. 6b. The first cube was moved to the buffer and can be seen in Fig. 6a. Manipulating the cube is done by using RML with the robot control GUI presented in Fig. 7b. In the robot control GUI, selection of a cube number on the left side of the interface and pressing the button 'Take the cube to the buffer table' makes the Baxter pick up the designated cube from the utility stand and place it on the buffer stand. Setting the rotation arguments in RML and pressing 'Rotate the cube' makes the robot to perform the manipulation on the buffer. Pressing 'Put the cube to the utility table' makes the robot move the cube to the utility stand.

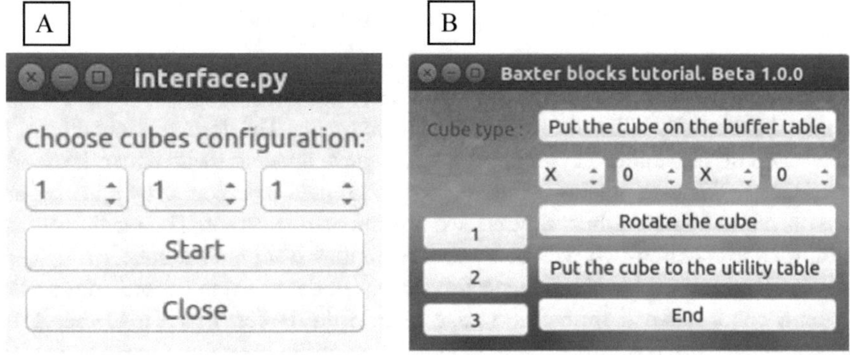

Fig. 7. The control interface: **A** Configuration GUI; **B** Robot control GUI.

3.2 Robotics Workshop

The 6-h workshop was delivered to 25 first-year students participated in the Introduction to Industrial Engineering and Management course. The workshop included a lecture, and two exercises in operating of two different digital twins of robot manipulators. The lecture exposed the students to industrial robotics and introduced the concepts of robot operation in a workspace. The capabilities of the 7 DOF mechanical arm of Baxter in manipulating objects were discussed through comparison with that of the conventional 5 DOF vertical articulated arm of Scorbot.

The first preparatory exercise (Verner and Gamer, 2015) was with the twin of Scorbot, operated in the robotic simulation environment RoboCell (Fig. 8a). In the second main exercise, which was developed in this study, the students operated the digital twin of Baxter in Gazebo (Fig. 8b). In both exercises the task was to manipulate three identical cubes with digits from 1 to 6 irregularly oriented on their faces, pick each cube in an initial orientation and place it in a desired orientation. The students ascertained that in some cases it is possible to find a suitable pick-and-place operation which implements the desired manipulation of the cube, while in other cases they need to plan a sequence of two operations.

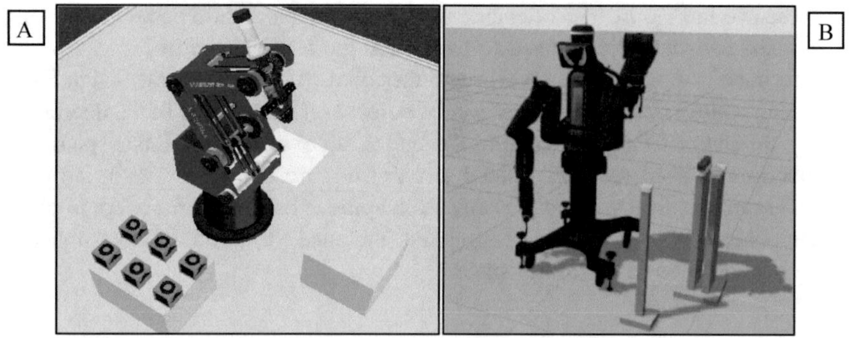

Fig. 8. **A** Scorbot RoboCell; **B** Baxter's virtual workcell; **C** Students perform the exercise.

The orientations of the cubes in the tasks were designed so that the students could evaluate the dexterity differences between the two robots. The desired orientation of the first cube could be achieved in a single pick-and-place operation with either the Scorbot or the Baxter. The manipulation of the second cube could be executed by one pick-and-place operation of Baxter but required two operations of Scorbot. The manipulation of the third cube required two pick-and-place operations from both robots.

In the Baxter exercise, the students used the GUI, described in the previous section, to pick a cube, move it from the storage area to the buffer, rotate it to the desired orientation, and place it in the destination position at the assembly area. The students worked in pairs and were given a time limit of 45 min to complete the exercise. During the exercise we noticed that some students took advantage of the opportunity to change workcell perspective as described in the previous section. Some students used the default perspective of an observer looking at the Baxter from the front (Fig. 5b), while some decided to change it and preferred Baxter's point of view (Fig. 5c). The students were offered a physical cube to help them in planning robot manipulations. Some of the students used the cube, while some others imitated movements of the robot gripper with their hand (Fig. 9). Some didn't use any physical objects and planned the rotations mentally in their heads.

Fig. 9. Students perform the exercise.

4 The Study

Our study aimed to evaluate learning outcomes of the workshop with special attention to the exercise with the Baxter's digital twin. We inquired how the workshop influenced students' understanding of the role of robotics in modern manufacturing, and their interest in the subject as future engineers. We asked whether and how the practice in operating the digital twin exposed the students to industrial robotics and contributed to their spatial awareness in robot operation. We also inquired into the spatial difficulties that the students faced when they performed rotation manipulations in the virtual robot cell.

4.1 Data Collection and Analysis

In this study, two post-workshop questionnaires were administered. The first one requested students' reflections on the workshop experience. We asked the students if the workshop exposed them to industrial robotics and raised their awareness of spatial problems in planning and operating robot manipulations. Responses were accepted from 25 students. Results of the questionnaires were analyzed and triangulated with the evidence elicited from the laboratory reports and observations.

The second questionnaire was about the spatial difficulties faced by the students when they performed the exercise with the digital twin of Baxter. The students were asked to evaluate the level of difficulties in using the rotation manipulation language of spatial codes to find optimal sequences of rotation manipulations to perform the robot task. An additional question was about the difficulties in applying the exocentric view of the workspace from the robot's view point. When performing the exercise, some students, who had spatial difficulties, upon request got physical cubes, same as that rotated by the robot, and rotated them with their hands, to make it easier to imagine manipulations by the robot. We asked those students, to what extent the manual rotations helped them to plan the robot manipulations.

4.2 Findings

4.2.1 Workshop Contribution

From the results of the first post-workshop questionnaire, most of the students did not study robotics and had no experience with virtual or real robots before the workshop. Their evaluation of the workshop's contribution was highly positive:

- 86% reported that the workshop exposed them to industrial robotics,
- 62% noted that the workshop was useful and relevant to their future profession,
- 43% pointed that the workshop contributed to their understanding of basic concepts in the field of industrial robotics.
- 71.5% reported that performing the exercise on a digital twin of Baxter contributed to their understanding about spatial skills required to operate robot-manipulator.

We found moderate Pearson correlations between the above-mentioned contribution factors. Statistically significant positive correlations were found for the awareness of spatial problems and for the exposure to industrial robotics $r = 0.52$ ($p < 0.05$) as well as between the relevance to future profession and the awareness of spatial problems $r = 0.8$ ($p < 0.0001$).

The highly positive evaluation of the learning practice is expressed also in students' reflections. The repeated reflections related to exposure to industrial robotics:

> *The lecture and the labs exposed me to industrial robotics, the subject I had never heard before. It gave me general knowledge of this profession and a taste of experience into subject. This exposed me also to a variety of robots and their capabilities.*
> *An industrial engineer needs to know new worlds because everything can be relevant to his field of competence, especially technology and robots. Exposing such subjects will help him in the future profession. I was interested in the robot-twin. The way to operate it related to different aspects such as angles, point of view, and rotation commands.*

The workshop helped students understand the value of industrial robotics for their future profession:

> *In my opinion as future industrial engineers, we had to know how to interact with machines and robots, and this was a learning experience.*

The students noted their progress in spatial skills for robot operation:

> *The experience with the robot contributed to my spatial skills and exposed me to the wide range of robot movements that can be performed with minimum number of commands.*

4.2.2 Spatial Difficulties in Manipulating Objects with the Digital twin

In the second post-workshop questionnaire the students noted spatial difficulties that they experienced when performing the exercise. In particular, 65% of them noted difficulties in using the rotation manipulation language, 61% in finding optimal sequences of rotation manipulations and in considering the workspace from the exo-centric point of view. 87% of the respondents noted that the opportunity to rotate the oriented cube by hand significantly helped them to overcome the spatial difficulties and successfully complete the exercise.

5 Conclusion

Our research explores ways to introduce industrial engineering students at the first year of their studies to the concepts and technologies of industrial robotics. In the previous stages of research, we developed and conducted a workshop in which the students controlled and operated a conventional robot manipulator Scorbot in physical, virtual and remote environments, to manipulate objects in the workspace.

In the current study we developed an exercise in control and operation of a new generation robot Baxter. We conducted a workshop which included a preparatory exercise with the digital twin of Scorbot and the new exercise with the digital twin of Baxter. In both exercises, the students controlled and operated robots to execute complex spatial manipulation tasks that require translation and rotation of oriented objects (cubes) in the workspace. We developed a language of spatial codes to describe pick-and-place manipulations providing different rotations of the cube. The students used the language for planning robot manipulations. The language also helped them to explore and compare the dexterity of the conventional and modern robots.

The experiments in using new generation industrial robots such as Baxter in education of novice students started very recently. For our knowledge, the developed exercise and the workshop for first-year industrial engineering students are the first of their kind. The workshop experience showed that despite the mechanical and software complexity of robot Baxter, its virtual twin can be used for hands-on experimentation of novice engineering students. Results of the educational study indicated that for students' opinion the workshop effectively exposed them to industrial robotics and was useful and relevant to their future profession. They noted that the exercise on a digital twin of Baxter challenged their spatial reasoning and contributed to the awareness about spatial skills required in operating industrial robots.

Acknowledgements. This study is supported by the Israel Science Foundation grant.

References

1. ISO 8373 (2012) https://www.iso.org/standard/55890.html. Retrieved October 28, 2018
2. Luo, R.C., Kuo, C.W.: Intelligent seven-DoF robot with dynamic obstacle avoidance and 3-D object recognition for industrial cyber–physical systems in manufacturing automation. Proc. IEEE **104**(5), 1102–1113 (2016)
3. Schuster, K., Groß, K., Vossen, R., Richert, A., Jeschke, S.: Preparing for industry 4.0–collaborative virtual learning environments in engineering education. In: Engineering Education 4.0, pp. 477–487. Springer, Cham (2016)

4. Verner, I., Gamer, S.: Robotics laboratory classes for spatial training of industrial engineering and vocational school students. Int. J. Eng. Educ. **31**, 1376–1388 (2015)
5. Fernández-Llamas, C., Conde, M.A., Rodríguez-Lera, F.J., Rodríguez-Sedano, F.J., García, F.: May I teach you? Students' behavior when lectured by robotic versus human teachers. Comput. Hum. Behav. **80**, 460–469 (2018)
6. Jones, J.L., Lozano-Perez, T.: Planning two-fingered grasps for pick-and-place operations on polyhedra. In: Robotics and Automation, 1990. Proceedings., 1990 IEEE International Conference on, pp. 683–688. IEEE (1990)
7. Amir, M., Ghalamzan, E., Mavrakis, N., Stolkin, R.: Grasp that optimises objectives along post-grasp trajectories. In: 2017 5th RSI International Conference on Robotics and Mechatronics (ICRoM), pp. 51–56. IEEE (2017)
8. Mavrakis, N., Ghalamzan, E.A.M., Stolkin, R.: Safe robotic grasping: Minimum impact-force grasp selection. In: Intelligent Robots and Systems (IROS), 2017 IEEE/RSJ International Conference on, pp. 4034–4041. IEEE (2017)
9. Verner, I.M., Gamer, S., Polishuk, A.: Development of spatial awareness and operation skills in a remote robot laboratory. In: Global Engineering Education Conference (EDUCON), 2018 IEEE, pp. 389–393. IEEE (2018)
10. Kumar, R.R., Chand, P.: Inverse kinematics solution for trajectory tracking using artificial neural networks for SCORBOT ER-4u. In: Proceedings of the 6th International Conference on Automation, Robotics and Applications (ICARA), pp. 364–369. IEEE (2015)
11. Baxter workspace guidelines: http://sdk.rethinkrobotics.com/wiki/Workspace_Guidelines. Retrieved 31 Oct 2018
12. Baxter arms: http://sdk.rethinkrobotics.com/wiki/Arms. Retrieved 31 Oct 2018

Part II
Remote Engineering

Development of a Remote Tube Bending Lab to Illustrate Springback and Determine Process Limits

Siddharth Upadhya(✉), Alessandro Selvaggio, Tobias R. Ortelt, Joshua Grodotzki, and Erman A. Tekkaya

IUL, TU Dortmund University, Dortmund, Germany
{Siddharth.Upadhya, Alessandro.Selvaggio, Tobias.Ortelt, Joshua.Grodtzki, Erman.Tekkaya}@iul.tu-dortmund.de

Abstract. The benefits of remote labs in overcoming the time and accessibility constraints of conventional labs is well known. To complement the already existing labs at the IUL, a remote tube bending lab incorporating an industry-relevant process was developed to aid in the understanding of important process limits and features of profile bending. The remote lab was developed in a bottom-to-top approach where the underlying tube bending lab was conceptualized and developed keeping in mind the best way to educate the students. Based on a problem based learning approach, the lab gives the students the freedom to explore and discover the field of tube bending on their own. The state of the art bending machine, with open platform communication (OPC) capabilities, coupled with a dedicated specimen handling robot, allows for easy automation of the lab while the use of augmented reality and web cameras facilitate easy visualization of the process and the product.

Keywords: Remote laboratory · Engineering education · Forming technology · Laboratory learning

1 Introduction

Metal forming, an isochoric manufacturing process, where metal is plastically deformed to a desired shape is a key manufacturing method, with the contribution to the automotive industry alone being valued at $221.2 billion [1]. Pushing the frontiers in this field through research and dissemination of this knowledge are the cornerstones of the work being done at the Institute of Forming Technology and Lightweight Components (IUL) of the TU Dortmund University. Realizing early on the need for a change in the existing education paradigm, the TU Dortmund University partnered with the RWTH Aachen University and the Ruhr University Bochum and worked under the "ELLI—Excellent Teaching and Learning in Engineering Education" banner with a collective goal of improving the engineering education in Germany. The benefits of remote labs, which allows students time- and location- independent access to labs, while allowing them to learn and explore at their own pace and convenience are well established [2, 3]. To this end, more than 10 different remote labs were developed by

M. E. Auer and K. Ram B. (Eds.): REV2019 2019, LNNS 80, pp. 95–106, 2020.
https://doi.org/10.1007/978-3-030-23162-0_10

the different universities with the contribution of the IUL being the development of a completely automated tele-operated testing cell for material characterization, a significant progress in comparison with the already prevalent FEM based mechanical engineering virtual labs, since remote labs allow for the unpredictability and inherent scatter in real experiments to be observed [4]. Through the material characterization lab, the students gained key insights into the link between theoretical hypotheses/models and reality. To build up on the fundamental knowledge obtained through the material characterization lab and to serve as a platform for the students to apply the knowledge and skills that they have acquired through previous courses, a remote lab, which incorporates a standard industry machine was envisioned as part of "ELLI2". The reasoning being that the majority of the students move on to work in the industry once they graduate. Therefore, it is not sufficient to acquire just theoretical knowledge rather the students must also understand manufacturing processes and their limit in depth and develop professionally relevant skills during their studies and this lead to the development of the remote tube bending lab.

2 The Process and Process Limits

Tube bending is an important manufacturing process and examples of its uses can be found all around, from staircase railings to automobile roll cages to aircraft bodies. Rotary draw bending (RDB), a form closed bending process, is one of the most widely used tube bending processes owing to its ability to bend thin-walled and tight bending radii tubes with a high accuracy. In RDB, a tube is pulled and bent around a bend die while generally being supported on the inside by a mandrel. The schematic of the process is shown in Fig. 1a. In the developed remote tube bending lab, the process is carried out on a Transfluid DB-2060 CNC rotary draw bender, as seen in Fig. 1b. The CNC, as well as OPC capabilities, allows for easy automation and application in an "Engineering Education 4.0" [5] use case.

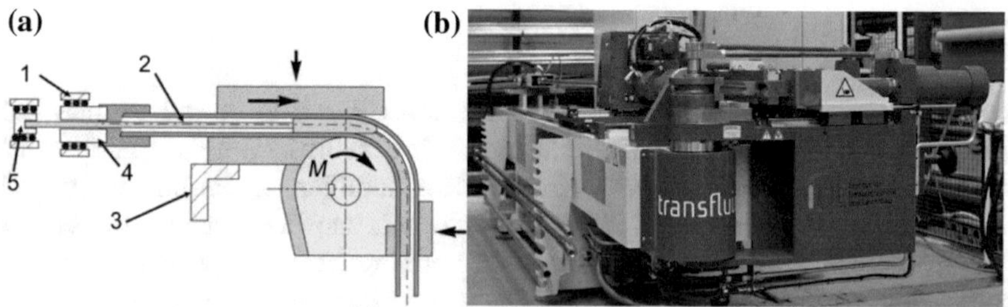

Fig. 1. **a** Schematic of RDB process [6] **b** DB 2060 CNC RDB machine

Although, the rotary draw bending process is more robust in comparison with the other bending processes, it too has its own set of limitations. Springback, a problem omnipresent in metal forming, plagues the RDB process as well. Springback, graphically represented in Fig. 2, is the geometrical change observed at the end of a forming process as the workpiece tries to go back to its original state due to the internal elastic stresses. It is critical that the students have a good understanding of this concept since springback calculation and compensation are an integral part of most product and process development cycles. The students need to understand the effect, its various influencing parameters (such as Young's modulus, yield strength, wall thickness etc.) and how to calculate and compensate for it to prevent errors or failures in the final bent tube.

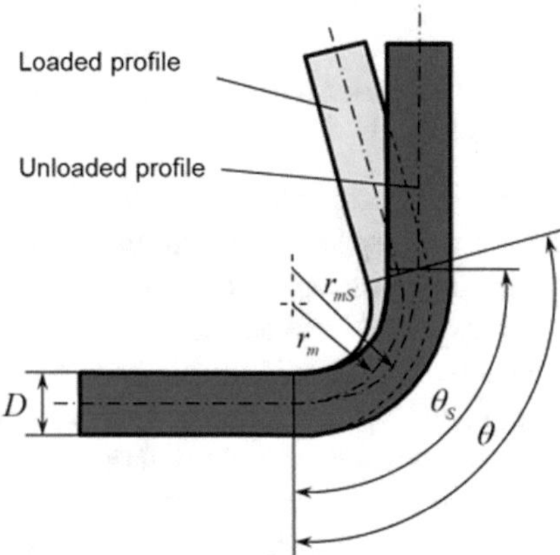

Fig. 2. Springback of a bent profile once the load is removed.

Apart from springback, process limits due to cracking (Fig. 3a) and wrinkling (Fig. 3b) are commonly observed in rotary draw bending. These failures lead to functional, as well aesthetical, unsuitability of the parts. There is a similar necessity to understand the effects, the influence of the material, as well as machine, properties on its intensity and the remedial measures to tackle them.

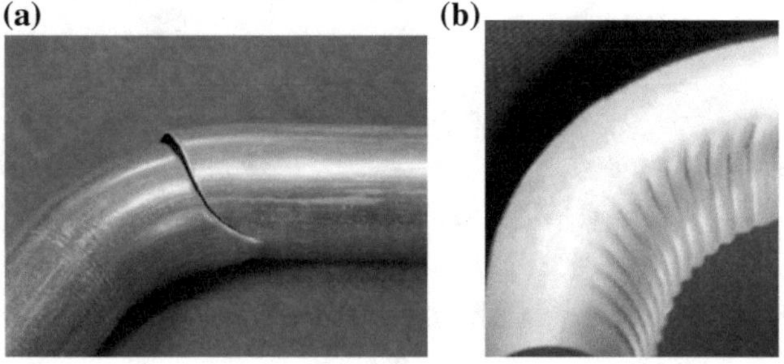

Fig. 3. **a** cracking, **b** wrinkling [7]

3 Design of the (Remote) Tube Bending Lab

The most straightforward approach to develop a remote lab is to automate an already existing conventional lab and offer it remotely with no or few minor changes to the underlying lab. With regards to the remote tube bending lab, there was no already existing tube bending lab and thus, it was decided to take a step back and tackle this with a bottom-to-top approach where the underlying lab was developed first, keeping in mind that the efficacy of this lab is what makes or breaks the subsequent remote lab.

Working under the "ELLI" banner allowed the authors, who all have an engineering background, to collaborate with didactic experts and to develop the lab in such a manner that it would benefit the students the maximum. Three key didactic tools were used during the design of the lab:

Constructive Alignment: is a principle used in designing teaching and learning activities. The 'constructive' aspect refers to the idea that students construct meaning through relevant learning activities. The idea being that meaning is not something imparted or transmitted from teacher to learner, but is something learners have to create for themselves. The 'alignment' aspect refers to what the teacher does, which is to set up a learning environment that supports the learning activities appropriate to achieving the desired learning outcomes. In an aligned system, one specifies the desired outcomes of teaching in terms not only of topic content, but in the level of understanding one wants students to achieve. Then an environment is set up that maximizes the likelihood that students will engage in the activities designed to achieve the intended outcomes. Finally, assessment tasks are chosen that will quantify how well individual students have attained these outcomes [8]. Figure 4 shows the graphical representation of the process.

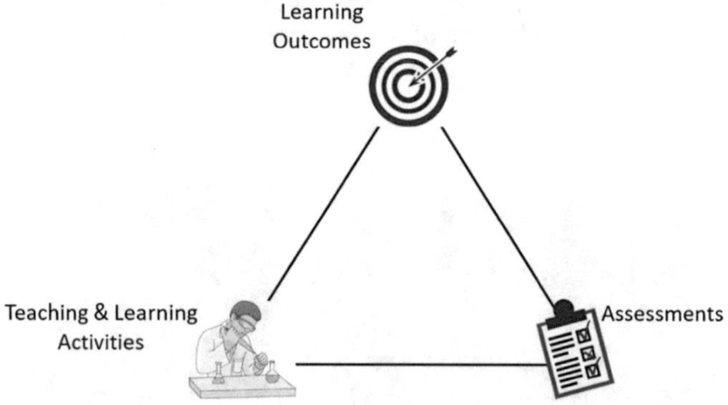

Fig. 4. The constructive alignment approach to course development.

Kolb's Experiential Learning Cycle: It is concerned with the learner's internal cognitive processes. It provides a holistic model of the learning process is called 'Experiential Learning' to emphasize the central role that experience plays in the learning process [9]. There are four stages in the cycle, as seen in Fig. 5:

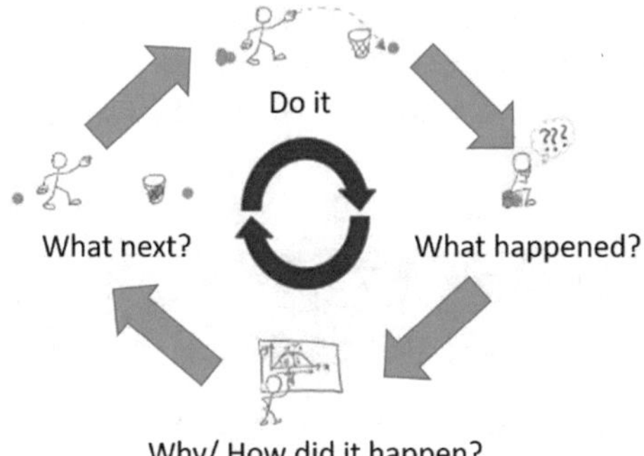

Fig. 5. Experiential learning cycle as postulated by Kolb.

(a) *Concrete Experience*: A new experience or situation is encountered.
(b) *Reflective Observation*: The experience is reviewed. Questions are raised and discussions are carried out.
(c) *Abstract Conceptualisation*: The process of making sense of what has happened. Involves the interpretation of the events and understanding the relationships between them and previously acquired knowledge or observations.
(d) *Active Experimentation*: The ideas born from conceptualisation and contemplation and put into practice and a new experience is encountered and thus the process continues.

Problem Based Learning (PBL): is a teaching method where real-world problems are used as a vehicle to promote learning as opposed to direct presentation of facts and concepts. In PBL, a student is presented with the problem and he or she determines the necessary skillset required to solve this problem and goes about acquiring it before eventually using it and solving the problem. PBL promotes the development of critical thinking skills, problem-solving abilities and communication skills [10].

The learning outcomes as mentioned in constructive alignment were used to define the aims of the lab while the learning activities were designed based on a PBL approach and in alignment with the Kolb's learning cycle. Details of the same will be discussed in the following chapter.

4 The Tube Bending Lab

The aims of the lab or learning outcomes are as follows:

- To motivate the students and stimulate their interest in the field of forming technology, especially tube bending.
- To deepen their understanding of the different bending processes, the fundamental process characteristics and the process limits.
- To provide students an opportunity to work together.
- To provide an opportunity for the development of professionally relevant skill and attitudes.

The learning activities, echoing a PBL approach, were modelled along the lines of a simplified product manufacture cycle (Fig. 6), with the intention of giving the students a taste of how things work in the industry while they learn new concepts as well as apply the already learnt theory.

Fig. 6. Design of lab using a simplified product manufacture cycle as reference

The flow of the lab is as follows:

4.1 Problem

It is literally the 'Problem' in PBL. A problem which needs to be solved and at the same time an opportunity for the students to learn and grow. The problem can be either an actual problem or derived from one. A possible sample problem for the tube bending lab would be the design of a pipeline connecting two points while fulfilling certain boundary conditions as seen in Fig. 7. This is a typical scenario observed in industrial planning as such pipelines are needed to transport fluids in various industries.

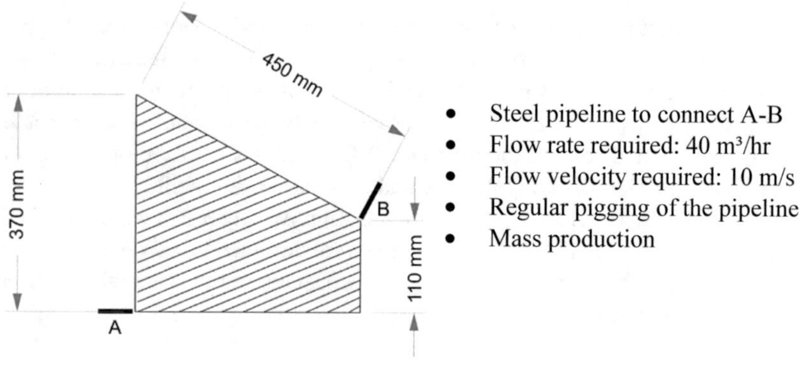

Fig. 7. Problem statement and the boundary conditions.

4.2 Product Design

The solution to the above problem is in the form of a bent tube as seen in Fig. 8. The students need to fall back on their previous knowledge to develop the part that will fulfill the requirements. The problems are constructed in such a way that there a limited number of solutions which fulfills the requirements and all of them can be manufactured by the available machine and the tools on hand. In a sense, it is an autocratic democracy, where the students feel they have complete freedom but in actuality they are guided down a specific path due to the constraints. This is necessary due to limits of the machine, available specimens, safety of the machine and other equipment. For example, the requirement of pigging, which is the act of sending an inspection probe through the pipeline, was included so that the students design a bent part with a radius of 200 mm since if pigging is required, the bend radius should be 5 times the tube diameter and R200 is a tool which is available at the lab.

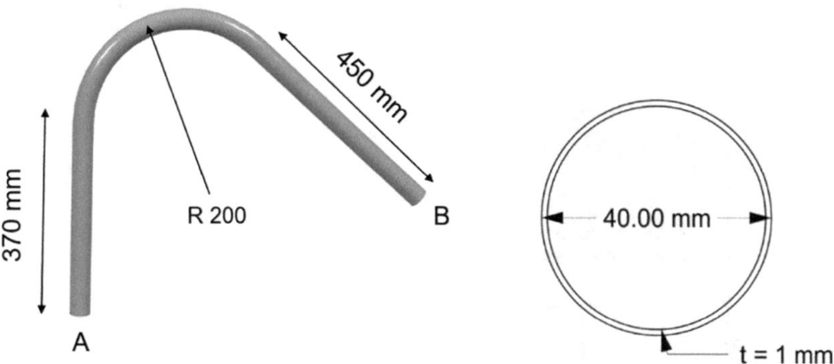

Fig. 8. Product geometry

4.3 Process Planning

Once a solution for the problem in the form of a product is determined. The next step in the sequence is to choose how exactly should the product be manufactured. In a traditional metal forming lab, the activities would begin from this point, but to give the students a broader perspective and a context to their actions, the authors chose a more holistic problem based approach as seen in the previous two stages. In the process planning stage, a strong link exists between the theory learnt in the lectures and the activities in the lab. The students have to fall back on their theoretical knowledge to compare the different bending processes, evaluate the pros and cons of each and finalize upon a suitable bending process. Autocratic democracy is incorporated here as well. To satisfy the requirement of mass producing a bent thin walled (1 mm) tube, the best possible method to bend the tube would be using the RDB process instead of a conventional 3-roll bending process.

Once the process is chosen, the students must identify the different tools required as well as the tool dimensions. Throughout the entire process planning phase, the students need to scientifically justify their actions and pen down the same in their report.

4.4 Validation and Testing

The validation and testing, a critical phase in the product manufacture cycle, is represented in Fig. 9 for our use case.

The validation and testing loop is similar to the Kolb's experiential learning cycle. An initial bend is carried out according to the machine parameters which are automatically generated based upon the input geometry. This is the 'new experience' or the activity.

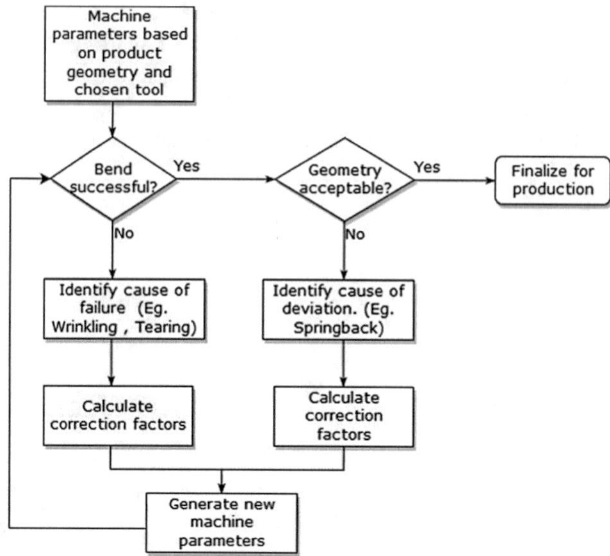

Fig. 9. Validation and testing loop based on Kolb's cycle

Next, the students reflect upon their observations. Questions are asked, Is the bend successful? Is the geometry in accordance with the target geometry? Are the boundary conditions met? etc.

Then they try to make sense of their observations. They fall back on their theory knowledge to understand what is happening or predict what will happen and how to rectify it. Remedial measures and compensation factors are calculated and a further iteration through the loop is carried out.

The AHA-effect [11], integral to learning is also experienced in this phase. Assume 2 students A and B are participating in the lab. Student A does not consider springback in his initial design while student B anticipates it and calculates the expected springback according to theoretical formulas. But at the end of the 1st iteration loop, both observe that the actual geometry is not same as the target geometry and the AHA moment occurs. Student A learns about the springback while B learns that theory is not always the same as reality and is in certain cases just an approximation. Here, the relevance and influence of the assumptions that go into the analytical description are better understood.

Once the validation and testing is completed, in reality the mass production of the product will begin. In our case, this marks the end of the lab and the students would be required to document their activities in the form of a report.

5 Structure of the Remote Tube Bending Lab

The following section details the structure of the remote lab based on the above mentioned.

5.1 Infrastructure and Layout

Machines:

To enable the complete automation of the bending process, a 6 axis KUKA KR90 robot works in tandem with the rotary draw bending machine seen in Fig. 1b. The robot deals with the loading and unloading of the specimens. A graphical representation of the layout is seen in Fig. 10.

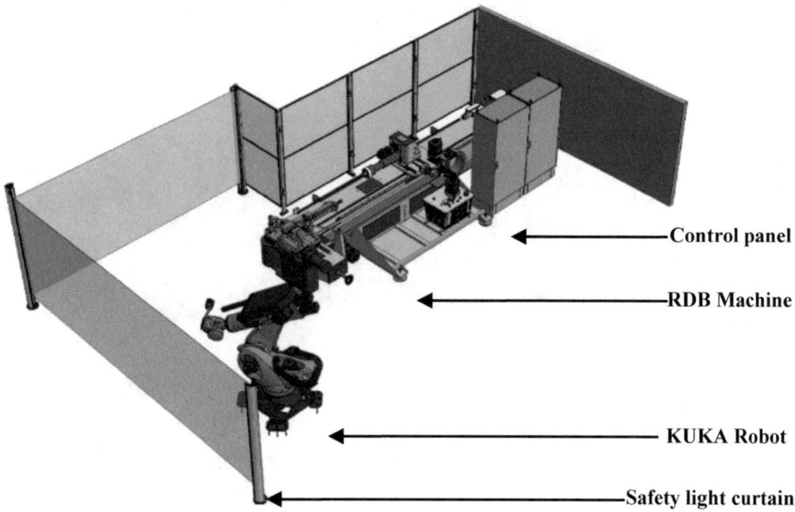

Fig. 10. Layout of the remote tube bending cell

IT Components:

- *Real-time control system*: The high level supervision of the bending machine and the robot is carried out using a PXI system from National Instruments and hence the main control software, as well as the user interface, is developed in LabVIEW.
- *Safety system*: To ensure the safety of the users, as well as the machines, light curtains and laser scanners are deployed which ensures automatic machine shutdown in case of movement within a defined danger zone.
- *Camera system*: Multiple strategically placed IP- cameras allow for the observation of the entire process from different perspectives.

5.2 Communication

The benefits of open platform communication, a standard feature of "Industry 4.0", is that the communication between the machines becomes more straightforward and thus automation is easier. The machine was developed with the intention of working in tandem with a robot and there is a dedicated interface which allows for the communication with the robot.

The main control system which is developed in LabVIEW and hosted on the PXI communicates with the machine and the robot using PROFINET, an industry technical standard for data communication over Ethernet, which is especially suited for real time applications.

The main control system takes inputs from the test manager hosted on the main experimental server and carries out the test in coordination with the machine, robot, cameras and the safety system.

This test manager is mirrored to the user using the 'remote front panel' function of LabVIEW and this is the user interface through which the user carries out the tests. An overview of the entire system is seen in Fig. 11.

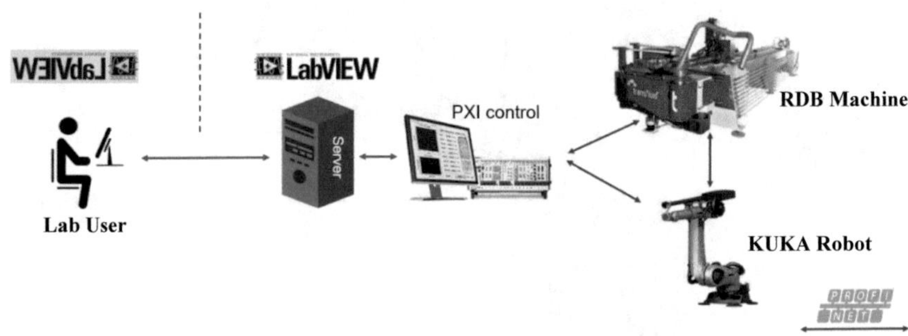

Fig. 11. Network topology of the remote tube bending cell

5.3 User Interface

Initially, the user interface is developed such that only the validation and testing loop, i.e. the actual bending process, can be conducted remotely. In the future, this will be expanded so that the entire lab can be done remotely. The functionalities of the user interface are as follows:

- Geometry data creation
  - User inputs the required geometry
  - Verify whether data is within the safety limit
  - Display the tube in 2D (Fig. 12)
- Machine data generation and transfer
  - The machine data is automatically generated based on geometry
  - Transfer of this data to the main controller
- Tools for observation
  - Camera systems to observe the process from different perspectives
  - Additional augmented reality tool for springback visualization and measurement. A prototype is seen in Fig. 13.
- Multiple Choice Question (MCQ) checkpoints
  - Checkpoints to check whether students have learnt the required information before moving on to the next stage

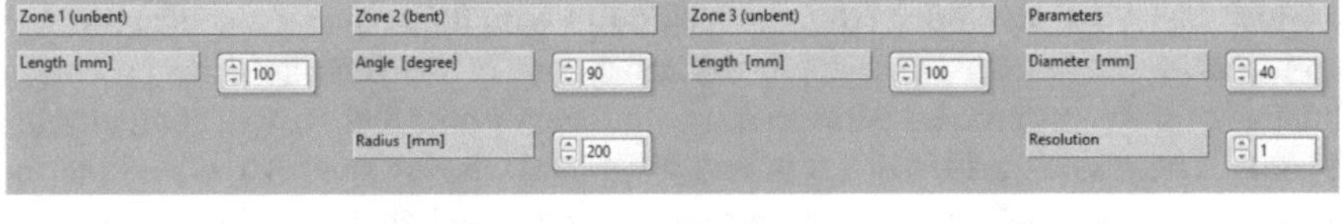

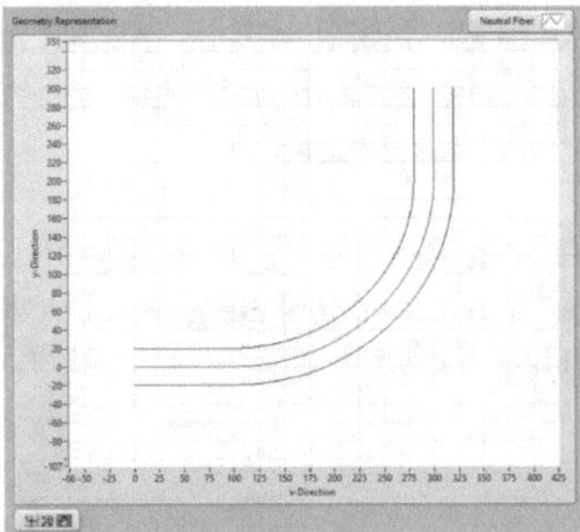

Fig. 12. 2D Geometry visualization based on user inputs.

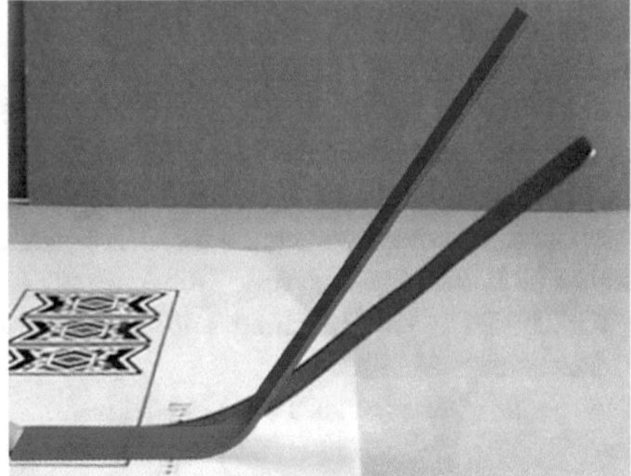

Fig. 13. Simple prototype of an augmented reality tool to visualize springback

  - Acts as a safety against improper or dangerous usage
  - Enables students to learn from their mistakes
- Data logging
  - Template using which the students record the necessary experimental data
  - Conscious decision to not offer automatic data logging as data recording is a key professionally relevant skill.

6 Conclusion and Outlook

Once the remote tube bending lab is fully functional, it will be offered to the students of the TU Dortmund university as an optional lab course. The access to the lab, just like in the case of the other labs at the department, will be gained by reserving a time slot. The lab which has been developed from the inside-out with the help of didactical tools, will help the students to gain an in-depth understanding of the tube bending process, the influence of different process parameters, the process limits and how to compensate for the same. These skills are essential in today's manufacturing scenario, where the focus is one "batch size one" production and there is only one chance to get it right. After evaluation of the effectiveness of the lab, it will be offered to students around the world from other partner universities. The desired outcome being a global increased interest in, and improvement of, engineering education.

Acknowledgements. The work was done as part of the "ELLI2—Excellent Teaching and Learning in Engineering Science" and the authors are grateful to the German Federal Ministry of Education and Research for funding the work (project no.: 01PL16082C).

References

1. Global Metal Forming Market for Automotives 2018-2025
2. Corter, J.E., Nickerson, J.V., Esche, S.K., Chassapis, C., Im, S., Ma, J.: Constructing reality: A study of remote, hands-on, and simulated laboratories. ACM Trans Comput-Human Interaction **14.2** (2 2007), 1–27 (2007)
3. Araujo, A.S., Araújo, A.S., Cardoso, A.M., Cardoso, A.M.: Pedagogical effectiveness of a remote lab for experimentation in Industrial Electronics. In: 2009 3rd IEEE International Conference on E-Learning in Industrial Electronics (ICELIE), pp. 104–108. Porto (2009)
4. Ortelt, T.R., Sadiki, A., Pleul, C., Becker, C., Chatti, S., Tekkaya, A.E.: Development of a tele-operative testing cell as a remote lab for material characterization. In: 2014 International Conference on Interactive Collaborative Learning (ICL), pp. 977–982. Dubai (2014)
5. Engineering Education 4.0—Excellent Teaching and Learning in Engineering Sciences, Springer International Publishing AG (2016)
6. VDI 3430: Rotary draw bending of profiles June (2014)
7. Li, H., Yang, H., Zhang, Z., Wang, Z.: Size effect related bending formability of thin walled aluminum alloy tube. Chin J Aeronaut **26**(1), 230–241 (2013)
8. Biggs, J.: Enhancing teaching through constructive alignment. High. Educ. **32**(3), 347–364 (1996)
9. Kolb, A.D.: Experiential Learning: Experience as the source of Learning and Development (2014)
10. Duch, B.J., Groh, S.E., Allen, D.E. (eds.): The Power of Problem-Based Learning. Stylus, Sterling, VA (2001)
11. Wills, T.W., Estow, S., Soraci, S.A., Garcia, J.: The AHA effect in groups and other dynamic learning contexts. J General Psychol **133**(3), 221–236 (2006)

Micro-modular myDAQ Labs

Doru Ursutiu[1(✉)], Andrei Neagu[2], and Cornel Samoila[3]

[1] Transylvania University of Braşov—AOSR Academy, Braşov, Romania
udoru@unitbv.ro
[2] Transylvania University of Braşov, Braşov, Romania
andrei.c.neagu@gmail.com
[3] Transylvania University of Braşov—ASTR Academy, Braşov, Romania
csam@unitbv.ro

Abstract. Many actual developments in mobile learning deal with "Remote Laboratories". The development of different electronics kits and the decrease in their price lead directly and easily to the idea of "student home labs" and empowered the idea of "learning by doing". Following discussions with many students we reached the idea of developing myDAQ Micro-Modular units able to be implemented in university laboratories or even by students in their homes. We started to develop and implement some easy to build and low price Micro-Modular devices for myDAQ (and/or myRIO) controlled by simple LabVIEW applications. These devices will be tested and developed during the next semester at the 2nd year bachelor students within the Electronics Measurement Laboratories.

Keywords: LabVIEW · Modular systems · MicroLabs · Mobile learning · Voltera · myDAQ · myRIO · etc.

1 Introduction

We present in this work a number of activities that can be implemented very fast and easy by using this systems: myDAQ/myRIO, student developed modules and Micro-Lab preliminary kits [1, 2].

Educators around the world can use the National Instruments Educational Laboratory Virtual Instrumentation Suite (NI ELVIS) and LabVIEW to teach concepts in circuit design, instrumentation, physics, etc. In the last years NI launched this small devices myDAQ (www.ni.com/myDAQ) and myRIO well suited for hands-on training [3–5].

One first successful development in this direction was done by our StudentEDEA module developed based on the PSoC1 programmable system on chip from Cypress. The PSoC1 embedded system is the first programmable, single-chip in the world. It integrates in a single chip: configurable analog and digital peripheral functions, analog and digital buss, memory and a micro-controller [6].

Last year, one of our PhD students presented a first attempt and was able to finish the laboratory. All 52 students prepared a final work on one small device, wired on the NI ELVIS system and controlled in LabVIEW. These students invited the Creativity

M. E. Auer and K. Ram B. (Eds.): REV2019 2019, LNNS 80, pp. 107–112, 2020.
https://doi.org/10.1007/978-3-030-23162-0_11

Laboratory industrial partners and concluded the Electronics Measurement Laboratories with a festive presentation of all this 52 modules.

Now we intend to move to the next level and build together with our students more elaborated and software assisted devices Micro-Modular systems and provide our students the necessary skills to create similar systems in their homes—useful for their personal training (learning by doing).

We started to do some simple specifications and select some small groups of students interested in the project design and development, starting from simple prototypes, up to a small mobile laboratory module.

Our main idea was linked to the decrease in dimension of components, thus obtaining both a minimum in energy consumption and a maximum of interactivity and mobility.

In some cases our students can quickly add remote control facility and thus use these systems in a collaborative manner of personal training and easy interactions with other colleagues.

2 First Developments

In one primary step, we have encouraged students to get involve in creating several sensors (Fig. 1) to test the concepts and ideas. As a first phase of the development we have been able to better understand the perspectives and also to better understand how we could integrate the concept.

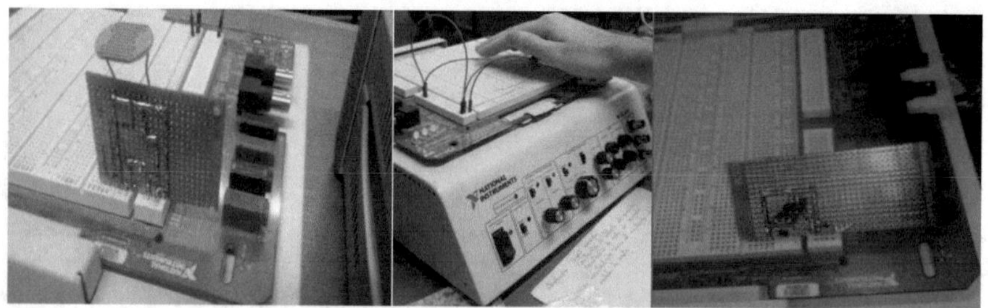

Fig. 1. One of 52 experiment NI ELVIS implemented by students

By all this approaches and preliminary steps we want to:

- Increase student involvement in learning
- Increase student—student interactions
- Increase student—professors interactions
- Increase student creativity
- Induce the first notions of student developed factories, etc.

3 Simple myDAQ Tools for Module Learning

After the first step we push students to develop one multi-use tool that connects to National Instruments myDAQ board, easy to use with different SAMPLE boards that are free (or at low price) in Creativity Laboratory for students training and development.

Here, in this category, we enumerate some of this simple boards: "Tiny low-power ADCs" from MAXIM (MAXDAClite+), TS1002 DEMO BOARD from Touchstone Semiconductor, LPC800-Mini-Kit manufactured by NXG Technologies for NXP, a lot of DIGILEND PMOD sensors and actuators, etc.

We like to present this prototyping system using two examples, how in Electronics Laboratory and in home experiments, based on myDAQ board and our add-on developed board, students can learn and understand, steps by step, electronics.

In Fig. 2 we present this add-on for myDAQ and one LabVIEW application developed to test and learn how to use the Touchstone Semiconductors TS1002 (and/or the TS1004) boards—with the industry's first and only dual or quad single supply, precision CMOS operational amplifiers fully specified to operate at 0.8 V while consuming less than 0.6 μA supply current per amplifier. These amplifiers are optimized for ultra-long-life battery-powered applications and of great interest now for all wearable devices.

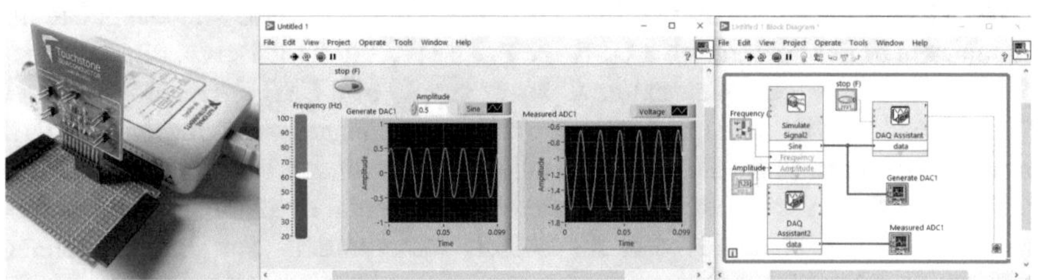

Fig. 2. Testing TS1002 with myDAQ and a simple LabVIEW application

The LabVIEW application was developed with express VI's and generate desired signal with DAC1 and receive amplified signal with ADC1. Students can change the frequency and for a deeper test they can change the levels of signals. This TS1002 board was powered from the myDAQ board and in the next steps students can measure also the low level consumption of this board or other parameters (for example to change the level of power in single-supply operation from 0.65 to 2.5 V).

The second example (presented in Fig. 3) offers the possibility to test the MAXIM kit MAXADClite+ using signal generated/measured directly by NI MAX and compared with the information's received directly from the Graphical User interface GUI developed by MAXIM for testing this kit.

The MAXADClite+ Rev-D evaluation kit (EV kit) evaluates the MAX11645, Maxim's smallest, very-low-power, 12-bit, 2-channel analog-to-digital converter (ADC). The EV kit allows usage of the ADC's internal reference or an externally applied reference voltage.

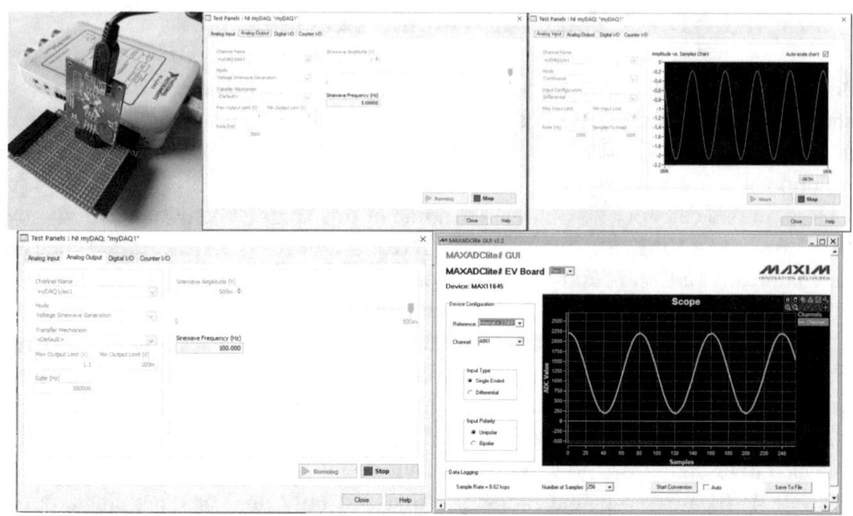

Fig. 3. Test MAXADClite + with NI MAX generated signal and MAX GUI monitor

The EV kit includes Windows compatible software GUI for data acquisition through a USB cable.

We selected this evaluation kit/board because it is a plug-n-play data-acquisition kit that connects to the PC through a USB cable. The EV kit digitises two different analog inputs and does not require an external power supply or a USB device driver. The EV kit is preloaded with default firmware that communicates with the MAXADClite evaluation software. Software can be installed and run on any Windows-based system. The ADC on the EV kit board can be selected for internal or external reference mode. The internal 2.048 V reference is the default configuration. The external reference mode can be selected through software.

Now we increased the myDAQ/myRIO add-on board complexity (especially for myRIO applications) adding one another extension connector useful for extended functionality with other boards but especial important for Digilent PMOD boards [7].

Also we add (see Fig. 4) two BNC for more easy connection within university laboratory or at home for signal generators and oscilloscopes/analysers.

4 Developments of MicroLab Final Modules

After this steps we need to teach students how to arrive at the final products and maybe to push them to develop some "student factories" and start selling their products.

In Creativity Laboratory of "Transilvania" University of Brasov we decided to use the new release of Voltera system (https://www.voltera.io). We invested in one Voltera V-One circuit printing system and his drilling head [8].

Voltera is an all-in-one PCB printer that is revolutionising the way that PCB prototyping is done. We selected Voltera, based in Canada, because we understand that the product they manufacture could be revolutionary to PCB prototyping in the same

Fig. 4. Add-on board with PMOD and BNC connectors used with myRIO

way that 3D printers changed the traditional prototyping industry. Thousands of businesses now print prototypes in plastic in-house on 3D printers instead of shipping CAD files off to machine shops for prototypes. We think the V-One will revolutionise PCB prototyping for electronics designers, university researchers, and electronics institutions in much the same way.

From faster prototyping and quick iterations, to new conductive materials research, to testing out new flexible and thin sheets for traces the possibilities for this versatile and accurate solder dispensing printer are seemingly limitless. The V-One can already be found at many of the leading research institutes and universities across the world.

In this manner we can teach students: circuit simulation (in our case using Multisim from National Instruments NI or any other), and from here move to developing directly the corresponding PCB (we use Ultiboard from NI but can be any other software). Voltera's software is designed to be able to import your Gerber files to the moment you press print, the software walks you through each step with built in videos and tutorials. Compatible with EAGLE, Altium, KiCad, Mentor Graphics, Cadence, DipTrace and Upverter.

In Fig. 5 we present the Voltera system and its ability to develop normal PCB's and flexible electronic devices, well-targeted for sensor and IoT device developments and in our case the optimal device for fast small MicroLab components development and test for final implementation.

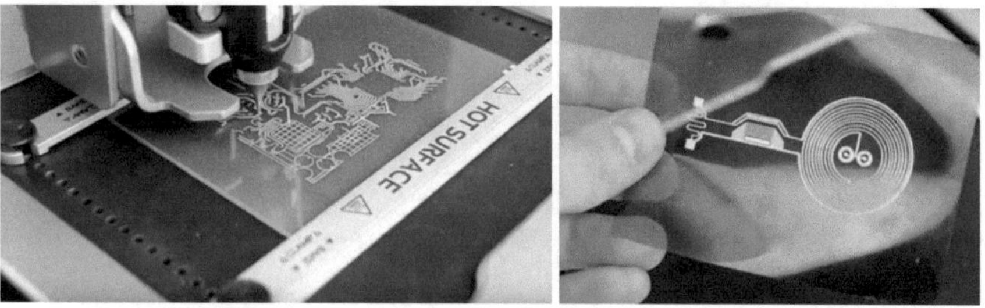

Fig. 5. Voltera printing one PCB and one flexible PCB

5 Conclusions

The MicroLab components presented in this paper was tested in the 2nd year of bachelor degree in Applied Electronics and will enter in final test next semester.

Till now we have successfully involved more than 50 students who developed some preliminary models, especially with sensors.

Now we developed some intermediate boards for myDAQ and integrated this student developed modules and some special "samples" received from different companies. At this point, we are ready to develop hardware and software for the new concept of MicroLab using at the entry level knowledge the myDAQ and at more advanced level the myRIO National Instruments systems.

The binder of the two technologies and the final implementation is facilitated by the new acquired VOLTERA system that must be put into direct use in the curriculum specialisation of the Applied Electronics bachelor studies.

Acknowledgements. We like to thanks to Voltera for their support and fruitful discussions and to National Instruments for great and permanent support offered to Creativity Laboratory of CVTC from Transilvania University of Brasov.

References

1. Presentation of myDAQ device: http://www.ni.com/pdf/manuals/373061f.pdf
2. Presentation of myRIO device: http://www.ni.com/pdf/manuals/376047c.pdf
3. Toribio Destro, F.H., Costa, R., Iaione, F.: A low-cost system for experiments with digital circuits. In: Frontiers in Education Conference (FIE) 2015 IEEE, pp. 1–6 (2015)
4. Essick, J.: Hands-on Introduction to LabVIEW for Scientists and Engineers. Oxford University Press, 2015, ISBN 019021189X, 688p
5. Ehsani, B.: Data Acquisition Using LabVIEW. Packt publishing, 2016, ISBN 9781782172161, 122p
6. Ursutiu, D. et al.: StudentEDEA and myDAQ in education & research laboratory. In: 2014 11th International Conference on Remote Engineering and Virtual Instrumentation (REV) (2014), pp. 67–69
7. Digilent PMOD IO Peripheral Modules: https://store.digilentinc.com/pmod-modules
8. Voltera system review: https://techcrunch.com/2015/01/06/the-voltera-v-one-makes-circuit-boards-in-minutes

Interactive Lab for Electronic Circuits

K. C. Narasimhamurthy(✉), S. R. Kavyashree, K. Kavana, G. Maithri Vaidya, and D. K. Kumuda

Siddaganga Institute of Technology, Tumkur, India
kcnmurthy@sit.ac.in, {kcnmurthy, srkavyashree061997, kavana.k46, maithrivaidyag, kumudamurthy}@gmail.com

Abstract. This paper describes an innovative method of self Learning methodology for Analog Electronic circuits. In this self learning methodology, Common Emitter (CE) amplifier is considered for demonstration. It explains every step involved in, how a user accessing the CE amplifier circuit remotely can interact with the circuit and verify own circuit design by conducting an experiment. National Instrument's LabVIEW, Analog Discovery-2 NI Edition, Arduino, and Digilent Pmod Digipot are used. This is a first step towards interactive lab (*i*Lab).

Keywords: Interactive lab · LabVIEW · Analog electronic circuits · Analog Discovery-2 · Digipot

1 Introduction

Engineering education is all about "Do Engineering" rather than "Read and Write Engineering". To enable this Do Engineering student need to be guided while conducting experiments to understand the concepts. However, it's not possible to have a professor/instructor with the student whenever he feels to conduct the experiments from anywhere at any time. The best solution is to have remote lab access in which student can conduct experiments from anywhere at any time. This remote lab must also have the capability of allowing student to verify their own design and must also provide assistance to get the desired output.

Many techniques have been introduced to improve the learning ability of students. These include use of ICT tools in classroom teaching, Flipped classroom, portable lab etc. Virtual/Remote labs have been augmenting the hands-on physical laboratories in several ways by allowing students to learn most elements of the experiment prior to physically seeing or experimenting with the instrumentation. Presently, there are many remote/virtual access are there for students for "Do Engineering" [1–8]. In νLabs [1] user can simulate analog experiments; however the user will not feel the real hardware challenges and real time signals. In VISIR [2], user can build basic analog circuits and conduct the experiments, however there no guidelines for effective conduction of experiments. vLabs [3] allows students to access the software anytime from anywhere. In all existing remote labs, if a student's fails to get the output for their own design, there is no online help available to get the desired output.

M. E. Auer and K. Ram B. (Eds.): REV2019 2019, LNNS 80, pp. 113–122, 2020.
https://doi.org/10.1007/978-3-030-23162-0_12

This paper describes, how circuit will communicate with the user to get the required output by providing guidance through proper instructions throughout the conduction of experiment using Interactive lab (iLab) platform. User while performing experiments using *i*Lab will have the similar experience of conducting the experiments under the supervision of an instructor. Measurements at the key test points are made using National Instruments (NI) Analog Discovery-2 NI Edition and Arduino. This *i*Lab gives a platform for the students to solidify their concepts learned in the classroom by repeatedly doing experiments with different designs. As first step towards *i*Lab, experiments on CE amplifier is considered as a proof of concepts.

*i*Lab is also used as platform for experiential learning in Analog Electronic circuit courses, in which the verification of most of the circuit concepts can be demonstrated in the classroom itself by remote access.

The *i*Lab will provide crucial guidance during conduction of experiment, provides waveforms at important test points and their voltage values. At present, Artificial Intelligence has become a fascinating technology in designing human robots. Similarly this project is designed in such a way that, a system could able to detect the faults in the circuit which normally requires human intelligence. This enables the user to perform experiments without any difficulties.

2 Overview of *i*Lab

*i*Lab can be implemented to conduct experiments of various electronic circuits. It helps students to verify their own paper design by conducting the experiments whenever they want to solidify their concepts.

Figure 1 depicts conduction of experiments in the conventional laboratory and using *i*Lab platform.

In conventional Lab, student has the support of an instructor to debug the circuit to verify the paper design. In *i*Lab, circuit with the help of LabVIEW gives the necessary instructions to the user for verifying the paper design. Here the circuit itself will interact with the student.

3 Realization of iLab

*i*Lab is implemented using LabVIEW, Arduino. Analog Discovery-2 NI Edition, Digilent PmodDPOT [4]. Analog Discovery-2 is like a portable lab with features like oscilloscope, power supply, waveform generator, network analyzer etc., which are essential for conduction of any electronic experiments. Waveforms 2015 is the virtual instrument suite for Analog Discovery-2. LabVIEW with necessary drivers can communicate to Arduino and Analog Discovery. Arduino is used as a backend tool to provide options for the user to set the desired component values. SPI protocol is used to establish the communication between Arduino and LabVIEW.

CE amplifier circuit shown in Fig. 2 is used for demonstrating *i*Lab implementation. The design equations of the amplifier are listed below

Fig. 1. Overview of conventional lab and *i*Lab

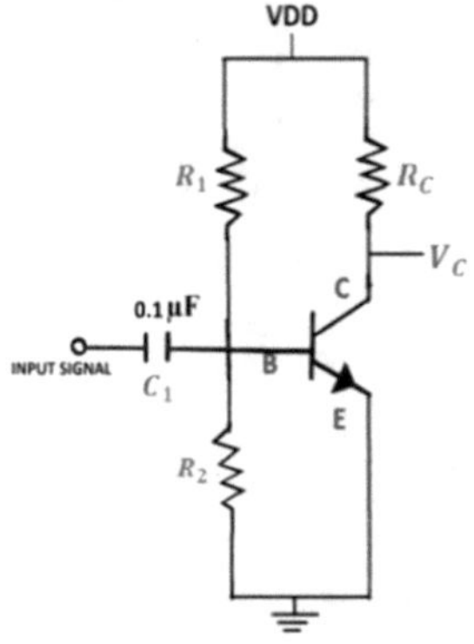

Fig. 2. CE amplifier

$$\text{Voltage gain } A_v = -g_m R_C \tag{1}$$

$$\text{Trans conductance } g_m = I_c V_T \tag{2}$$

$$I_c = \frac{V_{DD} - V_C}{R_c} \; or \, I_c = I_{CBO} \, e^{-V_{BE}/V_T} \tag{3}$$

$$V_{BE} = \frac{V_{DD} R_2}{R_1 + R_2} \tag{4}$$

Design equations relieves that, the voltage gain of the CE amplifier (A_v) can be varied by changing R_2 or R_c. User can do any number of paper design for CE amplifier to amplify ac signal, by choosing any value of R_2 and/or R_c. The same can be verified using *i*Lab, during the experiment conduction steps, the *i*Lab system will guide the student to change the biasing resistor values if it from the desired DC operating conditions. Resistors R_2 and R_c are realized using Digital potentiometer as shown in Fig. 3 so that user can modify the value remotely. The digital potentiometer is implemented using Digilent PmodDPOT, a digitally-controlled electronic component that mimics the analog functions of potentiometer. The Pmod DPOT utilizes Analog Devices AD5160 to digitally set a desired resistance between two terminals. With 256 possible step values, a resistance between 60 Ω and 9961 Ω can be programmed. User can change the range of resistance values by changing the design configuration with respect to paper design.

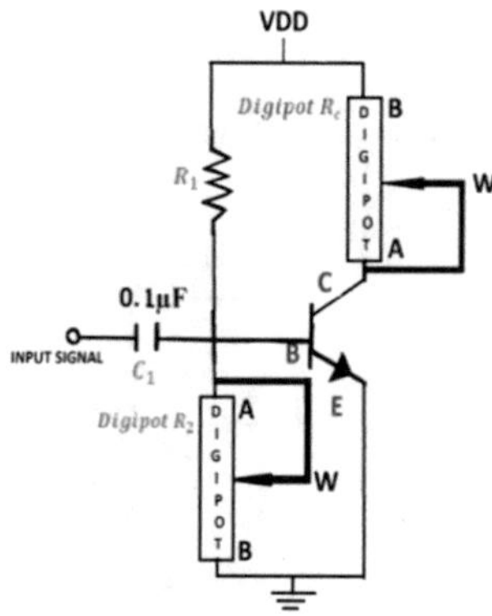

Fig. 3. CE amplifier with R_C & R_2 is realized by Digipot

The block diagram of *i*Lab is shown in Fig. 4. LabVIEW is used as a frontend tool in *i*Lab. LabVIEW having Front panel and Block Diagram providing the user to choose options for different parameters (like resistance values). It enables to set the desired resistance (Pmod Dpot) value using Arduino. Analog Discovery 2 is used to measure, visualize, generate, record, and control mixed-signal circuits of all kinds. Arduino also is used to read the analog voltages from desired nodes of the circuit.

4 Conduction of CE Amplifier Experiment Using iLab

User Block Diagram and Front Panel of LabVIEW developed for *i*Lab to perform CE amplifier are shown in Figs. 5 and 6.

Interactive experiment concept using LabVIEW programming for CE amplifier is shown in Fig. 5. The graphical representation programme is carefully coded by considering all key aspects of the CE amplifier such that, it can provide guidance to the user even at extreme failure conditions. Front panel of the Interactive lab is designed to

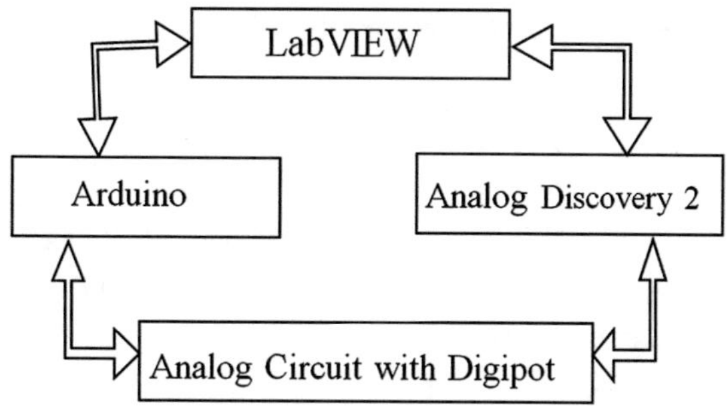

Fig. 4. Block diagram of interactive lab

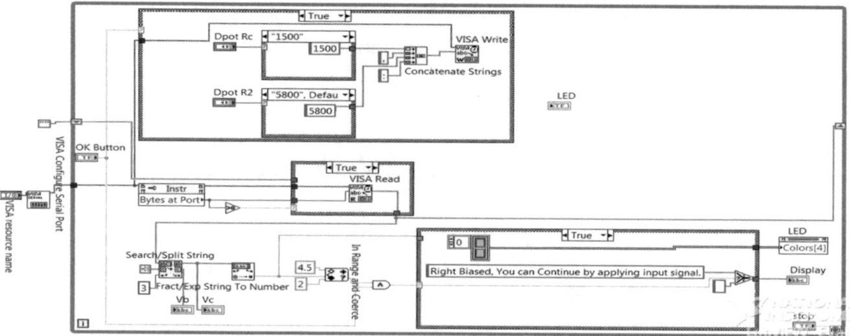

Fig. 5. LabVIEW block diagram of interactive lab

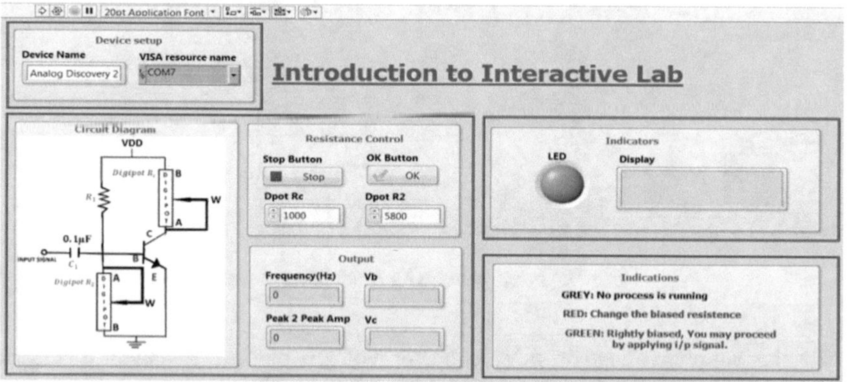

Fig. 6. LabVIEW front panel of interactive lab

provide best possible interaction between the circuit and user in absence of the instructor.

User should follow the steps given below to conduct the experiment.

- User must run the program by clicking Run button.
- In order to proceed with the experiment, one should click OK button.
- Choose the Digipot resistance of desired value from the drop-down menu of resistors R_c and R_2.
- Observe the DC voltages at collector and base of the circuit at V_c and V_c respectively
- If DC bias conditions are not satisfactory, LED colour will be **Red** and it will be **Green** if the DC biasing is satisfactory.

Detailed analysis of the CE amplifier circuit at various operating conditions is discussed in next section.

5 Analysis of CE Amplifier Using Interactive Lab

Once the user is ready with the paper design, the designed value of biasing resistors may select using the front panel as shown Fig. 7. The DC bias voltages at base and collector are 0.67 and 4.67 V respectively as shown in Fig. 7.

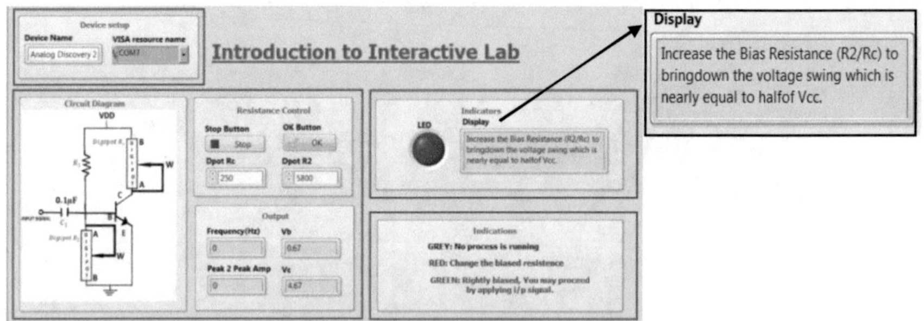

Fig. 7. Front panel with guidelines to user

As the supply voltage is 5 V, collector voltage of 4.67 V leaves a very less headroom (330 mV) for the output signal to swing (on the +ve half cycle). This DC condition is not satisfactory for amplification and needs modification. This will be conveyed to the user by the circuit itself by RED LED and displaying suggestions to vary the bias resistors as shown in Fig. 7.

If the user modifies the biasing resistors and selects resistors value as shown in Fig. 8 and press OK button. The circuit uses the new value of resistor and reads the latest node voltages at base and collector as 0.67 and 0.12 V respectively. As the collector voltage is 120 mV, this leaves less than 50 mV swing for the output signal (on the −ve half cycle). As this is also not a suitable DC conditions for amplification again RED LED and display suggesting change in the biasing resistor values as shown in Fig. 8.

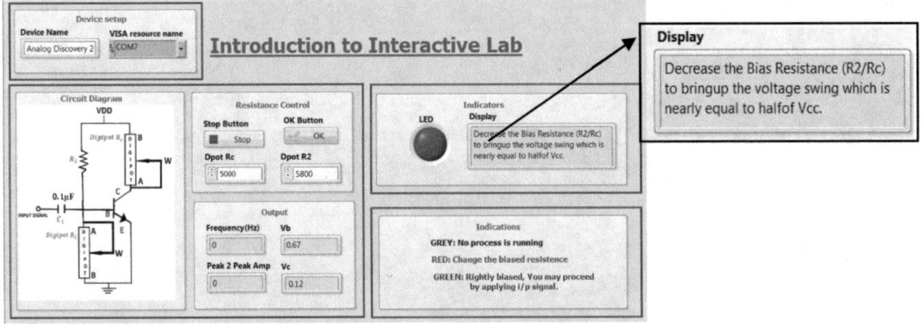

Fig. 8. Front panel with guidelines to user

In both the cases the DC voltage at the collector is nowhere close to 2.5 V ($1/2\ V_{cc}$), now user should do redo his paper design and estimate proper value of resistors that may provide satisfactory DC bias conditions for amplification.

Assume user has selected biasing resistors as shown in Fig. 9 and node voltages at base and collector are 0.67 and 3.84 V respectively. This collector voltage will provides at least 1.16 V swing for the output signal (on the +ve half cycle) and maybe considered as suitable DC bias conditions even though not the best one (2.5 V). This will be indicated by a GREEN LED and display suggesting the user to apply input ac signal for amplification.

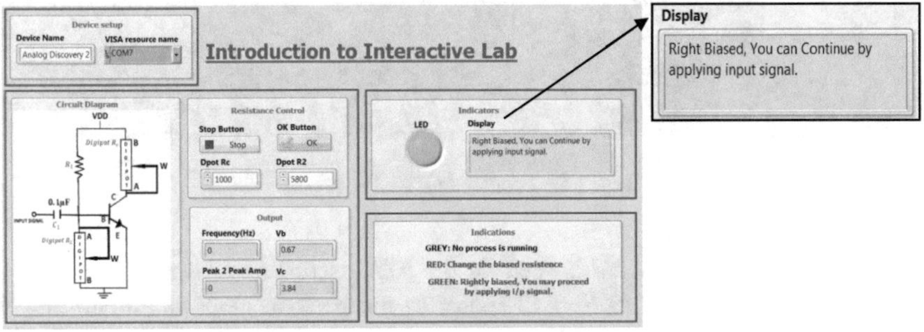

Fig. 9. Front panel with guidelines to user

DC voltages at base and collector nodes of the circuit can also be displayed on two-channel oscilloscope as shown in Fig. 10.

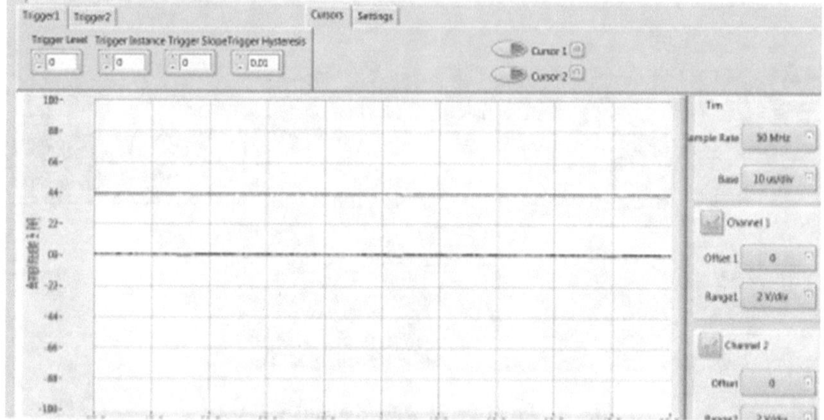

Fig. 10. Front panel showing scope

As the DC conditions are satisfactory for amplification, using wavegen option user can apply sine wave signal of suitable amplitude and frequency for amplification as shown in Fig. 11

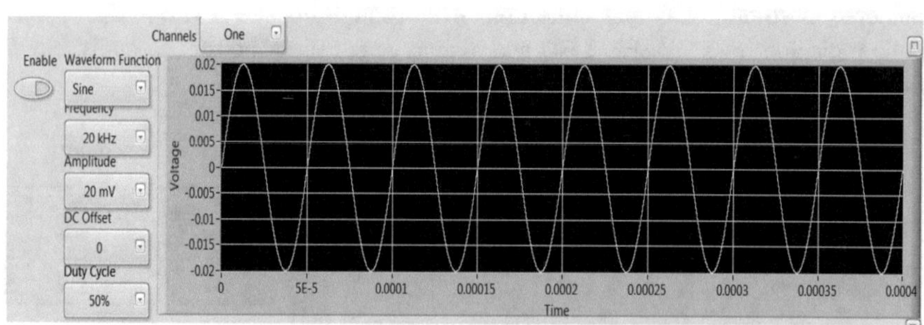

Fig. 11. Front panel of Wavegen

The amplified output signal is seen on the oscilloscope as shown in Fig. 12.

The front panel shown in Fig. 13 indicates the peak to peak amplitude and frequency of the output signal as 621 mV and 20 kHz respectively, for input signal of amplitude 40 mV. Table 1 shows the both theoretical and practical voltage amplification for various R_c values and with other biasing resistors as R_1 = 33 kΩ and R_2 = 5.8 kΩ. It is also possible to vary R_2 value and do verification of voltage gain, however variation in gain is not appreciable as in case of R_C.

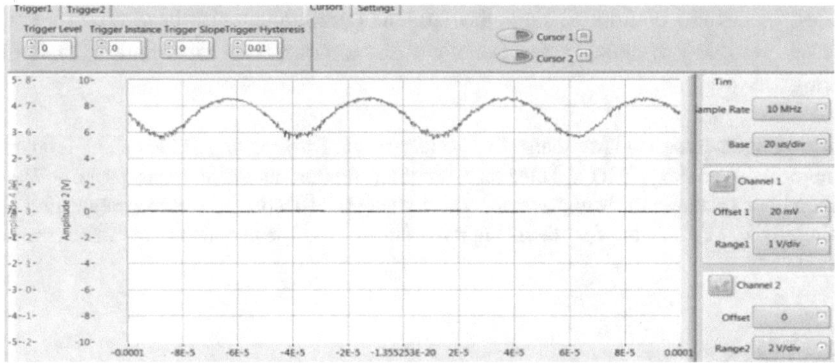

Fig. 12. Amplified output waveform on the scope

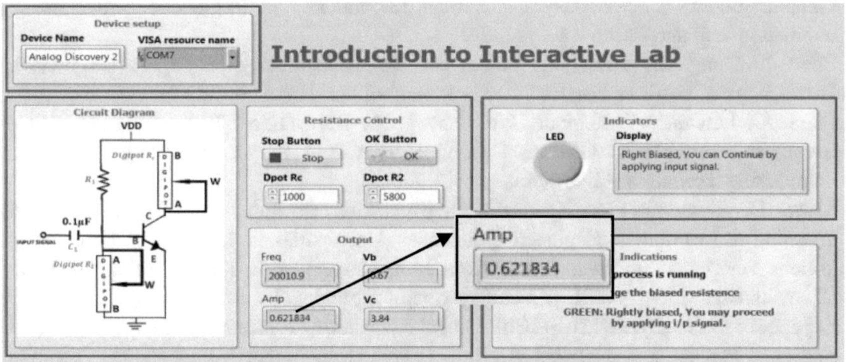

Fig. 13. Front panel of interactive lab

Table 1. Comparison of CE amplifier theoretical and practical voltage gain for different values of R_C

Resistance (R_C) (Ω)	Theoretical gain (A_v)	Practical gain (A_v)
1000	49.82	44.6153
1500	74.7	65
2200	109.5	94.2307

6 Conclusion

Paper titled "Interactive Lab for Analog Circuits" has described the Interaction between the CE amplifier circuit and the user to obtain the desired output. The circuit has instructed user to change the design to get the intended result. This iLab is a proof of concept using this methodology it's possible to develop similar systems for many

Analog electronic circuits to help the student community. As these circuits can be accessed remotely it makes more beneficial for anyone from anywhere to "Do Engineering" at anytime.

Acknowledgements. Authors thank the Management, Director and Principal of Siddaganga Institute of Technology (SIT), Tumakuru for their support in making this project. We also extended our gratitude to faculty members and students of Dept. Of Telecommunication Engineering, SIT, Tumakuru for their support during the development of this innovative methodology.

References

1. http://vlab.co.in/
2. http://ohm.ieec.uned.es/portal/?page_id=76
3. http://www.uml.edu/IT/Services/vLabs/
4. Balamuralithara, B., Woods, P.C.: Virtual Laboratories in Engineering Education: The Simulation Lab and Remote Lab (2007)
5. Chen, X., Song, G., Zhang, Y.: Virtual and remote laboratory development: a review. In: Proceedings of Earth and Space 2010, pp. 3843–3852, Honolulu (2010)
6. Chen, X., Lawrence O Kehinde, P.E., Zhang, Y., Darayan, S., Texas Southern. David O. Olowokere, Mr. Daniel Osakue, USING Virtual And Remote Laboratory To Enhance Engineering Technology Education
7. Kruse, D.: Sulamith Frerich from Ruhr-Universität Bochum, Project ELLI - Excellent Teaching & Learning in Engineering Sciences, Universitätsstr. 150 | 44801 | Germany and Marcus Petermann, Andreas Kilzer from Ruhr-Universität Bochum, Particle Technology, Universitätsstr. 150 | 44801 | Germany titled "Virtual Labs And Remote Labs: Practical Experience For Everyone" 2014 IEEE Global Engineering Education Conference (EDUCON)
8. Nedic, Z., Machotka, J., Nafalski, A.: Remote laboratories versus virtual and real laboratories, 33rd Annual Frontiers in Education, 2003. FIE (2003)
9. https://reference.digilentinc.com/reference/pmod/pmoddpot/reference-manual

Design and Implementation of an Architecture for Hybrid Labs

Lucas Mellos Carlos[1,2(✉)], José Pedro Schardosim Simão[3], Hamadou Saliah-Hassane[2], Juarez Bento da Silva[1], and João Bosco da Mota Alves[1]

[1] Federal University of Santa Catarina, Araranguá/SC, Brazil
lucas.mellos@posgrad.ufsc.br,
{juarez.silva;joao.bosco.mota.alves}@ufsc.br
[2] TELUQ University, Montreal/QC, Canada
{lmellos},hamadou.saliah-hassane@teluq.ca
[3] Federal Institute of Santa Catarina - Campus Tubarão, Tubarão/SC, Brazil
jose.simao@ifsc.edu.br

Abstract. This work aims to describe the design and implementation of an architecture for hybrid labs focused in the integration of remote and hands-on laboratories. The research was conducted following four different steps and based on applied procedures. The first step describes a systematic and exploratory literature review, in which the results were used as the basis for this work following by applied procedures for the next two steps. For the evaluation of the architecture, was adopted a hypothetical-deductive methodology that brings an analysis of two different learning scenarios. The application was designed to support two different interaction formats - following a generic-based and project-based use. These formats were designed in order to interact on the collaborative learning environment application adopting the concept of mobile laboratories. The findings revealed that the application might be addressed in different approaches adopting different learning methodologies, being the solution developed able to be used in consonance with hands-on and remote activities.

Keywords: Hybrid labs · Online labs · Learning scenarios

1 Introduction

Science laboratories are indispensable in education. Experimentation is critical factor in engineering courses [1]. In this sense, hands-on laboratories play a crucial role when knowledge is constructed, these laboratories are part of the real world, where students will act as professionals after graduation [2].

Online laboratories allow laboratory practices using electronic devices outside classroom [3,4]. They are divided by their nature as remotes virtual and hybrids. These laboratories encompass a significant number of characteristics related to their capacity to provide services, aligned with interoperability, quality and expandability factors from different natures [5].

M. E. Auer and K. Ram B. (Eds.): REV2019 2019, LNNS 80, pp. 123–142, 2020.
https://doi.org/10.1007/978-3-030-23162-0_13

Online and hands-on laboratories might be used as a complement to each other in order to favor the learning process. They have higher availability, 24h/7, enabling students to perform their experiences at any time and anywhere, since they have an Internet connection.

Hands-on labs encompass a significant number of limiting factors, related to their acquisition cost, maintenance, and availability to a limited number of users in a limited period of time [6].

Considering remote labs, they have limiting factors related to the format which an experience can be performed on it. In this sense the user is not totally free to interact physically with the laboratory, only setting inputs and receiving outputs from the laboratory.

Hybrid labs that merge remote and hands-on laboratories aim to join the qualities from both, the capability of build an own laboratory and share with others over the Internet adopting the concept of the remote lab [7].

Considering this new approach, it is necessary to adopt different learning scenarios, considering each laboratory model to favor the learning process regarding the different interactions models in each model of the online lab [8].

Hybrid labs that combine remote and hands-on are closer to traditional labs, once students can act in developing their practices and share with others using the Internet.

Regarding this statement, this project was developed according to the following research question: "How to design learning scenarios and implement an architecture for hybrid labs based in remote and hands-on labs?".

Based in the research question, this paper aims to describe the design and the implementation of an architecture for hybrid labs focused in the interaction of remote and hands-on laboratories at the same time. The architecture enables the design of different learning scenarios allowing users to share their labs with others students or professors in the same collaborative environment.

This paper was structured in five following sections. The first and current section describes the primary elements in this paper. The Sect. 2 describes the theoretical background related to online labs and their correspondent fields in this paper. The Sect. 3 describes the methodology adopted among the development of this paper. The Sect. 4 presents and discuss the results and findings in this paper. The last section shows the final remarks in this research.

2 Theoretical Background

Regarding to the theoretical background basis from this work, this section aims to describe online laboratories and their evolution among the years related to learning scenarios.

2.1 Online Laboratories

Online laboratories are considered a powerful tool that enable students to perform activities repeated times breaking problems associated with theoretical and practical activities [3].

In the start of the 2000s, new visions were aggregated in learning when adopting online laboratories. In this advent, new concepts like Mobile Learning (M-Learning), Mobile Remote Experimentation (MRE) and others began to appear.

One of the most difficult challenges in this context of remote instrumentation in engineering courses is promoting the most realistic environment. Online laboratories are divided into three main areas, based in their nature in remote, virtual and hybrid labs, considering the vision adopted by Balamuralithara and Woods [9], Zutin et al. [10].

Those labs are divided considering their nature and their location, in this sense, they can be from nature virtual or real, and the location can be local or remote.

Authors like [7] purposed a matrix to explain the combination between location and nature and the union of more than one category, resulting in hybrid models. The Fig. 1 shows the matrix developed by the authors.

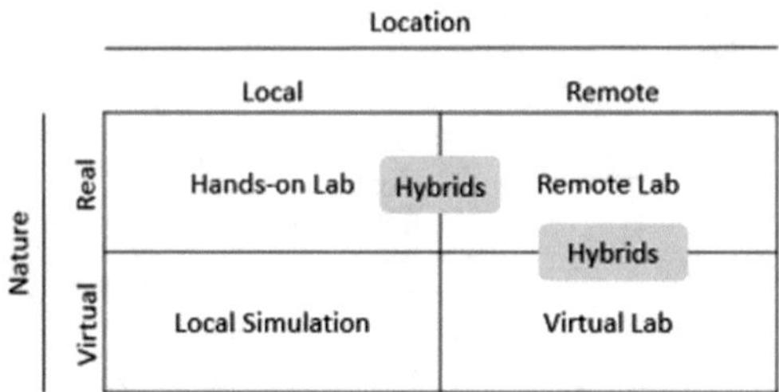

Fig. 1. Labs classification [7].

2.2 Remote Laboratories

Remote laboratories are characterized by a set of hardware that passes for an adaptation process using specific software architecture to enable students in anywhere and any time to perform activities via the Internet [11,12].

The primary characteristic of remote labs is given by the definition of *inputs* values in actuators to receive *outputs* that can be retrieved using sensors or visualization mechanisms.

A generic architecture model was described by [5], in which defined layers, communications protocols and critical elements in the development of a generic architecture for remote laboratories.

The architecture was based in three different layers, being: (1) laboratory consumer; (2) laboratory provider; and (3) physical laboratory.

The first layer, related to the users, enables access to all contents provided for them over an RLMS, LMS (Learning Management System) or others. The second layer, related the laboratory provider, describes the connectives elements, such as the Graphical User Interface (GUI).

The third and last layer, describes the physical lab. It details the communications protocols, elements, and format that they communicate with each other physically. The Fig. 2 shows the diagram developed by the authors and their connections between layers.

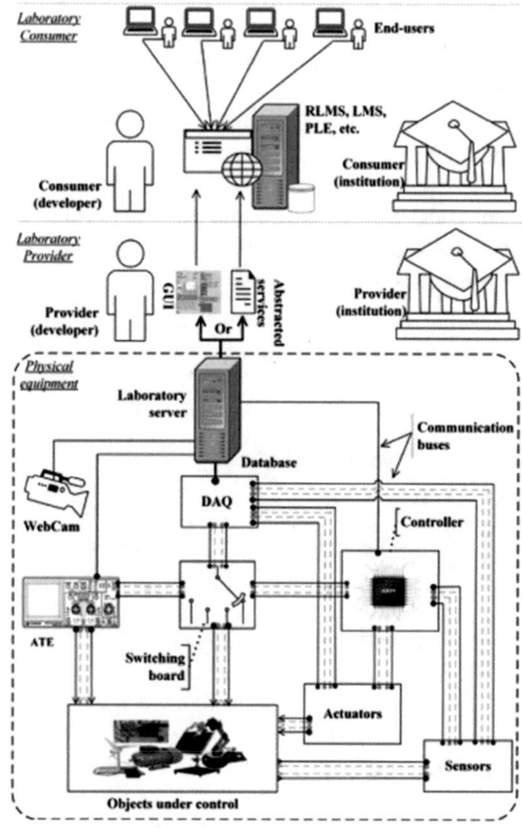

Fig. 2. Generic architecture for remote laboratories developed by [5].

2.3 Virtual Laboratories

Virtual laboratories have started to be discussed in the XX Century, when authors like [13] proposed the use of virtual laboratories in engineering courses.

The first ideas were to offer introductory concepts in practical courses. This type of laboratory has been commonly associated in the field of simulations, being used as an auxiliary tool [14].

The impact caused by those laboratories are highlighted in how activities are performed on it. Commonly, it has a computational device processing information and accessing Web-Services via Internet connections.

Simulations can be executed in standalone mode when it has a computational program with all elements prepared to execute a specific application.

2.4 Hybrid Laboratories

Hybrid laboratories can appear from two different derivations, from the union of remote and virtual or hands-on and remote [10,14].

Authors like [15,16] present some examples of hybrid laboratories based on remote experimentation using virtual technologies. This model can be built adopting virtual worlds, argument reality, exploring these technologies although serious games. The same lab might be applied in different teaching levels using different approached.

This integration is indicated when is desired a higher number of details in experimentation. The student is able to verify non-visual elements, such as electrical conduction in laboratories related to physics studies.

The union of remote and hands-on is highlighted in works of [17]. The authors describe a model that join both natures in the same architecture adopting the concept of *flipped* lab.

Practices that are developed using this type of laboratory may combine different numbers of elements, how for example, elements of sensors network. It can help in practices that are performed on online collaborative environments, encompassing the scalability and interactivity increment.

2.5 Online Collaborative Environments

Authors like [18] describe the traditional context of online collaborative environments where commonly they have computers' connected via the Internet, in a classes propitious to read.

In studies made by Curtis and Lawson [19], the authors report that is necessary to analyze the user behavior to discover which types of knowledge's are shared and how - via e-mail, chat or others. In another way, [20] affirms that in these activities students will be engaged if evaluated.

These models of environments have a direct inference in learning methods in order to build knowledge. In those collaborative environments, the effectiveness will be related in features developed or relevant functions in learning spaces through users behavior measurement [21].

However, taking in consideration online labs, more precisely remote labs, the most part of RLMS were developed to be used in individual mode, this fact in the most of times reduce scalability factors.

Others methods that aims to reduce this factor is the use of non-interactive labs, where the user watches records of laboratory execution and not the execution in real-time [22].

3 Methodological Approach

The research followed a qualitative approach, which according to [23] can be related to an attempt to explain more deeper the results and their characteristics.

This research was conducted over four different steps. The first step, a bibliographical review was made to compose the theoretical basis about architectures for hybrid laboratories in the context of remote laboratories.

The next steps were the project pilot and the development, respectively. The four and last step was an analysis of the learning scenarios and their limiting factors.

For the first step, about the bibliographical analysis, was performed an exploratory search on the literature about online laboratories followed by a systematic search on IEEE Xplore® and Scopus®, regarding to architectures for hybrid laboratories, centered in remote and hands-on laboratories.

The search was query was build based in an adaptation of [24,25] search model.

The query used is presented by the Fig. 3. The search resulted in 85 works, being 80 on IEEE Xplore® and five on Scopus®. These works were grouped on a bibliographic manager software, and after removed duplicates, the final result was the number of 84 works.

```
TOPIC = (
( "remote lab*" AND hands*)
AND ( "hybrid lab*" OR mixed OR flip* )
AND ( "online learning" OR education OR learning )
)
```

Fig. 3. Systematic search query adopted.

The second step was generated from the previous one. The most relevant works were analyzed in the fields of this research. This step, referent to the project development, was conducted using a task stage model in a task management tool to group elements necessaries among the pilot development and evaluation.

The architecture in three main layers was developed in the third step. The architecture was composed by the *lab layer*, in consonance with a *smart gateway* and an online collaborative environment.

The fourth and last step, refers to an analysis of the learning scenarios. The analysis was performed following an analysis about features and tools available to enhance the learning on each scenario based on [26] precepts.

4 Results and Discussion

This section aims to describe the findings and discuss them in four following sub-sections.

The first will describe three different architectures of remote labs and determine their particularities following the literature review.

The second sub-section will present the architecture developed, supported by a discussion about the learning scenarios based on the architecture implemented, described in the final sub-section, respectively.

4.1 Management Environments and Architectures for Remote Laboratories

RLMS is an acronym for *Remote Laboratories Management System*, an environment that storage communication interfaces and information related to laboratories and their users [11].

The development of an RLMS can vary according to the architecture adopted to communicate with the laboratories. Different layers can be added, such as access queue, booking, authentication, and others.

The next three sub-sections will be destined to discuss and analyze three different architectures and RLMS designed for remote laboratories.

WebLab-Deusto & LabsLand The WebLab-Desuto is an RLMS developed at Deusto University in Spain, released in 2000. How others RLMSs, the WebLab-Deusto has some particularities, how for example a specific module to integrate the system OpenLabs Electronics Laboratory, powered by VISIR remote laboratory [27].

The Fig. 4 shows the architecture adopted by WebLab-Deusto. The laboratories are hosted in rooms, making connections be forwarded to the laboratory room server.

Another WebLab-Deusto RLMS characteristic is that it provides support for Gateway4labs - a smart gateway that manages connections although different institutions and environments. The use of Gateway4labs enables different institutions to share their resources based on the same level using specific plug-ins of support [28] enhancing scalability factors.

iLab Shared Architecture Developed at Massachusetts Institute of Technology (MIT), the iLab Shared Architecture (ISA) has a shared architecture that provides different functionality' although an unified framework [29].

The ISA integrates three different models of laboratories:

- **Batch**: Laboratories where the user can configure and send inputs to the server to execute when it be possible, for example, when the lab be available;
- **Interactive**: Laboratories where the user interact in real-time;
- **Sensor**: Laboratories where the user can only measure sensor values.

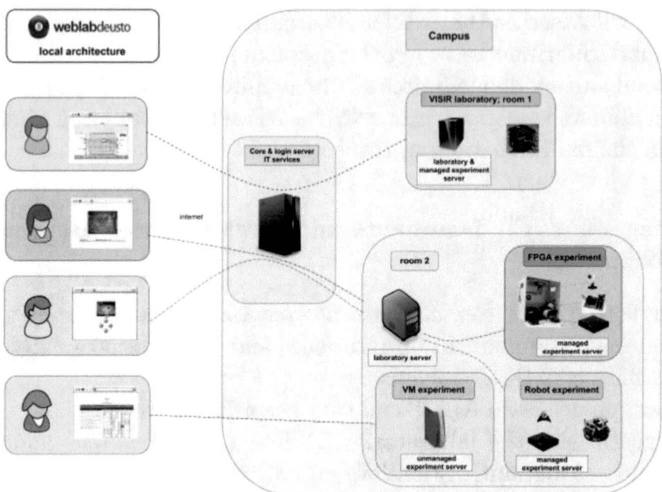

Fig. 4. WebLab-Desuto architecture [27].

The most common model, for interactive laboratories, uses an Interactive Service Broker (ISB) in order to congregate different elements, such as scheduling services, user management, logs storage service and LabServer communication interfaces although the Interactive Lab Server (ITS) model [30,31].

The Fig. 5 shows how the communication works between all the elements using the ISB model.

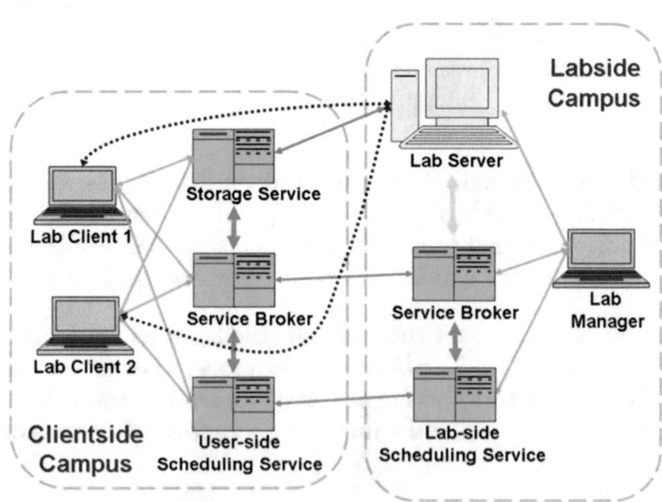

Fig. 5. iLab shared architecture for interactive service broker [30].

RELLE The *Remote Labs Learning Environment* (RELLE) was developed by the Mobile Remote Experimentation Work-group (GT-MRE) in Brazil in 2015. The RLMS is focused in group laboratories from different areas, such as physics, biology, robotics and others [32].

The labs might come from different partners, adopting a model based on independent labs, where these laboratories are connected on the Internet and must have a public IP address.

The RELLE architecture was developed based in three layers. The first one, describes the RLMS by itself, the second the scheduling service that manages connections between the first and the third layer (physical lab layer).

The RLMS compose the first RELLE layer, in this sense are allocated common functions to the user, laboratory, booking, and resource management e.g. (videos, documents, tutorials). The second layer, called RELLE-LIS describes the *Labs Instance Scheduling Service*, a service that is the responsible for managing labs connections and making authorization between users and labs.

The final layer describes the laboratory supported by the concept of instances (the same type of laboratory may have multiples instances). The generic model for the lab layer is based on a lab server that manage connections and a (Control and Acquisition Board) CAB that retrieve values from sensors and send actions to actuators [33].

The Fig. 6 reports the architecture and the connections between their elements.

Labs hosted on RELLE have individual access, and some have a queue in their own LabServer, not on the RLMS. The related change break the RELLE-LIS use by adding different connections methods, additionally being more scalable, once the student can use the interface simultaneously to others students.

4.2 Architecture Developed

The architecture developed aims to integrate models using hands-on and remote laboratories in order to build a hybrid and mobile model following the concept of flipped lab.

The feature of mobile laboratories make a laboratory capable of moving readily geographically. In this sense a student can perform activities at home, at university or anywhere since they have an Internet connection.

One of the barriers in connecting mobile labs anywhere is that this device must use a valid IP address. In this way, there is two assumption, the user and lab must be connected in client mode.

It was necessary to develop a smart gateway to connect labs and users to share messages between them, supporting the communication between both elements.

The architecture was developed in three layers. Each one of these layers is represented by the Fig. 7 that shows the modules and their connections.

The first layer, called *Lab@Home*, is the responsible for controlling the collaborative learning environment. The application contains services related to users and projects management and uses more than a single user role.

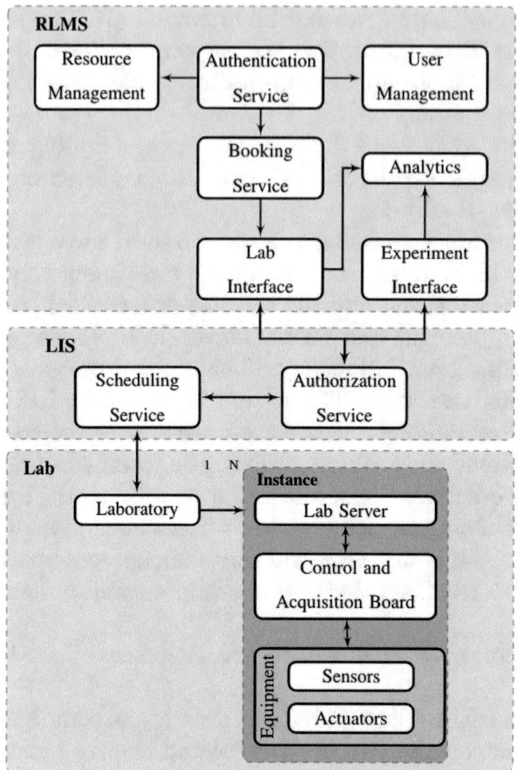

Fig. 6. RELLE architecture.

The services *Lab Interface* and *UI Composition* are responsible for managing the User Interface (UI). JSON files are exchanged between the modules containing UI preferences.

Those modules are interconnected by the *Smart Device Agent (SDA)* module to receive data from the Lab Layer. This integration transforms the communication in a unique and single format to the labs, and *UI Composition*, supporting Smart Device methods and delivery format [34].

These data that are received for the SDA and managed for the *UI Composition* according to the user preferences. The outputs are exhibited to the final user thought a *Lab Interface* module.

The second layer, named as *Gateway layer*, is the responsible for integrating both layers. Their function is receive and forward messages using the HTTP protocol.

The *Communicator Service* make requests to the *Authorization Service* using a hash password to enable the communication between them.

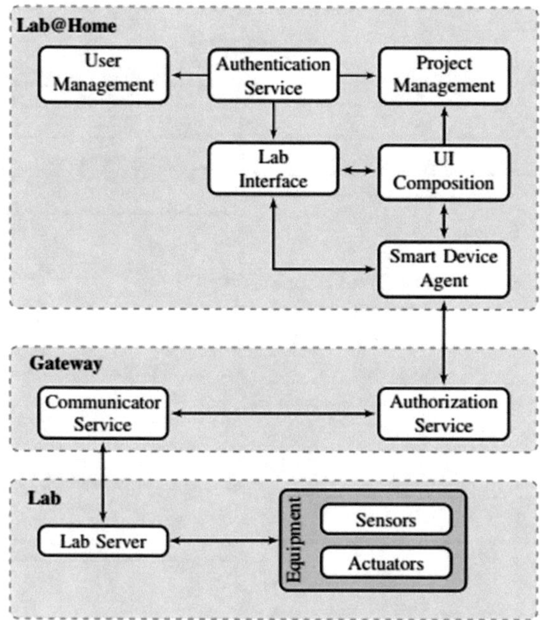

Fig. 7. Architecture developed.

The gateway layer fills the gap that all IP addresses must be public. In this sense, breaking the client/server communication, mobile lab and user are connected in the gateway server as client applications.

The *Lab layer* is the responsible for executing tasks related to laboratory management. This layer is composed by a generic Single Board Computer (SBC).

The SBC contains a *Lab Server* application, that communicates with sensors forwarding messages formatted through Smart Device methods and receive messages from actuators inputs (i.e. values that were defined by user).

The Lab Layer layer can be considered a generic layer. The *equipment* module can be built using any model of sensor or actuator or external equipment.

The equipment generic format is only supported because the data delivery format is encapsulated by Smart Device methods which manages values from the sensors in order to delivery custom outputs.

The Fig. 8 details the communication messages in order to present how users, labs and gateway share messages between them supported by authorization mechanisms and Smart Device methods.

To start the communication between user and lab, both services must authenticate in the gateway service (i.e., gateway server), if authenticated, they can send and receive messages using the Smart Device methods. The Smart Device methods implemented were *getSensorMetadata()*, *getActuatorMetadata()*, *getSensorData()* and *sendActuatorData()*.

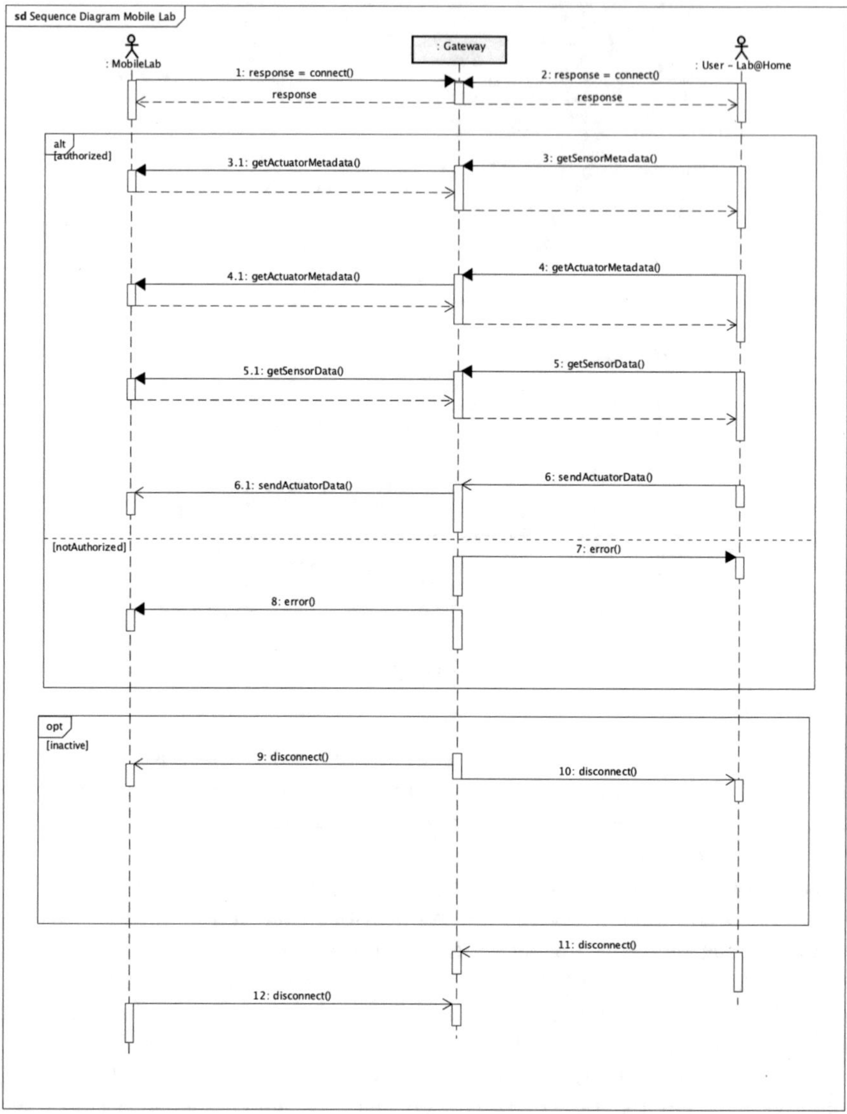

Fig. 8. Sequence diagram.

The gateway service works receiving and forwarding the messages as a *chat room*. In case of inactivity, the service is autonomous to disconnect user and lab and end the communication.

The Fig. 9 shows an use case diagram, considering the possibles operations for each user role. The user roles implemented were student, professor, and administrator. Administrator role has access to administrative functionality', such as create and edit user profiles and view usage information.

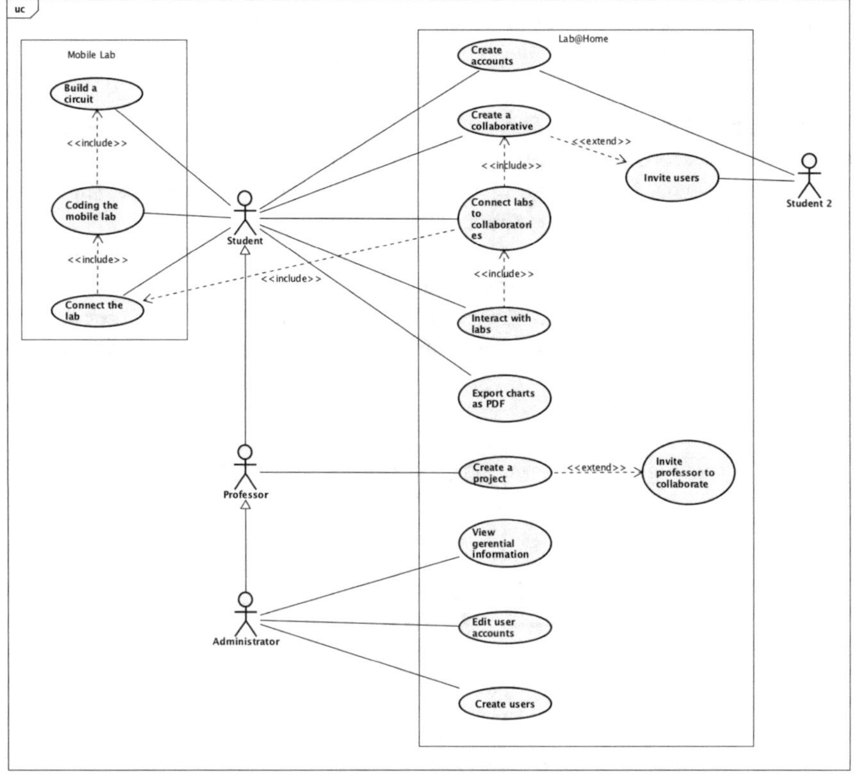

Fig. 9. Use case diagram.

Users with professors role are able to create projects in collaboration with others professors and share examples of collaboratories on it.

Students' can create, connect and manage labs in different collaborative environments. Labs might be created from a generic model or based on a project, previously shared by a professor, additionally students are able to invite others users to collaborate in their collaboratories.

4.3 Learning Scenarios

Lab@Home was developed as a collaborative learning environment to support two different learning scenarios for different user roles. The application considers that users can act as administrator, professors, and students.

The aggregated value in online labs describes how these labs will be helpful in the learning process. This value that aims aims to help students to understand general subjects adopting practice activities of experimentation.

In this sense, were developed two different learning scenarios supported by Lab@Home. The analysis aimed to describe the integration into students activities in classroom. The first one describing a generic model and the second based in the project model.

Generic-Based Scenario The generic based is the most basic and starting scenario. It describes a student that is free to create, build and share their lab to any user to access and interact with a lab according to the way that it was built during the development process.

Following this statement, students have in their flow of tasks the following steps: creating a collaboratory and connecting to a mobile lab, after this step, the system checks if this laboratory is valid, after validated, the student can start the communication and interact with them. The Fig. 10 describes this scenario in a flow chart.

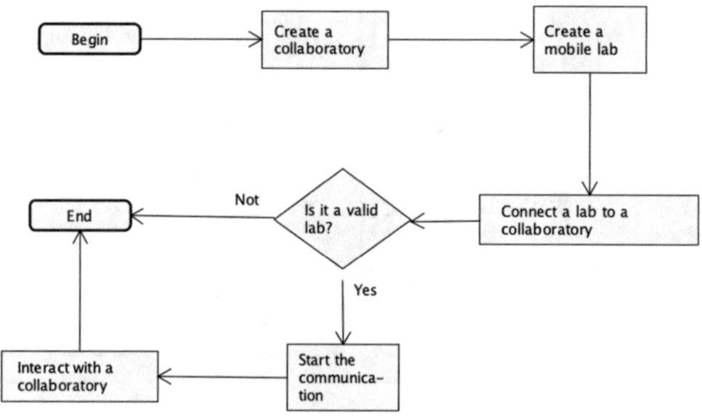

Fig. 10. Generic-based scenario.

Even if this scenario was developed using individual approach, users can share their labs with others in order to interact and prove from the same experience in a collaborative model. This model might undergo variations, following the interface and the starting point.

Project-Based Scenario Professors congregate students role, and additionally, they can create projects in collaboratively with others professors. Furthermore, managing their students, been able to see student progress in single or collaborative mode.

The project-based scenario describes a scenario where professors can create their labs from generic models and share with students inside their projects.

The projects might be aligned with tasks related to the labs. Students will be able to see all content (e.g. supporting or technical documentation) and configurations such as user interface and component specifications.

In this scenario, students have an interface that was already built. It these cases their main concern is in develop the physical lab, following the previous specifications that were given by the project creator.

The Fig. 11 shows a flowchart describing the students' steps in this scenario. Students must enroll in a project to have access to all content that had been made available by the project leaders.

The students can access all examples available and create different copies creating a lab and connecting it to a collaboratory using the same authentication method as previously described.

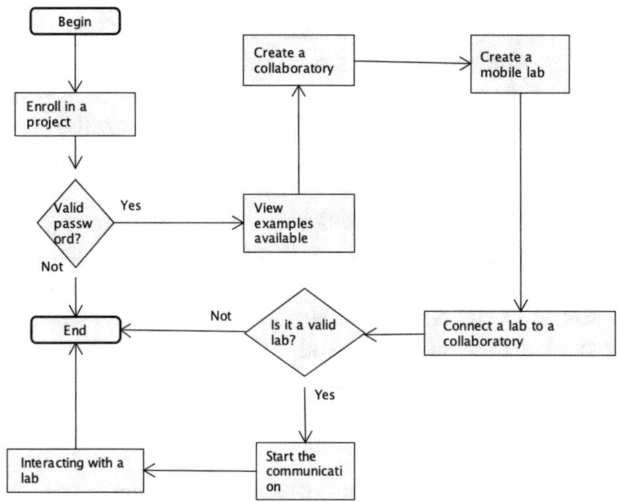

Fig. 11. Project-based scenario.

The communication methods provide enable users to share different' labs and connect them in different environments (i.e., collaboratories).

Those labs can be addressed in different purposes. The components can be shared in different aspects (e.g., different configurations, different tasks).

4.4 Learning Scenarios Evaluation

In both learning scenarios will be adopted the same problem situation that describes studies related to physics in the subject of teaching circuits in series, parallel and mixed resistors association.

The mail goal is to define how these studies can be tested and analyzed in engineering courses, considering different approaches that can be addressed adopting the solution developed.

Considering studies of series, parallel and mixed resistors association it has different' approaches in each scenario, in which will be described in the following paragraphs.

In a generic-based scenario, the student starts his experience with a new collaboratory, without any step already filled out. In this sense the student must define their experience goal, create the lab and define how will be interface behavior.

This approach considers that the student can measure values in series and parallel circuits, but does not mean that the value measured will be in accordance from what is expected (i.e., the student built a circuit, but not as expected).

Commons mistakes in this scenario refers to sensors positioned in wrong points, the student will receive value, but not as expected and will not receive support from a specialist directly.

In another hand, project-based scenario has support from professors. Those professors that were the responsible for previously develop examples.

The activity focuses on the lab development in order to fill tasks from a previous experimentation activity. Considering series and parallel studies, students have to adjust resistors in accordance to combine sensors following the UI that was already built by the project creator (i.e., professor).

In this scenario, the student might share the collaboratory to any student enrolled in the same project without restrictions. These students are able to see each collaboratory and cooperate with each one in order to prove and compare the same experience in the collaborative mode.

Regarding the learning scenarios evaluation, the Table 1 summarize the features that were described above in each scenario, considering how students can act in each one.

Table 1. Learning scenarios evaluation.

Characteristics	Generic-based	Project-based
Have to develop the lab	X	X
Have to develop the UI	X	
Export report in PDF	X	X
Modifications without disconnect	X	X
Invite users to collaborate	X	X
Users automatically invited		X
Focus on lab development		X
Tasks associated		X
Evaluation support		X

5 Final Remarks

The problem related that a student can not use equipment or a component, being electronic or not physically in an ideal way considering a lot of limiting factors, how for example time spent in class, is considered a barrier on experimentation.

Even if, the high cost to maintain a physical lab be a limiting factor, it has on another side, online labs that support this problem promoting an environment with full-time access with a reduced cost.

These experiences that can be addressed on Lab@Home using a generic Single-Board Computer (SBC) or embedded computer are only supported because the same transportation method that was adopted over HTTP supported by the Smart Device methods implanted.

Initiatives regarding to improve scalability in remote laboratories have made some advancements, but some initiatives have lost some crucial elements considering interaction.

How for example, non-interactive labs in remote experimentation. In this example users have video records of a laboratory execution, not allowing a complete personalized experience.

In this sense, the development of an application based on a hybrid model bring some non-conventional concerns on remote laboratories applications.

The fact that the laboratory must be mobile and that these labs need to move quickly without connectivity problems related to the Internet and their sensors and actuators.

Approaches that bring collaborative interaction on online laboratories are not common among the literature. The collaborative approach on online laboratories can be exploited towards to verify how students learn with others in order to improve their knowledge.

The project-based approach brings to students a scenario focused on learning specif concepts that had been given by professors, in this sense this scenario is mainly focused learning, being the professor able to verify how the students are developing their labs and collecting data' among the user, collaboratory and physical laboratory interaction.

Collaborative' experiences bring opportunities to explore different teaching methods related to adopting concepts that can be shared in the same project, adopting methodologies such as blended learning or inverted classroom.

Regarding to the outcomes generated by this work, future works can explore analysis about scalability supported by the application presented, being measured values when users share their labs in projects.

Acknowledgments. The authors would like to thank the Foundation for Research and Innovation Support of the State of Santa Catarina (FAPESC) and the Brazilian Coordination for the Improvement of Higher Education Personnel (CAPES) for the master's scholarships, and Global Affairs Canada for the mobility fellowship granted through the Emerging Leaders in the Americas Program (ELAP).

References

1. Jara, C.A., Candelas, F.A., Puente, S.T., Torres, F.: Hands-on experiences of undergraduate students in automatics and robotics using a virtual and remote laboratory. Comput. Educ. **57**(4), 2451–2461 (2011). https://doi.org/10.1016/j.compedu.2011.07.003, http://www.sciencedirect.com/science/article/pii/S0360131511001515
2. Ackovska, N., Kirandziska, V.: The importance of hands-on experiences in robotics courses. In: IEEE EUROCON 2017—17th International Conference on Smart Technologies, pp. 56–61 (2017). https://doi.org/10.1109/EUROCON.2017.8011077
3. Mujkanovic, A., Zutin, D.G., Schellander, M., Oberlercher, G., Vormaier, M.: Impact of students' preferences on the design of online laboratories. In: 2015 IEEE Global Engineering Education Conference (EDUCON), pp. 823–826 (2015). https://doi.org/10.1109/EDUCON.2015.7096067
4. Tawfik, M., Sancristobal, E., Martin, S., Gil, R., Diaz, G., Colmenar, A., Peire, J., Castro, M., Nilsson, K., Zackrisson, J., Hakansson, L., Gustavsson, I.: Virtual instrument systems in reality (visir) for remote wiring and measurement of electronic circuits on breadboard. IEEE Trans. Learn. Technol. **6**(1), 60–72 (2013). https://doi.org/10.1109/TLT.2012.20
5. Tawfik, M., Lowe, D., Salzmann, C., Gillet, D., Sancristobal, E., Castro, M.: Defining the critical factors in the architectural design of remote laboratories. IEEE Rev. Iberoamericana de Tecnologias del Aprendizaje **10**(4), 269–279 (2015). https://doi.org/10.1109/RITA.2015.2486388
6. Ramos, S., Pimentel, E.P., Marietto das Gracas, M.B., Botelho, W.T.: Hands-on and virtual laboratories to undergraduate chemistry education: toward a pedagogical integration. In: 2016 IEEE Frontiers in Education Conference (FIE), pp. 1–8 (2016). https://doi.org/10.1109/FIE.2016.7757580
7. Saliah-Hassane, H., Simao, J.P.S., Lima, J.P.C., Alves, G.R.C., Silva, J.B., Alves, J.B.M.: Mobile laboratories as an alternative to conventional remote laboratories. In: 15th LACCEI International Multi-Conference for Engineering, Education, and Technology (2017)
8. Saliah-Hassane, H., Kourri, A., Teja, I.D.L.: Building a repository for online laboratory learning scenarios. In: Proceedings of 36th Annual Conference Frontiers in Education, pp. 19–22 (2006). https://doi.org/10.1109/FIE.2006.322603
9. Balamuralithara, B., Woods, P.C.: Virtual laboratories in engineering education: the simulation lab and remote lab. Comput. Appl. Eng. Educ. **17**(1), 108–118 (2009). https://doi.org/10.1002/cae.20186, http://dx.doi.org/10.1002/cae.20186
10. Zutin, D.G., Auer, M.E., Maier, C., Niederstätter, M.: Lab2go—a repository to locate educational online laboratories. In: IEEE EDUCON 2010 Conference, pp. 1741–1746 (2010). https://doi.org/10.1109/EDUCON.2010.5492412
11. Orduña, P., Rodriguez-Gil, L., Garcia-Zubia, J., Angulo, I., Hernandez, U., Azcuenaga, E.: Labsland.: a sharing economy platform to promote educational remote laboratories maintainability, sustainability and adoption. In: 2016 IEEE Frontiers in Education Conference (FIE), pp. 1–6 (2016). https://doi.org/10.1109/FIE.2016.7757579
12. Gustavsson, I.: Remote laboratory experiments in electrical engineering education. In: Proceedings of the Fourth IEEE International Caracas Conference on Devices, Circuits and Systems (Cat. No.02TH8611), pp. I025–1–I025–5 (2002). https://doi.org/10.1109/ICCDCS.2002.1004082

13. Mosterman, P.J., Dorlandt, M.A., Campbell, J.O., Burow, C., Bouw, R., Brodersen, A.J., Bourne, J.R.: Virtual engineering laboratories: design and experiments. J. Eng. Educ. **83**(3), 279–285 (1994). https://doi.org/10.1002/j.2168-9830.1994.tb01116.x, http://dx.doi.org/10.1002/j.2168-9830.1994.tb01116.x
14. Gomes, L., Bogosyan, S.: Current trends in remote laboratories. IEEE Trans. Ind. Electron. **56**(12), 4744–4756 (2009). https://doi.org/10.1109/TIE.2009.2033293
15. Rodriguez-Gil, L., Garcia-Zubia, J., Orduna, P., de Ipina, D.L.: Towards new multiplatform hybrid online laboratory models. IEEE Trans. Learn. Technol. **PP**(99), 1–1 (2017). https://doi.org/10.1109/TLT.2016.2591953
16. Callaghan, M.J., McCusker, K., Losada, J.L., Harkin, J., Wilson, S.: Using game-based learning in virtual worlds to teach electronic and electrical engineering. IEEE Trans. Ind. Inf. **9**(1), 575–584 (2013). https://doi.org/10.1109/TII.2012.2221133
17. Chacón, J., Saenz, J., de la Torre, L., Sánchez, J.: Flipping the remote lab with low cost rapid prototyping technologies. In: Auer, M.E., Zutin, D.G. (eds.) Online Engineering & Internet of Things, pp. 250–257. Springer International Publishing, Cham (2018)
18. Scott, K., Benlamri, R.: Context-aware services for smart learning spaces. IEEE Trans. Learn. Technol. **3**(3), 214–227 (2010). https://doi.org/10.1109/TLT.2010.12
19. Curtis, D., Lawson, M.: Exploring collaborative online learning. J. Asynchronous Learn. Netw. **5**(1) (2001) (cited By 225)
20. Macdonald, J.: Assessing online collaborative learning: process and product. Comput. Educ. **40**(4), 377–391 (2003). https://doi.org/10.1016/S0360-1315(02)00168-9, http://www.sciencedirect.com/science/article/pii/S0360131502001689
21. Ryu, H., Parsons, D.: Risky business or sharing the load? Social flow in collaborative mobile learning. Comput. Educ. **58**(2), 707–720 (2012). https://doi.org/10.1016/j.compedu.2011.09.019, http://www.sciencedirect.com/science/article/pii/S0360131511002363
22. García-Zubía, J., Angulo, I., Martínez-Pieper, G., de Ipiña, D.L., Hernández, U., Orduna, P., Dziabenko, O., Rodríguez-Gil, L., van Riesen, S.A.N., Anjewierden, A., Kamp, E.T., de Jong, T.: Archimedes remote lab for secondary schools. In: 2015 3rd Experiment International Conference (exp.at'15), pp. 60–64 (2015). https://doi.org/10.1109/EXPAT.2015.7463215
23. De Oliveira, M.: Como Fazer Pesquisa Qualitativa. Marly DE Oliveira (2007)
24. Heradio, R., de la Torre, L., Galan, D., Cabrerizo, F.J., Herrera-Viedma, E., Dormido, S.: Virtual and remote labs in education: a bibliometric analysis. Comput. Educ. **98**, 14–38 (2016). https://doi.org/10.1016/j.compedu.2016.03.010, http://www.sciencedirect.com/science/article/pii/S0360131516300677
25. Ma, J., Nickerson, J.V.: Hands-on, simulated, and remote laboratories: a comparative literature review. ACM Comput. Surv. **38**(3) (2006). https://doi.org/10.1145/1132960.1132961, http://doi.acm.org/10.1145/1132960.1132961
26. Bryman, A.: Social Research Methods. Oxford University Press, Oxford (2012)
27. Rodriguez-Gil, L., Orduña, P., Garcĩa-Zubia, J., de Ipiña, D.L.: Advanced integration of openlabs visir (virtual instrument systems in reality) with weblab-deusto. In: 2012 9th International Conference on Remote Engineering and Virtual Instrumentation (REV), pp. 1–7 (2012). https://doi.org/10.1109/REV.2012.6293150
28. Orduña, P., Uribe, S.B., Isaza, N.H., Sancristobal, E., Emaldi, M., Martin, A.P., DeLong, K., Bailey, P., de Ipiña, D.L., Castro, M., Garcia-Zubia, J.: Generic integration of remote laboratories in learning and content management systems through federation protocols. In: 2013 IEEE Frontiers in Education Conference (FIE), pp. 1372–1378 (2013). https://doi.org/10.1109/FIE.2013.6685057

29. Mendes, L.A., Li, L., Bailey, P.H., DeLong, K.R., del Alamo, J.A.: Experiment lab server architecture: a web services approach to supporting interactive labview-based remote experiments under mit's ilab shared architecture. In: 2016 13th International Conference on Remote Engineering and Virtual Instrumentation (REV), pp. 293–305 (2016). https://doi.org/10.1109/REV.2016.7444486
30. Harward, V.J., del Alamo, J.A., Lerman, S.R., Bailey, P.H., Carpenter, J., DeLong, K., Felknor, C., Hardison, J., Harrison, B., Jabbour, I., Long, P.D., Mao, T., Naamani, L., Northridge, J., Schulz, M., Talavera, D., Varadharajan, C., Wang, S., Yehia, K., Zbib, R., Zych, D.: The ilab shared architecture: a web services infrastructure to build communities of internet accessible laboratories. Proc. IEEE **96**(6), 931–950 (2008). https://doi.org/10.1109/JPROC.2008.921607
31. Hardison, J.L., DeLong, K., Bailey, P.H., Harward, V.J.: Deploying interactive remote labs using the ilab shared architecture. In: 2008 38th Annual Frontiers in Education Conference, pp. S2A-1–S2A-6 (2008). https://doi.org/10.1109/FIE.2008.4720536
32. Mellos Carlos, L., Cardoso de Lima, J.P., Schardosim Simão, J.P., Bento da Silva, J., Sommer Bilessimo, S.: Estratégias de integração de tecnologia no ensino: Uma solução baseada em experimentação remota móvel. In: Libro de Actas Tical (2017)
33. de Lima, J.P.C.: Desenvolvimento De Servidores Para LaboratÓrios Remotos Baseado No Paradigma De Dispositivos Inteligentes. Universidade Federal de Santa Catarina, Trabalho de conclusão de curso (2016)
34. Salzmann, C., Govaerts, S., Halimi, W., Gillet, D.: The smart device specification for remote labs. In: Proceedings of 2015 12th International Conference on Remote Engineering and Virtual Instrumentation (REV), pp. 199–208 (2015). https://doi.org/10.1109/REV.2015.7087292

Quality and Efficiency Indicators of Remote Laboratories

Mykhailo Poliakov[1], Heinz-Dietrich Wuttke[2(✉)], and Karsten Henke[2]

[1] Zaporizhzhya National Technical University, Zaporizhzhya 69063, Ukraine
polyakov@zntu.edu.ua

[2] Ilmenau University of Technology, TU Ilmenau, 98693 Ilmenau, Germany
{Dieter.Wuttke,Karsten.Henke}@tu-ilmenau.de

Abstract. In the course of digitization of educational processes, the importance of using a complex of remote laboratories in the processes of distance learning programs in engineering specialties is increasing. Indicators are proposed for assessing the quality of remote laboratory applications in hybrid learning processes in which laboratory work is performed remotely and in class. These indicators assess the extent to which curricula and courses are covered by remote laboratory work, the applicability of a remote laboratory of one or a set of such laboratories to use in a training course or program, indicators of the diversity of experiments and the loading of a remote laboratory.

Keywords: Remote laboratory · Indicators of applicability and quality

1 Introduction

In the process of creating and using remote laboratories in engineering education, the stages of creating a scientific and technical base for remote experiments and remote laboratories that are effective for individual experiments, training topics and courses started mostly from 2000 to the present [1–3]. The urgent task of the near future is the creation and use in the educational process of complexes of remote laboratories that will be effective in the framework of training programs (BS, MS, and Ph.D.), specialties and structural divisions of universities. The practice of using remote laboratories has shown their advantages, but has not supplanted the traditional forms of laboratory work. The educational process in which laboratory work is performed both remotely and in classes at the university will be called hybrid.

Currently, a remote laboratory is considered effective if with its help a student can carry out one or several remote experiments included in the program of the training course, the experiments are available 24/7 and are positively evaluated by the student during the survey. But if remote experiments cover only a few percent of the volume of laboratory work on the curriculum, then it will be premature to talk about effective remote training.

The following sections of the article provide an overview of the literature on the selected topic, describe the interaction of people and remote laboratories, describe the quality and efficiency indicators of remote laboratories, as well as the peculiarities of determining these indicators for experiments with physical models of control objects.

M. E. Auer and K. Ram B. (Eds.): REV2019 2019, LNNS 80, pp. 143–154, 2020.
https://doi.org/10.1007/978-3-030-23162-0_14

2 State of the Art: Quality Indicators of a Remote Laboratory

In well-known publications, remote laboratories are evaluated in terms of the internal structure [4–6], descriptions of the experiments conducted in them [7–9], pedagogical training scenarios, and the benefits of distance learning [10, 11].

Practically in all the works, various quantitative and qualitative indicators are described to substantiate the effectiveness of the proposed solutions, but the laboratory indicators are not transformed into indicators of the effectiveness of the use of these laboratories. This makes it difficult to design, assess the significance of the achieved or planned level of distance education. The relevance of indicators for these assessments will increase in the process of further development of distance learning tools. People and Remote Laboratories Interaction.

Different groups of people interact with remote laboratories with different roles in the life cycle. These categories of their role and current indicators of remote laboratories are given in Table 1.

Table 1. Who and how interacts with remote laboratories: categories, roles and relevant indicators

Category	The role in the life cycle of a remote laboratory	Actual indicators
Designers	Selection and implementation of remote experiments	Themes and variety of experiments
Operating personnel	Maintenance of laboratory equipment, setting tasks to designers, storing and transferring the results of experiments to students, teachers and administrators	Reliability, performance/throughput, maintainability
Teachers	Task setting for designers, analysis of experimental results	Covering training courses and programs with remote experiments
Students	Use to gain knowledge and skills in the process of distance learning	Waiting time, availability of methodological documentation, quality and clarity of experiments, and others
Administrators of the educational process and university departments	Decision on use, selection of laboratories, setting targets for their improvement and cooperation with other laboratories	Coverage of training courses and programs with remote experiments, level of use, performance/throughput

Normally, specialists in computer science, who do not always understand the subtleties of conducting experiments in other areas, do designing remote laboratories. Therefore, probably, remote laboratories on the subject of “computer scene” are more than in other areas. In the selection of subjects of experiments, the designers should work in contact with teachers, and in questions of increasing reliability—with those who operate the laboratory.

Teachers are customers in the formulation of remote experiments. Their task is to form a variety of learning tasks in the framework of experiments with a single physical model of the object of study/control and/or experiment control technology. By the course coverage of a training course/program by remote experiments, we mean the conjunction of the set of laboratory work, included in this course/program and the set of laboratory work, performed by a remote laboratory or complex of such laboratories.

Students are end-users of remote laboratories, consumers of educational services. The maximum student requirements for laboratories are access mode—24/7; the waiting time of the experiment is 0 min; the duration of the experiment is unlimited; a property of physical models of the object of study—full compliance with real objects. At the same time, the performance of real laboratories in real educational processes does not always justify the student's expectations.

The task of administrators is to determine the optimal level of remote experiments in a hybrid educational process at a university/department, examination and organizational and financial support for proposals from other participants in the laboratory's life cycle.

Thus, the diversity of participants and their roles gives rise to a variety of indicators related to the quality of the laboratory and the effectiveness of its use, which are discussed in the following sections.

3 Quality and Efficiency Indicators of Remote Laboratories

The space of indicators determines the ternary relation between the parameters of the educational process, the capabilities of the laboratory and the flow of trainees. The indicators proposed below estimate one element of this relationship relative to one or two others in areas such as the nomenclature, the duration of use.

Coverage indicators describe the nomenclature. For example, the laboratory coverage indicator for a nomenclature for some academic discipline assesses the proportion of laboratory work in this discipline that can be performed using this remote laboratory. And the coverage of the discipline/program is an indicator of its level in terms of the implementation of distance learning. We will assume that this level increases with the share of remotely performed laboratory work.

Indicators of the duration of use specify the indicators on the nomenclature in the case when the remotely performed laboratory work differs significantly in the time that is scheduled for their implementation. For example, the indicator of the duration of the use of a laboratory for some academic discipline estimates the proportion of time, spent on performing laboratory work in this discipline, which can be performed using this remote laboratory.

The general problem of using remote laboratories in some educational process is that they do not provide full coverage of laboratory work in all academic disciplines. A typical example of a coverage map is shown in Fig. 1.

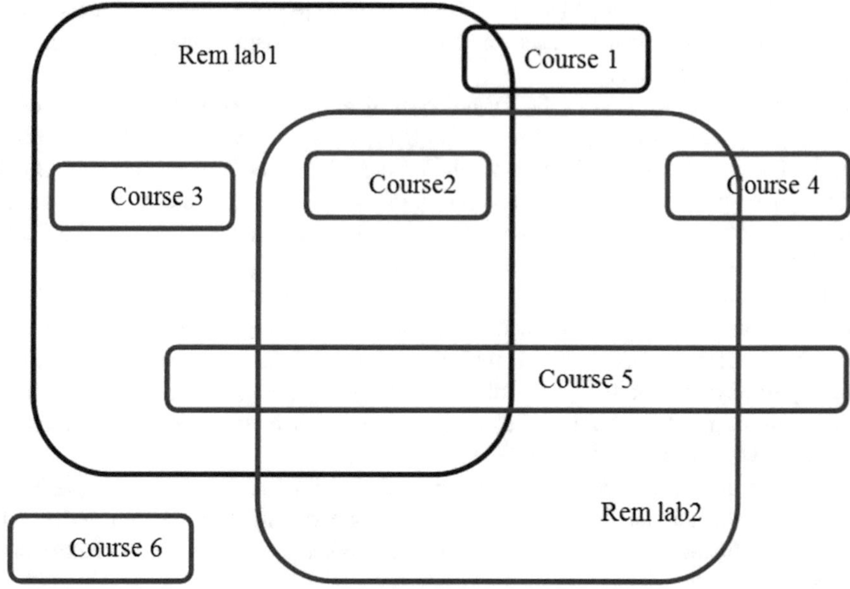

Fig. 1. Coverage of laboratory work for curriculum courses using remote laboratories 1 and 2

The coverage map in Fig. 1 shows the following options for covering laboratory work for curriculum courses, consisting of courses 1–6 using remote laboratories Rem lab 1 and 2. On the map, we will highlight typical coverage variants:

- Partial coverage of one of Rem lab—Course 1 partially covered by Rem lab 1.
- Full coverage of all Rem lab—Course 2 fully covered and Rem lab 1 & Rem lab 2.
- Full coverage of one (several, not all) Rem lab—Course 3 is fully covered by Rem lab 1, and Course 4 is fully covered by Rem lab 2.
- Partial coverage of each Rem lab and all—Course 5 is covered partially by Rem lab 1, partially by Rem lab 2, partially by both Rem lab, and partly not covered by any of the Rem Labs.
- A course that is not covered by any Rem Labs (Course 6).

To formalize the indicators we introduce the following notation:

- p is the number of the curriculum at the university;
- n is the number of the training course in the program;
- k is the number of the remote laboratory;
- m is the number of laboratory work from the set, performed in this training course;
- j is the number of laboratory work from the set, performed in this laboratory;
- t_{nm} is the time allotted by the curriculum for the m-th laboratory work in the n-th course;
- c_{nmk} is the coverage of the k-th remote laboratory of the m-th laboratory work in the n-th training course:

$$c_{nmk} = \begin{cases} 1, & \text{if it covers} \\ 0, & \text{if it does not} \end{cases};$$

- P: is the set and number of programs for which training is conducted;
- N: is the set and number of courses in the program;
- K: is the set and number of available remote laboratories;
- M_n: is the set and number of laboratory works in the *n*-th course,
- R_{kn}: is the set and number of laboratory works that are performed remotely using the *k*-th laboratory for the *n*-th course;
- R_n: is the set and number of laboratory works that are performed remotely using at least one remote laboratory for the *n*-th course;
- L_n: is the set and number of laboratory works that you complete in the classes for the *n*-th course.

Suggested indicators are listed in Table 2.

These indicators are calculated by the formulas:

$$CCn = Ln, \tag{1}$$

$$CCn(\%) = \left(\frac{Ln}{Mn}\right) * 100\%, \tag{2}$$

$$CCp = \sum_{n=1}^{Np} Ln, \tag{3}$$

$$CCp(\%) = \left(\frac{CCp}{\sum_{n=1}^{Np} Mn}\right) * 100\%, \tag{4}$$

$$CnR = Mn - Ln, \tag{5}$$

$$CnR(\%) = \left(1 - \frac{CnR}{Mn}\right) * 100\%, \tag{6}$$

$$CpR = \sum_{n=1}^{Np} Mn - \sum_{n=1}^{Np} CnR, \tag{7}$$

$$CpR(\%) = \left(CpR/\sum_{n=1}^{Np} Mn\right)100\% \tag{8}$$

$$C_nR_{kn} = R_{kn}, \tag{9}$$

Table 2. Indicators of remote laboratories and the level of distance education

Indicator	Name	No.
Coverage of the classroom lab works for the n-th training course	CC_n	1
Coverage of the classroom lab works for the n-th training course (percentages)	$CC_n(\%)$	2
Coverage of the classroom lab works for the p-th curriculum	CC_p	3
Coverage of the classroom lab works for the p-th curriculum (percentages)	$CC_p(\%)$	4
Coverage of the Rem Lab works for the n-th training course	C_nR	5
Coverage of the Rem Lab works for the n-th training course (percentages)	$C_nR(\%)$	6
Coverage of the Rem Lab works for the n-th training course for the p-th curriculum	C_pR	7
Coverage of the Rem Lab works for the n-th training course for the p-th curriculum (percentages)	$C_pR(\%)$	8
Coverage of the n-th training course in the k-th Rem Lab	C_nR_{kn}	9
Coverage of the n-th training course in the k-th Rem Lab (percentages)	$C_nR_{kn}(\%)$	10
Coverage of the p-th curriculum in the k-th Rem Lab	C_pR_k	11
Coverage of the p-th curriculum with work in the k-th Rem Lab (percentages)	$C_pR_k(\%)$	12
The duration of the work performed in the classes for the n-th course	TCC_n	13
The duration of the work performed in the classes for the n-th course (percentages)	$TCC_n(\%)$	14
The duration of work performed in Rem Lab for the n-th course	TRC_n	15
The duration of work performed in Rem Lab for the n-th course (percentages)	$TRC_n(\%)$	16
Duration of work performed in Rem Lab for the p-th curriculum	TRC_p	17
Duration of work performed in Rem Labs for the p-th curriculum (percentages)	$TRC_p(\%)$	18
The duration of the work performed in the k-th Rem Lab for the n-th training course	TR_kC_n	19
The duration of the work performed in the k-th Rem Lab for the n-th training course (percentages)	$TR_kC_n(\%)$	20
The duration of the work performed in the k-th Rem Lab for the p-th curriculum	TR_kC_p	21
The duration of the work performed in the k-th Rem Lab for the p-th curriculum (percentages)	$TR_kC_p(\%)$	22

$$C_nR_{kn}(\%) = (R_{kn}/M_n) * 100\%, \tag{10}$$

$$C_pR_k = \sum_{n=1}^{N} C_nR_{nk}, \tag{11}$$

$$C_pR_k(\%) = \left(\sum_{n=1}^{N}\left(\frac{C_nR_{nk}}{M_n}\right)\right)100\% \tag{12}$$

$$TCC_n = \sum_{m \in L_n} t_{nm}, \tag{13}$$

$$TCC_n(\%) = \left(\frac{\left(\sum_{m \in L_n} t_{nm}\right)}{\sum_{m=1}^{M_n} t_{nm}}\right) * 100\%, \tag{14}$$

$$TRC_n = \sum_{m=1}^{M_n} t_{nm} - \sum_{m \in L_n} t_{nm}, \tag{15}$$

$$TRC_n(\%) = \left(1 - \frac{\sum_{m \in L_n} t_{nm}}{\sum_{m=1}^{M_n} t_{nm}}\right)100\% \tag{16}$$

$$TRC_p = \sum_{n \in N_p} TRC_n, \tag{17}$$

$$TRC_p(\%) = \left(\frac{TRCp}{\sum_{m=1}^{M_n} t_{nm}}\right) * 100\% \tag{18}$$

$$TR_kC_n = \sum_{m=1}^{M_k} t_{nm} * c_{nmk} \tag{19}$$

$$R_kC_n(\%) = \left(\frac{TR_kC_n}{\sum_{m=1}^{M_n} t_{nm}}\right) * 100\%, \tag{20}$$

$$TR_kC_p = \sum_{n \in R_{kn}} TR_kC_n, \tag{21}$$

$$TR_kC_p(\%) = \left(\frac{TR_kC_p}{\sum_{m=1}^{M_p} t_{nm}}\right) * 100\%. \tag{22}$$

The application of the proposed indicators considers an example. The initial data for the calculation are given in Tables 3 and 4.

As can be seen from Table 4, the educational process for this program is a hybrid: 40% of laboratory work is performed in classes, and the rest in remote laboratories. These laboratories have different coverage for a variety of laboratory work programs. Seven variants of complexes of remote laboratories are possible in this example.

Table 3. Curriculum parameters

Course number	Lab number	Number of works in the course	Duration, hour (s)	Duration at the training course, hours
1	1	5	2	13
	2		4	
	3		2	
	4		4	
	5		1	
2	6	5	2	10
	7		2	
	8		2	
	9		2	
	10		2	
3	11	5	4	14
	12		6	
	13		1	
	14		1	
	15		2	
Total		15	37	37

As a result of calculation by formulas (1)–(22), the following values of indicators are obtained:

CC1 = 1 h; CC2 = 3; CC3 = 2; CC1(%) = 20; CC2(%) = 60; CC3(%) = 40; CCp = 6; CCp(%) = 40; C1R = 4; C2R = 2; C3R = 3; C1R(%) = 80; C2R(%) = 40; C3R(%) = 60; CpR = 9; CpR(%) = 60; C1Rk1(k = No. 1) = 3; C2Rk2(k = No. 1) = 2; C3Rk3(k = No. 1) = 3; C1Rk1(%)(k = No. 1) = 60; C2Rk2(%)(k = No. 1) = 40; C3Rk3(%)(k = No. 1) = 60; CpRk(k = No. 1) = 8; CpRk(%)(k = No. 1) = 53.3; CpRk(%)(k = No. 2) = 40; CpRk(%)(k = No. 3) = 26.6; CpRk(%)(k = No. 1 + No. 2) = 53; CpRk(%)(k = No. 1 + No. 3) = 60; CpRk(%)(k = No. 2 + No. 3) = 53.3; CpRk(%)(k = No. 1 + No. 2 + No. 3) = 60; TCC1 = 2; TCC2 = 6; TCC3 = 7; TCC1 (%) = 15.4; TCC2(%) = 60.0; TCC3(%) = 50.0; TRC1 = 11; TRC2 = 4; TRC3 = 7; TRC1(%) = 84.6; TRC2(%) = 40; TRC3(%) = 50; TRCp = 22; TRCp(%) = 59.5; TRkC1 = 4; TRkC2 = 8; TRkC3 = 3; TRkC1(%) = 30.8; TRkC2(%) = 61.5; TRkC3 (%) = 23.1; TRkCp(k = No. 1) = 15; TRkCp (k = No. 1)(%) = 40.5.

These indicators can be used to analyze the level of distance education for the curriculum. As a result of the analysis of these indicators it is possible to draw, for example, the following conclusions:

- The educational process for the program as a whole and for each training course included in it is a hybrid one. In the classes, 40% of laboratory work is performed on the number envisaged by the program.
- The duration of work performed in classes is 15.4, 60 and 50% for courses 1, 2, 3, respectively.

Table 4. The possibility of using remote laboratories

Course number	Lab number	Tool to perform (1—applies; 0—does not apply)							
		In classes	No 1	No. 2	No. 3	No. 1 + No. 2	No. 1 + No. 3	No. 2 + No. 3	No. 1 + No. 2 + No. 3
1	1	1	0	0	0	0	0	0	0
	2	0	1	1	1	1	1	1	1
	3	0	1	0	1	1	1	1	1
	4	0	0	0	1	0	1	1	1
	5	0	1	1	0	1	1	1	1
2	6	1	0	0	0	0	0	0	0
	7	1	0	0	0	0	0	0	0
	8	1	0	0	0	0	0	0	0
	9	0	1	1	1	1	1	1	1
	10	0	1	1	0	1	1	1	1
3	11	0	1	1	0	1	1	1	1
	12	1	0	0	0	0	0	0	0
	13	1	0	0	0	0	0	0	0
	14	0	1	1	0	1	1	1	1
	15	0	1	0	0	1	1	0	1
Total		6	8	6	4	8	9	8	9
Total in percentages		40	53.3	40	26.6	53	60	53.3	60

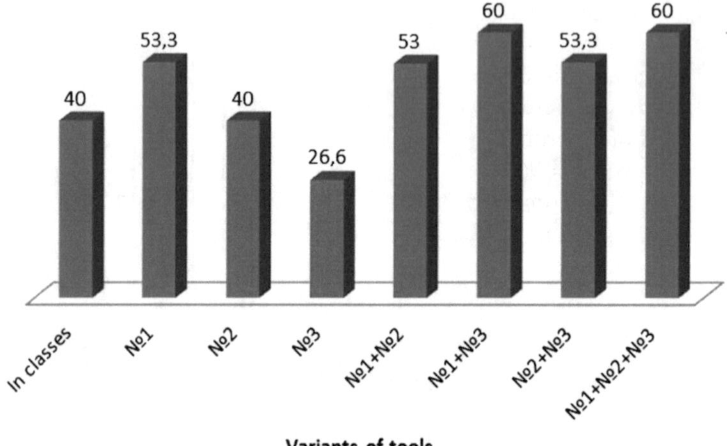

Fig. 2. Coverage of laboratory works on the curriculum using various tools (percentages)

- None of the three remote laboratories available provides full coverage of laboratory work on the program that is not performed in classes. As can be seen from Fig. 2, their coverage by nomenclature ranges from 53.3 to 26%.

In terms of the amount of coverage of laboratory work on the curriculum, Rem Lab3 makes the smallest contribution (26%). Only this laboratory allows you to perform laboratory work number 4 remotely and therefore it is included in the complex Rem Labs number 1 and number 3. Such a complex covers all remaining laboratory work that is not performed in classes.

To automate the determination of indicators by the formulas (1)–(22) it is advisable to use the computer program of the table processor.

4 Indicators of the Number of Experiments with Models in a Remote Laboratory

The number variety (diversity) of experiments with models in a Rem Lab is one of the types of quality indicators for it. In which case (under what conditions) can two experiments be considered different? This of course depends on the purpose of the laboratory.

For Rem Labs, designed to study digital control systems, the main condition for diversity is the difference in the applied control devices (FSM formalism, microprocessor board, programmable logic controller, FPGA) and structures of the projected behavior. If we restrict ourselves to the control structures of the electro-mechanics of physical models, then the number of experiments will be small. For example, the control structures of electric lift mechanics set the cyclical movement of the cabin between floors with stops and opening/closing of doors on certain floors. If you include an elevator in the passenger delivery system, the number of experiments with different options for processing the flow of service requests or with different goals for the operation of such a system will be much higher. For example, suppose that in some Rem Lab there are four physical models, the number of possible control structures for each model is six, and these structures can be designed using one of 4 control devices. Then the number of experiments in such a laboratory will be determined by multiplying these figures 4 * 6 * 4 = 96.

The number of experiments at Rem Lab should be correlated with the number of students undergoing training. If the curriculum provides for each student to conduct one laboratory work with each physical model, then in the above conditions it is possible to simultaneously train six students.

5 Conclusion

Today, distance-engineering education in terms of laboratory performance is still hybrid, that is, it combines classes in classes and remotely with the help of remote laboratories. A significant number of these laboratories are available on the Internet 24/7. However, the significance of the use of remote laboratories is not quantified due to the lack of a set of indicators of their quality and effectiveness of application.

To determine these indicators, categories and roles of people who interact with a remote laboratory are considered: projectors, operational personnel, teachers, students, administrators of the educational process, and university departments.

The ternary relationship between the parameters of the educational process, the capabilities of the laboratory and the flow of students determines the space of actual indicators of quality and effectiveness of the use of remote laboratories. The parameters of the educational process are set by the nomenclature and duration of laboratory work on training courses, programs and departments of the university. The number of experiments that can be performed determines the capability of a remote lab and the degree of coverage of all experiments required in a course. The significance of the use of a remote laboratory is determined by the proportion of work performed in it regarding the needs of a training course or program. The parameters of the flow of students set the requirements for the number of channels of simultaneous access to a remote laboratory.

The proposed indicators allow you to choose a rational composition of a complex of remote laboratories and to assess the level of application/significance of distance learning tools for the selected training course and program. The selection process is illustrated by an example.

To assess the quality of remote experiments with physical models of objects of study, an indicator of the variety of experiments is proposed, which depends on the number of options for constructing control devices, the number of available physical models, and the number of control tasks.

The results of the work will be used in the design of training courses using remote laboratories GOLDi [12].

References

1. Azad, A.K.M., Auer, M.E., Harward, V.J. (eds.): Internet accessible remote laboratories: scalable e-Learning tools for engineering and science disciplines. In: Engineering Science Reference, 645p (2012)
2. Gravier, C., et al.: State of the art about remote laboratories paradigms—foundations of ongoing mutations. Int. J. Online Eng. (iJOE) **4**(1), 1–9 (2008)
3. Gomes, L., Bogosyan, S.: Current trends in remote laboratories. IEEE Trans. Ind. Electron. **56**(12), 4744–4756 (2009). https://doi.org/10.1109/TIE.2009.2033293
4. Gonzalez, J.L.V., Aviles, J.B., Muñoz, A.R., Palomares, R.A.: An industrial automation course: common infrastructure for physical, virtual and remote laboratories for PLC programming. IJOE **14**(08), 4–19 (2018). https://doi.org/10.3991/ijoe.v14i08.8758
5. Remote and virtual tools in engineering: monograph/general editorship Karsten Henke, Dike Pole, Zaporizhzhya, Ukraine, 250p (2015). ISBN 978–966–2752–74–8
6. Dintsios, N., Artemi, S., Polatoglou, H.: Acceptance of remote experiments in secondary students. IJOE **14**(05), 4–19 (2018). https://doi.org/10.3991/ijoe.v14i05.8678
7. Chevalier, A., Copot, C., Hegedus, A., De Keyser, R.: Remote laboratory as a novel tool for control engineering studies: a feedback study. In: Processing of 14th International Conference on Optimization of Electrical and Electronic Equipment OPTIM 2014 May 22–24, 2014, Brasov, Romania
8. Gomes, L., Garsia-Zubia, J.: Advances on Remote Laboratories and e-learning Experiences, 310p. Deusto, Bilbao (2007)
9. Odeh, S.: A web-based remote lab platform with reusability for electronic experiments in engineering education. iJOE **10**(4), 40–45 (2014)
10. Tho, S.W., Yeung, Y.Y.: Technology-enhanced science learning through remote laboratory: System design and pilot implementation in tertiary education. Australas. J. Educ. Technol. **32**(3), 96–111 (2016)
11. Chandre, B.R., Geevarghese, K.P., Gangadhara, K.V.: Design and implementation of remote mechatronics laboratory for eLearning using LabVIEW and smartphone and cross-platform communication toolkit (SCCT). Procedia Technol. **14**, 108–115 (2014)
12. GOLDi-labs cloud Website: http://goldi-labs.net

Design of a Low Cost Switching Matrix for Electronics Remote Laboratory

Abderrahmane Adda Benattia[1(✉)], Abdelhalim Benachenhou[2], Mohammed Moussa[2], and Abdelhamid Mebrouka[2]

[1] University Ibn Khaldoun of Tiaret, Tiaret, Algeria
abderrahmane.addabenattia@univ-mosta.dz
[2] Université de Mostaganem, Mostaganem, Algeria
{abdelhalim.benachenhou,mohamed.moussa,
abdelhamid.mebrouka}@univ-mosta.dz

Abstract. Most of currently remote laboratories implementations include interactive experimentation. In this case, students use real devices and equipment to perform real experiments, which need some flexibility of interaction with the hardware platform. The hardware platform is composed of raspberry pi as lab server, a switches matrix, a practical work circuit, and some measurement instruments. The switches matrix is used to make configuration of experimentation by establishing connection between the practical work circuit and measurement instruments. During the experimentation process, students manipulate the setup using a web page. In the background, the hardware configuration is realized using switches matrix, which is controlled by the manipulation server. The purpose of this work is to develop a new switches matrix in order to provide more possibilities, interaction flexibility with the hardware platform, and ease of use, and finally to improve performance in response time. Our approach aims to improve the flexibility, performance and ease of use. We use 'spst' switches instead of relays because of their less consumption in current and less response time; by addition, we design our matrix so that it can be inserted directly on raspberry pi to facilitate the assembly, and incorporate SPI bus in order to control some electronic components using SPI bus communication such as potentiometer. Considering the new enhancement in our switches matrix, we have obtained better control of practical work circuits and more flexibility and performance when using the existing experimentation such as Filter circuit, AOP circuit, dipole circuit or when developing new ones. Results of experimentation are presented and discussed.

Keywords: Remote laboratory · Switching matrix · Reusability

1 Introduction

Laboratories are generally classified into two categories: Virtual Laboratories and Real Laboratories. In virtual laboratories, the student interacts with a mathematical model that simulates a physical phenomenon [1]. They can be accessed locally or remotely. In the latter, the student interacts with real equipment either locally (these are the classical

M. E. Auer and K. Ram B. (Eds.): REV2019 2019, LNNS 80, pp. 155–164, 2020.
https://doi.org/10.1007/978-3-030-23162-0_15

laboratories) or remotely via the Internet. Remote real laboratories require hardware and software development [2, 3].

MOSTALAB is a remotely accessible electronic laboratory, developed within the framework of a call for projects of the "Agence Universitaire de la Francophonie" (AUF) in order to pool the laboratories within a network [4] and the ERASMUS+ e-LIVES project which aims to promote the concept of e-engineering which is the combination of remote lab and e-learning [5].

Maghreb universities are facing massification of students in the early cycles. Meeting the needs of practical work sessions by the multiplication of measurement benches and premises is very expensive. By bringing the traditional laboratories remotely, we meet the needs of premises; the solution discussed in this paper meets the need for multiplication of measurement benches.

The switching matrix presented makes it possible to control reconfigurable circuits and the sharing of a single measurement bench between several simultaneous users.

Section 2 of this paper describes the remote laboratory MOSTA-LAB and the place of the switching matrix, its environment and software configurations. Section 3 presents a case study of remote experiment.

2 Remote Lab Architecture

This section contains a brief overview of our system; this one is composed of a Lab server hosted on a Single-Board Computer Raspberry pi [6], a switching matrix (SM), a practical work circuit, and some measurement instruments (Fig. 1).

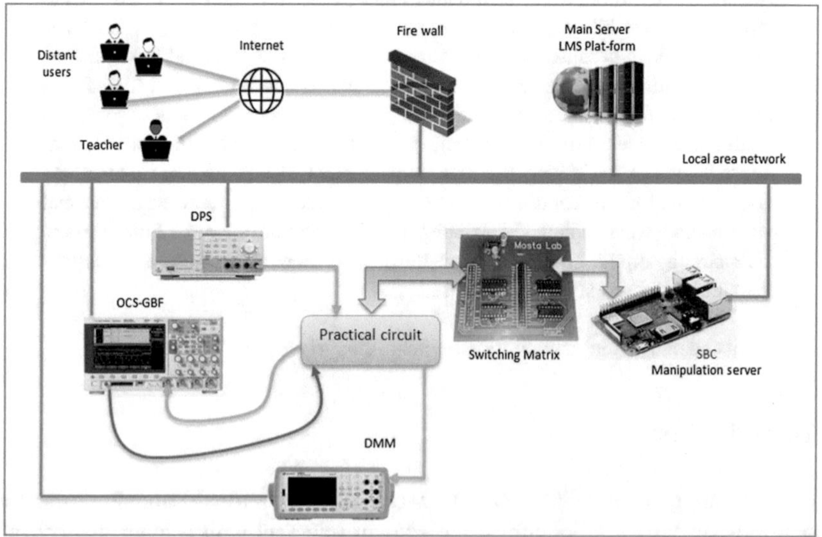

Fig. 1. Remote laboratory architecture

The switching matrix is used to select a configuration of a desired experiment by establishing connection in one hand, between the practical work circuit components, in the other hand with measurement instruments [7, 8].

During a lab session, students select an experience using a web interface. In the background, the switching matrix, controlled by the lab server configures the desired hardware.

2.1 Place of the Switching Matrix in the Hardware Architecture

In this section, we outline the hardware architecture of our system and describe the key mechanisms responsible for an efficient and reliable execution of remote experimentation.

Our system is designed to deliver multiple online accesses to reusable experimentation on Electronics. The hardware setup is composed of a set of measurement instruments, a practical work circuit, a lab server as a single board computer (raspberry pi) and a switching matrix, the entire system is accessible remotely via the main server.

With the developed switching matrix (SM), it is possible to reconfigure and control remotely the same setup used for multiple users and for personalized experiments.

In fact, it became possible to use the same park of measurement instrument such as function generator, digital scope, digital multimeter, for a multitude of experiments over the same practical work circuit, for a high number of students. This makes costs very lower, with the proliferation of instruments for experimentation. Another improvement of this SM leads in the fact that it is directly plugged to the raspberry pi, so it's designed as a shield, as described in Figs. 2 and 5.

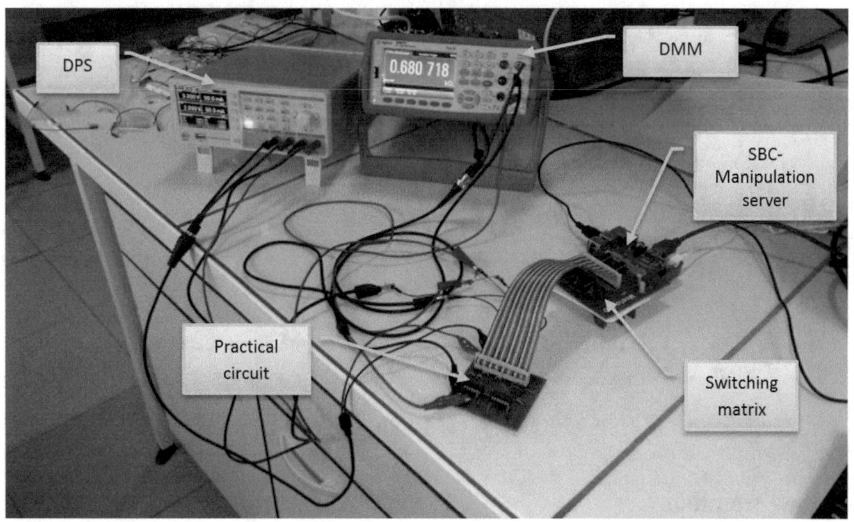

Fig. 2. Real hardware setup

The practical work circuit is designed with multiple combinations of connection, which are established using the flexible switching matrix; this one is managed by the lab server. A variant of experimentation consists of establishment of connection between components of the experimental circuit and appropriate measurement instruments as shown in Fig. 2.

2.2 Evolution in the Developed Switching Matrix

Our approach aims to improve the flexibility, performance and ease of use. Our matrix has been improved even from the hardware and management software.

Hardware evolution
Initially, we have used relays to establish connections in the designed hardware set-up, recently, we use 'spst' switches in our new matrix instead in order to ensure a less consumption in the current, a shorten response time (Figs. 3, 4).

By addition, our matrix incorporates more switches than the previous one, in order to maximize the possibilities of configurations for the practical circuit. To facilitate the assembly, the SM is designed with reduced size and to be inserted directly on raspberry pi board. Finally, it incorporates SPI bus in order to control some components on the practical circuit using SPI bus such as potentiometer Fig. 5).

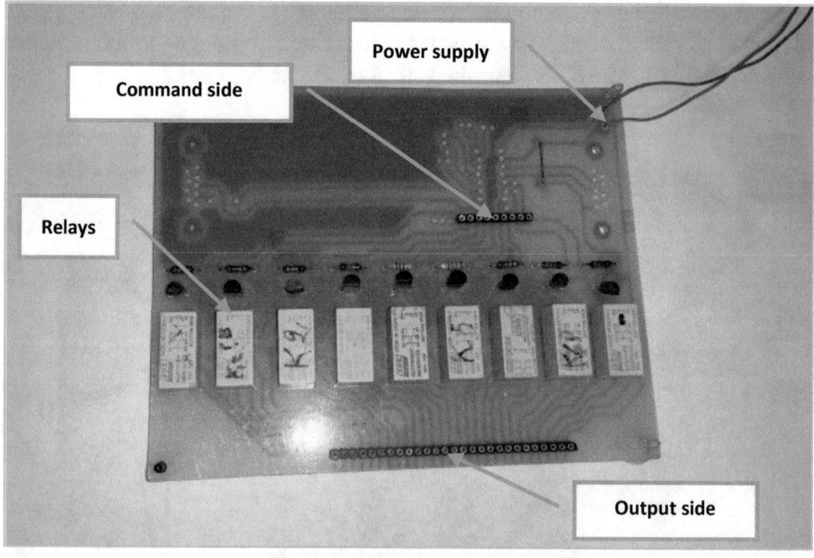

Fig. 3. SM with relays

Software evolution
In a remote experimentation, student control the state of the lab with a web page which is the front end, the user interface sends the parameters remotely to the backend as the stat of the connection of the switching matrix and values of different components

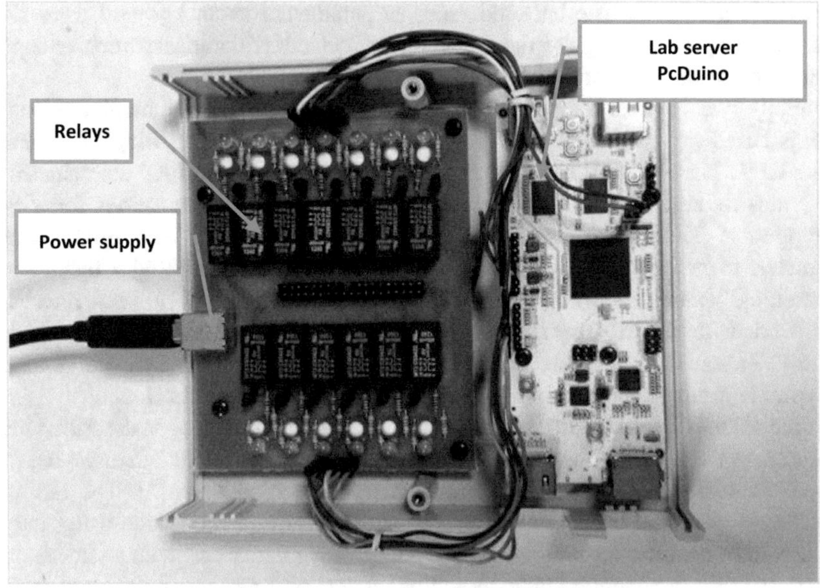

Fig. 4. SM with relays and PcDuino board [9]

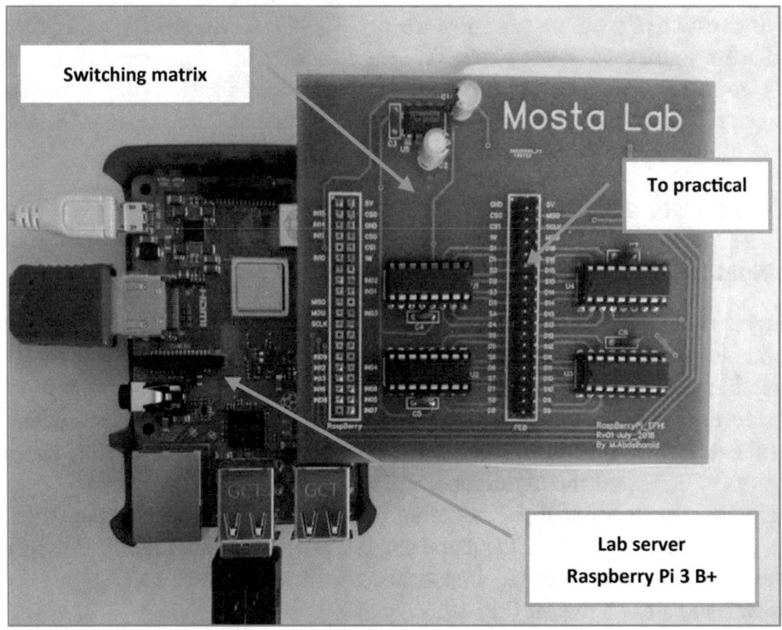

Fig. 5. SM with "*SPST*" switches plugged on Raspberry Pi 3B+

included in the setup. The backend receives parameters in background from a high level interface, when doing manipulation such as selecting a practical work, changing a slider value or click on submit button.

Initially, we have developed the backend which is the server manipulation with Node.js based on JavaScript modules [10], we have used the "Wiring Pi" library to access GPIO pins on the Raspberry Pi 1 model B [11]. We controlled the values of the GPIO pins to activate relays of the matrix in order to make connections for a circuit configuration. The lab server is designed to receive remotely the GPIO values as parameters from the client side via a user interface. We have used Ajax technique for transmitting parameters from user interface to backend [12]. The user interface is developed with HTML/PHP/JavaScript languages [13]. The lab server and the experimentation web page are implemented on the same single board computer, to minimize equipment, but it was a limited solution for future extension.

In the recent version of SM, there is some notable improvement as described below. To optimize the code of the lab server, we have used the "onoff" library [10] under Node.js to access GPIO pins on the Raspberry Pi 3 model B+ [6]. The lab server receives remotely the GPIO values as parameters from the client side using the web socket technique, thus we minimize response time of interactions with switches on the matrix [14]. We use Python language to implement control of some component on the practical circuit connected to the SM using SPI protocol of communication.

The user interface is developed with HTML5/JavaScript/CSS languages. It's composed of input fields; submit buttons, results display as text or images and iframe for displaying results of measurement instruments. The lab server and the experimentation web page are implemented separately. We implemented the experimentation web page under Moodle LMS to allow some evolution of the system, such as duplication of rigs, or to manage the sustained pedagogical control of student.

3 SM in Use, a Case Study

3.1 Dipole Characterization

This experimentation aims to study a dipole circuit. After authentication, user access to the online experimentation through Moodle LMS, it is provided as a test including embedded answers using "cloze" questions. The aim of this manipulation is to characterize a dipole circuit as shown in Fig. 8. Thanks to the SM, each student get a personalized variant of experimentation from the same practical circuit, this makes costs very lower regard the high number of student (Fig. 6).

To realize this manipulation according to a pedagogical scenario, the distant user wires a dipole circuit to an oscilloscope and a wave generator (Fig. 8), then performs measurements and calculates (Z) impedance matrix elements, finally submits results from a web page as shown in Fig. 10.

$$\begin{bmatrix} v_1(\omega) \\ v_2(\omega) \end{bmatrix} = \begin{bmatrix} z_{11}(\omega) & z_{12}(\omega) \\ z_{21}(\omega) & z_{22}(\omega) \end{bmatrix} * \begin{bmatrix} i_1(\omega) \\ i_2(\omega) \end{bmatrix} \quad (1)$$

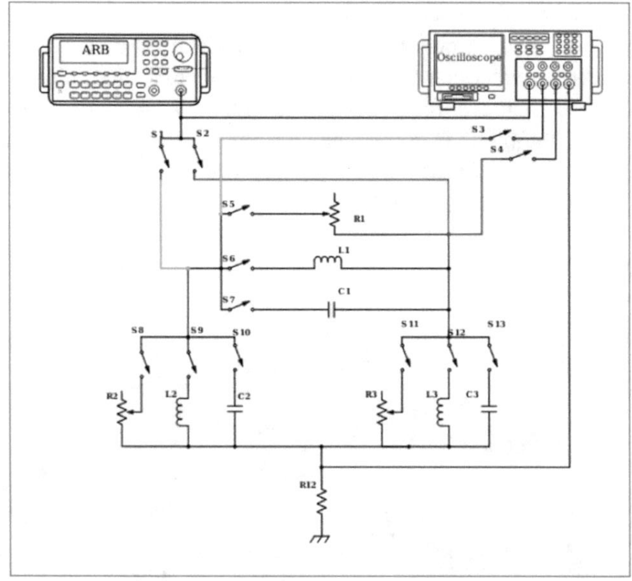

Fig. 6. Multi-variant setup of experimentation for dipole study

Theoretically, Eq. (1) is the base of all mathematical calculation, it's the Frequency-Domain impedance matrix; student has to determine "z_{ij}" elements by changing wires of dipole circuit (Fig. 8), observe and note measurements displayed on scope interface as shown in Fig. 9. In this case study capacitor = 4.7 μF, resistor R = 1 kΩ, R′ = 200 Ω, and frequency value f = 1 kH.

$$z_{11} = \frac{v_1}{i_1}\bigg|_{i_2=0} \tag{2}$$

Student calculates z_{ij} using measurements, for example, Eq. (2) determines the impedance z_{11} which is the ratio of the voltage at port1 to the current in the same port when port 2 is open. Student performs the same calculations for other z_{ij} values.

3.2 Transmission Technique

From a technical side, in background, the front-end software sends automatically the configuration code of the practical circuit to the flexible switching matrix via the lab server using "web socket" technique. This configuration is a string of binary code to control GPIO status on the lab server, these GPIO pins activate switches on the SM so that to establish connection between components on the practical circuit and measurement instruments. The designer of experimentation specifies this code of configuration according to the set of components and measurement instruments includes in the practical circuit.

Figure 7 illustrates the transmission technique, the submission of data input for the appropriate practical work circuit as resistor values are transmitted to the server using SPI bus transmission technique to control potentiometers, this allows measurement instrument software to make customized measurements, send back and display the results. Web socket technique is implemented in JavaScript, where transmission by SPI bus is implemented using Python language. We use "iframe" to access and display scope web page using appropriate URL.

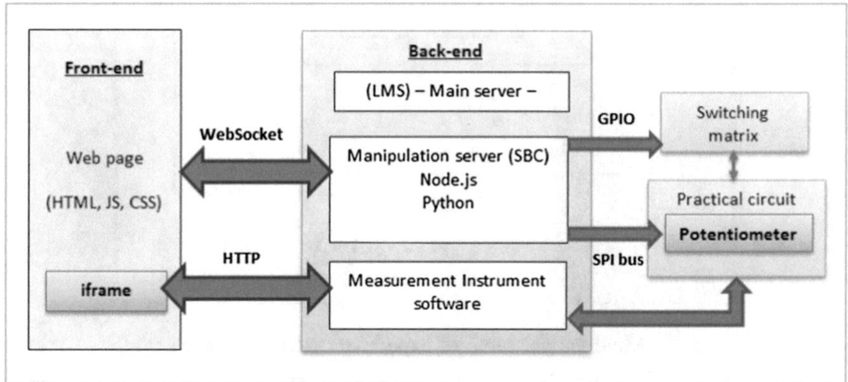

Fig. 7. Interaction technique

Figure 8 shows the user interface web page; it consists of the practical circuit components for the dipole study such as resistors and capacitor, a wave generator and an oscilloscope. Each student gets the same interface with appropriate resistor values, therefore, on the scope, it will be displayed a specific plotted graph different from each other's.

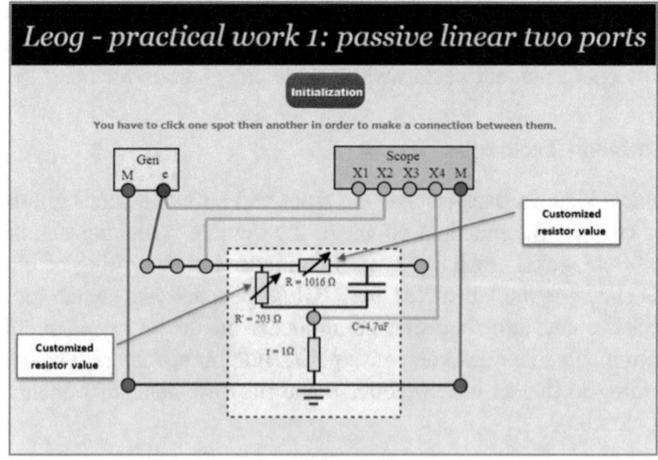

Fig. 8. User interface for dipole characterization

Figure 9 shows the scope display of output signals according to one variant of resistor value, for the same values of capacitor and frequency. This display shows a result of measurement of z_{12} and z_{22} impedances elements, according to resistance values $R = 1\ k\Omega$ and $R' = 200\ \Omega$, capacitor $C = 4.7\ \mu F$, and frequency value $f = 1\ kH$.

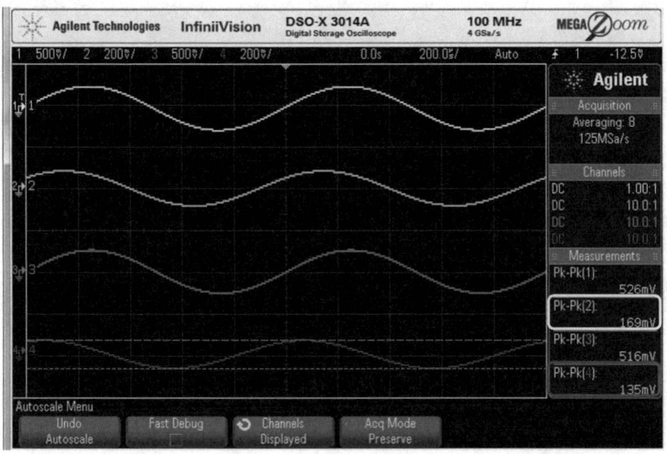

Fig. 9. Scope display, calculate impedances z_{12} and z_{22}

Using Moodle LMS, we have implemented the manipulation as test, in which students manipulate the dipole circuit and visualize the scope in “iframe”, complete mathematical calculations and thus inputs results as a 4 × 4 matrix of impedance (Z), as shown in Fig. 10.

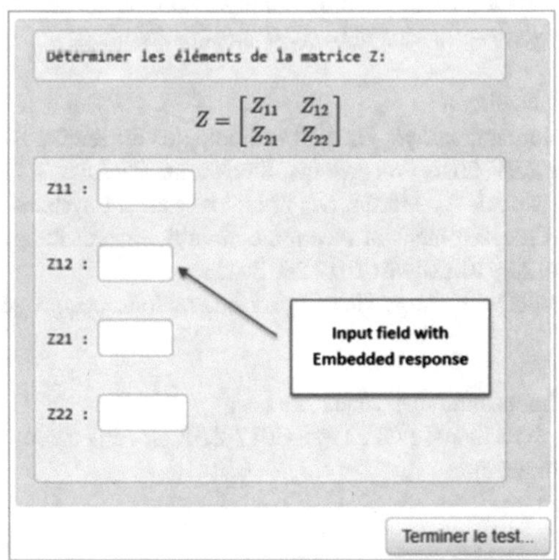

Fig. 10. Input field implementation for user response

4 Conclusions

Considering the new enhancement in our switches matrix, we have obtained better control of practical work circuits and more flexibility when developing new experimentation setup based on our new switches matrix, that combine performance and ease of use. Some existing experimentations are tested with our switching matrix, such as: study of fundamental theorem, filter circuit (RC, CR, RLC …), operational amplifier, and diode.

The use of switches matrix in a remote laboratory is very important to maximize configuration of manipulation and reuse of a same practical work circuit, this makes costs very lower regard the high number of student. In the developed remote laboratory, students manipulate experimentation by interaction with a switches matrix and some measurement instruments; the key of success of the manipulation is the performance in terms of response time, the reusability, the flexibility and ease of use.

Acknowledgements. This work is supported by the e-LIVES and ReLaTraP-AUF projects.

References

1. Gustavsson, I., et al.: On objectives of instructional laboratories, individual assessment, and use of collaborative remote laboratories. IEEE Trans. Learn. Technol. **2**(4) (2009). https://doi.org/10.1109/tlt.2009.42
2. Limpraptono, F.Y., Faradisa, I.S.: Development of the remote instrumentation systems based on embedded web to support remote laboratory. In: Proceedings of Second International Conference on Electrical Systems, Technology and Information 2015 (ICESTI 2015). Lecture Notes in Electrical Engineering 365. https://doi.org/10.1007/978-981-287-988-2_60
3. Farah, S., Benachenhou, A., Neveux, G., Barataud, D.: Design of a flexible hardware interface for multiple remote electronic practical experiments of virtual laboratory. Int. J. Online Eng. (iJOE) **8** (2012). http://dx.doi.org/10.3991/ijoe.v8iS2.2004
4. https://www.auf.org/asie-pacifique/nouvelles/actualites/resultats/
5. https://e-lives.eu/
6. https://www.raspberrypi.org/
7. García-Zubía, J., Hernández-Jayo, U.: LXI technologies for remote labs: an extension of the VISIR project. In: REV 2010 Proceedings, Stockholm, 29 June–2 July 2010
8. Tawfik, M., Sancristobal, E., Martin, S.: Virtual instrument systems in reality (VISIR) for remote wiring and measurement of electronic circuits on breadboard. IEEE Trans. Learn. Technol. https://doi.org/10.1109/tlt.2012.20, 2012
9. "programming the pcDuio", https://learn.sparkfun.com/tutorials/programming-the-pcduino/
10. https://nodejs.org/
11. http://wiringpi.com/
12. https://www.w3schools.com/jquery/ajax_ajax.asp
13. Achour, M., Betz, F.: Manuel PHP. 1997–2017 PHP Documentation Group. https://secure.php.net/manual/fr/index.php
14. https://www.python.org/

Development of Internet of Things Platform and Its Application in Remote Monitoring and Control of Transformer Operation

R. Venkataswamy[1(✉)], K. Uma Rao[2], and P. Meena[3]

[1] Department of Electrical & Electronics Engineering, Faculty of Engineering, Christ (Deemed to be University), Bangalore, India
venkataswamy.r@gmail.com

[2] Department of Electrical & Electronics Engineering, R. V. College of Engineering, Bangalore, India
umaraok@rvce.edu.in

[3] Department of Electrical & Electronics Engineering, B. M. S. College of Engineering, Bangalore, India
meenabms@gmail.com

Abstract. Internet of Things platforms deployed on the system will exhibit numerous benefits such as real time monitoring, faster operation and cost effectiveness. A system oriented IoT platform is developed which features database connotation, web services, setup portal, cloud hosting, drivers or listener for programming languages and hardware devices. The functional parameters of transformer in electrical power system vary around the limit and beyond, which is observed by the IoT platform for remote analysis and to report deformation in the winding. The frequency response measurement from the transformer terminal unit is send to cloud database which is then fetched to remote application through IoT client. At remote monitoring tool, the diagnostic algorithm is executed to estimate the location and extent of deformation. IoT based frequency response analyzer and transformer diagnostic tools developed reports the status of the transformer health condition. Depending upon the extent of deformation, the transformer is isolated from power system.

Keywords: Internet of things · Remote monitoring · Transformer winding deformation · Transformer health monitoring system

1 Introduction

The internet users across the globe has crossed 4 billion among the existing population of around 7.5 billion. The penetration and growth rate of internet user is increasing exponentially. The internet enabled devices are introduced effectively in every domain. The Internet of Things (IoT) products are meeting the current

M. E. Auer and K. Ram B. (Eds.): REV2019 2019, LNNS 80, pp. 165–183, 2020.
https://doi.org/10.1007/978-3-030-23162-0_16

demands of the industry and consumers. The market survey depicts the prediction of more than 100 billion IoT enabled devices into the consumer arena within a decade. IoT technology is creating a major revamp in the physical infrastructure and functional strategies by exhibiting inherent advantages. The adaptation of IoT spans across all the domains such as engineering, management, science and smart applications. Electrical energy is a backbone of digital and engineering system. The electrical system itself is undergoing migration from conventional grid to smart grid by incorporating information and communication technologies alongside. The IoT is being adapted in every possible level in the power system to realize the smart grid in many countries. Though, IoT platforms are generic in nature, there is a requirement of domain oriented IoT platform to furnish better performance, improve latency trade and to fetch mutual economic benefit. In this work, an inter-operatable IoT platform is developed which includes, database integration, web services creation, application portal, cloud server setup, driver services for programming languages and hardware devices. The cryptosystem is also provided for the secure communication between sensor and framework. Transformers are major component in electrical power system which is connotated with the IoT platform for remote analysis and to report deformation in the winding. The frequency response analysis (FRA) from the transformer terminal unit is acquired through IoT client. The diagnostic algorithm is executed to estimate the radial and axial deformation at the remote machine. It also reports the extent of deformations. IoT based frequency response analyser is developed which works with transformer diagnostics tool to report incipient, alert, abnormal and critical condition of the transformer. The deformation extent is used to open circuit breaker by comparing with allowable reference. The circuit breaker is closed after the maintenance or repair.

2 Research, Development and Products

An economical solution for developing energy utilization and environmental monitoring through IoT framework, focusing on educational buildings. In [1], hardware module is designed with analog sensors with conduction of a set of experiments which exhibiting the similar outcome compared to existing available frameworks. In [2], vibration model of transformer core and winding was developed. The transformer vibration THD was created in distribution transformers and tested for inter-turn and inter-disc fault. The vibration THD is descend under short circuit currents initiation. The signal stored and acquired using cloud environment.

In [3], cloud platform has been developed which combines IoT and cloud framework to facilitate all kind of users. Domain specific requests are generated by the IoT devices which is sensed by the IoT platform and processes the request in real time with the stored data depending upon the user requirements. Virtual environment is created upon the physical layer to accelerate the request. Virtual docker container is implemented to facilitate cloud services such as PaaS, SaaS. Request distribution scheme is proposed for faster operation and improve the quality of service. The scheme is also optimizes the communication cost involving in the bidirectional dataflow.

In [4], detailed summary of the user-centric cloud enabled model along with sufficient flexibility to meet the diverse and many a times facilitating the needs of various domains were drafted. The platform activated by a scalable cloud to provide the capability to adapt the IoT. The platform offers communication, runtime, database and visualization applications, thereby focusing complementary individual growth in all the sector by shared hosts. The future challenges have been discussed ranging from data analytics and visualization of the big data The international initiatives were consolidated wherein adaptation of IoT in smart application, transportation and plug and play devices were presented.

In [7], IoT testbench is implemented to provide technical assistance for control operation which including communication architecture and database on cloud. A prototype development of color sensor is presented to validated the work.In [8], developed a GSM/GPRS based transformer health monitoring which captures the current, temperature, oil level, vibrations and humidity. Microcontroller based monitoring node is developed which acquires data and transmismits through GSM system. The status is communicated under abnormal or unstable situations.

The major IoT platforms available in market are Eclipse IoT, OpenHab, Cisco IoT, Industry 4.0, Google IoT Services, SAP IoT and OpenIoE platform. The AWS released IoT related products such as Core, FreeRTOS, greengrass, 1-click, IoT analytics, IoT buttons and Device Management tools [9]. Heroku provides the containers with lightweight, isolated dynos that provide compute, storage, operating system and file system which are typically run on a shared host in cloud framework [10]. Google cloud platform offers products such as computing engine, cloud storage, databases, Cloud IoT Core, Cloud IoT Edge, Analytics, Machine learning, Identity and security applications to integrate with physical system to realize IoT enable system [11].

3 Internet of Things

3.1 Contextual Background

In the current scenario, more than 51% of the world population using internet with the growth rate being around 1000%. Continent-wise percentage of internet users are shown in Fig. 1. Statistics reveals the high penetration or faster adaptability of IoT in various domain such as home, industry, education, management, power, transportation and all other engineering. Almost 20–25 billion IoT devices are at consumer end which is effectively playing its role and attracting manufacturer to consumers.

3.2 Architecture

The general architecture of IoT can be visualized as shown in Fig. 2. The physical parameters are acquired from sensor and transducers. Sensor value then fed to

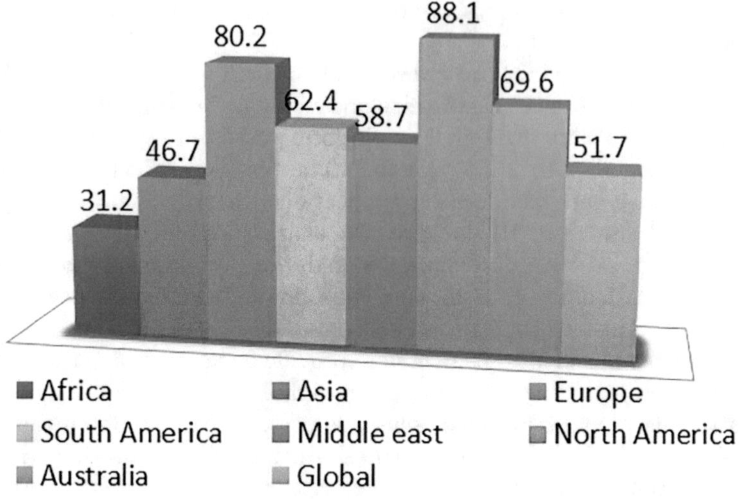

Fig. 1. Internet users

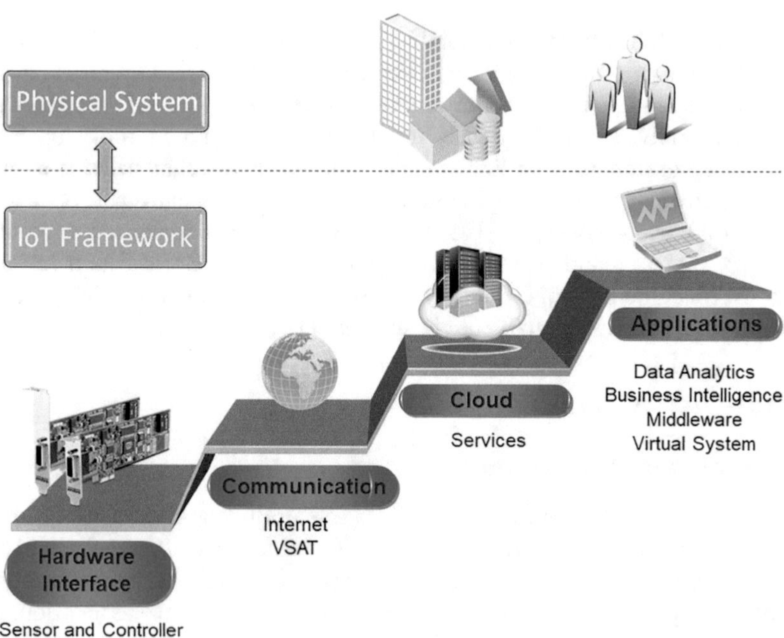

Fig. 2. IoT architecture

the digital devices. Digital device data then passed to cloud through internet or VSAT communication medium. The control signal for the physical system is given by IoT platform using the same infrastructure.

3.3 Future of IoT

There will be a significant growth in IoT domain with respect to investment, deployment, data, life style, migration and technologies. The approximate number of IoT devices penetrating into the market year-wise is shown in Fig. 3. More than 75 billion devices are expected in the year 2025 [13]. The adaptability of IoT seems to be faster in industry, education, health care, transportation, banking, administration and every possible entities. It has been predicted that 47% and 35% of IoT deployment will happen in industry and health care respectively in the year 2025. Around 3 trillion dollars investment is planned on IoT market cumulatively by various companies. The data traffic expected in the year 2020 would be 14,000 exabytes [14].

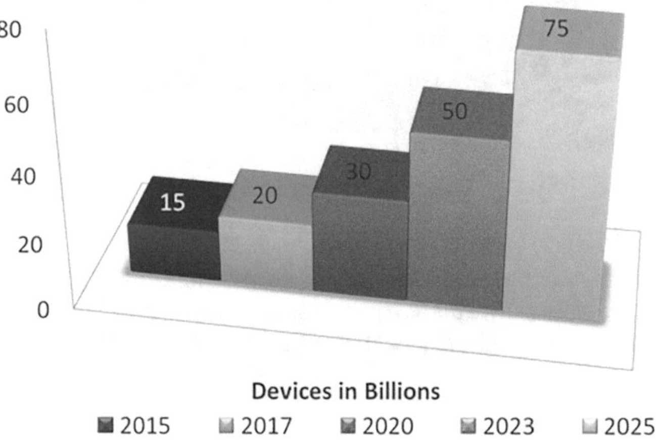

Fig. 3. IoT devices projection

3.4 Hardware Ecosystem

The continuous design, fabrication, manufacturing and deployment of IoT devices on the physical system consuming material, time and intellectual power of man. The rigid or proprietary devices may not be recycled or reused for advanced or modified IoT platform or system. Therefore, there is a need of sustainable software and hardware interoperatable model which surely addresses the societal or holistic ecosystem. The Fig. 4 shows how the e-Waste management and open source platform play vital role in green ecosystem.

3.5 Cloud Environment

Cloud services are the backbone of the IoT framework. Cloud services such as HaaS, PaaS, SaaS etc are used by IoT wherever applicable. The physical system,

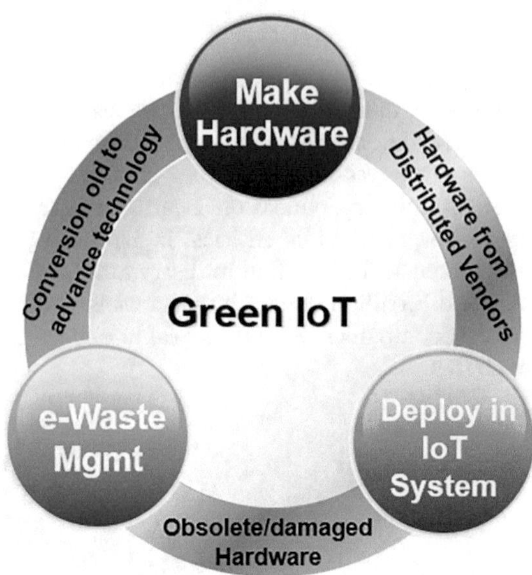

Fig. 4. Hardware ecosystem

interfacing hardware and cloud services together constitute IoT platform. There are countless vendors are available in market who provide different cloud services. The key providers being Amazon web services, Cisco IoT, IBM and Google cloud platform.

4 Transformer Diagnostics

The transformer is a essential component in the power system. The unavailability of transformers due to damages in any of its parts will lead to the outages. The consumers will be affected until transformers are restored. Therefore, diagnostics and control of transformer is very important aspects.

4.1 Winding Forces

The major issue in the transformer is winding deformation which occurs due to the unstable load currents. The current density will result in thermal change, axial and radial movements.

Thermal Change The An approximate expression for the temperature rise of the conductor after ΔT seconds [15] is given by,

$$\Delta\theta = \frac{\Delta T(1 + W_e)D^2\rho\alpha^2}{dh} \tag{1}$$

where W_e is unit eddy current loss, ρ is resistivity of the conductor material, α is factor for short-circuit current as multiple of rated full-load current. d is density of the conductor material, h is the specific heat of the conductor material.

Radial Force The average radial force per mm length of conductor of a circular coil is given by the expression,

$$F_R = \frac{0.031\sigma A_c}{W_D} \tag{2}$$

where σ is the peak value of mean hoop stress in kN/mm^2, A_c is the cross-sectional area of the conductor by which the force is required in mm^2. W_D is the mean diameter of winding in mm.

Axial Force The maximum axial force inner and outer turn is given by,

$$F_A = \frac{510U}{(z_t + z_s)fh} \tag{3}$$

where U is rated kVA, z_t is per unit impedance of transformer, z_s is per unit impedance of supply on the basis of tranformer rating, f is the frequency, h is the axial height of the windings in mm.

4.2 Frequency Response Analysis

A FRA is generally a plot of transfer function and frequency typically from 20 Hz to 2 MHz usually measured at very low voltages. Typically winding admittances, voltage ratio between windings or the DPI are the well known transfer function representation. One such FRA transfer function with impedance as the measured quantity has been shown in Fig. 5 which is obtained using SpiceOpus circuit simulator by substituting circuit elements in the four section ladder circuit model of

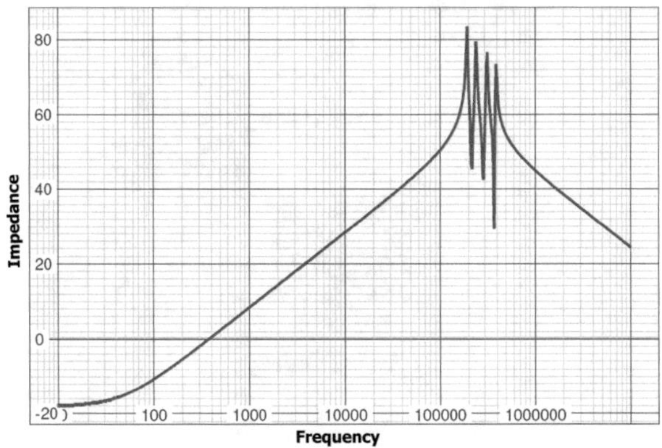

Fig. 5. FRA of transformer model (Four section)

the transformer. At low frequencies, winding exhibit an inductive nature. In the intermediate-frequency range, the combination of winding leakage inductance and series capacitance results in a paralleled LC. At low frequencies, winding exhibit an inductive nature. Therefore, physical changes will result in the variation in the Intermediate and high frequency range of FRA.

5 Objectives

The main objectives of the work undertaken are,

- Development of IoT platform: It includes the creation of database schema, web service creation, drivers and web portal for user activities.
- Transformer winding diagnostics: It includes development of IoT based Frequency response Analyzer, integration of transformer diagnostic tool.

6 Development

6.1 System Overview

The system includes transformer winding attached with frequency response measurement unit interfaced to local and remote transformer diagnostics tools using IoT platform. The diagnostic algorithms are executed either in cloud or at remote base unit or at local unit to know about the winding health status. The pictorial representation of the system is shown in Fig. 6.

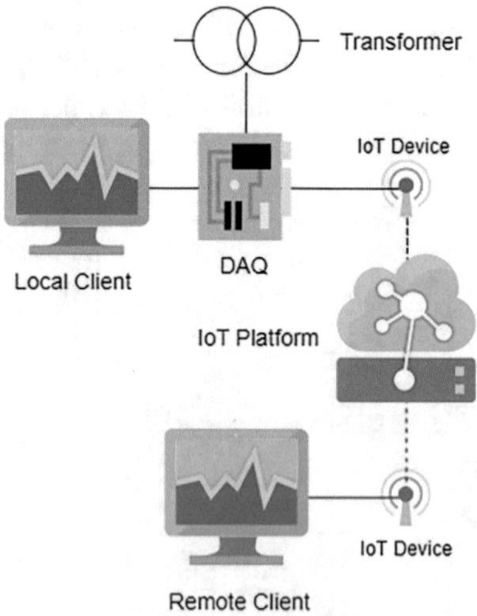

Fig. 6. System overview

6.2 Development of Internet of Things

Architecture the architecture of the IoT platform and the technologies used are in brief is depicted in Fig. 7.

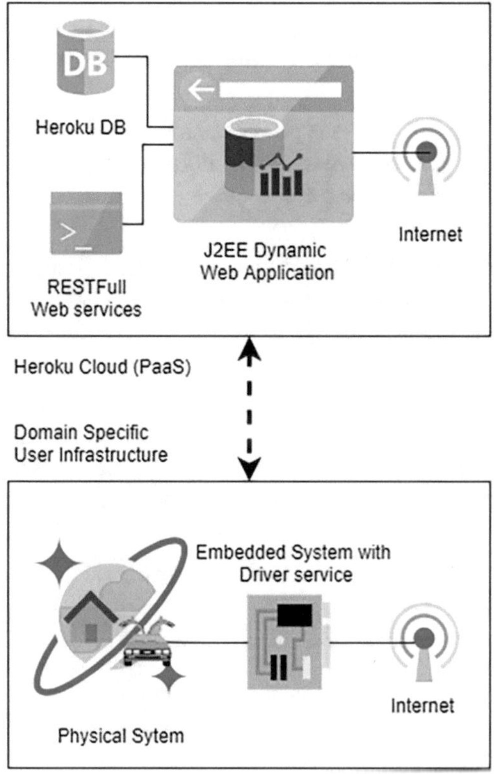

Fig. 7. Architecture

Features The features of the IoT platform are,

- Creation of identity for the hardware device.
- Creation of the required database schema at the back-end.
- Bi-directional access, that is hardware device to IoT and
- IoT to hardware device via webservices.
- Execution of the webservices in the console.
- Provision of data hosting in the form of XML, JSON and scalar value.
- Driver programs for the Matlab, Scilab, Python programs. Supports all the hardware such as Raspberry Pi, Arduino, Beagle bone, NodeMCUs etc.

Device Selection One can connect almost all the electronics boards directly or indirectly into the network. The pin details, specifications and operating modes are different for different embedded boards. Depending upon the hardware, default data will be loaded into the database.

Device Identification The hardware section is used to create the identity for the hardware devices. It gives sequential 4 digit user id and 3 digit random pin which could used for bi-directional communication between the ad-hoc nodes.

Data Formats There are three types of data support is given. User can choose any one among xml, json or scalar value. It depends on the programming languages and library available for parse.

Cryptosystem The system stores the data securely for the user opted for cryptosystem. All the data is encrypted and keys are generated and given to the user. It can be used in any level of data transfer or communication.

Web Services The web services are exposed for retrieval, storage and computing operations, few among them are shown in Table 1. A generic framework has been created to add Restfull web services. Web services acts as back-end engine which could be consumed by any hardware or software applications.

Table 1. Web services

Web service	Description
showjson/userid/userpin	To get hardware JSON from webservicefor a given user id and pin
showvalue/userid/userpin	To get hardware value from webservice for a given user id and pin
showxml/userid/userpin	To get hardware XML file from webservice for a given user id and pin
updatejson/userid/userpin/json	Update the hardware JSON passed as parameter against the user id and pin in cloud database
updatevalue/userid/userpin/value	Update the hardware value passed as parameter against the user id and pin in cloud database
updatexml/userid/userpin/xml	Update the hardware XML passed as parameter against the user id and pin in cloud database

Console Service The console service is to testing web services in the web portal by using browser as a client. URI is constructed with the credentials and request is sent to the server. The response from the server is captured and displayed in the browser window.

Hardware Drivers Hardware driver is a service written in any programming languages which has run-time environment on the hardware. Many hardware drivers were written using matlab, python, scilab and java for personal computer, raspberry pi, arduino and data acquisition cards. The hardware drivers fetches the data from database through web service. The data is parsed and flushed onto the signal ports or pins. It also scans the signal from all the ports or specific ports, merges and uploads to the database through web services.

6.3 Development of Frequency Response Analyzer

Frequency Response Analyzer is a instrument which measures gain and phase variations with respect to the variation in the frequency. This FRA is virtual instrument which connects IoT and data acquisition unit with the element under test.

Features : The features of IoT based FRA are,

- Measurement of FRA using NI DAQ
- Measurement of FRA from IoT connected device.
- Upload the FRA to cloud.
- Remote and local FRA measurement.

Data Acquisition System A high sampling rate data acquisition can be connected across the winding. The channels and voltages are configured. Using current and voltage sensor values, DPI is calculated.

Data Persistence Instantaneous values from the measurement unit is packed as XML and uploaded into the cloud. A comma separated variable file is stored in the application folder.

Operating Modes Two modes incorporated are,

- DAQ Mode: In this mode, the DAQ card connected locally is used measure FRA and perform diagnostics in the system.
- IoT Mode: In this mode, FRA is fetched from remote DAQ card and diagnostics of transformer is performed at base unit.

TDT Interface The FRA is available in cloud as well as in local application folder. It can be imported manually and diagnostic algorithm is executed. It can also be executed by cloud FRA.

6.4 Transformer Winding Diagnostics

Algorithm The steps followed for the winding diagnostic using FRA is shown in the Fig. 8.

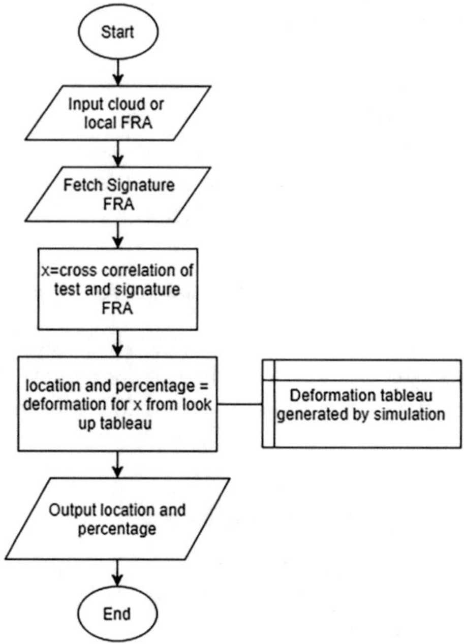

Fig. 8. Algorithm

Application A Matlab graphical user interface is prepared for the diagnostic algorithm. An integration service is optional for the user which fetches data if the user opt for it.

6.5 Control of Transformer Operation

Transformer is isolated from the power network by operating circuit breaker as shown in Fig. 9. The isolation is depends on the extent, types, part and location of deformation. The reference values are stored and categorized as normal, incipient, abnormal or critical problem in the transformer.

7 Results and Discussion

7.1 FRA Measurement Using IoT Platform

The FRA measured using IoT is shown in Table 2. FRA data obtained by the application or file will be stored and served from IoT platform. The first URI data is FRA setting at IoT. Second data is FRA consisting of magnitude and phase values for all frequencies.

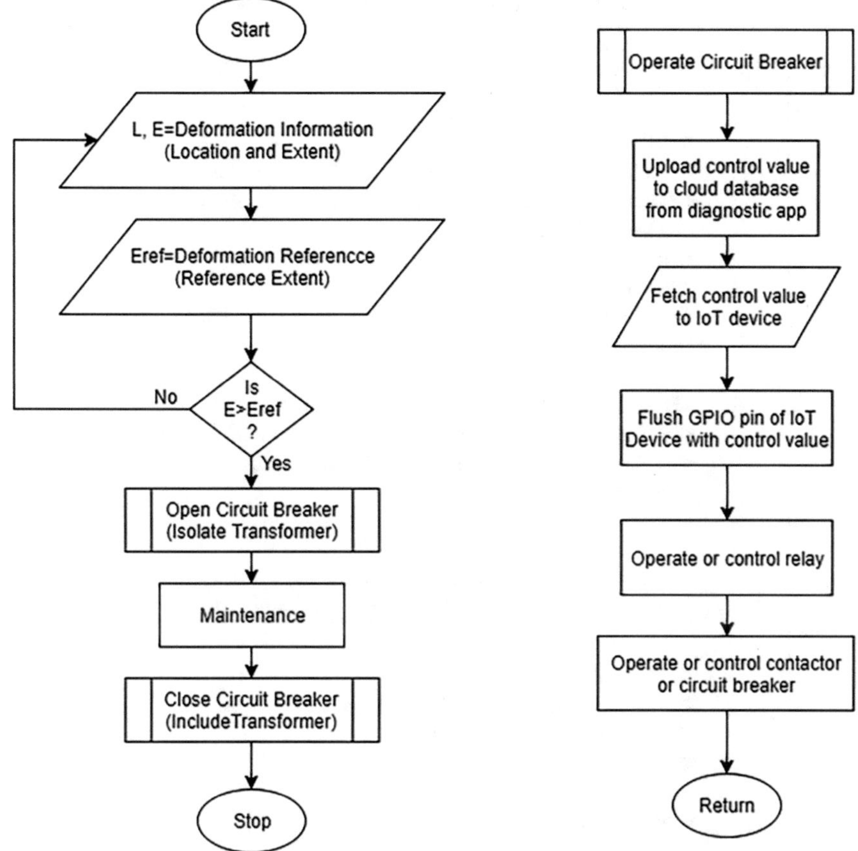

Fig. 9. Control of circuit breaker at primary and secondary side of transformer

7.2 FRA Measurement Using IoT Frequency Response Analyzer

The application is developed using Matlab app designer. The database schema and identity were created. The application checks for internet and IoT connection. It retrieves the FRA uploaded by transformer remote terminal unit. One such case is shown in Fig. 10.

7.3 Transformer Diagnostics

FRA readings are obtained in ambient condition. A deformation of 3% introduced in 2rd section and FRA readings are taken again. From Fig. 11, it can be proved that at 2nd section 3% of radial deformation is present as differential cross correlation value is tending to zero. This validates the precision and functioning of proposed algorithm.

Table 2. FRA measurement using IoT platform

URI	https://openioe.herokuapp.com/api/showxml/1357/73
DATA	<xml> <FRAName>Transformer Winding FRA</FRAName> <FRADesc>Transformer Winding FRA</FRADesc> <Date>28-9-2018</Date> <FRADataIDDAQ>1358</FRADataIDDAQ> <FRADataPINDAQ>3</FRADataPINDAQ> <FRADataIDDAQSetting>1360</FRADataIDDAQSetting> <FRADataPINDAQSetting>673</FRADataPINDAQSetting> <FRADataIDIoT>1359</FRADataIDIoT> <FRADataPINIoT>615</FRADataPINIoT> <FRADataIDIoTSetting>1361</FRADataIDIoTSetting> <FRADataPINIoTSetting>549</FRADataPINIoTSetting> <Check>1</Check> </xml>
URI	https://openioe.herokuapp.com/api/showxml/1359/615
DATA	<FRAData> <Date>September 28 2018 19:24:41</Date> <F>50</F> <M>1.5708</M> <P>88.182</P> <F>51.1</F> <M>1.6048</M> <P>88.201</P> <F>52.2</F> <M>1.6388</M> .. </FRAData>

7.4 Control of Transformer Operation

A deformation of 3% introduced in 2rd section and 2% is set as reference. The trip signal is generated and triggered the excitation coil of contactors to isolate primary and secondary side of the transformer.

8 Conclusion and Future Scope

Internet of Things platform is developed using open source tools provided by various vendors. Driver services are written ensuring the suitability to fifth generation programming languages and popular hardware modules. IoT based Frequency Response Analyzer is developed to enable data acquisition from remote

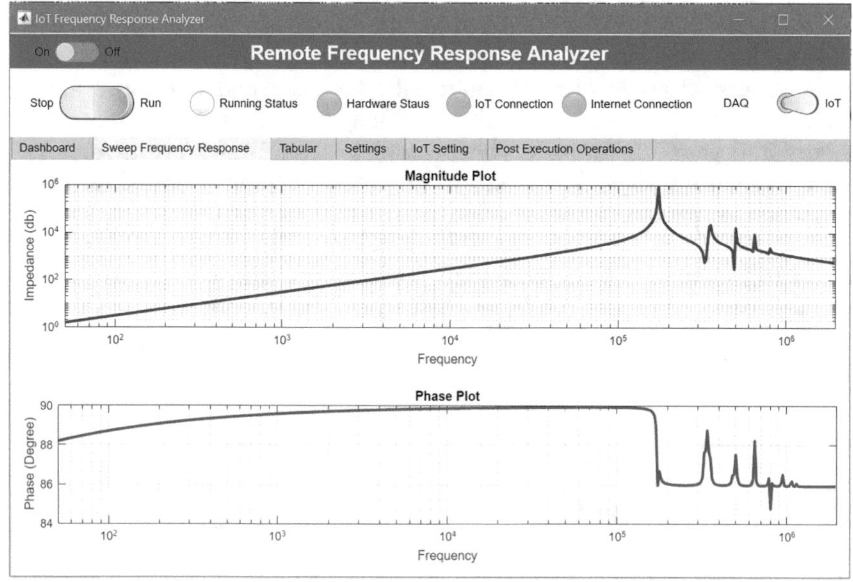

Fig. 10. FRA measurement using frequency response analyzer

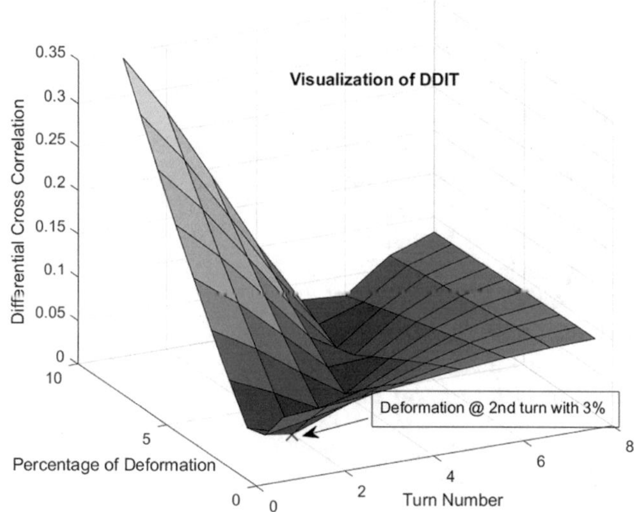

Fig. 11. Cross correlation visualization

client that being transformer terminal unit. The transformer diagnostic tool is integrated with base or coordinator program to identify the local and extent of deformation of transformer winding.

The methodology can be extended for the diagnostics of other parts of transformer such as core, insulation or any other ancillary modules. The conceptualization can also be applied to perform specific operation and monitoring activities by using functional parameters such as voltage, current, temperature, vibration, pressure and gases. In broader view, complete transformer health monitoring and maintenance task could be used to improve the stability and controllability of smart power grid.

Acknowledgements. The authors would like to acknowledge and express the deepest gratitude to management of Faculty of Engineering, Christ(Deemed to be University); BMS College of Engineering, and RV College of Engineering for providing freedom to work in the laboratory.

Appendices

Appendix 1: Glossary

IoT Internet of Things
SFRA Sweep Frequency Response Analysis
FRA Frequency Response Analysis
THD Total Harmonic Distortion
AWS Amazon Web Services
GCP Google Cloud Platform
PaaS Platform as a Service
SaaS Software as a Service
DPI Driving Point Impedance
TDT Transformer Diagnostic Tool
VSAT Very Small Aperture Terminal
GSM Global System for Mobile
DAQ Data acquisition
GPRS General Packet Radio Service
RTOS Real-Time Operating System
DIT Deformation Information Tableau
DDIT Differential Deformation Information Tableau

Appendix 2: IoT Platform

2.1 IoT Details

The URLs of the application are shown in Table 3.

2.2 Hardware Selection and Identity

The user should select the hardware, cryptosystem type and data format. The database infrastructure is created against the user id and pin. The sample option and generated credentials are shown in Fig. 12.

Table 3. IoT details

Main application	https://openioe.herokuapp.com
Hardware	https://openioe.herokuapp.com /hardware.jsp
Console	https://openioe.herokuapp.com /console.jsp

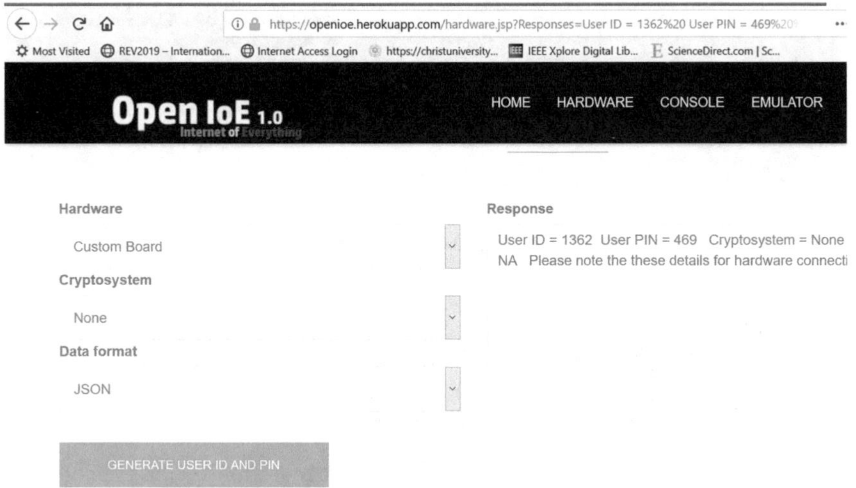

Fig. 12. Hardware selection and identity

2.3 Console

Web service get and post operation can be perform in the console without programming language or service program. The user should enter service name without the application context. The FRA IoT setting retrieved using console is shown in Fig. 13.

Appendix 3: Frequency Response Analyzer

The dashboard shown in Fig. 14 will display status of DAQ, internet, IoT connection, instantaneous values of frequency, magnitude and phase. The provisions given the tool are DAQ settings, IoT settings and Upload to cloud.

Appendix 4: Transformer Diagnostic Tool

The cross correlation of signature and deformed FRA is calculated. The differential deformation information tableau is generated by comparing the value with DIT. The graphical window is shown in Fig. 15.

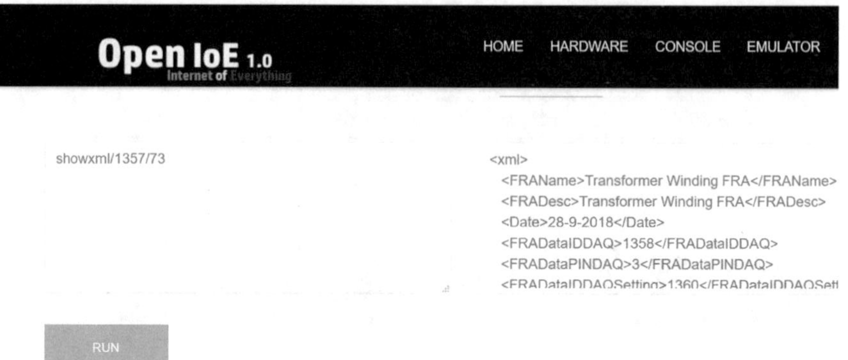

Fig. 13. Console

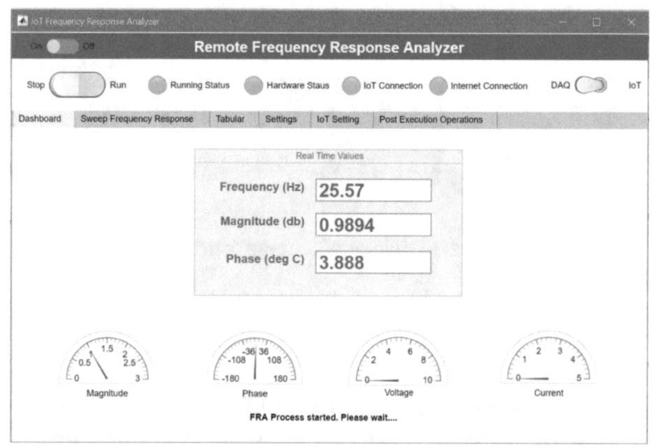

Fig. 14. Dashboard

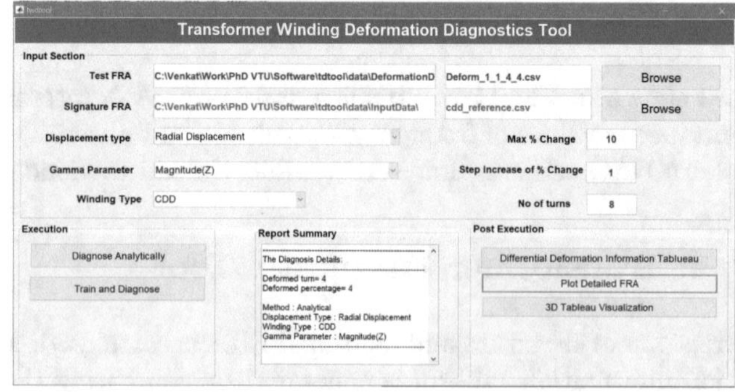

Fig. 15. Transformer winding diagnostics tool window

References

1. Pocero, L., Amaxilatis, D., Mylonas, G., Chatzigiannakis, I.: Open source IoT meter devices for smart and energy-efficient school buildings. In: Elsevier Licensed Under CC BY-NC-ND, vol. 1, pp. 54–67, Greece, Italy (2017)
2. Bagheria, M., Nezhivenkoa, S., Naderib, M.S., Zollanvaria, A.: A new vibration analysis approach for transformer fault prognosis over cloud environment. Int. J. Electr. Power Energy Syst. **100**, 104–115 (2018)
3. Dehury, C.K., Sahoo, P.K.: Design and implementation of a novel service management framework for IoT devices in cloud. J. Syst. Softw. **119**, 149–161 (2016)
4. Gubbi, J., Buyya, R., Marusic, S., Palaniswami, M.: Internet of Things (IoT): a vision, architectural elements, and future directions. Future Gener. Comput. Syst. **29**(7), 1645–1660 (2013)
5. Kamlaesan, B., Kumar, K.A., David, S.A.: Analysis of transformer faults using IOT. In: IEEE International Conference on Smart Technologies and Management for Computing. Communication, Controls, Energy and Materials (ICSTM), vol. 1, pp. 239–241. Tamil Nadu, India (2017)
6. Oh, S., Kim, Y.: Development of IoT security component for interoperability. In: 13th International Computer Engineering Conference (ICENCO), pp. 41–44. Cairo, Egypt (2017)
7. Okuda, M., Mizuya, T., Nagao, T.: Development of IoT testbed using OPC UA and database on cloud. In: 56th Annual Conference of the Society of Instrument and Control Engineers of Japan (SICE), pp. 607–610. Kanazawa, Japan (2017)
8. Pawar, R.R., Deosarkar, S.B.: Health condition monitoring system for distribution transformer using Internet of Things (IoT). In: International Conference on Computing Methodologies and Communication (ICCMC), pp. 117–122. Erode, India (2017)
9. Amazon Web Services (AWS)—Cloud Computing Services. https://aws.amazon.com. Last Accessed 26 Oct. 2018
10. Heroku-Cloud Application Platform. https://www.heroku.com. Last Accessed 26 Oct. 2018
11. Google Cloud Including GCP & G Suite, -Google Cloud. https://cloud.google.com. Last Accessed 26 Oct. 2018
12. V. R—Open IoE. https://openioe.herokuapp.com. Last Accessed 26 Oct. 2018
13. IoT: Number of Connected Devices Worldwide 2012–2025. https://www.statista.com/statistics/471264/iot-number-of-connected-devices-worldwide. Last Accessed 29 Oct. 2018
14. IoT White Papers and Case Studies—Cisco. https://www.cisco.com/c/en_in/solutions/internet-of-things/resources/case-studies.html. Last Accessed 29 Oct. 2018
15. Heathcote, M.: J & P Transformer Book, 13th edn. Elsevier Science, Great Britain (2011)

Verification of Zener Voltage Regulation Phenomenon Using Remote Engineering

Rayudu Chakravarthy[1], Sanku Supriya[1], Kavirayani Srikanth[1(✉)], Divya Sree Uddandapu[1], Ajit Kumar Rout[2], Asapu Mohan Krishna[2], and Srinivasu Tangudu[2]

[1] GVP College of Engineering (A), Visakhapatnam, India
{rayuduchakravarthy, Supriyasameera}@gmail.com, {kavirayanisrikanth, divyasree}@gvpce.ac.in

[2] GMR Institute of Technology, Srikakulam District, India
ajitkumar.rout@gmrit.org, {asapu.mohankrishna, tangudusrinivasu17}@gmail.com

Abstract. Voltage regulators play an important role in maintaining constant output voltage across electronic circuits. They are also used for impedance matching. Zener diode is most commonly used semiconductor device in designing voltage regulator. Stabilizing circuits whose essential parts contain Zener diodes are used for protection of circuits from over voltages. It is very important to study the characteristics of Zener diode for the designing the stabilizing circuits and voltage regulators as well as clippers. In this paper Zener voltage regulation phenomenon is verified using the Remote Engineering. The entire setup of the voltage regulator circuit is controlled remotely through mobile or personal computer. This can be achieved with the help of Arduino. Arduino is a traditional tool used in many electronics circuits mainly involving Embedded systems, Internet of things and Remote Engineering as well. The voltage regulation phenomenon of the Zener diode has been a very important experiment for the beginners in electronics engineering. However this phenomenon can also be verified using the remote engineering. The student can go through the circuit using his/her mobile or laptop. The student can watch the entire setup in web camera and variation in the output voltage is displayed. The student also has the tendency to control the input voltage and observe the constant output voltage.

Keywords: Zener diode · Voltage regulation · Remote engineering · Digital voltmeter · MOSFET · Wemose d1 mini wifi controller module

1 Introduction

Remote engineering and virtual labs are an emerging arena having good scope for student research and advanced studies. This research will help the society in terms of parameters like cost, space and mainly time and the liberty of the students to work at their own space. There is lot of work that has been already done in this arena. Arduino being the state of art technology has been used referring to model circuits in [1, 2] and [4]. The basic Zener regulator circuit has been adopted from [3]. Hashemian et al. [5]

M. E. Auer and K. Ram B. (Eds.): REV2019 2019, LNNS 80, pp. 184–192, 2020.
https://doi.org/10.1007/978-3-030-23162-0_17

discusses a method of the cloud deployment of a reconfigurable interface for remote engineering. Salzmann et al. [6] discusses how web 2.0 technology can be used for development and deployment approaches in tackling challenges in remote and virtual laboratories. Fayolle et al. [7] elaborates on the distinction among virtual, remote and hybrid labs and elaborates on issues in collaborative remote laboratories in successful recreation of classroom. Heradio et al. [8] discusses about remote labs for control education as a survey paper based on local area and remote area configurations. This present work focuses on the idea of applying the logic to a simple remote lab for the phenomenal study of zener voltage regulation.

1.1 Zener Voltage Regulation

Zener diode is similar to a normal p-n junction diode except its reverse characteristics. When a large negative voltage is applied across the normal p-n junction diode, the voltage across the diode does not remain constant whereas in case of the Zener diode, the voltage across it remains constant once it enters into the breakdown region, even by varying input voltage. This phenomenon is called voltage regulation. There are two types voltage regulation. They are line regulation and load regulation. Achievement of constant Zener voltage regulation is done by keeping the load resistance constant by varying the input voltage is called line regulation. On the other hand if the Zener voltage regulation is achieved by keeping the input voltage as constant by varying the load resistance. Then it is called load regulation. This paper mainly focuses on the line regulation of a Zener diode i.e. varying the input voltage by keeping the load resistance constant.

1.2 Aim of the Work

The voltage regulation phenomenon has very extensive applications in designing voltage regulators, stabilizers, clippers. A voltage regulator circuit can be designed using a Zener diode due to its unique reverse characteristics. However once the voltage regulator circuit is built, it should be tested in the laboratory. The same can be done using the Remote engineering. The student will be able to go through the circuit even if he is not present in the laboratory. The student can conduct the experiment i.e. Zener voltage regulation phenomenon from a distant place, using his mobile or PC. He/she will be able to watch the output voltage and the entire circuitry through a web camera installed in the laboratory. The student can also control the input voltage and observe the output readings. In order to observe the output readings, a digital voltmeter is connected to the output such that the display is visible through the web camera.

1.3 Implementation Requirements

The main aspects of this work are divided into the following and are discussed in the next section in detail.

1. Voltage regulator circuit
2. Variable voltage supply

3. Digital voltmeter
4. Arduino web interfacing using Wifi module.

The implementation block diagram as shown in Fig. 1 is as follows and the details of each module will be explained in the next section.

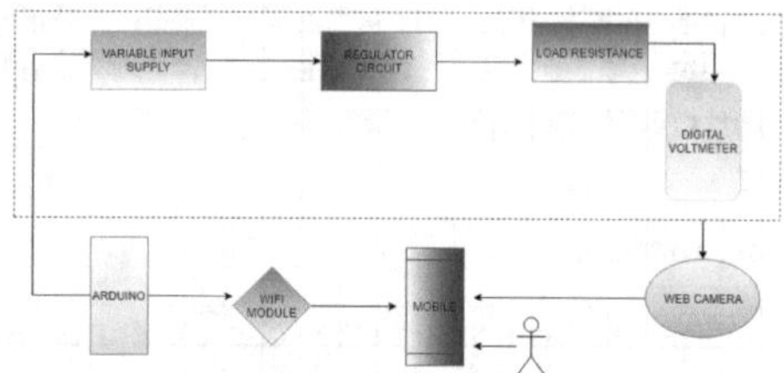

Fig. 1. Block diagram

2 Setup of the Remote Lab

2.1 Voltage Regulator Circuit

A simple voltage regulator circuit can be designed by using a Zener diode in series with a known resistance. When the input voltage is increased, the output voltage is same as that of input voltage. If the input is further increased, the diode enters into the breakdown region and the digital voltmeter reads the Zener breakdown voltage. The output voltage remains constant even if the input voltage is increased beyond the breakdown voltage. Figure 2 shows how Zener diode can be used to design a voltage regulator circuit.

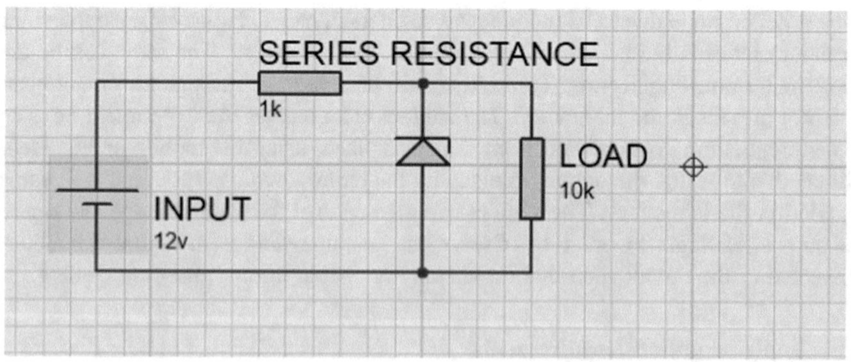

Fig. 2. Voltage regulator circuit

2.2 Variable Voltage Source

A variable voltage source is designed using a field effect transistor i.e. MOSFET (metal oxide semiconductor field effect transistor). MOSFET is a voltage controlled device. Figure 3 is the circuit acting as a variable voltage source. The circuit is connected as an input voltage source to the regulator circuit.

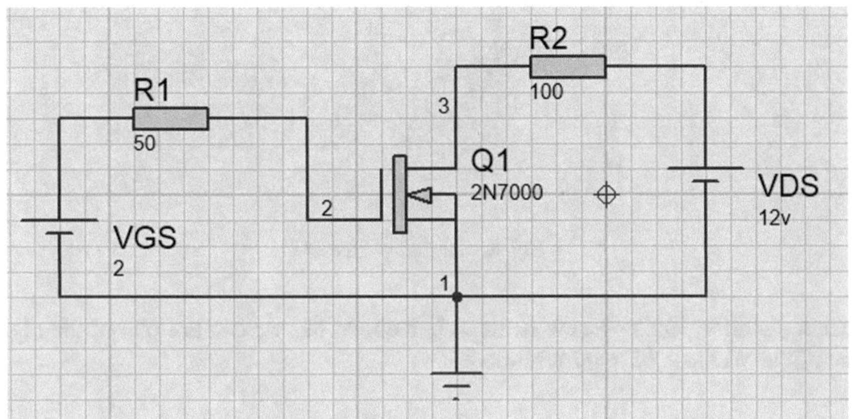

Fig. 3. Variable voltage source

By varying the gate source voltage, the drain source voltage varies accordingly input to the regulator circuit varies. This can be achieved using an Arduino software. The gate source voltage is supplied using the Arduino board. The output of a Arduino board is capable of generating voltage to 5 v. So in order to meet this requirement the gate source voltage required to vary the input voltage supply to the regulator circuit should not exceed 5 v. Hence considering this limit of operation code for the Arduino software is developed accordingly. Not only MOSFET a Bipolar junction transistor can also be used to develop a variable voltage source.

2.3 Digital Voltmeter

Digital voltmeter is the most essential instrument in electronics engineering. It is used to measure the voltage across two points and display the output digitally. A digital voltmeter can be developed using Arduino as shown in Fig. 4.

A digital voltmeter essentially contains the above mentioned components. As stated before, the maximum voltage which can Arduino withstand is 5 v but the voltage to be measured might be greater than 5 v. So in order to limit the voltage, the input voltage is fed to a potential divider which is used to step down the voltage to measure below 5 v. After stepping down the voltage below 5 v, this voltage is processed according to the code and the measured voltage is displayed on the LCD screen as shown in Fig. 4. This digital voltmeter is fit into the casing along with the whole circuit on the PCB (printed circuit board). This helps in reducing the risk of damage of the whole setup. In order to

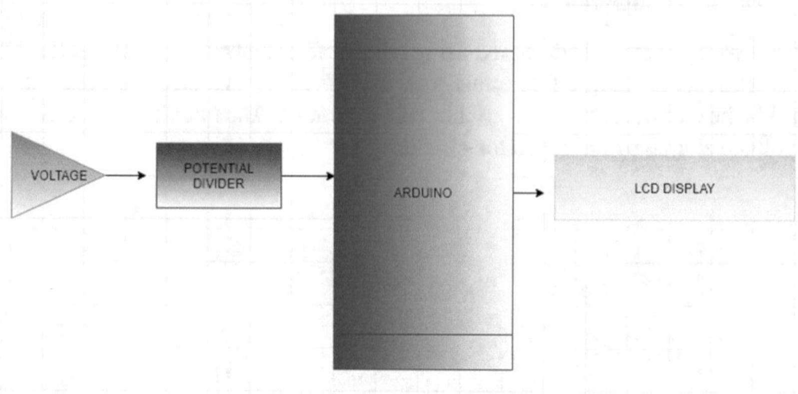

Fig. 4. Digital voltmeter

make a digital voltmeter a new Arduino is required but we can use one of the digital pins of the Wemose d1 mini wifi module.

2.4 Arduino Web Interfacing

Generally, Arduino board connected to a wifi module is not economical So a special type of Arduino board with inbuilt wifi module is used. This module is known as Wemose ESP8266 which helps in reducing the cost. It is same as Arduino i.e. it uses a similar software and is robust in performance. The entire code for the module is prepared in a software tool called Arduino IDE 1.8 ('IDE' stands for integrated development environment) and after successful compilation of the code it is uploaded into the Wemose module.

The main objective of this project is to perform the experiment online. In order to control the circuit through mobile or a computer, it is essential to create an interface. When the student enters the IP address a web page appears. This web page can be created using Javascript and the code is written in the programmable memory of the Wemose module. The Wemose wifi module has a specific IP address. It is difficult for one to remember the IP address so a domain name is created. One can also create apps to perform the experiment in mobile. The web page acts as an interface between the Arduino and the computer. One can alter the circuit operations using this interface. The web page allows the student to vary the voltage and observe the live streaming video of the experimental setup. The digital voltmeter displays the output voltage which is observed through the web camera.

The web page consists of a range slider. The input voltage to the circuit is varied by moving the position of the slider on the web page. The position of the slider decides the value of voltage applied the circuit. Once the slider position is changed the Wifimodule receives a request from the web page it responds by producing an input voltage to the circuit from one of its analog pins. The Arduino board used is capable of producing only 5 v, but there is a necessity of voltage levels below 5 v. This is obtained by using

PWM i.e. pulse width modulation. Using PWM one can acquire voltages below 5 v by varying the duty cycle. The required range of voltage levels are obtained by varying the duty cycle.

3 Expected Output

The Zener voltage regulator circuit is therefore designed and simulated using Proteus software and the expected outputs are as shown in Fig. 5.

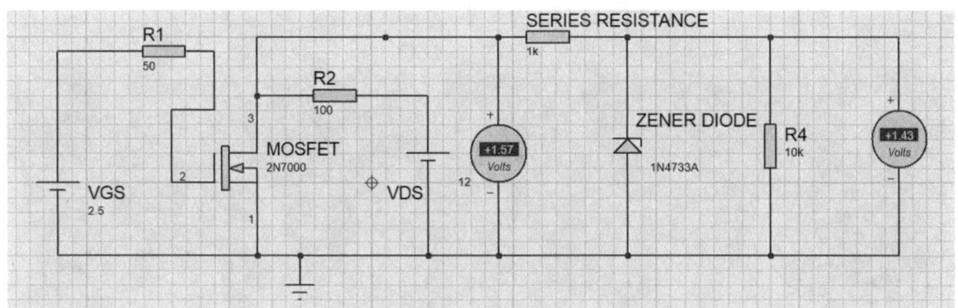

Fig. 5. Simulation result-1

When the input voltage is zero the output is also zero. The input voltage is further increased the output voltage is same as that of input as shown in Figs. 5 and 6.

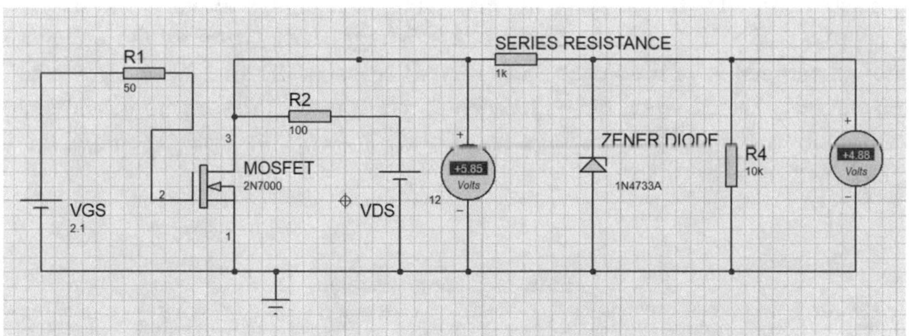

Fig. 6. Simulation result-2

The output follows the input only up to certain voltage i.e. Zener breakdown voltage. Once the input is increased beyond the Zener breakdown voltage the output remains constant and displays approximately 5 v as shown in Figs. 7 and 8.

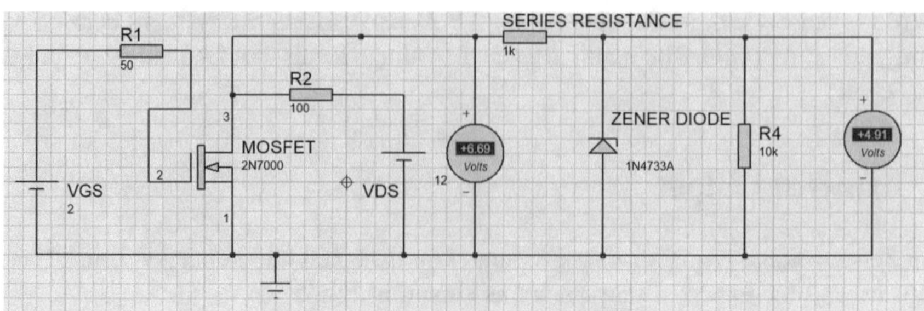

Fig. 7. Simulation result-3

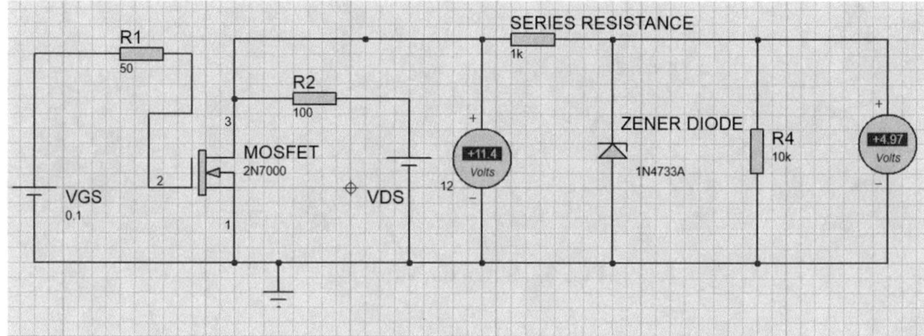

Fig. 8. Simulation result-4

When the input is increased to the maximum voltage, the output is still displaying 5 v approximately. Hence the output voltage is regulated. Therefore the voltage regulation phenomenon of the Zener diode is verified using the simulation software. The same can be obtained when the experiment is performed offline i.e. in the laboratory. Table 1 shows the sample readings and Graph 1 shows the sample graph for line regulation of a Zener diode.

Table 1. Sample readings

Reading No.	Input voltage (volts) (voltage across the input terminals of Zener regulator circuit)	Output voltage (volts) (voltage across the load resistance of the Zener diode)
1	0.83	0.75
2	1.57	1.43
3	5.85	4.88
4	6.69	4.91
5	11.4	4.97

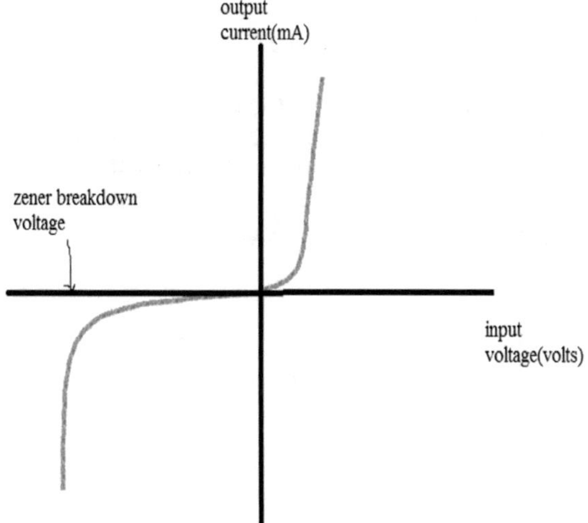

Graph 1. V–I characteristics of Zener diode

4 Conclusion

Remote labs play a very important role in performing the experiments through online. One can perform the experiment without visiting the lab and the only requirement is the user should login to the respective created web page.

The Zener voltage regulation phenomenon is verified using remote engineering. This work finds its application in designing a clipping circuit also. They can be used as shunt regulators in many electronic circuits and stabilizing circuits. This Remote Engineering technique can be used to access the experimental setup from home and various kinds of experiments related to Electrical and Electronics engineering. The Remote engineering is used not only for performing experiments but also to control certain systems whether they may be electrical systems or mechanical systems.

References

1. Arduino UNO: forum.arduino.cc
2. Digital arduino voltmeter: ElectroSchematics.com
3. Boylestad, R.: Electronic Devices and Circuits, 10th edn.
4. ESP8266 core for Arduino: github.com
5. Hashemian, R., Pearson, T.R.: Teaching hardware design with online laboratories. In: Internet Accessible Remote Laboratories: Scalable E-Learning Tools for Engineering and Science Disciplines. IGI Global, 18–39. Web. 12 Nov. 2018 (2012). https://doi.org/10.4018/978-1-61350-186-3.ch002
6. Salzmann, C., Gillet, D., Esquembre, F., Vargas, H., Sánchez, J., Dormido, S.: Web 2.0 open remote and virtual laboratories in engineering education. In: Cyber Behavior: Concepts,

Methodologies, Tools, and Applications. IGI Global, 559–580. Web. 12 Nov. 2018 (2014). https://doi.org/10.4018/978-1-4666-5942-1.ch028
7. Fayolle, J., Callaghan, M., Gravier, C., Harkin, J., Jailly, B.: Stakes and issues for collaborative remote laboratories in virtual environments. In: Internet Accessible Remote Laboratories: Scalable E-Learning Tools for Engineering and Science Disciplines. IGI Global, 2012, 529–542 (2018). https://doi.org/10.4018/978-1-61350-186-3.ch027
8. Heradio, R., de la Torre, L., Dormido, S.: Virtual and remote labs in control education: a survey. Ann. Rev. Control **42**, 1–10. Web (2016)

Remote Laboratory for Implementation of Various Applications Using Linear and Digital IC's

D. Bindu Sri(✉), M. Srividya, and Y. Syamala

Gudlavalleru Engineering College, Vijayawada, Andhra Pradesh, India
{bindusri606, sriviyasharma2499, coolsyamu}@gmail.com

Abstract. In the present scenario people are moving towards to the technology life at the same time there has been a move towards using remote laboratories in engineering education. So this is one idea to present a new hardware system that lets develop labs of linear and Digital ICs. This is the system used to verify the output functions of linear and digital, so teachers can create new experiments by modifying the problem proposed, within the range of input variables and output functions that system supports. The students have to solve the problems finding and simplifying the functions. Then, they must implement the results in the proposed system. This laboratory is the extension of the existing lab, fully functional remote lab, Net Lab, which can be used in teaching student's on-campus as well as off-shore. This enables students from different places to access equipment at any time, enabling students to experiment on it to meet their needs. Though installing remote lab is costly affair, it revolutionizes the current era of educational system and brings a change for betterment. This facilitates good practical exposure to different kinds of experimentation. In this work, different linear and digital circuits such as digital to analog convertor and 4-bit arithmetic and logic unit will be implemented using IC's. Further it is interfaced with remote lab website using scripts to enable users to access remotely.

Keywords: Linear design-DAC · Digital design-ALU · Remote laboratories

1 Introduction to Remote Laboratories

Engineers act as trouble shooter in solving many practical problems. Thus it's crucial for transferring theoretical knowledge to obtain practical applications. To realize this, Remote Laboratories are designed to allow a student or researcher to remotely conduct real experiments across the Internet. Remote laboratories allow students to understand scientific principles and resulting phenomena through hands-on laboratory experience, "learning by doing," from wherever they are located. Asking questions, systematically and objectively gathering data, and testing hypotheses to find answers is the essence of scientific inquiry. Web-based labs make possible to illustrate scientific phenomena that require costly or difficult-to-assemble equipment. There are two complementary approaches for web-based labs: Virtual Labs or Experiments provide computer based simulations which offer similar views and ways of work to their traditional counterparts. Now a days, simulations have evolved into interactive graphical user interfaces

M. E. Auer and K. Ram B. (Eds.): REV2019 2019, LNNS 80, pp. 193–202, 2020.
https://doi.org/10.1007/978-3-030-23162-0_18

where students can manipulate the experiment parameters and explore its evolution. The experimental laboratory space, materials and operating equipment are in one geographical location, while the researcher is controlling experiments from a different, sometimes very distant location. Remote Labs or Experiments use real plants and physical devices which are mobile operated in real time.

Figure 1 illustrates basic block diagram of a remotely accessible laboratory.

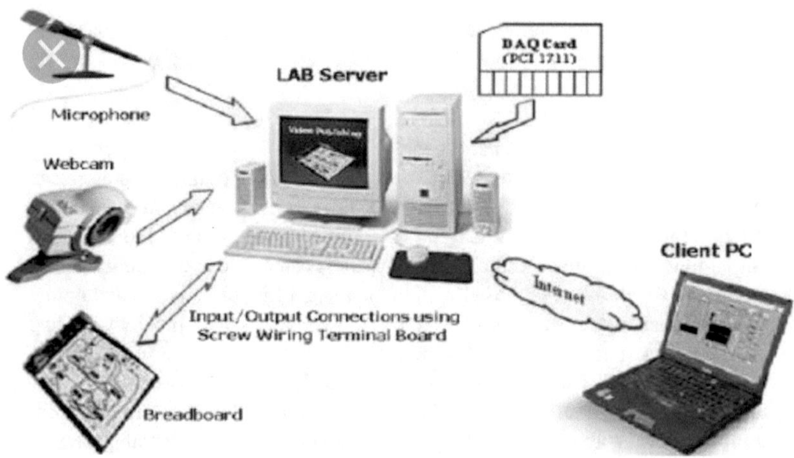

Fig. 1. Block diagram of remotely accessible laboratory

The first important component of a remote lab is its experimentation circuit or kit. It can also involve microphone so that communication can be set, display to read values and etc.

Another important component of remote laboratory is webcam. A webcam is a video camera that feeds or streams its image in real time to or through a computer to a computer network. When "captured" by the computer, the video stream may be saved, viewed or sent on to other networks travelling through systems such as the internet, and e-mailed as an attachment. When sent to a remote location, the video stream may be saved, viewed or on sent there.

There are many physical parameters or values that are involved while conducting an experiment. These can be measured by physical contact or using devices physically. To overcome this, Data acquisition (DAQ) is involved in process. DAQ is the process of measuring an electrical or physical phenomenon such as voltage, current, temperature, pressure, or sound with a computer. A DAQ system consists of sensors, DAQ measurement hardware, and a computer with programmable software. Compared to traditional measurement systems, PC-based DAQ systems exploit the processing power, productivity, display, and connectivity capabilities of industry-standard computers providing a more powerful, flexible, and cost-effective measurement solution.

Thought all the data is set, it should be transmitted to remote places. For this Cloud computing can be used to provide the remote access of this circuit or lab equipment. The software usually used to connect devices to cloud are VMware and XenServer.

It is commonly accepted that digital media (such as simulations, videos, interactive screen experiments or web labs) can positively impact student knowledge, skills and attitudes. Consequently, tools such as Moodle and web-based labs have become widespread in distance education in the last decade. Moodle supports the administration, documentation, tracking, and reporting of training programs, classroom and online events.

However, remote laboratory applications in teaching are newer, and are given significant impetus by growing research in robotics. Increasing demand for online learning similarly drives development, especially with the potential of reaching learners of all ages who are remotely located and without laboratory resources.

1.1 Programmable Logic Based Experimentation

The basic system consists the hardware components that are required to perform the experiment as in the laboratory. The hardware connections corresponding to the specific experiment will set before hand and the system will be focused and analyzed by using a web camera. The functionality of the circuit is observed from the webpage that will be interfaced to the kit and also to the user so that they can actually interact with the circuit by changing the voltage or current levels and also can check its functionality using graphs.

1.2 Practical Approach

As explained the experimental setup required in order to find how this process works let's consider the experimental setup for the experiments Linear (weighted resistors and R-2R ladder) and digital (ALU) integrated circuits.

The components that are required for the development of the circuitry are as follows for the linear experiment.

2 Linear and Digital Integrated Circuits

2.1 Digital to Analog Converter

Digital systems are used in ever more applications, because of their increasingly efficient, reliable, and economical operation with the development of the microprocessor, data processing has become an integral part of various systems. Data processing involves transfer of data to and from the microcomputer via input/output devices. Since digital systems such as micro computers use a binary system of ones and zeros, the data to be put into the micro computer must be converted from analog to digital form. On the other hand, a digital-to-analog converter is used when a binary output from a digital system must be converted to some equivalent analog voltage or current. The function of DAC is exactly opposite to that of an ADC.

Digital to Analog Converters (DAC): The process of converting digital signal into equivalent analog signal is called D/A conversion. The electronics circuit, which does this process, is called D/A converter. The circuit has 'n' number of digital data inputs with only one output. Basically there are three types of D/A converter circuits: Weighted resistors DAC, R-2R ladder DAC and Inverted R-2R ladder DAC.

The binary-weighted-resistor DAC employs the characteristics of the inverting summer Op Amp circuit. In this type of DAC, the output voltage is the inverted sum of all the input voltages. If the input resistor values are set to multiples of two: 1R, 2R and 4R. The output voltage would be equal to the sum of V1, V2/2 and V3/4.

Figure 2 circuit consists the op-amp of IC number 741 of specifications. The Input impedance of the IC 741 op amp is above 100 kΩ. The output impedance of the 741 IC op amp is below 100 Ω. The frequency range of amplifier signals for IC 741 op amp is from 0 Hz to 1 MHz. Offset current and offset voltage of the IC 741 op amp is low. Figure 3 gives the pin description of the IC 741 that being used in the design of the experiment.

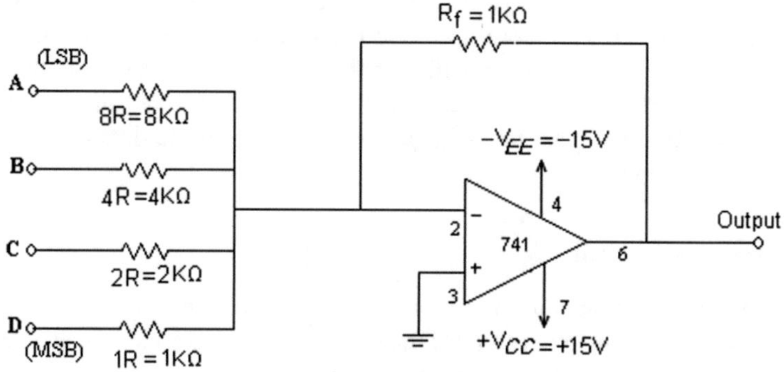

Fig. 2. Binary weighted resistor type DAC

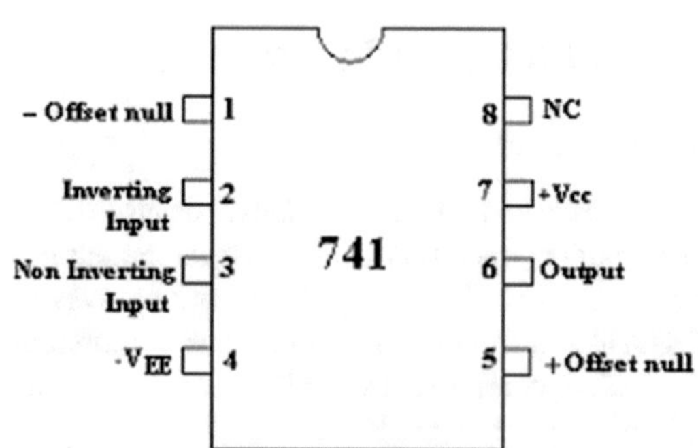

Fig. 3. Pin description of IC 741

Figure 4 represents the schematic diagram of the R-2R ladder type of DAC. In IC 741 is used as a primary component.

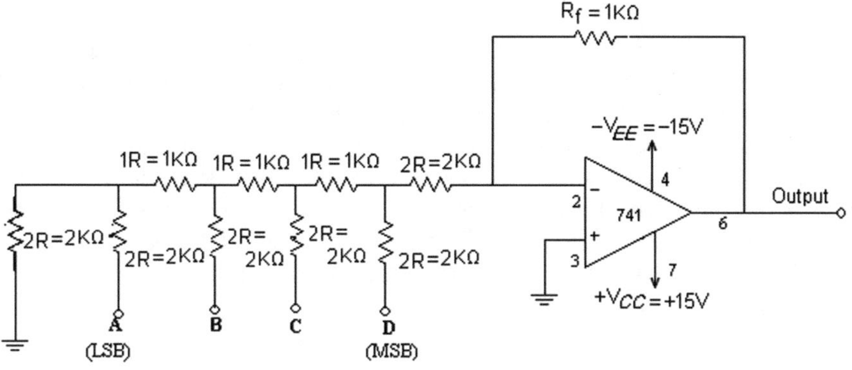

Fig. 4. R-2R ladder DAC

Calculations

(i) Weighted Resistor DAC

$$\mathrm{V_o} = -\mathrm{R_f}\left[\frac{b_A}{8R} + \frac{b_B}{4R} + \frac{b_c}{2R} + \frac{b_D}{R}\right] \tag{1}$$

For a digital sequence—1111

$$\mathrm{V_0} = \left[\frac{1}{8} + \frac{1}{4} + \frac{1}{2} + 1\right]\frac{R_f}{R} \times 5 \tag{2}$$

$\mathrm{V_o} = -9.375$ V

(ii) R-2R Ladder Type DAC

$$\mathrm{V_o} = -\mathrm{R_f}\left[\frac{b_A}{16R} + \frac{b_B}{8R} + \frac{b_c}{4R} + \frac{b_D}{2R}\right] \times 5 \tag{3}$$

2.2 Arithmetic Logical Unit

The arithmetic logic unit (ALU) is fundamental building block of a central processing unit (CPU) of a computer. Even one of the simplest processor has one ALU unit for purposes such as maintaining the timers. The logic circuitry in this units is entirely combinational (i.e. consists of gates with no feedback and no flip-flops). The ALU is an extremely versatile and useful device since, it makes available, in single package, facility for performing many different logical and arithmetic operations. ALU is a core component of all central processing unit within in a computer and is an integral part of execution unit. The ALU takes as input the data to be operated on (called operand) and the code from the control unit indicating which operation to perform. Basically as this is 4 bit ALU, 16 different arithmetic operations, of them we have chosen few operations and the required operation can be designed by the respective code logic which is used to interact with the device so that we can get the required output. Arithmetic operation includes (Addition, Subtraction, Increment, Decrement, and Transfer). Logical operations (AND, OR, XOR, NOR, EX-NOR, EX-OR). Figure 5 is the basic diagram of 4-bit ALU that represents the number of inputs that are given and the number of outputs that are taken from it.

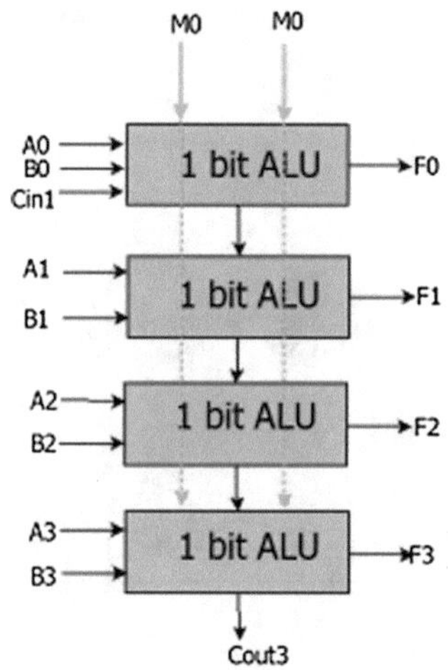

Fig. 5. 4-bit arithmetic logic unit

Program:

```
Library IEEE;
Use IEEE.std_logic_1164.all;
Use IEEE.std_logic_unsigned.all;
Use IEEE.std_logic_arith.all;
Entity LC2_ALU is
port ( A: in std_logic_vector (3 downto 0);
B: in std_logic_vector (3 downto 0);
S: in std_logic_vector (3 downto 0);
Y: out std_logic_vector (3 downto 0));
End LC2_ALU;
Architecture bhv of LC2_ALU is
Begin
process (A, B, S)
begin
case S is
   when "0000" => Y <= A or B;
   when "0001" => Y<= A and B;
   when "0010" => Y <= A;
   when "0011" => Y <= not A;
   when "0100" => Y <= A xor B;
   when "0101" => Y <= A nand B;
   when "0110" => Y <= A + B;
   when "0111" => Y <= A - B;
   when "1000" => Y <= A*B;
   when "1001" => Y <= A/B;
   when others => null;
end case;
end process;
end bhv;
```

3 Results

3.1 Weighted Resistor DAC

The following results are obtained by testing it for the weighted type of digital to analog converter (Table 1).

Table 1. Observations of weighted resistors type of DAC

D	C	B	A	Theoretical voltage (V)	Practical voltage (V)
0	0	0	0	0	0
0	0	0	1	−0.62	−0.66
0	0	1	0	−1.25	−1.02
0	0	1	1	−1.87	−1.74
0	1	0	0	−2.5	−2.36
0	1	0	1	−3.12	−3.08
0	1	1	0	−3.75	−3.44
0	1	1	1	−4.37	−4.16
1	0	0	0	−5	−4.95
1	0	0	1	−5.62	−5.66
1	0	1	0	−6.25	−6.02
1	0	1	1	−6.87	−6.73
1	1	0	0	−7.5	−7.35
1	1	0	1	−8.12	−8.07
1	1	1	0	−8.75	−8.43

3.2 R-2R Ladder Type DAC

The following results are obtained by testing it for the weighted type of digital to analog converter (Table 2).

3.3 Simulation Waveforms of ALU

The following simulated waveforms are the result of the arithmetic logic unit (Fig. 6).

4 Conclusion

The main motto to opt the remote labs is to provide the students, the accessibility of the labs from rural location students. These can be used to provide quality education to people in rural places, making them understand how the devices work practically without actually spending on the equipment. This is hence beneficial in present education system, as practicality is being given the highest preference in the pyramid of

Table 2. Observations of R-2R type of DAC

D	C	B	A	Theoretical voltage (V)	Practical voltage (V)
0	0	0	0	−0.31	−0.05
0	0	0	1	−0.62	−0.6
0	0	1	0	−0.93	−0.7
0	0	1	1	−1.25	−1.22
0	1	0	0	−1.56	−1.27
0	1	0	1	−1.87	−1.91
0	1	1	0	−2.18	−1.96
0	1	1	1	−2.5	−2.41
1	0	0	0	−2.81	−2.52
1	0	0	1	−3.12	−3.06
1	0	1	0	−3.41	−3.11
1	0	1	1	−3.75	−3.63
1	1	0	0	−4.06	−3.69
1	1	0	1	−4.2	−3.7
1	1	1	0	−4.37	−4.32
1	1	1	1	−4.68	−4.38

Fig. 6. Simulation waveforms for arithmetic logic unit

learning. The applications doesn't end here, this can be used in solving many real time complications in human life. This actually makes the way of life a bit easier and saves time for human beings.

References

1. Palloff, R., Pratt, K.: The Virtual Student—A Profile and Guide to Working with Online Learners, Jossey-Bass (2003). ISBN 0- 7879-6474-3
2. Lee, R.C.T., Lang, C.L.: Some properties of fuzzy logic. Info. Control **9**, 413–431 (1971)
3. Daponte, P.: Guest Editor Special Issue on DAC modelling and testing. Measurement, vol. 31, No. 3 (2002)
4. Ursulet, S., Gillet, D.: Introducing flexibility in traditional engineering education by providing dedicated on-line experimentation and tutoring resources. In: The International Conference on Engineering Education, Manchester, UK, 18–21 Aug. 2002
5. Wagner, B.: From computer-based teaching to virtual laboratories in automatic control. In: The Twenty Ninth ASEE/IEEE Frontiers in Education Conference, San Juan, Puerto Rico, 10–13 Nov. 1999

6. Buitrago, G.C.: Simulation et Controle Pedagogique: Architectures Logicielles Reutilisables. Ph.D. thesis, Universite Joseph Fourier–Grenoble (1999)
7. Stalling, W.: Computer Organisation and Architecture: Designing for Performance, 7th ed., p. 13. Pearson Prentice Hall (2006)
8. Vento, J.A.: Application of labview in higher education laboratories. In: Proceedings of Frontiers In Education Conference, pp. 444–447 (1998)
9. Gomes, L., Patricio, G., Ferreira, R., Costa, A.: Remote experimentation for introductory digital logic course. In: ELearning in Industrial Electronics, pp. 98–103 (2009)
10. Datta, K., Sass, R.: RBoot: software infrastructure for a remote FPGA laboratory. In: Annual IEEE Symposium Field-Programmable Custom Computing Machines, pp. 343–344 (2007)

Design of Single Patient Care Monitoring System and Robot

M. N. Mamatha(✉)

Electronics and Instrumentation Engineering Department,
BMS College of Engineering, Basavanagudi, Bangalore 560019, India
mamathamn.bms.intn@bmsce.ac.in

Abstract. The advent of VLSI technology brings in its wake huge benefits to patients, doctors and designers, undreamt before, by bringing in automation and raising the level of care nurses are able to provide. Instead of walking to each patient's bedside to collect vital information, nurses can collect it in one location and concentrate their efforts on caring for those who need them the most. As an outcome of these requirements, a prospect arose to design a highly efficient patient care and monitoring system which can handle patients and multiple parametric measurements from every single patient in real time. Patient care and monitoring embedded systems are proposed in this work to acquire, store, display physiological information obtained from multi patients and are vital in operating rooms, emergency rooms, intensive care units, ambulances as well as at homes. The design exploits development of Novel, Fast Algorithms, Design of Architectures using labVIEW implementations.

Keywords: Partially paralytic · Patient care and monitoring · ASM chart · Finite state machine

1 Introduction

Paralytic Patients need supportive services for executing the tasks in their daily activities. These services may not only be rendered to people with cognitive impairments but also be extended to the aged. Supportive services are to be designed which ensures an acceptable level of autonomy to the above mentioned category of patients [1–4]. One such supportive device for paralytic patients has been designed and presented in this chapter.

Single Patient Care and Monitoring System is proposed to be implemented using a Microcontroller. Before undertaking such a time consuming Design and Implementation, development of Finite State Machine (FSM) and Simulation of the same using software such as the National Instruments 'labVIEW' will prove the concepts and ideas envisaged for the implementation. In the next section, the design of the PCM System using labVIEW is elaborated. This will be followed by the detailed design and implementation using Microcontroller.

M. E. Auer and K. Ram B. (Eds.): REV2019 2019, LNNS 80, pp. 203–216, 2020.
https://doi.org/10.1007/978-3-030-23162-0_19

2 Purpose or Goal

2.1 labVIEW Simulation for Single Patient Care and Monitoring System

Sequential control is crucial in many control applications as in the design of Patient Care Monitoring System. A sequential control procedure can be represented graphically by one of the following two methods:

- A State diagram
- An Algorithmic State Machine (ASM) Chart

When the number of inputs and outputs of a system increases, then ASM chart based design will be more convenient than the state diagram approach. For this reason, ASM Chart has been used in the design and simulation using National labVIEW [5–8]. The same chart will also be used as a model later in the Microcontroller based PCM System.

3 Development of ASM Chart for Single PCM System

The proposed design is for controlling the movements of a robotic arm for single patient care and monitoring. In the designed system, the user's EEG, eyeball, eye blink and flex movements are sensed and interpreted in order to control a robot to pick and place small objects as per the wishes of the patient. Before developing the ASM Chart, we need to identify the Inputs and Outputs of the System.

Various signals used in the system are conceived in Table 1. Inputs such as EEG, Eye blink, Eye ball (Left, Right) and Flex are respectively connected to the IEEG [7:0], IEbL, IEba [7:0], IFlex_0, IFlex_1. OEEG_0 [7:0] and OEEG_1 [7:0] are sensor outputs indicating awake and asleep modes respectively. OEba_L [7:0] and OEba_R [7:0] are the sensor outputs corresponding to the eyeballs moving towards the left and the right respectively. The signal OEba_st [7:0] is for the detection of the eye balls looking straight which arises if the patient is neither looking left nor right.

OEbl_1 is output signal of eye blink when the eyes are closed. OFlex_0 and OFlex_1 are the two signals from touch (flex) sensor to make the robotic arm to open and close. This helps the robot arm to grab an object.

The ASM Chart presented in Fig. 1 describes the step by step functioning of the system.

The system works as follows:

(1) Initially, the patient is required to wear the sensors whenever it is required to operate the robot.
(2) The System On/Off condition is dependent on the frequency of alpha and beta signal to initiate its operation. System is turned "ON" if EEG frequency is above 15 Hz, thus indicating that the patient is "awake" and also to avoid any mishap when the patient is asleep or drowsy. This operation is sensed in S1.

Table 1. Signal description of the PCM system

Signal	Input/output	Description	
Clk	Input	System clock	
IEEG [7:0]	Input	Patient status from headband through probe at P1.0 of μc	
IEbL [7:0]	Input	Eye blink input from probe at P3.7 of μc	
Iflex_0	Input	Flex 1 input	
Iflex_1	Input	Flex 2 input	
IEba [7:0]	Input	Eyeball input from specks	Left move at P1.2 of μc
			Right move at P1.3 of μc
			Straight move at P1.4 of μc
EOC	Input	ADC to controller, End of conversion	
OEba_L [7:0]	Output	Eyeball—Left	
OEba_R [7:0]	Output	Eyeball—Right	
OEba_st [7:0]	Output	Eyeball position—Straight	
OEbl_0	Output	Eye blink—Open	
OEbl_1	Output	Eye—Closed	
OFlex_0	Output	Gripper-robot open	
OFlex_1	Output	Gripper-robot closed	
OEEG_0 [7:0]	Output	Patient status—awake	
OEEG_1 [7:0]	Output	Patient status—asleep	

(3) If PB_1 is pressed in S2 state, then S3 to S6 are executed sequentially. In these states, the voltages corresponding to the Left/Right eye ball movements of the patient are set.

(4) If PB_2 pressed in S6 state, then move the robot corresponding to the movements of eye ball of the patient else wait in the same state as controlled by S7 to S11 states. Robot has now reached where the object has to be grabbed or dropped.

(5) In state S11, if PB_3 is pressed, move the robot arm up and down as in states S12 to S14. By moving the eye ball "left", robot arm can be moved "up" and by moving it "Right" the arm moves "down".

(6) If touch (flex) sensor is operated, then grabbing or dropping an object is executed as in states S15 to S17.

(7) At any point in the process of operation, if the patient wishes any assistance, closing the eyes for about 15 s would activate a voice chip calling for "Help". Note: Press PB_4 will reset the system

4 Implementation of Patient Care Monitoring System Using labVIEW

Sequential control can be implemented using graphical programming of state diagrams in labVIEW. The finite state machine implementation of the Single PCM System in labVIEW is shown in Fig. 2 and is implemented based on the ASM Chart shown in

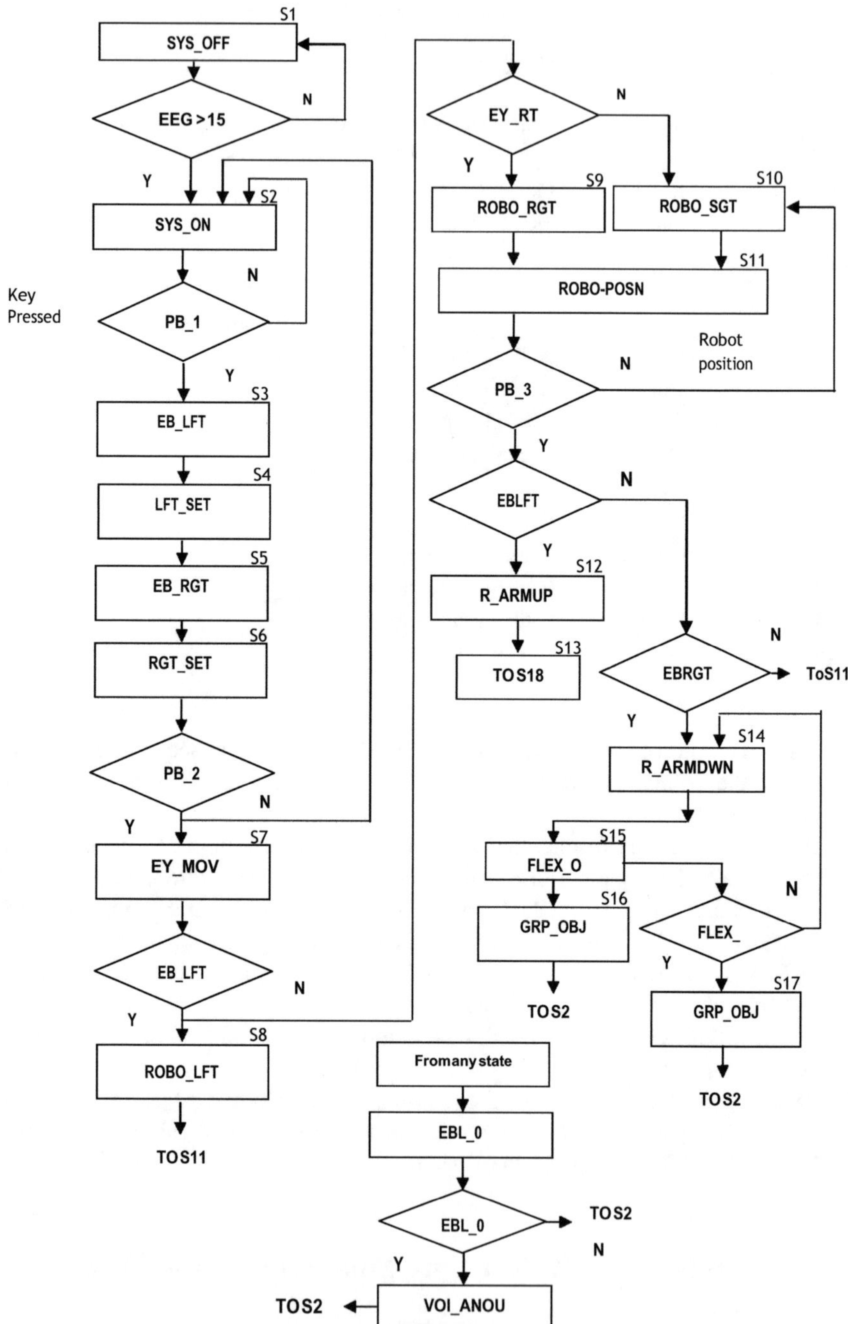

Fig. 1. ASM chart to control the robot using bio potentials

Fig. 1. It shows the Front Panel controls as well as the state chart sequencing through various states. State charts have states with actions to be executed whenever a state is active. They may have transitions from one active state to another when the transition condition is logically true. The state machine always starts in a particular state defined as the initial state as shown.

The entire design is simulated using state chart feature of labVIEW. Earlier in Table 1, the system I/Os were listed. The same has been changed slightly in the labVIEW implementation and is presented in Table 3.

The complete operation of the robot movement is subdivided into 5 substates as mentioned in Table 2, based on the functionality of the robot in response to the acquired biopotentials.

Table 2. Sub state description of state chart

Sl No	State	State name		Description
1	S0	SYS_OFF		System is powered OFF
	S1	SYS_ON		System is powered ON
2	S2	INITIALIZE	LOOK_LEFT2	Eye to left
			INIT_L2	Set voltage when eyes turned left
			LOOK_RGT2	Eye to right
			INIT_R2	Set voltage when eyes turned right
3	S3	ROBO_MOV	ROBO_STR	Robot moving straight
			ROBO_R	Robot moving right
			ROBO_L	Robot moving left
4	S4	ROBO_ARM	ARM_ROT	Robo arm activated
			ARM_DOWN	Robo arm moves down
			ARM_UP	Robo arm moves up
5	S5	PICK/DROP	PICK/DROP	Flex sensor activated
			PICK	Object gripped
			DROP	Object ungripped

The front panel of the state chart as implemented in this work, consists of two clusters—inputs and outputs. Figure 2 represents the design implementation of front panel and state chart.

When the patient is awake and active, the EEG frequency ranges between 15 and 30 Hz. The front panel shows switches "Sw1–Sw4" corresponds to "PB1–PB4" in the actual system (Table 3).

"SW1_INIT" is for setting the voltages corresponding to the patient's eye ball voltages when the sensors are worn by the patient. Once the voltages are set, "SW2_R/L" is used for moving robot in right and left directions. The robot is made to navigate to the desired position. "SW3_U/D" is pressed to operate the arm either up or down. The input settings for all these operations are simulated from the front panel. One more switch "SW4_P/D" is used to simulate "Pick or Drop" of an object. To set

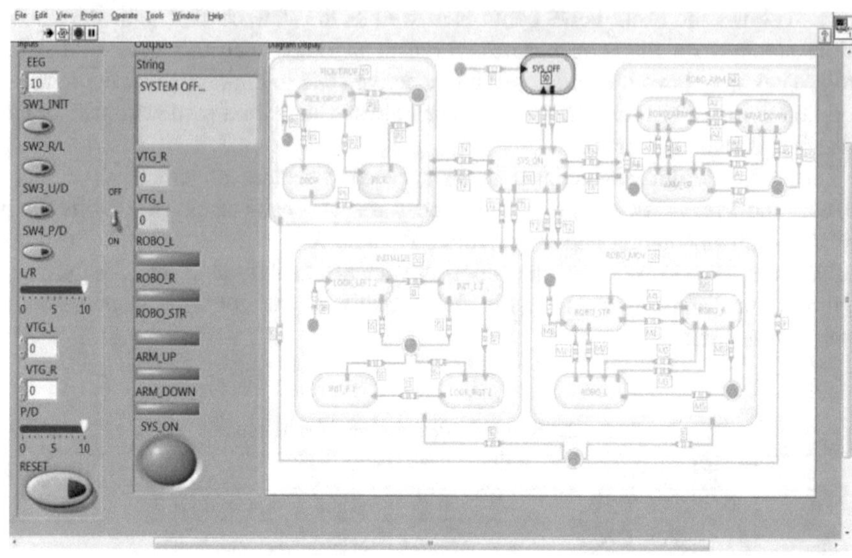

a

b

Fig. 2. labVIEW simulation of single patient PCM system. **a** Front panel design. **b** State chart realization (Sw 1 to Sw 4—PB1 to PB4, L/R—Left/Right, P/D—Pick/Drop)

Table 3. Patient care and monitoring system simulation in labVIEW

Inputs sensor	Description
EEG (Hz)	Senses the EEG frequncies from the EEG sensor
SW1_INIT	Switch for initialising the voltages for the left and right movement of the eyeball
SW2_R/L	Switch to move the robot forward, backward or straight depending on the position of the eyeball
SW3_U/D	Switch to move the robot arm up or down by the lateral movement of the eyeball
SW4_P/D	Switch that activates the flex in the robot to either pick or drop the object
L/R slide button	This is the simulated eyeball position recorder Slide range – Eyeball position 0–4 – Left 4–6 – Straight 6–10 – Right
VTG_L	It is the numeric control that records the voltage of the eye of a particular patient when he looks left
VTG_R	It is the numeric control that records the voltage of the eye of a particular patient when he looks right
P/D	This slide bar replicated the flex motion of the patient Slide bar position – Action of End effectors 0–4 – Grip 6–10 – Ungrip
RESET	Boolean button that helps in resetting the system for reuse

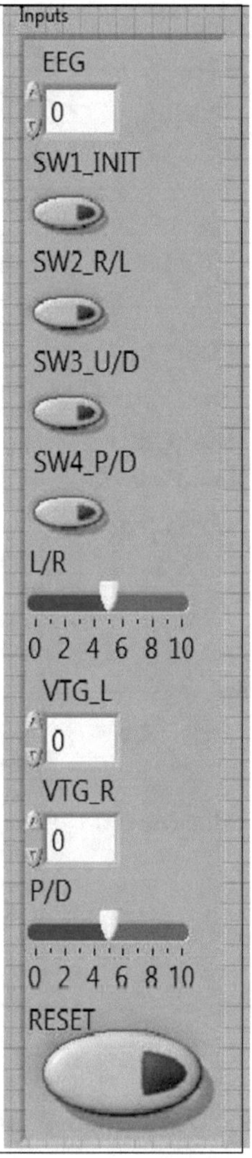

the ranges for all the above said operations, a slider ranging on a scale of "one to ten" is provided. The working of the simulated system will follow in the succeeding sections.

Figure 3 shows the initial state when the system is turned "ON". The voltage settings corresponding to positions of the eyeball is shown in Fig. 4a. The slider movements corresponding to "Right" and "Left" eye movements are depicted in Fig. 4b and c respectively.

The next task is to move the robot according to the wishes of the patient. This is the sub-state named "Robo_Mov" as shown in Table 2. The following section deals with an in-depth analysis of the substates as it involves major operations.

Output displays	Description
String	A display where actions are described to make them easily understandable
VTG_R	Numeric display to display the voltage produced by the EOG when looked right at the time of initialisation and retains it till the system is reset
VTG_L	Numeric display to display the voltage produced by the EOG when looked left at the time of initialisation and retains it till the system is reset
ROBO_L	It is a boolean LED that is switched on when the robot gets the command to move to its left
ROBO_R	It is a boolean LED that is switched on when the robot gets the command to move to its right
ROBO_STR	It is a boolean LED that is switched on when the robot gets the command to move straight
ARM_UP	It is a boolean LED that is switched on when the robot gets the command to move to its arm up
ARM_DOWN	It is a boolean LED that is Switched on when the robot gets the command move to its arm down
SYS_ON	It is a boolean LED that is switched on when the patient is awake and active.(i.e. EEG is in the range of 15Hz to 30Hz)

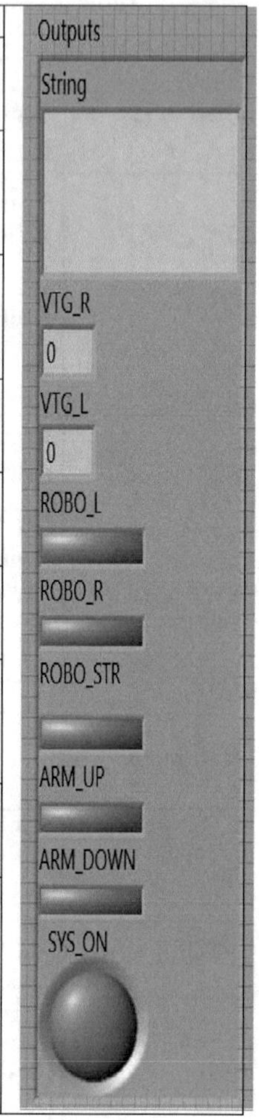

4.1 Description of Robot Movement (Robo_Mov)

The control is at the state S1 from either S0 or S2. Now, to move the robot around, instructions about the directions are given by the eyeball movements by the patients. To move the robot, key2 (SW2_R/L) must be pressed. This key is checked in the transition T2 along with the status of the patient corresponding to EEG frequency.

The transitions from the ROBO_STR to ROBO_L and ROBO_R to ROBO_L are checked by the transition node M2 and M3. The condition of the transition is true if the slide bar range is less than 4, in real-time, the voltage would be compared to be less than VTG_L and robot is moved to its left.

From ROBO_L state transition to ROBO_R is through transition M3″ and, to ROBO_STR, it is through transition M2″. Transition M3″ is similar to M1″ in that it changes the control from states ROBO_L and ROBO_STR respectively to ROBO_R.

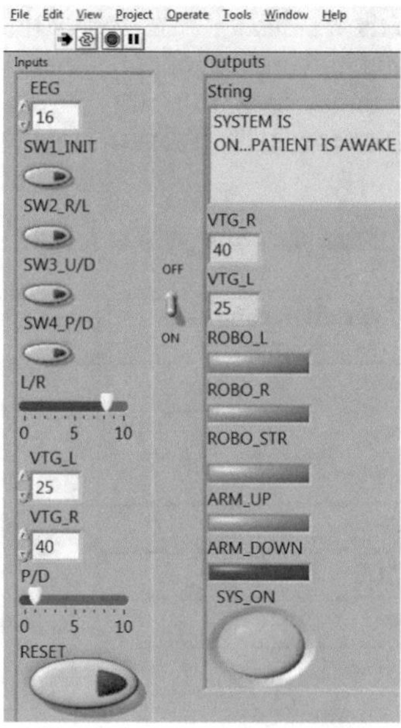

Fig. 3. State showing system ON for EEG beyond 15 Hz

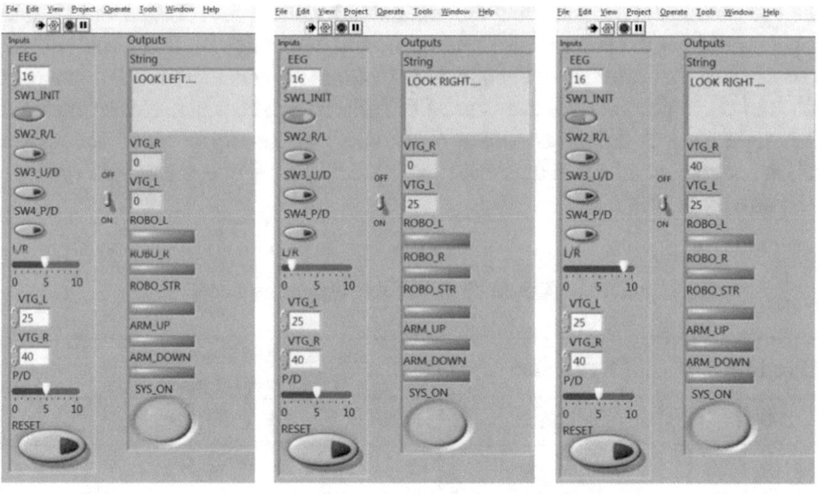

a b c

Fig. 4. Voltage settings of the patient for robot movements. **a** Initialise. **b** Slider less than 4. **c** Slider greater than 6

When the robot reaches the destination from where the object is to be picked or to be placed, the patient can stop moving the robot by releasing key 2, SW2R/L. Release of switch would pass the control back to state1, SYS_ON, from state S3 through transition T2″. These transitions are as in Table 4.

Table 4. Description of signals in robotic movement

Sl. No.	Transitio	Transition condition (Slider)	State diagram
1	MB	–	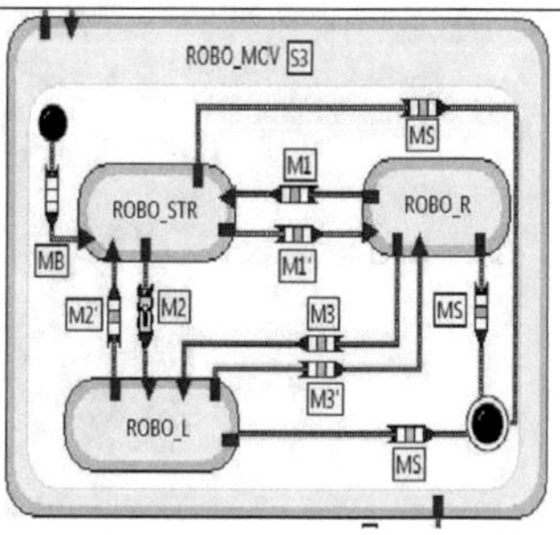
2	M1	range 4–6V	
3	M1''	range 6–10V	
4	M2	range 0–4	
5	M2''	range 6–10V	
6	M3	range 0–4V	
7	M3''	range 6–10V	
8	MS	Switch 2 asserted LOW	

The eye movements to the robot are simulated by slider movements as in Table 5. The robot would get the commands to move around depending on the eyeball position. If the position of the slider is between “0 and 4”, then it is considered as left gaze of the eyeballs and corresponding voltage is set. Similarly if the slider is between “6 and 10”, then it is right gaze. A default condition of the slider in the region “4–6” is considered as straight movement of robot as shown in Table 5. The above mentioned operations are summarized in Table 6.

Table 5. Operation of robot to Eye movements

Position	Eye slider	Record the voltage (VTG) level operation
0–4	Left gaze	VTG_L
4–6	Straight gaze	VTG_S
6–10	Right gaze	VTG_R

Table 6. State chart representation for robot movements

Slidebar	Movement of robot	
0–4	Robot moves left	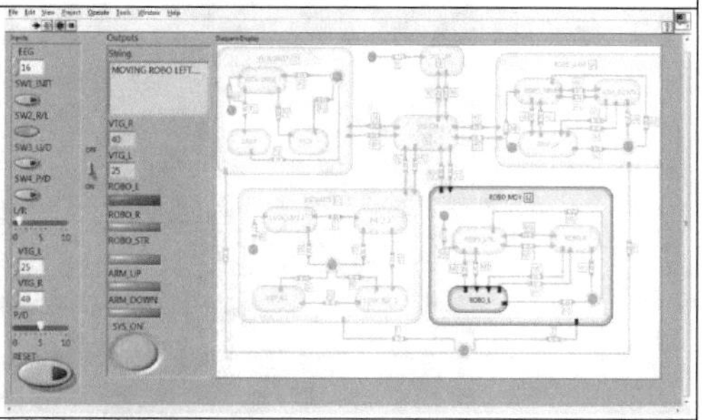
4–6	Robot moves forward/straight	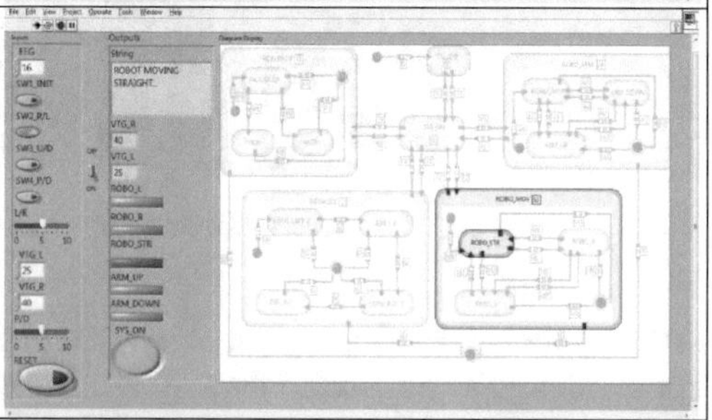
6–10	Robot moves right	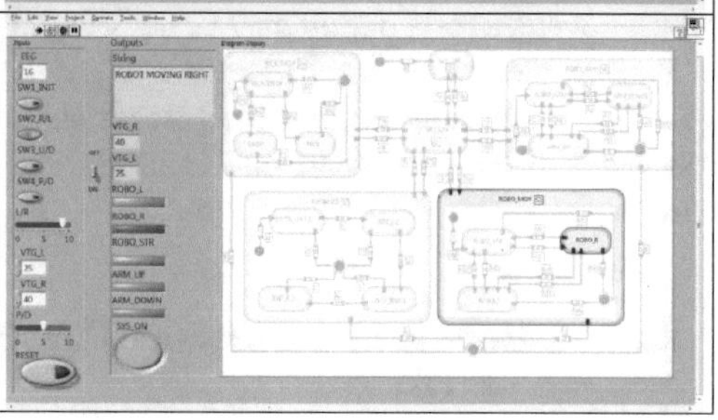

After the robot has reached its position, it has to perform the arm movement. The eyeball movement as simulated for the movement of the arm of the robot is given in Table 7.

Table 7. Robotic Arm Movements

Slidebar range	Robot Arm Movement
0–4 (left)	Arm moves up
6–10 (right)	Arm moves down

The robot arm is operated to "Pick" or "Drop" the desired object. These operations are in Fig. 5 and Table 8.

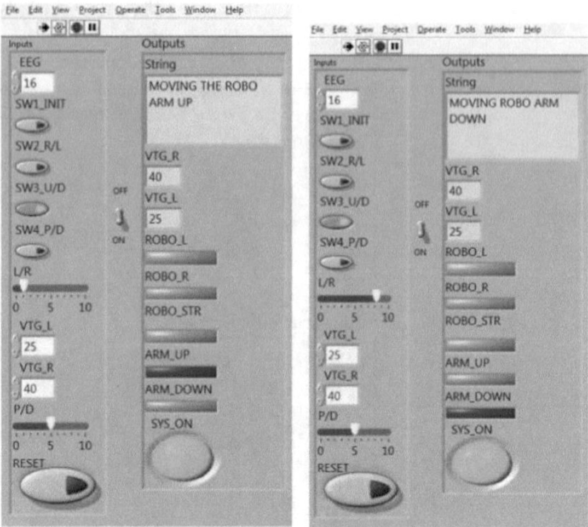

Fig. 5. Move robot arm UP or DOWN to pick or place

After grabbing or releasing the object, the patient releases the key 4, and hence the sub states should be terminated through PS. At any point of time if the patient wants to reset the device, a reset button may be activated. Whenever the device is to be used by a new patient, the system has to be reset to get the new voltages for both right and left gaze. The system is reset after the present state is completed. This is checked in the transition E of the state chart that connects each state with the terminal node of the program.

5 Role of Robots in a Partially Paralytic Patient

Several technologies are useful in providing supportive services for physically or mentally disabled people [9–12]. Robots can be used to perform a fixed sequence of actions. A robot interacts with the environment through its actuators and sensors. Many research efforts have been undertaken in this direction; few of them have focused on devising robots for disabled or elderly people [13, 14].

The following section presents the design of microcontroller based patient care system and a robot which can do mundane work based on the acquired bio potentials from a patient. The design of sensors for EEG, EOG and EMG bio potential acquisition has already been presented in the previous Chapter. These acquired signals will be used to operate a robot in real time.

5.1 Actual or Anticipated Outcomes

Single Patient Care and Monitoring System is proposed to be implemented using a Microcontroller. Before undertaking such a time consuming Design and

Table 8. State Chart for Pick and Place Operation

Range in slidebar	**Operation of robot**	**State Chart**
4–6	Robot arm ready to Pick/Drop	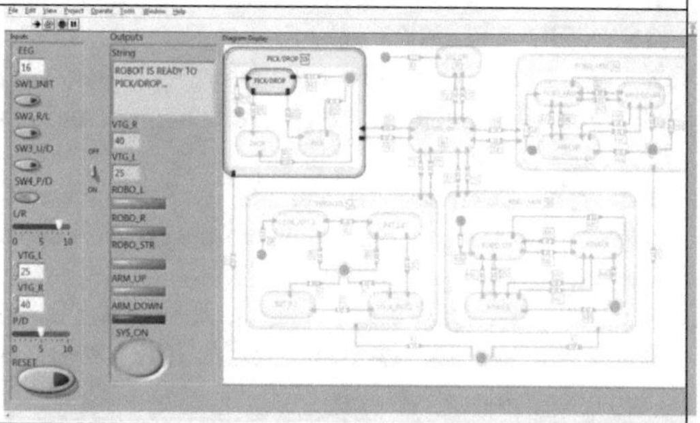
0–4	Pick/Grip the object	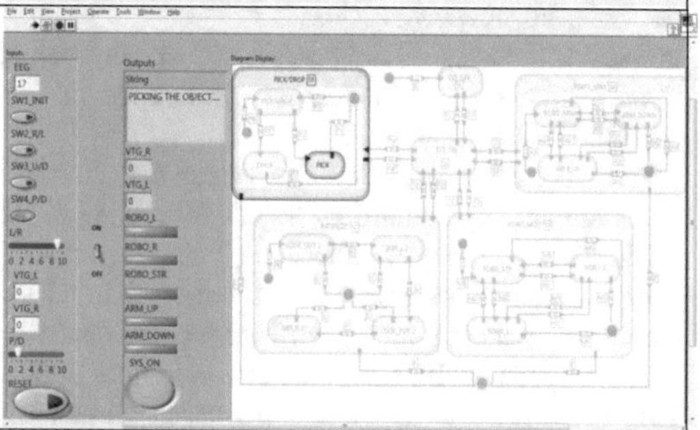
6–10	Drop the object	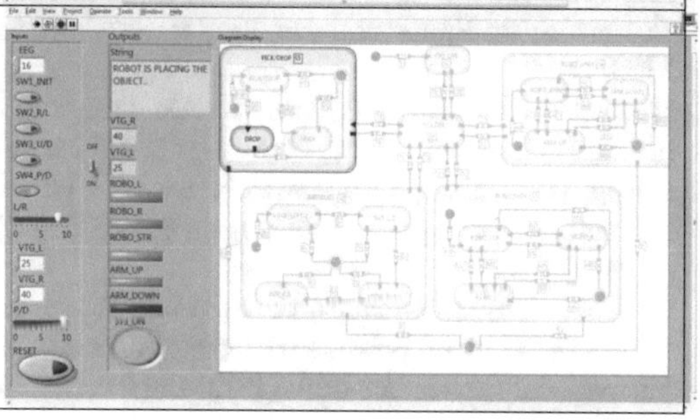

Implementation, development of Finite State Machine (FSM) and Simulation of the same using software such as the National Instruments "labVIEW" will prove the concepts and ideas envisaged for the implementation.

6 Conclusions

Algorithmic State Machine Charts are good design tools that aid in designing architectures efficiently and expeditiously. This paper dealt with the development of ASM Charts for Bio signal Data Acquisition System.

References

1. Ives, T.G., Kampmann, J.R., Pastor, F., Pascual-Leone, M.A.: A new device and protocol for combining TMS and online recordings of EEG and evoked potentials. J. Neurosci. Methods **141**(2), 207–217 (2005 Feb)
2. Di, W., Zhu, D.: Study on brain fag based on EEG signal analysis. In: 2009 ETP International Conference on Future Computer and Communication, Wuhan, China, pp. 134–137, June 2006
3. Zhuang, X., Sekiyama, K., Fukuda, T.: Evaluation of human sense by biological information analysis. In: International symposium on Micro-NanoMechatronics and Human Science, Nagoya, Japan, pp. 74–79, Nov 2009
4. Guçluturk, Y., Guçlu, U., Samraj, A.: An online single trial analysis of the P300 event related potential for the disabled. In: IEEE 26th Convention of Electrical and Electronics Engineers, Eliat, Israel, pp. 338–341, Nov 2010
5. Website: http://www.NI.Com
6. Kaper, M., Ritter, H.: Generalizing to new subjects in brain-computer interfacing. In: Proceedings of the 26 Annual Conference of the IEEE EMBS, San Francisco, USA, pp 4363–4366, Sept 2004
7. Harun, H., Mansor, W.: EOG signal detection for home appliances activation. In: 5th International Colloquium on Signal Processing & Its Applications (CSPA), Kuala Lumpur, Malaysia, pp. 195–197, Mar 2009
8. Kirbis, M., Kramberger, I.: Multi channel EOG signal recognition for an embedded eye movement tracking device. In: 16th International conference on Systems, Signals and Image Processing, IWSSIP, Chalkida, pp. 1–4, June 2009
9. Lin, M., Li, B.: A wireless EOG-based human computer interface. In: 3rd International Conference on Biomedical Engineering and Informatics, Yantai, Japan, vol. 5, pp. 1794–1796, Oct 2010
10. Matsuoka, Y., Afshar, P., Oh, M.: On the design of robotic hands for brain–machine interface. Neurosurg. Focus **20**(5), 1–9 (2006 May)
11. Huang, H.-P., Hsu, L.-P.: Development of a wearable biomedical heath-care system. In: International Conference on Intelligent Robots and System, Taipei, Taiwan, pp. 1760–1765, August 2005
12. Isais, R., Nguyen, K., Perez, G., Rubio, R., Nazeran, H.: A low-cost microcontroller-based wireless ECG-blood pressure telemonitor for home care. In: Proceedings of the 25th Annual International Conference of the IEEE Engineering in Medicine and Biology Society IEEE, vol. 4, pp. 3157–3160, Sept 2003
13. Bobbie, P.O., Chaudhari, H., Ari, C.-Z.: Homecare telemedicine: analysis and diagnosis of tachycardia condition in an M8051 microcontroller. In: IEEE/EMBS International Summer School on Medical Devices and Biosensors (ISSS-MD), Hongkong, pp. 47–52, Sept 2006
14. Milea, P.L., Stefan, G., Moga, M., Barbilian, A., Mitulescu, S., Cernat, E., Oltu, O., Moldovan, C., Pompilian, S.: Hardware and software package for locomotory disabled patients training. Int. J. Syst. Appl. Eng. Dev. **5**(3) (2011)

Portable Digital Remote Labs Designed for the Students Using Inexpensive Hardware and Open Source Prototyping

H. O. Darshan, I. Pooja, J. Gaurav(✉), and J. Nikhil

Cymbeline Innovations Pvt Ltd., Bengaluru, India
{darshanhol6, poojairannal, gauravprasad96, nikhil.janardhana}@gmail.com

Abstract. The concepts of engineering education is mainly restricted to the prescribed syllabus for a period of 4 years. Due to this, the growth of practical knowledge among the students for conducting and analysing a particular experiment in a lab duration of 3 h is fragile, depending on his or her understanding capabilities. In order to bring out a change in education pattern this paper throws light on the introduction of PORTABLE REMOTE LABORATORIES which can be easily built by the students to perform digital experiments such as logic gates. This pays a path for developing practical knowledge which often leads to deep understanding of a theoretical concept through the act of performing experiments and adds personal approach in order to gain practicality over the subjects that is prescribed to them. The overall setup highlighted in the paper is less expensive.

Keywords: Portable remote labs · Logic gates · Practicality · Less expensive

1 Introduction

Presently in India, there are 1.7 million engineering graduates, out of which maximum number of students lack practical approach towards the concepts that are taught in the classrooms. One of the major reason for this kind of draw-backs is non availability of laboratories for students, more over these conventional laboratories works only for 6 h per week i.e.., they work only for 2 days this factor highlights the gap between the students to gain practical knowledge on daily basis and exposures to trending technologies.

According to the reports of IAMAI and Kantar IMRB, the number of internet users in India would reach up to 500 million by 2018 therefore, depending on this fact it provides a strong basis to setup portable remote laboratories using microcontrollers such as Wemos and arduino IDE as a open source software at educational institutions. These portable remote laboratories form the key factor for students to understand the concepts practically at their own phase as the remote laboratory is created by themselves.

Students can better learn the prescribed set of digital experiments such as verification of some logic gates in the curriculum remotely through less sophisticated remote infrastructure in real time and also can be accessed from any geographical locations,

M. E. Auer and K. Ram B. (Eds.): REV2019 2019, LNNS 80, pp. 217–228, 2020.
https://doi.org/10.1007/978-3-030-23162-0_20

they also have a revision of these experiments during their examinations and also they can perform the experiments repeatedly as per their convenience. The key feature of this paper is to educate rural students who are underprivileged of complex laboratory equipments through building cost effective portable remote lab development kit which can be understood with minimum practical knowledge. Therefore, this step induces a sense of curiosity to develop a strong base of practical knowledge among the students in rural areas.

The paper introduces an idea of application for the development kit as a trigger circuit in order to have an access and control over light bulbs, fans and other household appliances serving the purpose of home automation (Fig. 1).

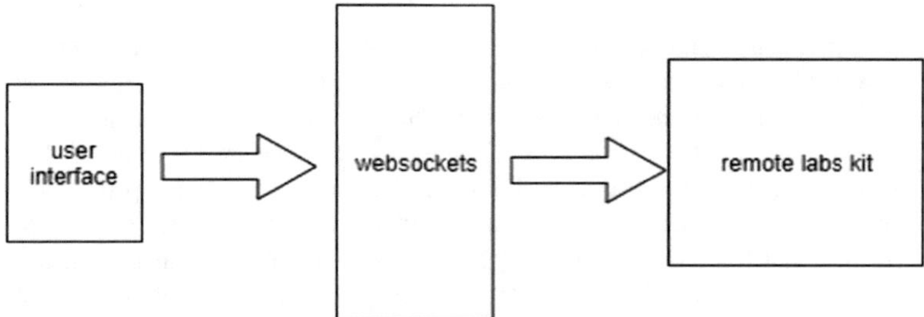

Fig. 1. Basic architecture

2 Drawbacks of Present System

The existing education pattern in all engineering colleges lacks to deploy practical knowledge to students due to confined practical hours. The deep understanding of the theoretical concepts can be achieved through the act of conducting experiments and gaining personal experiences on a day to day basis in order to have hands on experiences. The usual situation in the lab comprises of three to four sets of apparatus to conduct experiments among thirty students in a batch. This ultimately makes one student to conduct the experiment for the entire three hours of lab duration. This restricts other students to develop and gain practical knowledge to conduct experiments and gain hands on experience. The restricted time allocation for a lab during the entire year of syllabus, constraints the students approach towards practicality of analysing and understanding the theoretical concepts that are taught in classrooms.

3 Proposed System

The portable remote labs proposed under this paper can purely be built by the students by using low cost hardwares such as wemos d1 mini microcontroller, few 220 Ω resistors, LED's and general purpose pcb board as shown in Fig. 2. The firmware is burnt and updated over the air (OTA), this firmware is free of cost i.e., a open source prototyping platform with large free library packages which can be accessed by the students to write a code.

Fig. 2. Hardware placement

The completely built kit can be carried by the students at any point of time to conduct experiments related to digital labs such as verification of logic gates with its truth table. Therefore this newly built systems overcomes the drawbacks of the outdated education pattern and provides flexible lab durations to carryout experiments individually thereby emphasizing on individual growth of practical knowledge.

4 Implementation Details

The block diagram of portable remote laboratory in order to conduct digital experiments is as shown in Fig. 3. The setup is less complex in nature and inculcates low power consuming components as it's hardware requirement such as:

- Voltage Regulator: It is a step down power supply module which can supply 5 V DC from 230 V AC supply.
- ESP8266 WiFi Module: It is a low cost miniature wireless 802.11 microcontroller full fledged development board.
- LED: It is used to verify truth tables of logic gates through ON/OFF state of the LED.
- Resistors: Forms voltage biasing circuit.

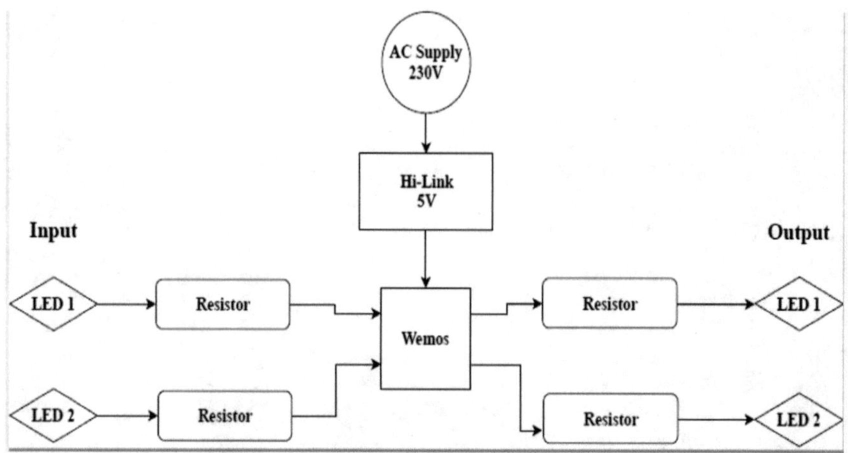

Fig. 3. Block diagram of portable remote labs

The circuit diagram is as shown in Fig. 4. Once the components are connected as per the circuit diagram, coding is done in Arduino IDE along with the help of Spiffs, Web-Sockets, Mdns. When the ESP8266 wifi module is connected to internet, it generates an IP address, further this IP address is hosted over the webpage in order to acquire remote access to the experiment setup.

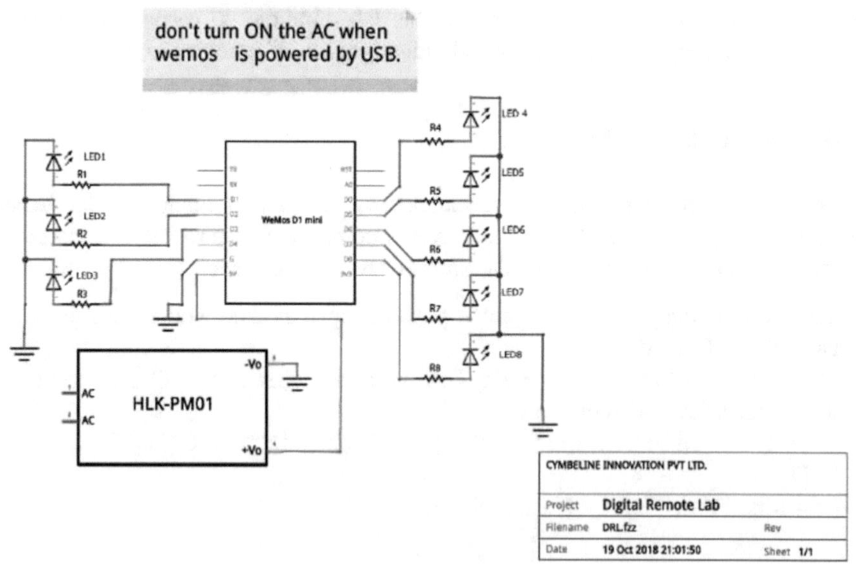

Fig. 4. Circuit diagram

5 Hardware Details

The components used to build this portable remote lab kit is chosen by keeping the cost and efficiency has the constraints and we have selected the low cost components which are efficient for our usage.

The list of hardware used is:

5.1 **Wemos d1 Mini**: The WeMos D1 min is a miniature wireless 802.11 (Wifi) microcontroller development board. It turns the very popular ESP8266 wireless microcontroller module into a fully fledged development board. It's a mini wifi board with 16 MB flash, external antenna connector and built-in ceramic antenna based on ESP-8266EX. It has 11 digital input/output pins, Interrupt/pwm/I2C and 1 analog input (Fig. 5; Table 1).

Fig. 5. ESP8266

Table 1. Specification of microcontroller

Sl no.	Specification	Range
1	Power supply	5 V
2	Operating voltage	3.3 V
3	Digital, I/O pins	11
4	Analog input	1
5	Clock speed	240 MHz
6	PSRAM	4 M bytes

5.2. **LED's**: LED'S are light emitting pn junction diodes which are often used to indicate the states of logic gates in a digital experiments. They are cheap in cost and comes in different colors such as red, green, blue and so on (Fig. 6).

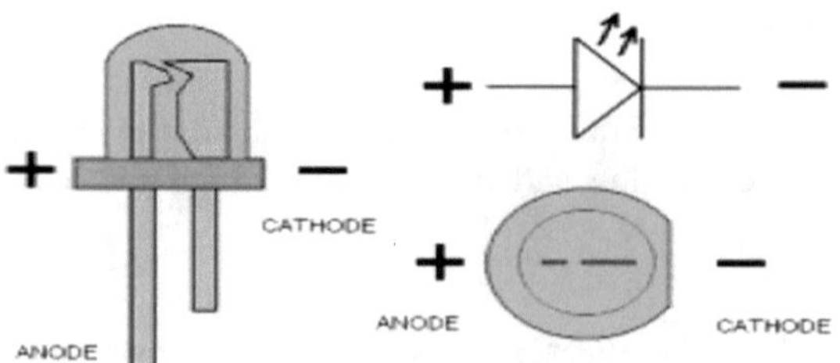

Fig. 6. LED

5.3. **Resistors**: Resistors are generally used for voltage dividing purpose and also to limit the rate of current flow to the LED's ensuring their safety while the system is put in use (Fig. 7).

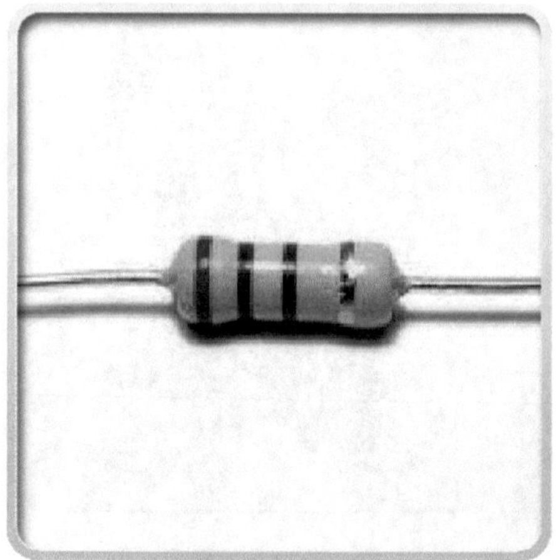

Fig. 7. Resistors

5.4. **General Purpose PCB Boards**: These boards helps in building the initial stage of the portable remote labs, where all the hardware components inculcated are soldered to form a circuit which is low of cost (Fig. 8).
5.5. **Voltage Regulator**: Voltage regulators such as Hi-Link, comes as an pcb mounted module which converts 230 V input to 5 V DC output supporting small projects which are often used to power up microcontrollers and other low voltage components. They are highly efficient, precise as well as highly secured in terms of safety (Fig. 9).
5.6. **2-Pin Power Cable**: These are the power cords which supply 230 V AC power supply to the input of the voltage regulator in order to get %V DC output power supply.
5.7. **Connecting Wires**: These wires forms the connection paths between different components that are used in the circuit.

Fig. 8. GCB

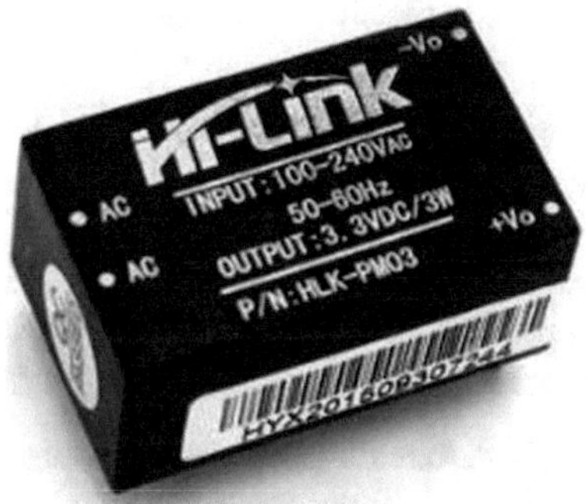

Fig. 9. Voltage regulator

6 Software

The portable remote labs makes use of Arduino IDE as its software due its open source in nature as well as availability in large libraries which are free of cost. Then for device to connect over the internet the system integrates other softwares such as Web-Sockets, SPIFFS, mDNS as its supporting system.

6.1 **ARDUINO IDE**: It makes it easier to write a code and upload it to the board. IDE comes with a software library called "Wiring" which makes students to perform input or output operations in an easier way by just defining two functions such as Setup(); and Loop();.

6.2. **Web-Sockets**: An inter-active open communication session between the user's browser and a server is only possible through advanced technology such as WEB SOCKETS. With the help of this API, messages can be sent to the server and can receive event driven responses without polling to the server for a reply.
6.3. **SPIFFS**: Serial Peripheral Interface Flash File System is mainly used to integrate HTML and CSS files to be stored in a ESP8266 File system in order to build a web server. HTML and CSS files are separately created in Arduino IDE, therefore, building controls such as ON or OFF at the ESP8266 output (Fig. 10).

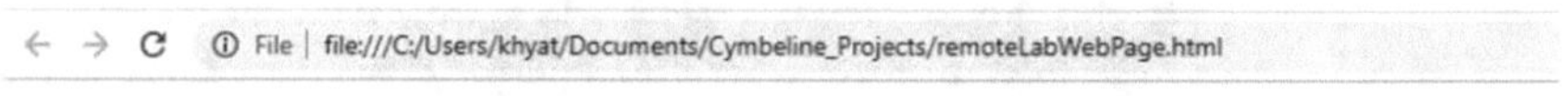

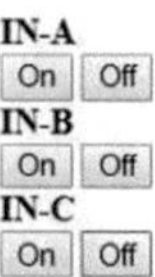

Fig. 10. Web page

6.4. **mDNS**: DNS stands for Domain Name System which mainly facilitates naming system for the connectivity of resources such as computers, laptops etc., over the Internet. The primary function of the DNS is to memorize the domain names to the IP addresses for locating and identifying computer based services with the underlying network protocols. It specifies the technical functionality of data base services which are present at its core, thus forming a internet protocol suit.

For small projects which requires small networks adapts mDNS protocol which resolves host name to IP address without including local name server. It has similar interfaces, packet formats and operations as that of DNS system (Fig. 11).

7 Working

The hardware components were taken and were placed on the general purpose pcb board the connections were made by soldering and connecting all the components as shown in the circuit diagram. A fine casing is put inorder to protect the circuit from external factors. Once the hardware is built the code is dumped into the wemos D1 mini and the built-in led glows twice and goes off this indicating the code is dumped successfully. The ESP8266 module will scan for access points and it will searches for the priority number one hotspot and will connect to it. If the hotspot is unavailable connect it to the next hotspot on the list till the board is connected. After the

```
#include <ESP8266WiFi.h>
#include <WebSocketsServer.h>
#include <Hash.h>
#include <ESP8266WebServer.h>
#include <ESP8266mDNS.h>


static const char ssid[] = ".........";// Your WiFi SSID comes here
static const char password[] = "........";//Your WiFi password comes here

MDNSResponder mdns;
int i;
static void writeLED1(bool);
static void writeLED2(bool);
static void writeLED3(bool);
// GPIO#0 is for Adafruit ESP8266 HUZZAH board. Your board LED might be on 13.

const int LEDPIN[3] = {D3,D2,D1};//Input LED pins to indicate inputs
// Current LED status
bool LEDStatus[3];

//Default server port is 80
ESP8266WebServer server(80);
//Web socket port at 81
WebSocketsServer webSocket = WebSocketsServer(81);

//User Auth
const char* www_username = "USER";
const char* www_password = "USERPASS";

//Index web page for connected client
static const char PROGMEM INDEX_HTML[] = R"rawliteral(
```

Fig. 11. Glimpse of the code

connection, start the TCP server on the IP and enter the port number as specified. Host the webpage stored in SPIFFS and wait for the client to respond. When the client is connected, open sockets for communication for various buttons. The data is received through sockets and is passed to microcontroller for computation. The computation of different gates takes place in microcontroller and outputs are sent to the respective leds. So we can take the respective truth tables of the gates and verify them (Fig. 12).

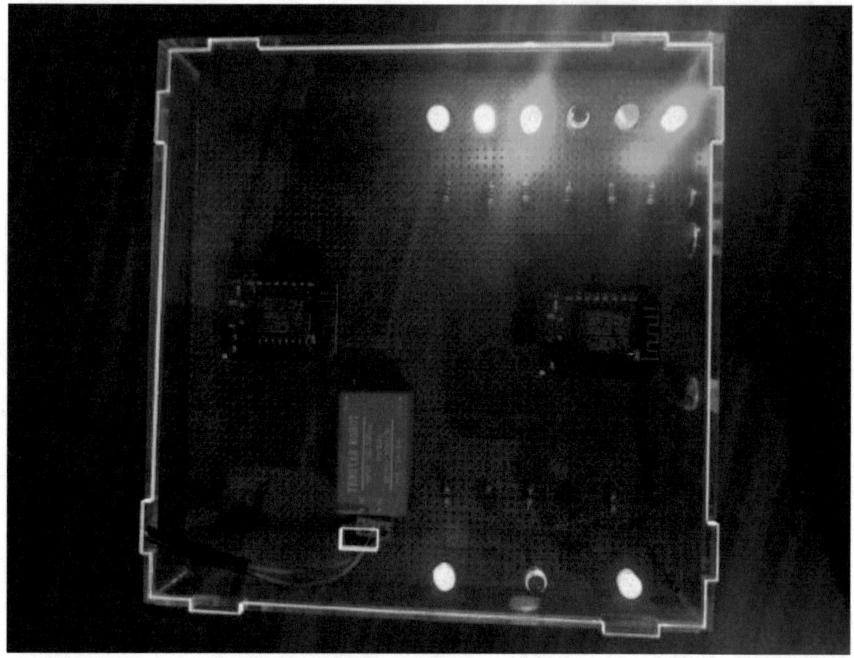

Fig. 12. Portable remote lab

8 Outcomes

The proposed portable remote lab under this paper was successfully able to develop practical knowledge among the students and was also able to provide the knowledge of cutting edge technologies of the future. Students from various engineering colleges in Bangalore, Vijayawada as well as Madurai have undergone hands on training on developing these portable labs in a one day workshop that was conducted with help of college assistances. Students were able to learn the technology from scratch to fully fledged operating system. The workshop also was able to convey students to learn how a application was useful in controlling any devices remotely (Fig. 13).

9 Conclusions

Over all this paper was indeed able to arise the need among students to get hands on experiences over the theoretical concepts that were taught in class hours. Also helped in gaining practical knowledge and a sense of approach to learn future remote technologies. The proposed system if at all built on a large scale with different type of experiments, students from any remote locations can be able to conduct experiments at duration of time.

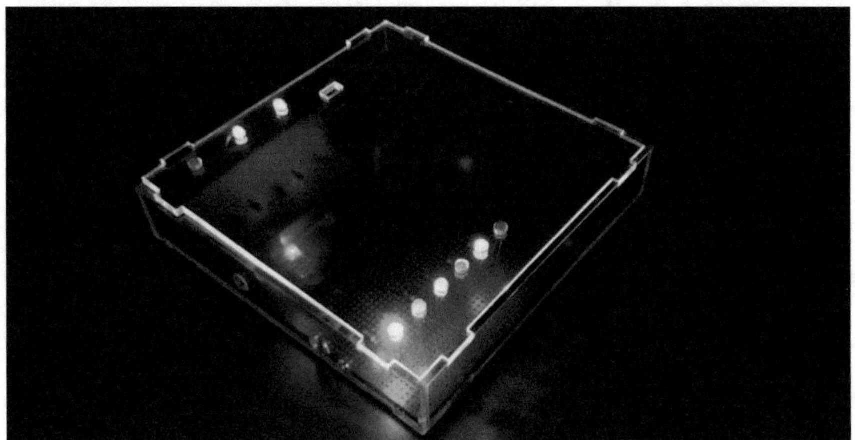

Fig. 13. Final product

References

1. Auer, Michael E.: Virtual lab versus remote lab. In: 20th World Conference on Open Learning and Distance Education (2001)
2. Ram, B.K., Kumar, S.A., Sarma, B.M., Mahesh, B., Kulkarni, C.S.: Remote software laboratories: Facilitating access to engineering softwares online. In: 2016 13th International Conference on Remote Engineering and Virtual Instrumentation (REV), pp. 409–413. IEEE (2016)
3. Pruthvi, P., Jackson, D., Hegde, S.R., Hiremath, P.S., Kumar, S.A.: A distinctive approach to enhance the utility of laboratories in Indian academia. In 2015 12th International Conference on Remote Engineering and Virtual Instrumentation (REV), pp. 238–241. IEEE (2015)
4. Esche, S.K., Chassapis, C., Nazalewicz, J.W., Hromin, D.J.: Architecture for multi-user remote laboratories. Dynamics (with a typical class size of 20 students), 5, 6 (2003)
5. Outram, J.D., Outram, R.G.: Adaptive data logger. U.S. Patent No. 4,910,692. 20 Mar. 1990
6. Yunlong, F., Anping, F., Ning, L.: Cortex-M0 processor: an initial survey. Microcontrollers Embed. Syst. **6**, 033 (2010)
7. D'Ausilio, A.: Arduino: a low-cost multipurpose lab equipment. Behav. Res. Methods **44**(2), 305–313 (2012)
8. Gontean, A., Szabó, R., Lie, I.: LabVIEW powered remote lab. In: 2009 15th International Symposium for Design and Technology of Electronics Packages (SIITME). IEEE (2009)
9. Auer, M., Pester, A., Ursutiu, D., Samoila, C.: Distributed virtual and remote labs in engineering. In: 2003 IEEE international conference on Industrial technology, vol. 2, pp. 1208–1213. IEEE (2003)
10. Aloni, E., Arev, A.: System and method for notification of an event. U.S. Patent No. 6,965,917. 15 Nov. 2005
11. Shnayder, V., Hempstead, M., Chen, B.R., Allen, G.W., Welsh, M.: Simulating the power consumption of large-scale sensor network applications. In Proceeding of the 2nd International Conference on Embedded Networked Sensor Systems, pp. 188–200. ACM (2004)
12. Tinga, T.: Application of physical failure models to enable usage and load based maintenance. Reliab. Eng. Syst. Saf. **95**(10), 1061–1075 (2010)
13. Vuletid, M., Pozzi, L., Ienne, P.: Seamless hardware-software

14. Nikhil, J., Pavan, J., Darshan, H. O., Kumar, G.A., Gaurav, J., Devi, C.Y.: Digital remote labs built by the students and for the students. In: International Conference on Remote Engineering and Virtual Instrumentation, pp. 261–268. Springer, Cham (2018)
15. Gowripeddi, V.V., Bhimavaram, K.R., Pavan, J., Janardhan, N., Desai, A., Mohapatra, S., Devi, C.Y.: Study of remote lab growth to facilitate smart education in Indian Academia. In: International Conference on Remote Engineering and Virtual Instrumentation pp. 694–700. Springer, Cham (2018)
16. Murthi, M., Guio, A.C., Dreze, J.: Mortality, fertility, and gender bias in India: a district-level analysis. Popul. Dev. Rev. **21**(4), 745–782 (1995)
17. Premawardhena, N.C.: http://mhrd.gov.in/sites/upload_files/mhrd/files/statistics/AISHE 2015–16.pdf (2012)
18. Cheney, G.R., Ruzzi, B.B., Muralidharan, K.: A profile of the Indian education system. Prepared for the New Commission on the Skills of the American Workforce (2005)
19. Dunnette, M.D., Campbell, J.P.: Laboratory education: impact on people and organizations. Ind. Relat. J. Econ. Soc. **8**(1), 1–27 (1968)
20. Argyris, C.: On the future of laboratory education. J. Appl. Behav. Sci. **3**(2), 153–183 (1967)
21. Pruthvi, P., Jackson, D., Hegde, S.R., Hiremath, P.S., Kumar, S.A.: A distinctive approach to enhance the utility of laboratories in Indian academia. In: 2015 12th International Conference on Remote Engineering and Virtual Instrumentation (REV), pp. 238–241, IEEE (2015)
22. Natarajan, R.: Emerging trends in engineering education-Indian perspectives. In: Proceedings of 16th Australian International Education Conference, vol. 30 (2002)
23. Kozma, R.B.: National policies that connect ICT-based education reform to economic and social development. Hum. Technol. Interdisc. J. Hum. ICT Environ. **1**(2), 117–156 (2005)
24. Ferrero, A., Salicone, S., Bonora, C., Parmigiani, M.: ReMLab: a Java-based remote, didactic measurement laboratory. IEEE Trans. Instrum. Meas. **52**(3), 710–715 (2003)
25. Nordhaug, O.: Human Capital in Organizations: Competence, Training, and Learning. Scandinavian University Press, Oslo (1993)
26. Nickerson, J.V., Corter, J.E., Esche, S.K., Chassapis, C.: A model for evaluating the effectiveness of remote engineering laboratories and simulations in education. Comput. Educ. **49**(3), 708–725 (2007)
27. Diwakar, S., Kumar, D., Radhamani, R., Sasidharakurup, H., Nizar, N., Achuthan, K., et al.: Complementing education via virtual labs: implementation and deployment of remote laboratories and usage analysis in South Indian villages. Int. J. Online Eng. **12**(3), 8–13 (2016)
28. Achuthan, K., Sreelatha, K.S., Surendran, S., Diwakar, S., Nedungadi, P., Humphreys, S., Sreekala S., et al.: The VALUE@ Amrita virtual labs project: using web technology to provide virtual laboratory access to students. In: 2011 IEEE Global Humanitarian Technology Conference (GHTC), pp. 117–121. IEEE (2011)
29. van Joolingen, W.R., de Jong, T., Lazonder, A.W., Savelsbergh, E.R., Manlove, S.: Co-Lab: research and development of an online learning environment for collaborative scientific discovery learning. Comput. Hum. Behav. **21**(4), 671–688 (2005)
30. Kraut, R.E., Fussell, S.R., Brennan, S.E., Siegel, J.: Understanding effects of proximity on collaboration: implications for technologies to support remote collaborative work. In: Distributed Work, pp. 137–162 (2002)

Internet of Things Based Autonomous Borewell Management System

R. Venkataswamy(✉), Rinika Paul, and Jinu Jogy

Department of Electrical & Electronics Engineering, Faculty of Engineering, Christ (Deemed to be University), Bangalore, India
venkataswamy.r@christuniversity.in, rinika@btech.christuniversity.in, jinu@btech.christuniversity.in

Abstract. Water is a basic need for all living beings. At present, due to a large population, water level is getting depleted at an alarming rate particularly in urban region. During summer season, there is no continuous flow of water or availability of water. In electrical contingency situations, bore-wells are prone to damages. The utilization of power at dry run condition affects the economy of the consumers. Despite having no water in the bore-well, if the motor runs, the motor windings may burdened and gives rise to unnecessary power loss. In the present scenario, conservation of energy is a major concern. The conservation of energy as a whole will take place when an individual take an active part by using autonomous and effective methodologies or controllers. The issue is solved by managing the borewell using Internet of Things (IoT) as a platform to automate and manage. The IoT based borewell management system is designed to provision scheduling, manual operation, avoidance of borewell motor running at dry run condition and also nullifies energy loss. The automated borewell operations can be executed from a remote control and measurement unit by the measurement of electrical parameters and analytics. The proposed system minimizes man power, saves time and conserve energy loss. The paper presents operating the conventional borewell by deployment of smart controller which handles the information and communication technology at client and base units.

Keywords: Internet of things · IoT controller · Dry run · Bore-well centrifugal pump · Energy conversation

1 Introduction

Our planet is witnessing excessive water usage from different types of consumers. It also include usage and waste or not properly utilized. In long run, there will sustainability problem arises. Water scarcity at many part of the globe is caused by the climate transients, pollution content, and excessive consumption by people and also use of water more than the requirement. Due to inadequacy of proper protection system, around 15% of the water pump failures are due to dry run. This failure occurs when the motor is in running condition but does not pump

M. E. Auer and K. Ram B. (Eds.): REV2019 2019, LNNS 80, pp. 229–243, 2020.
https://doi.org/10.1007/978-3-030-23162-0_21

water. This leads to voltage fluctuations but heavily loaded condition which heats up the motor and cause heavy power loss. So to avoid the situation, we provide the motor with some intelligent controllers. The motor hence can detect running and dry run condition by measuring the current value and hence can turn on and off to avoid damage. In this way, we prevent power loss and hence conserve energy. Internet of Things (IoT) is emerged, adapted and associated with almost all the physical system. The multi-domain integration and digital twins are exposing the possibility of prosumer profit. Bore-well is basically hydro system driven by electrical supply. Integration of IoT devices at the required places will makes it completely autonomous and intelligent.

2 Current Research and Developments

Gunde et al. [1] has proposed automated water management system for water tank. The water level is detected by the ultrasonic sensor and informed to the user. The embedded system is connected with GSM module to notify the alert information through the SMS. The Monitored sensor data is uploaded to the cloud database for the future use.

Predescu et al. [2] presented online development of IoT client module for pumping water in a water contained in the distribution system. The hardware and software framework is implemented in generic manner so as to suit any system with least portability configuration. The remote sensor and actuator data is processed using open source hardware nd software and its integration with highest flexibility.

Getu et al. [3] has designed an embedded system to manage the level of water in a tank or water container using ultrasonic sensor information. The embedded system is designed in such a way so that it can display at a scale value from 0 to 9. The procedure used for the design is generic in nature and can be adapted by any other physical system where water level indicator and controlling are required.

Gurguiatu et al. [4] presented the mechanism of improving the power factor using active power filters considering the water pump load as nonlinear. The power electronic drive is implemented to trigger active filters. The energy saving hence proved by the installation of active filters.

There are numerous IoT platforms are available in market. Free and open source platforms are Eclipse IoT, OpenHab, Heroku and OpenIoE platform. The AWS released IoT related products such as Core, analytics, buttons and Device Management applications [5]. Heroku provides the dockers with light, isolated cartridges that facilitate computing engine, file space, database space and operating system which are typically run on a connected terminals in cross cloud framework [6]. Google cloud platform offers products such as computing engine, cloud space, databases, IoT Core, Edge, IoT Analytics, Machine learning, Identity and security tools to connect with physical system to realize IoT enable ecosystem [7].

3 Modeling of System

3.1 Architecture

The system consists of device, sensor, transmitter, receiver and user display device. The operating device is the borewell. There are two sensors- current sensor and voltage sensor that detects current and voltage values; and sends it to the communication devices. The setup consists of arduino and ZigBee at the control end of the borewell. The data received by raspberry pi via ZigBee communication is updated into the cloud. The remote terminal unit processes the data and transmits the data to the cloud. The receiver consists of the control box along with sensors and arduino and ZigBee. Control action takes place at relay unit and consequently operation of the bore-well is controlled. The state of the bore-well is displayed to the user in the display device such as computer, tablet and mobile (Fig. 1).

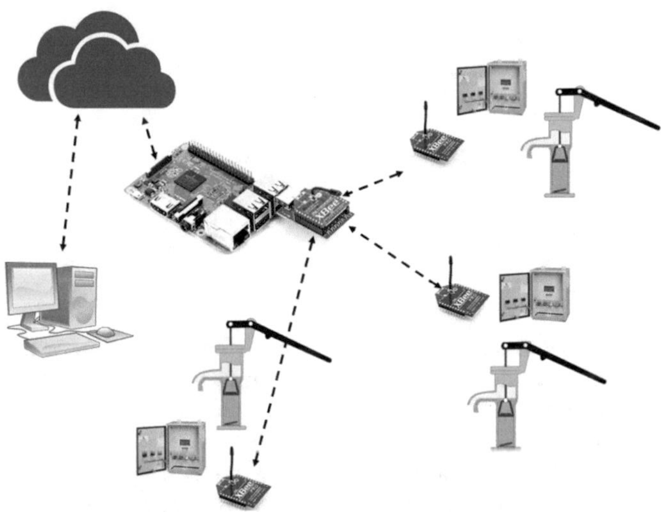

Fig. 1. Pictorial view of hardware model

3.2 Functional Model

The IoT technology has been incorporated in the functional model. Basically there are two ends namely base coordinator node and IoT client. The analog data is sent from the IoT client to the base coordinator and the values are sent to cloud for processing and calibrating sensor values to water availability and electric energy for a particular period. The most important aspects of proposed work are, it can be controlled manually or by scheduling method from anywhere, anytime. There will one base coordinator known as master and the IoT client termed as slaves, which functions in broadcast mode. Hence control and monitoring of the multiple number of bore-wells could be done with great ease (Fig. 2).

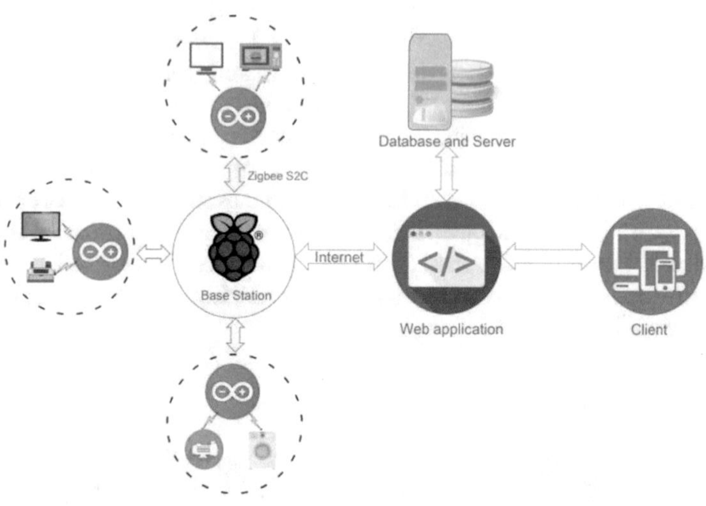

Fig. 2. Functional model

3.3 Circuit Model

The model consists of mainly two parts named as base coordinator node and IoT client. The IoT client node controls the relay simultaneously fetches the real time sensor data. The transmitter node or base coordinator consists of a Raspberry Pi, single board computer which sends the control values and receives the sensor values. Zigbee is used to transfer data from transmitter end to receiver end and vice versa. The IoT client consists of an Arduino microcontroller with current sensor, relay module, voltage sensor, OLED display, SD card module and ZigBee for data transmission. The IoT client reads the current and voltage values, sends it to the base coordinator and also to cloud for future reference, the current values obtained is used to control the functioning of a motor. If water is present then load is more, current values will be high. When no water is present load will be less. The load difference is monitored and a current limit is referred, so when current values drops beyond a certain limit, it is postulated as no water is present and hence the motor gets off. This helps to prevent dry running of motor and hence conserve water and electric energy (Fig. 3).

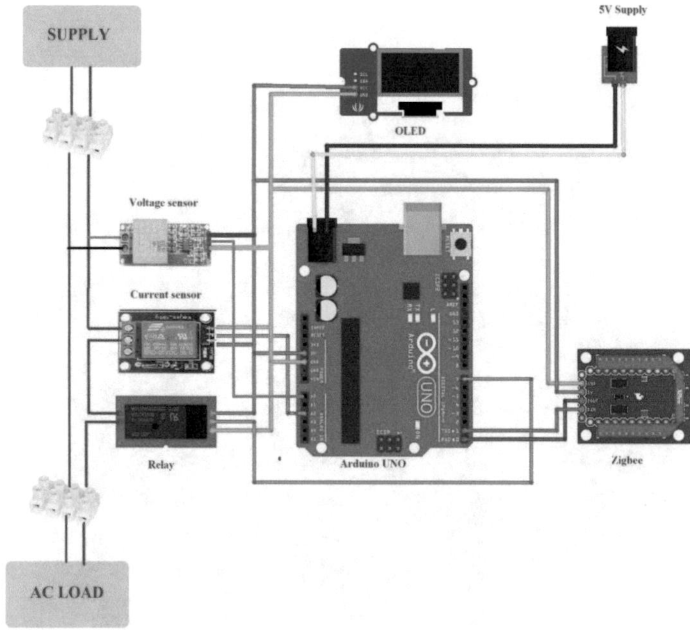

Fig. 3. Circuit model

3.4 Hardware Model—IoT Client

The hardware model shown in Fig. 4 is comprises of current sensor, voltage sensor, relay, arduino, ZigBee. The current sensor detects and converts current to an voltage type, which is proportional to the current through the measured path. Current sensor is connected in series with the line. The voltage sensor is connected across the line and neutral. A NC of relay and common terminal of relay is connected in series to the line. Relay is used as it is necessary to control a power circuit by a separate low-power signal. It is a switching device to switch on and off the moderate power circuit. Arduino, a single board computer is used to take digital decisions and interactive objects that can sense and control objects in the physical and digital world. Zigbee is used for high-level communication protocols to create personal area networks between transmitter and different receiver nodes.

3.5 Base Coordinator

Base coordinators shown in Fig. 5 works like a master controller and parent hub for all routers and nodes in the network. They control the managerial function of the network with proper security. Normally, single board computers are used as the Base coordinator. It acts as the internet gateway to sensor nodes. The Raspberry Pi uses GET and POST request library to communicate with the

Fig. 4. Hardware model

Fig. 5. Base coordinator

Table 1. Heroku user ID and pin

Parameter	User ID	Pin
V1	1285	829
I1	1286	600
V2	1241	608
I2	1242	680
Relay1	1243	232
Relay2	1287	354

cloud server. Rapberry Pi3 is used which has ARM cortex A7 CPU 900 MHz, Four USB ports, an ethernet port and 40GPIO pins for a GUI though monitor. The Raspberry Pi GPIO serial port is referenced as '/dev/ttyUSB0' read the incoming value and upload the values to the cloud. At the same the it fetch the relay status and control the relay.

3.6 IoT Platform

Open IoE is an open source IoT platform. It uses Heroku as Cloud Paas(platform as a service) which enables us to build, run and operate fully in the cloud platform itself allied with PostgreSQL database. The cloud is accessed from the Raspberry Pi using WiFi. To Create infrastructure in the IoT platform, each variable is assigned with specific user id and pin. The credentials can be generated in https://openioe.herokuapp.com. Dataformat and cryptography is selected accordingly for every variable and generated the id and pin are noted as shown in the Table 1.

To see the value : https://openioe.herokuapp.com/Console/showvalue/ID/Pin

To update the value https://openioe.herokuapp.com/Console/updatevalue/ID/Pin/value.

4 Methodology

4.1 Objectives

The objectives of the proposed work includes,

- Conservation of energy, water and time.
- IoT based system that can be accessed from anywhere in the world.
- The system can be useful to compare the previous values of current, voltage and power consumption for better management.
- IoT based scheduling for water pumping and current, voltage and power consumption monitoring.

4.2 Features

The features of the developed system includes,

- IoT based Manual Control
- IoT based Automation and Scheduling
- Dry run identification and curtailment
- Centralized monitoring and control system
- Peak loading and consumption issues
- Energy and bill calculation.

4.3 Working Principle

At control centre, there exists ZigBee connected to the Raspberry Pi, which receives the information send by the motor and gives the control command to the motor. The motor runs if the current is near to full load current. At 70% of the full load current, the motor is considered as running in dry and automatically switches off. The physical system connected and controlled through cloud using Matlab application as an interface.

At each bore-well location there exists control box consisting of current sensors, relays and a ZigBee module connected to Arduino, which is connected to the motor. The motor senses the load current using current sensor. It is considered that if there is a reduction in 30% of the full load current, the motor goes to dry run condition. So the control box measures the current value at each interval and sends it to the cloud. The pictorial representation is shown in Fig. 6.

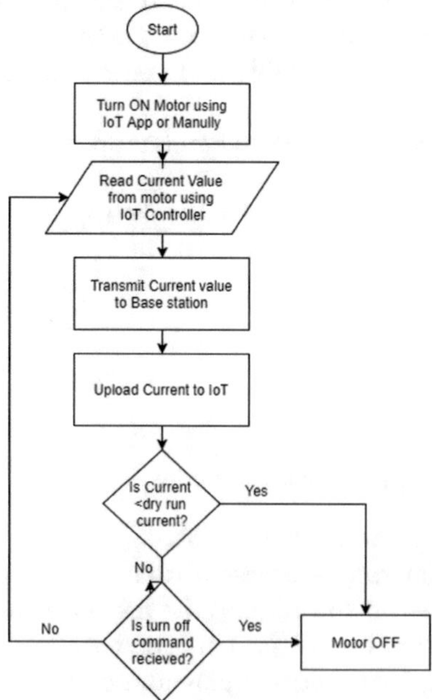

Fig. 6. Circuit model

4.4 System Configuration

Configuration 1: The base coordinator and IoT client ZigBee should be configured as shown in Table 2. Basically setting for source, destination, system and security identities.
Configuration 2: Calibrate current sensor with resistive, inductive and capacitive loads.
Configuration 3: Create user id and pin for all the variables as shown in Table 1 and update the placeholders in the program.
Configuration 4: The setting values of all the bore-well are recorded in remote application.

Table 2. Xbee configuration parameter

Setting	Acronym	Coordinator	Node 1	Node 2
Channel	CH	C	C	C
PAN ID	ID	1234	1234	1234
Destination address high	DH	0	0	0
Destination address low	DL	33	44	44
16-bit source address	MY	44	33	33

4.5 Sensor Data Acquisition

The current sensor reads the current values from the loaded and the unloaded pump condition. As the motor is pumping water to the overhead tank, the current sensor senses the current value which is greater than 0.4 A and is considered as loaded condition. When there is no water and the motor runs without pumping water, the current value is very small (about 0.2 A) and the motor is said to be in dry run (unloaded condition). The motor is given a value below which it will turn off. This value is the 30% reduction in the current value. So when the current value of 70% of its full load current or less, the motor turns off. The sensor data is displayed using Arduino and transmitted to Raspberry Pi through ZigBee. At the control centre, MATLAB is used as interface to control the motor through cloud computation. Data is displayed in the form of a GUI on the cloud, which is visible to the end user.

4.6 Bore-Well Management System

The software is developed using Matlab consisting of the following modules,

1. BMS Dashboard: This tool is used to display all the information of bore-wells connected in campus or in the cluster.
2. BMS Controller: Individual bore-well operation and monitoring can be performed. The scheduling job can be defined using the navigation menu.
3. Event Logger: Information is persisted for analysis.

5 Case Study

Centrifugal universal motor with dual valve arrangement rated at 50 W is connected through IoT based controller as shown in Fig. 7. Dry run and loaded conditions are emulated using valve arrangements.

Fig. 7. Prototype for PoC

6 Results and Discussions

The experiment is conducted by measuring current values under loaded and unloaded conditions which is shown in Table. 3.

The experiment is further iteratively carried out to identify the percentage reduction of current value from loaded condition to unloaded condition.

It has been proved for the binary logic of bore-well operation, that 30% reduction from the loaded current would be the appropriate value to map the dry run situation.

Table 3. Current measurements

Condition	Current (A)
Loaded	0.385
Dry run	0.220

Theorem 1. *The dry run condition of the bore-well could empirically be considered as 30% reduction of current value from its full load value.*

6.1 Cluster Statistics

Real time bore-well cluster information will be displayed for administrator as shown in Fig. 8.

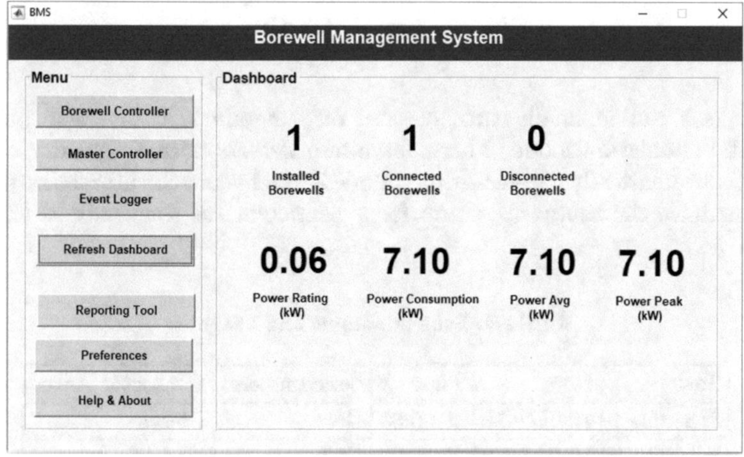

Fig. 8. BMS dashboard

6.2 Normal Operation

The bore-well would operate under normal current which reflects in BMS as shown in Fig. 9.

The controller board displays all the control operations as well as the current, voltage and power values of the motor being operated. There are 3 different modes,

- Manual mode
- Automatic simulation
- Scheduler.

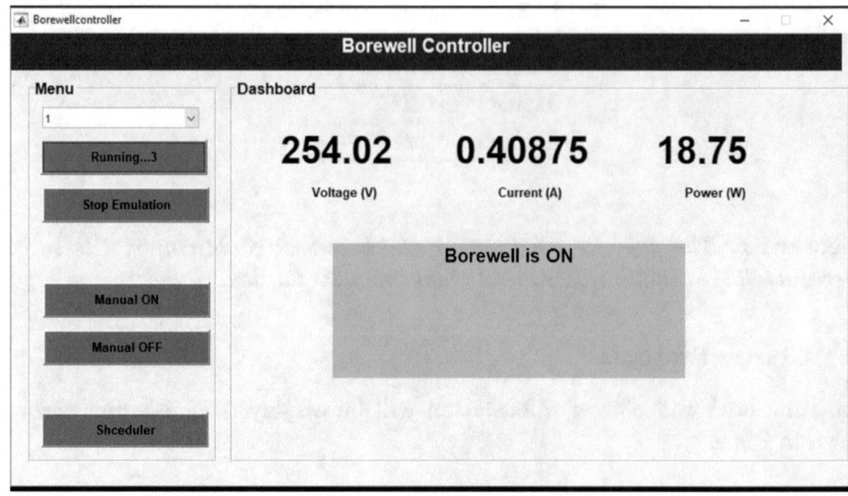

Fig. 9. Normal operation

The user can manually turn on and off the pump. It can also put it in automatic simulation mode, where the pump detects normal and dry run and turns off automatically. The scheduler mode is used when the user is not present. This schedules the pump operation for a particular set time only as shown in Table 4.

Table 4. Schedule input and output

Task	Date	Time	Set command	Contactor status
Schedule 1	10-08-2018	09:00:00	ON	Closed
Schedule 2	10-08-2018	09:15:00	OFF	Opened

6.3 Dry Run Operation

The bore-well would issue trip command on retrieval of dry run current which reflects in BMS as shown in Fig. 10.

6.4 Manual Operation

The bore-well would issue turn on and turn off command on retrieval from IoT platform shown in Fig. 11.

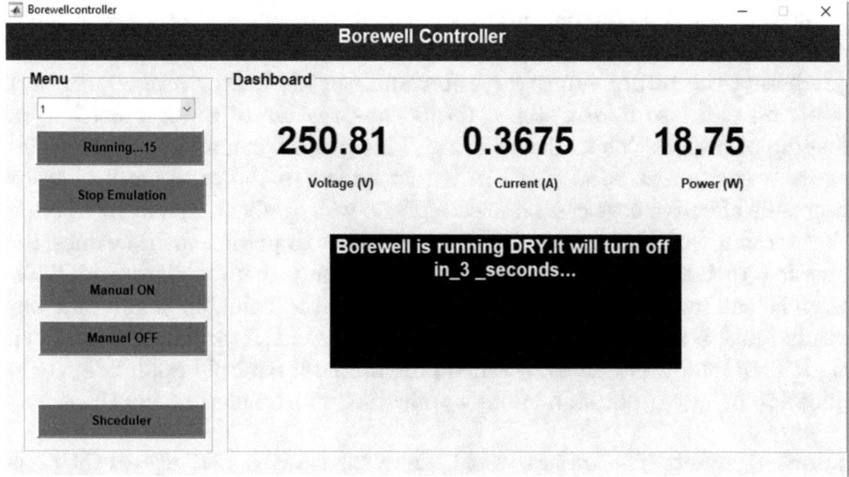

Fig. 10. Dry run operation

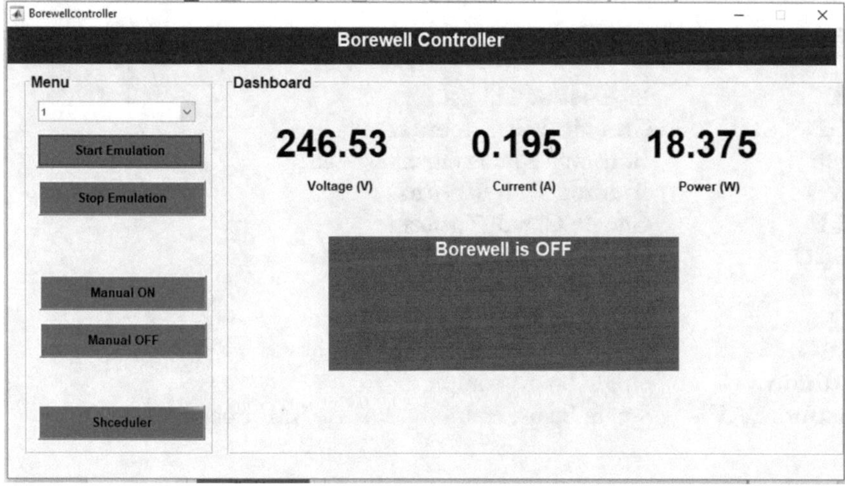

Fig. 11. Manual turn off operation

7 Conclusion and Future Work

In this work, implementation of the bore-well management system is validated which will conserve energy as well as protect the pump merely by preventing dry run of the centrifugal pump by using IoT based communication system. This paper proposes a proof of concept of system design, implementation and description of required tools and technologies to develop IoT based bore-well management system along with scheduling.

The work could be further improved using the results and observations of the current project. The monitored values of current, voltage and power are helpful in predicting the future values for water and electric energy usage for a particular period and also take a step towards conservation of water, electricity and damaging of motors due to dry running. This project can added to the existing bore wells with great ease. Since it can be added to the pre-existing bore-well system, the effective cost can be lowered. The web application can be developed for interaction with the end user and help the user to predict future values. Since it uses IoT technology it can be controlled from anywhere in the world. This as a product can surely save money and moreover the valuable time. This paper presents Matlab application as an interface between the communication devices using IoT manually as well as by scheduling method. Android application or web application or any other standalone application can be used as interface.

Acknowledgment. The authors would like to acknowledge and express the deepest gratitude to management of Faculty of Engineering, Christ (Deemed to be University) for providing freedom to work in the laboratory.

Appendix: Glossary

IoT	Internet of Things
GUI	Graphical User Interface
BMS	Bore-well Management System
AWS	Amazon Web Services
GCP	Google Cloud Platform
OLED	Organic Light-Emitting Diode
NC	Normally Open
NO	Normally Closed
GPIO	General Purpose Input Output
Ardiuno	Single board computer
Raspberry Pi-	Single board computer with ARM processor

References

1. Gunde, S., Chikaraddi, A.K., Baligar, V.P.: IoT based flow control system using Raspberry PI. In: International Conference on Energy Communication, Data Analytics and Soft Computing (ICECDS), pp. 1386–1390. Chennai, India (2017)
2. Predescu, A., Mocanu, M., Lupu, C.: Real time implementation of IoT structure for pumping stations in a water distribution system. In: 21st International Conference on System Theory. Control and Computing (ICSTCC), pp. 529–534. Sinaia, Romania (2017)
3. Getu, B.N., Attia, H.A.: Automatic water level sensor and controller system. In: 5th International Conference on Electronic Devices Systems and Applications (ICEDSA), pp. 1–4. Ras Al Khaimah, United Arab Emirates (2016)

4. Gurguiatu, G., Balanuta, C. D., Munteanu, T., Gaiceanu, M.: Energy savings generated by installing active power filters in water pumping stations. In: 4th International Symposium on Electrical and Electronics Engineering (ISEEE), pp. 1-5. Galati, Romania (2013)
5. Amazon Web Services (AWS)—Cloud Computing Services. https://aws.amazon.com. Last Accessed 26 Oct. 2018
6. Heroku-Cloud Application Platform. https://www.heroku.com. Last Accessed 26 Oct. 2018
7. Google Cloud including GCP & G Suite—Google Cloud. https://cloud.google.com. Last Accessed 26 Oct. 2018
8. V. R - Open IoE. https://openioe.herokuapp.com. Last Accessed 26 Oct. 2018

Work-in-Progress: Enhancing Collaboration Using Augmented Reality Design Reviews for Product Validation on the Example of Additive Manufacturing

Opportunities and Challenges

Daniel Eckertz(✉), Jan Berssenbrügge, Harald Anacker, and Roman Dumitrescu

Fraunhofer Institute for Mechatronic Systems Design IEM, Paderborn, Germany
{daniel.eckertz, jan.berssenbruegge, Harald.anacker, roman.dumitrescu}@iem.fraunhofer.de

Abstract. The "DigiKAM" project aims to create a digital collaboration network that enables the efficient exploitation of Additive Manufacturing (AM). Via a scalable platform solution, spatially separated AM customers and experts are interconnected throughout the entire development process, i.e. from potential identification to development and production. The digital collaboration will be supported by Augmented Reality (AR) Design Reviews in which the realistic visualization and shared viewing of components at the later operation site enables a faster and more meaningful validation. However, there is no integrated software solution that meets all requirements. In addition, AR know-how is necessary to prepare and execute AR Design Reviews, which the collaboration partners usually do not have. Therefore, the goal of this project is a set of tools and methods that enables AR Design Reviews to be conducted without AR know-how. Application scenarios for AR Design Reviews are identified and analyzed in order to derive rules for the requirements definition and the selection of AR hardware and software. Furthermore automated tools are developed to provide support for the complex parameterization of the conversion process from CAD to AR-capable data. In addition, a software will be developed and prototypically implemented to execute collaborative AR Design Reviews.

Keywords: Augmented reality · Remote engineering · Collaboration · Virtual environments · Design review

1 Digital Collaboration Network for the Exploitation of Additive Manufacturing

The digitalization of the industry is revolutionizing products and production systems, which is leading to a fundamental change in the value creation of product manufacturers. The combination of local information processing with global communication capabilities in tomorrow's products and production systems opens up new

M. E. Auer and K. Ram B. (Eds.): REV2019 2019, LNNS 80, pp. 244–254, 2020.
https://doi.org/10.1007/978-3-030-23162-0_22

opportunities for collaboration in an increasingly inter-organizational value chain. Technologies whose potential can only be fully exploited through digital collaborations in the shortest possible time are gaining considerable importance. One such technology is Additive Manufacturing (AM). It makes it possible to manufacture customized products with complex component structures and a high degree of functional integration in small quantities while maintaining a high level of economic efficiency. However, the development of the technology has so far been reserved primarily for large industrial groups. SMEs in particular lack the resources to independently build up and profitably apply the necessary knowledge about possible applications and benefits as well as the new product design requirements associated with AM (e.g. design guidelines, solution knowledge, cost/effort estimation methods). Companies are therefore relying on collaborations with AM know-how providers. However, there is currently no suitable contact point (e.g. Internet platform) that takes care of this mutual interaction between demanders (customers) and know-how carriers (AM experts).

The joint project "DigiKAM"[1] aims to create a digital collaboration network that enables the exploitation of Additive Manufacturing along the entire value creation process. The network is aimed in particular at small and medium-sized enterprises (SMEs) who, due to limited resources, are not able to build up the necessary AM know-how in the shortest possible time. By means of a scalable platform solution, AM customers and AM experts, most likely geographically separated, are efficiently networked throughout the entire AM development process, i.e. from potential identification to development and production. This is the basis for fully exploiting the existing potential of the technology with regard to efficient and fast production of components.

2 Enhancing Collaboration Through Augmented Reality

Augmented Reality (AR) is an innovative technology that allows virtual content to be visualized embedded in real environments. AR devices can be used to interconnect participants and enable visual immersion in the situation of the communication partner. In addition, the communication of complex facts is considerably simplified by viewing interactive AR scenes together. During the AM development process, the AM expert designs three-dimensional components with complex functional integrations and AM-specific design options. This must be coordinated with and checked and approved by the customer before the manufacturing. To support the collaboration between AM experts and AM customers, AR is integrated into the platform-based and inter-organizationally distributed development process. With AR for example product and component validation can be significantly improved. This leads to more efficient product development, faster availability of components and ultimately higher customer satisfaction.

[1] Funded by the Federal Ministry of Economics and Energy (BMWi) as part of the "Digital Technologies for Business (PAiCE)" funding programme.

2.1 Reference Process of the Development of AM Components

In order to integrate AR into the development process and to be able to use the existing potentials, it first had to be understood how the cooperation of an AM expert and the AM customer usually takes place and which work steps are to be performed during the AM development. A corresponding process model did not exist until now. Therefore, the AM development process was analyzed in cooperation with a company specialized on AM and documented in its entirety as an AM reference process including the interfaces between AM expert and customer (Fig. 1). The modeling was carried out with OMEGA [1] and includes interactions and synchronization points between the process steps including necessary data formats. The process covers all relevant process steps from potential identification, development and validation to the production of the component and is applicable for an initial concept model as well as for serial parts.

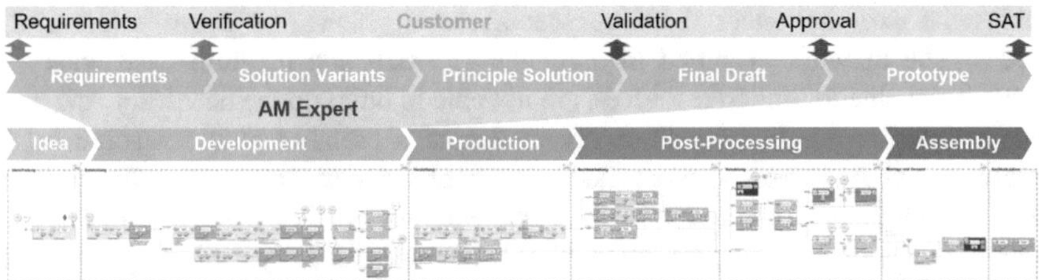

Fig. 1. Additive manufacturing reference process (bottom: detailed description using OMEGA process modeler, top: simplified description in functional phases and detailed development phase with interconnections between AM expert and customer)

For a simplified representation, the process was subdivided into individual superordinate function modules or phases. After an initial ideation phase the actual development follows. In this phase the knowledge-intensive and collaborative work steps take place, so it is considered in more detail. The development is essentially carried out by the AM expert, but there are collaborative connections to the customer (see arrows in Fig. 1). The first step is to define the requirements, which incorporate the customer's application-specific requirements and the expert's know-how. The customer has to verify the resulting requirement specification before the expert starts with the development. After various solution variants have been evaluated, first a principle solution is developed followed by the final design and a physical prototype. Between these steps the customer needs to validate and approve the 3D construction data and use the prototype for a site acceptance test (SAT). Only after that the actual production takes place. Following the production, the component must be post-processed and, if necessary, refined. At the end of the process, it is finally assembled and put into operation.

2.2 Possible Interconnections of the Reference Process with Augmented Reality

Subsequently, in cooperation with the AM company and application companies of the project consortium, it was investigated at which interfaces in the AM reference process AR can be integrated in a supportive way. The focus was on the development phase, as the production and subsequent post-processing steps are essentially carried out by AM experts and do not require coordination with the customer. The interfaces validation, approval and SAT (red arrows in Fig. 1) were identified. The validation and the approval process involve checking and confirming the 3D construction data. Here the component geometry and dimensions are of interest as well as the design and aspects made possible by AM such as functional integrations, component miniaturization and weight savings. All these aspects can best be checked and discussed by a 3D visualization, ideally directly at the future installation site. Collaboration can be significantly simplified by direct communication and joint viewing of the 3D construction data. The same applies to the SAT interface, where AR can be used on the prototype to explain final properties and functionalities of the component through animations.

3 Augmented Reality Design Reviews

All three identified AR interfaces are about the validation of three-dimensional construction data by the customer. Accordingly, AR Design Reviews can be integrated into the AM reference process. The aim is for the customer to check the component and give the expert feedback so that the component is revised as required. 3D construction data is used as input data, the production of the constructed component following the Design Review (DR) is equally based on the revised 3D data. First, the process of an AR Design Review is analyzed in more detail and modelled. On this basis, the difficulties associated with the process extension and the integration of an additional technology are examined and a short state of the art as well as existing solutions are presented.

3.1 AR Design Review Process Within the AM Reference Process

In the current AM reference process, CAD data is used consistently. Accordingly, CAD data is the input data for the Design Review phase that is considered here in detail and CAD data is also used after the phase. Figure 2 shows the detailed process of an AR Design Review in the context of collaboration via the platform. As a participant in the collaborative process, on the one hand there is the customer who provides the use case and has the need for the component. On the other side there is the AM expert who holds the necessary AM know-how and has created the component construction and wants to present it to the customer via Augmented Reality. In order to bring the two participants together, the Design Review is conducted between them, which should be collaborative and enable the validation of the component.

The DR is divided into three phases. During the **planning** phase, the special properties that characterize the component and are to be communicated to the customer

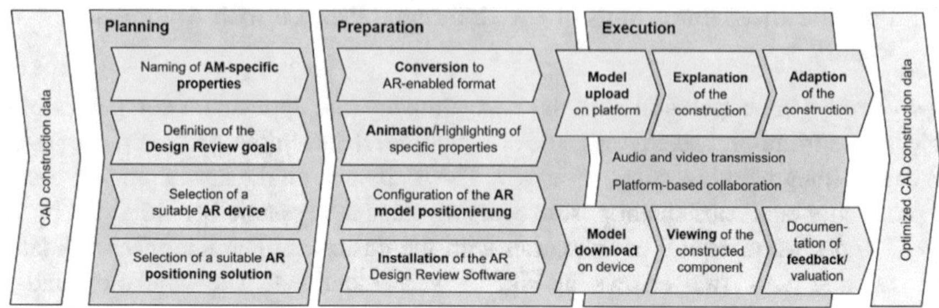

Fig. 2. AR design review process with the three phases planning, preparation and execution. Orange tasks are done by the customer and blue tasks by the AM expert.

must first be named. This can be, for example, an integrated functionality or an AM-specific construction method. Affected by this the goals of the DR are to be determined. From these in turn the requirements for the used hardware and the AR tracking and positioning result, that likewise must be defined. During the **preparation**, the DR is configured according to the planning. First, CAD data must be converted into an AR-capable format. The AR scene must also be created. This includes positioning the virtual component for later correct visualization at the customer's desired location as well as creating animations to illustrate functions or visually highlight special parts. In addition, the customer must install AR Design Review software on his AR device. At the beginning of the DR **execution**, the expert uploads the AR scene to the platform and the customer synchronizes it to his device and starts the visualization. Then the expert and the customer can talk and discuss the component. The expert sees what the customer sees and can point out special aspects through animations and highlights. The customer must be able to give feedback on the virtual component, for example through visual annotations (arrows, circles, etc.) or through three-dimensionally positioned voice or text messages. The feedback must be documented on the platform so that the expert can use it to conduct an optimization of the component before manufacturing takes place.

3.2 Problems and Challenges

The objective is for the AM expert and the customer to be able to carry out the AR Design Review independently. However, it can be assumed that both have no special AR know-how, neither in the use nor in the programming of AR applications.

First problems arise in the planning phase. The objectives of the Design Review are influenced by the AM-specific properties of the component, which are to be communicated and illustrated to the customer. If, for example, the component has special material or surface properties, the aim is to visualize the optical design. A construction approach specifically adapted to a machine is more likely to involve checking the installation space. Accordingly, the requirements for the Design Review are very different. For example, a particularly detailed visualization is necessary to illustrate the design. In contrast, high-precision positioning is absolutely essential when checking the installation space. On the market, there are a large number of devices of different device categories, some of which differ greatly in their properties and functionalities. For example, AR can be realized via standard smartphones and tablets, however only via the screen. In order to achieve a three-dimensional perception in the real environment, glasses with binocular displays are necessary. Here again there are differences in the display resolution and the tracking of the device, which is necessary for a positionally accurate visualization. When positioning the components, a distinction is made between marker-based AR and dynamic positioning through software-based reconstruction of the environment. On the one hand there are different forms of positioning solutions and on the other hand there are compatibility problems between different software tools and the possible hardware. Finding an optimal choice of the hardware and the positioning solution is therefore not possible without AR know-how.

In the preparation phase, the biggest problem is the conversion of CAD data, which is a parametric representation of the geometry. AR formats, on the other hand, are polygon-based. A conversion is therefore highly complex and requires complex software which is additionally subject to a licence fee. Like the hardware and the software the requirements to the conversion are dependent on the DR goals. For example, the level of detail can be configured using a large number of parameters. The correct and efficient use of the conversion tools requires appropriate knowledge and experience. In some cases, common CAD tools allow manual conversion, but this is very time-consuming and work intensive. An automatic conversion is possible in part, but must be parameterized very precisely. The configuration of the AR scene, i.e. the positioning of the component and the creation of animations and highlights must currently be done manually, for example in the Unity 3D software, and requires programming knowledge.

The Design Review should be carried out collaboratively via the DigiKAM platform and enable the expert and the customer to communicate directly via audio and video. The collaboration should aim at a level where both participants can interact in the AR scene. The customer should be able to give feedback on the component and, for example, be able to add three-dimensional annotations anchored to the component in text or speech form as well as icons such as arrows or circles. At the same time, the expert must be able to start the animations in order to draw the customer's attention to special functionalities or to highlight special areas of the component. Depending on the device, different solutions are available for the interaction, each of which requires individual user interfaces. The entire Design Review including feedback and comments must be documented and stored on the platform so that the expert can modify the component accordingly. For minor adjustments, the expert must be able to provide a modified component as an AR scene while the DR is still running.

Overall, the Design Review involves problems that are difficult to solve without AR know-how. In addition, software is required that offers all the necessary functionalities and interfaces. In any case, the expert and the customer need information about the procedure and work steps of an AR Design Review.

3.3 Existing AR Design Review Tools and Software Solutions

There are already a number of software tools available on the market that enable collaborative viewing of CAD and 3D models. With the Thingworx platform [2],

PTC enables visualization via Augmented Reality directly from the in-house CAD tool Creo. However, this only enables positioning via marker tracking and local visualization via smartphones or tablets without the involvement of a second person. The software is also subject to a charge and restricted to PTC software. With Worksense [3], Daqri offers CAD visualization for the company's own AR eyewear, but no direct platform connection or collaboration. Re'flekt enables remote AR [4] to connect experts. Although the corresponding application is independent of operating systems and devices, it does not allow collaborative DR of 3D data. In addition, there are service providers on the market who implement individual solutions for customers, but do not offer a generally applicable platform-based solution. EngineeringPeople, for example, has developed a HoloLens application that can be used for Design Reviews of crane cab [5].

There are various other tools, but each of them only covers partial aspects of the AR Design Review process necessary for efficient collaboration. In research, collaborative AR is mainly researched in the field of education [6, 7]. There are also various papers on AR as a remote support for service activities such as maintenance and repairs [8, 9]. Multimodal interaction is also still an area of research, especially the effectiveness and intuitiveness of solutions [10, 11]. The topic of collaborative Augmented Reality Design Reviews for product validations as a whole cannot be found in research.

4 Support in the Preparation and Execution of Augmented Reality Design Reviews

Basically it can be said that Augmented Reality can be used to improve distributed collaboration in a value chain. At present, however, there is no software available that enables Augmented Reality Design Reviews to be embedded in a platform-based value creation process. Furthermore, due to the variety of possible AR devices and software and the complexity of the necessary tasks, it is not possible to plan, prepare and conduct AR Design Reviews without AR know-how for a specific application case.

For this reason, a set of instruments is necessary that supports the planning, preparation and execution of AR Design Reviews and makes them possible without AR know-how. This set of instruments is to be carried out using the value-added system of Additive Manufacturing as an example and embedded in the DigiKAM platform. The requirements for an AR Design Review depend on its objectives. For the development of the set of instruments different scenarios for AR Design Reviews are analyzed and defined, on the basis of which insights and solutions for the support during planning and preparation are expected. As part of the instruments, on the one hand tools and methods are to be developed that support planning and preparation, and on the other hand software that enables the actual execution. Additionally a comprehensive systematic is to be developed. This should guide the use of the tools and methods as well as the software and the collaboration of the expert and the customer throughout the entire Design Review process on the basis of a procedure model. In the following, the individual aspects and present findings are presented in more detail.

4.1 Augmented Reality Design Review Scenarios

When it comes to the visualization of three-dimensional construction data, AR will make sense in most cases. Nevertheless, one should always consider the effort of preparation and execution in comparison to the benefit and see whether in some cases it might be more efficient to review a component using a 2D sketch or 3D rendering. The corresponding data could easily be provided via the platform. Accordingly, there may be application scenarios in which an AR Design Review would not be useful. In cases where it is useful, the selection of devices and software is of crucial importance. Different device classes could be used to meet different requirements. The devices differ in price, functionalities and supported interaction options. Different positioning solutions can also be used to achieve different accuracies and different flexibility.

The goal of future research work is to define application scenarios that are linked to different objectives and thus different requirements. The scenarios can be used for planning and preparation, by assigning a given use case to a scenario and thus deriving the requirements and supporting the selection of hardware and software. Figure 3 shows two examples of scenarios, an ergonomics test (left) and a installation space check (right). An ergonomics test is about checking the ergonomic properties of a component, in this case a control device. Accordingly, the hands must be free and the customer must perceive the component three-dimensionally in order to get a realistic impression. This is why binocular AR glasses are necessary and freehand gesture control. The position of the virtual component must be dynamic and adapt to the position and orientation of the customer's hand. When checking the installation space, a more cost-effective tablet is sufficient to visualize the size of the component in the installation space. However, very precise positioning using a marker is necessary here. In this case, a touch user interface via the display serves as interaction.

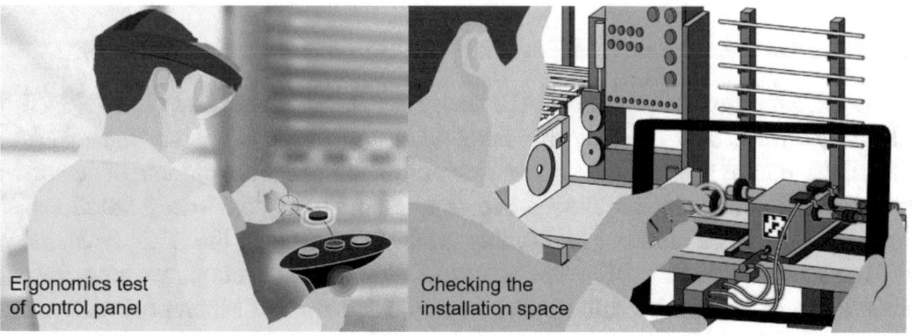

Fig. 3. Two possible AR design review scenarios.

4.2 Tools and Methods for the Planning and Preparation of AR Design Reviews

When planning and preparing an AR Design Review, various decisions have to be made that actually require AR know-how. Based on the identified scenarios and their analysis rules and conditions can be derived, which in turn can be used for the development of supporting software tools. Existing knowledge and experience as well as research results on existing technologies and solutions will be analyzed and formalized.

Based on this, a pre-defined digital catalogue of requirements for the planning of AR Design Reviews is targeted. Through a tool, the user is asked specific questions that directly or indirectly relate to the objectives of the Design Review. The characteristics of individual aspects can already be predefined, so that the user of the tool only has to select from a list of possible answers. The requirements are automatically derived from the answers and inputs of the user. For the selection of the device and the positioning solution, a selection tool is developed that generates suggestions and alternatives based on the requirements. By integrating aggregated knowledge into a tool, the user no longer needs specific AR know-how to plan an AR Design Review. In addition to the tools, the AM expert and the customer need instructions on how to use the tools. This is to be realized through a process model for the joint preparation of the Design Review, which will be provided interactively via the platform.

When preparing the Design Review, the conversion of the CAD data is crucial. Due to the high complexity of the conversion process, it is not the aim of the project to develop a new software solution integrated into the platform. Rather, the existing well-engineered tools should be used as a basis. What is needed for the preparation is experience knowledge and knowledge about the parameterization of the conversion. Since the properties of the AR model and thus the parameters of the conversion depend on the goals and requirements of the Design Review, a tool will be developed that utilizes the defined requirements and automatically suggests optimal parameters through additional specific questions. For this purpose, the conversion process must first be analyzed more precisely and rules and conditions be derived. The creation of animations and highlighting as part of the preparation is shifted into the integrated software solution, as this functionality can be used more flexibly and effortlessly.

4.3 Integrated Software Tool for Platform-Based AR Design Reviews

For the execution of collaborative AR Design Reviews, a software is developed that will be integrated into the platform functionality. This includes the transmission of audio and video as well as the shared viewing of 3D construction data. The software will run on different device classes and support different positioning solutions. The customer needs to be able to synchronize available models on the AR device and call the expert. In order to enable intuitive and efficient interaction, various forms of interaction will be conceptualized and implemented for the different device classes. This allows the expert to point out special aspects of the component and to execute animations. In contrast to the configuration of animations and highlighting in the preparation phase, these functionalities can then be used more dynamically according to requirements and the course of the Design Review. On the other hand, the customer is given the opportunity to give feedback on the component via interaction and to anchor it three-dimensionally to the component. In addition, the documentation of the feedback and the storage on the platform is foreseen. Initially, a prototypical implementation of this software to evaluate and illustrate the possibilities and benefits will be realized.

5 Conclusion

The ongoing research project DigiKAM is investigating to what extent digital collaboration can be simplified and made more efficient through the use of the innovative Augmented Reality technology. Based on the technological experience and the current project results, it can be said that there is an enormous potential to support product development processes through collaborative Augmented Reality Design Reviews and to enable a much faster and more meaningful validation of products and components through realistic visualizations at the future site of use. However, the necessary AR know-how for the planning, preparation and execution of AR Design Reviews is problematic. There is no ready-to-use solution for an AR Design Review within a platform-based value creation network and no known specific research intentions. The goal of this project is therefore a set of instruments in the form of tools and methods that enable experts and customers to conduct AR Design Reviews without AR know-how. The next steps are to identify and analyze application scenarios for AR Design Reviews in order to derive rules and conditions for the definition of requirements and the selection of AR hardware and software. In addition, the conversion process from CAD to AR-capable data will be examined in order to provide support for parameterization. In complement to automated tools and procedures, software for collaborative Design Reviews is currently being developed and prototypically implemented.

References

1. Gausemeier, J., Lewandowski, A., Siebe, A., Soetebeer, M.: OMEGA: object-oriented method strategic redesign of business processes. In: IiM 98—Changing the Ways We Work, Göteborg (1998)
2. PTC: Thingworx. https://www.ptc.com/de/products/iot (2018). Accessed 30 Oct 2018
3. Daqri: Worksense. https://daqri.com/worksense/ (2018). Accessed 30 Oct 2018
4. Re'flekt: Remote. https://www.re-flekt.com/de/rf-remote (2018). Accessed 30 Oct 2018
5. EngineeringPeople: Augmented Reality Design Review für die Kress Corporation. https://www.youtube.com/watch?v=ySQBmiD0YUc (2018). Accessed 30 Oct 2018
6. Phon, D.N.E., Ali, M.B., Halim, N.D.A.: Collaborative augmented reality in education: a review. In: International Conference on Teaching and Learning in Computing and Engineering, Kuching, pp. 78–83 (2014). https://doi.org/10.1109/latice.2014.23
7. Martín-Gutiérrez, J., Fabiani, P., Benesova, W., Meneses, M.D., Mora, C.E.: Augmented reality to promote collaborative and autonomous learning in higher education. Comput. Human Behav. Part B **51**, 752–776 (2015). https://doi.org/10.1016/j.chb.2014.11.093
8. Abramovici, M., Wolf, M., Adwernat, S., Neges, M.: Context-aware maintenance support for Augmented Reality assistance and synchronous multi-user collaboration. Proc. CIRP **59**, 18–22 (2017). ISSN 2212-8271. https://doi.org/10.1016/j.procir.2016.09.042
9. Mourtzis, D., Vlachou, A., Zogopoulos. V.: Cloud-based augmented reality remote maintenance through shop-floor monitoring: a product-service system approach. ASME J. Manuf. Sci. Eng. (2017). https://doi.org/10.1115/1.4035721

10. Chen, Z., Li, J., Hua, Y., Shen, R., Basu, A.: Multimodal interaction in augmented reality. In: IEEE International Conference on Systems, Man, and Cybernetics (SMC), Banff, AB, pp. 206–209 (2017). https://doi.org/10.1109/smc.2017.8122603
11. Nizam, S.S.M., Abidin, R.Z., Hashim, N.C., Lam, M.C., Arshad, H., Majid, N.A.A.: A review of multimodal interaction technique in augmented reality environment. Int. J. Adv. Sci. Eng. Inf. Technol. **8**(4–2). ISSN: 2088-5334. https://doi.org/10.18517/ijaseit.8.4-2.6824

Part III

Internet of Things and Industrial IoT

Work-in-Progress: 'ParKnow'—A System for Smart Parking Management

Medha Bharadwaj[1], S. Mahalakshmi[1], and Veena N. Hegde[2(✉)]

[1] Department of ISE, B.M.S. College of Engineering, Bangalore, Karnataka, India
medhavkrishna@gmail.com, mahalakshmi.ise@bmsce.ac.in

[2] Department of EIE, B.M.S. College of Engineering, Bangalore, Karnataka, India
veenahegdebms.intn@bmsce.ac.in

Abstract. Cities are changing because of movement of people out from the centre, increased suburbanization and desire for spacious residences and job opportunities. This change has created an impact on journey-to-work by choosing a public mode of transport system to convenient to travel, affordable travel by car. There is a demand for parking spaces as a part of space management. It is found that lack of knowledge of motorists regarding valid parking spaces cause increased traffic congestion. As we know, awareness of any system is necessary for efficient utilization of that system, the knowledge of infrastructure availability of parking avenues is first step towards facilitating parking space. The paper presents a solution to the driver, which is capable of analyzing the availability of vacant slots in a given parking space based on digital image processing techniques combined with an Internet of Things (IoT) technology. Further analytics of the solution are used to create interfaces that can make its users aware of the vacant parking slots in and around their area of travel.

Keywords: Image processing · ThingSpeak · Smart parking · IoT

1 Introduction

1.1 Overview of Traffic in Metropolitan Cities

A survey of automated solutions for parking management, 30% of cars circling a city at any time do so as the drivers are searching for a parking vacancy [1]. This may seem inconsequential but actually this adds to the traffic congestion and is a very significant reason for overcrowded roads. The slow moving vehicles in search of parking vacancies decrease the mobility of vehicles in the road and cause long chains of traffic congestions in the connected road network. According to Dr. Jean-Paul Rodrigues of Hofstra University's Department of Global Studies and Geography in central areas of large cities, cruising accounts for more than 10% of the local circulation as drivers spend minimum of 20 min looking for a parking spot. Study from environmental standpoint this contributes to incalculable amounts of wasted fuel and hazardous carbon emission.

M. E. Auer and K. Ram B. (Eds.): REV2019 2019, LNNS 80, pp. 257–268, 2020.
https://doi.org/10.1007/978-3-030-23162-0_23

On one side cruising in search of parking places causes traffic congestion and on the other parking vehicles in unauthorized parts of the road due to unavailability of parking adds to the woes of commuters as stagnant vehicles parked wrongly on the roads are more difficult to maneuver through than cruising vehicles. Keeping in mind the above details this paper aims to address the issues caused by cruising vehicles in search of parking spaces based on the principle that "No system succeeds without the users being aware of it". The existing system to indicate the validity of parking spaces is by using sign boards. But this system is highly inefficient. Hence there is a severe need for a robust system capable of keeping motorists informed of valid parking spaces available near their places of travel. Even if a robust system is designed to locate valid parking places in city the next issue in line is how to ensure uniform distribution of vehicles in these parking lots. If uniform distribution is not ensured it again results in invalid parking of vehicles or cruising of vehicles in search of parking places. Hence the need of the hour is a well-designed technology friendly way to ensure that motorists are aware of valid parking places in and around their area of travel and also know if there are vacant slots in the respective parking places. This way the motorists can decide their route in advance. Based on the real time availability of parking spaces higher percentage of motorists finding a vacant parking slot.

Several technological means have been used in developed countries to achieve the above. Countries like United States have taken the help of the trending mobile and cloud technology to design applications capable of informing its users about the existing parking places and their real time availability. But no such successful efforts have been made in India. The work in this project aims to provide a solution capable of analyzing the number of vacant parking slots available in a parking space, so that this data can be used to guide motorists to an assured parking slot in and around their area of travel. There have been many technologies developed in past decade to find out ease mechanisms for traffic management and vehicular management. The image based systems can be classified into two categories while detecting the car parking. One is car driven and another is space driven, as explained by Hang and Wang, 2010 [1]. In this work algorithms are developed to detect cars, which are object of interest. There are other algorithms developed which detect, presence of car as an object [2]. But the object identification alone for car detection suffers from perspective distortion, as car image taken from a distance may look that it occupies a small space. It is found from the review that the image acquisition of cars considered for the processing, depends on position of the camera [3]. In algorithms where the analysis is based on empty spaces detection, that assumes the background of the space is stationary and its range of pixel values remain the same [4].

The image acquisition, irrespective of the space or the presence of car detection, is susceptible to season changes. The images acquired during summer and rainy season for the same parking slots, vary in their pixel values due to visibility [5]. For that the robust filters such as Gaussian filters have been used [6]. It has been observed that, though there have been enough research and development and classification of parking slots based on presence and absence object using SVM and other classification algorithms, collections of images and creating a data set to try different techniques has been a concern. Hang, in their paper have provided the method of collection of images for processing also have given enough image repository. Other researchers can use this baseline database as reference when testing their own algorithms on this dataset. Karuna Moorthy et al. [3] brought out a new algorithm for car parking using an integrated ANN with the image to count the number of cars. The researchers designed the parking area with four parking lots to be parked in the specified slots. The images of various combinations that are possible for vehicles to be parked were stored in advance in a database. Whenever an input image is fed to the system, the system compares the fed images with the database and classifies the image as valid or not. As an alternate approach in [4] RF transmitter was installed to the number plate of registered users, giving any user a unique id such that whenever the registered users entered the parking lot. Unlike NN based system, this system does not need training; however user has to register himself/herself to the automated system. Sayanti Banerjee et al. [5] came up with a car parking system in which a camera was installed at the entry point and the sequence of images produced is processed at every 2 s. These images were compared with the reference images, Gamma correction was used to enhance the image depending upon the brightness in the image [4]. Prewitt edge detection operator was used for feature extraction. The prototype proposed was capable of having 20 cars. The disadvantage of this method is cars having similar geometry and color can be easily confused by the processor as cars manufactured by a same company are identical. The template used for identify the car was number plate stored priory in the database. This is more effective as far as the identification of the slot and cars are concerned, but the system is costlier and only one car can be parked at a time.

Though there have been several attempts to address the issue of identifying the valid parking slot while at the same time, designing an effective system that gives the feedback of the parking space surroundings to the user is rare. The technique needs to be integration of video as well as image processing to produce more reliable result. The autonomous device has to be easy to use, lightweight and the user should learn how to use this device with minimal training. Auditory feedback systems would hamper the impaired people, and visualization is one of the senses that many can rely most upon. The system designed should not mistake it for an object in the parking slot and should also be able to detect obstacles at the floor level and overhanging obstacles. The paper is arranged in different sections. Methodology has been provided in Sect. 2. Section 3 gives the algorithm used Sect. 3 provides the result and paper is concluded in Sect. 4.

2 Methodology

2.1 Proposed Block Schematic

The overall methodology for building the parking management system has been represented in the subsequent sections.

2.2 Stages of Implementation

The overall block diagram for implementation is shown in Fig. 1. As presented in the block diagram the image acquired of parking space is analysed for empty and filled slots using image processing algorithms. The chosen image processing algorithms are

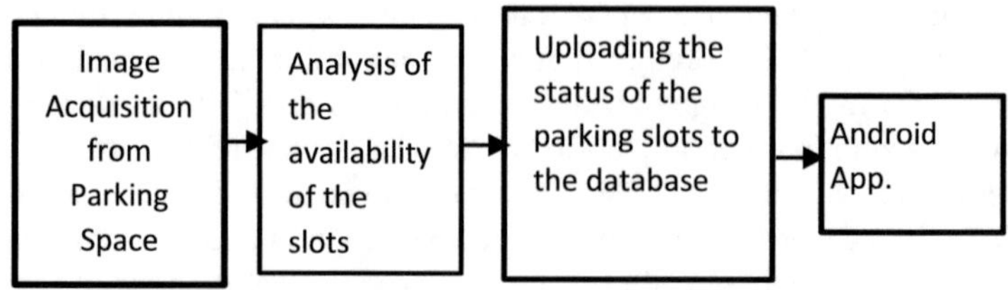

Fig. 1. Overall block diagram

experimented for maximum efficiency and robustness. The module is trained with exhaustive datasets available such that it is made ready with best accuracy for any test image input. The analysis of the parking space for empty slots every screening is uploaded in the central database. That also helps in creating the inventory of database for developing periodical trend analysis of utilization of space for parking. The status of the parking space will be made available on the cell phone of the driver.

Stage I: The proposed method was experimented by building a prototype of sensory section, communication section and a cloud section as shown in Fig. 3.

Figure 2 represents the cycle of periodical processing of image of the parking space and sequence of operations that take place while processing the frames. As indicated, every time a new image frame of parking slot is captured after receiving an acknowledgement from the central database to the processor for completion of information transmit operation. A prototype of the design procedure for image processing

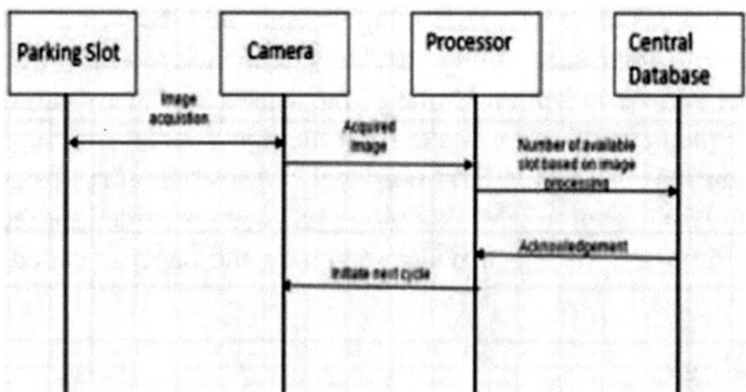

Fig. 2. Cycle of processing the image of parking space

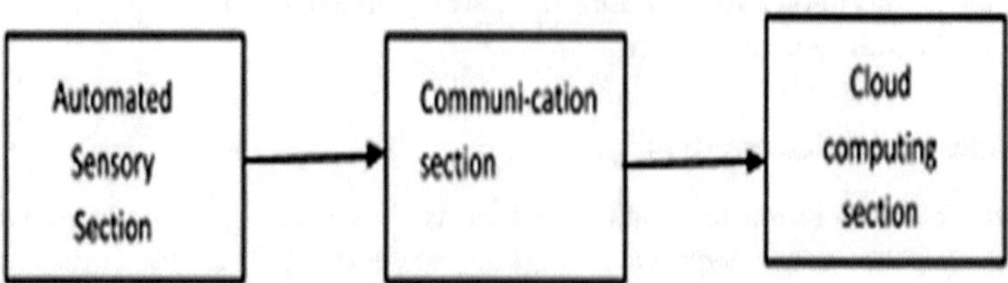

Fig. 3. Prototype block diagram

Detailed design

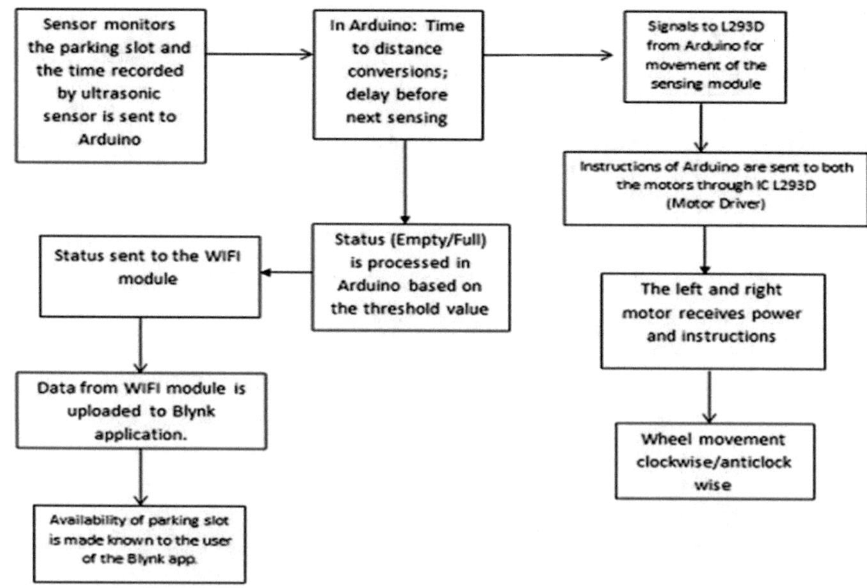

Fig. 4. Detailed design of proposed implementation

was tested with the help of simple Arduino board with a ultrasonic sensor and motors as shown in Fig. 4. The block diagram indicates the hardware circuit built to test the working of sensing, commutation and cloud section. The same design procedure was extended with image processing techniques later as shown in Fig. 5.

Stage II: The stage II implementation is proposed with image acquisition in sensor section, Image processing and shape recognition in processing section and ThingSpeak implementation in communication and cloud section.

Figure 5 represents the block schematic of data processing and data analytics being transferred to cloud. The process involves processing of image frames for slot identification in which the presence of car indicates the occupancy of the parking space. The non-occupancy of the parking space will be analysed to be empty space. Identification of the empty spaces is done using image processing techniques, in order to have count of number of empty spaces for parking. The number of parking slots and their location in a given parking space will be communicated to the android application of the driver through cloud. The flowchart (Fig. 9) gives the steps followed to extract the features for classification.

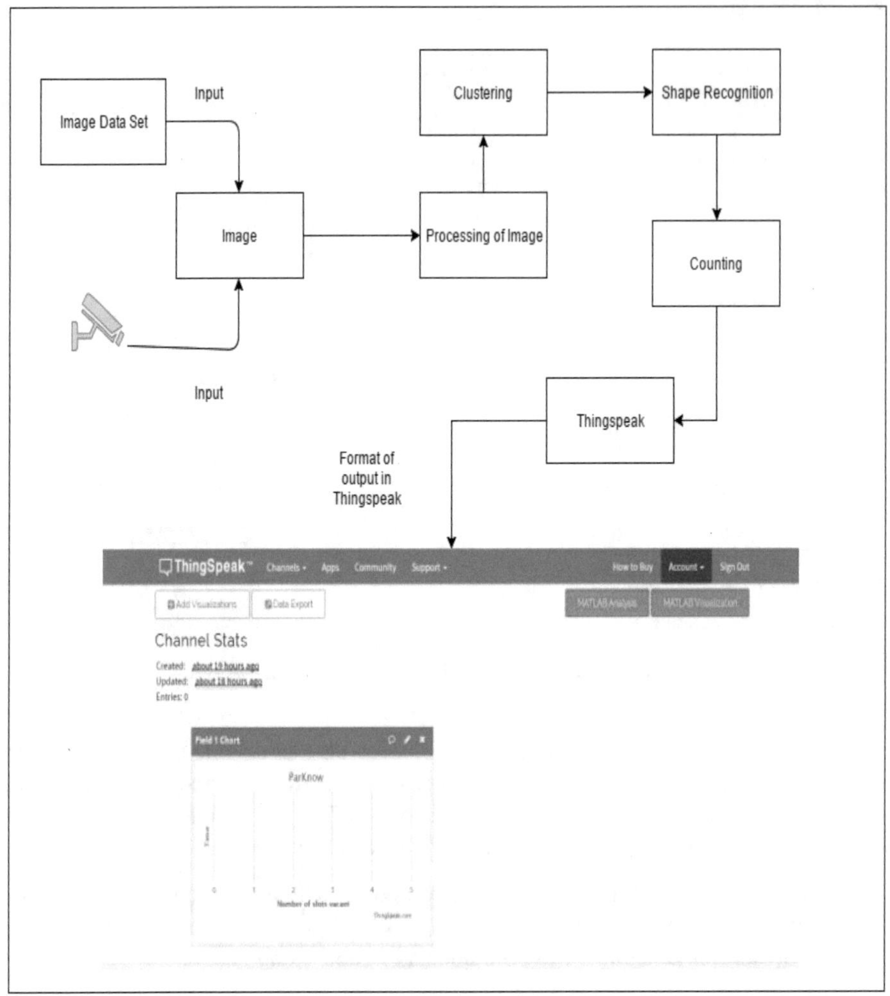

Fig. 5. The schematic representation of data processing and to cloud

2.3 Algorithms

The main two algorithms used for the work are for object detection and classification.

The acquired image of the parking lot goes through, object feature identification and classification. The shape estimation algorithm is used to extract the car feature and Vector Quantization is carried out using Linde Buzo and Grey (LBG) algorithm for classification. The flow charts of the two algorithms are shown in Figs. 6 and 7.

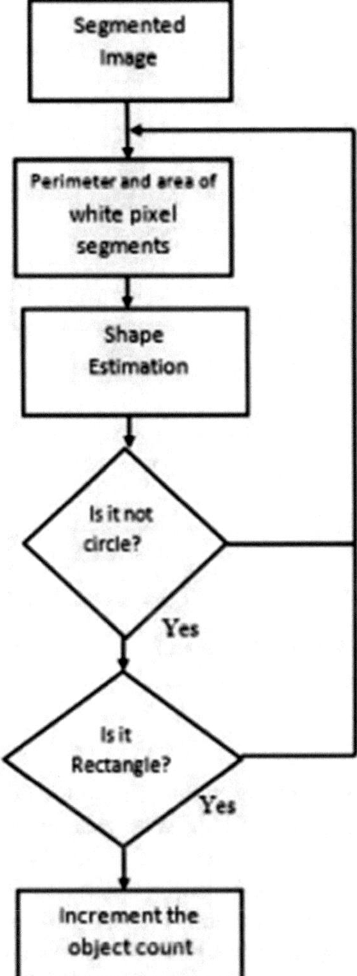

Fig. 6. Feature extraction

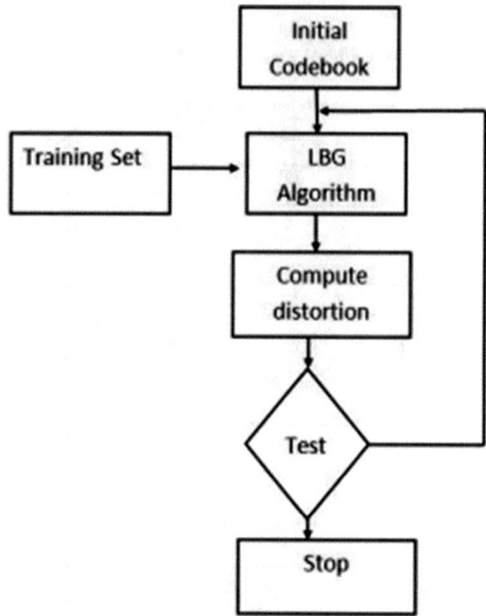

Fig. 7. Algorithm for classification

3 Results

The proposed method was tested by building a prototype using Aurdino board and Ultrasonic sensors for sensory section. Wifi was used for communication with ThinkSpeak for cloud computing. The count of vacant slots is uploaded to a channel in ThingSpeak that can save the count based on timestamp. ThingSpeak is open Source software for Iota that allows storage and retrieval of data and also provides several functions and API's for Analytics. The count of vacant slot against timestamp is plotted in the Channel. This channel can be constantly updated in real time. Data from the ThingSpeak channel can be used by any user interface applications to make its users.

Algorithm running for identification of an empty slot in real time operation was captured and Fig. 8 provides the moving prototype searching for empty slot.

Figure 9 shows two cases, when a car is partially or completely inside a slot, then the LED Widget is ON in the Blynk application implying that the slot is full (Case 1), When a car is a the edge of a slot as LED Widget is ON in the Blynk application implying that the slot is full (Case 2). The same was extended to Image processing technique where images were captured to identify the empty, full and outside the vehicle space when parked using clustering and classification algorithms. There were two sets of image data collected, one for training and another set for testing. Figures 10, 11 and 12 show the results of classification of parking slots with/without cars.

Fig. 8. Prototype that senses the slot state and communicate to cloud aware if the number of parking slots vacant in a parking space at any given time

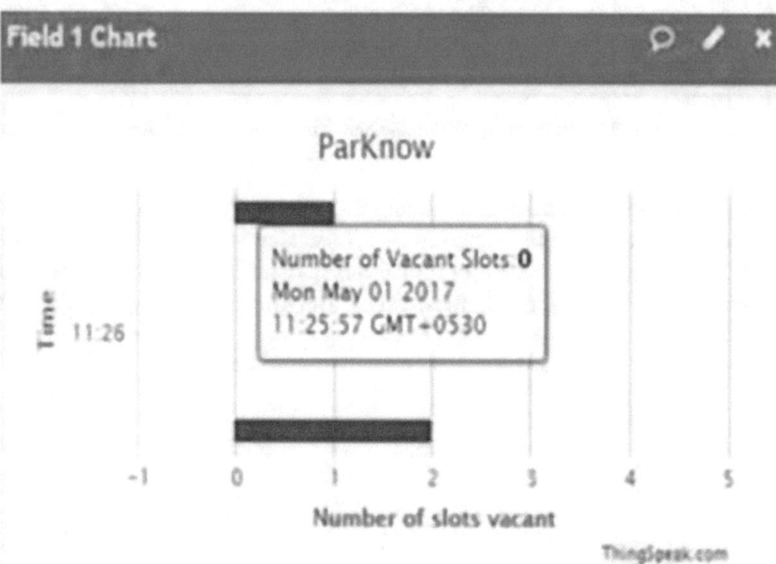

Fig. 9. Communication to ThingSpeak via Wi-Fi

Case 2: When a car is at the edge of the slot

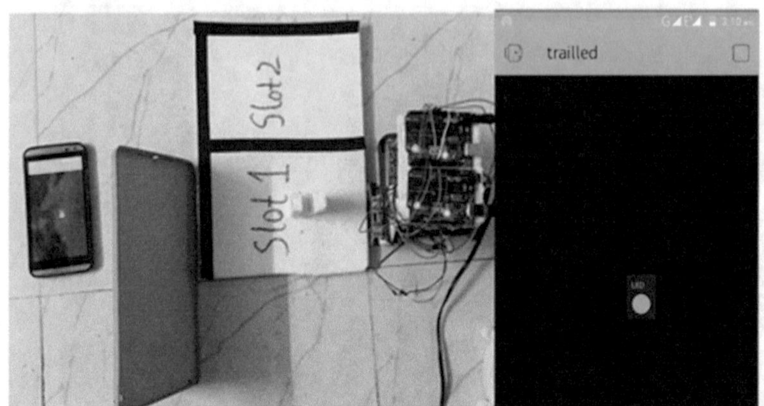

Case 1: When a car is inside a slot completely or partially

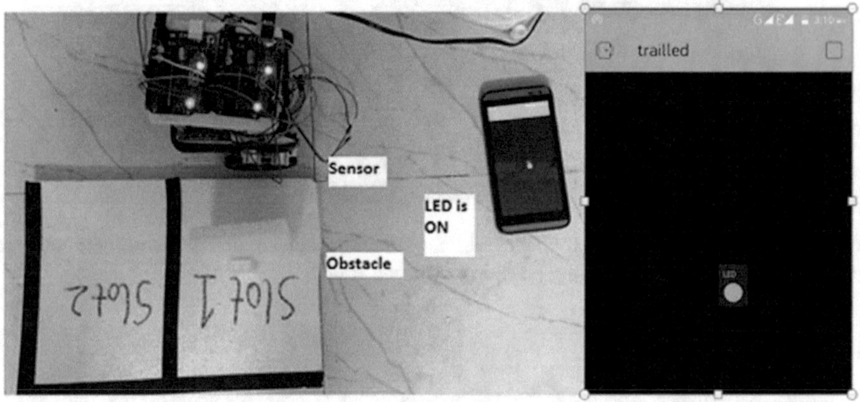

Fig. 10. Parking space image with different combinations

Table 1 provides the classification using the Vector Quantization (VQ) classification method as explained in Flowchart in Fig. 7. As shown in Figs. 10 and 11 the slots indicate that the objects have been occupied the space and their analytics are obtained for user using an IoT analytics platform called ThinkSpeak as shown in Fig. 12.

Fig. 11. Object recognition and clustering. **a** Original image. **b** Cluster 1 car present. **c** Cluster 2 outside space of car

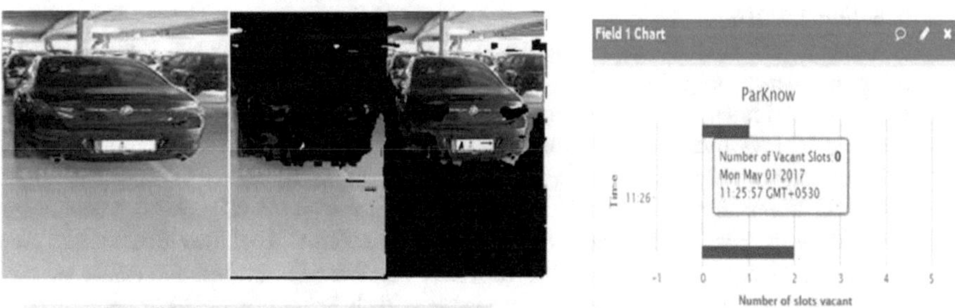

Fig. 12. Object detection through clustering and communication to cloud

Table 1. Classification

Class	Indicators	Interpretation
1	The occupied slots	No. of cars available in parking
2	The occupied slots are cars	No. of free slots available in the parking space
3	The surroundings	The space without car slots marked

4 Conclusion

The proposed work in this paper, "ParKnow "has been successfully implemented with the tested data samples of images. The algorithms proposed for image clustering and classification using Vector Quantization has worked effectively in identifying the valid parking slots for cars and also is able to provide the number of occupied parking slots in a structured parking space. The classifier is able to provide a good accuracy up to 90% in most of the cases. Further the algorithms of classification and segmented image processing to extract the numerical details of the parking slots, have been experimented on different seasonal images collected, such as Rainy, cloudy, sunny etc. It is proved that, the simplest K-Means classifier, using Vector quantization can be applied for the classification purpose. The classifier results also have been tested with single car, parked in a particular slot, by taking images from different directions and hence has proved robust. The analytics derived out of the interpretations of the segmented images have been successfully represented using Think Space on cloud, which can be made readily available for IoT based vehicular technology.

References

1. Gonzalez, R.C., Woods, R.E.: Digital Image Processing. Addison-Wiley Publishing (1993). ISBN 0-201-50803-6
2. Ahrnbom, M., Åström, K., Nilsson, M.: Fast classification of empty and occupied parking spaces using integral channel features. In: 2016 IEEE Conference on Computer Vision and Pattern Recognition Workshops (CVPRW) (2016), Las Vegas, NV, United States, June 26, 2016–July 1, 2016
3. Chowdhury, Md. T. H., Zahir, E.: Automotive parking lot and theft detection through image processing. Am. J. Eng. Res. **2**(10)
4. Scalise, L., Primiani, V. L., Russo, P., Shahu, D., Mattia, V.D., Leo, A.D., Cerri, G.: Experimental investigation of electromagnetic obstacle detection for visually impaired users: a comparison with ultrasonic sensing. IEEE Trans. Instrum. Meas. **61**(11) (2012 Nov)
5. Shoval, S., Borenstein, J., Koren, Y.: Auditory guidance with the Navbelt—a computerized travel aid for the blind. IEEE Trans. Syst. Man. Cybern. Part C Appl. Rev. **28**(3) (1998 Aug)
6. Strakowski, M.R., Kosmowski, B.B., Kowalik, R., Wierzba, P.: An ultrasonic obstacle detector based on phase beam forming principles. IEEE Sens. J. **6**(1) (2006 Feb)
7. Arora, S., Acharya, J., Verma, A., et al.: Multilevel thresholding for image segmentation through a fast statistical recursive algorithm. Pattern Recogn. Lett. **29**, 119–125 (2008). Elsevier 2007
8. Wang, L., Bai, J.: Threshold clustering by grey levels of boundary. Pattern Recogn. Lett. **24**, 1983–1999. Elsevier Science
9. Yusnita, R., Norbaya, F., et al.: Intelligent parking space detection system based on image processing. Int. J. Innov. Manag. Technol. **3**(3) (2012 June)
10. Linde, Y., Buzo, A., Gray, R.M.: An algorithm for vector quantizer design. IEEE Trans. Commun. **COM-28**(1) (1980 Jan)
11. Gray, R.M.: Vector quantization. IEEE ASSP Mag. pp. 4–29 (1984 Apr)

IoT Remote Laboratory Based on ARM Device Extension of VISIR Remote Laboratories to Include IoT Support

Pablo Baizan, Alejandro Macho, Manuel Blazquez, Felix Garcia-Loro, Clara Perez, Gabriel Diaz, Elio Sancristobal, Rosario Gil, and Manuel Castro(✉)

Spanish University for Distance Education (UNED), Madrid, Spain
{pbaizan, amacho, mblazquez, fgarcialoro, clarapm, gdiaz, elio, rgil, mcastro}@ieec.uned.es

Abstract. This paper presents an extension of the VISIR Remote Laboratory to support the IoT Technologies. By means of this new extension, the users of the remote laboratory can experiment by programming an ARM device (a raspberry Pi) and interact with sensors and actuators. Furthermore, all the characteristics of VISIR are still preserved. Therefore, the user can wire in the breadboard their circuits including sensors and actuators and interact with the instrumentation. On the other hand, a Raspberry Pi OS (Raspbian) complete remote desktop is available as an instrument more of VISIR.

Keywords: VISIR · Remote laboratory · Internet of Things · Raspberry pi · ARM · IoT · Python · Virtual instrumentation · Virtual box

1 Introduction

The Internet of Things (IoT) is an emerging technology, which is increasingly present in many sectors, such as industry automation, home appliances, mobile devices, agriculture, medicine, security, etc. Although these are some examples of sectors where IoT is present, new applications for this technology appear every day (as a rule, everything that can be actuated, sensing and/or integrated on the Internet). The rise of this technology implies an increase in the demand of qualified professionals. Therefore, the educational institutions must incorporate in their areas of knowledge a learning of the technologies of IoT, if they do not want to remain obsolete.

For students to gain the skills of a professional in an efficient manner, it is necessary to consolidate knowledge by practicing. To this end, educational institutions must have a practice laboratory with the necessary instrumentation and equipment. Due to the high cost of this equipment and instrumentation, in many cases, the institutions cannot acquire as many as they are needed. This implies that students must share equipment and instrumentation during the practice session. On the other hand, these practice sessions are usually carried out in the buildings of the educational institution at a certain timetable, forcing the students to move and adapt to the timetables.

M. E. Auer and K. Ram B. (Eds.): REV2019 2019, LNNS 80, pp. 269–279, 2020.
https://doi.org/10.1007/978-3-030-23162-0_24

Thanks to the advances in Internet technologies (where IoT is included), a new type of laboratory, the Remote Laboratory, has appeared. In these, all the instrumentation and equipment are connected to the Internet, presenting many advantages, among the main ones are:

1. Availability of 24 h and 365 days, the laboratory is not limited to the hours of the institution.
2. Accessible from a device with an Internet connection, avoiding travel to the institution.
3. Better support for people with disabilities or sick, thanks to the possibility of connection when and where they want (whenever they have compatible device and with Internet connection).
4. Lower costs, more efficient use of resources, opportunity to share access and resources with other institutions.
5. Scalable, it is easy to increase the number of practices, equipment, instrumentation and laboratories.
6. Robust, the laboratory is always connected, if a fault occurs the experiment can be transferred to another remote laboratory automatically.

This scenario has increased the interest in the use of a remote laboratory as an alternative to on-site laboratories by educational institutions (especially those related to science, engineering and Technology). An interesting remote laboratory project related to electronic engineering is the Virtual Instruments System In Reality VISIR [1]. The VISIR remote laboratory project began as a research project at the Blekinge Institute of Technology, Sweden by Gustavsson and others [1] in 1999. Today, VISIR has the support of a consortium of universities in Sweden, Spain, Portugal, Austria, Georgia and India. VISIR is one of the most popular analog electronic remote laboratories and possibly also one of the most replicated. There are numerous research projects on new educational technologies and remote electronic labs based on VISIR [2–5]. These characteristics make the VISIR a project base ideal for the development of a laboratory of remote IoT.

Many batch projects are based on the embedded arm, such as [6–8], these devices being the most used for the development of batch technologies. Of the ARM devices analyzed for the development of the remote laboratory of IoT, highlights the Raspberry PI for various reasons:

1. Many IoT projects include raspberry PI, such as [6, 9, 10].
2. The raspberry PI has been developed to promote the teaching of electronic and informatics. So, it includes a very good learning support [8].
3. There are a lot of examples of open source code applications and increasing every day.
4. The raspberry pi includes numerous interconnectivity ports, such as USB, Ethernet, I2C, SPI, GPIOS, and the last versions WIFI and Bluetooth.
5. It is a low-cost environment, which makes it easy to replicate. Also, on the market there are many development and expansion boards also low-cost.

This paper describes research around an IoT ARM embedded devices remote laboratory based on VISIR. Section II provides a general description of VISIR remote laboratory. Section III describes the requirements and problems to be faced when implementing an extension IoT for VISIR. The Section IV analyzes the architecture used to implement the remote laboratory of IoT.

2 VISIR, Features and Architecture

This section provides a brief description of the characteristics and architecture of VISIR, which is important because it is important for a better understanding of the following sections. VISIR is a remote laboratory for wiring and measuring electronics circuits, the user designs an electronic circuit by web browser on a virtual breadboard, sends the circuit to the server and once verified, it is built by a relay switching matrix board (RSM) and the measurements are returned to the user. The instrumentation is based on National Instrument (NI), PCI eXtensions for Instrumentation (PXI), this include an embedded PC controller board and following instrument boards: Digital multi-meter (DMM), DC power supply (PS), Digital oscilloscope (DO) and Analog function generator (FGENA). The instrumentations are connected to RSM and this communicates with the controller by USB.

The RSM is a stack of "PCI/104" sized boards, the circuits are built by opening and closing relays with the received circuit design. The instrumentation and the electronic components necessary for the construction of the circuit are connected to one of the 10 nodes available (A–I) of the RSM (Figs. 1 and 2) by means of relays of the components and instruments boards.

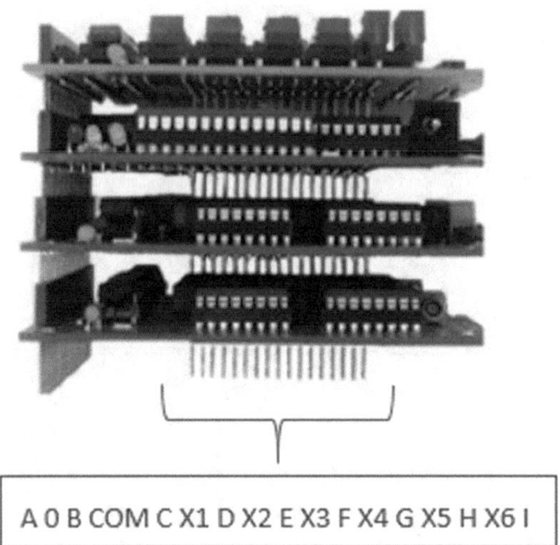

Fig. 1. RSM available nodes

Fig. 2. VISIR instruments and RSM boards assembly

The RSM form at least three instrument boards and can include up to sixteen components boards (see Fig. 2). The component boards contain 10 sockets connected to a Double-Pole Single-Throw (DPST) relays (4 of these sockets can be connected instead to 2 Single-Pole Single-Throw (SPST) relays).

The DO and DMM channels can be connected to any node depending on the design, instead, the channels FGENA, +6 V, +25 V, −25 V and AUX are connected to A, X1, X2, X3 and X4 nodes respectively (Fig. 3).

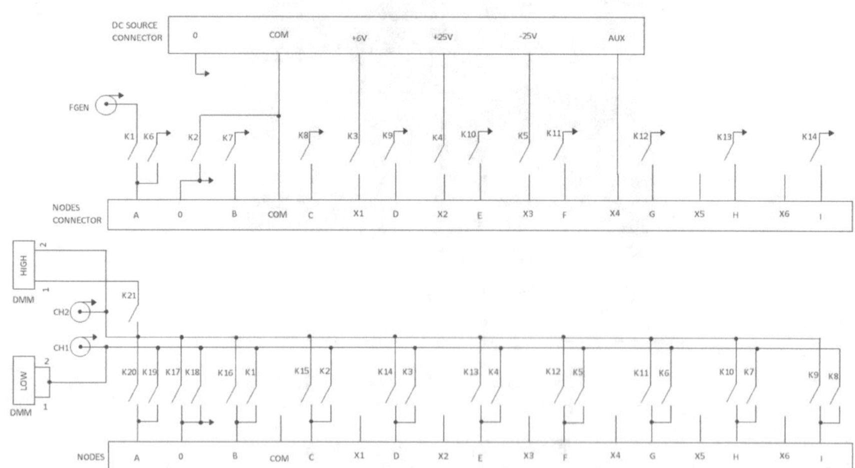

Fig. 3. VISIR instrumentation board schematic

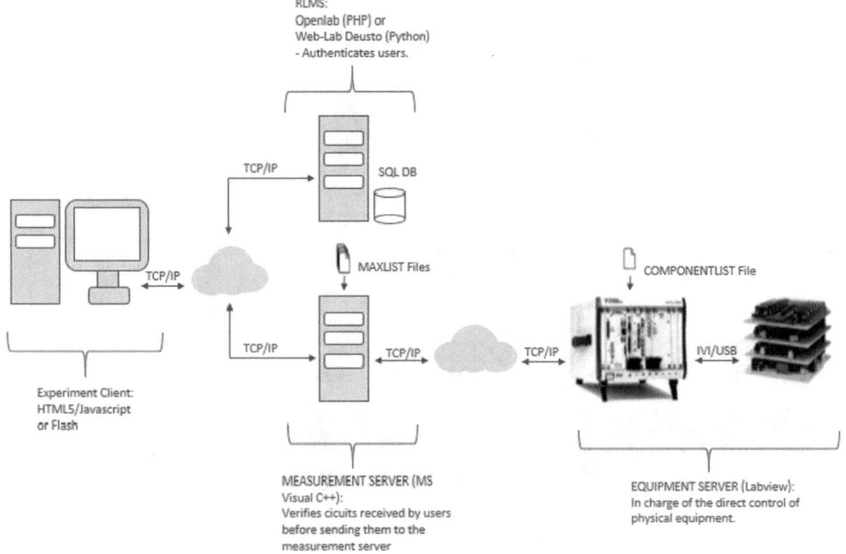

Fig. 4. VISIR architecture

VISIR include an open source software under the GPL license. In the current version its architecture (Fig. 4) consists of four independent components, which are:

1. Experiment client.
2. Measurement server.
3. Equipment server.
4. Remote Laboratory Management Systems (RLMS).

2.1 Experiment Client

The Experiment client is a JavaScript embedded code into html5 user interface that simulate an analog electronic experiment laboratory. The user wires the circuit on a simulated breadboard by clicking and dragging the components and wires, as shown in Fig. 5. Once the circuit has been designed, the user can select the virtual instrument with which to measure and configure it. The interfaces simulate real instruments that the user could handle in an on-site laboratory (Fig. 5) these are the following:

1. A HP 33120A Analog Function Generator.
2. A Fluke 23 Digital Multimeter.
3. An Agilent 54622A Digital Oscilloscope.
4. A E3631A DC power supply.

When the user has finished configuring the instrumentation can press the “perform experiment” button to perform the measurement on the physical equipment. The Experiment client send the experiment to Equipment server by XML protocol [10]. The XML protocol is also used to receive the results of the experiment.

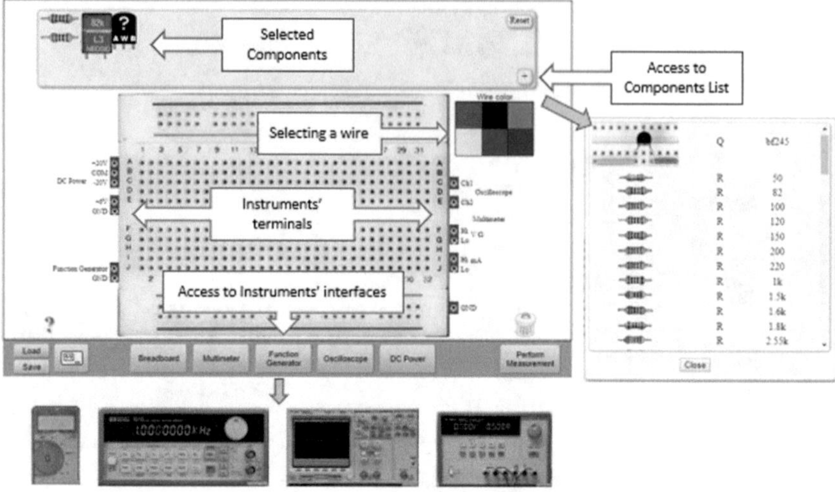

Fig. 5. VISIR client interface

2.2 Measurement Server

The measuring server is the VISIR engine, it is a software application developed in Microsoft Visual C++ and can be run on a separate server or in the same as the rest of the VISIR software. The main functions of the measurement server are as follows:

1. Ease of use, to this contributes besides other things: automatic memory management, simple reading and writing of operations or being a dynamic typed language. Reducing programming times versus other languages.
2. XML Server, which receives the XML requests from the experimental client, parses and returns the XML response to the client.
3. Authentication Module, verifies the credentials and permissions of the users, assigns them a single session key and the valid one in each request of the client.
4. EqCom, its functions are, communicating via TCP/IP with the Equipment Server, sending the circuit to build, configuring the instrumentation and receiving the answers from this. All this through a protocol defined for that purpose [11].

2.3 Equipment Server

The Equipment Server software is a LABVIEW development. This application controls the instrumentation and matrix switching board. The Equipment Server receives the validated experiment from Measurement server by TCP/IP, executes it and returns the results.

2.4 RLMS

The RLMS provides the initial and config web pages, user authentication and authorization methods, management of access to the different laboratories and a database (used to store users, circuits and laboratory information). There are several options of RMLS, the original of VISIR is OpenLabsweb [12] and is programmed in the PHP language. Another interesting RMLS alternative option is Weblab-Deusto [13].

3 IOT Remote Laboratory

For the experiments of an IoT remote laboratory is interesting the concepts of VISIR exposed in section II. Where the user can drag and drop sensors, manipulators and communication devices to a breadboard and wire them virtually. Another interesting option of VISIR for an IoT remote laboratory is to be able to interact with instrumentation. This way the user can validate if the reading of the programmable device corresponds to the value of the sensor, can generate signals identical to those of a sensor to validate a software, check if a manipulator is active, etc.

Therefore, it has been considered interesting to take as the base of a remote VISIR laboratory. But adding access to the programming of an ARM device as if it were an additional instrument. As has been advanced in the introduction, the ARM device chosen is the Raspberry PI. The Raspberry PI Foundation has created a Debian-specific distribution for this device called Raspbian. This operating system includes an integrated development and learning environment (IDLE) for Python. Idle is destined to be an integrated development environment (IDE) simple and suitable for beginners, especially in an educational environment.

But Raspbian presents two major problems as a remote laboratory development environment:

1. There is a need for strong access policies, there is a great danger that users, either intentionally or by accident, can leave the system inoperable.
2. The VISIR Multiplex (where several users can perform their experiments at the same time) is very complex to implement, in addition, the greater the number of users who access at the same time to the Raspberry PI the performance of the same decreases considerably.

Following the principles of VISIR, the solution to these problems is in the virtualization of the operating system. In the next section, the architecture implemented to achieve this is exposed in greater detail.

4 Implementation

The programming language chosen for the IoT arm devices remote lab implementation has been Python. This language is ideal compared to others thanks to its multi-platform support, this is a great advantage because the code can be used on different types of servers and is not limited to a single operating system. Other advantages that this language offers are:

1. Ease of use, to this contributes besides other things: automatic memory management, simple reading and writing of operations or being a dynamic typed language. Reducing programming times versus other languages.
2. Readability of the code, the structure of the code promotes a way of writing that facilitates its reading later. This gives a great advantage to the maintenance or extensions of the code, that allows to incorporate new technologies to our laboratory in a simple way and to avoid that it becomes obsolete.
3. Abundance of libraries. There are many libraries available to extend the basic Python functionality to any field. For example, SQLAlchemy, bottle, PYPDF2, etc.

As the main disadvantage of this its Runtime versus other languages, an identical code in Python and C runs in less time in the latter. This disadvantage is due to two of the advantages of Python indicated above, language typed dynamically and interpreted (the latter favors the multiplatform). Therefore, this disadvantage and more considering the continuous advance of the processors every day faster does not limit the use of Python for remote labs.

The architecture of the remote laboratory Fig. 6 is based on the VISIR but has been rewritten the entire measurement server in Python. The biggest problem that this type of labs presents is to protect the operating system of the ARM device, as already mentioned. To do so, it has decided to virtualize the OS Raspbian in Virtual Box and make use of "snapshot" of this, allowing us to freeze in a default state the OS. Because Raspbian is a Debian-based distribution, there is a version of the first for devices other

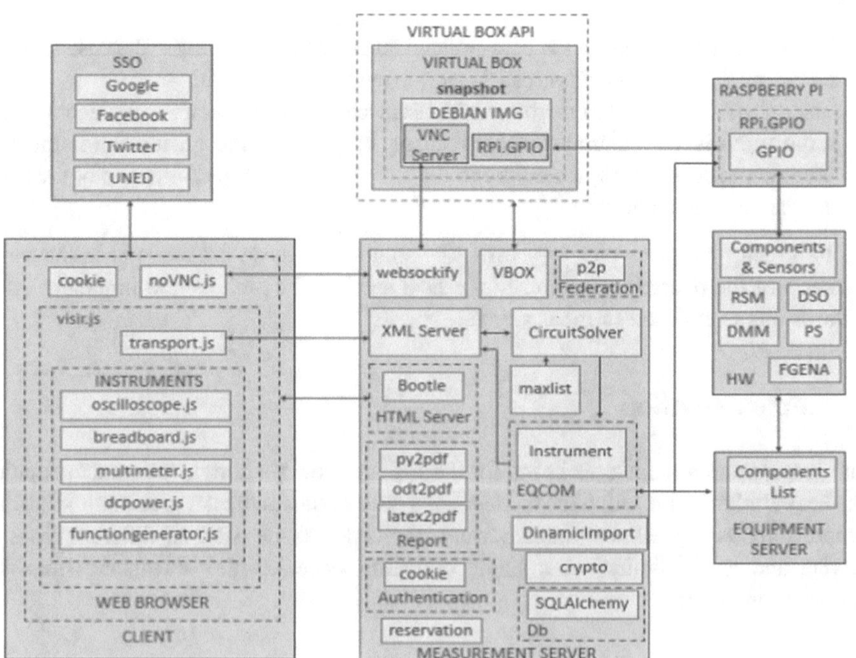

Fig. 6. IoT remote laboratory architecture

than the Raspberry Pi (Raspbian desktop). This Raspbian desktop is identical to the raspberry, but does not count as it is logical, with access to the GPIOs. To fix this, the RPI library is modified. GPIO in such a way that it is accessed virtually by an Ethernet socket to the GPIOs of a Raspberry Pi.

On the side, the Raspberry Pi runs an application that includes a socket server. This server manages the queue of requests to access the GPIOs (allows several users to access the GPIOs with little latency in the replies) and making changes in the GPIOs at the request of the libraries of the Raspbian Desktop.

To manage the virtual box (manage, destroy, boot images of OS, etc.) is used the Python library that communicates with this through the Virtual Box API. On the other hand, Raspbian images run a VNC server (Virtual Computing network) allowing access to the desktop remotely.

As the client must access through his web browser, on the client side is the remote desktop of the raspberry PI as if it were one more instrument (Fig. 7). This is done thanks to the JavaScript library NoVNC.js on the client side and to the library in Python websockify [14].

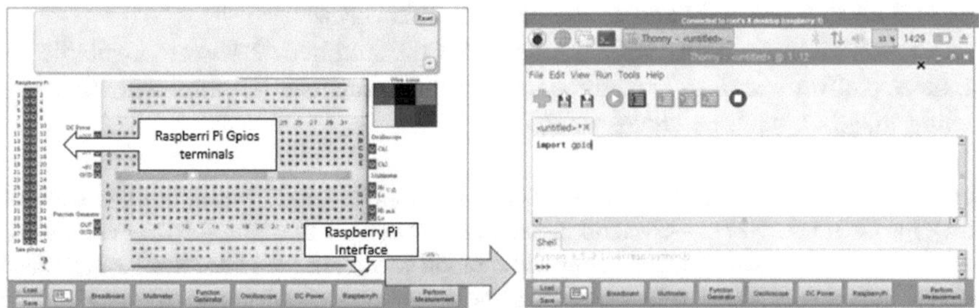

Fig. 7. New client interface

To build the circuits and manage the instrumentation connected to the GPIOs of the raspberry PI, our measurement server follows the architecture of VISIR. To do this, a XML server has been written (which processes the requests of the clients) a circuit solver module (that does the functions of virtual tutor, validating the circuits) and the EQCom (in charge of communicating with the Equipment Server and with the Raspberry Pi, helping the latter with the management of the GPIOs). This XML server is listening to client requests and creates a thread per request (the maximum number of sessions to handle can be configured). Parses these requests to obtain the circuit and the configuration of the instrumentation and pass this data to the circuit solver. This module, circuit Solver, analyzes the circuit and compares it (as in VISIR) with the list of permitted circuits (Maxlist) previously created by qualified personnel to not compromise the system. If the circuit is validated, it passes to the stack of the module EQCom, which converts the circuit and the instrumentation configurations requested by the client into commands (commands defined in the communications protocol of the VISIR Equipment server) and that will be sent to the equipment server and the Raspberry Pi by Ethernet. The values measured by the instrumentation are sent from the Equipment Server to the EQcom module in the form of commands, which are processed. The data obtained by the EQcom are forwarded to the XML server that parses them and sends them as a response to the client.

On the server of measure has embedded the HTTP server, this is based on the Python web framework Bottle. This allows us to include web dynamics in a simpler way giving the server new services. Some of these features are:

1. SSO-based authentication capability, Google, Facebook.
2. Generation of reports with results and sending these automatically to the email of the client and/or instructor.
3. Cookie-based federation and management reserves.

5 Conclusions

The system presented extends the teaching capacities of VISIR, as well as continuing to be fully compatible. Providing the ability to make the programming of ARM devices accessible and allowing the device to interact with sensors and communication interfaces and all this remotely and available 24 h, 365 days.

As future improvements will be to expand the number of sensors available from this base platform and to be complemented by practical training with theoretical teaching material, such as videos or PDFs.

Acknowledgements. The authors acknowledge the support of the Escuela de Doctorado de la UNED, the S2013/ICE-2715—eMadrid project, PILAR project Erasmus + Strategic Partnership no 2016-1-ES01-KA203-025327 (Platform Integration of Laboratories based on the Architecture of visiR). And the e-LIVES. e-Learning InnoVative Engineering Solutions-Erasmus+ Capacity Building in Higher Education 2017—585938-EPP-12017-1-FR-EPPKA2-CBHE-J, the Education Innovation Project (PIE) of UNED, "Desarrollos Avanzados Multi-Objetivo de Laboratorios Remotos para Actividades Educativas – DAMO-LRAE", from the Academic and Quality Vicerectorate and the IUED of the UNED and to the project 2018-IEQ18 from the Industrial Engineering Technical School of UNED.

References

1. Gustavsson, I., Zackrisson, J., Lundberg, J.: VISIR work in progress. In: 2014 IEEE Global Engineering Education Conference (EDUCON), Istanbul, pp. 1139–1148 (2014). https://doi.org/10.1109/educon.2014.6826253
2. Soria, M.F., et al.: First practical steps on the educational activities using VISIR and remote laboratories at UNSE in partnership with UNED inside the VISIR + Project. In: 2018 IEEE World Engineering Education Conference (EDUNINE), Buenos Aires, pp. 1–5 (2018). https://doi.org/10.1109/edunine.2018.8450995
3. Fischer, T., Scheidinger, J.: VISIR—microcontroller extensions. In: Proceedings of 2015 12th International Conference on Remote Engineering and Virtual Instrumentation (REV), Bangkok, pp. 177–179 (2015). https://doi.org/10.1109/rev.2015.7087287

4. Tawfik, M., et al.: Online experiments with DC/DC converters using the VISIR remote laboratory. In: IEEE Rev. Iberoam. Tecnol. Aprendiz. **10**(4), 310–318 (2015 Nov). https://doi.org/10.1109/rita.2015.2486459
5. Garcia-Zubia, J., et al.: Empirical analysis of the use of the VISIR remote lab in teaching analog electronics. IEEE Trans. Educ. **60**(2), 149–156 (2017 May). https://doi.org/10.1109/te.2016.2608790
6. Pardeshi, V., Sagar, S., Murmurwar, S., Hage, P.: Health monitoring systems using IoT and Raspberry Pi—a review. In: 2017 International Conference on Innovative Mechanisms for Industry Applications (ICIMIA), Bangalore, pp. 134–137 (2017). https://doi.org/10.1109/icimia.2017.7975587
7. Beltran, V., Martinez, J.A., Skarmeta, A., Martinez-Julia, P.: An ARM-compliant IoT platform: security by design for the smart home. In: 2016 IEEE 5th Global Conference on Consumer Electronics, Kyoto, pp. 1–2 (2016). https://doi.org/10.1109/gcce.2016.7800512
8. Shang-Fu, G., Xiao-Qing, Y.: Solution of home security based on ARM and ZigBee. In: 2016 International Symposium on Computer, Consumer and Control (IS3C), Xi'an, pp. 89–91 (2016). https://doi.org/10.1109/is3c.2016.33
9. Basil, E., Sawant, S.D.: IoT based traffic light control system using Raspberry Pi. In: 2017 International Conference on Energy, Communication, Data Analytics and Soft Computing (ICECDS), Chennai, pp. 1078–1081 (2017). https://doi.org/10.1109/icecds.2017.8389604
10. URL:https://github.com/VISIRServer/Measurement/blob/master/src/docs/Interface.doc, Website title: GitHub/VISIRServer/Measurement. Date Published: May 3, 2016, Date Accessed: October 07, 2018
11. URL https://github.com/VISIRServer/Equipment/blob/master/Docs/Protv4_1.pdf, Website title: GitHub/VISIRServer/Equipment. Date Published: May 3, 2016, Date Accessed: October 07, 2018
12. Gourmaj, M., Naddami, A., Fahli, A., Moussetad, M.: Integration of virtual instrument systems in reality (VISIR) OpenLabs with Khouribga OnlineLab. In: 2015 International Conference on Interactive Collaborative Learning (ICL), Florence, pp. 793–797 (2015). https://doi.org/10.1109/icl.2015.7318129
13. Rodriguez-Gil, L., Orduña, P., García-Zubia, J., López-de-Ipiña, D.: Advanced integration of OpenLabs VISIR (Virtual Instrument Systems in Reality) with Weblab-Deusto. In: 2012 9th International Conference on Remote Engineering and Virtual Instrumentation (REV), Bilbao, pp. 1–7 (2012). https://doi.org/10.1109/rev.2012.6293150
14. URL:https://github.com/novnc/websockify, Website Title: GitHub/novnc/websockify. Date Published: September 10, 2018, Date Accessed: October 07, 2018

Locker Security System Using Matlab

L. R. Karl Marx(✉) and R. Pradheep Kumar

Department of Mechatronics, Thiagarajar College of Engineering, Thirupparankundram 625015, India
lrkarlmarx@tce.edu, pradheepkumarr@yahoo.com

Abstract. The project proposed an effective monitoring and controlling system for security locker rooms, homes, bank lockers, jewellery outlets etc., which is completely autonomous. The security system is designed to detect the illegal entrance in the locker room area that commonly happens in the case of robberies. The major concern with the current manually supervised security system is that if the robbery occurs then the banks are not able to identify the robber due to lack of proof. The system will focus on the safety of locker rooms in an effective way by detecting and controlling unauthorized motion. The proposed system will save the images that can be used for further investigation. The proposed security system will detect the motion through PIR sensor and provide security by three different preventing actions as taking photos, alarm signal, spraying chloroform and can send the taken photo, warning e-mail to the owner using MATLAB IOT.

Keywords: Security system · IOT · Embedded system

1 Introduction

For a common man, the locker means a place with top level of security where he can store his invaluable like jewellery, documents, cash etc. To secure our expensive jewellery, important documents or cash, we need to use the locker rooms. To survive in his competitive world and to achieve continuous growth, the banking [1] industry needs to provide a high degree of security [2]. Video surveillance in moving areas have become a current topic of interest of computer vision. These CCTV Camera are used to monitor unauthorized motion activity. It needs to be monitored continuously. The alarm switch also needs to be pressed manually. Hence a more advanced system can be developed which will detect unauthorized motion and inform to the security officials by different ways without the need of a human being. A prototype of this security system has been designed in the dissertation to increase the level of security in locker rooms effectively.

The motion detection can be done through the Passive InfraRed (PIR) sensor itself and the hardware associated with it will provide 3 different ways to inform the security officials i.e., using alarm system, a warning Email and capture an image and that image will be automatically be sent through an E-mail which can downloaded anywhere. For emailing, MATLAB is used. So the important objectives of the bank security system

M. E. Auer and K. Ram B. (Eds.): REV2019 2019, LNNS 80, pp. 280–288, 2020.
https://doi.org/10.1007/978-3-030-23162-0_25

are tracking the bank locker room areas, detection of motion and taking the necessary control action. The further sections will describe how the objectives are attained.

The term Internet of Things (IoT) has become a buzzword [3] in the recent times. It has been gaining a lot of attention and development with the advancement and development of wireless technologies. The main idea behind Internet of Things (IoT) is the diverse objects such as mobiles, sensor, actuator, and communication devices such as Bluetooth, RFID, and Wi-Fi etc. to be integrated and interconnected to function together as a system and communicate among one another with a distinct physical address.

Earlier, the IoT systems were based on Radio Frequency Identification (RFID) system. But now various technologies have been developed in the market such Near Field Communication (NFC), Machine to Machine (M2M) communication and Vehicle to vehicle (V2V) communication etc. Further the use of IoT have enabled to individual components such as sensors, actuators, and other components into smart systems, which can communicate with each other and can even take decisions on their own [4].

Machine to Machine (M2M) communication essentially refers to the communication and interaction that occurs between smart devices [5], embedded controllers, sensors and other mobile platforms and devices [6]. In today's industrial environment. With the advancement of sensors and communication technologies, machines have become highly sophisticated and the process, control and monitoring of systems in industry have become more complex and difficult. By using IoT [7, 8] and machine-to-machine communication technologies, such processes have become easier to control and monitor. By the use of Wireless Sensor Nodes, remote units can be connected to a central controller and control be performed. With the advancement and development of cloud technology and cloud systems, data sharing, storage and analysis can be performed with the huge amount of Big Data produced by the sensor and smart machines. Hence Industrial Internet of Things (IIoT) is gaining popularity and is being implemented in Industries [4].

Industry 4.0 has been for emerged which is based on different types of data and from different devices even from different manufacturers. Ii is based on development of Cyber Physical Systems (CPS), which includes an embedded system that is connected to networks and can pass information between the nodes on another system and with the main application. The concept of Industry 4.0 has brought the idea of Open Platform Communication (OPC) Architecture that facilitates different software to integrate with machine of different Vendors, thus bringing integration over the industrial scale environment. Thus Internet of Things (IoT), Machine-to-Machine (M2M) communication, cloud based systems and other technologies play an important role in Industry 4.0 [9].

The first part of this paper gives an introduction about the work. The second part defines the existing problem at hand. The third part gives the block diagram and a generalised idea of the work. The fourth part describes the experimental setup used. The fifth part deals with algorithm used. The sixth part summarises the result.

2 Problem Statement

For the view of common man, lockers are safe houses where he stores his jewels, invaluable and other important things in his life. Hence the security of the locker system is an important part that needs to be addressed to. In existing system, it send the warning text short message service (SMS) to the operator using GSM technique [10, 11–13]. This system is highly cost effective, some disadvantages are there and during investigation there is no proof but in our implementation there is a low cost, intelligent and secured system for the locker room areas. The system will focus on the safety of the locker rooms in an effective way by detecting and alerting unauthorized motion. The proposed security system will detect motion through PIR sensor and provides security by taking three different preventing actions. In this system MATLAB based process is used to send a snapshot via email. The unauthorized image detection signal is also online communicated to the microcontroller that sets the warning alarm nearby also send the warning email to the owner using IOT technique for taking necessary control actions. The proposed intelligent system has advantages over the current security system like low cost, large coverage area, and better communication cost.

3 Proposed Model

The aim of the system is to provide a low cost, intelligent and secured system for the locker room areas. The microcontroller gets inputs from the PIR, Light Dependent Resistor (LDR) and metal detector and gives a suitable output response through Alarm, Web Camera, Bulb and Solenoid Valve. The microcontroller also gives a suitable email through MATLAB software based on the input. The following image 1 shows the block diagram of the experimental setup

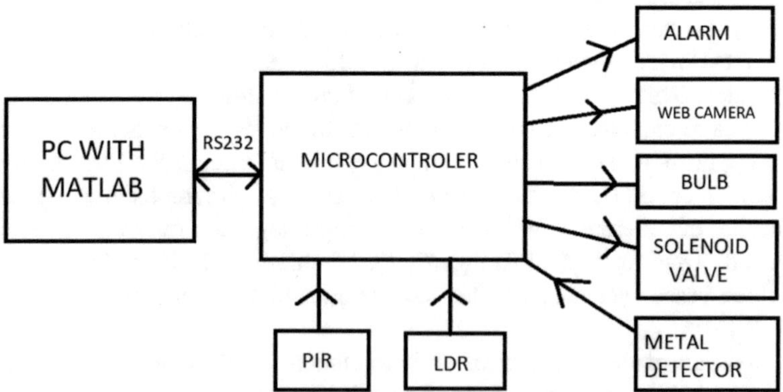

Fig. 1. Block diagram of the experiment

When a normal person enters the locker room, the Passive InfraRed (PIR) sensor is disturbed. Then the LDR is also disturbed. Here the person will have the key to the locker and hence will not inflict damage to the locker. Hence metal detector will not be disturbed. Hence only a mail will be sent as "someone has entered' and no photograph will be taken. From the time of the mail, we can check the working of the system. However if a thief enter, he will try to break the locker, hence the metal detector will also be disturbed. Thus from the LDR value and time of the mail we can confirm its day or night. A mail will also be sent to the user along with the photo of the thief.

4 Experimental Setup

The main objective of the project is finding the theft contro and monitoring the other peripherals that are connected to the Atmega 8 board. Atmega8 is manufactured by AVR microcontroller, based on Harvard architecture and works rapidly with RISC. These microcontrollers are very fast and utilize lower power in power saving mode. It is programmed using WINAVR software development tool.

PIR sensor allow us to detect motion, mostly used to detect whether a human has moved in or out of the sensors range. They are small, inexpensive, low-power, and easy to use. PIRs are mostly made of a pyroelectric sensor, which can detect infrared radiation. Every object emits some low level radiation, and the hotter the object is, the more amount of radiation is emitted. The sensor in a PIR is actually split in two halves. The reason for that is that we are looking to detect motion (change) not average IR levels. The two halves are wired in a manner so as to cancel each other out. If one half sees more or less IR radiation than the other, the output will swing high or low. We here use a PIR module built based on BISS0001 chip which gets input from the IR sensor.

The PIR and LDR sensor are connected to the microcontroller through analog-to-digital converters (ADC).The metal detector is directly connected to the microcontroller's I/O port. The alarm, camera, bulb and solenoid are connected to the microntroller's I/O port. The alarm is actually a buzzer which is an active buzzer, which means that it produces sound by itself, without needing an external frequency generator. Similarly the bulb too is connected to the microcontroller through a relay. The solenoid valve is actuated by using a 24 V relay which is controlled by the input ports from the microcontroller. The microcontroller communicates with the MATLAB in PC through Serial port through Rs. 232. If the PIR sensor detect a unauthorized motion, then camera are rotated along with the PIR to get a clear picture of the thief.

The system has three cases. In first case, if LDR and PIR are disturbed the mail is sent through MATLAB as 'Someone is trying to thrash the locker'. In second case, if the metal detector and PIR are disturbed, the solenoid is opened to spray chloroform and buzzer will ring and snap of thief will be taken and mail is sent with it as 'Locker thrashed without light'. In third case, LDR, PIR sensor and metal detector are disturbed, the solenoid valve will be opened and spray the chloroform to the thief and the buzzer will be rung after that bulb will be ON and snap will be taken and mail will be sent to user with image. If Both LDR and metal detector are not disturbed, then the PIR values are checked again continuously.

The experimental setup used has been shown in Fig. 2.

5 Flow Chart

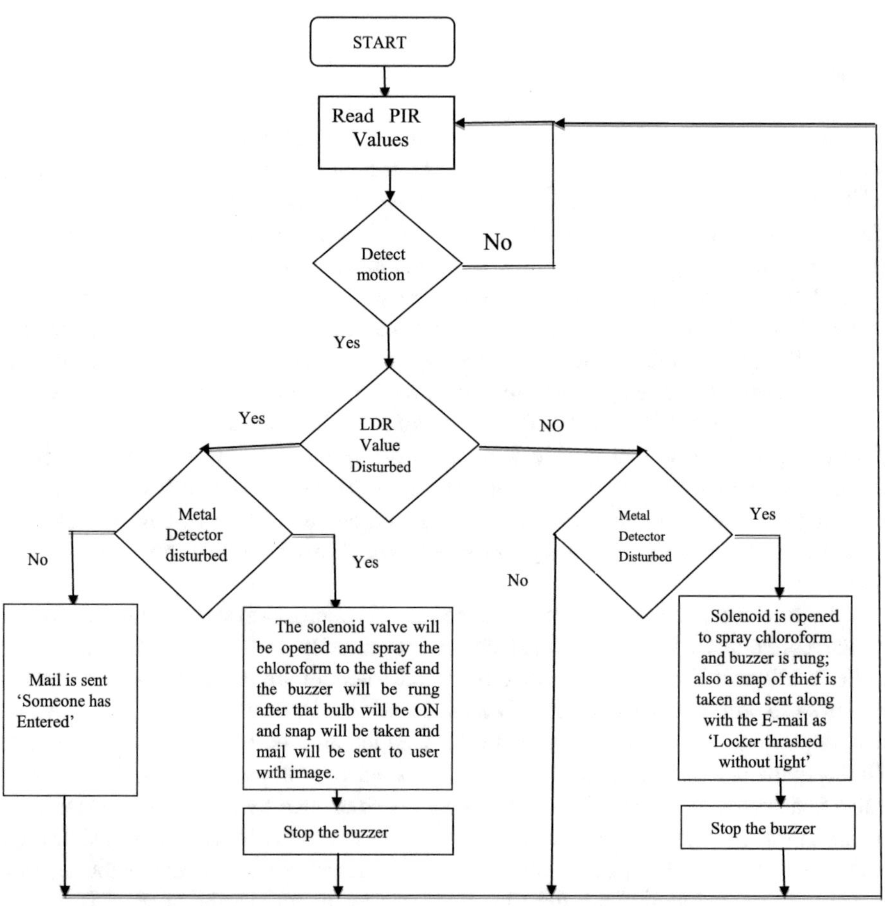

The algorithm is as follows:

1. Start the program.
2. Continuously monitoring the PIR values for human motion.
3. If any human detected move to step 5 or check the values.
4. If no human detected, go to step 2.
5. Check if LDR is disturbed. If Yes, move to step 6 else go to step 9.
6. Check if metal detector has been disturbed, if no go to step 7 else go to step 8.

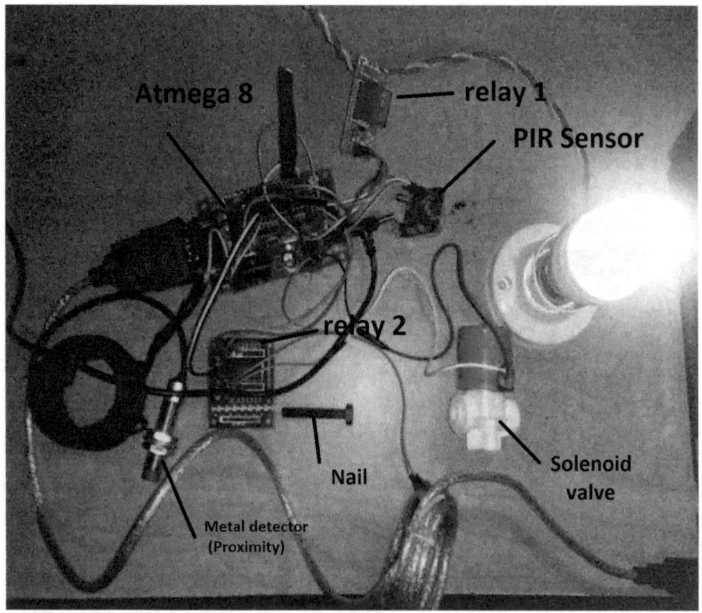

Fig. 2. The experimental setup

7. As PIR and LDR have been disturbed, a mail will be sent as 'someone has entered'. After this move to step 12.
8. As PIR, LDR and metal detector have been disturbed, the solenoid valve will be opened and spray the chloroform to the thief and the buzzer will be rung after that bulb will be ON and snap will be taken and mail will be sent to user with image. The buzzer rings until stop button is pressed. After this move to step 12.
9. Check if metal detector has been disturbed, if no go to step 10 else go to step 11.
10. As PIR sensor alone has been disturbed, no action is taken, move to step 12.
11. As PIR and metal detector have been disturbed, Solenoid is opened to spray chloroform and buzzer is rung; also a snap of thief is taken and sent along with the E-mail as 'Locker thrashed without light'. The buzzer rings until stop button is pressed. After this move to step 12.
12. Move to step number 2 to repeat the process.

6 Result

Hence by using the PIR, LDR and metal detector, a security system has been developed for Lockers using Atmega8 microcontroller and the program are loaded into it. Thus a low cost, highly secure security system for locker has been developed using Microcontroller and the warning is sent to the user through E-mail along with a photograph as shown in Fig. 3.

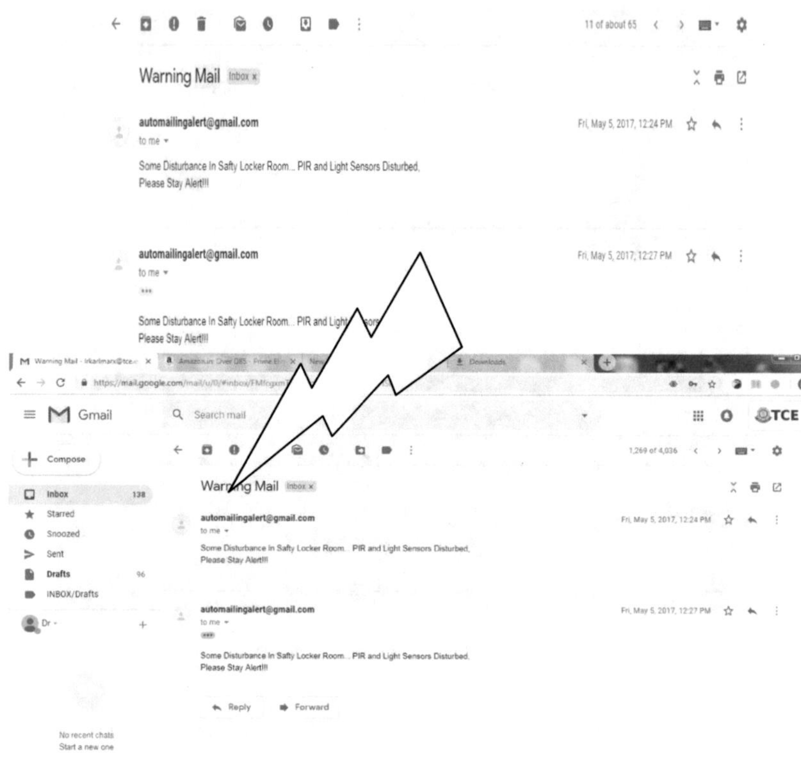

Fig. 3. Image of mail received

7 Bill of Materials

See Table 1.

Table 1. Bill of materials

Material	Cost (in Rs.)
Atmega 8 microcontroller kit	1000
Proximity sensor(inductive)	300
Buzzer	300
PIR sensor	160
Bulb + holder	150
Camera	1500
Wifi dongle	800
PC + Matlab(Home DTP)	5000
Total	9210

8 Conclusion and Future Works

Thus a low cost, highly secure security system for locker has been developed using Microcontroller and the warning is sent to the user through E-mail along with a photograph. The system can be further enhanced using more sophisticated sensors and a completely secure locker system can be developed. In case of robberies, most thieves are masked, however the reveal a lot of biological information such as iris, body form, hair colour etc. By making using of proper iris scanning algorithms on the image obtained from the camera, the thief can be searched from a database. Hence the system can be further enhanced by using recognition systems and Artificial intelligence. Further thermal sensor can also be added to the locker to check for flame cutting sensing. Hence with the addition of more sensors, the locker system can be made to be more secure.

References

1. Tejesvi, S.V., et al.: Implementation of embedded based bank security system using knock-out gas. Int. J. Eng. Res. Appl. **6**(2), 31–34 (Part-2) (2016 Feb). ISSN: 2248-9622
2. Khera, N., Verma, A.: Development of an intelligent system for bank security. In: 2014 5th International Conference—Confluence the Next Generation Information Technology Summit (Confluence), Noida, pp. 319–322 (2014). https://doi.org/10.1109/confluence.2014.6949339k
3. Ogri, U., Donatus, B., Etim, A.: Design and construction of door locking security system using GSM. Int. J. Eng. Comput. Sci. **2**, 2235–2257 (2013)
4. Shah, S.H., Yaqoob, I.: A survey: internet of things (IOT) technologies, applications and challenges. In: 2016 IEEE Smart Energy Grid Engineering (SEGE), Oshawa, ON, pp. 381–385 (2016). https://doi.org/10.1109/sege.2016.7589556
5. Zuo, F., de With, P.H.N.: Real-time embedded face recognition for smart home. IEEE Trans. Consum. Electron. **51**(1), 183–190 (2005 Feb)
6. Dhondge, K., Ayinala, K., Choi, B., Song, S.: Infrared optical wireless communication for smart door locks using smartphones. In: 2016 12th International Conference on Mobile Ad-Hoc and Sensor Networks (MSN), Hefei, pp. 251–257 (2016). https://doi.org/10.1109/msn.2016.047
7. Baidya, J., Saha, T., Moyashir, R., Palit, R.: Design and implementation of a fingerprint based lock system for shared access. In: 2017 IEEE 7th Annual Computing and Communication Workshop and Conference (CCWC), Las Vegas, NV, pp. 1–6 (2017). https://doi.org/10.1109/ccwc.2017.7868448
8. Anu, Bhatia, D.: A smart door access system using finger print biometric system. Int. J. Med. Eng. Inf. **6**, 274–280 (2014). https://doi.org/10.1504/IJMEI.2014.063175
9. Lu, Y.: Industry 4.0: a survey on technologies, applications and open research issues. J. Ind. Inf. Integr. **6** (2017) https://doi.org/10.1016/j.jii.2017.04.005
10. Ramani, R., Selvaraju, S., Valarmathy, S., Niranjan, P.: Bank locker security system based on RFID and GSM technology. Int. J. Comput. Appl. **57**(18):15–20 (2012 Nov)

11. Cortez, C., Badwal, S.J., Hipolito, R.J., Astillero, J.C.D., Dela Cruz, M.S., Inalao, J.C.: Development of microcontroller-based biometric locker system with short message service. Lecture Notes on Software Engineering, vol. 4, pp. 103–106 (2016). https://doi.org/10.7763/lnse.2016.v4.233
12. Adamu, M., Susan, A.A., Ijemaru, G., Oluwatosin, O.B., Habibu, H.: Design of a GSM-based biometric access control system (2014)
13. Gayathri, M., Selvakumari, Brindha, R.: Fingerprint and GSM based security system. Int. J. Eng. Sci. Res. Technol. **3**(4). ISSN: 2277-9655 (2014 Apr)

Digital Twins in Remote Labs

Heinz-Dietrich Wuttke(✉), Karsten Henke, and René Hutschenreuter

TU Ilmenau, 98693 Ilmenau, Germany
Heinz-Dietrich.Wuttke@tu-ilmenau.de

Abstract. Accelerated by current research on the Internet of Things, working with virtual models and things has taken a significant step forward. The term digital twins refers to the development of virtual models of processes, products or services that allow the simulation of system behavior and the early detection of problems before the real system exists. Simulations have been state of the art for a long time, but with the service-oriented possibilities of the Internet a new quality of simulation results. In the meantime, it has become possible, for example, to configure a virtual computer on the Internet and run algorithms on it without this computer in fact existing, see e.g. AWS (AMAZON web services). This paper discusses possibilities that the digital twin concept can offer to educational purposes in remote and virtual laboratories.

Keywords: Digital twin · Remote lab · Learning scenario

1 Introduction

Marr in [1] points out the importance of digital twins as follows "While the concept of a digital twin has been around since 2002, it's only thanks to the Internet of Things (IoT) that it has become cost-effective to implement. And, it is so imperative to business today, it was named one of Gartner's Top 10 Strategic Technology Trends for 2017 [2]. Quite simply, a digital twin is a virtual model of a process, product or service. This pairing of the virtual and physical worlds allows analysis of data and monitoring of systems to head off problems before they even occur, prevent downtime, develop new opportunities and even plan for the future by using simulations. Thomas Kaiser, SAP Senior Vice President of IoT, put it this way: 'Digital twins are becoming a business imperative, covering the entire lifecycle of an asset or process and forming the foundation for connected products and services. Companies that fail to respond will be left behind.'

As digital twins become that important for the industry [3] we as University teachers should think about, how to use this technology in classrooms. In [4], Y. Sulema et al. presents three models for multimodal data representation that are suited for the concept of digital twins. The authors consider that the concept of digital twins is a possible use case in future education of engineers. Following this idea, we will discuss in this paper, how we use virtual and real components in our hybrid online lab. The rest of the paper is organized as follows.

First, we give an overview of the structure of our lab from the digital twin perspective. In section three we discuss learning scenarios, were the use of digital twins

M. E. Auer and K. Ram B. (Eds.): REV2019 2019, LNNS 80, pp. 289–297, 2020.
https://doi.org/10.1007/978-3-030-23162-0_26

shows its big potentials and in the fourth section we scrutinizes the actual implementation of our lab to describe, how to extent experiments of the lab.

2 Structure of the Hybrid Lab

Labs are divided into remote, virtual and hybrid laboratories [5]. Remote Labs allow remotely controlled experiments. Virtual labs work in virtual artificial worlds. Hybrid labs combine both approaches by allowing work with simulations in addition to remote experimentation.

The GOLDi laboratory flexibly realizes all variants. Thus it is possible to carry out all experiments either completely virtually or on real devices or in a combination of both. In the experiments, the students have the task of designing a control algorithm (CA) for a certain control object (CO) running on certain control-units (CU) [6].

Before starting an experiment, the student must decide, how to carry out the experiment: with real or virtual devices. Figure 1 shows the configuration panel.

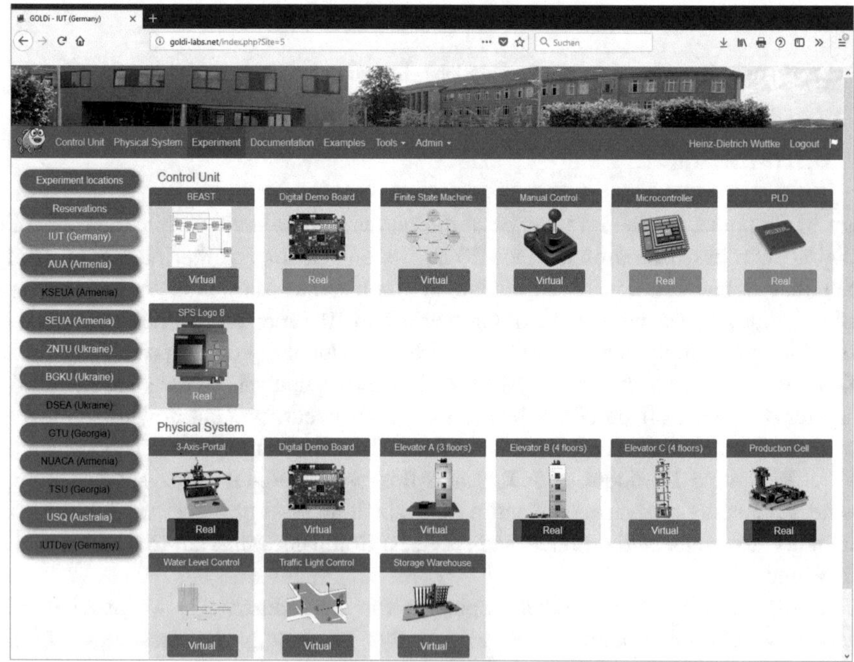

Fig. 1. Configuration of experiments

User choses from the available sets of devices by mouse-click a real or virtual control unit (first two lines) and a real or virtual control object (last two lines, in Fig. 1 called "Physical System"). Afterwards, he/she can start the experiment, see [7].

Microcontrollers, FPGAs, PLCs are available as real control-units and an interpreter is available as virtual control unit (vCU). As control-objects, there are physical systems (electromechanical models of an elevator, a production cell, a pump station and further objects). A digital twin exists for each control object as virtual control object (vCO).

Control units are either virtual either real. A virtual CU is an interpreter of an abstract CA description, based on finite state machines (FSMs) and independent from a concrete implementation. That way, the CA can be tested in experiments without implementing it on a real CU.

The architecture of the GOLDi-lab allows any combination of real and virtual devices. While virtual control units and control objects are JAVA-scripts and run in the clients' browser, the real devices are hardware in the lab. Figure 2 gives an overview of these devices.

Control Units		Control Objects	
virtual	real	virtual	real
Finite state machine interpreter	Microcontroller FPGA Siemens PLC	Digital twin of each real CO	3-achsis portal, Elevator, ...
Browser based	Hardware devices	Browser based	Hardware devices
Sensor → CU → actuator		Actuator → CO → Sensor	

Fig. 2. Virtual and real parts of the GOLDi lab

As depicted in Fig. 2, control-units receive sensor signals and send out actuator signals. Opposite, control-objects receive actuator signals and send out sensor signals.

Virtual devices are realized as JAVA-scripts and run on a browser. The real devices are located in the lab and connected to the lab server via a local Ethernet. The server realizes the communication between the users' browser on the client devices (PCs, laptops or mobile devices) via Web-sockets. Figure 3 shows the structure of this connection.

Abbreviations in Fig. 3: **CA**: control algorithm, **CH**: control hardware, **CO**: control Object, **CU:** control unit, **vCA**: virtual (abstract) control algorithm, **vCH** virtual control hardware (client device).

By configuring the experiment, the student decides between four different modes of experiments:

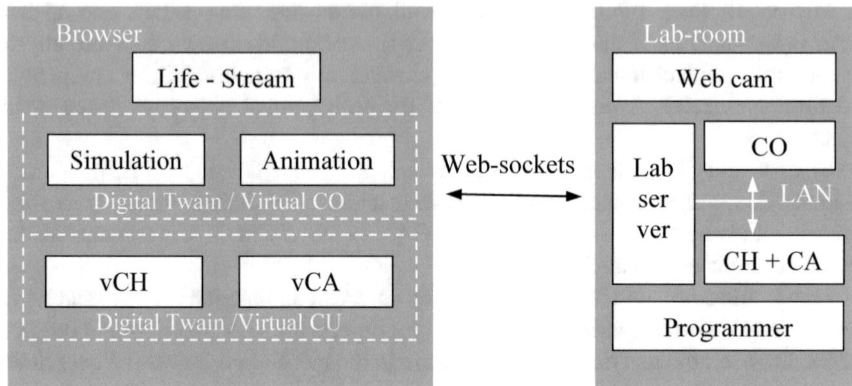

Fig. 3. Connection between real and virtual devices

(A) Virtual experiments (CU virtual, CO virtual, CA abstract)
(B) Abstract experiments (CU virtual, CO real, CA abstract)
(C) Real remote experiments (CU real, CO real, CA implemented) and
(D) Implementation test experiments (CU real, CO virtual, CA implemented)

Figures 4, 5, 6 and 7 show a deeper insight of the necessary connections between the lab devices and the users' browser to realize the different modes.

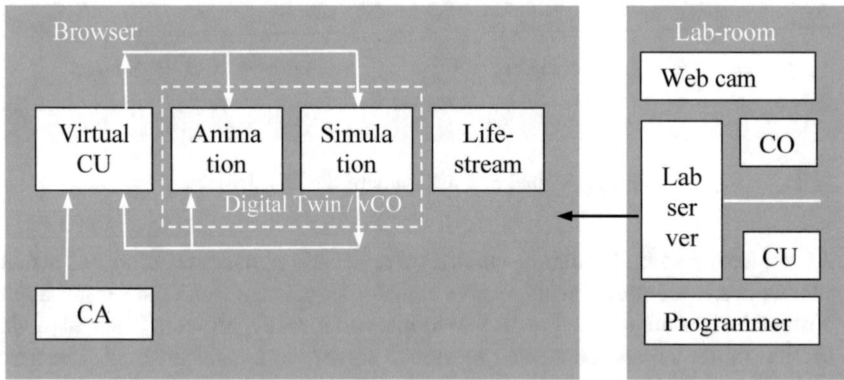

Fig. 4. Mode (A): virtual CU, virtual CO

In mode (A), if after configuration all necessary JAVA-scripts are downloaded from the lab-server, the virtual CU interprets the abstract FSM-based control algorithm (CA). The digital twin receives the output signals of the virtual CU and calculates, based on a behavioral model of the real CO, the actuator signals. The animation part of the digital twin receives the input/output signals and animates the virtual CO.

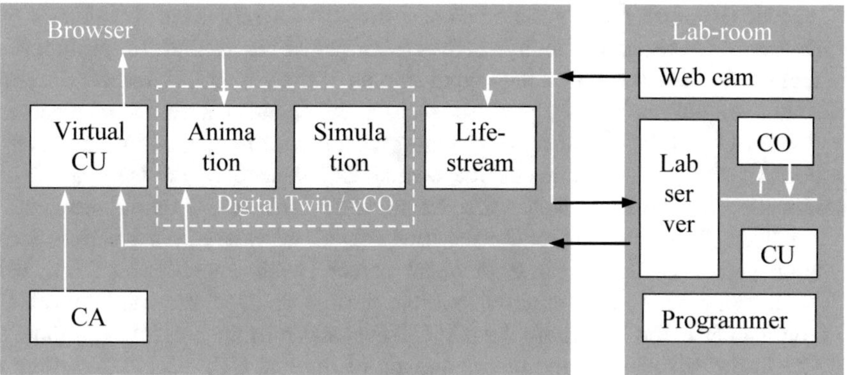

Fig. 5. Mode (B): virtual CU, real CO

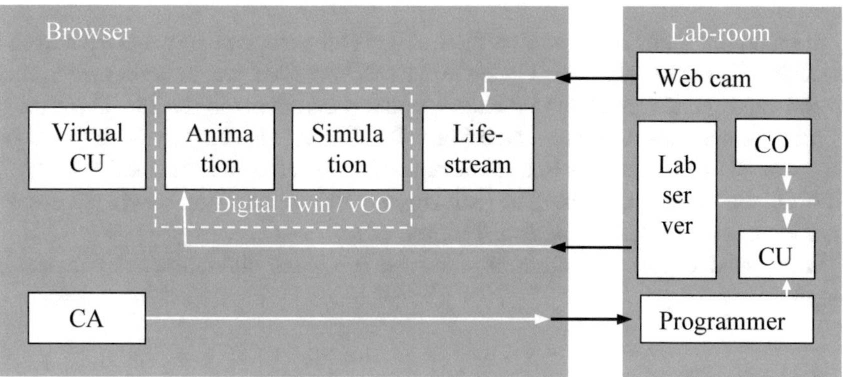

Fig. 6. Mode (C): real CU, real CO

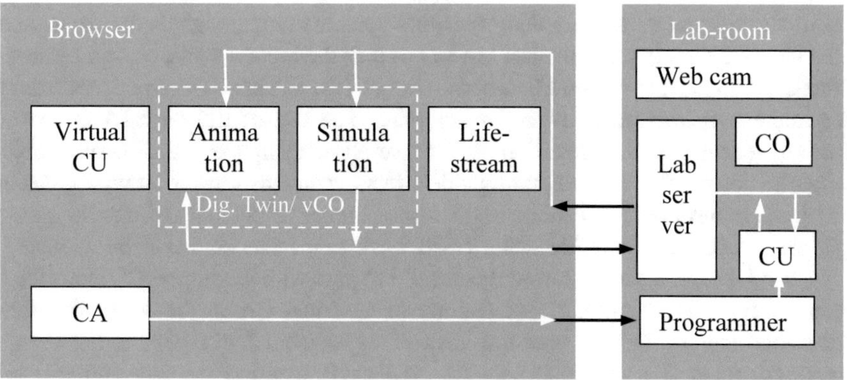

Fig. 7. Mode (D): real CU, virtual CO

In mode (B), if after configuration all necessary JAVA-scripts are downloaded from the lab-server, the virtual CU interprets the abstract FSM-based control algorithm (CA). The real CO in the lab receives the output signals of the virtual CU, reacts concerning the actuator signals and produces sensor signals, which are sent via the web-server to the virtual CU and the digital twin. The animation part of the digital twin receives all input/output signals and animates the virtual CO depending on the real signals. Additionally, the user can watch a life stream of the moving CO on the browser.

In mode (C), after configuration the user gets all necessary JAVA-scripts for the animation part of the digital twin from the lab-server. He/she translates his CA by using a CU-conform programming environment into the appropriate format and uploads it to the programmer in the lab. There the real CU gets and runs the control algorithm. The real CO in the lab receives the output signals of the real CU, reacts concerning the actuator signals and produces sensor signals. The animation part of the digital twin receives all input/output signals and animates the virtual CO dependent on the real signals. Additionally, the user can watch a life stream of the moving CO on the browser.

In mode (D), after configuration the user gets all necessary JAVA-scripts from the lab-server. He/she translates his CA by using a CU-conform programming environment into the appropriate format and uploads it to the programmer in the lab. There the real CU receives and runs the control algorithm. The digital twin receives the output signals of the real CU and calculates in its simulation part, based on a behavioral model of the real CO, the sensor signals. The animation part of the digital twin receives all input/output signals and animates the virtual CO.

In the next chapter we describe, how we use these different modes in learning scenarios.

3 Learning Scenarios

The task in the experiment is to control the CO with an own designed control algorithm, following a given movement scenario. To do so, the student has develop a control algorithm, i.e., to formulate rules that generate output signals for the actuators of the CO and keep in mind the last state as well as the input signals, caused by sensors of the CO. This is a so-called FSM-based design of a CA and is independent from the later implementation in hardware or software. The use of the digital twin concept allows us teaching in the classroom, where we allow using the virtual control objects during the design phase for exploring side effects, which can occur on wrong designs. The virtual control objects behave in the same manner as the real objects. So once the algorithm works on the digital twin, it will work in the real environment as well.

Figure 8 shows a screenshot of the users' browser with the virtual CO (left) the life stream of the real CO (middle) and parts of a virtual CA in the form of Boolean expressions (right). This is how the abstract experiment looks like, if the user has configured mode (B). In mode (A), the web cam stream is missing and in modes (C) and (D), the CA-part is absent.

The following examples discuss which mode is useful for what learning scenario.

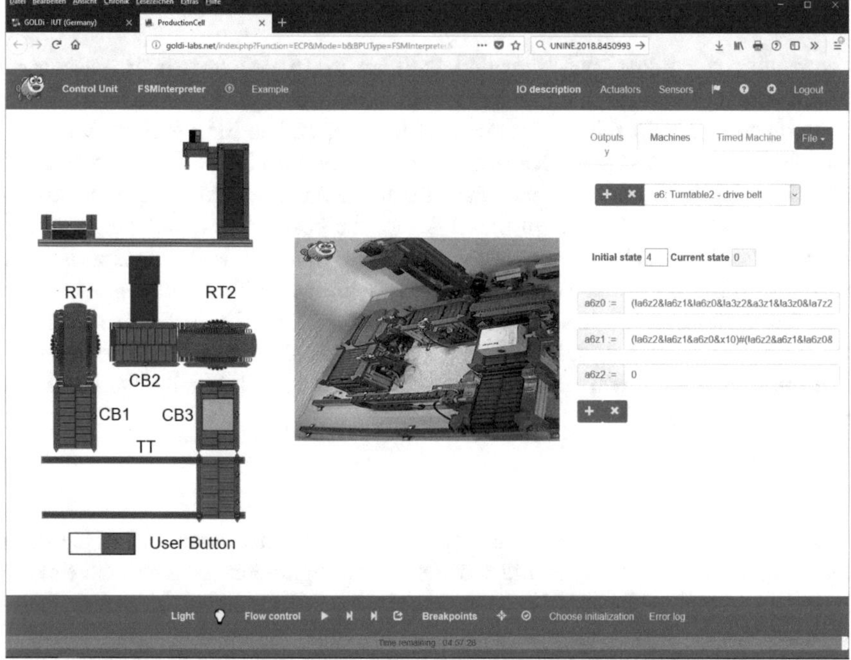

Fig. 8. Digital twin of a production cell (Mode B)

3.1 Lecture Hall

In the lecture hall, mode (B) is suitable, because with **abstract experiments** the design process of an FSM-based CA can be demonstrated. We use this mode to explain the concepts of Boolean constants, variables and expressions as well as the concept of finite state machines in a first year basic course in computer science.

We can insert constant Boolean values (e.g. set the actuator y_7 to logical "1") or variables or expressions (e.g. $x_1\&!x_2$, if sensor x_1 is true and x_2 is not true) to the virtual CU and demonstrate immediately its effects on the experiment. For first year students it is very motivating to see the application of what they are enforced to learn.

3.2 Reflection/Flipped Class Room

For flipped classroom scenarios as well as for self-studies **virtual experiments** (mode A) are beneficially. These kinds of experiments run offline on the browser once they were configured and started. Therefor the students can make their experiments also independent from the Internet-connectivity and prepare questions for the discussion with the coach in the seminars. They can try different variants of CAs and explore the differences or repeat the experiments that were demonstrated in the lecture hall. It requires no maintenance and reservation of lab equipment and therefor many students can benefit from the experiments at the same time.

3.3 Guided Design

In workshops for vocational education with up to 15 people, we first use **virtual experiments** (mode A) and let the learners retrace each step, we demonstrate, on their own computer. We do this on the abstract level of FSMs, independent from a later implementation. That way the learners get familiarly with the design process and the lab work. Afterwards we give a new task that the learners should solve by themselves. We discuss different solutions on the desk and the best solution can switch to mode (B) and control the real CO. Our experience shows that this is motivating the participants and enforces them to do their best in concurrency with the others.

In further lessons, we teach the implementation of the CA in hardware and software. To be sure, that the implementation was successful, first we let the learners work in the **implementation test mode** (D) were the CO is virtual. Finally, we use the **real remote experiment** (C).

3.4 Lab Exercises

To replace real hands on experiments by using remote labs, mode (C) is the choice. A booking system allows reserving a dedicated configuration for a defined time slot. The architecture of the lab enables an easy extension of COs and CUs because they are connected to a LAN and scanned permanently. In case, a new device is connected to the LAN, it is immediately accessible for experiments. For example, if there are three instances of the same CO installed, for the user this is transparent. He/she will not see which concrete device is connected to the experiment. Only if all devices are occupied, the user has to look for another time slot.

4 Actual Implementation

In the actual implementation, we have two virtual and four real CUs. One virtual CU allows manually controlling a CO, the other virtual CU is supported by two design tools, allowing to design CAs based on FSMs. Tools for designing the FSM- functions (next state function and output function) are BEAST (Block diagram Editing and Simulation Tool) [8] and GIFT (Graphical Interactive Finite State Machine Tool) [9]. The output, they produce, is the file-format for the interpreter of the virtual CU. Real CUs are a microcontroller for software-oriented implementation and an FPGA for hardware-oriented implementation, a CPLD demo-board and a Siemens-ECU are available as well.

Digital twins exit for an elevator, a 3-axis-portal, a production cell, a storage warehouse, a traffic light, and a water level control. If one of the real COs is defect, it will be disconnected from the LAN and only its digital twin is available. Thus, we can guarantee that experiments are running at any time in mode (A), even if parts of the hardware of the lab are under construction or maintenance.

5 Conclusions

The paper describes the architecture and usage of the digital twin concept in a hybrid remote laboratory. Main advances of the concept are the different use cases in learning scenarios, checking of abstract control algorithms independent from a later implementation, the test of implementations on digital twins of the real control objects and the possibility to run experiments even if a group of students wants to do the same experiment. It is possible, testing and correcting the students design on digital twins before it affects the real physical systems in the lab.

Four variants of experiments are available: virtual experiments, abstract experiments, real remote experiments, and implementation test experiments.

Acknowledgements. We want to thank our colleagues Tobias Fäth, Felix Seibel and Johannes Nau for the hardware design and maintenance of the GOLDi-lab.

References

1. Marr, B.: What is digital twin technology—and why is it so important? [Online]. Available: https://www.forbes.com/sites/bernardmarr/2017/03/06/what-is-digital-twin-technology-and-why-is-it-so-important/#22384ad22e2a. Accessed 25.10.2018
2. Top trends in the Gartner hype cycle for emerging technologies, 2017. [Online]. Available: https://www.gartner.com/smarterwithgartner/top-trends-in-the-gartner-hype-cycle-for-emerging-technologies-2017/. Accessed 6.12.2018
3. Haag, S., Anderl, R.: Digital twin—proof of concept. Manuf. Lett. **15**, 64–66 (2018)
4. Sulema, Y., Dychka, I., Sulema, O., Sikorsky, I.: Multimodal Data Representation Models for Virtual, Remote, and Mixed Laboratories Development, pp. 560–569. Springer, Cham, Switzerland (2019)
5. Pop, D., Zutin, D.G., Auer, M.E., Henke, K., Wuttke, H.-D.: An Online Lab to Support a Master Program in Remote Engineering, pp. 1–6. IEEE, Piscataway, NJ (2011)
6. Henke, K., Vietzke, T., Hutschenreuter, R., Wuttke, H.-D.: The remote lab cloud "GOLDi-labs.net", pp. 37–42. IEEE, Piscataway, NJ (2016)
7. Remote Lab GOLDi, TU Ilmenau, 15.8.2017. [Online]. Available: https://www.goldi-labs.net/. Accessed 15.8.2017
8. Knüpper, J., Seeber, M.: Browser Based Digital Circuit Simulation. Ilmenau
9. Henke, K., Fäth, T., Hutschenreuter, R., Wuttke, H.-D.: GIFT—an integrated development and training system for finite state machine based approaches. In: Online Engineering & Internet of Things, vol. 22, pp. 743–757. Springer International Publishing, Cham (2018)

Fog Based Architectures for IoT: Survey on Motivations, Challenges and Solution Perspectives

S. Thiruchadai Pandeeswari(✉) and S. Padmavathi

Department of IT, Thiagarajar College of Engineering, Madurai, India
eshwarimsp@tce.edu

Abstract. Advent of New age applications encompassing IoT, 5G wireless systems, Artificial Intelligence, real time analytics and stream mining has added new challenges to how cloud computing paradigm handles computing, storage and bandwidth. Many of the applications from the aforesaid domains are time critical and hence have stringent latency requirements and Network bandwidth constraints. It is impractical to depend on the cloud for processing the voluminous amount of data produced by these applications. Hence change in the computing- storage and Networking architectures of these cloud based systems is inevitable. The emerging "Fog Computing" paradigm promises to address the challenges specified above. Fog moves significant amount of computing, storage, communication and Networking away from cloud and performs the same in proximity to end user, thereby addressing the latency and bandwidth requirements related issues. Hence, Fog architecture enhances the distributed nature of an application and adds agility and efficiency to the applications by reducing the dependency on Cloud. However, this emerging Fog computing paradigm also poses its own set of challenges which must be understood and addressed in order to provide a viable framework of computing. This paper presents a thorough study on motivations behind the fog computing paradigm and various challenges in proposing a viable fog computing architecture for the resource intensive and time critical applications. Number of existing solution perspectives to address these challenges is studied and their strengths, benefits and drawbacks are analyzed. Also, scope for future research in this domain are highlighted.

Keywords: Fog · IoT · Cloud · Software defined networks (SDN) · Docker containers · Virtualization

1 Introduction

In this era of IoT, interaction between physical world and digital world is achieved through sensors and actuators together with networking and computing capabilities. More and more devices become connected, talk to each other over IP, takes decisions and produce results, thereby making human life easy like never before. Automated cars, Smart Homes, Smart cities and Smart farming are few examples from the family of new age modern IoT applications. These applications are supported by real time data analytics, stream mining and quick communication. These IoT applications produce

M. E. Auer and K. Ram B. (Eds.): REV2019 2019, LNNS 80, pp. 298–305, 2020.
https://doi.org/10.1007/978-3-030-23162-0_27

huge amount of data every second, all of which needs to be processed and analyzed in short span of time to produce desirable results.

The computing architecture of IoT systems basically contains three layers viz. Perception layer—which contains the sensors that sniff the ground and produce data, Network layer—which concerns with transporting the data and Application layer—where the data is processed and rendered as desired. This layered IoT architecture can also be extended to a five layer architecture as given in [1] to include transport layer and Processing layer in place of Network layer. The transport layer is responsible for transporting the data for processing while the processing layer is responsible for processing the huge amount of data produced at the perception layer. It is in the processing layer, number of technologies such as cloud computing and big data processing are applied.

Though lot of emphasis is given to the processing layer as it is the one that actually churns out the voluminous data, due importance must be given to the transport layer as well, as it is the one that is responsible for delivering the data to the processing layer. Hence lot of research is being carried out in this area that has resulted novel proposals on architectures that make use of number of technologies such as Fog computing, Software defined networks and Information centric networks to enhance quick transporting and processing of data. This paper presents a study on such research works especially on fog based architectures for IoT systems and further analyzes their viability.

2 Requirements and Challenges of IoT Systems

Most of the IoT applications are time critical and resource intensive requiring real time analytics and advanced processing such as stream mining. Not to forget, the amount of data generated at the perception layer is huge. For example, an application of automated cars produces roughly about one gigabyte of data per second [2]. Such voluminous amount of data needs to be processed in seconds to support real time decision making. Such applications have stringent latency requirements and huge network bandwidth requirements. Also, the IoT end user devices are small, hand-held and battery powered with limited resources. Hence confining the processing to local devices may not be feasible. Thus these systems require novel computing architecture and processing mechanisms to meet the aforementioned requirements. Sethi and Sarangi [1] presents a research taxonomy based on the five layer IoT architecture. The taxonomy prescribes research in three major areas pertaining to processing and transport layers. The areas under focus are Networking and middleware. While [1] prescribes WSN, Near field Communication and IP based communication for Networking, new mechanisms and protocols suitable for IoT devices may be incorporated here. Lot of options on Middleware services suitable for IoT systems that bridge the back end services with the application are available. However the applicability, feasibility and efficiency of these Networking and middleware technologies to IoT systems are subjective and require research to establish a well rounded framework.

Number of frameworks for IoT systems has been proposed in the past. However none of the frameworks is considered as the de facto standard like What OSI model is for the Data Communication Networks. Initially, Cloud based architectures for IoT looked promising as cloud caters to the flexibility, mobility and intensive computing requirements of IoT applications. However, with increase in the rate at which the data is being produced and scale of such applications, it is almost impossible to transport all of the data to cloud and process them because of network bandwidth and latency constraints. These drawbacks of cloud based IoT architecture lead to the development of Fog Computing. The needs, motivations and challenges in fog based IoT architectures are discussed in the following sections.

3 Fog Based Architectures for IoT Systems

3.1 Needs and Motivations of Fog Computing

As stated in [3], Fog computing refers to bringing networking resources close to the underlying network. Fog extends cloud and brings the services closer to the Data sources to eliminate unwanted latency and hence support time critical applications. As explained in the above section, it is highly taxing on resources to transport all the data to cloud—a centralized infrastructure system to do the processing and analysis. Instead of transporting all of the data from the perception layer to cloud, a discretion at the middle based on the importance and timeliness of the data could be done. Also, data pre-processing, data trimming and time critical operations could be performed next to perception layer. This will reduce latency and Network bandwidth consumption.

According to Smart Gateway Architecture highlighted in [3], the Network gateway, which is responsible for delivering data from the underlying network to cloud should be provided with capability to perform functionalities like data trimming, pre-processing and simple processing. The operations that must be carried out by Fog and When those operations must be carried out is decided by pre-programming the rules or dynamically setting up the rules based on given situation. This in itself opens up huge scope for research on computational offloading between cloud and fog.

It must be noted that the smart gateway architecture proposed in [1] is not sufficient to support mobility as the sources generating data is not static. This creates a multi-hop communication scenario. There arises a need for another layer of sink nodes where the data generated by various source nodes are aggregated. The smart gateway must pick up data from sink nodes that has aggregated data from multiple heterogeneous sources which necessitates the interoperability among smart gateways.

Fog computing, which mimics the cloud architecture at edge of the network presents a highly virtualized environment suitable for supporting mobility and interoperability. Thus fog can be defined as a processing entity in the network layer which may be realized either on network gateways or general purpose physical servers connected to a local network device. Since fog is located between the end user devices that are the sources of data generation and the cloud, fog can perform time sensitive control and analytics operations on behalf of cloud and also resource intensive tasks of end user devices may be offloaded to fog nodes. Thereby, fog addresses the major issues such as latency limitations, Network bandwidth constraints and resource constraints of end user devices. Chiang and Zjang [2] explains in detail the various advantages of fog computing especially in the context of IoT applications.

3.2 Challenges of Fog Computing

As mentioned in [3], Resource management, Service creation, Service discovery, Service management, data storage and power management are important aspects of any IoT systems. Considering the five layer architecture of IoT system and employment of fogs in the Network layer, number of questions arise on the fog service model, establishment of infrastructure for fog nodes, framework for communication among fellow fog nodes and cloud. Though the fog computing paradigm promises to be good complement to cloud to support IoT applications, it poses number of feasibility challenges and mandate changes in networking landscape. From the literature survey carried out, the following are identified as major challenges of fog computing paradigm

- Standard framework for dynamic relocation of computing, storage and networking tasks among Cloud, fog and Data sources.
- Orchestration of fog nodes (Placement of fog nodes, workload sharing and communication among fog nodes).
- High costs involved in replacement of existing Networking equipments to enable fog computing.
- Feasibility of creating infrastructure required for realizing fog computing and the ambiguity on primary realization agent (Whether the cloud service provider or IoT service provider).
- Lack of well evolved fog service model considering resource requirements of fog services, service usage by the users, and the profitability of the intermediation agent.
- Lack of standard, highly customizable, light weight, time sensitive, wireless enabled communication protocols suitable for fog.

Rest of the paper explains each of the aforementioned challenges briefly and highlights the solutions proposed in literature to address these challenges, excluding the resource allocation and computational offloading issues.

3.3 Related Works to Address Challenges of Fog Computing

Reference [4] highlights the various feasibility issues in establishing Fog Computing environment. It focuses on the lack of well-defined service model for fog i.e. ambiguities in the benefits for the users who participated in building the infrastructure for fog, service usage metering and billing. Since, fog computing architecture necessitates the network gateways and other network equipments to be smart with optimum processing capability to carry out the pre-processing and time critical processing tasks, high costs are involved. Reference [1] also highlights this high CAPEX requirement, as there is a need for replacement of existing network equipments to enable fog computing. A user participation based model is proposed in [4] to address this challenge. This research work proposes a novel idea for construction of fog nodes. In order to manage the usage billing and incentives to the users participating in the construction of infrastructure, an independent management entity called Fog Portal is proposed. In the existing model, anybody such as general users, network administrators or public agencies could construct the infrastructure for fog and service operators who would use the fog must enter into contract with the one who built it. In the proposed model, the user has to build the required fog nodes and services, declare them in the fog portal and delegate the operational authority to portal. Other users who have registered their nodes and services in the portal may use the services of other fog nodes as and when required. When a service of a user's fog node is used, the user would be given incentives in terms of revenue or other resources. Kim and Chung [4] has used Docker container technology and SDN to implement fog portal, service delegation and discovery.

Reference [5] presents taxonomy on important aspects of fog computing such as Fog nodes configuration, Communication among fog nodes, Resource/Service provisioning, techniques to achieve SLOs, Networking and Security. The taxonomy provides clarity on the implementation of these aspects and highlights various options available to realize the fog computing environment. Further the paper explores the existing research on these aspects of fog computing. As per the taxonomy proposed in [5], numbers of options are available for configuring the fog nodes. While the existing network equipments such as gateways, servers and other networking devices may be enhanced as fog nodes, fog nodes may also be realized on vehicles, ad hoc points and base stations. Cloudlets may also be leveraged to act as fog nodes. However their relative performance and suitability is still an open question as the performance depends on multiple factors such as execution environment, resource provisioning schemes, underlying networking capabilities and limitations and importantly the context and characteristics of application.

Another very important aspect of fog computing is communication and collaboration among the fog nodes. It must be noted that Placement and orchestration of fog nodes itself has attracted extensive research and lot of contemporary research is being carried out in this area especially in the context of IoT. Conceptually the Orchestration of fog nodes is directly related to the service provisioning. Existing research works such as [6] proposes fog based orchestration framework that leverages micro-services and Container technology. Software Defined Networking and Virtualization could also be thought as technology elevators for fog nodes orchestration.

Fog nodes orchestration plays a vital role in accomplishing the Service level objectives as well. Latency, Cost, Network, Data and Computation are noted as important factors that impact the service level objectives in [5]. In the existing research works, Proper placement of fog nodes and appropriate nodal collaboration are indicated as solutions for latency management. Many solutions provided in literature leverages Virtualization and SDN concepts to reduce CAPEX (Capital expenses) involved in establishing fog nodes and helps in cost management.

4 Discussions

4.1 Enabling Technologies

From the survey carried out on the various aspects, requirements and challenges of fog computing paradigm in the context of IoT applications, the following are identified as important technologies that would enable the realization of fog based architectures for IoT applications.

SDN
This networking paradigm introduces programming and flexibility in network behavior by separating the communication network into three different planes viz control plane, Data Plane and Management plane. This paradigm also opens up the networking landscape to open source switches and general purpose processors rather than proprietary networking equipments. SDN has centralized controller in the control plane that controls less expensive, less functional, open standard based switches in the data plane that simply forwards the packets based on the flow rules programmed. Since SDN introduces programmability into network, it enables Context aware, QoS aware Service Provisioning and Service discovery. Establishment and Configuration of fog nodes become simpler and less expensive with SDN, since networking equipments need to be updated with certain computing capabilities to act as fog nodes. SDN enhances nodal collaboration and orchestration inherently with its centralized control concept, programmability and flexibility.

Virtualization
Virtualization was the basic technology enhancer for Cloud computing. Similarly, Virtualization has an important role in fog computing also. Given the context of IoT and fog, network resources requirement, traffic characteristics, network configuration are dynamic in nature. The Service provisioning, latency and bandwidth requirements are not static. Network reconfiguration is needed more often than not. This is catered with the help of Network function virtualization where the network functionalities such as load balancing and firewall are softwarized. Network function virtualization helps to isolate the network functions from network infrastructure and thereby makes reconfiguration much simpler.

Containers
Container technology is a linux based technology that acts as an light weight alternative to virtualization. Containers make application execution independent of environment by packaging code and other dependencies into a single package that could easily be configured, run and ported. Containers also provide almost all the benefits of virtualization while being light weight, more portable and efficient. Many of the fog implementations in the existing researches such as [7, 6] have made use of container technology.

4.2 Taxonomy of Enabling Technologies for Implementing Fog Computing Architecture

From the literature survey carried out, it was observed that, combination of the above mentioned three technologies are leveraged in many of the fog based computing architecture proposals. Taxonomy of the applicability of these enabling technologies over the important aspects of the fog based computing architectures is given in Fig. 1. The following are identified as important aspects of any fog based computing architecture, to make the implementation absolute.

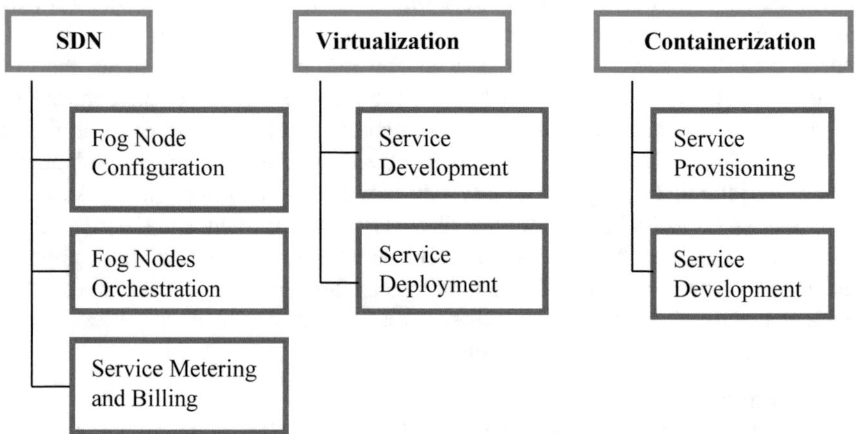

Fig. 1. Taxonomy of enabling technologies over the important aspects of fog computing architecture

- Fog Node Configuration—deals with configuring a network node such as a Physical server, networking devices, base stations etc. with fog computing capabilities.
- Fog Nodes Orchestration—deals with communication and collaboration among the fog nodes. This is required to ensure service continuity and Quality of Experience to the end users of the application.
- Service Development—deals with development of a functionality as a service that is used either by the end users or the intermediate nodes to accomplish the computing mechanism.
- Service Provisioning—deals with making the service available to the target users in such a way that service level objectives (SLOs) are attained.
- Service Metering and billing—deals with monitoring the usage of service by the end users, so that the usage can be metered and billed.

5 Conclusion

Internet of Things has disrupted the existing business models and application architectures. However technologies and architectures that accentuate the benefits of IoT need to be evolved and grown well rounded. Fog is one such paradigm that enables the growth and impact of IoT applications. Our survey on the existing Fog based IoT architectures has helped us to understand the many aspects of Fog computing that must be considered and provisioned for effective implementation of IoT applications. It must be noted that any fog based architecture for IoT must address the concepts of fog nodes configuration, placement, orchestration, Service provisioning and discovery. Though many solutions have been proposed in the past for optimized placement of fog nodes, their orchestration and service provisioning, a unified framework that addresses all these aspects with context-aware and QoS aware services is yet to be evolved. It is noteworthy that application of SDN, NFV and Container technology in the framework for fog based IoT applications is much open for exploration.

References

1. Sethi, P., Sarangi, S.R.: Internet of things: architectures, protocols, and applications. J. Electr. Comput. Eng. (2017)
2. Chiang, M., Zhang, T.: Fog and IoT: an overview of research opportunities. IEEE Internet Things J. **3**(6), 854–864 (2016)
3. Aazam, M., Huh, E.N.: Fog computing and smart gateway based communication for cloud of things. In: 2014 International Conference on Future Internet of Things and Cloud (FiCloud), pp. 464–470. IEEE (2014 Aug)
4. Kim, W.S., Chung, S.H.: User incentive model and its optimization scheme in user-participatory fog computing environment. Comput. Netw. **145**, 76–88 (2018)
5. Mahmud, R., Kotagiri, R., Buyya, R.: Fog computing: a taxonomy, survey and future directions. In: Internet of Everything, pp. 103–130. Springer, Singapore (2018)
6. Donassolo, B., Fajjari, I., Legrand, A., Mertikopoulos, P.: Fog based framework for IoT service provisioning. In: IEEE Consumer Communications and Networking Conference (2019)
7. Salman, O., Elhajji, I., Chehab, A., Kayssi, A.: IoT survey: an SDN and fog computing perspective. Comput. Netw. (2018)
8. Liu, Y., Fieldsend, J.E., Min, G.: A framework of fog computing: architecture, challenges, and optimization. IEEE Access **5**, 25445–25454 (2017)
9. Cai, H., Xu, B., Jiang, L., Vasilakos, A.V.: IoT-based big data storage systems in cloud computing: perspectives and challenges. IEEE Internet Things J. **4**(1), 75–87 (2017)
10. El-Sayed, H., Sankar, S., Prasad, M., Puthal, D., Gupta, A., Mohanty, M., Lin, C.T.: Edge of things: the big picture on the integration of edge, IoT and the cloud in a distributed computing environment. IEEE Access **6**, 1706–1717 (2018)

A Review of Techniques in Practice for Sensing Ground Vibration Due to Blasting in Open Cast Mining

Surajit Mohanty[1], Jyotiranjan Sahoo[2], Jitendra Pramanik[3], Abhaya Kumar Samal[2], and Singam Jayanhu[4(✉)]

[1] Department of CSE, DRIEMS, Cuttack, India
mohanty.surajit@gmail.com
[2] Department of CSE, Trident Academy of Technology, Bhubaneswar, India
{jyoti.sahoo,abhaya}@tat.ac.in
[3] Centurion University, Bhubaneswar, India
jitendra.pramanik@cutm.ac.in
[4] Department of Mining Engineering, National Institute of Technology, Rourkela, India
sjayanthu@nitrkl.ac.in

Abstract. Blasting is one of the most popular and a common technique widely used in quarries and mining production processes. It is the standard industrial cost-effective methodology that provides achievement of expected results quickly in a short period of time with involvement of relatively low cost of expenditure. Usage of Explosive generates ground vibration which is undoubtedly influence on surrounding structure. The basic purpose of this paper is to present a review of various sensing tools and technique to detect, monitor and measure the intensity and extent of vibration that has occurred due to blasting in open cast mining operation.

Keywords: Opencast mine · Sensors · Blasting · Ground vibration · Internet of things · Wireless sensor network

1 Introduction

Blasting is an important process step for mining operation that makes use of a lot of explosives for this purpose. The blasting process and usage of explosives, however, remain a potential source of numerous human and environmental hazards. Various studies indicate that fragmentation accounts for only 20–30% of the total amount of explosive energy used while the rest of the energy is lost in the form of ground vibration, fly rock, air overpressure and noise. The specific problem associated with ground vibrations represents the human response to them. Blasting vibrations may also cause a significant damage to nearby buildings or various structures.

Despite, the drilling and blasting combination is still an economical and viable method for rock excavation and displacement in mining as well as in civil construction works. The ill effects of blasting, i.e., ground vibrations, air blasts, fly rocks, back breaks, noises, etc. are unavoidable and cannot be completely eliminated but certainly

M. E. Auer and K. Ram B. (Eds.): REV2019 2019, LNNS 80, pp. 306–314, 2020.
https://doi.org/10.1007/978-3-030-23162-0_28

be minimized up to a permissible level to avoid possible damages to the surrounding environment with the existing structures.

Among all the ill effects, ground vibration is a major concern to the planners, designers and environmentalists. A number of researchers have suggested various methods to minimize the ground vibration level during the blasting. Ground vibration is directly related to the quantity of explosive used and distance between blast face to monitoring point as well as the geological and geotechnical conditions of the rock units in excavation area.

The application of proper field controls during all steps of the drilling and blasting operation will help to minimize the adverse impacts of ground vibrations, providing a well-designed blast plan that has been engineered.

The design would consider the proper hole diameter and pattern that would reflect the efficient utilization and distribution the explosives energy loaded into the blast hole. It would also provide for the appropriate amount of time between adjacent holes in a blast to provide the explosive the optimum level of energy confinement.

The key motivation behind this paper is to serve as the pre-cursor to the research leading to the design of a Wireless Sensor Network (WSN) in Internet of Things (IoT) framework to design a cost effective solution for sensing of ground vibration due to blasting in the opencast mining site.

Rest of the paper is organized into five sections. In Sect. 2, a short review of literature is presented to throw light on related works going on in the domain while in Sect. 3, we present basic concepts and various terminology and parameters associated with of the opencast mining activity related to the work presented in this paper. In Sect. 4, we present a study of geophone based vibration monitoring systems, followed by an overview of WSN based ground vibration monitoring system in Sect. 5. And finally, we present concluding remarks in Sect. 6.

2 Literature Survey

In this section we present few noteworthy reviews of previous works carried out by various researchers in international and national context through study of literatures from various sources.

Gui et al. [1] have designed a numerical model to understand the impact of the attenuation law and the geological features on the blast wave propagation. The field test results are then used to calibrate the numerical model. From the calibration, the parameters involved in the general form of peak particle velocity have been determined. It is demonstrated that the blast wave propagation in the free field is significantly governed by the field geological conditions, especially the interface between rock and soil layers.

Ragam et al. [2] have suggested wireless sensor network to monitor the effect of blast induced vibrations on structures. The paper explains a frame work of the process of monitoring and transferring sensed ground vibration data from a blast site to monitor site while blasting in mines.

Nishikawa et al. [3] have proposed an ANN based approach to predict blast-induced ground vibration of Gol-E-Gohar iron ore mine. To demonstrate the supremacy of ANN approach, the same 69 data sets were used for the prediction of PPV with four common empirical models as well as multiple linear regression (MLR) analysis. The results revealed that the proposed ANN approach performs better than empirical and MLR models.

Kumar et al. [4] proposes a wireless control and monitoring system for an induction motor based on Zigbee communication protocol for safe and economic data communication in industrial fields, where the wired communication is more expensive or impossible due to physical conditions. This system monitors the parameters of induction machine and transmit the data. A microcontroller based system is used for collecting and storing data and accordingly generating a control signal to stop or start the induction machine wireless through a computer interface developed with Zigbee.

3 Background Concepts

3.1 Opencast Mining

Opencast mining is a surface mining technique of extracting rock or minerals from the earth by their removal from an open pit or borrow. This form of mining differs from extractive methods that require tunneling into the earth such as long wall mining. Civil Engineers play a crucial role in mining as they are extensively involved in the design and building of the infrastructure for the mines, including the entire civil infrastructure and the processing facilities etc. In mining Civil Engineers will perform engineering duties very specific to mining operations where they will be involved in planning, designing and overseeing the construction and maintenance of various structures for both open-pit and underground mines (Fig. 1).

Fig. 1. Opencast mining

3.2 Blasting Practices at Open Cast Mines

The most important part in extraction from an open cast mine is blasting. Following are some important facts related to blasting practices in open cast mines:

- **Drilling**: There are two major types of drilling techniques being used Down the Hole (DTH): It is used when the blast hole diameter varies from 100 to 380 mm and the depth of blast hole is 12–15 m.
- **Rotary**: When the blast hole diameter >400 mm and the depth is about 20 m, rotary drilling is used.
- **Bench Height**: This depends on the excavator employed in the operation i.e. Shovel (4.6, 6, 8 m^3): 10–12 m Dragline (24 m^3/96 m): 30 m.
- **Blast Hole Diameter**: Up to bench height of 8 m, 100 mm,, 150 mm, 163 mm of blast hole diameter is preferred but for bench height >8 m, a minimum of 250 mm diameter of blast hole is efficient.
- **Spacing and Burden**: In coal measure rocks spacing varies from 3 to 8 m whereas burden is selected about 0.8 times the spacing.
- **Depth of Sub-grade Drilling**: In coal mines the length of sub-grade drilling for a bench height 8 m is generally 0.3 m.
- **Explosives**: In Indian coal mines nitro-glycerine based explosives, slurries, Ammonium Nitrate and Fuel Oil Mixture (ANFO) and Liquid oxygen type Explosive (LOX) are the main explosives which are used [5].

3.3 Ground Vibration

The movement of any particle in the ground can be described in three ways; displacement, velocity and acceleration. Velocity transducers (geophones) produce a voltage which is proportional the velocity of movement, and can be easily measured and recorded. They are robust and relatively inexpensive and so are most frequently used for monitoring. It has been shown in many studies, most notably by USBM that it is velocity which is most closely related to the onset of damage, and so it is velocity which is almost always measured [6, 7]. If necessary, the velocity recording can be converted to obtain displacement or acceleration. Each trace has a point where the velocity is a maximum (+ve or −ve) and this is known as the Peak Particle Velocity (or PPV) which has units of mm/s.

3.4 Waveforms of Blast Vibrations

The ground vibration wave motion consists of different kinds of waves:

- Compression (or P) waves.
- Shear (or S or secondary) waves.
- Rayleigh (or R) waves.

The Compression or P wave is the fastest wave through the ground. The P wave moves radially from the blast hole in all directions at velocities characteristic of the material being travelled through (approximately 2200 m/s). The Shear or S wave travels at approximately 1200 m/s (50–60% of the velocity of the P wave). The P waves and S waves are sometimes referred to as—body waves because they travel through the body of the rock in three dimensions. The Rayleigh or R wave is a surface wave, which fades rapidly with depth and propagates more slowly (750 m/s) than the other two waves.

3.5 Parameters Influencing Propagation and Intensity of Ground Vibration: *Reduction of Ground Vibration*

The parameters, which exhibit control on the amplitude, frequency and duration of the control vibration, are divided in two groups as follows:

A. Non-controllable parameters
B. Controllable parameters

The Non-controllable parameters are those, over which the blasting engineer does not have any control [8]. The local geology, rock characteristic and distances of the structure from blast site is non-controllable parameters. However, the control on the ground vibrations can be established with the help of controllable parameters. The same have been reproduced below:

- Charge weight
- Delay interval
- Type of explosive
- Direction of blast progression
- Coupling
- Confinement
- Spatial distribution of charges
- Burden, spacing and specification and specific charge.

3.6 Reduction of Ground Vibration

So as to protect a structure, it is necessary to diminish the ground vibrations from the blast [9]. The techniques which are acceptable for reduction and control of vibrations are:

(a) *Reduce the charge per delay*:
This is the most important measure for the purpose. Charge per delay can be controlled by:

- Reducing the hole depth
- Using small diameter holes
- Delay initiation of deck charge in the blast holes
- Using more numbers of delay detonators series
- Using sequential blasting machine

(b) *Reduce explosive confinement by*:

- Reducing excessive burden and spacing
- Removing buffers in front of the holes
- Reducing stemming but not to the degree of increasing air blast and fly rock

4 Geo Phone Based Vibration Monitoring Systems

As per the existing practice of ground vibration monitoring, geophone based vibration monitoring systems are in use. Instruments based on this techniques are the systems like Instantel Minimate plus, Micromate, Blastmate-II, Blastmate-III, etc. are generally used to monitor ground vibrations in mines [10–12]. The existing blast-induced ground vibration monitoring systems are of wire-based systems for acquiring and transferring the field data. In addition, the wire-based systems can only be applied to a limited area and the system becomes incapable if damages occur to the wires, which happens frequently [13].

4.1 Blasmate

A number of manufacturers have developed different models of blasting seismograph with different features and modified to reflect advancing technology. Most commonly used Instrument in India is Instantel—Blastmate-II and Blastmate-III. Few of these conventional vibration monitoring systems are shown in Fig. 2.

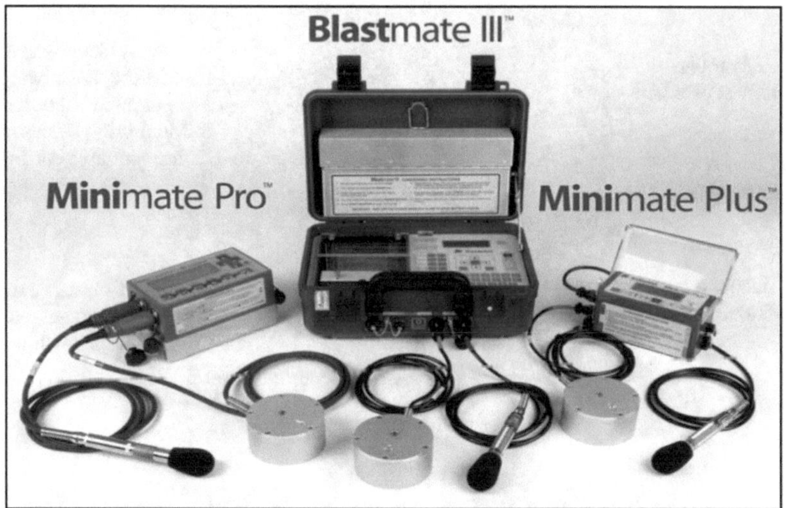

Fig. 2. Minimate Pro, Blastmate-III and Minimate Plus

So, the problem with conventional vibration sensors is that if vibrations need to be measured at several points (different positions), requires additional geophone sensors, which are expensive or if single geophone has considered, data collection was tedious and time-consuming.

4.2 Minimate Plus

The Instantel Minimate Blaster is a vibration and air overpressure monitor, a simple and economical package, with all the reliable operation that maybe demanded. It is lighter, more compact and easier to use than any monitor previously available. Blastware, comprehensive vibration analysis software, is provided as standard. Use the Minimate Plus monitor with an Instantel Standard Triaxial Geophone and an overpressure microphone to provide a rugged, reliable compliance monitoring system. Available Instantel 8-channel option allows for two standard geophones and two microphones to be operated from one Minimate Plus monitor.

4.3 Blasting Seismographs

Blasting seismograph consist of a transducer (generally a geophone, although an accelerometer may be used) connected to a processor to collect any analyze the signals. The monitor incorporates an eight-key tactile keypad and on-board LCD, with a clearly structured, menu-driven interface (Fig. 3).

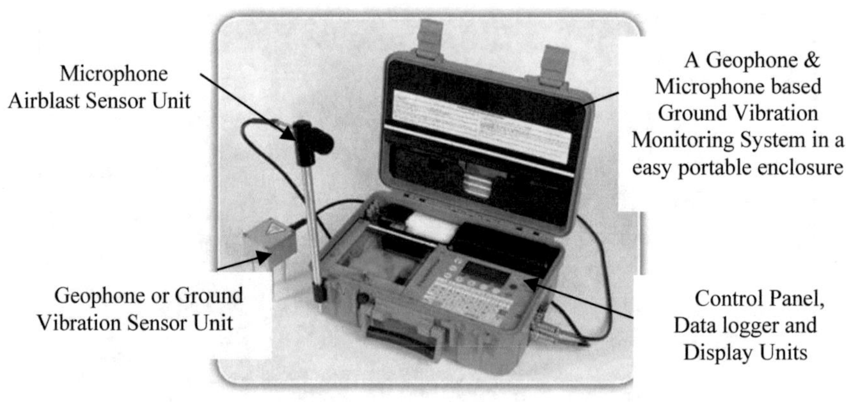

Fig. 3. Parts of a seismograph

It measures both ground vibrations and air pressure. The monitor measures transverse, vertical and longitudinal ground vibrations. Transverse ground vibrations agitate particle in a side to side motion. Vertical ground vibration agitate particle in an up and down motion progressing motion. Longitudinal ground vibration agitates particles in a forward and back motion progressing outwards from the event site. By measuring air pressure, we can determine the effect of air blast energy on structures.

Geophone amplitudes do not directly represent the actual magnitude of ground velocity. It is well known in seismic exploration that geophone amplitudes do not directly represent the actual magnitude of ground velocity. Although a geophone shows a fairly flat frequency response to ground velocity above the resonant frequency, it cannot be considered a "true velocity sensor" because low frequencies (below resonance) are recorded with small amplitudes relative to the amplitudes at high frequencies, and there is varying phase rotation up to 100 Hz or more.

5 Multi-sensor Based Wireless Monitoring System

A WSN consisting of multiple sensor nodes that are connected wirelessly to a sink node (central node). Kim et al. developed a vibrating wire WSN to monitor the tunnel construction with ZigBee protocol [8]. The proposed system suggested in the paper is equipped with a 3-axis acceleration sensor, an A/D converter, Atmega microcontroller and ZigBee-based communication system, Fig. 4. Sensor has a ZigBee based wireless sensor network in the PCB board so that the data can transmit wirelessly send and transmit the sensed data between each other.

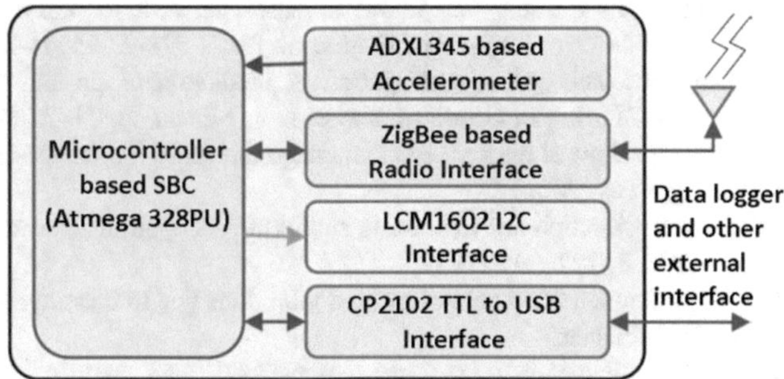

Fig. 4. Sensor node in the developed WSN system

6 Conclusion

Vibration monitoring can ascertain motion levels of natural as well as seismic activity. This can provide objective information about seismic effects, in particular when seismic operations are undertaken over sensitive and restricted areas. The monitoring results suggest that three components of ground motion need to be measured, since the PPV in the longitudinal and transverse components could be larger than the PPV measured in the vertical direction. Vibration measurements are fairly simple and straightforward in terms of peak values of particle velocity and air-overpressure. However, waveform analysis of ground motion and air vibration may reveal some further interesting features in the data.

References

1. Gui, Y.L., Zhao, Z.Y., Zhou, H.Y., GohL, A.T.C., Jayasinghe, B.: Numerical simulation of rock blasting induced free field vibration, Procedia Eng. **191**, 451–457 (2017)
2. Ragam, P., Jayanthu, S., Karthik, G.: Wireless sensor network for monitoring of blast-induced ground vibration. IEEE **8**(2), 14–21 (2016)
3. Nishikawa, Y., et al.: Design of stable wireless sensor network for slope monitoring. IEEE (2018)
4. Kumar, M., Sharma, M., Narayan, R., Joshi, S., Kumar, S.: Zigbee based parameter monitoring and controlling system for induction machine (2013)
5. Kwon, S.W., Kim, J.Y., Yoo, H.S., Cho, M.Y.: Wireless vibration sensor for tunnel construction. In: Proceeding of the 23rd International Symposium on Automation and Robotics in Construction (2006)
6. Kim, J.Y., Kwon, S.W., Cho, M.Y.: Development of wireless module for tunnel vibrating wire type sensor. In: Proceeding of the 24th International Symposium on Automation and Robotics in Construction Automation Group (2007)
7. Ragam, P., Jayanthu, S., Karthik, G.: Wireless sensor network for monitoring of blast-induced ground vibration. The Institute of Engineers. ISBN 978-93-86171-07-8
8. Singh, B., PalRoy, P.: Generation, propagation & prediction of ground vibration from opencast blasting. In: Blasting in Ground Excavation & Mines, pp. 21–35 (1993)
9. Singh, P.K.: Standardisation of blast vibration damage threshold for the safety of residential structures in mining areas (2006)
10. Crandell, F.J.: Ground vibration due to blasting and its effects upon structures. J. Boston Soc. Civ. Eng. Sect. ASCE **3**, 222–245 (1949)
11. Singam, J.: Study of frequency content of ground vibrations due to blasting. M.Tech. thesis, 1989, IIT BHU (Unpublished)
12. Singam, J.: Study of influence of operational parameters on performance of large diameter blast hole drills at Block II and Kusunda OCP, BCCL, 1989, 45 p. (Unpublished Report)
13. Singam, J.: Evaluation of blasting operations and suggestions on safe blasting to limit ground vibrations at Jindal Power Open Cast Coal Mines, Tamnar, Raigarh, 2011, 40 p. (Unpublished Report)

Improve VLC LiFi Performance for V2V Communication

Cristian-Ovidiu Ivascu(✉), Doru Ursutiu, and Cornel Samoila

Universitatea Transilvania, Brasov, Romania
cristian.ivascu@unitbv.ro

Abstract. LiFi VLC Communication for V2V Purposes. The goal of V2V communication is to prevent accidents by allowing vehicles in transit to send position and speed data to one another over an ad hoc mesh network. Depending upon how the technology is implemented, the vehicle's driver may simply receive a warning should there be a risk of an accident or the vehicle itself may take pre-emptive actions such as braking to slow down.

Keywords: LiFi · VLC · V2V · V2I · Vehicular networks · Visible light communication · Vehicle-to-Vehicle · Vehicle-to-Infrastructure · Vehicle-to-Internet

1 Introduction

Li-Fi (short for light fidelity) is a technology for wireless communication between devices using light to transmit data and position. In its present state only LED lamps can be used for the transmission of visible light [1]. The term was first introduced by Harald Haas during a 2011 TEDGlobal talk in Edinburgh. In technical terms, Li-Fi is a visible light communications system that is capable of transmitting data at high speeds over the visible light spectrum, ultraviolet and infrared radiation [1].

In terms of its end use the technology is similar to Wi-Fi. The key technical difference is that Wi-Fi uses radio frequency to transmit data. Using light to transmit data allows Li-Fi to offer several advantages like working across higher bandwidth working in areas susceptible to electromagnetic interference (e.g., aircraft cabins, hospitals) and offering higher transmission speeds.

V2V communication, vehicle-to-vehicle communication, is the wireless transmission of data between motor vehicles.

Motivated by the looming radio frequency (RF) spectrum crisis, this paper aims at demonstrating that optical wireless communication (OWC) has now reached a state where it can demonstrate that it is a viable and matured solution to this fundamental problem. Light fidelity (Li-Fi) which is related to visible light communication (VLC) offers many key advantages, and effective solutions to the issues that have been posed in the last decade [2].

M. E. Auer and K. Ram B. (Eds.): REV2019 2019, LNNS 80, pp. 315–329, 2020.
https://doi.org/10.1007/978-3-030-23162-0_29

2 Goal

The goal of V2V communication is to prevent accidents by allowing vehicles in transit to send position and speed data to one another over an ad hoc mesh network. Depending upon how the technology is implemented, the vehicle's driver may simply receive a warning should there be a risk of an accident or the vehicle itself may take pre-emptive actions such as braking to slow down.

V2V communication is expected to be more effective than current automotive original equipment manufacturer (OEM) embedded systems for lane departure, adaptive cruise control, blind spot detection, rear parking sonar and backup camera because V2V technology enables an ubiquitous 360-degree awareness of surrounding threats. V2V communication is part of the growing trend towards pervasive computing, a concept known as the Internet of Things (IoT).

Connected vehicles use wireless technology to connect vehicle information and location to other vehicles (V2V); to infrastructure (V2I); or to other modes, such as internet clouds, pedestrians, and bicyclists (V2X). The wireless technology typically used for connected vehicles is DSRC, Dedicated short-range communications, but some functions may use cellular or other types of communication.

The goal is to use LiFi in order to communicate from one vehicle to another one and from vehicle to infrastructure or from vehicle to internet. Connected vehicles offer additional functions related to roadside devices and fleet-level information. Connected vehicles bring additional mobility and environmental benefits that cannot be achieved through automation alone.

V2V technology represents the next great advance in saving lives. This technology could move us from helping people survive crashes to helping them avoid crashes altogether—saving lives, saving money, and even saving fuel thanks to the widespread benefits it offers [3].

3 Approach

With the omnipresence of light-emitting diodes (LEDs) in outdoor and automotive lightings, VLC, Visible Light Communication, emerges as a natural candidate for vehicle-to-vehicle (V2V) and vehicle-to-infrastructure (V2I) communications. We first provide an overview of this emerging research area highlighting recent advances and identifying open problems for further research. Then, we present the performance evaluation of a typical V2V VLC system with realistic automotive light sources. Our evaluation takes into account the measured headlamp beam pattern and the impact of road reflected light. We demonstrate that depending on the photodetector (PD) position above the ground level, a data rate of 50 Mb/s can be achieved at 70 m.

There has been a growing interest in the field of intelligent transportation systems (ITSs) in an effort to improve road safety and traffic flow and to address environmental concerns. The ITS involves the application of the advanced information processing, control technologies, sensors, and communications in an integrated approach to improve the functioning of the road transportation systems. Considerable efforts have been made in the last decade by researchers from both academia and industry to enable the cooperative ITS, which is seen as the next generation of ITSs and it is enabled by V2V and V2I communications.

We consider VLC as a complementary and/or an alternative technology to RF-based systems. An alternative communication means can turn out to be useful to offload the RF channel. VLC refers to the use of optical radiation at the visible wavelengths to transmit data in an unguided medium. Since the human eye perceives only the average intensity when light is switched on and off fast enough, then it is possible to transmit information data using LEDs without a notable effect on the light illumination level and the human eye. Recent advances in materials and solid-state technologies have enabled the development of highly efficient LEDs that are now being widely used in outdoor lighting, traffic signs, and advertising displays. Furthermore, many automotive manufacturers have started to employ LEDs due to their high resistance to vibration, improved safety performance, and long life span. LEDs can be now found in brake lights, turn signals, and headlamps in most new vehicles. The outdoor and on-vehicle omnipresence of LEDs makes the use of VLC for V2V and V2I communications possible [4].

As illustrated in Fig. 1, a vehicular VLC network consists of on-board units, i.e., automobiles, motorcycles, other type of vehicles, and RSUs, i.e., traffic lights, street-lamps and digital signage. In addition, RSUs are connected to the backbone network via the roadside infrastructure (RSI) network. Vehicles fitted with LED based front and back lights can communicate with each other and with the RSUs through the VLC technology. In a vehicle, a VLC transceiver is connected to both headlamps and taillights, which serve as “optical transmit antennas.” As “optical receive antennas,” PDs are placed next to each of these four lights and connected to the VLC transceiver. Additional PDs can be installed on the sides of vehicles (such as the back sides of rearview mirrors) to enhance coverage. As an alternative to PDs, an image sensor (IS) (i.e., a camera), which is now available in many cars, could be also used for reception.

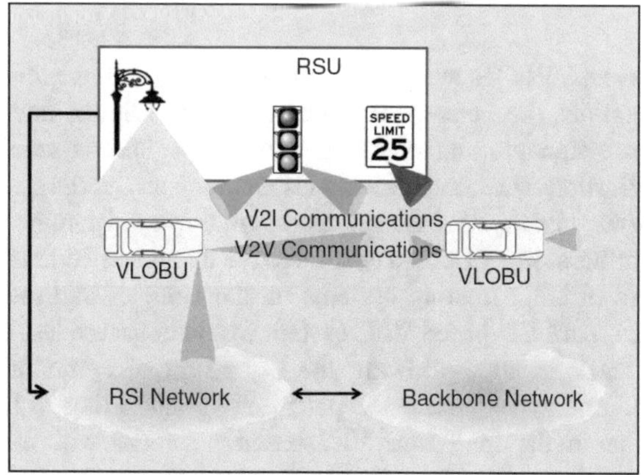

Fig. 1. Vehicular VLC network. VLOBU: VLC-based on-board unit

4 Comparison of RF and VLC for Vehicular Networking

The unique properties of optical propagation and VLC offer many advantages compared with those of the RF communications including immunity to the electromagnetic interference, operation in unlicensed bands, inherent safety and security, and a high degree of spatial confinement allowing a high reuse factor. A comparison of the main features of both VLC- and RF-based (Specifically 802.11p [5]) vehicular networking is provided in Table 1.

Table 1. A comparison of the VLC and IEEE 802.11p performance characteristics

Type	VLC	802.11.p
Communication mode	Point-to-point (LOS or diffuse)	Point-to-multipoint/broadcasting
Latency[a]	Very low	150 ms
Data rate[b]	Up to 400 Mb/s	Up to 54 Mb/s
Range	Up to 100 m (single hop)	Up to 1 km
Frequency band	400–790 THz (390–750 nm)	5.8–5.9 GHz
License	Unlicensed	Licensed
Cost	Low	High
Mobility	Medium	High
EMI	No	Yes
Power consumption	Relatively low	Medium
Coverage	Narrow	Wide
Weather conditions	Sensitive	Robust
Ambient light	Sensitive	Not affected

[a]The latency is defined as the maximum time span during which information is successfully delivered to targeted vehicles

[b]Range refers to the distance that a signal can propagate delivering the required quality of service

As Table 1 reveals, VLC is well positioned to address both the low latency required in safety functionalities (i.e., emergency electronic brake lights, intersection collision warning, in-vehicle signage, and platooning) and high speeds required in so-called infotainment applications (i.e., map downloads and updates, media downloads, point of interest notifications, high-speed Internet access, multiplayer gaming, and cooperative downloading). Furthermore, VLC is a cost-effective and green communication solution since the dual use of LED lighting systems on the vehicles and the roadside infrastructure is targeted. An LED-based VLC system would consume less energy compared with the RF technology, thus allowing the expansion of communication networks without the added energy requirements, potentially contributing to the global carbon emission reduction in the long run. VLC-based networks will further offer better scalability in scenarios with high vehicle density, where the RF-based vehicular communications experience longer delays and lower packet rate because of the channel congestion. The directionality of optical propagation provides an inherent advantage in vehicular VL since only a small number of neighboring vehicles, typically within the direct line of sight (LOS) of the receiver (Rx), are in the same contention domain, thereby significantly lowering collision probability and increasing scalability. However, this advantage is at the cost of reduced coverage compared with RF technologies and requires judiciously designed upper-layer protocols to handle more frequent handovers. VLC is also appealing for vehicular scenarios in which the use of a RF band is restricted or banned due to the safety regulations, e.g., industrial parks such as in oil/gas/mining industries, and military vehicle platoons, to name a few. Another attractive feature of VLC is the positioning and navigation capabilities. Although the global positioning system (GPS) is widely used today, it fails to provide a sufficient accuracy in environments where there is no LOS paths such as tunnels, indoor parking lots, and some urban canyons. For such cases, VLC-based positioning technology could be used to complement the accuracy of other localization systems, knowing that the lighting fixtures offer very high accuracy, up to of tens of centimeters, which is much more suitable for vehicle safety applications, compared with a typical positioning error of up to 10 m associated with the GPS. VLC also presents some challenges due to operation in outdoor environments such as severe weather conditions, sunlight, and ambient light. Visibility-limiting conditions such as heavy fog or snow could decrease the operation range. These degrading effects can be kept at minimum with highly sensitive RXs. Another potential concern is direct sunlight or strong ambient light, which could saturate the VLC Rx. This is usually addressed by utilizing proper optical filtering.

5 Vehicular VLC Research and Open Problems

Early research efforts on the use of LEDs for V2I communications can be traced back to a U.S. patent issued in 1997 (Patent 5 633 629), which utilizes traffic lights to optically transmit information to a vehicle. In initial performance evaluation studies [6, 7], the basic performance metrics such as the received optical power and the signal-to-noise ratio (SNR) have been derived for different road surfaces (e.g., asphalt versus cement or concrete) for a V2I communication system between a street/traffic light and a vehicle. More recent experimental works [8–10] have used off-the-shelf LEDs and PDs to investigate the throughput and the error rate performance of VLC-based V2V and V2I links. Such works have demonstrated the feasibility of the vehicular VLC. However, in many respects, this technology is in its infancy and requires further research efforts in several areas including channel modeling, physical layer design, and upper-layer protocols.

6 Channel Modeling Aspects

Most works on the propagation modeling and characterization of VLC channels are mainly limited to the indoor environments. For the outdoor environment, as in the case of vehicular networking, additional noise sources such as the ambient interference due to the background solar radiation and artificial light sources from cars and streetlights must be taken into consideration as mentioned earlier. Furthermore, the visibility limiting weather and environmental conditions (such as rain, snow, fog, and car fumes) and heat-induced turbulence (i.e., scintillation effects) need to be further considered. The outdoor VLC channel modeling has been investigated only in sporadic works [11], where the effects of solar irradiance and artificial light sources are modeled and incorporated in the channel model. While these initial works point out the striking difference between the indoor and outdoor VLC channels, systematic modeling and characterization of outdoor VLC channels particularly considering that the ITS environments and scenarios does not exist yet.

7 Physical Layer Design Issues

VLC channels are of multipath nature and exhibit frequency selectivity, which results in inter symbol interference (ISI), and leads to a reduced data rate. The conventional solution for the ISI mitigation in a single carrier system is to adopt the time-domain equalization (TDE) and use modulation formats with wide-enough pulse duration. Although nonlinear TDEs are particularly effective in handling ISI, the number of operations per signaling interval grows linearly or exponentially with the ISI span, or, equivalently, with the data rates. This results in an excessive complexity and makes TDEs unfeasible for VLC systems where data rates of several hundreds of megabits per second are targeted. A powerful alternative to the single-carrier TDE would be the deployment of multicarrier communication techniques. The most popular form of multicarrier communications is the orthogonal frequency-division multiplexing (OFDM), which has been already adopted in various wireless RF and wireline standards and recently applied to VLC systems [12]. Different from the RF approach, in OFDM for optical systems, the dc biasing or asymmetrical clipping have been introduced to ensure the non-negativity of the intensity-modulated optical signal. Multiple-input, multiple-output (MIMO) communication, another innovative technique that was also originally proposed for RF systems, has been further applied to VLC systems in recent studies. Since a number of LEDs are used to achieve the required intensity in a typical lighting application, the MIMO techniques emerge as a natural candidate in VLC systems to boost the data rate. The combination of OFDM and MIMO is considered a powerful physical layer solution for high-speed vehicular VLC systems to support bandwidth-hungry infotainment applications. Another key component to enable connectivity in vehicular VLC networks is the concept of multi hop transmission. There are only sporadic works that have addressed multi hop VLC transmission. For example, researchers from the Disney Research Center have demonstrated toy-to-toy car communications where messages sent via VLC are passed from one toy car to another in a multi hop mode. Performance analysis of a multi hop transmission system is presented in [13] for an indoor scenario where the light signal emitted from the ceiling is relayed through a desk lamp. The current works on multi hop VLC are limited to indoor environments and there is an obvious need for a thorough performance evaluation of multi hop vehicular VLC networks.

8 Imaging Sensor Communications

ISs can be also used as optical detectors. This type of VLC is also referred to as camera communication (CamCom). In recent years, we have seen extensive use of smartphones, tablets, and event data recorders for car navigation, in car entertainment, and information related to vehicle crashes or accidents recording. All these devices have at least one camera and some of them even have the ability to perform video signal processing. As in an RF MIMO channel, the LED lights function as the transmit antennas and each camera pixel element as a receiving antenna in CamCom. In RF MIMO schemes, the signal quality is mainly affected by the path loss, multipath induced fading, and co channel interference. However, in CamCom, the channel experiences negligible multipath fading but suffers from path loss and interference from other light-emitting sources. The latter manifest itself as visual distortions on the camera output (i.e., the image), which can be modeled using a traditional camera imaging theory. The channel noise also affects the signal quality at the camera Rx. Under a high level of ambient light, the noise in a pixel can be considered signal independent. In typical V2V CamCom applications, headlamps are used to transmit traffic-related information such as the vehicle's speed, position, status of brakes, and changing lanes and statues of the road to the nearby vehicles. Within the receiving vehicles, a low-frame rate camera is used to capture the transmitted signal as a video stream. The imaging lens and an IS-based Rx have the ability to spatially separate light signal from different directions, thus giving CamCom a unique advantage of interference-free communication based on space-division multiple access and MIMO, not found in the traditional single PD-based VLC systems. Recent experimental results in this area [14] demonstrate that various vehicle internal data (such as speed) and image data (320 # 240, color) were transmitted successfully in challenging outdoor conditions.

9 Upper-Layer Protocols

In addition to physical layer issues discussed previously, there are also other design considerations in the upper layers that require further attention to realize a fully functional vehicular VLC network. For example, the medium access control (MAC) protocols have been widely investigated in the literature assuming isotropic radiation of RF systems. VLC systems with their inherent directionality render conventional MAC schemes practically useless. There have been recent efforts on the design of MAC protocols taking into account this directionality feature as reported in [15]. It should be further noted that a similar line of research on the MAC and upper layers is also conducted in the context of sub terahertz communications (i.e., 60 GHz for short range wireless personal area networks, and 30–40 and 70–90 GHz for long-range wireless applications). The similarities in these works can be leveraged to some extent in MAC protocol designs for vehicular VLC networks. In contrast to RF links with the isotropic coverage, neighbor discovery and link establishments impose further challenges in vehicular VLC networks, which have not yet been addressed fully.

10 V2V VLC Evaluation Scenario

We consider a typical V2V VLC scenario as depicted in Fig. 2. Since both the left and right headlamps have almost identical output light distribution.

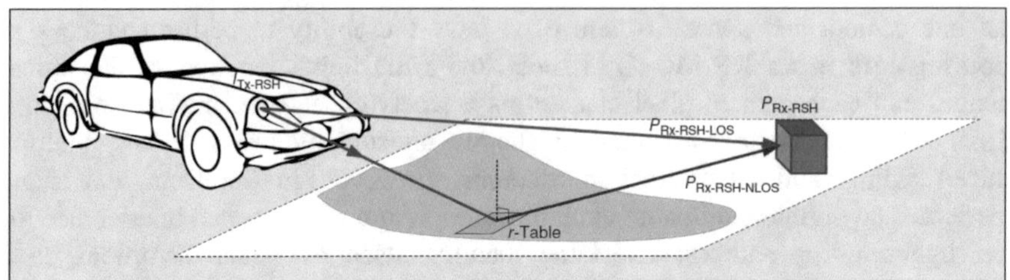

Fig. 2. V2V VLC system configuration

We consider both the LOS and the non-LOS (NLOS) components due to reflections from the road surface, which are assumed to have Lambertian profiles. In the following, we discuss the modeling of vehicle headlamps and the channel (i.e., the road surface reflection). The main purpose of both high- and low-beam vehicle's headlamps is to provide a safe and comfortable lighting environment for the drivers and other road users, during the day and night times and in all weather conditions. In detail, the high beams are used for long distance visibility with no oncoming cars, whereas the low beams with an asymmetrical light pattern provide maximum forward and lateral illuminations while minimizing the glare toward oncoming cars and other road users. In this article, we have adopted the widely used mathematical model for the headlamp beam pattern model (developed by the Transportation Research Institute of the University of Michigan).

Experimental data evaluated of 25 tungsten halogen headlamps. Following photometric data measurement using a gonio photometer, the collected data were weighted by the current sales figure for the corresponding vehicles and integrated into the mathematical model.

Figure 3 shows the isocandela and isoilluminance diagrams of the road surface from pair of high beam (luminous intensities at the 50th percentile, a lamp mounting height of 0.62 m, and a lamp separation of 1.12 m) and the low-beam headlamps (luminous intensities at the 50th percentile, a lamp mounting height of 0.66 m, and a lamp separation of 1.20 m), respectively. As can be seen for high-beam headlamps, a narrow flat beam is projected in a horizontal direction a few degrees to the left, thus providing a symmetrical illumination pattern on the road. However, the low beam headlamps provide an asymmetrical pattern designed to offer adequate forward and lateral illuminations, in addition to controlling glare by limiting light being directed toward the eyes of other drivers. The reflectance from the road surface depends on the materials used, the road surface, time of the year, and the weather conditions. For wet or moist road surfaces, a large amount of specular reflection may occur, thus resulting in greater luminance nonuniformity reflections. In the ideal case, the reflected radiant intensity is proportional to the cosine value of polar angle. However, for the arbitrary surface, both diffuse and specular reflections will take place, and the reflected light will be emitted in all directions of the upper hemisphere, which is challenging to describe mathematically. Therefore, a table of luminance coefficients (known as a r-table) has been developed for different road surface classifications based on a large number of photometric measurements as given in [16].

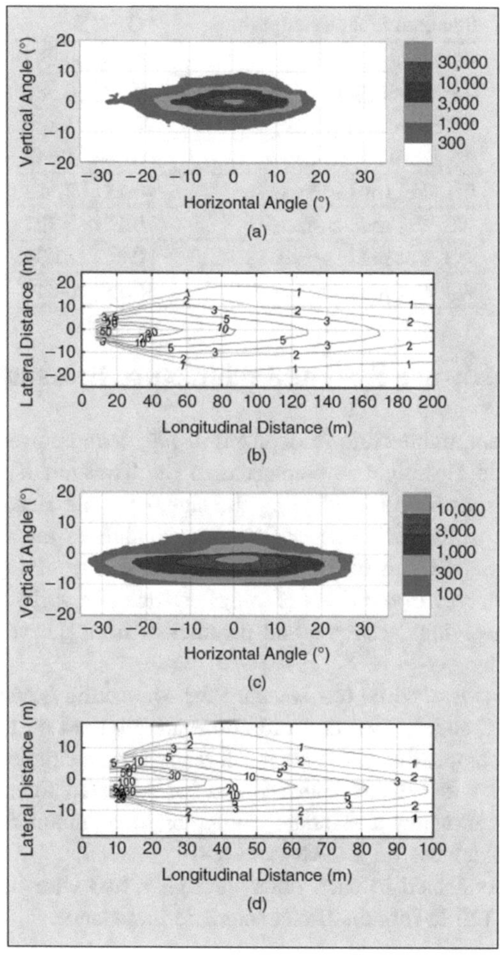

Fig. 3. The isocandela and isoilluminance diagrams of the road surface. **a**, **b** high beam, **c**, **d** low beam

In addition to the detailed reflective property for a range of incidence angle, the average luminance coefficient (Q) and the specular factor (S) should also be defined to describe the general reflection properties of a road surface. In general, the larger the Q is, the more reflective the road surface is. The larger the S is, the more the road surface looks like a mirror. The International Commission on Illumination has classified the road surfaces into different categories for a range of road surface materials and wet conditions based on the values of Q and S. Table 2 shows eight road-surface classifications defined in terms of the r-table, Q, and S. Note that R1 is the most diffused surface, while W4 is the most specular surface.

Table 2. Road-surface classifications

Standard table description	Q	S
R1 Mostly diffused	0.10	0.247
R2 Mixed (diffused and specular)	0.07	0.582
R3 Slightly specular	0.07	1.109
R4 Mostly specular	0.08	1.549
W1 Wet road surface	0.11	3.152
W2 Wet road surface	0.15	5.722
W3 Wet road surface	0.2	8.633
W4 Wet road surface	0.25	10.842

11 Vehicle Hardware System Architecture Schematic

The Hardware system architecture in depicted in Fig. 1 and consists of the following components: Vehicle Headlight is connected to the Transmit Driver Amplifier (TX Driver) which is responsible for adapting the signal for the Headlight LED, the TX Driver is connected to the LiFi Access Point that modulates the signal which is connected to the Ethernet Switch in the same network with the NI sbRIO Device with CAN Interface connected to the CAN (Controller Area Network) Bus Network of the car, sending and receiving messages and parameters from the vehicles ECUs (Electronic Control Unit).

CAN is a multi-master serial bus standard for connecting Electronic Control Units ECUs also known as nodes. Two or more nodes are required on the CAN network to communicate. The complexity of the node can range from a simple I/O device up to an embedded computer with a CAN interface and sophisticated software. The node may also be a gateway allowing a standard computer to communicate over a USB or Ethernet port to the devices on a CAN network.

All nodes are connected to each other through a two wire bus. The wires are a twisted pair with a 120 Ω (nominal) characteristic impedance.

12 Software System Architecture Schematic

The software architecture is depicted in Fig. 3, data acquired from the vehicle CAN Bus is analyzed and calculated trough decision algorithms programmed in NI sbRIO CAN Interface after processing, data I transferred through Ethernet to the LiFi Access point, data coming from other vehicles is acquired through the receiver module, then sent to the sbRIO CAN Interface for processing though the LiFi AP and Ethernet Connection.

13 Performance Results

For the performance evaluation, we assume the deployment of low-beam headlamps and consider R2 and W3 road surfaces. First, the received optical power of two types of road surfaces on a vertical plane at different transmission spans should be calculated followed by the noise variance on the same vertical plane. For the LOS path, the provided PD's position is fixed, the distance between the transmitter (Tx) and the Rx and the direction from Tx to Rx (in relation to the headlamp axis) can be obtained. Therefore, by checking and interpolating the low-beam headlamp model, the luminous intensity at that particular direction and the illuminance at the PD's surface as well as the received optical power (P_r) can be determined. For the NLOS path, the illuminance value at a small area A on the road surface should be determined, which acts as a secondary light source. As the reflected light follows the road-surface reflection properties, the luminous intensity of the reflected light in the direction from A to the Rx can be predicted and Pr from A can be determined. The total received optical power from the NLOS path is determined by integrating all reflected lights from the entire road surface. Additive white Gaussian noise is assumed in our work. This is a good approximation of the shot noise caused by the background radiation when the intensity of the light incident on the PD is sufficiently high, which is easily justified in practice. Based on the received optical power, we obtain SNR and bit error rate (BER). Figure 4 demonstrates the relationship between the communication range and the BER performance of V2V VLC system at a data rate of 50 Mb/s on a vertical plane for three different transmission spans of 20, 40, and 70 m. It is notable that as the distance

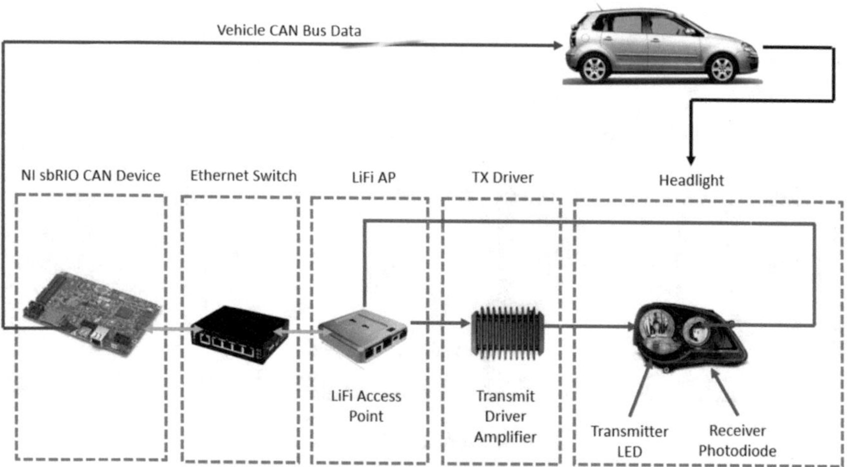

Fig. 4. Hardware schematic

increases, the BER performance decreases. We can observe from Fig. 4 that the lowest contour lines for the BER degrade from 10–80 at 20 m to 10–4 at 70 m. This is because when the link span changes from 20 to 70 m, the received optical power reduces by more than ten times. It can also be observed that the zones with the lowest BER tend to be smaller and shorter as the distance increases, and their positions are inclined to be more skewed to the right. This becomes more apparent when examining lines for the BER of 10–30 in Fig. 4, since the adopted low-beam headlamp model is designed for U.S. cars with the driving on the right W3 has higher Q and S compared with R2. Thus, the W3 road surface reflects more light than the R2 road surface, which increases the received light from the NLOS link. From the BER profiles of two road

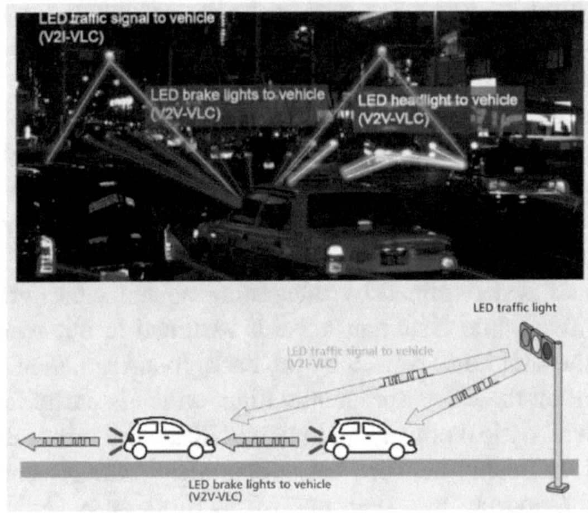

Fig. 5. Vehicle-to-Vehicle V2V and Vehicle-to-Infrastructure V2I schematic

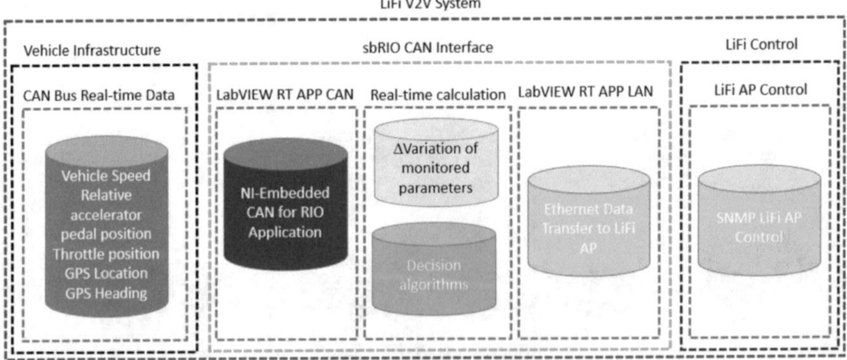

Fig. 6. Software architecture

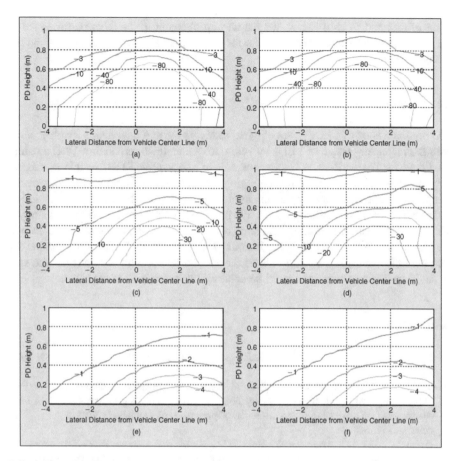

Fig. 7. The BER distribution on a vertical plane at three different distances with clean low-beam amps. **a** log10 (BER) on a vertical plane (20 m from Tx and R2). **b** log10 (BER) on a vertical plane (20 m from Tx and W3). **c** log10 (BER) on a vertical plane (40 m from Tx and R2). **d** log10 (BER) on a vertical plane (40 m from Tx and W3). **e** log10 (BER) on a vertical plane (70 m from Tx and R2). **f** log10 (BER) on a vertical plane (70 m from Tx and W3).

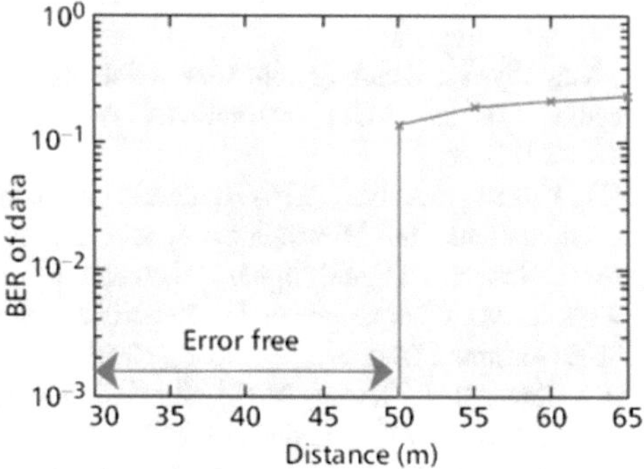

Fig. 8. BER over distance

surfaces (R2 and W3) in Fig. 4, we see an improvement in the BER performance for the W3 road surface compared with that for the R2 road surface, especially for the link span of 40 m.

14 Conclusion

The omnipresence of LEDs in streetlights, traffic signs, and advertising displays along with the increasing usage of LEDs by automotive manufacturers makes VLC a natural candidate for the wide-scale implementation of cooperative ITSs. In this article, we have provided an overview of this emerging research area by highlighting some recent results and identifying open problems in channel modeling, the physical layer design, and upper layers to be further pursued. We presented a performance evaluation study of a V2V system using a measured headlamp beam pattern model and taking into account the impact of road reflected light. Results demonstrated that depending on the PD location in the car (i.e., its height above the ground level), a higher transmission span up to 70 m between two cars could be achieved at a data rate of 50 Mb/s.

References

1. Islim, M.S., Haas, H.: Modulation Techniques for Li-Fi, vol. 02, 004 (2016). ISSN 1673-5188
2. Tsonev, D., Videv, S., Haas, H.: Light Fidelity (Li-Fi): Towards All-Optical Networking
3. U.S. Transportation Secretary Anthony Foxx, http://www.nhtsa.gov/About+NHTSA/Press+Releases/NHTSA-issues-advanced-notice-of-proposed-rulemaking-on-V2V-communications
4. Uysal, M., (Fary) Ghassemlooy, Z., Bekkali, A., Kadri, A., Menouar, H.: Performance study of a V2V system using a measured headlamp beam pattern model. https://doi.org/10.1109/mvt.2015.2481561
5. Morgan, Y.L.: Notes on DSRC & WAVE standards suite: Its architecture, design, and characteristics. IEEE Commun. Surv. Tut. **12**(4), 504–518 (2010)
6. Akanegawa, M., Tanaka, Y., Nakagawa, M.: Basic study on traffic information system using LED traffic lights. IEEE Trans. Intell. Transport. Syst. **2**(4), 197–203 (2001)
7. Kitano, S., Haruyama, S., Nakagawa, M.: LED road illumination communications system. In: Proceeding of IEEE 58th Vehicular Technology Conference Fall, vol. 5, pp. 3346–3350 (2003)
8. Liu, C., Sadeghi, B., Knightly, E.: Enabling vehicular visible light communication (V2LC) networks. In: Proceedings of 8th ACM International Workshop on Vehicular Inter-Networking, pp. 41–50 (2011)
9. Lourenco, N., Terra, D., Kumar, N., Alves, L.N., Aguiar, R.L.: Visible light communication system for outdoor applications. In: Proceedings of 8th International Symposium on Communication Systems, Networks Digital Signal Processing, pp. 1–6 (2012)
10. Yu, S.-H., Shih, O., Tsai, H.-M., Wisitpongphan, N., Roberts, R.: Smart automotive lighting for vehicle safety. IEEE Commun. Mag. **51**(12), 50–59 (2013)
11. Lee, S.J., Kwon, J.K., Jung, S.Y., Kwon, Y.H.: Simulation modeling of visible light communication channel for automotive applications. In: Proceedings of 15th International IEEE Conference on Intelligent Transportation Systems, pp. 463–468 (2012)

12. Mesleh, R., Elgala, H., Haas, H.: On the performance of different OFDM based optical wireless communication systems. J. Opt. Commun. Network. **3**(8), 620–628 (2011)
13. Kizilirmak, R.C., Uysal, M.: Relay-assisted OFDM transmission for indoor visible light communication. In Proceedings of IEEE International Black Sea Conference on Communications Networking, pp. 11–15 (2014)
14. Takai, I., Harada, T., Andoh, M., Yasutomi, K., Kagawa, K., Kawahito, S.: Optical vehicle-to-vehicle communication system using LED transmitter and camera receiver. IEEE Photon. J. **6**(5) (2014)
15. Tomas, B., Tsai, H.-M., Boban, M.: Simulating vehicular visible light communication: physical radio and MAC modeling. In: Proceedings of IEEE Vehicular Networking Conference, pp. 222–225 (2014)
16. Luo, P., Ghassemlooy, Z., Minh, H.L., Tang, X., Tsai, H.-M.: Undersampled phase shift ON-OFF keying for camera communication. In: Proceedings of 6th International Conference on Wireless Communications Signal Processing, 2014, pp. 1–6
17. https://purelifi.com/lifi-products/

Real Time Web Enabled Smart Energy Monitoring System Using Low Cost IoT Devices

J. Gaurav(✉), I. Pooja, C. R. Yamuna Devi, and C.R. Prashanth

Dr. Ambedkar Institute of Technology, Bangalore 560056, Karnataka, India
{gauravprasad96, poojairannal, yamuna.devicr, Prashanthcr.Ujjaini}@gmail.com

Abstract. The renewable energy is one of the best options to meet the modern energy requirements but today, renewable energy resources account for about 33% of India's primary energy consumptions. India's population is more than 1028 million, growing at an annual rate of 1.58% every year for this reason, modern forms of energy management and diagnostics acts as one of the most important drivers in economic growth and providing access to cheap reliable energy to consumers, which can be achieved by integrating IoT system—a fast booming technology, which facilitates energy monitoring and smart metering in real time. This paper is an application based low cost IoT design consisting of a powerful microcontroller, Sensors that can measure or sense current and voltage levels in a house hold appliances, integration of web interfaces that are capable of keeping analytical and graphical track reports of daily power consumptions.

Keywords: Renewable energy · IoT systems · Smart energy monitoring · Sensors · Web interfaces

1 Introduction

India being a developing country, has its traditional grid which meets the demand of electricity with maximum losses hence [1], it becomes inadequate and exorbitant to adopt the upcoming technologies like Smart Grids and Smart Metering system in its present established grid network [2, 3].

Generally, these new technologies creates an important innovation for the modern world by delivering reliable power supply [4] to the consumers across the country with a feature of two way communication between the consumers and the utilities [2]. The attempts made to evolve the existing grid system into smart grids, along with the deployment of smart meters which results in improved efficiency, reliability and increased optimal use of available power resources [5] which requires a major wide scale investment at that instant, which would directly disturb the economy of the nation [6]. Thus bearing this financial crisis of the nation this paper emphasizes Internet of Things as one of the important technical innovation with its vast reliable features like intelligence [2, 7], energy monitoring, wide connectivity secured features [8] and sensing which

M. E. Auer and K. Ram B. (Eds.): REV2019 2019, LNNS 80, pp. 330–342, 2020.
https://doi.org/10.1007/978-3-030-23162-0_30

considerably reduces regular human interventions for performing operations. The proposed system specified in this paper predominantly makes use of particle photon [9]—which is a scalable IoT device, along with high precision, low cost sensors such as a current sensor WCS-2702, single phase AC voltage sensor ZMPT101B [10] which contributes on building a cost effective smart energy monitoring system which is both scalable as well as accessible to industrial standards. This system makes use of open source software such as particle photon dashboard that provides a secured connection for hardware over Internet, integrating web interfaces which are capable of analytical, graphical and digital displays like ThingSpeak and Blynk App that provides historical reports on daily power consumption with high accuracy [11]. Further more, Fig. 1 demonstrates the basic idea of sensors that can be connected to the energy resources thus, enabling two way communication between grids and utilities [12] further the sensors' output data that are continuously stored in the cloud platform for monitoring purposes [13, 14].

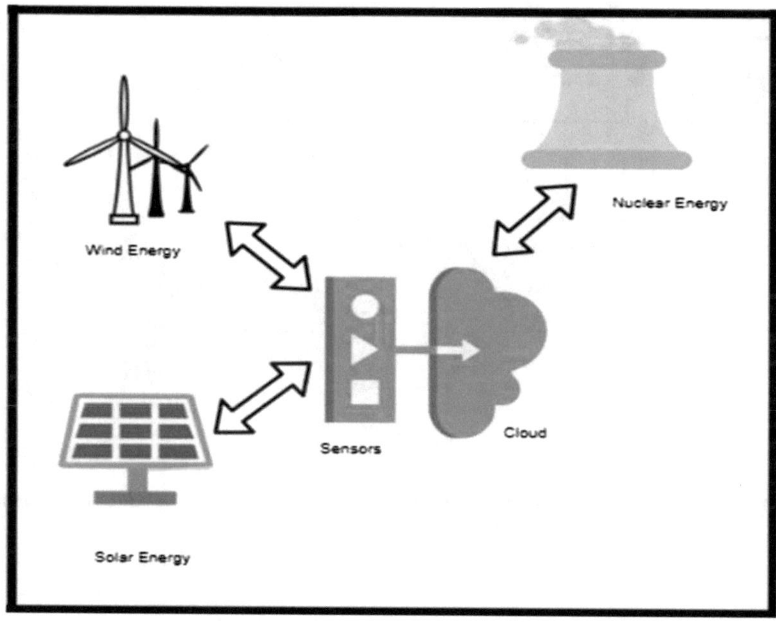

Fig. 1. Generic idea of IoT system in energy monitoring.

2 Architecture

The interpretation of the proposed concept under this paper is briefly described in the architecture as shown in Fig. 2. The AC power supply i.e., 230 V from the mains are supplied to 5 A or 15 A switch board that handles various types of loads such as fans, light bulbs, washing machine, fridge and many more house hold appliances [15] that are connected to it, for which smart energy monitoring operation has to be performed [10, 15]. In the block diagram two inexpensive, highly accurate sensors are being used such as WCS-2702 current sensor, single-phase AC voltage sensor-ZMPT101B that are connected to the load, for example we have considered a room heater. The output of these sensors are integrated to powerful ARM Cortex 3 microcontroller [9] such as a particle photon which is an IoT device that helps in performing accurate energy monitoring operations for smart electrical appliances through it's highly flexible firmware platform, secured connectivity to other cloud platforms over the internet (OTA) such as ThingSpeak, Blynk App where in live data consisting of analytical, historical and digital forms of power consumption reports [16, 17] are updated on daily basis to the user thus, preventing over estimation of power consumptions by the household appliances [11].

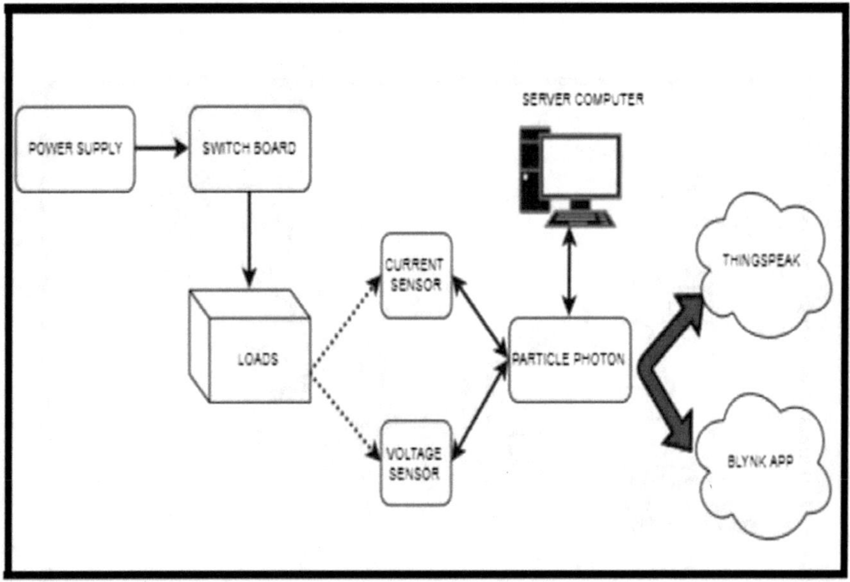

Fig. 2. Architecture

3 Implementation

The limelight of carrying out the implementation process is the selection of hardware components due to their important features of operations that are scalable from low to high standards for meeting industrial requirements. In order to build cost effective [6, 15] smart energy monitoring system proposed under this paper, for homes or industries the following technical aspects of hardware components are as listed in Table 1.

Bearing in mind, the critical financial dynamics of INDIA [1, 4], the proposed paper aims at building a smart energy monitoring system by using a truncated hardware

Table 1. Important features of hardware components

Components	Features	Expenses (INR)	Expenses (USD)
Particle photon	1. Open source design	3000–4000 Rs.	$40.53–$54.04
	2. Soft AP setup		
	3. 12 bit A/D inputs		
	4. 120 MHz frequency		
	5. Arm Cortex M3		
Current sensor WCS 2702	1. Low noise analog signal path	180 Rs.	$2.43
	2. Min. sensing current (0–5) A		
	3. High sensitivity 250 mV/A		
	4. Zero magnetic hysteresis		
	5. 23 kHz bandwidth		
Voltage sensor ZMPT101B	1.250 AC voltage measuring	480 Rs.	$6.48
	2. Micro-precision-voltage transformer		
	3. Accuracy class of 0.2		
	4. Frequency 50–60 Hz		
	5. Phase ≤ 20		
	Total cost	3660–4660Rs.	$49.44–$62.96

circuits which are easily programmed and upgraded using a reliable open source software [18, 19]. An IoT cloud platform along with IoT dashboards forms a vibrant backbone for accumulating, scrutinizing and logging the data in real time [20]. Therefore, the implementation process is realized in three major steps which comprises of STEP-1, STEP-2, and STEP-3 which are as follows:

3.1 STEP-1: Input Connections of Current and Voltage Sensors w.r.t. the Load

Initially, the input of the current sensor WCS 2702 i.e., +ip and −ip terminals are connected in series to the load and voltage sensor ZMPT101B terminals such as Line (L) and Neutral (N) are connected parallel to the load respectively [10], through intermediate 15 A switch board whose pin configuration is shown in Fig. 3. After this initial connections being made to the sensors, their outputs are monitored through STEPS 2 and 3.

3.2 STEP-2: Connecting Sensors' Output Pins to 12 Bit ADC

In order to find out the nature of output pins i.e.., Vout signals from both the sensors, the connections are being made to 12 bit ADC for testing purpose. The outputs of the sensors are connected to Analog Input Channels 0 and 1, for the purpose of monitoring over current and voltage consumed by a smart appliance such as a room heater [7, 10].

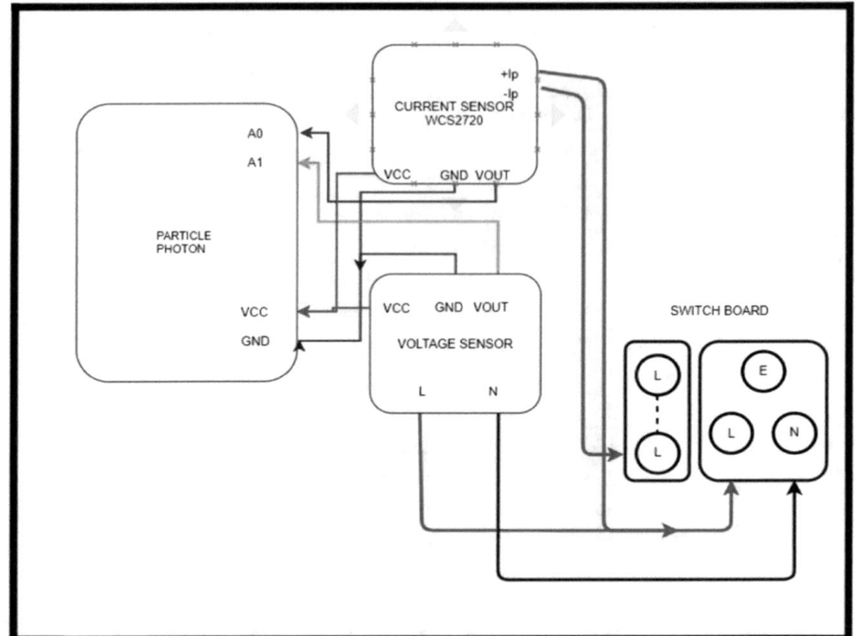

Fig. 3. Pin configuration of switch board with sensors

LabVIEW is used to design a code for overall working of the system and the resulting waveforms such as current and voltage values are noted during ON or OFF state of the appliance as shown in Fig. 4, for the purpose of programming in IoT dashboards.

3.3 STEP-3: Connecting Sensors' Output Pins to Particle Photon

To ensure the safety of Particle Photon, STEP-2 is followed first and then proceeded to connect the sensor's output pins i.e.., Vout to analog input pins—A0 and A1 of Particle Photon as shown in Fig. 5. After, the complete integration of these sensors along with a load, the firmware is burnt and updated via Wi-Fi [9, 20]. The signals from the pin Vout of respective sensors are monitored and the parameters inculcating current and voltage are derived by programming. Further these data are published to Blynk App as well as ThingSpeak in-order to have remote monitoring capabilities [21]. The Blynk App is installed in mobile phones and is synced to particle photon in-order to have complete observations over the phones [22].

Initially the first prototype of the system is carried out by soldering the hardware components as a plug and play system over a GPB board as shown in Fig. 6. Remaining connections are being made as depicted in STEPS 1, 2 and 3.

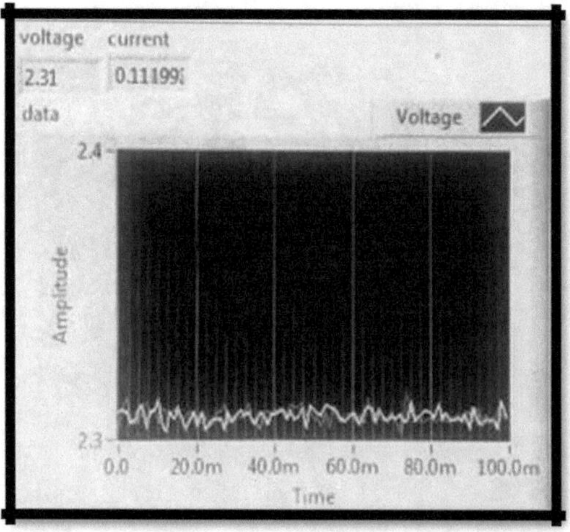

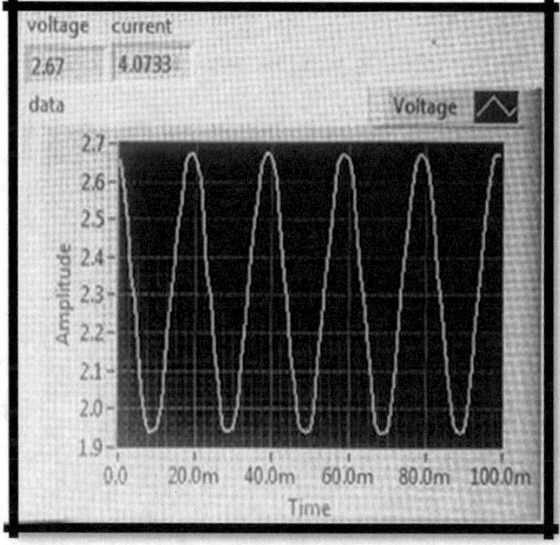

Fig. 4. Sensor outputs during ON or OFF state of the appliance.

4 Software

As stated earlier particle photon is a powerful STM32 ARM Cortex M3 microcontroller with Broadcom BCM43362 Wi-Fi chip as its connectivity to the internet with 12 bit Analog to Digital inputs ranging from 0 to 4095 [9], which makes it easy for creating own Internet of Things prototype which can be scaled on a larger aspects. It has an open source web based IDE, where the photon code is written, edited and shared for developing products with cloud based messaging platform that allows the sensors to communicate with each other in a secured way [8], which flashes the built code to the photon over the air, thus connecting the device at much higher rates over the internet [9]. The code for the proposed system is based on the linear equation **y = mx + c**,

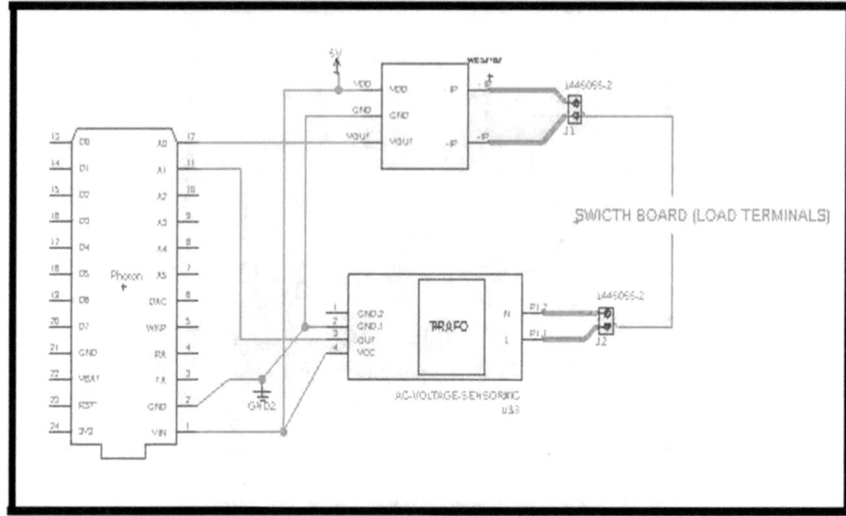

Fig. 5. Circuit connection with particle photon

Fig. 6. Hardware assembly.

power **P = V** × **I**; with voltage and current values as a function during their On and Off state conditions over a definite period when the load is put in use as shown in Fig. 7.

Command statement such as **analogRead()** reads the output values of both the sensors from A0 and A1 pins from the microcontroller and maps it to integer value

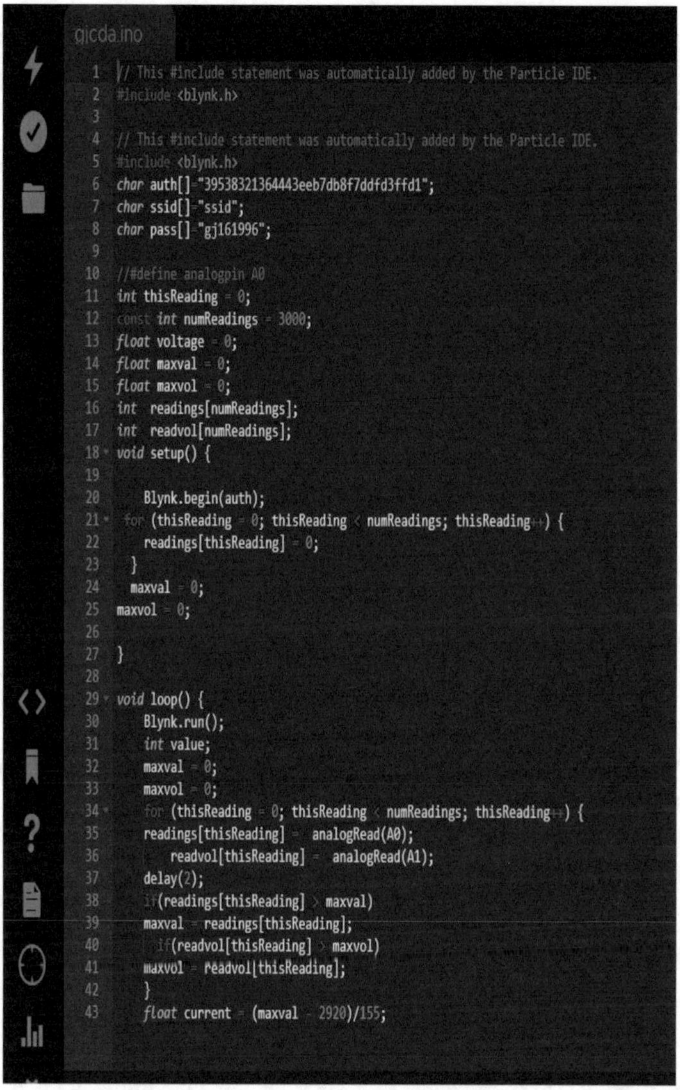

```
gicda.ino
// This #include statement was automatically added by the Particle IDE.
#include <blynk.h>

// This #include statement was automatically added by the Particle IDE.
#include <blynk.h>
char auth[]="39538321364443eeb7db8f7ddfd3ffd1";
char ssid[]="ssid";
char pass[]="gj161996";

//#define analogpin A0
int thisReading = 0;
const int numReadings = 3000;
float voltage = 0;
float maxval = 0;
float maxvol = 0;
int  readings[numReadings];
int  readvol[numReadings];
void setup() {

    Blynk.begin(auth);
  for (thisReading = 0; thisReading < numReadings; thisReading++) {
    readings[thisReading] = 0;
  }
  maxval = 0;
maxvol = 0;

}

void loop() {
    Blynk.run();
    int value;
    maxval = 0;
    maxvol = 0;
    for (thisReading = 0; thisReading < numReadings; thisReading++) {
    readings[thisReading] = analogRead(A0);
        readvol[thisReading] = analogRead(A1);
    delay(2);
    if(readings[thisReading] > maxval)
    maxval = readings[thisReading];
      if(readvol[thisReading] > maxvol)
    maxvol = readvol[thisReading];
    }
    float current = (maxval - 2920)/155;
```

Fig. 7. Photon code

between 0 and 4095, another statement such as **particle.publish("voltage", voltage)** with similar statements for current and power sends message to the cloud that some event has occurred letting to know the states of the device, this event is realized in console part of the IDE. Photon's web IDE supports large number of libraries in order to view the results from a virtual places, two such libraries are **#Thingspeak** which enables the hardware to write or read data to or from thingspeak which is an open data platform for IoT with MATLAB analytics, visualizations and provides historical data of daily power consumption through secured channel and API keys [17], secondly,

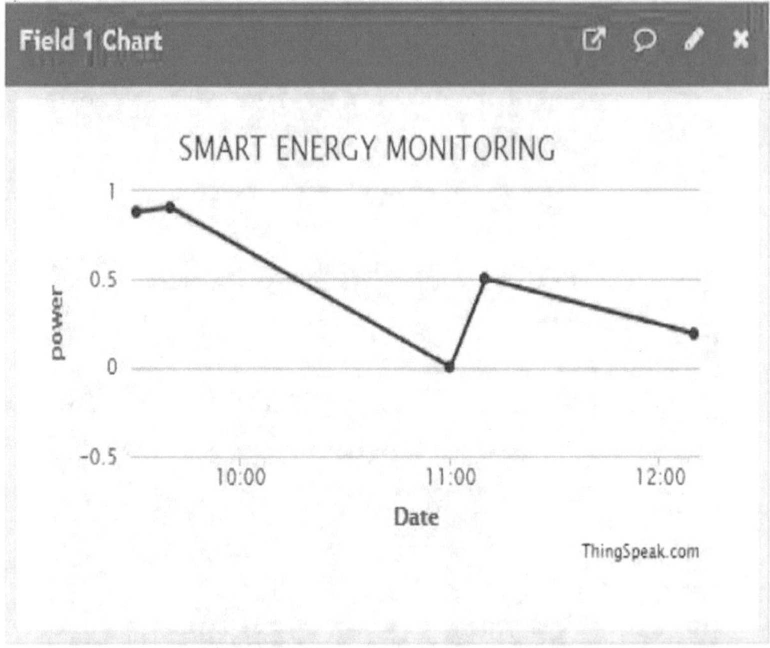

Fig. 8. ThingSpeak data

#include "blynk.h" creates graphical user interface through AUTH key mailed to the registered account in mobile phone where user can create pins and widgets in the app installed from the play-store to have digital display of output values as shown in Figs. 9 and 10.

5 Outcomes

The proposed smart energy monitoring system was able to successfully monitor 1000 W room heater operating for a particular duration of time, during which the energy or power consumed by the appliance was continuously monitored by publishing the sensor output values to ThingSpeak and Blynk app platforms [23]. These published sensor output values to cloud platforms have a consistent record or history of power consumed by the smart electrical appliance connected to the switch board which are held by Fields in a ThingSpeak channel.

Figure 8 shows the curve of power consumed in Kilo Watts ranging from 0 KW during OFF state and 0.99–1 kW during ON state for a time period of 15–20 min when the room heater was put to use. The graph also gives complete information of time duration for days, weeks, months, for a year and also the current dates along with the historical data regarding the use of appliance through which one can avoid peak hours of electricity demand, black hours by scheduling the use of devices when the power is

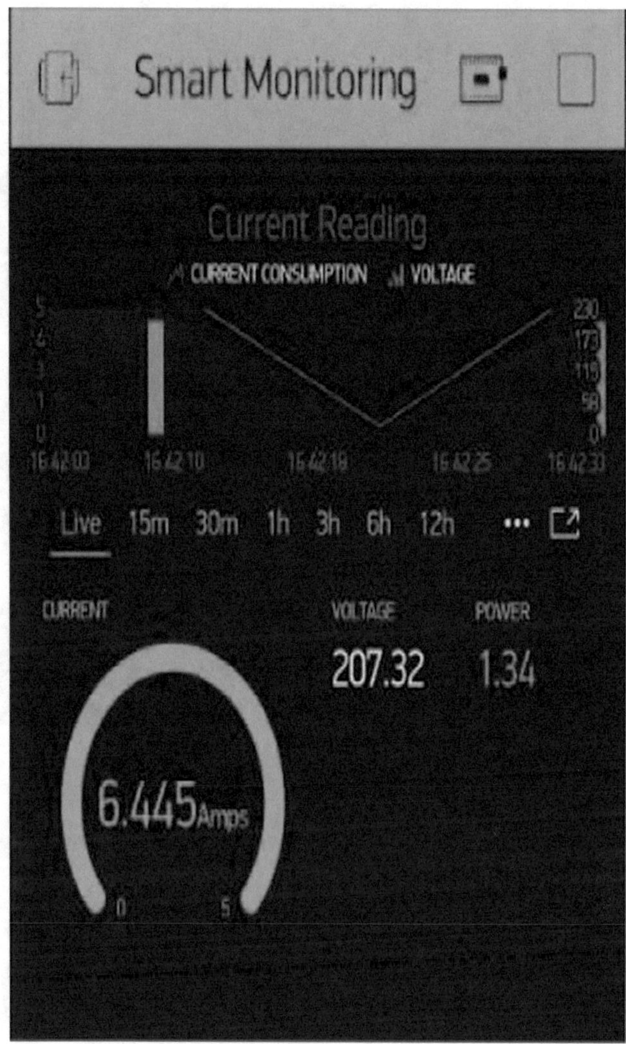

Fig. 9. Blynk app voltage and current widgets

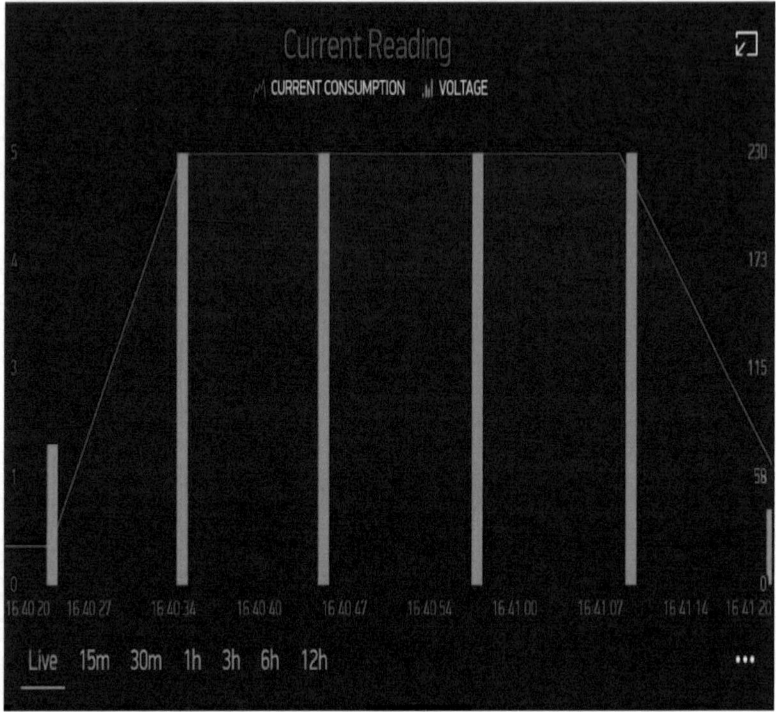

Fig. 10. Super chart display of live current and voltage consumptions

cheap, thereby having control over electricity bill at the end of each month or for a year.

Similar data is obtained by Blynk app in the mobile phone through which one can view the current and voltage widgets from virtual place which shows current consumption of 6.445 A and voltage of 207.32 V as shown in Fig. 6, as well as live stream of current and voltage consumptions in super chart display as shown in Fig. 7 for a time duration of 15, 30 m and so on.

6 Conclusions

Overall this paper substantially highlights the hands on experience gained by the students in creating and innovating cost effective smart energy monitoring systems with the most evolving technologies like IoT [20], Cloud Computing and other convenient forms of web interfaces which would reduce the cost of introduction of smart meters over the existing grids on a wide scale. It also serves as a proof of concept for Internet of Energy-IoE [21].

Acknowledgements. The associated authors of this paper would like to express deep sense of gratitude to the respected **Principal Dr. C. Nanjundaswamy** and our respected **Head of the Department (TCE), Dr. Yamuna Devi C. R** and other teaching staffs for their unending support for this research paper.

References

1. Saini, S.: Evolution of Indian power sector at a glance. Natl. J. Multidiscip. Res. Manag. Accepted (2018)
2. Lloret, J., Tomas, J., Canovas, A., Parra, L.: An integrated IoT architecture for smart metering. IEEE Commun. Mag. **54**(12), 50–57 (2016)
3. Kappagantu, R., Daniel, S.A.: Challenges & Issues of smart grid implementation: a case of Indian scenario. J. Electr. Syst. Inf. Technol. (2018)
4. Tripathy, B.R., Sajjad, H., Elvidge, C.D., Ting, Y., Pandey, P.C., Rani, M., Kumar, P.: Modeling of electric demand for sustainable energy and management in India using spatio-temporal DMSP-OLS night-time data. Environ. Manag. **61**(4), 615–623 (2018)
5. Jithin Jose, K., Mohan, L., Nijeesh, U.K., Benny, T.C.: Smart energy meter
6. Fankhauser, S., Jotzo, F.: Economic growth and development with low-carbon energy. Wiley Interdiscip. Rev. Climate Change **9**(1) (2018)
7. Imran, M.I.: Intelligent home control and monitoring system via internet. Int. J. Sci. Dev. Res. (IJSDR) **1**(4), 82–87 (2016)
8. Jain, S., Kumar, V., Paventhan, A., Chinnaiyan, V.K., Arnachalam, V., Pradish, M.: Survey on smart grid technologies-smart metering, IoT and EMS. In: 2014 IEEE Students' Conference on Electrical, Electronics and Computer Science (SCEECS), pp. 1–6. IEEE (2014 Mar)
9. Bista, A.: Smart home using particle photon. Technical Report (2017)
10. Abubakar, I., Khalid, S.N., Mustafa, M.W., Shareef, H., Mustapha, M.: Calibration of ZMPT101B voltage sensor module using polynomial regression for accurate load monitoring (2006)
11. Prasad, A., Chawda, P.: Power management factors and techniques for IoT design devices. In: 2018 19th International Symposium on Quality Electronic Design (ISQED), pp. 364–369. IEEE (2018 Mar)
12. Perez, N., Luis, H., Vega, D.: State of the art and trends review of smart metering in electricity grids (2016)
13. Chakraborty, A., Sharma, N.: Advance metering infrastructure: technology and challenges (2016)
14. IoT Based Smart Energy Meter. Int. Res. J. Eng. Technol. (IRJET) **04**(04) (2017 Apr). e-ISSN: 2395-0056
15. Hiremth, M., Kumar, M.: Internet of things for energy management in the home power supply
16. Naveenkumar, J.S.: Smart energy meter
17. Maureira, M.A.G., Oldenhof, D., Teernstra, L.: ThingSpeak—an API and web service for the internet of things. Retrieved7/11/15World WideWeb, http://www.Mediatechnology.leiden.edu/images/uploads/docs/wt2014_thingspeak.pdf (2011)

18. Barua, A.M.: Smart Metering System for Energy Conservation and its Implementation in Assam (2018)
19. India Smart Grid Forum (ISGF) Newsletter, Smart Grid Bulletin (2017 Apr)
20. Singh, K.J., Kapoor, D.S.: Create your own internet of things: a survey of IoT platforms. IEEE Consum. Electron. Mag. **6**(2), 57–68 (2017)
21. Jung, A.: Building the internet of energy supply. Spiegel Online International, Dec 2010. Available: http://www.spiegel.de/international/business/0,1518,694287,00.html
22. Bressan, N., et al.: The deployment of a smart monitoring system using wireless sensor and actuator networks. In: Proceedings of IEEE SmartGridComm, Oct 2010
23. Castellani, A.P., et al.: Web services for the internet of things through CoAP and EXI. In: Proceedings of IEEE ICC RWFI workshop, Kyoto, Japan, June 2011
24. Rahman, M.M., Islam, M.O., Salakin, M.S.: Arduino and GSM based smart energy meter for advanced metering and billing system. In: 2015 International Conference on Electrical Engineering and Information Communication Technology (ICEEICT), pp. 1–6. IEEE (2015 May)
25. The Internet of Energy: A Web-Enabled Smart Grid System. IEEE Netw. (2012 July)

Smart Attendance System Using Deep Learning Convolutional Neural Network

I. Pooja, J. Gaurav(✉), C. R. Yamuna Devi, H. L. Aravindha, and M. Sowmya

Dr. Ambedkar Institute of Technology, Bangalore 560056, Karnataka, India
{poojairannal, gauravprasad96, Yamuna.devicr, arvindhlait, sowmya.mallik}@gmail.com

Abstract. Image recognition has been playing an increasingly larger role in the modern life like driver assistance systems, medical imaging system, quality control system to name a few. Artificial Neural Network models are extensively used for the above purposes due to their reliable success. One such update used here is the convolutional neural network (CNN, or ConvNet). This paper highlights the importance of pre-trained neural networks as well as the significance of Deep Learning used in the field of Academics and Advancement which is implemented in MATLAB Software. Smart Attendance Systems involves the image (face) detection and analyzes the data accurately. This approach solves the time consuming traditional method of attendance system and paves way for new advanced technologies.

Keywords: Convolutional neural network (CNN) · Deep learning · AlexNet · Transfer learning

1 Introduction

In earlier days supervision system was cumbersome to detect the location of student and it is very tedious task to take attendance manually. As technology has advanced, integrating the monitoring system with an automation technology will provide more convenient way in monitoring the student. The traditional method of attendance system is time and energy consuming. As technology took its shape, techniques like barcode system, RFID systems evolved eventually. The attendance maintaining system is difficult process if it is done manually. The smart and automated attendance system for managing the attendance can be implemented using the various ways of biometrics. Face recognition is one of them. By using this system, the issue of fake attendance and proxies can be solved. In this paper we throw light upon much more advanced and reliable way of smart attendance system i.e., use of artificial intelligence (Convolutional Neural Networks) which does such compassionate task more effectively and quickly.

Machine-learning technology powers many aspects of modern society: from web searches to content filtering on social networks to recommendations on e-commerce websites, and it is increasingly present in consumer products such as cameras and smartphones. Machine-learning systems are used to identify objects in images, transcribe speech into text, match news items, posts or products with users' interests, and

M. E. Auer and K. Ram B. (Eds.): REV2019 2019, LNNS 80, pp. 343–356, 2020.
https://doi.org/10.1007/978-3-030-23162-0_31

select relevant results of search. Increasingly, these applications make use of a class of techniques called deep learning. The concept of Deep Learning is integrated onto the attendance systems. This approach eliminates the time consuming traditional approach.

Deep learning allows computational models that are composed of multiple processing layers to learn representations of data with multiple levels of abstraction. These methods have dramatically improved the state-of-the-art in speech recognition, visual object recognition, object detection and many other domains such as drug discovery and genomics. Deep convolutional nets have brought about breakthroughs in processing images, video, speech and audio, whereas recurrent nets have shone light on sequential data such as text and speech.

Neural Networks are implemented to solve scientific and engineering challenges. The network upholds several pre-trained networks that are very useful in providing reliable solutions to the network. The layers are trained individually to provide probabilities of the solution. Because the computer gathers knowledge from experience, there is no need for a human computer operator formally to specify all of the knowledge needed by the computer. The hierarchy of concepts allows the computer to learn complicated concepts by building them out of simpler ones; a graph of these hierarchies would be many layers deep.

Deep learning is especially suited for image recognition, which is important for solving problems such as facial recognition, motion detection, and many advanced driver assistance technologies such as autonomous driving, lane detection, pedestrian detection, and autonomous parking.

2 History

(a) In 1960s, the first semi-automated system for facial recognition to locate the features such as eyes, ears, nose and mouth on the photographs.
(b) In 1970s, Goldstein and Harmon used 21 specific subjective markers such as hair colour and lip thickness to automate the recognition.
(c) In 1988, Kirby and Sirovich used standard linear algebra technique to the face recognition.

3 Existing System

3.1 Biometric (Fingerprint) System

- The biometric system is one such approach which has existed from past few years. This approach is fairly acceptable due to its added disadvantages.
- The error rates in biometric systems are the False Acceptance Rate (FAR) and False Rejection Rate (FRR).
- The other factors include cost, delay, complexity, scanning difficulty for physically challenged individuals etc. (Fig. 1).

Fig. 1. Biometric system

3.2 RFID System

- The Radio Frequency Identification (RFID) technology is one of an automation technology that is proved to be beneficial than the traditional barcode system.
- This system uses individual RFID tags for each user identification.
- This approach cannot be reliable as the card can be misused by unauthorized identities (Fig. 2).

Fig. 2. RFID system

4 Proposed System

- In our approach we use Deep Learning technique (Neural Networks) which identify the individuals based on stored Image data sets [Image Recognition].
- Convolution Neural Networks pose a great deal of advanced technology. CNN is a form of machine learning that enables computers to learn from experience and understand the world in terms of a hierarchy of concepts.

Fig. 3. Convolutional neural network

- These networks use features to classify images. The network learns these features itself during the training process without the involvement of human (Man Power) (Fig. 3).

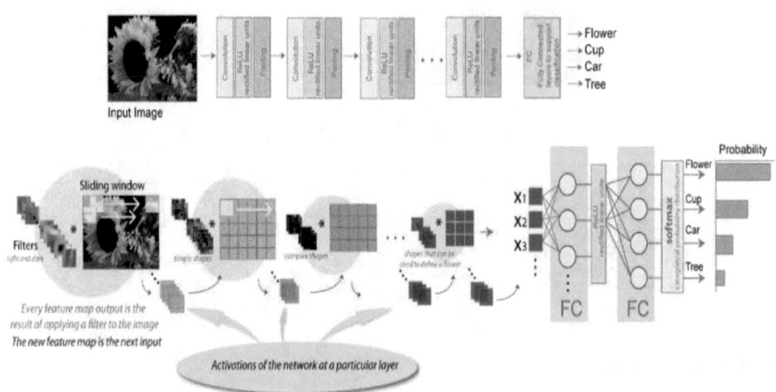

Fig. 4. Pre-trained network

5 What Is Pre-trained Convolutional Neural Network?

As the name suggest, these networks are previously trained with large data sets such as image database, pattern database etc. Each such pre-trained neural networks will have layers in which the data sets are embedded accordingly. These neural networks with inbuilt data sets are further trained and fine tuned for classification, transfer learning, pattern recognition, feature extraction and many such practices (Fig. 4).

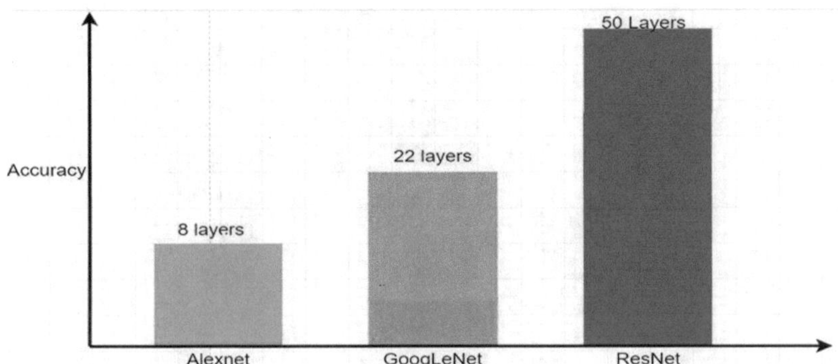

Fig. 5. Types with accuracy

Pretrained networks have learned rich feature representations for a wide range of natural images. You can apply these learned features to a wide range of image classification problems using transfer learning and feature extraction. The pretrained networks are trained on more than a million images and can classify images into 1000 object categories, such as keyboard, coffee mug, pencil, and many animals. The training images are a subset of the ImageNet database, which is used in ImageNet Large-Scale Visual Recognition Challenge (ILSVRC).

5.1 Following Are the List of Few Popular Pre-trained Networks

See Fig. 5.

1. **AlexNet**

AlexNet is 8 layers deep with 5 convolutional layers and 3 fully connected layers. AlexNet is fast for retraining and classifying new images.

2. **GoogLeNet**

GoogLeNet uses a typical inception module which consists of multiple-convolutional layers whose outputs will be concatenated to obtain best results in the system. Using the inception module, GoogLeNet is able to achieve high accuracy using limited computational cost. GoogLeNet is 22 layers deep.

3. **ResNet-50 and ResNet-101**

ResNet involves much more deeper layers and hence more accurate the result. As the names suggest, ResNet-50 is 50 layers deep and ResNet-101 is 101 layers deep.

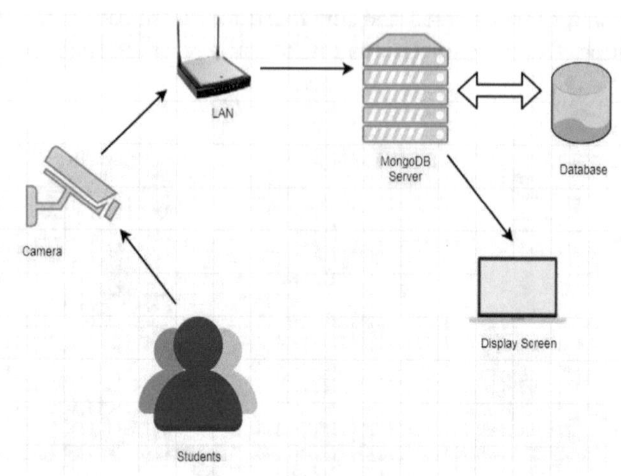

Fig. 6. Block diagram

6 Block Diagram

In Fig. 6, we show the system architecture in which the camera detects and recognize students' faces then sent the captured pictures to the system that compare the sent pictures with retrieved images from the database which contains students' information and images. Once the recognized face match a retrieved image, the attendance is marked for that person and the attendance sheet is updated. If there are missed faces while detecting and recognizing the process, the process will work repeatedly until all faces recognized. Finally, the result will be shown at the display board attached with the camera.

7 Implementation

In this paper, the implementation of smart attendance system uses MATLAB for training the neural networks. Deep learning is a branch of machine learning that teaches computers to do what comes naturally to humans: learn from experience. Machine learning algorithms use computational methods to "learn" information directly from data without relying on a predetermined equation as a model.

A Convolution Neural Network is composed of a series of layers, where each layer defines a specific computation. Training the network requires only training the last few layers of the pre-trained network instead of training from scratch.

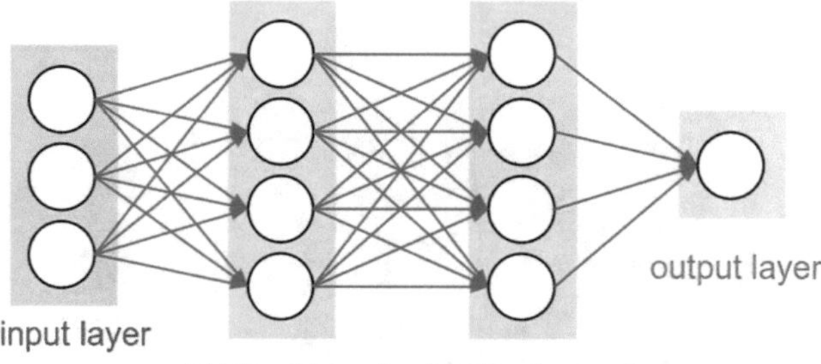

Fig. 7. Neural network process

- imageInputLayer—Image input layer
- convolution2dLayer—2D convolution layer for Convolutional Neural Networks
- reluLayer—Rectified linear unit (ReLU) layer
- maxPooling2dLayer—Max pooling layer
- fullyConnectedLayer—Fully connected layer
- softmaxLayer—Softmax layer
- classificationLayer—Classification output layer for a neural network (Fig. 7).

7.1 Step-1: Labelling of Images

The images are imported onto a single database or a single variable called the "**groundTruth**" variable. The groundTruth variable stores all the image data set that are labelled and will be further used for training purposes.

Fig. 8. Anitha

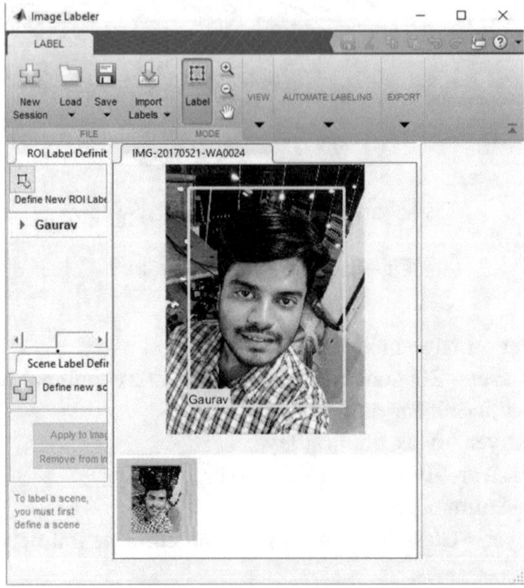

Fig. 9. Gaurav

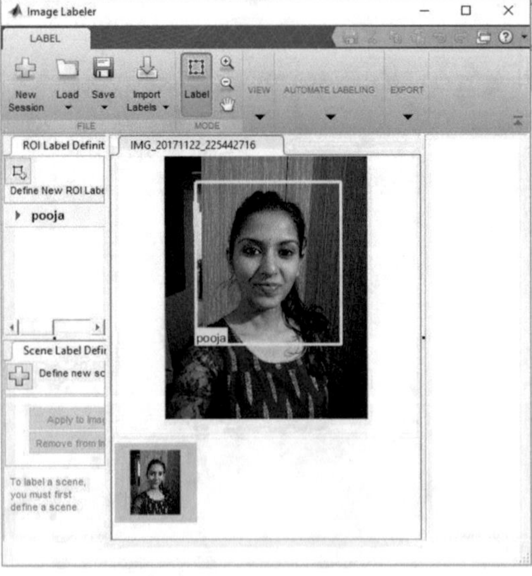

Fig. 10. Pooja

In order to create a groundTruth variable, one must define ROI (Region of Interest) i.e., the face of a student in this case. The **Imagelabeller** app used in MATLAB is exclusively used to create a rectangular **ROI label** (Region Of Interest) and Scene labels for conditions such as lightning and weather conditions for object detection,

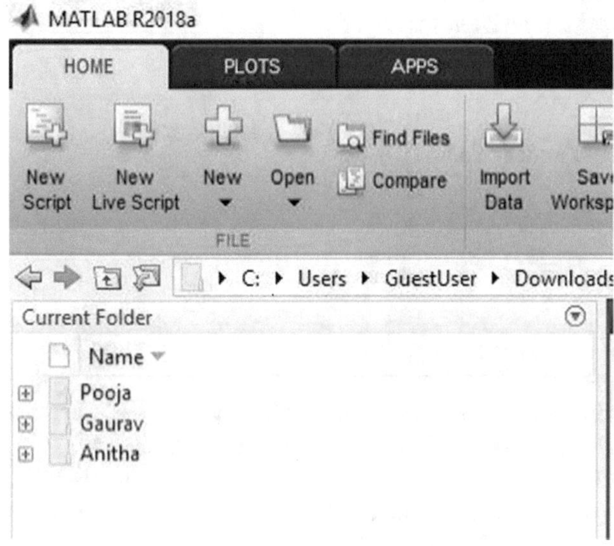

Fig. 11. Labelling of images

pixels for semantic segmentation, and scenes for image classification. In this case, we label the face of each student with his/her respective names.

In this session, we label 3 students images and store them onto a single folder or database. The names are as follows:

- Anitha (Fig. 8)
- Gaurav (Fig. 9)
- Pooja (Fig. 10)

7.2 Step 2

Once the images are labelled these are stored in a single database (or) a single folder as shown in Fig. 11. Each folder consists of 30 images of the following 3 students.

7.3 Step 3: Training Algorithm

```
>> dir = ('C:\Users\GuestUser\Downloads\ICD');
imds=imageDatastore(dir,'IncludeSubfolders',true,'LabelSource',
'foldernames'); [imdsTrain,imdsTest] = splitEachLabel(imds,0.7,'randomized');
numTrainImages = numel(imdsTrain.Labels);
idx = randperm(numTrainImages,16);
figure
for i = 1:16
subplot(4,4,i)
I = readimage(imdsTrain,idx(i));
Imshow (I)
end
net =
alexnet;
net.Layers
inputSize = net.Layers(1).InputSize
augimdsTrain =
augmentedImageDatastore(inputSize(1:2),imdsTrain); augimdsTest =
augmentedImageDatastore(inputSize(1:2),imdsTest); layer = 'fc7';
featuresTrain =
activations(net,augimdsTrain,layer,'OutputAs','rows'); featuresTest =
activations(net,augimdsTest,layer,'OutputAs','rows'); YTrain =
imdsTrain.Labels;
YTest = imdsTest.Labels;
classifier =
fitcecoc(featuresTrain,YTrain); YPred =
predict(classifier,featuresTest); idx = [1 5
10 15];
figure
for i =
1:numel(idx)
subplot(2,2,i)
I =
readimage(imdsTest,idx(i));
label = YPred(idx(i));
imshow(I) title(char(label))
end
accuracy = mean(YPred == YTest)
```

7.4 Step 4

In the training algorithm, one can select any random number of images to classify. In this segment the algorithm randomly selects 16 images for classification (Fig. 12).

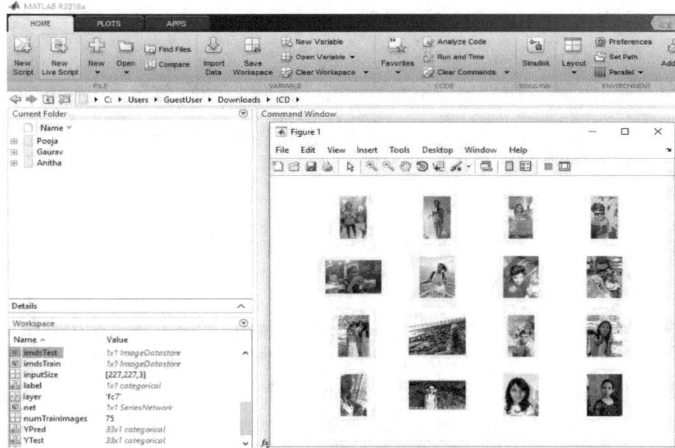

Fig. 12. Classifying images

8 Software Required

MATLAB Learning Tool box is used to collect the images and detect them according to the training model as mentioned earlier.

This is sufficient for a set of students or faces. The image recognition will get complicated as it involves more students, in such case one has to use a reliable database system which can incorporate such a tedious task. The trained model in MATLAB Tool Box which consists of several students' faces need to be securely stored and be easily accessible. Hence in this paper we use MongoDB which is the most popular and Document oriented database system (Fig. 13).

8.1 MongoDB

As the experiment involves huge collection of students data and information, one must ensure that the database used, needs to be much more comfortable and easy in order to make amendments to the student's profile whenever it is necessary. Therefore the software presented in this paper is MongoDB.

MongoDB is one of the most popular NoSQL type of database engine. It has gained attraction amongst the database designer communities over the relational databases.

It is a type of Non-Relational and document oriented NoSQL database system. The data entered is non relational i.e., the data need not be related to the other documents. It is a database which consists of collections which are preferably known as tables in the traditional relational type of database. The rows or tuples in MongoDB are known as documents.

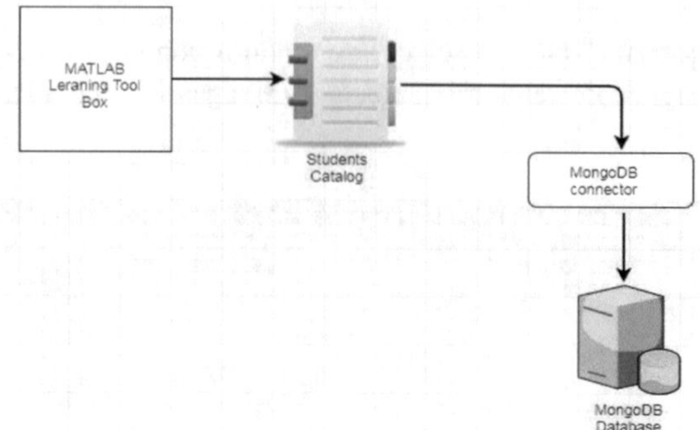

Fig. 13. Software structure

Basically, the usage of MongoDB can simply host multiple databases and it serves much functionality such as providing high performance, flexibility, availability, simplicity and easy scalability. Due to that, MongoDB is well-suited to be utilized for any system that requires large and dynamic database experience.

This type of database system is an open source system and uses Simple Query language.

9 Result

The images are classified according to the labelled images (Fig. 14).

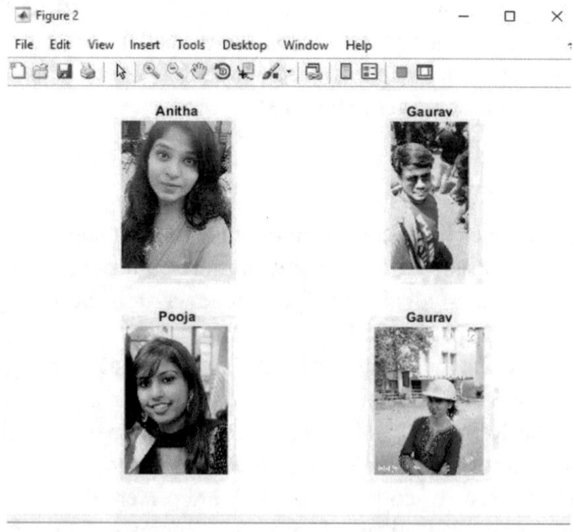

Fig. 14. Output figure

9.1 Accuracy

The training process eventually reaches maximum accuracy as it is trained and tested. The initial accuracy is 84.85% which is quite reliable (Fig. 15).

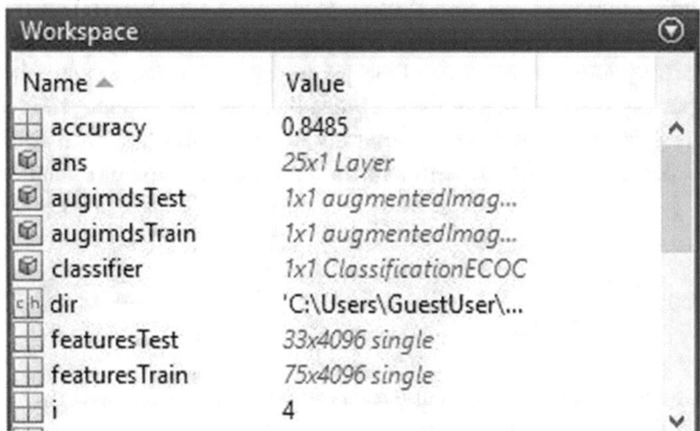

Fig. 15. Accuracy estimate

9.2 Advantages Over Traditional Method

- Provides a valuable attendance service for both teachers and students.
- Reduce manual process errors by providing automated and a reliable attendance system that uses face recognition technology.
- Increase privacy and security.
- Ease of use.
- Multiple face detection.
- Flexibility.

9.3 Applications

- Neural Networks (Image Recognition) finds its utmost use in medical imaging system that can give a preliminary diagnosis from the image. In detecting the sensitive images/medical disorders of the human body.
- The system is used in automatic driver assistance systems which indicates the presence of tree, pedestrian etc.
- It is also used in Quality Control Systems that can examine images of objects coming off a production line and flag the ones with defect.

10 Future Development

So far we have been able to develop a system that recognizes images and classify them accordingly for the attendance system using MATLAB Learning Tool Box and MongoDB software. The system can be further developed to incorporate few additional features. The future scope of this project paves way for determining the students' record furthermore to identify students with less percentage of attendance.

The future system can set a deadline percentage or a break point. If a student's attendance per say is to fall below the set threshold, then the system can be triggered to automatically send the details and information to their respective parents/guardians. In this way the performance of each student can be regularly monitored and improved in this regard.

References

1. LeCun, Y., Bengio, Y., Hinton, G.: Nature (2015)
2. Goodfellow, I., Bengio, Y., Courville, A.: The MIT Press, Cambridge, MA, USA (2016)
3. ImageNet. http://www.image-net.org
4. Russakovsky, O., Deng, J., Su, H., et al.: ImageNet large scale visual recognition challenge. Int. J. Comput. Vis. (IJCV) **115**(3), 211–252 (2015)
5. Gagare, P.S., Sathe, P.A., Pawaskar, V.T., Bhave, S.S.: Int. J. Recent Innov. Trends Comput. Commun. **2** (2014 Jan). ISSN: 2321-8169
6. Published in: 2015 International Conference on Green Computing and Internet of Things (ICGCIoT), 14 Jan 2016
7. Krizhevsky, A., Sutskever, I., Hinton, G. E.: ImageNet classification with deep convolutional neural networks. Adv. Neural Inf. Process. Syst. (2012)
8. BVLC-AlexNet-Model. https://github.com/BVLC/caffe/tree/master/models/bvlc_alexnet
9. Face Recognition Attendance System. Essays, UK, 26th July, 2017
10. Zhu, W.P., Xin, L.M., Chen, H.: Using MongoDB to implement textbook management system instead of MySQL. In: Proceedings of 3rd International Conference on Communication Software and Networks (ICCSN), 27–29 May 2011, pp. 27–29
11. Rahman, M.N.A., Seyal, A.H., Tajuddin, S.T., Azmi, H.M.: Int. J. Comput. Electr. Autom. Control Inf. Eng. **10**(5) (2016)

Work-in-Progress: Challenges in IoT Security

Sreelatha Malempati(✉) and V. S. J. R. K. Padminivalli(✉)

Department of Computer Science & Engineering, R.V.R & J.C. College of Engineering, Guntur, Andhra Pradesh, India
{lathamoturicse, srivallivasantham}@gmail.com

Abstract. Many IoT devices lack basic security requirements. The small size and limited processing power of many connected devices could inhibit encryption and other robust security measures. People need understanding of limitations of devices due to their size and the approaches for providing security. The challenges for implementing the security of embedded devices and providing end-to-end security are the outcomes of this study.

Keywords: Internet of Things · IoT security · Embedded devices · Hardware security

1 Introduction

1.1 Internet of Things

The Internet of things (IoT) is the network of physical devices, vehicles, home appliances, and other items embedded with electronics, software, sensors, actuators, and connectivity which enables these things to connect, collect and exchange data. The number of IoT devices increased 31% year-over-year to 8.4 billion in the year 2017 and it is estimated that there will be 30 billion devices by 2020. Security is essential for the safe and reliable operation of IoT connected devices.

1.2 Applications

Today, Internet of Things is used in many applications like Home automation, Personal Health Monitoring, Building automation, Industrial automation and Smart cities. The first and most obvious advantage of Smart Homes is comfort and convenience, as more gadgets can deal with more operations which in turn frees up the resident to perform other tasks. In Personal health monitoring, it increase the relationship between consumer/patient and healthcare providers and payers. Patient engagement and consumer consciousness play an important role here and in the relationship with healthcare payers. Building Automation processes related to energy efficiency, temperature control, security, and even sanitation can improve operations in ways that directly impact the production cost and maintenance cost. In Industrial automation the Internet of Things (IoT) helps to create new technologies to solve problems and increase productivity. Lastly in smart cities it helps in managing of city wastes, energy efficient lightening system, environment monitoring.

M. E. Auer and K. Ram B. (Eds.): REV2019 2019, LNNS 80, pp. 357–364, 2020.
https://doi.org/10.1007/978-3-030-23162-0_32

1.3 Working of an IoT End Device

Consider the following IoT device. This is a small micro controller which is connected with a sensor, power supply, actuator and RF transceiver. All these micro controllers will form a network and will be connected to the internet with the help of a gateway as shown in Fig. 1.

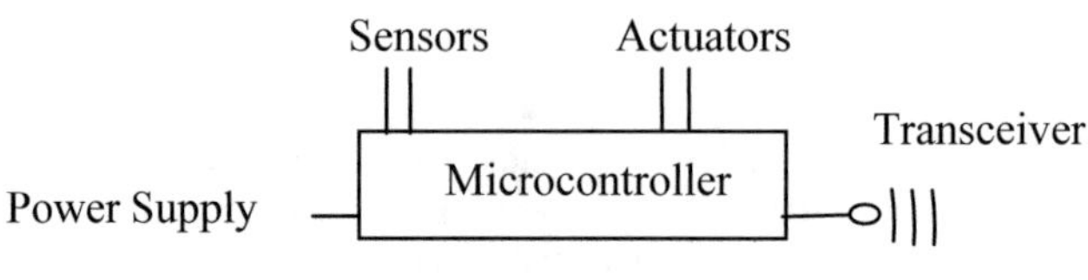

Fig. 1. IoT device

1.4 IoT Elements

IoT devices can communicate with the Internet. The End device send information to the gateway with the help of communication protocol. The information then send to a cloud where the information will be processed and the respective actions will be sent to the receiver device. Figure 2 [1] illustrates the how IoT devices connect to the internet.

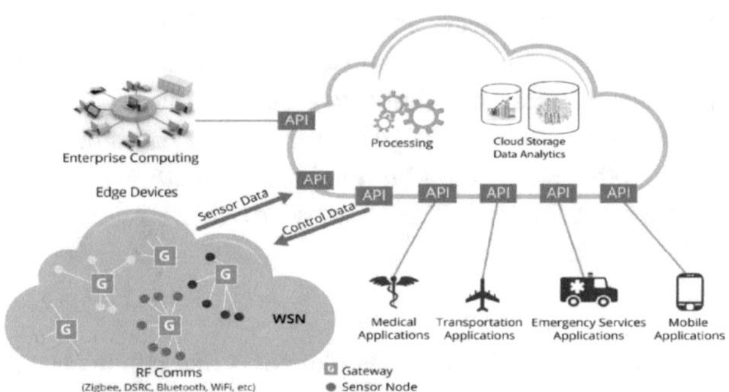

Fig. 2. IoT connecting different platforms

1.5 Communication Models

There are four models [2] to provide communication between devices and the Internet. One of them is Device-to-Device Communications [3]. In this the devices will communicate autonomously without centralized control and collaborate to gather, share and forward information. The information will be transformed into intelligence and create a intelligent environment. The quality of information gathered is depend upon the protocol we are using in getting the information. Example is the home automation system. In this small data packets flow in low data rate. Secondly, Device-to-Cloud

Communications [4]. Here IoT device connect to application service provider for exchanging data and other control information. The service provider offer cloud services which makes the data storage not a big problem. The application in IoT device has a remote access to the cloud service provider and also can be updated easily. This is useful applications which require remote monitoring. The main limitation is the interoperability between vendor of the device and cloud service provider. Device-to-Gateway Model [5] is another in which the IoT device connect to the application service provider with intermediary device. This gateway provide security, data or protocol translation. It also pairs IoT device and communicates with a cloud service. For example health monitoring app in the smart mobiles will come under this model. The app continuously monitors the respective health parameter and send to the service provider. From there the analysis will be performed and an health alert will be send to the end-user. Lastly, Back-End Data-Sharing Model. Here, the data obtained from the devices are shared between different application service providers. This help in extending the services to the end-users. For example a corporate company easily analyze the data in the cloud produced by the devices. This an extension to Device-cloud communication. Here also interoperability is a major concern.

2 Methodology

IPv6 with 2 to the 128th power addresses, is for all practical purposes inexhaustible. This represents about 340 undecillion addresses, which is more than the demand of the estimated 100 billion IoT devices going into service in the coming decades. It enables direct connection of physical objects to the Internet using microcontrollers which are constrained in computational power, memory and in power consumption. The main goals are Confidentiality, Integrity and availability. The main threats are Snooping, Traffic Analysis, Spoofing, Replaying, Repudiation, modification, Denial of service.

IoT involves extending Internet connectivity beyond standard devices, such as desktops, laptops, smart phones and tablets, to any range of traditionally dumb or non-internet-enabled physical devices and everyday objects. Embedded devices are designed for low power consumption, and have limited connectivity. They typically have only as much processing capacity and memory as needed for their tasks. Generally people overlook the risks of internet connected devices without taking proper security measures. The level of security required for an embedded device varies depending upon the function of the device and the vulnerabilities of the device. Security requirements must take into consideration the cost of a security failure, the risk of attack, available attack vectors, and the cost of implementing a security solution.

IoT security requires an end-to-end approach. Developing secure end-to-end IoT solutions involves multiple levels of security across devices, communications and Cloud. Smart devices is about giving your device the power to evolve, making it more powerful/useful/helpful over time. This work deals with securing IoT devices using proper filters, firewalls, configurations, software updates, OS patches, IoT authentication, encryption and hardware security.

2.1 Addressing of IoT Devices

There are many Issues in addressing. They are Compound Object in which an object consists of many objects. The Object Lifetime might range from years or decades down to days or minutes. The ownership and identity relationship between objects. The authentication and authorization procedure in use. The support for mobility in which dynamic objects connect from one network to another. There are many security issues will be there. The main concern is on Protection of Devices, Protection of Data, Secure communication, Secure applications. The Secure communication require use of confidentiality and data integrity mechanisms. To protect the devices against various attacks we have to use a secure operating system environment. We have to choose only those applications where security is a major concern. If we use all of them then the

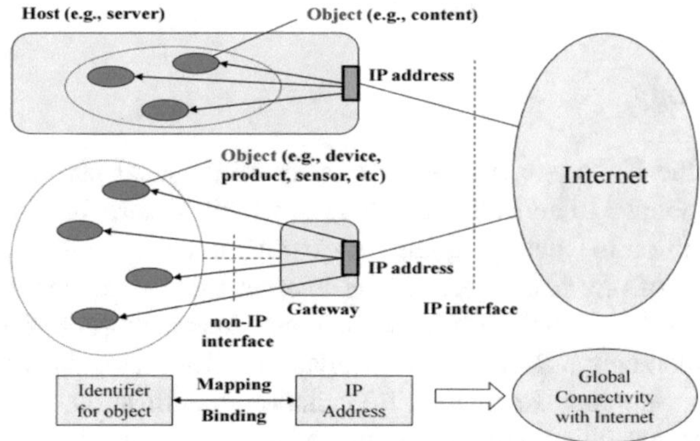

Fig. 3. Addressing scheme

Internet of Things will a good solution for many problems. For addressing we may use IPv6 as it can connect up to 26 billion devices. The non IP interfaces can also be connected to the Internet with the help of the gateways (Fig. 3).

2.2 Lifecycle of IoT Security

To protect the devices we can use Code signing and run-time protection. Cryptographically ensure code hasn't been tampered after being "signed" as safe for the device, it can be done at "application" and "firmware" levels. All critical devices should be configured to only run signed code and never run unsigned code. Be sure malicious attacks don't overwrite code after it is loaded. OS hardening, lockdown, white listing, sandboxing, network facing intrusion prevention, behavioral and reputation based security, including blocking, logging, and alerting. Many chipmakers already build "secure boot" capabilities into their chips. Open-source, and client-side libraries like OpenSSL can be used to check signatures of code. Challenge is "managing the keys," and "controlling access to the keys" for code signing and protection of embedded software. In some cases they offer hosted services that make it easy to safely and securely administer code-signing. Sign and update individual blocks or chunks of updates and not force anyone to sign entire monolithic images, or even an entire binary file. Software signed at the block or chunk levels can enable updates to be done with much finer granularity without sacrificing security and without having to sacrifice the battery for security. When the devices are reverse engineered, vulnerabilities are discovered and exploited, they need to be patched as quickly as possible. Code obfuscation and code encryption can considerably slow down the reverse engi-

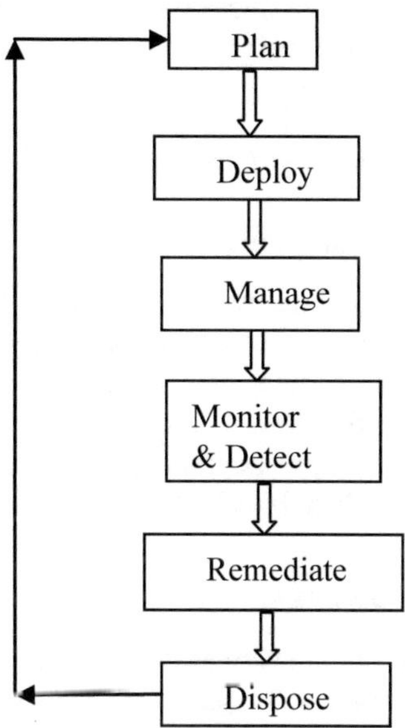

Fig. 4. Lifecycle of IoT security

neering process, but not entirely prevent reverse engineering. Over-the air (OTA) manageability must be built into devices before they ship for software/firmware security patches and functionality updates. Threats can defeat all of those countermeasures, best understand your system, identify anomalies that might be suspicious or dangerous, malicious or not, diagnostics and remediation (Fig. 4).

For Data at Rest Encryption, protect information in case of device theft/loss. For Data in Transit Confidentiality, Integrity, Authentication is required. For Data in Use Trusted execution environment, Trust Zone-ARM is used. For Data Loss Prevention Sensitive data not to be distributed outside of the user base or network.

2.3 Secure Communications

The IoT devices are having very less memory requirements. Using of encryption algorithms such as DES, AES etc. makes impossible. So better algorithm is elliptic curve cryptography for providing confidentiality for the messages. But this requires key certificates to be securely exchange between the devices. A number of certificate authorities are there to provide the embedded "device certificates". By using this certificates we can exchange the securely between the devices. Certificates are transferred with the help of the protocols [6] such as Simple Certificate Enrollment Protocol (SCEP), Enrollment over Secure Transport (EST), and Online Certificate Status Protocol (OCSP). With a strong CA helping to handle certificates, keys and credentials, authentication can easily be done by standards like Transport Layer Security (TLS) and Datagram TLS (DTLS). Each browser has a few "roots" of trust against which all certificates are evaluated. Embedding these roots [7] into browsers enabled security to scale to millions of servers on the web.

3 Other Security Considerations

3.1 Application Level Considerations

Applications (could be web, mobile, cloud, etc.) must be developed with industry standard secure coding practices such as OWASP, SAFECode, SANS/CWE, etc. to minimize the risk of application related attacks. E.g. preventing SQLi, XSS, data leakage, session replay, buffer overflow attacks, etc. Leveraging best practices such as file restrictions (e.g. type, size), input validation, etc. Scanning/testing the applications (dynamic, static, hybrid) for vulnerabilities and taking corrective measures.

3.2 Device and Gateway Security

The device security is also major concern as they are also vulnerable to do the attacks. So mechanisms such as disabling external device connectivity e.g. USB drives, disabling direct internet access from sensitive devices/endpoints if not required, disabling or blocking of unused services such as open ports, insecure protocols, Secure booting (using keys) and Secure firmware, authentication of devices during connection, applying regular patches on device OS, etc., Secure and authenticated firmware [8] upgrades, establishing connections with white listing instead of blacklisting devices, using of secure key exchange protocols can be implemented in the IoT devices to enhance the device security. Another important place where security can be considered is gateway security. Intruders can enter into the network wit the help of gateway. So ensure that the IoT/M2M gateway is secured from intrusions and malware by using appropriate mechanisms such as ACLs, IPS, filtering, etc.

3.3 Device Constraints

Facilities should have adequate physical security such as security guards, access cards, visitor logs, CCTV cameras, secure zones, etc. for preventing unauthorized access.

Appropriate security mechanisms [9] should be leveraged for isolating sensitive information bearing segments such as IDS/IPS, firewalls, network ACLs, etc. Service provider should obtain and produce assurance certifications such as ISO 27,001 SSAE/ISAE SOC reports, privacy seals, etc. Allow only strong authentication (e.g. MFA) for remote access to privileged users like administrators, clinicians, maintenance personnel for logging in securely from outside the company network. Usage of secure communication channels [10] such as VPNs-S2S, C2S for regular employees accessing the company network from branch offices or outside locations and disabling that access when no longer needed.

3.4 Challenges to Be Considered

The cryptographic algorithms should work on tiny, low-power devices. Lightweight, fast and secure communication protocols are to be used in IoT devices. Hardened operating systems and secure applications are to be used across all devices. Usability concerns and privacy issues are to be.

4 Conclusion and Future Work

IoT security is complicated by the fact that many 'things' use simple processors and operating systems that may not support sophisticated security approaches. Awareness is required in the public about IoT security challenges and the proposed solutions. We are thinking to enhance the security policies for the IoT communication so that an attacker may not have a chance to do attacks.

References

1. De Poorter, E., Moerman, I., Demeester, P.: Enabling direct connectivity between heterogeneous objects in the internet of things through a network-service-oriented architecture. EURASIP J. Wirel. Commun. Netw. (2011)
2. Kulkarni, S., Kulkarni, S.: Communication models in internet of things: a survey. Int. J. Sci. Technol. Eng. **3**(11), 88–91 (2017)
3. Bello, O., Zeadally, S.: Intelligent device-to-device communication in the internet-of-things. IEEE J., pp 1172–1182 (2016)
4. Biswas, A.R., Giaffrede, R.: IoT and cloud convergence. In: IEEE World Forum on Internet of Things (2014)
5. Altamimi, A.B., Ramadan, R.A.: Towards internet of things modelling: a gateway approach. Complex Adapt. Syst. Model. (2016)
6. Ezema, E., Abdullah, A., Sani, N.F.M.: A Comprehensive survey of security related challenges in internet of things. Int. J. New Comput. Archit. Their Appl. **8**(3), 160–167 (2018)
7. Yoon, S., Kim, J.: Remote security management server for IoT devices. In: International Conference on Information Communication Technology Convergence, Jeju, South Korea (2017)

8. Oh, S., Kim, Y.: Development of IoT security component for interoperability. In: 13th International Computer Engineering Conference, Ciaro Egypt (2017)
9. Suresh Babu, G.N.K., Kumaraswamy, M.: Security considerations for IoT technologies. Int. J. Adv. Eng. Res. Dev. **5**(4), 1592–1599 (2018)
10. Tiburski, R.T., Amaral, L.A., de Matos, E., de Azevedo, D.F.G., Hessel, F.: The role of lightweight approaches towards the standardization of a security architecture for IoT middleware systems. IEEE Commun. Mag. **54**(34), 56–62 (2016)

Part IV

Networking and Grid Technologies

Design and Development of Solar Electric Hybrid Heated Bed Smart Electric Stove

C. Lakshminarayana[1(✉)], Mohammed Arfan[1], and Desh Deepak Sharma[2]

[1] Department of Electrical and Electronics Engineering, BMS College of Engineering, Bangalore, India
lngp.eee@bmsce.ac.in, mdarfaneee@gmail.com
[2] Electrical Engineering Department, MJP Rohilkhand University, Bareilly, India
deshdeepak101@gmail.com

Abstract. India's overwhelming energy challenges are likely to increase in upcoming years. Looking at the great potential for renewable energy and energy storage, India will meet its unprecedented energy demands. Thus the goal of this work is to design and develop a solar-powered electric stove, the conventional energy shortage in India is motivating for development of solar electric stoves. The developed system consists of a heating bed which gets power from solar panels providing power of 300 W. There is a 12 V lead acid battery which supplies power in the absence of sunlight. While the stove is not in use batteries are charged in day light, a controller is designed and mounted onto the setup which allows to regulate heat. The stove was used in real time, various food was cooked and the data such as cooking time and power was recorded and analyzed. The results obtained were satisfactory in terms of cooking time and energy consumed. Which concludes the effectiveness of the solar stove in rural areas and remote locations.

Keywords: Heating bed · Solar electric stove · Heat controller

1 Introduction

Large scale energy crises have made India one of its victims, but this has led interest in discovering various ways of harnessing renewable energy [1]. The country is splendidly gifted with renewable energy sources, there is abundance of sunlight throughout the year in semitropical region. India is one of its benefice, considering these facts, this paper presents a new technique of cooking which could take the full advantage of renewable energy, that is, solar photovoltaic energy. Our goal is to create a solar electric stove which is cost effective as well as similar in temperament with the typical cylinder and gas stoves.

Solar cooker is a device that cook's food using only sun energy in the form of solar radiation. The solar cooking saves a significant amount of conventional fuels. The solar cooking is the simplest, safest, clean, environment friendly, and most convenient way to cook food without consuming fuels or heating up the kitchen.

M. E. Auer and K. Ram B. (Eds.): REV2019 2019, LNNS 80, pp. 367–378, 2020.
https://doi.org/10.1007/978-3-030-23162-0_33

A major concern of today is the rapidly depleting natural resources. So it is the urgent need of time to reduce the dependency on non-renewable sources, judiciously using the remaining sources and at the same time switching to new and better alternatives and renewable source of energy [2, 3]. In most parts of India, solar energy is available almost throughout the year and can be used as alternate input to meet out energy needs. Solar energy is the cheapest, inexhaustible and can be used for various domestic and agricultural requirements including cooking, drying, dehydration, heating, cooling and solar power generation [4].

2 Literature Review

Solar cooking is not a new technology; it is estimated to be in use since 1767 [5]. In the last twenty years there has been tremendous advancement in this solar cooking [6]. A French–Swiss scientist in 1767 built a greenhouse with five boxes made up of glass with one inside the other and set on a black table top [6]. The food placed in the innermost box was cooked thoroughly, this was probably the beginning of solar cooking. Barbara Kerr et all were developing solar cooker models in 1980s, solar cooker international was later founded in July 1987 [7].

3 Present Scenario of Electricity in India

The most common form of power, Electricity, is an issue for economic expansion for any country. The average maximum demand was 525,672 GWh in 2007 which then increased to 785,194 GWh in 2012 with a rate of 51,904.4 GWh per annum [8]. Assuming the same increase in demand rate, the average demand would stand at 1,300,000 GWh in 2018. While the average generation was 771.551 BU in 2009 which then increased to 811.143 BU in 2010 with increase rate of 5.8%. Which is comparatively low when compared to the demand. While the government is taking various measures and initiatives to solve the energy crises but still the crises preserves due to the unprecedented growth in demand and population. Which is due to the large scale industrial development in recent years and population growth, more the population more the number of consumers.

4 High Coal Dependency

Coal happens to be another major source of energy in our country, almost 75% of commercial energy is from coal [8]. Presently about 60% of natural gas production is used for electricity generation of country. Looking at the present consumption rate and taking into account a probable 10% growth rate in consumption, the remaining recoverable gases would only be sufficient for next 10 years from (2018 to 2028) [8].

If the capacity doesn't increase to the estimated, then probably after 2028 the supply would not be able to meet the demand and there would be large scale energy crises.

5 The Opportunity of Solar Energy in India

India is one of the major markets for solar, due to its geographical location, it happens to have a great solar energy potential. 5000 trillion KWh of energy per year is incident over the India's land, with 5–7 Kwh per Sq. per day. Hence there is great scope for technology to harness solar energy for both commercial as well as domestic consumption such as solar thermal and solar photovoltaic can be scaled easily in a country like India. Therefore, a solar business in India would be great to start in a country like India.

6 Design of the Proposed System

For the developed solution, 300 W solar panel was used. This panel of 300 W fulfils the requirement of the desired current and voltage. The developed system is 12 V and has a current rating of 15 A. Therefore, a single 12 V 20 Ah lead acid battery is used. a charge controller of 12 V 20 A is used to charge the battery. The system is setup in such a way that the grid will supply power for the stove when both solar and battery are unavailable.

The following block diagram briefly describes the setup (Fig. 1).

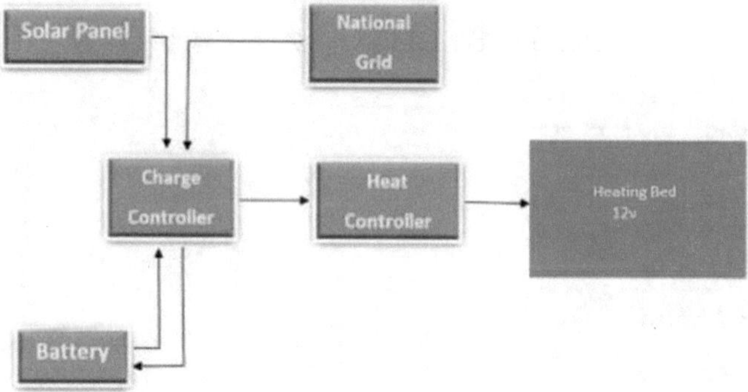

Fig. 1. Overview

A. *Configuration of the heating coil*

The coils that has been used is in the form of a flat plate used mostly in 3D printing. This is the first time heating bed is proposed as a solution for cooking application, they are very economical and can be designed for any shape and wattage. For the proposed solution the coil has been designed for wattages between 150 and 200 W.

The heating bed works on the principle that when a high current is passed through a low resistance thin copper trace, the copper trace gets heated due to the heating effect of electricity (i.e. I^2R).

The design and development of the coil involves the calculation of width and length of the copper trace to obtain the required wattage and temperature.

Electrical Equations:

$$\text{Resistance} = \text{Resistivity} * \text{Length}/\text{Area} * (1 + (\text{Temp\_Co} * (\text{Temp} - 25))$$

where,

Area = Thickness*Width
A copper Thickness of 1 Oz/ft^2 = 0.0035 cm
Copper Resistivity = 1.7E−6 Ω cm
Copper Temp_Co = 3.9E−3 Ω/Ω/C.

Thermal Equations:

$$\text{Thermal Resistance} = \text{Thermal Resistivity} * \text{Length}/\text{Area}$$
$$\text{Copper Thermal Resistivity} = 0.249\,\text{Cm K/W}\,(\text{at}\,300\,\text{K})$$

From the above equations a heating bed to have a resistance of around 1 Ω, so that at 12 V it draws a current of 12 A.

$$\text{Wattage of heated bed} = 12^2 \times 1 = 144\,\text{W}$$

From the equations at 25 °C

Width = 2 mm

Thickness = 4 oz/ft^2

Length = 16.5 m

Resistance = 1 Ω

So, from the above data a coil was designed using a PCB designing software and etched using ferrous chloride solution.

The developed PCB heated bed had a resistance slightly greater than theoretical value (Fig. 2).

B. ***Heat controller***

In order to control the heat of the stove to obtain various heating ranges for cooking sundry items, a microcontroller based control circuit was developed. Since, the current drawn by the heating bed was huge, a high current rating relay (30 A) was used. This relay cuts off the power to the heating bed when the temperature of the heating bed goes beyond the set temperature.

Fig. 2. Heating bed

There are two temperature sensors in the stove, one monitors the heated bed temperature and the other monitors the temperature inside the stove. So that, when the circuitry inside becomes too hot it turns on the cooling system.

The block diagram of the heat controller is shown in Fig. 3. The structure of the whole body is shown in Fig. 4.

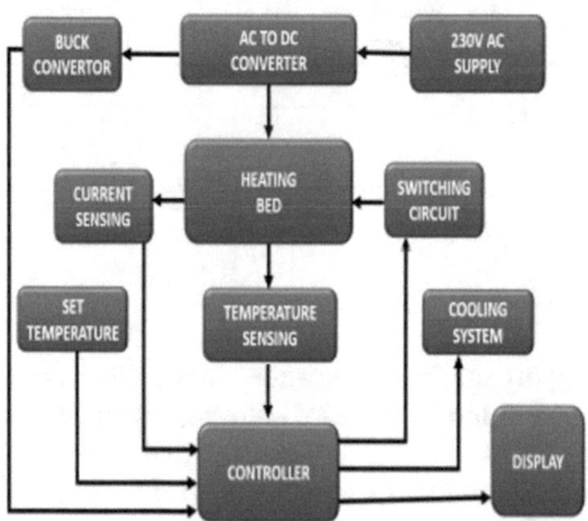

Fig. 3. Block diagram of the heat controller

Fig. 4. Body construction

7 Real Time Testing and Analysis

When the stove is not in use during the day light the batteries are charged through the charge controller from the PV panels while at the time of cooking the power is shared by the panel and the battery. The stove was built and setup, to check the effectiveness of the stove different food items were cooked and data was recorded.

The power sharing between the PV panels and the batteries depends on the sunlight, when there is good amount of sunlight there no need of power from the battery, the PV panels can alone supply the power.

The cooking time of the food items were recorded and compared with the gas stove cooking time (Table 1).

Table 1. Comparison of cooking time

No.	Food item	Time required on gas stove (in minutes)	Time required on solar stove (in minutes)
1	Bread Toast	4	7
2	Egg omelet	2.5	6
3	Chapatti	5	10

8 Future Development

The stove doesn't support boiling applications since there is only heating from the bottom, and the wattage is too low thus would take a lot of time to boil water. The future development would be to add side heaters so that water can be boiled. Right now the system is open and there is lot of heat losses to the surrounding and thus causing power loss and increase in cooking time. Another future development would be to find new ways to prevent heat to flow out of the system so that it is contained inside and utilized to cook food even faster.

9 Payback Calculation

To carry out the payback calculation firstly the normal expenses the people has to bear for cooking when the use gas cylinders, which the most common form on which the people are depended was calculated and analyzed. On an average in India each cylinder costs about 700 Rs, Table 2 shows the expenses while using cylinder gas.

Table 2. Yearly cost of gas cylinder

Family size	Cylinder consumption	Per cylinder cost	Total cost per month	Total cost per year
Small (2–4 person)	1	560	560	6720

Now the cost involved in development of a solar electric hybrid stove that has been proposed in this paper was calculated (Table 3).

Table 3. Development cost

Material	Description	Cost	Total cost (cost in INR)
Panel	250 W	24 Rs/W	6000
Batteries	1 pc (12 V, 30 A)	3500 Rs	3500
Charge controller	1	1500 Rs	1500
Heat controller	1	1000 Rs	1000
Stove body	–	–	500
Wires and others	–	–	500

Payback calculation was done for two different cases.

For case 1 the system cost was calculated including the battery and charge controller cost. For case 2 it was calculated excluding the battery cost. The reason behind the case 2 is, the stove can work independently without the battery on a bright sunny day as the panel installed is 250 W which is 1.6 times the stove consumption thus we don't need to spend for the batteries and charge controller as the consumer can get the power from the installed PV panels or in case of shortage of power he can draw it from the grid.

Case 1 (including the Battery): Total cost for the solar electric hybrid stove: 6000 Rs + 3500 Rs + 3500 = 13,000 Rs. If a family uses the solar stove instead of the cylinder gas, the total money can be paid back in less than 2 years.

Case 2 (excluding the PV panel): Total cost for the solar electric hybrid stove: 9500 Rs.

If a family uses the solar stove instead of the cylinder gas, the total money can be paid back in approximately 15 months for small family.

10 Case Study and Analysis

The energy calculations were performed for following specifications:

Installed Solar panel: 250 W
Battery: 12 V 30 Ah
Stove consumption: 150 W
Number of cooking hours: 3-h daily

Energy calculation:

Assuming the fact that in India PV panels get proper sunlight for approximately 6 h.

The total power of the PV panel was 250 W

$$E = P * t * \text{delta} = 250 * 6 * 0.41 = 615\,\text{Wh}$$

where,

E = PV panel supply in Wh
P = Power rating of PV panel
t = time of solar incidence per day
Delta = average co-efficient of irradiance (as obtained from SWERA)

So, the energy, the PV panels can provide is 615Wh

Assuming stove runs for 3 h a day and Stove has the power rating of 150 W

$$E^1 = P^{1*} t = 150 * 3 = 450\,\text{Wh}$$

where,

E^1 = Energy consumed in KWh
P^1 = power rating of stove
t = time for stove is in use

So, the energy, consumed by the stove is 450 Wh.

The energy supplied form the PV panels is 615 KWh and the consumed energy by the stove is 450 KWh. Therefore the rest of the power is stored in the battery.

$$E^2 = E - E^1 = 615 - 450 = 175\,\text{Wh}$$

where,

E^2 = energy flowing into the battery
E = energy from the panels
E^1 = energy consumed by stove

So, the stored energy is 175 Wh.

Assuming 75% discharge. Only 75% of this stored energy can be supplied by the battery later,

$$E^3 = E^2 * 75\% = 175 * 75\% = 131.25\,\text{Wh}$$

where,

E^3 = Energy available for use
E^2 = Energy stored in battery

So, the battery can supply 131.25 Wh later.

The outlook of the stove is shown in Figs. 5 and 6.

Fig. 5. Physical look of the stove

Fig. 6. Inside the stove

The cooking experience on the stove is shown in Figs. 7, 8 and 9.

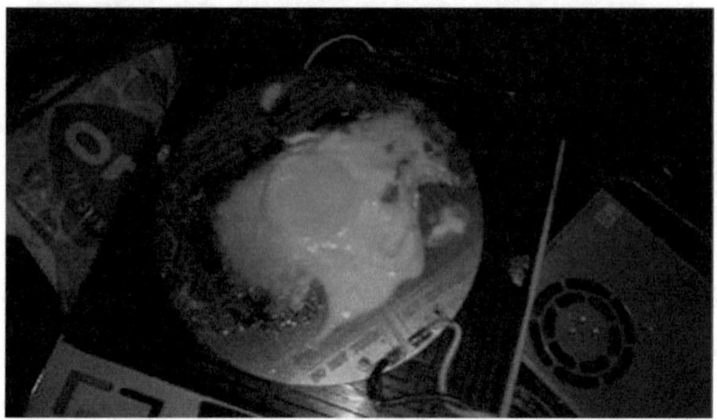

Fig. 7. Making omelet

Fig. 8. Making chapatti

Fig. 9. Bread toast

11 Feasibility

The main reason for this project is to develop a stove that harnesses solar energy so as to replace conventional gas stoves. In recent days there is an enormous increase in solar panels and decrease in cost of the panels as well. Earlier around three decades back the solar panels costed USD 15/W peak. But since then the price of the panels has been continuously being falling and it has also been expected that the prices will still fall currently at USD 0.6/W peak. It is estimated that prices will fall even further and hit USD 0.3/W peak in upcoming years. From the year 2006 to 2014 a span of 8 years the world module prices have dropped by about 78% from $3.25 per watt to about $0.72 per watt. The system proposed requires a 250 W panel which may happen to be costly now but in the upcoming days with the fall of prices on PV the it would be cheap and affordable. Thus the current moderately expensive solution would look attractive and economic solution. In upcoming days there is a probability that Indian government may make it mandatory for new high rise building to install solar panels on the rooftop. Solar stove can serve as a great energy efficient and economical alternative in rural and remote areas. So, analyzing the overall situation of India it can be said that, the implementation of the solar electric hybrid stove is feasible in this country.

12 Conclusion

This project has a great potential to revolutionize clean cooking. The project is mainly aiming at finding an alternative to conventional methods so as to reduce the gas and electricity shortage of the country. While making use of cleaner, safer and more economic ways of cooking such as proposed in this paper.

The solar stove developed would be easy to sue with the controller provided with it, the coils are designed so as to produce maximum heat and provide uniform cooking. The controller can help save power as well as regulate heat. Batteries serve as a backup for solar as well as grid. From various tests that were carried on the stove it has been found that this stove can serve as efficient, economic, safe and easy to use where time being the only issue.

Acknowledgements. We are thankful to Department of Electrical and Electronics Engineering B.M.S College of Engineering for assisting us in conducting this study. We are also thankful to all the faculties of Department of EEE for their valuable suggestions.

References

1. History of Solar Cooking [online]. Available: http://www.solarcooker-at-cantinawest.com/solarcooking-history.html. Accessed 21 Mar 2016
2. Kundapur, A., Sudhir, C.V.: Proposal for new world standard for testing solar cookers. In: International Solar Cooker Conference at Granada, Spain in June 2007 (also http://solcooker.tripod.com)

3. Kalbande, S.R., Mathur, A.N., Kothari, S., Pawar, S.N.: Design, development and testing of paraboloidal solar cooker. Karnataka J. Agric. Sci. **20**(3), 571–574 (2007)
4. Rajamohan, P., Shanmugan, S., Ramanathan, K.: Performance analysis of solar parabolic concentrator for cooking applications. In: International Solar Food Processing Conference (2009)
5. State of the art of solar cooking
6. Early uses of the sun to serve humanity [Online]. http://vignette2.wikia.nocookie.net/solarcooking/images/5/51/Sam.pdf/revision/latest?cb=20070122015559. Accessed 21 Mar 2016
7. History of solar cooking [Online]. Available: http://solarcooking.wikia.com/wiki/History_of_solar_cooking. Last Modified: 1 Jan 2016
8. Cantina West Parabolic Solar Burner [online]. Available: http://www.solarcooker-at-cantinawest.com/parabolic_solar_cooker_solar_burner.html. Accessed 2 Aug 2016
9. Siddiqua, S., Firuz, S., Nur, B.M., Shaon, R.J., Chowdary, S.J., Azad, A.: Development of double burner smart electric stove powered by solar photovoltaic energy. In: IEEE Conference on Global Humanitarian Technology Conference, 16 Feb 2017. https://doi.org/10.1109/ghtc.2016.7857319
10. Parabolic solar cooker designs [Online]. Available: http://solarcooking.wikia.com/wiki/Parabolic_solar_cooker_designs. Last Modified: 14 July 2016
11. Solar Electricity Costs [Online]. Available: http://solarcellcentral.com/cost_page.html. Accessed 2 May 2016
12. Patil, R.C., Rathore, M.M., Chopra, M.: An overview of solar cookers. In: 1st International Conference on Recent Trends in Engineering & Technology, Mar 2012 Special Issue of International Journal of electronics, Communication & Soft Computing Science & Engineering. ISSN: 2277–9477

Remotely Operated Distribution System

U. Kamal Kumar[(✉)], Suresh Babu Daram, D. Sreenivasulu Reddy, I. Kumara Swamy, and T. Nageswara Prasad

Department of EEE, Sree Vidyanikethan Engineering College, Tirupati, India
{kumarkamal41, sureshbabudaram, seenu.d7}@gmail.com, np_thunga@yahoo.com

Abstract. Fault location, fast service restoration and controlling of abnormal loadings are the key challenges in electrical distribution system. Under normal operating conditions, the switches are electrically separated. The conventional methods used to detect the fault location; power thefting and control of overloading are time-consuming and costly, disturbing the entire distribution system. The need of automation of the distribution system plays a vital role for the proper detection, isolation and detection of the fault. In this paper, the distribution automation system is developed for the rapid fault detection, isolation, and service restoration to reduce time and maintenance costs comparatively. Initially, the fault and theft is detected and smart energy meters are used for communication purpose through GSM. The information will be sent to the customers remotely through the network to the utility providers.

Keywords: Arduino controller · GSM module · Mobile · Smart energy meter

1 Introduction

Distribution systems carry a large amount of power as compared to earlier era because of increase in per capital consumption of electricity. It leads to the substantive increase in complexity and diversity of the consuming market, the Electric Power System (EPS) has been demanding a considerable updating and also a significant improvement of the monitoring, control and protection equipment's. Faults in power distribution systems cause supply interruptions being responsible of process disturbances, information and economic loss and equipment damage among others. The distribution companies are unable to keep track of the changing maximum demand. The consumers are affected with high billing and bills that have already been paid as well as poor reliability of electric supply and quality even if bills are paid regularly.

Fault location includes the determination of the physical location of the fault. About 80% of interruptions are caused by faults in distribution networks and the application of fault location algorithms developed for transmission system is not an easy task due to the topology and operating principles of the first (i.e., non-homogeneous feeders, load taps, laterals, radial operation and the available measuring equipment). Two foremost

M. E. Auer and K. Ram B. (Eds.): REV2019 2019, LNNS 80, pp. 379–389, 2020.
https://doi.org/10.1007/978-3-030-23162-0_34

things which are required for quick restoration of the faulty part are fault location and type of fault. Similarly, in digital distance protection system the appropriate operation of protective device and accurate classification of the fault are necessary.

Although, a large number of techniques are available for fault identification and classification, some of them are based upon continuous monitoring of voltages, currents and impedances, etc. All these techniques have their own advantages and disadvantages. Some intelligent techniques, (generally known as knowledge based techniques) of the fault classification in transmission line are based upon Neural Network, Fuzzy Logic and Fuzzy Neural Network and knowledge system based approach. All these techniques suffer from a major drawback that a proper training is required for neural network and these are not susceptible to high impedance faults [1–5].

The electricity is needed to be protected for efficient power delivery to the consumer because electricity is indispensable to domestic and industrial development activity. They have technical losses and Non-technical losses, T&D losses have been a concern for the Indian electricity sector. When the discussions about T&D losses it also includes the theft of electricity, although it is the part of commercial loss but there is no way to segregate theft from the T&D losses. With a technical view, "Power Theft" is a non-ignorable crime that is highly prevalent, and at the same time it directly affects the economy of a nation.

There exist a variety of approaches for locating faults and power theft in power distribution systems individually. In this work a design is proposed for detecting the location of fault, its type and power theft with an Arduino controller based and disseminating the information to concerned authorities using GSM module. The limitations in fault controlling devices and the issues related to power theft had been discussed taking into account of the non-technical losses.

A microcontroller is set to compare essential line parameters with preset value stored in the monitor. Whenever the preset threshold is crossed, the microcontroller instantly initiates a message to be sent to the authorities and utilities stating the value of the increased or decreased parameter, so that the type of fault, theft and its location can be properly predicted. This compared value is transmitted to electricity board, this value display in LCD. The information will then be quickly processed by the microcontroller and a SMS will be send through the GSM technology [10–14].

In the present billing system issues, the remedies for all these problems is to keep track of the consumers load on timely basis, which will help to assure accurate billing, track maximum demand and to detect threshold value. These are all the features to be taken into account for designing an efficient energy billing system.

The above problems can be solved with the help of a Smart Energy Meter. A Smart Meter is a new kind of electricity meter that can digitally send meter readings to the energy supplier for more accurate energy bills. They are a replacement for standard meters, which use technology created decades ago and require households track their own readings and submit them to suppliers if they want accurate bills. Smart Meters come with an in-home display screen that shows us exactly how much energy the consumer is using in real time and will bring an end to estimated bills (Fig. 1).

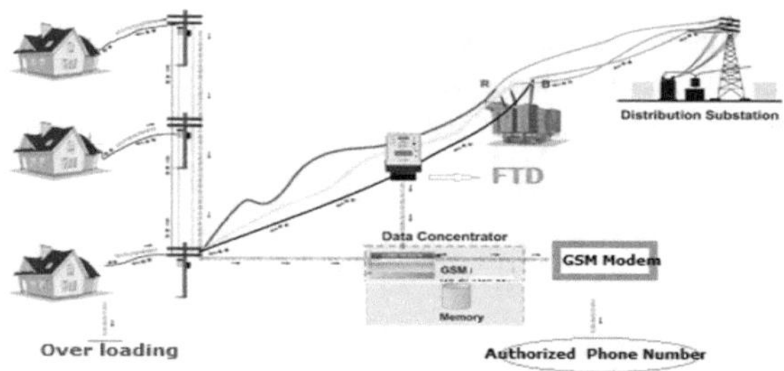

Fig. 1. Smart distribution system with fault theft detector

2 Faults in Distribution System

Faults can be defined as the flow of a massive current through an improper path which could cause enormous equipment damage which will lead to interruption of power, personal injury, or death. In addition, the voltage level will alternate which can affect the equipment insulation in case of an increase or could cause a failure of equipment start up if the voltage is below a minimum level. As a result, the electrical potential difference of the system neutral will increase. The severity of the fault depends on the short-circuit location, the path taken by fault current, the system impedance and its voltage level. In order to maintain the continuation of power supply to all customers which is the core purpose of the power system existence, all faulted parts must be isolated from the system temporary by the protection schemes.

Fault Current Calculations:

Line-to-ground fault Current,

$$I_f = \frac{3V_f}{Z_1 + Z_2 + Z_0 + Zf + Zn} \tag{1}$$

Line-to-line fault Current,

$$I_f = \frac{3V_f}{Z_1 + Z_2 + Zn} \tag{2}$$

Double line-to-ground Current,

$$I_f = \frac{3V_f}{Z_1 + Z_2 + Zf + Zn} \tag{3}$$

Three phase fault Current,

$$I_f = \frac{3V_f}{Z_1 + + Zf} \tag{4}$$

where

Z_1 = Positive Sequence Impedance
Z_2 = Negative Sequence Impedance
Z_3 = Zero Sequence Impedance
Z_n = Neutral Impedance
Z_f = Fault Impedance

Types of Fault Detection:

The fault occurring in the power lines and cable can be classified into four main categories: short circuit in the cable or transmission line, short circuit to earth, high resistance to earth and open circuit. Four method that are mostly used in detecting fault location are described as follow [7].

1. A frame
2. Thumper
3. Time domain reflectometer (TDR)
4. Bridge method.

The fault identification system is composed by five modules:

(i) Data acquisition,
(ii) Data preprocessing,
(iii) Transient identification,
(iv) Participant phases detection,
(v) Participant phases classification.

Power Thefting:

There are many reasons that encourage people to steal electricity. Among them the socioeconomic reasons induces people to a great extent in electricity theft. A common opinion amongst many people is that, it is fraudulent to steal something from their neighbour but not from the government and government aided private utility company. In addition, other reasons that provoke illegal consumptions are:

(a) Due to high amount of tariff collected from consumers, which leads them to steal electricity to escape from vast electricity consumption bills among most of the peoples.
(b) Poor financial condition in various countries.

There are various types of electricity power theft in meters, cables, transformers and overhead lines. Some of them are listed below:

- Tampering with meters and seals.
- By-passing the meters.
- Damaging or removing meters wires.
- Illegal tapping to bare wires or underground cables.

- Illegal terminal taps of overhead lines on the low voltage side of the transformer.
- Billing irregularities made by meter readers.
- Other unpaid bills by individual customers, government and private organizations (Fig. 2).

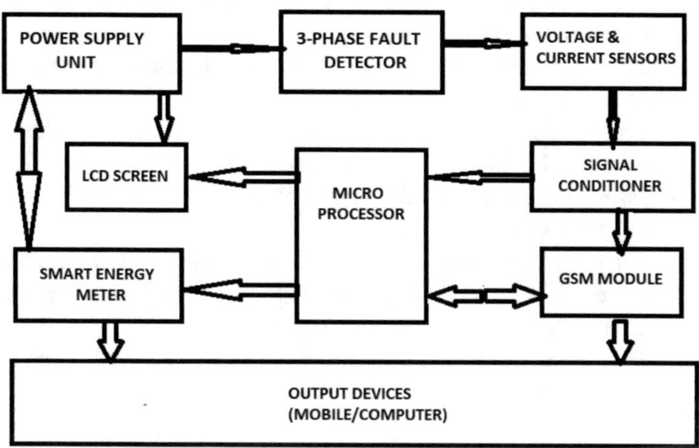

Fig. 2. Block diagram of remote distribution system

Fault detection and identifying type of fault:

The single phase supply is given to three transformers and voltage is stepped down from 230 to 12 V. The supply is converted from ac to dc through full wave rectifier and supply is driven to fault circuit. In the fault detection through manual operation the switches are operated. Three types of faults (L, L-G, L-L, L-L-L, L-L-L-G) is switched manually. Any fault is happened in fault circuit then supply is fed towards transistor and it is triggered and supply is driven through op-amp. From the op-amp it is send to perspective relay then the relay gets operated and type of fault is detected by relay operation. Then the corresponding relay will activate and signal will be given to Arduino mega 2560. From Arduino the signal will be send to the GSM and from the GSM the type of fault and in which phases it is occurred will be sent to operator of the substation.

The components are requiring to build the fault and theft detector with smart energy meter are as follows:

- Current sensor:
- Driver circuit:
- Microcontroller (Arduino MEGA 2560)
  - Communication
  - Programming
  - Automatic software reset
  - Physical characteristics

Microcontroller (Arduino MEGA) algorithm:

Step 1: Start.
Step 2: Initialize base current value of the system.
Step 3: Check for the actual current.
Step 4: The controller reads that if actual current exceeds the base current under any faults and overloading condition.
Step 5: If fault occurs, the microcontroller initiates and sends a message to along with alarm and the LCD display.
Step 6: In feeders, the controllers sense the abnormal current values with base values and identified its location based on feeder current. If there is no exceeding current then no theft takes place and LCD displays the power and goes to next step.
Step 7: Similarly under overloading the microcontroller senses the abnormal currents and initiates, sends a text message which also mentions that "there is power thefting occurs in distribution lines".
Step 8: If there are no faults then the normal power is displayed in the LCD screen and goes to step3.
Step 9: Stop.

- GSM Module;
- Energy Meter.

Energy meter is a device that measures the amount of electrical energy consumed by a residence, a business, or an electrically powered device. They are typically calibrated in billing units, the most common one being the kilowatt hour (kWh). They are usually read once in each billing period.

When energy savings during certain periods are desired, some meters may measure demand, the maximum use of power in some interval. "Time of day" metering allows electric rates to be changed during a day, to record usage during peak high-cost periods and off-peak, lower-cost, periods (Fig. 3).

3 Technical Details of Hardware Setup

1. Resistor of 1kohms is connected in series across the base of the transistor to avoid over currents and for protection of the transistor.
2. From the collector and emitter it is connected to op-amp inputs and a normal and variable resistor of 1kohms is connected to the op-amp for the protection of op-amp.
3. In between the relay and op-amp a resistor is connected (Fig. 4).

Working

At starting from single phase ac supply the current sensor is connected across the phase and from that point it is connected to lamps in the circuit. The single phase supply is

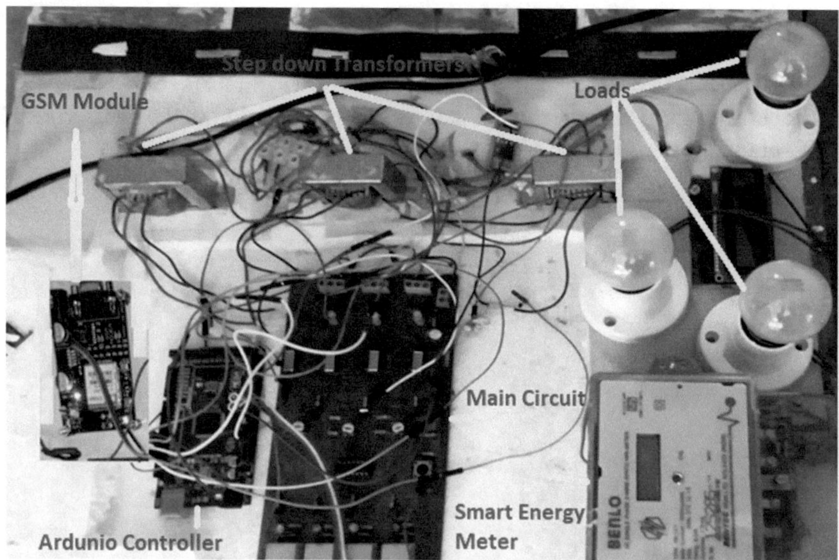

Fig. 3. Hardware setup of the remotely distribution system

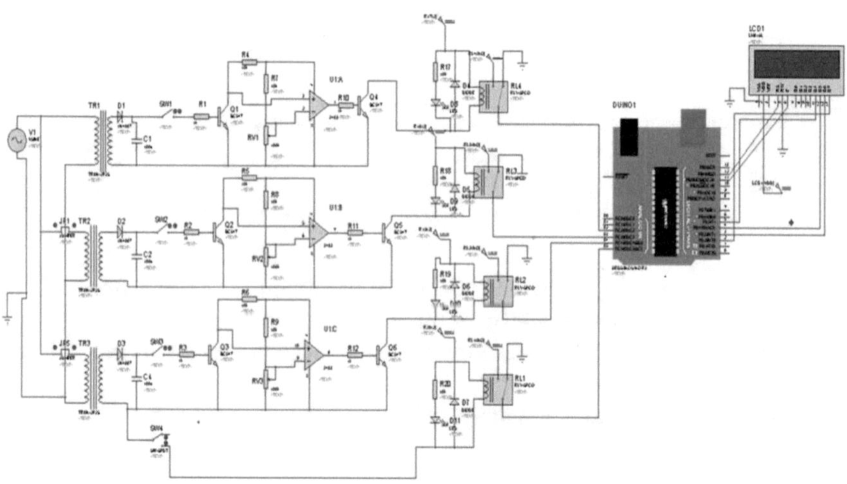

Fig. 4. Connection diagram of fault circuit

converted to three phase supply by bus then across each three wires is taken and it is connected to the bus with phase wire of single phase supply. The neutral wire is connected separately to the neutral of the bus primary their by the connections of the bus primary is completed. From bus secondary the three phase wires is connected to three primary transformer phases and neutral of bus is connected to the neutrals of the primary of the transformer. Then secondary of the transformers is connected to the full wave rectifier circuit which is in series with the transformer. From the rectifier circuit through resistance it is connected to diver circuit which consist of transistor. From the transistor across collector through a resistor is connected to the op-amp and from the emitter it is connected to other terminal of the op-amp. Similarly rest connection of transformer is connected in the fault circuit except the neutral from the transformer. The end driver circuit is connected to relay unit (Fig. 5).

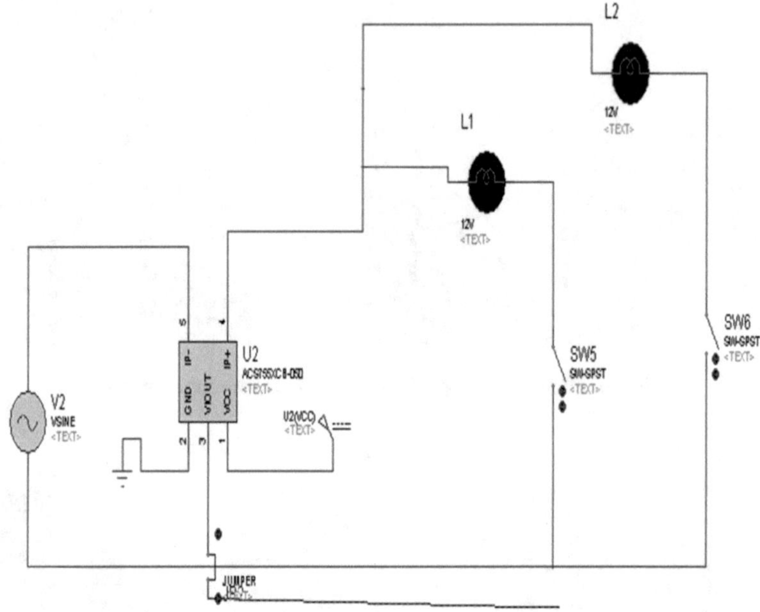

Fig. 5. Connection diagram of power theft circuit

- A phase from the single phase is connected to current sensor and from it is connected to the lamp holders which are in parallel connection through two switches.
- Then three from the current sensor the analog input from the current sensor is given to the Arduino
- Another one is given to 5 V supply and another one is grounded.

Theoretical calculations

See Tables 1 and 2.

Table 1. Fault calculations

S. No.	Condition	Voltage (V)	Current (I)	Power (W)
1	L-G fault	12	2.91	35
2	L-L fault	12	3.25	39
3	L-L-L fault	12	3.25	39

Table 2. Power theft calculations

S. No.	Condition	Voltage	Current	Power
1	Normal	230	0.26	60
2	Power theft	230	1.09	260

Energy and power factor calculations:

The power factor and Energy calculations are as follows:

- The Energy E is given by,

$$E = P * t \quad \text{Watt-Hour}$$

  where E is the Energy consumed in a particular time, P is the Wattage of the load, and t is the time duration in which the energy is being calculated.

- The Power factor is calculated using the formula given below:

$$P = V * I * \text{Cos}\,\alpha$$
$$\text{Cos}\,\alpha = P/V * I$$

  where α is the power factor angle (Table 3).

Table 3. Energy Calculations at different loads

Number of lamps	Load (Watts)	Energy (kWh)	Amount (Rupees)	Power factor
1	200	0.2	0.29	1.00
2	400	0.4	0.58	1.00
3	600	0.6	0.87	1.00
4	800	0.8	1.16	1.00

Pulse Calculations in Smart Energy Meter:

Here meter is interfaced with microcontroller through the pulse that is always blinked on the meter. Further that pulse is calculated as per its blinking period, using this principle we calculated it for one unit and accordingly what charge will be for a unit. After 0.3125 W energy uses Meter LED (calibrate) blinks. Means if we use 100 W bulb for a minute then the pulse will blink 5.2 times in a minute. And this can be calculated using given formula.

$$\text{One Pulse} = (\text{Pulse rate of Meter} * \text{watt} * 60)/(1000 * 3600)$$

4 FTD Results with Faults

If any fault occurs in the system then the along with the alarm a text message is sent to the prescribed mobile number (Fig. 6).

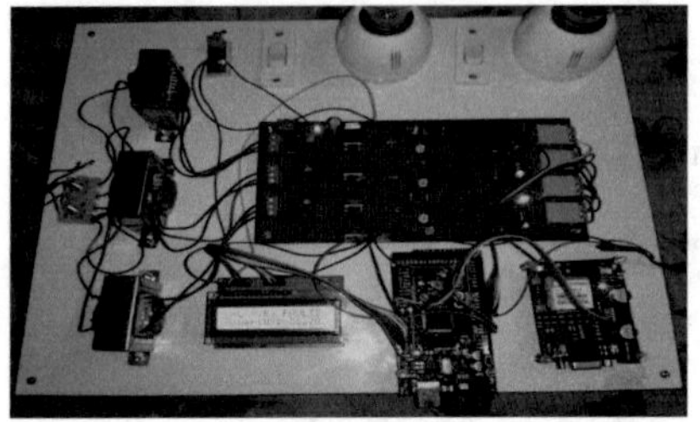

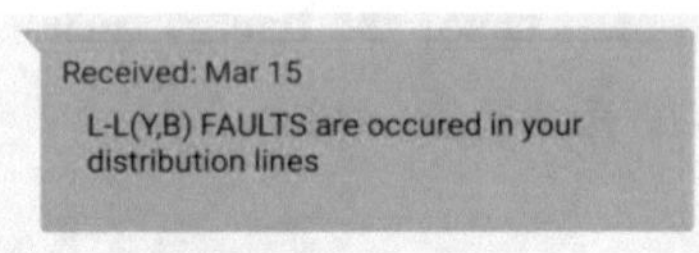

Received: Mar 15
L-L(Y,B) FAULTS are occured in your distribution lines

Fig. 6. During the occurence of fault and message received during fault

FTD during overloading

Whenever the system is overloaded, often treated as thefting, along with the alarm a text message is sent

The kit during overloading condition is as follows

Overall LCD displays during abnormal conditions

L(R) FAULT Power(W): 31.79	L-G(B,G) FAULTS Power(W): 39.74
L(Y) FAULT Power(W): 39.74	L-L(R,Y) FAULTS Power(W): 39.74
L(B) FAULT Power(W): 35.76	L-L(R,B) FAULTS Power(W): 39.74
G(G) FAULT Power(W): 35.76	L-L(Y,B) FAULTS Power(W): 35.76
L-G(R,G) FAULTS Power(W): 35.76	L-L-L(R,Y,B) FAU Power(W): 39.74
L-G(Y,G) FAULTS Power(W): 35.76	L-L-L-G(R,Y,B,G) Power(W): 31.79

5 Conclusion

This system is provided reduction in time requirement, to locate fault and power theft in the distribution system. It is also provided the information regarding the type of fault using GSM which leads to minimize power disruptions to distribution substations, continuity of supply and helps in saving expensive distribution transformers. In this system, smart energy meter is designed to eliminate third party involvement, tampering detection and supplier can disconnect service to the consumer in the event of meter tampering or unauthorized use of energy. It makes the relation between utility and user more transparent and reliable. Power saving is possible which contributes towards the minimization of the problem of energy crisis.

Acknowledgements. I express my heartfelt thanks to my co-Authors and Sree Vidyanikethan Engineering College for their moral support and good wishes.

References

1. Sonwane, N.S., Pable, S.D.: Fault detection and autoline distribution system with GSM module. IRJET 3(5) (May 2016)
2. Shunmugam, R., Ashok Kumar, K., Deebika Devi, A.R., Manoj Kumar, K., Mathivanan, A.M.: Distribution line fault detection and intimation using GSM. IJRTER **2**(4) (April 2016)
3. Tadadikar, N.S., Nerlekar, R.S., Kabade, R.M., Chormule, V.S., Gavali, D.D.: Three phase line fault detection on distribution lines using GSM technique. IJETIE **2**(3) (March 2016)
4. Kamal Kumar, U., Daram, S.B., Kumar, N.M.G.: Intelligent technique based fault analysis and location in a Distribution System. In: ICREU, Jan 2016
5. Kumar, N., Sharma, M., Sinha, A., Bhushan, I.: Fault detection on radial power distribution systems using fuzzy logic. IJEEE **7**(2) (July–Dec 2015)
6. Pandey, V., Gill, S.S., Sharma, A.: Wireless electricity theft detection system using Zigbee technology. IJRITCC **1**(4) (Mar 2013)
7. Mirzaei, M., Ab Kadir, M.Z.A., Moazami, E., Hizam, H.: Review of fault location methods for distribution power system. Aust. J. Basic Appl. Sci. **3**(3), 2670–2676 (2009)
8. Nareshkumar, K.: Application of Multi-Agents for Fault Detection and Reconfiguration of Power Distribution System. IEEE (2008)
9. Ziolkowski, V., da Silva, I.N., Flauzino, R., Ulson, J.A.: Fault identification in distribution lines using intelligent systems and statistical methods. In: IEEE MELECON 2006, Benalmadena (Malaga), Spain, 16–19 May
10. Mohite, N., Ranaware, R., Kakade, P.: GSM based electricity theft detection. IJSEAS **2**(2) (Feb 2016)
11. Monisha, N.N., Monica, R., Satishkumar, V.S.: Power theft identification using GSM technology. IJORAT **1**(5) (May 2016)
12. Prakash, T.: Power theft prevention in distribution system using smart devices. Int. J. Appl. Eng. Res. **10**(42) (2015), © Research India Publications
13. Kalaivani, R., Gowthami, M., Savitha, S., Karthick, N., Mohanvel, S.: GSM based electricity theft identification in distribution systems. IJETT **8**(10) (Feb 2014)
14. Executive Summary Power Sector, Government of India, Ministry of Power Central Electricity Authority New Delhi, July 2014. Available online http://cea.nic.in/reports/monthly/executive_r ep/jul14.pdf
15. Rastogi, S., Sharma, M., Varshney, P.: Internet of Things based smart electricity meters. Int. J. Comput. Appl. **133**(8) (Jan 2016)
16. Pandit, S., Mandhre, S., Nichal, M.: Smart energy meter using internet of Things (IoT). VJER Vishwakarma J. Eng. Res. **1**(2) (June 2017)
17. Jithin Jose, K., Mohan, L., Nijeesh, U.K., Benny, T.C.: Smart energy meter. Int. J. Eng. Trends Technol. (IJETT) **22**(4) (Apr 2015)
18. Li, L., Hu, X., Zhang, W.: Design of an ARM-based power meter having WIFI wireless communication module. In: Proceedings of IEEE 4th International Conference on Industrial Electronics and Applications, Xi'an, pp. 403–407, May 2009
19. Rahman, M.M., Noor-E-Jannat, Islam, M.O., Salakin, M.S.: Arduino and GSM based smart energy meter for advanced metering and billing system. In: 2nd International Conference on Electrical Engineering and Information & Communication Technology (ICEEICT), May 2015

Design of Energy Efficient ALU Using Clock Gating for a Sensor Node

M. Sharath(✉) and G. Poornima

B.M.S College of Engineering, Bengaluru, India
sharath.ithal@gmail.com, gpoornima.ece@bmsce.ac.in

Abstract. Recent growth in the need of fast devices and complex designs on a single System on Chip (SoC) mainly requires a low power and a highly efficient design in terms of speed and area. Moreover the Wireless Sensor Nodes which performs on the node processing are mainly powered by battery thereby requiring the processing element to consume less power for computations. The two ways of reducing power consumption is by either reducing static power consumption or by reducing dynamic power consumption. This paper has concentrated on reducing the dynamic power consumption using clock gating scheme. This scheme is applied to a 16-bit Arithmetic and Logic Unit with two distinct approaches namely Block Enabled Clock Gating and Functional Unit Enabled Clock Gating. A Comparative analysis is done for ALU without and with two distinct clock gating schemes. The reduction in clock power and dynamic power consumption is achieved with higher frequencies. Simulations and power analysis is done using Xilinx Vivado Simulator. With Functional unit enabled clock gating we could get more than 70% reduction in power for higher frequencies like 1 and 2 GHz.

Keywords: Functional unit enabled clock gating · Power reduction · Folded tree architecture

1 Introduction

The applications of Wireless Sensor Network (WSN) includes sensing of environmental parameters, inspection in industrial domain, surveillances in military domain and medical observing. WSN nodes fundamentally comprises of sensors, radio communication along with the microcontroller powered with restricted power source such as storage battery. Radio communication requires extravagant energy, the overall energy required for the node should be minimized for prolonging lifespan of sensor node. So, on node processing must be compromised for data communication which requires to convert large number of sensor readings into a smaller pool of beneficial data values.

Ever increasing complexity and speed required for contemporary designs suggests a greater upsurge in the energy consumption of very large-scale integration chips. To overcome the challenge, researchers have worked on several distinct design schemes to reduce the consumption of power. The beginning of the consumer era along with the admiration for mobile applications has made the power optimization one of the most

M. E. Auer and K. Ram B. (Eds.): REV2019 2019, LNNS 80, pp. 390–399, 2020.
https://doi.org/10.1007/978-3-030-23162-0_35

important design parameter. Designers would run through a number of iterations to optimize power with the purpose of accomplishing power savings. At each and every stage of design flow, the power consumption should be optimized. The optimizations in initial design phases is the chief influence in reducing overall power. Clock power is the component which consumes around 50–70% of total on chip power and it is expected it will grow ominously in succeeding design generations below 45 nm.

Power dissipation contains two important components namely static power and dynamic power. Static power can be because of mainly of the leakage current and switching activity results in dynamic power. Dynamic power can be defined as [10],

$$Pdynamic = \alpha * \mathrm{CL} * VDD^2 * f \tag{1}$$

where, α indicates switching activity, CL points to capacitance at load, VDD corresponds to voltage supply and f corresponds to frequency. This equation points to the fact that the dynamic power is directly proportional to the switching activity factor and frequency of operation. So, one should not reduce frequency of operation for reducing the overall power consumption as the devices should operate at high frequencies. So, one possible solution could be the reduction of switching activity.

Clock gating is the most common low power design technique which aids in reduction of power dissipation in VLSI design circuits. The objective of clock gating design scheme is to stop the clock transitions from propagating onto the clock path under some conditions influenced by gating circuits. Some power can be saved because of the reduction of loading unnecessary transitions whenever the functional unit is not active.

2 Literature Review

Walravens et al. [1] has described a new architecture for designing a processor for wireless sensor nodes which reduces the power consumption. Folding Tree Architecture uses the concept of component reuse to minimize the area consumed. In this work, the processing element considered was a Parallel prefix adder. The authors has illustrated a way of reducing the number of adders required for processing. Mrs. Jasmis C. S. et al., proposed a modification of Binary Tree Architecture into area and power effective Folded Tree Architecture (FTA) for comparator circuit. K. Hemapriya and R. Karthikeyan in their work have discussed about the low power design with the help of folded tree architecture and multibit flipflop merging procedure for processing of data on the node itself. The computation was carried out with the help of prefix operations that occur parallelly and vicinity of data in hardware. Parallel Prefix Adder employed in this paper is Ladner Fischer Adder because it had small amount of delay when compared to remaining parallel prefix adders [3]. Power consumption is reduced further by the usage of flip-flop merging technique in which smaller flipflops were exchanged by larger multi-bit flipflops. Le et al. [5] talks about 16-bit digital signal processor design. The architecture was built using Harvard architecture which consists of two buses meant for ALU along with multiply accumulator (MAC) with pipelining enabled. Sixteen general purpose 24-bit registers with forty-one 4-cycle instruction set constituted the design. The DSP processor was implemented on ASIC with the aid of "Silico on Thin Box" (SOTB) 65 nm process. Ferdous [6] has worked on design and implementation of a 32-bit DSP processor on FPGA in order to obtain high performance gain for RISC architecture-based DSP processors. This design included a pipelined architecture which was hazard optimized along with a dedicated MAC unit to boost the processing speed. The unique part in this design was hazard handling. The FSM was used to detect hazards and control the data path by resolving the hazard. Shaer et al. [8] have presented a DSP processor for hearing aid application which consumes low power. The design was implemented, and layout was designed using a 90 nm CMOS process. Shinde et al., have investigated numerous clock gating schemes which can be utilized for the optimization of power in the VLSI circuits at Register Transfer Level (RTL) in their work [9] and identified several concerns involved when applying these power optimization schemes at RTL level.

3 Implementation Details

The work done includes design of a 16-bit Arithmetic and Logic unit with clock gating. There are 2 16-bit inputs and result of the operation is of 32 bit. The ALU performs arithmetic operations such as addition, multiplication and subtraction and logical operations which comprises of AND operation, OR operation, NOT operation, NAND operation, NOR operation, XOR operation and XNOR operations. In this design there will be 3 selection lines which are represented by opcodes to select a particular operation the ALU must perform.

The three main components of this ALU design are:

- Ladner-Fischer Adder
- Multiplier using Folded Tree Architecture
- Clock Gating circuit.

3.1 Ladner-Fischer Adder

To accomplish addition operation in a swift manner, parallel prefix adder can be used. Ladner-Fischer adder is one of those adders which performs high performance parallel prefix addition operation. Ladner-Fischer adder has smaller delay in comparison with all other parallel prefix adders [4]. The construction of efficient Ladner-Fischer adder consists of three stages (Fig. 1).

In the Pre-Processing Stage, propagate and generate signals are calculated. The equations for the Pre-processing Stage are given by,

$$Pi = Ai\,XOR\,Bi \tag{2}$$

$$Pi = Ai\,AND\,Bi \tag{3}$$

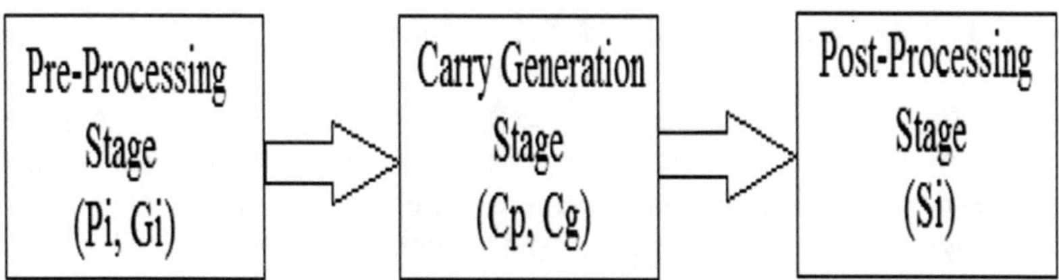

Fig. 1. Ladner-Fischer adder

In the Carry Generation Stage, equations for carry generate and carry propagate signals are given by,

$$Cp = P1\,AND\,P0 \tag{4}$$

$$Cg = G1\,OR\,(Pi\,AND\,G0) \tag{5}$$

The final stage of Ladner-Fischer Adder is post processing stage in which the carry of the first bit is XORed with the succeeding bit of propagate signal.

$$Si = Pi\,XOR\,Ci - 1 \tag{6}$$

3.2 Multiplier Using Folded Tree Architecture

In folded tree architecture, the adder is identified as the basic processing element which should occur more than once. Recursion process is performed in the adder with the aid of additional blocks such as finite state machine, iteration counter. The multiplier block diagram is as shown in Fig. 2. The 16-bit data input is represented by Data In block and it is given as an input to the adder. Finite state machine contains a counter which is used to increment after every iteration. Iteration count is a storage element which stores

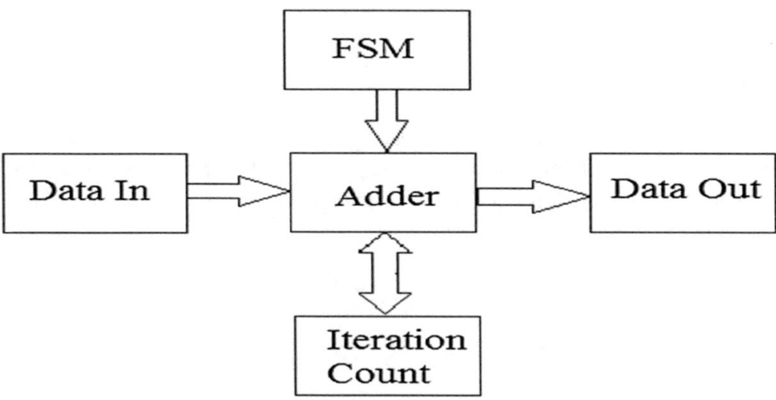

Fig. 2. Multiplier using recursion

the value until the recursion should happen. The final output is obtained in Data Out block.

In the array used for processing, for each iteration the preceding output value is used as a portion of input. The amount of occasions the operation of recursion needs to happen is stated in the iteration count i.e., the number of times the primary input must be added is given by the secondary input which will be stored in the iteration count. For each iteration, the counter value will be incremented, and it will be compared with the value stored in iteration count. When the counter outputs the same value as that of iteration count, the corresponding output of the adder indicates final output (Fig. 3).

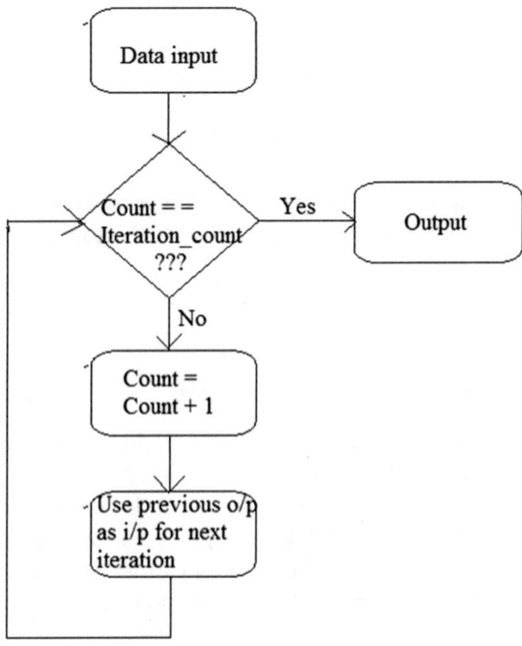

Fig. 3. Recursion part of the multiplication process

3.3 Clock Gating

Clock Gating mechanism works by capturing the enable conditions attached to the registers and utilizing them to gate the clock signal. The design should comprise of certain enable conditions to use and take advantage of clock gating scheme. The Clock gated signals are given by,

$$Gate\_mul = mul\_enable\,AND\,Clk \tag{7}$$

$$Gate\_add = add\_enable\,AND\,Clk \tag{8}$$

$$Gate\_sub = sub\_enable\,AND\,Clk \tag{9}$$

$$Gate\_and = and\_enable\,AND\,Clk \tag{10}$$

$$Gate\_or = or\_enable\,AND\,Clk \tag{11}$$

$$Gate\_xor = xor\_enable\,AND\,Clk \tag{12}$$

$$Gate\_nand = nand\_enable\,AND\,Clk \tag{13}$$

$$Gate\_nor = nor\_enable\,AND\,Clk \tag{14}$$

D flip-flop is a basic sequential circuit with which clock gating can be accomplished. The output of the D flip-flop is ANDed with the input clock and that signal is applied to ALU. Whenever the enable signal (EN) is becomes 1, a HIGH value is stored in the output (Q) of D flipflop which is subsequently applied to AND gate to generate gated clock signal to ALU (Fig. 4).

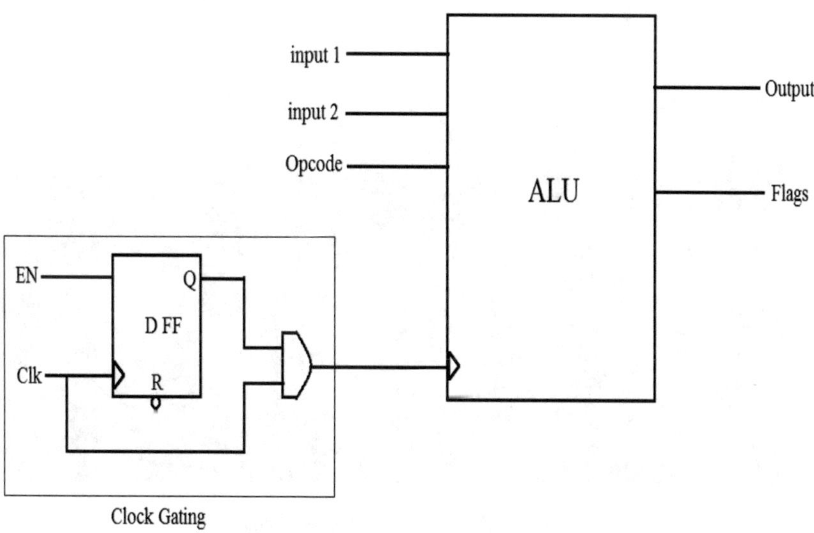

Fig. 4. Latch based clock gating circuits

The device utilization report for the ALU design without Clock Gating and with Clock Gating is tabulated in Table 1. Since additional circuits like registers and logic gates are needed, the area required for the design obviously increases. As the table shows the clock gating circuits consumes more components. This is the trade-off required to be satisfied for the reduction of power.

Table 1. Device Utilization Report with and without Clock Gating

Name of the resource	Resource used without clock gating	Resource used with clock gating
Slice LUTs	406	522
Slice Registers	362	310
F7 MUXes	16	66
F8 MUXes	0	1
Bonded IOB	69	69
BUFGCTRL	3	4

4 Simulation Results and Discussion

Verilog HDL is used to design 16-bit ALU with and without clock gating schemes and the simulations are performed on the design using Xilinx Vivado 14.4. The simulation waveform can be seen in Fig. 5. As the waveform shows, the appropriate output is obtained for the opcode. For adder, since there are three stages, we are getting the output after three cycles.

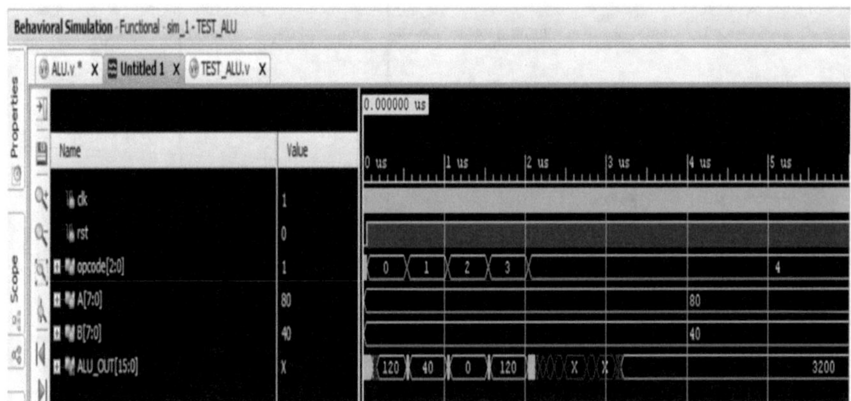

Fig. 5. ALU simulated waveform

4.1 Power Analysis Without Clock Gating Scheme

Power analysis is performed with the help of Xilinx Power analyzer. The clock input is applied to all units of the ALU and the inputs are sampled at every positive edge of clock signal, the power analysis results are tabulated in Table 2.

As the table shows, dynamic power is constituted mainly of clock power and IO power. Clock Power can be explained as the power consumed by the clock signal and IO power is the power constituted by IO signals of the design block. In the design of ALU without clock gating, clock signals and IO signals are loaded into all blocks of

Table 2. Power analysis without clock gating

F (MHz)	Clock power (mW)	Logic power (mW)	Signal power (mW)	IO power (mW)	Dynamic power (mW)	Total power (mW)
100	0.021	0.004	0.003	0.065	0.094	0.154
200	0.043	0.008	0.007	0.130	0.187	0.248
500	0.107	0.020	0.017	0.324	0.469	0.53
1000	0.215	0.041	0.033	0.648	0.937	0.999

design even though only one operation is executed at a moment of time. This contributes to the unnecessary power consumption.

4.2 Power Analysis with Block Enabled Clock Gating Scheme

With the application of gated clock to two blocks of ALU design, the clock power is reduced to some extent. Table 3 shows the power consumed for different clock frequencies.

Table 3. Power analysis with block enabled clock gating

F (MHz)	Clock power (mW)	Logic power (mW)	Signal power (mW)	IO power (mW)	Dynamic power (mW)	Total power (mW)
100	0.011	0.001	0.001	0.063	0.076	0.136
200	0.023	0.002	0.001	0.126	0.151	0.212
500	0.057	0.005	0.001	0.315	0.379	0.439
1000	0.113	0.010	0.003	0.631	0.757	0.819

As shown in Table 3, the power consumed by IO operations is reduced to lesser extent because if arithmetic block is enabled by the clock signal, all IO of the complete arithmetic blocks will be loaded hence contributing to additional IO power. Hence the power reduction will be less.

4.3 Power Analysis with Functional Unit Enabled Clock Gating Scheme

With the application of gated clock to each individual functional unit of ALU, the clock power is reduced to great degree. The table shows the power consumed for different clock frequencies.

In this clock gating design, the clock signal and IO signal will be loaded onto the functional units such as adder, multiplier only when a unit is selected to operate through opcode. Hence greater power reduction can be seen as shown in Table 4.

Table 4. Power analysis with functional unit enabled clock gating

F (MHz)	Clock power (mW)	Logic power (mW)	Signal power (mW)	IO power (mW)	Dynamic power (mW)	Total power (mW)
100	0.006	0.001	0.001	0.006	0.001	0.074
200	0.012	0.001	0.002	0.012	0.028	0.088
500	0.030	0.003	0.006	0.030	0.069	0.129
1000	0.060	0.006	0.011	0.061	0.138	0.199

The power reduction percentage can be calculated using the formula [10],

$$Power\,Reduction\,\% = \frac{\text{Power without CG} - \text{Power with CG}}{\text{Power without CG}} * 100 \tag{15}$$

By referring to Table 5, it can be inferred that with clock gating the dynamic power reduces drastically for higher frequencies. At higher frequencies, with functional unit enabled clock gating we can expect 75–80% of power reduction.

Table 5. Percentage power reduction with clock gating scheme

Frequency (MHz)	Power without CG (mW)	Power with BCG (mW)	Power with FCG (mW)	% Reduction with BCG (%)	% Reduction with FCG (%)
100	0.154	0.136	0.074	12	52
200	0.248	0.212	0.088	15	65
500	0.53	0.439	0.129	17	75
1000	0.999	0.819	0.199	18	80

5 Conclusion

In this work, clock gating scheme is applied to a 16-bit ALU and waveforms are simulated, and the power analysis is done using Xilinx Vivado Power Analyzer 14.4 Suite. The folded tree architecture is used for the design of a multiplier for low area and low power consumption. The two different techniques of Clock gating namely Block enabled and Functional Unit enabled clock gating techniques are applied. The power reduction is observed to a greater extent for the latter technique. For lower frequencies the power reduction is lesser as we observed that for 100 and 200 MHz the power reduction varies from 50 to 60% whereas for higher frequencies the power is reduced significantly like for 500 MHz and 1 GHz the power reduction is observed to be varying between 75 and 80%. The additional area utilized for clock gating circuits is compensated by the greater power reduction. The significant reduction in power persuades the use of clock gated ALU for wireless sensor applications where the power consumption is the critical issue.

Acknowledgements. I would like to thank my guide for the valuable guidance during the entire period of this work.

References

1. Walravens, C., Dehaene, W.: Low-power digital signal processor architecture for wireless sensor nodes. IEEE Trans. Very Large Scale Integr. (VLSI) Syst. **22**(2), 313–321 (2014)
2. Jasmin C.S., Mathew, A.P.: Power efficient comparator architecture for wireless sensor nodes. In: International Conference on Emerging Technological Trends (ICETT), p. 978-1-5090-3751 (2016)
3. Hemapriya, K., Karthikeyan, R.: Low-power folded tree architecture and multi-bit flip-flop merging technique for WSN nodes. In: International Conference on Information Communication and Embedded Systems, p. 978-1-4799-3834 (2014)
4. Ranjithkumar, K., Anandharajan, T.R.V.: DSP architecture with folded tree for power constraint devices. In: Sixth International Conference on Advanced Computing (ICoAC), p. 978-1-4799-81S9 (2014)
5. Le, D.-H., Sugii, N., Kamohara, S., Nguyen, X.-T., Ishibashi, K., Pham, C.-K.: Design of a Low-power fixed-point 16-bit digital signal processor using 65 nm SOTB process. In: 2015 International Conference on IC Design & Technology (ICICDT) (2015)
6. Ferdous, T.: Design and FPGA-based implementation of a high performance 32-bit DSP processor. In: 2012 15th International Conference on Computer and Information Technology (ICCIT), p. 978-1-4673-4836 (2012)
7. Nguyen, X.-T., Huu, T.H., Le, D.H.: An ASIC implementation of 16-bit fixed-point digital signal processor. J. Sci. Technol. **51**(4B), 282–289 (2013)
8. Shaer, L., Nahlus, I., Merhi, J., Kayssi, A., Chehab, A.: Low-power digital signal processor design for a hearing aid. In: 2013 4th Annual International Conference on Energy Aware Computing Systems and Applications (ICEAC), p. 978-1-4799-2543 (2013)
9. Shinde, J., Salankar, S.S.: Clock gating—a power optimizing technique for VLSI circuits. In: India Conference (INDICON), 2011
10. Kulkarni, R., Kulkarni, S.Y.: Power analysis and comparison of clock gated techniques implemented on a 16-bit ALU. In: Proceedings of International Conference on Circuits, Communication, Control and Computing (I4C 2014), p. 978-1-4799-5364 (2014)
11. Shrivastava, G., Singh, S.: Power optimization of sequential circuit based ALU using gated clock pulse enable logic. In: International Conference on Computational Intelligence and Communication Networks, p. 978-1-4799-6929-6 (2014)
12. Shanker, M., Bayoumi, M.A.: Clock gated FF for low power application in 90 nm CMOS. IEEE Trans. Circuits Syst. II, 978-1-429474-3 (2011)
13. Wimer, S., Koren, I.: The Optimal fan-out of clock network for power minimization by adaptive gating. IEEE Trans. Very Large Scale Integr. (VLSI) Syst. **20**(10), 1772–1780 (2012)
14. Oliver, J.P., Curto, J., Bouvier, D., Ramos, M., Boemo, E.: Clock gating and clock enable for FPGA power reduction. In: 8th Southern Conference on Programmable Logic (SPL), pp. 1–5 (2012)
15. Kamaraju, M., Lal Kishore, K., Tilak, A.V.N.: Power optimized ALU efficient datapath. Int. J. Comput. Appl. **11**(11) (2010)

Optimal Charging Strategy for Spatially Distributed Electric Vehicles in Power System by Remote Analyser

R. Venkataswamy(✉) and Teena M. Joseph

Department of Electrical & Electronics Engineering, Faculty of Engineering, Christ (Deemed to be University), Bangalore, India
venkataswamy.r@christuniversity.in, venkataswamy.r@gmail.com
teena.joseph@mtech.christuniversity.in

Abstract. The burden on the consumer for the price of fuel for classic vehicles is the root cause for the emergence of the fast growing trend in the power driven vehicles or electric vehicles. Less acceptance of electric vehicles by the customers and the hesitancy to replace traditional fuel powered vehicles by considering the economic factor is a major concern that existing in the current scenario. Therefore, for the proper balancing of the load with respect to the power available among different neighbouring charging stations in a given area, a load scheduling algorithm is used. The optimal route planner for the electric vehicles reaching the charging station is identified and then the power carried by each feeder is calculated by cumulative power of all the charging stations. The identification of the possible route is performed by the spatial network analysis which will be executing at remote analyzer. The location, state of charge, and other details of the electric vehicle through telemetry is used to find the best charging station for the particular vehicle in view of the cost, distance and the time. The performance of the technique is evaluated with and without optimization by considering the logical constraints; and the results are presented.

Keywords: Google maps · Load scheduling · Optimization · Optimal route planner

1 Introduction

One of the most relevant contribution to the electrical sector due to the upgradation of technology is the launch of electric vehicles. The upturn in the population cause to a gradual increase in the number of vehicles on the road. The traditional gas powered vehicles when handled or operated on lane will have the harmful gas emissions beyond the value which can be predicted. According to the surveys the major portion of the pollution that is, 75% is as a result of these poisonous carbon monoxide emissions expelling out from the automobiles. As part of a solution to this environmental pollution problem and for providing individual

M. E. Auer and K. Ram B. (Eds.): REV2019 2019, LNNS 80, pp. 400–412, 2020.
https://doi.org/10.1007/978-3-030-23162-0_36

benefits to customers more focus is given to EV's than the vehicles which run by using IC engines. Apart from all these, electric vehicles will be a solution for the expecting fossil fuel scarcity.

The time required to charge the battery fully hits a stumbling block in using EV for the long distance traveling. The customers are bothered about the SOC of the battery of electric vehicle whether they can reach the destination before the battery charge getting depleted. There should be enough CS available for charging since the number of power driven vehicles goes on increasing. Sometimes the consumers need to wait in queue to get their vehicles charged and in order to avoid such disturbances an optimal charging strategy required for EV which are distributed spatially.

Google maps is a widely used service by Google. Despite of where the driver is traveling, the application service required to locate the actual position of the user with reference to the co-ordinate axes. By using dynamic location the remote analyzer determines the distance between two charging stations and map to the center which is near by. To achieve the task, two set of data is created one which consists of the EV and its SOC, other including CS and the capacity of charging along with its per unit cost. This master and dynamic data will help to programmically map the appropriate point for charging station accordingly.

The impact which is given by the EV's to the power system is not comparable with respect to load for one residential building. The high EV load penetration may sometimes cause an unbalancing situation in the grid. This can be overcome by the proper planning and scheduling of the load. This paper also discuss about the scheduling of the load at the charging station level to the feeder level. By calculating all the power consumed by each EV entering the CS we can decide whether the existing system is profitable or not.

2 Current Research and Developments

Among the possible routes to a particular destination the selection of the best route to ensure the quicker travel time is important [1]. To provide the best route, the system will take the information regarding traffic, volume to the capacity ratio and vehicle queue length. The static and dynamic factors which affects the routing of vehicles is considered in [2]. The main consideration of this work is the SOC of the battery to determine the path through which it should travel. The developed model for efficient routing is compared and verified with the 306 km of actual driven data. Area of driving of the electric vehicle is divided into different sections and a RTOU price model is introduced in [3]. For optimizing the cost of electric vehicle charging schedule and also so as to bring down the difference in potential of the power system the node layer model and the regional layer model is developed.

Based on some optimal parameters, to make the load value constant at the peak hour and also to make the peak-valley difference to a lower value in the load curve of a residential area the charging control strategy is proposed in [4]. By selecting the effective interval for charging the vehicle the plan of charging and also the power grid load curve can be improved. The traditional random charging and the proposed method for charging is compared and analyzed. The strategies which used for the parking of EV integrated with the vehicle to grid capability such that the power exchange between the parking lot and the distribution system is considered [5]. The area of parking for the capacity of almost 100 vehicles is taken for analysis at the peak and off peak hours. The control and monitoring for the integration of renewable sources and the electric vehicle charging is discussed. The two models are proposed one is to shift the consumption time in order to get the lesser energy bill and second model is provided to make flexible services to the power system [6].

An optimized charging model is made in [7] with the intention of increasing the profit of charging station. The orderly charging and disorderly charging condition is compared taking the power consumed for both cases as a factor for analysis. The vehicle arrival characteristics at a particular period of time in each CS is obtained and from that data, charging demand of grid is calculated [12]. The uncontrolled charging may affects the power system when there is a high demand for energy. The optimal scheduling for charging leads to the stable system [9]. A two stage optimization model is proposed which meet the needs of both power grid and the users. The first stage optimization will focus on the economic requirements and second stage will coincide the stability of the grid as presented in [10].

3 Load Scheduling

The load scheduling is basically allocation of power to different feeder depends on the supplied power the region in the substation. A simple optimization problem is formulated for the validation of the proposed model. The formulation of load scheduling problem is shown below, The objective function can be complete utilization with profit for available power at substation,

$$\min\left(\sum_{i=1}^{k}\alpha_i \mathrm{P}_i\right) \tag{1}$$

Constraints are,

$$\sum_{i=1}^{k}\mathrm{P}_i = P_S \tag{2}$$

Feeder power limit are,

$$P_{lb} \leq P_i \leq P_{ub} \forall i \in R^k \tag{3}$$

where, α is cost per unit power, P_S is total power available at substation, $P_{lb}\ and\ P_{ub}$ are lower and upper bound of power allocation.

4 Google Maps

Google map APIs are used to get distance and time for given two location. In the study under consideration, the first point is dynamic EV location and second point is fixed CS location. The algorithm is executed every fixed time duration and distance and time tuples are formulated.

5 Remote Analyzer

Remote analyser is a IoT application where in it extracts dynamic location of EVs and charging station details. It uses google map APIs to generate distance and time vector in a cloud computing platform. The cost from database, distance and time are uploaded to cloud. An application is developed to send and fetch data from cloud to EVs. Load details are sent to substation for load scheduling. The working pictorial model is shown in Fig. 1.

6 Genetic Algorithm

Genetic algorithm which is one among the prominent evolutionary algorithm is adopted in this paper to obtain the expected results. The optimum value for the objective function is found out by giving some changes or alteration to the properties of individuals participating in the technique. The initial population is predetermined and encoded for the synthesis of the next generation results, then further undergone mutation and crossover to achieve the optimal solution. The flow chart shown in Fig. 2 depicts the same.

7 Mathematical Modeling

The mathematical model is basically the optimization problem. The type of optimization is inequality constrained binary optimization technique which optimizes multiple parameter namely time, distance and cost of charging. The mathematical formulation is as follows,
The objective function,

$$\min \sum_{j=1}^{m} \sum_{i=1}^{n} \beta_{ij} \Gamma_{ij} \tag{4}$$

Constraints are,

$$\overline{P_{EV}^{r}} = \overline{\beta \alpha}.\overline{P_{CS}} \tag{5}$$

$$\sum_{i=1}^{n} \beta_{ij} = 1 \forall j \in R^{m} \tag{6}$$

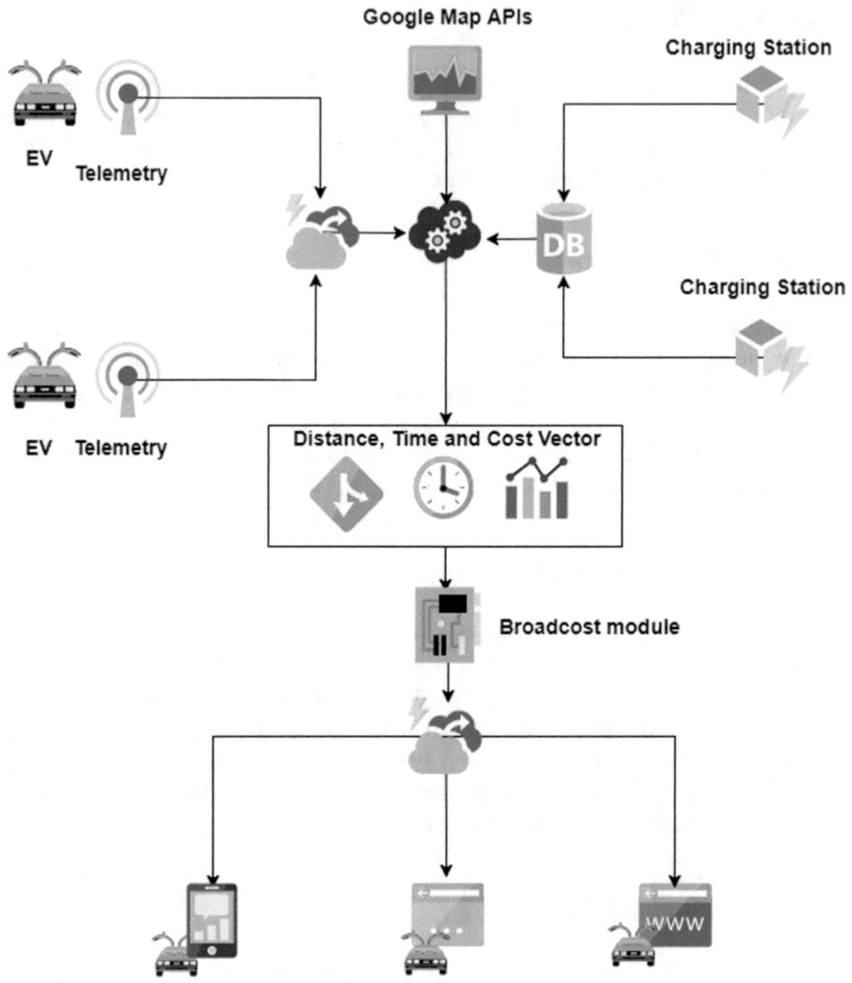

Fig. 1. Remote analyzer

$$\sum_{j=1}^{m} \beta_{ji} \leq m \forall i \in R^n \tag{7}$$

where,

$$\beta \in \{0, 1\} \tag{8}$$

$$\Gamma_{ij} = \zeta_{i1} c_{ij} + \zeta_{i2} d_{ij} + \zeta_{i3} t_{ij} \tag{9}$$

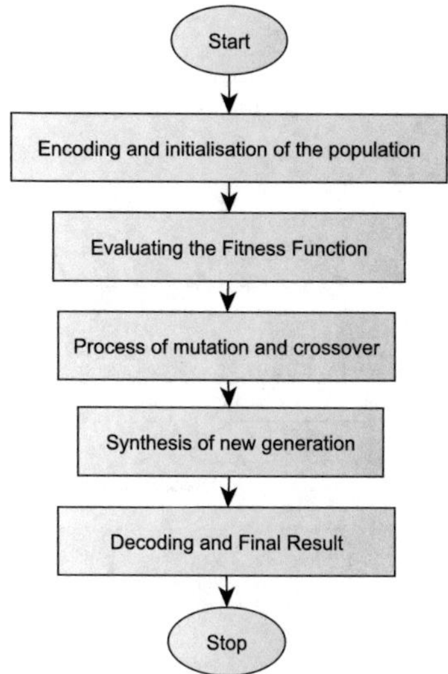

Fig. 2. Genetic algorithm

Equation 5 can be written as,

$$\begin{bmatrix} P^r_{EV1} \\ P^r_{EV2} \\ \vdots \\ P^r_{EVn} \end{bmatrix} = \begin{bmatrix} \beta_{11}\alpha_{11} & \beta_{12}\alpha_{12} & \cdots & \beta_{1m}\alpha_{1m} \\ \beta_{21}\alpha_{21} & \beta_{22}\alpha_{22} & \vdots & \beta_{2m}\alpha_{2m} \\ \vdots & \vdots & \vdots & \vdots \\ \beta_{n1}\alpha_{n1} & \beta_{n2}\alpha_{n2} & \cdots & \beta_{nm}\alpha_{nm} \end{bmatrix} \tag{10}$$

The vector (c, d, t) is the cost, distance and time required for EV for CS. ζ is the customer psychological behavior. This is an indicator whether driver is more oriented to cost, distance or time. α is the portion of power utilized from EV in the CS. n is the number of EVs. m is the total number of charging stations.

8 Methodology

The paper mainly deals with the mapping of electrical vehicles to the charging station by considering charging cost, distance of travel and time of charging. For the mapping ten EV's and CS are marked randomly at various points by using the cloud based application. The current latitude and longitude of the longitude of the location where the EV is at present is fetched from the telemetry system. The algorithm is executed on timely basis to know the CS from which they can get benefit. The multiple EV's and charging station are considered. The overall working of the model is shown in Fig. 3.

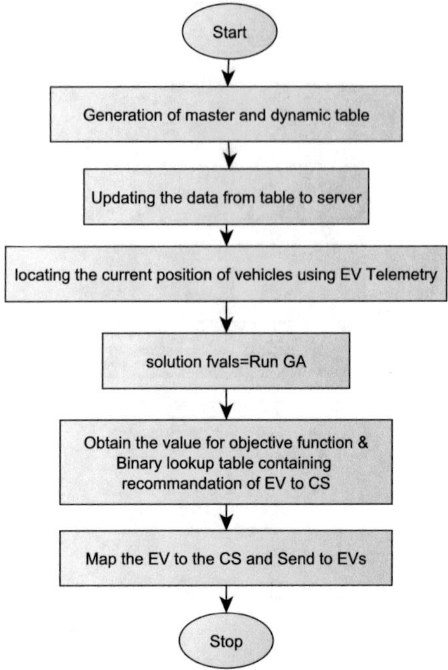

Fig. 3. Methodology

9 Results and Discussions

A case study or illustration is simulated or solved using the methodology proposed. The illustration considered is, three electrical vehicles that are placed around two charging stations. Unit driver psychological behavior is considered.

9.1 Remote Analyzer

The distance and duration matrices obtained using Google Map API is show in Tables 1 and 2 for every charging station for all electric vehicles. The cost assumed is given in Table 3.

Table 1. The distance matrix (values are in meters)

	EV1	EV2	EV3
CS1	1992	4422	5698
CS2	6302	4466	1385

Table 2. The duration matrix (values are in seconds)

	EV1	EV2	EV3
CS1	220	424	684
CS2	662	534	206

Table 3. The cost matrix (values are in rupees)

	EV1	EV2	EV3
CS1	10	10	10
CS2	5	5	5

Table 4. EV and CS mapping without optimization

Charging station	Electric vehicles	Objective function
CS1	2	13601
CS2	1,3	–

Table 5. EV and CS mapping with optimization

Charging station	Electric vehicles	Objective function
CS1	1,2	8899
CS2	3	–

The EV and CS mapping with optimization is shown in Table 4. The EV and CS mapping without optimization is shown in Table 5.

9.2 Load Scheduling

Three feeder substation is considered for the load allocation with maximum limit of 110, 100, 120 MW and minimum limit of 95, 40 and 40 MW respectively. The total input to the substation is considered as 200MW. First feeder is assumed to be having the power requirement of 70MW along with EV load of 20 MW. The load scheduling algorithm shown in Eq. 9 is executed with genetic algorithm to find the power allocation for other two feeder which found out to be 53.71 and 56.29 MW respectively by assuming unit cost.

10 Conclusion and Future Scope

The charging strategy of electrical vehicles implemented based on the optimal selection of charging cost, distance of travel and time to reach charging station. The charging station and electric vehicle dynamic location is passed to remote analyzer wherein the best charging station for driver psychological behavior is recommended through the IoT platform. The cumulative power at all the charging station is considered in the substation to schedule power. The comparative study is presented which validated using remote analytics.

The future scope of work would be considering cross constraints at load scheduling and optimal vector derivation. Considering waiting time at charging station, SoC variation for different types of vehicle would increase the performance of the proposed model. The load forecasting along with the prediction of electric vehicle would definitely add smartness in the complete ecosystem.

Acknowledgments. The authors would like to acknowledge and express the deepest gratitude to management of Faculty of Engineering, Christ (Deemed to be University) for providing freedom to work in the laboratory.

Appendices

A Glossary

EV	**E**lectric **V**ehicle
CS	**C**harging **S**tation
SOC	**S**tate **o**f **C**harge
GUI	**G**raphical **U**ser **I**nterface
GCP	**G**oogle **C**loud **P**latform
API	**A**pplication **P**rogram **I**nterface
IC	**I**nternal **C**ombustion
RTOU	**R**egional **T**ime **Of** **U**se

B Remote Analyzer Application

Remote Analyzer application is developed using Matlab app designer. The charging station map, charging station details and electric vehicle dynamic details are given in Figs. 4, 5, 6 and 7 respectively.

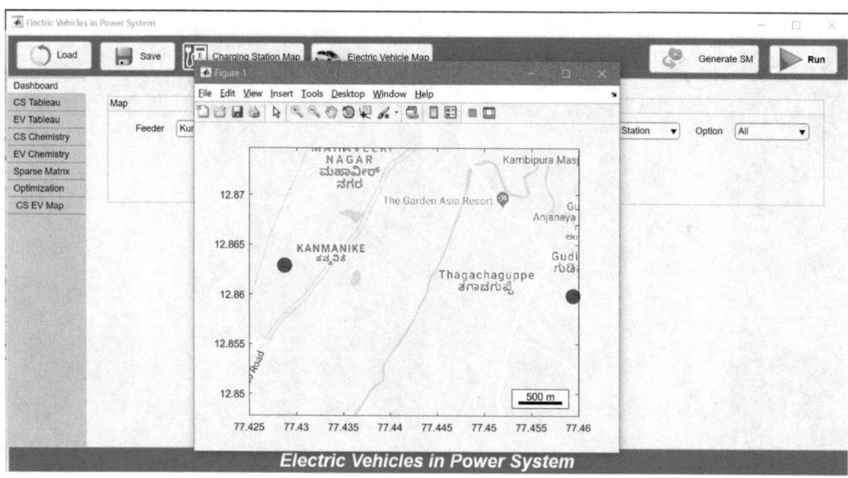

Fig. 4. Charging stations map

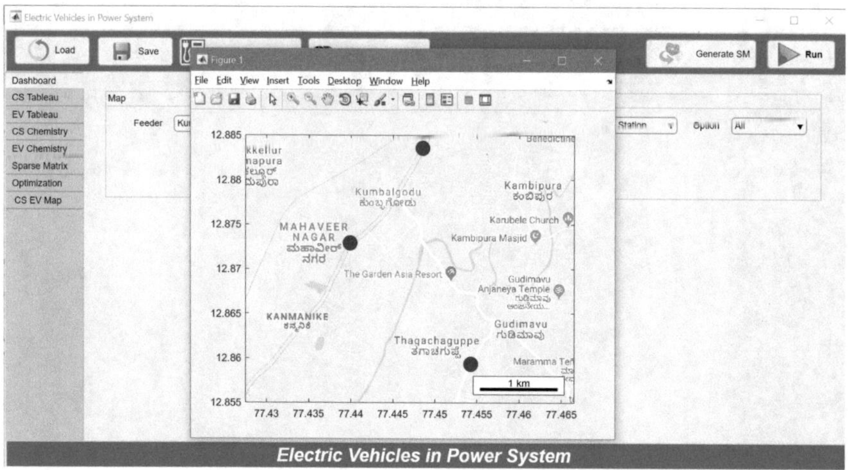

Fig. 5. Electric vehicles map

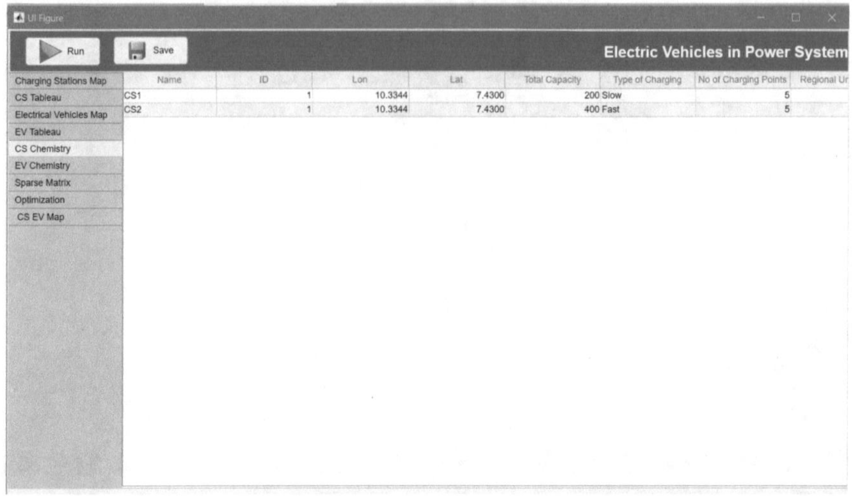

Fig. 6. Charging station details

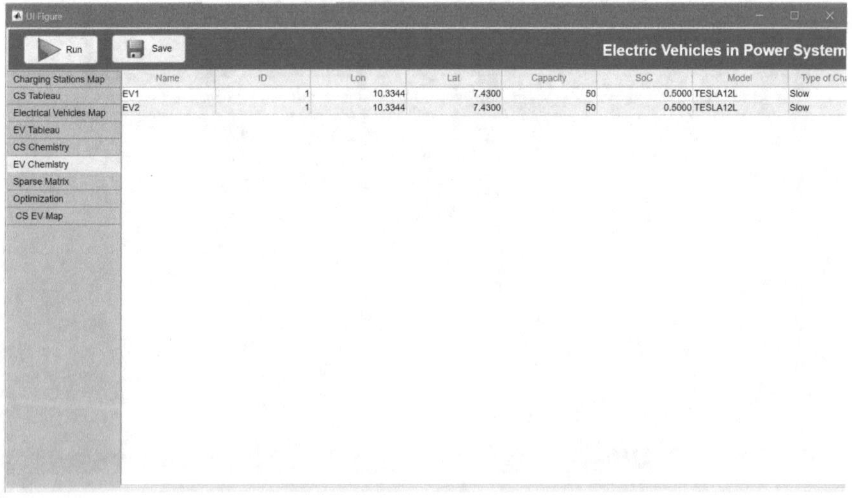

Fig. 7. Electric vehicle dynamic details

References

1. Abousleiman, R., Rawashdeh, O.: Electric vehicle modeling and energy-efficient routing using particle swarm optimisation. In: Research Article Published by Department of Electrical and Computer Engineering, Oakland University, IET, vol. 10, pp. 65–72 (2015)
2. Rahman, S., Yeasmin, N., Ahmmed, M.U., Kaiser, M.S.: Adaptive route selection support system based on road traffic information. In: 2nd International Conference on Electrical Engineering and Information and Communication Technology (lCEEICT), Dhaka, Bangladesh, pp. 1–6 (2015)
3. Chen, J., Yang, J., Lib, M., Zhoub, Q., Jiang, B.: An optimal charging strategy for electrical vehicles based on the electricity price with temporal and spatial characteristics. In: IEEE Power and Energy Society General Meeting (PESGM), Boston, MA, USA, pp. 1–5 (2016)
4. Boi, S., Zhongpei, S., Dajun, W., Li, Y.: Research on optimal control of electric vehicle charging in residential area. In: Proceedings of the 35th Chinese Control Conference, Chengdu, China, pp. 8617–8621 (2016)
5. Jozi, F., Mazlumi, K., Hosseini, H.: Charging and discharging coordination of electric vehicles in a parking lot considering the limitation of power exchange with the distribution system. In: IEEE 4th International Conference on Knowledge-Based Engineering and Innovation (KBEI), Tehran, Iran, pp. 937–941 (2017)
6. Georgiev, M., Stanev, R., Krusteva, A.: Flexible load control in electric power systems with distributed energy resources and electric vehicle charging. In: IEEE International Power Electronics and Motion Control Conference (PEMC), Varna, Bulgaria, pp. 1034–1040 (2016)
7. Liu, R., Zong, X., Mu, X.: Electric vehicle charging control system based on the characteristics of charging power. In: Chinese Automation Congress (CAC), Jinan, China, pp. 3860–3840 (2018)
8. Zhang, J., Zhou, H., Li, H., Liu, H., Li, B., Yan, C.L.: Multi-objective planning of charging stations considering vehicle arrival hot map. In: IEEE Conference on Energy Internet and Energy System Integration (EI2), Beijing, China, pp. 1–6 (2017)
9. Wang, Z., Paranjape, R.: Optimal scheduling algorithm for charging electric vehicle in a residential sector under demand response. In: IEEE Electrical Power and Energy Conference (EPEC), London, ON, Canada, pp. 45–49 (2015)
10. Jiang, R., Zhang, Li, J., Zhang, J., Huang, Q.: A coordinated charging strategy for electric vehicles based on multi-objective optimization. In: 2nd International Conference on Power and Renewable Energy, Chengdu, China, pp. 823–827 (2017)
11. Ceng, R.: Optimal charging/discharging control for electric vehicles considering power system constraints and operation costs. IEEE Trans. Power Syst. **31**, 1854–1860 (2016)
12. Huo, Y., Bouffard, F., Jos, G.: An energy management approach for electric vehicle fast charging station. In: IEEE Electrical Power and Energy Conference (EPEC), Saskatoon, SK, Canada, pp. 1–6 (2017)
13. Mistry, R.D., Eluyemi, F.T., Masaud, T.M.: Impact of aggregated EVs charging station on the optimal scheduling of battery storage system in islanded microgrid. In: North American Power Symposium (NAPS), Morgantown, WV, USA, pp. 1–5 (2017)

14. Hafez, O., Bhattacharya, K.: Integrating EV charging stations as smart loads for demand response provisions in distribution systems. IEEE Trans. Smart Grid **9**, 1096–1106 (2018)
15. Venkataswamy R.: Open IoE, https://openioe.herokuapp.com. Last accessed 26 Oct 2018

Switching Studies on $Ge_{15}Te_{70}In_5Ag_{10}$ Thin Films Device for Phase Change Memory Applications

Diptoshi Roy[1], Soniya Agrawal[1(✉)], G. Sreevidya Varma[2], and Chandasree Das[1]

[1] Department of Electrical & Electronics Engineering, BMS College of Engineering, Bangalore 560019, India
soniya.eee@bmsce.ac.in
[2] Department of Physics, Presidency University, Rajanakunte, Yelahanka, Bangalore 560064, India

Abstract. This paper describes the various studies performed on a representative quaternary thin film device of composition $Ge_{15}Te_{70}In_5Ag_{10}$, which could be an important material for phase change memory and data storage application. The I-V Characteristics (Switching characteristics), X-RAY Diffraction (XRD) and Energy-Dispersive X-ray analysis (EDAX) studies have been performed to show that the given material has some favorable characteristics as its application as phase change memory (PCM).

Keywords: Phase change memory (PCM) · Electrical switching · Thin film devices

1 Introduction

The property of the chalcogenide material to change its phase from amorphous to crystalline phase has been utilized from many years [1–3] for various applications. Phase Change Memory (PCM), is a currently developed semiconductor technology for non-volatile memory applications. Low programming energy, rapid switching among different phases and excellent scaling characteristics are the advantages that makes it promising candidate for non-volatile memory applications [4, 5]. On application of an apposite electric field (switching or threshold field) on chalcogenide glassy semiconductors, it switches from a high resistive "off" phase to high conductive "on" phase [6]. Threshold and memory switching are the two types of electrical switching manifested by chalcogenides. Glasses which exhibit threshold switching revert to their original state whereas glasses exhibiting memory switching retains its on state after removal of electrical field.

These PCM devices are based on the thermal transition of thin film chalcogenide material and its change in resistance to store data permanently. These types of devices

M. E. Auer and K. Ram B. (Eds.): REV2019 2019, LNNS 80, pp. 413–420, 2020.
https://doi.org/10.1007/978-3-030-23162-0_37

are essentially resistors made up of thin film chalcogenide materials, using Germanium (Ge)-Tellurium (Te) based alloys [7–9]. Change in resistance depends upon phase (Amorphous or crystalline) state of material of active region [10].

When an appropriate current or voltage pulse is given to the chalcogenide glassy semiconductors and if amorphous to crystalline phase change occurs in it, the phenomenon is called as SET process. Under the application of another sharp and short pulse of higher magnitude, the glassy semiconductor regains their amorphous phase this process is called RESET process as shown in Fig. 1 [11].

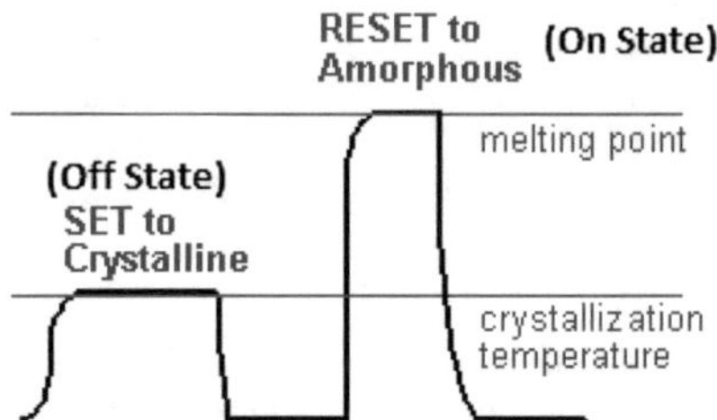

Fig. 1. Change in phase of the chalcogenide glassy semiconductors material at different temperature

For the investigation of SET-RESET phenomenon and their applicability in phase change memory applications, electrical switching studies on $Ge_{15}Te_{70}In_5Ag_{10}$ thin films materials has been performed.

2 Experimental Details

To Study the switching Characteristics of Ge-Te based thin film device a composition $Ge_{15}Te_{70}In_5Ag_{10}$ has been chosen. The composition has been fabricated on glass substrate in sandwiched geometry with aluminium as upper and lower electrodes.

The bulk composition has been prepared by melt quenching technique. The quartz ampoule inserted in the furnace is loaded with ±0.1 mg accurate and measured proportions of 99.999% pure elements and sealed under 10^{-5} mbar vacuum level. This sealed ampoule is placed in a high temperature rotary furnace and the temperature is increased by 100 °C every hour until it reaches to the temperature which is greater than the melting point of the composition, followed by the continuous rotation of the furnace containing the ampoule for 36 h at 10 RPM to maintain uniformity of the melt. The ampoule were quenched, to obtain bulk glassy sample in sodium hydroxide (NaOH) and ice water mixture, after finishing quenching the ampoule were broken and further it is coated in thin film form by adopting flash evaporation technique at a vacuum level of 10^{-6} mbar.

Electrical switching behaviour (I-V curve) of the sample is obtained by plotting the graph between the current (I) through device as a function of voltage (V) difference across the device. The above process was carried out with keithley (2410^c) source-measure unit using Labview 7. The source and measure unit is capable of sourcing current in the range of 50 pA to 1.05 A at a maximum adherence voltage of 1100 V. The sample is placed in a probe station which is proficient to move in X, Y, Z directions. Scanning electron micrograms (SEM) and Energy-Dispersive X-ray analysis (EDAX) studies are performed using VEGA3 TESCAN to observe the overall device characteristics. Optical absorption of the film is taken from a spectrophotometer (Specord S600). The amorphous nature of the thin film is studied by capturing X-Ray Diffraction (XRD) patterns with diffractometer (Bruker AXS D8 Advance).

3 Outcomes and Discussions

3.1 Electrical Switching Characterization

Electrical switching response of the $Ge_{15}Te_{70}In_5Ag_{10}$ composition of the thin film:

The electrical switching characteristics of $Ge_{15}Te_{70}In_5Ag_{10}$ composition has been done on devices which is coated on a single glass substrate using masks and I-V characteristics are performed on them, two of which has been shown in Fig. 2a, b. In this experiment, current is applied and the voltage across the sample is measured. When the voltage across the sample reached the threshold voltage the chalcogenide alloy switched from the high-resistive off state to the low-resistive state on state, and on the removal of current the material do not revert back to original amorphous state. Hence the material shows the memory type of electrical switching behavior. The threshold voltage is 4.82 V for first device and 2.85 V for second device. Figure 2a shows that when current is 0.09 mA the corresponding voltage drop across the sample is very less as it is in high conductive state, and resistance has dropped drastically. The same explanation holds for Fig. 2b also.

The low switching voltage of the composition confirms that material can be a suitable candidate for phase change memory applications. The close observation of I-V characteristic shows that data points are not obtained in negative resistance region during the transition from amorphous to crystalline phase which concludes that phase change is fast in case of SET process i.e. change in phase from amorphous to crystalline.

Figure 3 shows the amorphous nature of the thin film as characterized by XRD pattern.

Figure 4 shows the energy dispersive X-Ray (EDAX) pattern, which emphasizes the presence of different elements in the bulk as deposited thin film. The EDAX image of the sample confirms the presence of all elements which are present in bulk. Also to

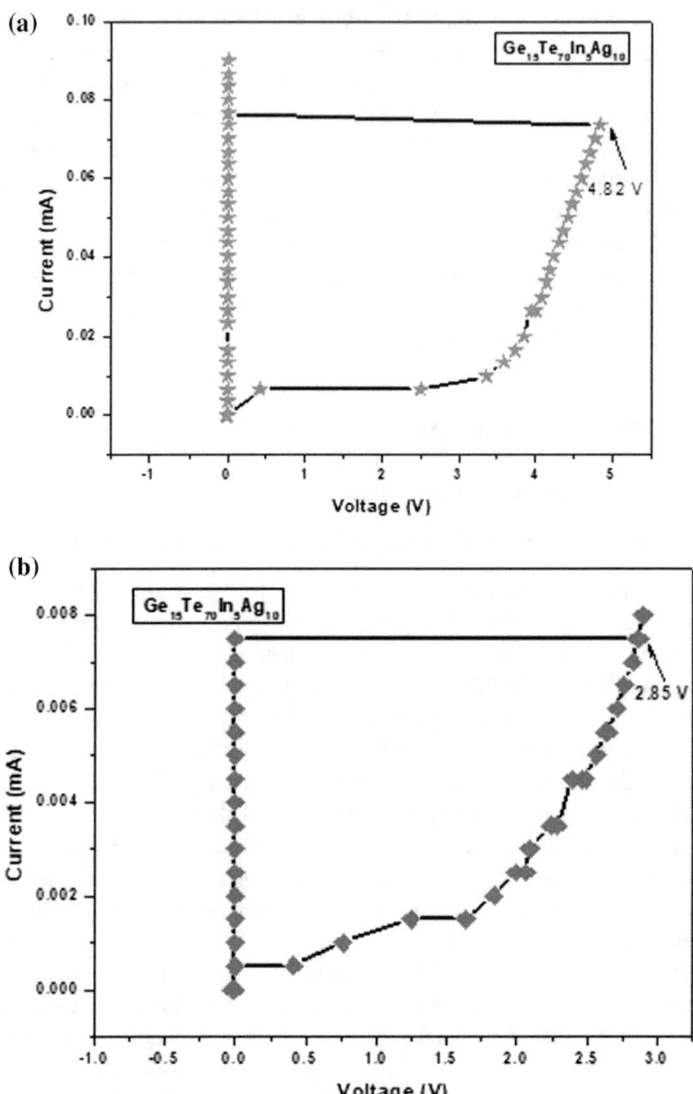

Fig. 2. **a** Switching characteristics (I–V curve) of $Ge_{15}Te_{70}In_5Ag_{10}$. **b** Switching characteristics (I–V curve) of $Ge_{15}Te_{70}In_5Ag_{10}$

verify applicability of quaternary compound in phase change memory application we have taken a quaternary compound (semiconductors composed from elements of the IIIrd and Vth column of the periodic system with four different elements) instead of ternary compound (Ternary compounds are composed of three different elements). Literature also shows that less number of devices have been fabricated with quaternary compounds.

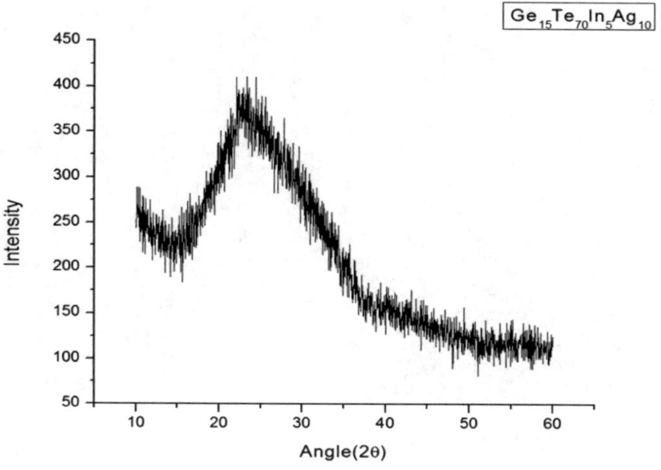

Fig. 3. X-ray diffraction graph (XRD) of $Ge_{15}Te_{70}In_5Ag_{10}$ thin films

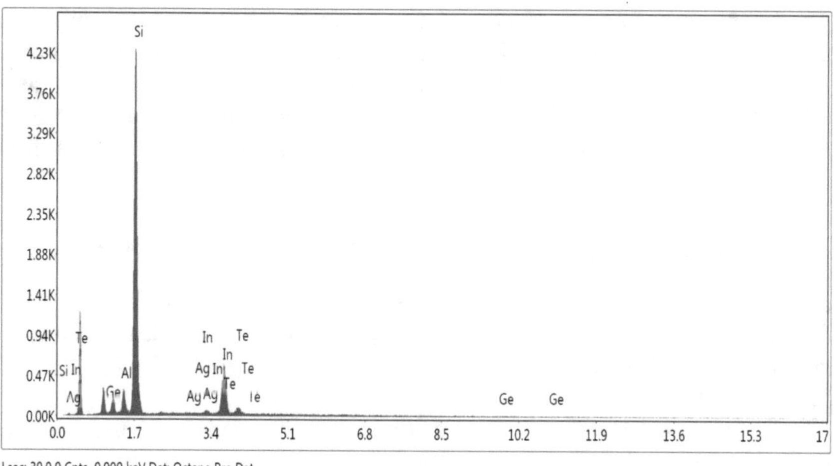

Fig. 4. EDAX analysis of $Ge_{15}Te_{70}In_5Ag_{10}$ thin film

3.2 Morphological Characterization of Amorphous $Ge_{15}Te_{70}In_5Ag_{10}$ Thin Films

The scanning electron micrograph (SEM) images of switched device of $Ge_{15}Te_{70}In_5Ag_{10}$ thin film sample can be seen in Fig. 5. The change in phase in the electrode region is understood by the image contrast in the SEM image. Morphological changes have also been observed in the sample at the place of switching. These changes could be due to any of the following reasons [12, 13]:

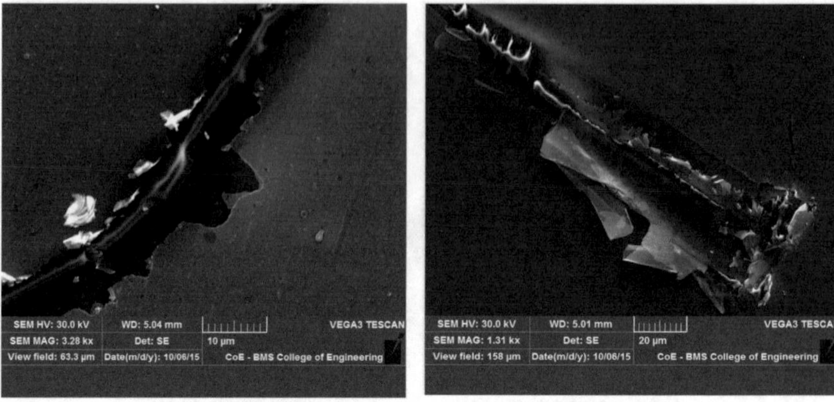

Fig. 5. SEM images of $Ge_{15}Te_{70}In_5Ag_{10}$ thin film

1. Flow of current during switching causes local melting and re-solidification of active material into crystalline state which results in surface relief causing image contrast.
2. Crystalline phase of a matter has a higher density compared to its amorphous state so glass to crystal phase transition amalgamated with local structural change and density solidification at the electrode region and this can show image contrast.
3. With the change in phase from amorphous to crystalline, the resistance also changes drastically by three orders of magnitude. The change in conductivity can lead to image contrast.

3.3 Optical Characteristics of Amorphous $Ge_{15}Te_{70}In_5Ag_{10}$ Thin Films

There are many models used to recognize the optical characteristics of amorphous thin films. The most frequently used model is the Tauc model [14] in which the square root of the product of the absorption coefficient and photon energy $(\alpha h\nu)^{1/2}$ is plotted versus photon energy $h\nu$.

The Tauc optical gap is described as occurrence of intercept of this linear estimation with the X-axis [15, 16]. The absorption coefficient α near the band edge in many amorphous materials shows an exponential relation with photon energy described by the following equation [17]:

$$\alpha h\upsilon = B\left(h\upsilon - E_g\right)^m$$

where $h\upsilon$ is the photon energy, B is a parameter that depends on the transition probability and E_g is the energy gap and m is a number characterizing the transition process which may take values 1/2, 1, 3/2 or 2, depending upon the nature of the electronic transitions responsible for the absorption [17, 18]. The optical band gap of amorphous $Ge_{15}Te_{70}In_5Ag_{10}$ thin films has been obtained from the Tauc plot. Figure 6 shows the plots of $(\alpha h\upsilon)^{1/2}$ versus $h\upsilon$ for $Ge_{15}Te_{70}In_5Ag_{10}$ thin film sample. This particular composition of thin film has a band gap of 1.16748 eV.

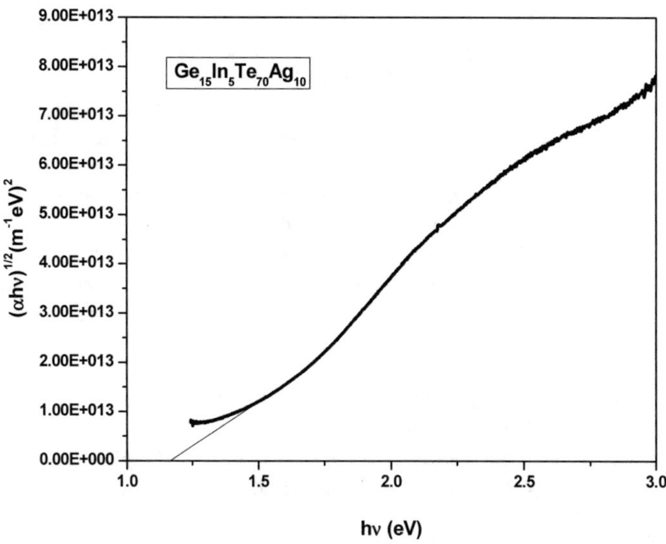

Fig. 6. Graph of $(\alpha h\upsilon)^{1/2}$ versus $h\upsilon$ for $Ge_{15}Te_{70}In_5Ag_{10}$ amorphous thin films

4 Conclusion

X-RAY Diffraction (XRD) has been utilized to determine the phases present in the thin film and result reveals that the composition is amorphous in nature and Energy-Dispersive X-ray (EDAX) patterns validates that the chemical composition of thin film is approximately same as the base bulk material. The I-V Characteristics shows that the memory type switching behavior is present in the composition and the electrical switching is faster in this particular composition. From Scanning Electron microscopy (SEM) which has been carried out to study the structural changes in the composition during switching. As per the studies done on the sample it could be concluded that the chosen composition $Ge_{15}Te_{70}In_5Ag_{10}$ may find an application in phase change memory (PCM) device.

Acknowledgements. Author, Chandasree Das acknowledges SERB for providing funding to establish the research facility to carry out the research works.

References

1. Ovshinsky, S.R.: Symmetrical current controlling device. U.S. Patent 3 271 591,1966
2. Kim, S.B., Lee, B., Asheghi, M., Hurkx, G.A.M., Reifenberg, J.P., Goodson, K.E., Wong, H.-S.P.: Thermal disturbance and its impact on reliability of phase-change memory studied by micro-thermal stage. In: IEEE International Reliability Physics Symposium, Anaheim, CA, 2–6 May 2010
3. Ovshinsky, S.R.: Reversible electrical switching phenomena in disordered structures. Phys. Rev. Lett. **21**, 1450–1453 (1968)

4. Das, C., Mahesha, M.G., Mohan Rao, G., Asokan, S.: Studies on electrical switching behavior and optical band gap of amorphous Ge–Te–Sn thin films. Appl. Phys. A **106**, 989–994 (2012). https://doi.org/10.1007/s00339-011-6726-0
5. Wang, L., Tu L., Wen, J.: Application of phase-change materials in memory taxonomy, pp. 406–429. Received 15 Mar 2017, Accepted 16 May 2017, Accepted author version posted online: 22 May 2017, Published online: 13 June 2017
6. Structural and thermoelectric properties of Se doped In 2 Te 3 thin films. AIP Adv. **8**(11), 115015 (Nov 2018). https://doi.org/10.1063/1.5057734
7. Laidani, N., Bartali, R., Gottardi, G., Anderle, M., Cheyssac, P.: Optical absorption parameters of amorphous carbon films from Forouhi-Bloomer and Tauc-Lorentz models: a comparative study. J. Phys. Condens. Matter. **20** (2008). https://doi.org/10.1088/0953-8984/20/01/015216
8. Waterman, A.: Phys. Rev. **21**, 540 (1923)
9. Vengel, T., Kolomiets, B.: Sov. Phys. Tech. Phys. **2**, 2314 (1957)
10. Wuttig, M., Yamada, N.: Nat. Mater. **6**, 824 (2007)
11. Electrical SET-RESET phenomenon in thallium doped Ge-Te glasses suitable for phase change memory applications. Chapter 12. https://shodhganga.inflibnet.ac.in/bitstream/10603/65492/12/12_chapter
12. Ali, M., Ürgen, M.: Switching dynamics of morphology–structure in chemically deposited carbon films—a new insight. Department of Physics, COMSATS Institute of Information Technology, Islamabad 45550, Pakistan
13. Das, C., Mohan Rao, G., Asokan, S.: Electrical switching behavior of amorphous Ge15Te85 xSix thin films with phase change memory applications. Mater. Res. Bull. **49**, 388–392 (2014)
14. Roy, D., Sreevidya Varma, G., Asokan, S., Das, C.: Switching studies on Ge15In5-Te56Ag24 thin films. World Academy of Sci. Eng. Technol. Int. J. Chem. Mol. Nucl. Mater. Metall. Eng. **10**(6) (2016)
15. Mok, T.M., O'Leary, S.K.: The dependence of the Tauc and Cody optical gaps associated with hydrogenated amorphous silicon on the film thickness: αl Experimental limitations and the impact of curvature in the Tauc and Cody plots. J. Appl. Phys. **102**, 113525:1–113525:9 (2007)
16. O'Learly, S.K., Lim, P.K.: On determining the optical GAP associated with an amorphous semiconductor: a generalization of the Tauc model. Solid State Commun. **104**, 17–21 (1997)
17. Urbach, F.: The long-wavelength edge of photographic sensitivity and of the electronic absorption of solids. Phys. Rev. **92**, 1324 (1953)
18. Adachi, S.: Optical Properties of Crystalline and Amorphous Semiconductors: Materials and Fundamental Principles (1st edn). Kluwer Academic Publishers, Boston, MA, USA, pp. 207–208 (1999)

IOT Implementation at Public Park to Monitor and Use Energy Efficiently

N. Elangovan[1(✉)], Preeti Biradar[1], P. Arun[1], Mahesh[1], U. Ajay[1], Anish[2], and Praveen Nair[2]

[1] Electrono Solutions Private Limited, Bengaluru, India
{elango.natraj3,preeti1905,ajayshankarusa7}@gmail.com, {arun,mahesh}@electronosolutions.com

[2] Bilva Infra, Bengaluru, India
{Anish.pandari,Praveen.nair}@bilvainfra.com

Abstract. Electricity is the mysterious force and power, without which we cannot think of modern life. Power is to be utilized effectively and it should be efficiently consumed without any outage. This project aims at efficient power consumption over a large area. So the continuous monitoring of power consumption and Environmental parameters in real time would help to develop a large area into a Smart Park. Real time data can be accessed Remotely and same would be displayed on the LED board and stored in Database.

Keywords: Remote laboratory · Environmental parameters · Smart street lights · RF communication

1 Introduction

There is a saying that goes as "money cannot buy you everything" and this particular phrase so true and evident in terms of power. Power which is generated once has to be used efficiently else it can never be recovered. Any small loss or outage in power would directly have adverse effects on our nation's wealth because in recent times most of the currency involved in Annual Budget is dedicated towards the upbringing and growth of power sector according numerous reports, surveys and statistics. It is such a vast operation from Generation to Distribution and the distance involved is many a miles for the same operation. Hence it is hard to keep tabs on the power transferred throughout the cycle.

Thus monitoring of power is must in the modern world. Physical monitoring can be acceptable and helpful up to a certain extent. Once there is no physical presence near the system it is hard to monitor. So in such cases we are blessed with a concept termed as "Remote Engineering" which involves monitoring a system remotely. This means the requirement of physical presence need not exist anymore and the same system can be monitored from any corner of the World. This unique way of monitoring will not only help us save power but also would indicate the exact reason for the losses and failure of any power system.

Having mentioned about the importance of power in the earlier wordings, it is necessary to implement the same. The idea or prototype was to monitor a number of

M. E. Auer and K. Ram B. (Eds.): REV2019 2019, LNNS 80, pp. 421–428, 2020.
https://doi.org/10.1007/978-3-030-23162-0_38

street lights and their power consumption. As we know, street lights are mainly required in the dark and there is no use of them being on throughout the day. This was the main objective the project and the spine of this paper. The normal street lights present needed to be turned off manually when not in use. So it was required to replace these lights with smart lights over a very large giving rise to a smart grid. So these smart street lights now would only glow on in the presence of a living being through pre-installed motion sensors. If the population is high then they would at the maximum and if vice versa then they would go dim. If no physical presence or motion is found then the lights would go off permanently till the detection of a new physical barrier. This would result in a lot of power being saved which was getting wasted unnecessarily.

2 Approach

As mentioned earlier, the aim of this project is smart street lighting. To go about the block diagram, it can be broadly divided into 2 parts. First part involves the smart street light and the latter includes Environmental sensor setup. Hence in smart street lights, the RADAR sensor is present to detect any human motion or appearance. If it does so, then it is communicated through RF communication by the RF trans-receiver. The street lights would taken in the same information and would dim and glow accordingly. If any human involvement is noticed only then the lights glow else they will be down. Another major part in this setup is the CCMS. CCMS can be thought to be a power monitoring device which is fed through the means of a microcontroller and can be obtained on a local server. The RF trans-receiver is also in contact with the microcontroller and through which CCMS is able to detect the amount of power loss and continuously monitors the power throughout the system. It not only helps in determining the power loss but also leads to the factor that had caused the loss (Figs. 1 and 2).

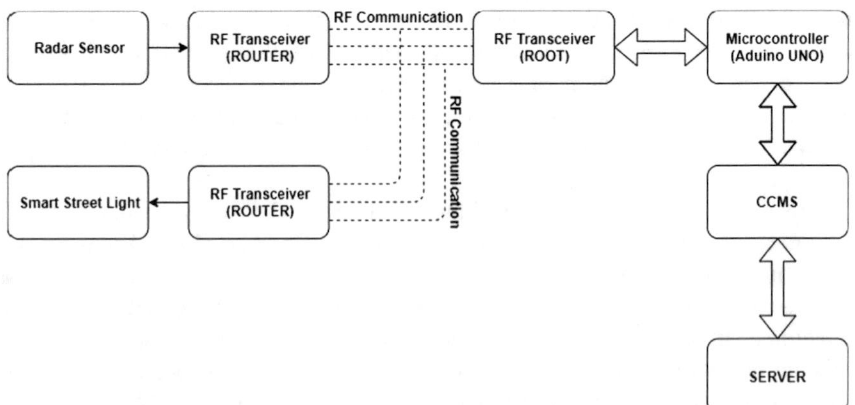

Fig. 1. Block diagram street light control

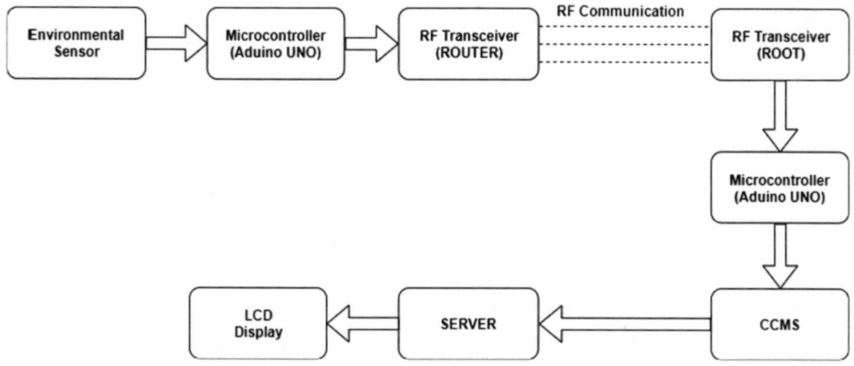

Fig. 2. Block diagram environmental sensor setup

The second part involves a slightly different set up to the earlier one. Instead of the RADAR sensor an Environmental sensor is made use of. This sensor would detect all the common surrounding atmospheric parameters like temperature, humidity, carbon di-oxide level and other similar factors. Once it has successfully detected it a micro-controller situated there can transmit the data through RF communication. The data sent by the router is gathered up by RF trans-receiver root. This is in contact with another microcontroller and then followed by the same CCMS system. It again monitors the power and other Environmental factors associated. These parameters through a particular server will be displayed on a LCD display projected in the park. The RF communication in both the setups works at 868 MHz.

3 Software Requirements

Software used:

[1] Arduino IDE is an easy to use software platform that has already built in code for various operations which can be easily modified based on the user's requirements.

[2] LabVIEW is an integrated development environment designed specifically for engineers and scientists building measurement and control systems.

ARDUINO IS CHOSEN FOR:

Simplicity,

Strong hardware—software interaction.

Code at an embedded C level.

Open source and a huge community for support.

Large database of libraries and binaries.

LABVIEW IS CHOSEN FOR:

Excellent design in form of front panel and block diagram.
Built in libraries and tools.
Precision measurement reading.
Highly customizable.

LINX: is a software upgrade developed by MakerHub to facilitate Arduino control through LabVIEW.

LINX provides an interface between LabVIEW and an Arduino.
LINX requires a data connection between LabVIEW and the Arduino always.
LabVIEW sends a packet to the Arduino.
The Arduino processes the packet and performs the specified operation (usually some I/O).
The Arduino sends a response packet back to LabVIEW.

4 Working

4.1 Working of Smart Street Light

Cubbon Park is 5 km in radius, we have installed 710 smart street lights in the park. There will be totally 50 Radar sensors in the park each sensor controls 15 street lights. The radar sensor is directly connected to the RF module-Router (Tarang UT-20) which transmits the radar sensor data to the receiver-Root which will be in CCMS (Centralised Control Monitoring System). From the CCMS the data is sent to the server through Arduino Ethernet, then in server LabVIEW process the data and sends the require command back to the Arduino which is present in the CCMS, from Arduino the is sent back to the Smart Street Light through RF which controls the Street Light.

For example: If Radar sensor-1 detects the human motion the data is transmitted from Router to the Root which is connected to Arduino which is in CCMS then the data is sent to the server through Ethernet then in server LabVIEW, depending on the data received LabVIEW sends corresponding commands to the Arduino then data is sent to the smart street light (Contains RF module which controls light) through RF module then it controls the 15 street lights which will be connected to sensor-1. After 1 min if the motion is not detected then the street lights goes on decreasing its intensity from 100 to 0%, if there is any motion in between the dimming of the light then it switch back to 100%.

Fig. 3. Environmental setup

4.2 Working of Environmental Sensors

There will be 50 environmental setup will be installed in the park which includes Atmospheric Pressure sensor, Noise sensor, Dust Particle sensor, Temperature sensor, Humidity sensor, Altitude sensor, Carbon-monoxide sensor. The environmental data will be captured by Micro-controller and the data is sent to the CCMS through RF communication from there the data sent to server through Ethernet then from the server the data is processed and displayed on the LED screens which will be present in the different parts of the park (Fig. 3).

CCMS BLOCKS:

5 Outcomes

1. Seamless monitoring of real time data related to power.
2. Automation indulging self turning on and reduction in intensity of installed street lights.
3. The data is getting stored in database which can be studied or reviewed for the research.
4. The user interface page is given for the Admin to make changes in settings.
5. RF communication has an upper hand when compared to other communications. Data communicated in **868** MHz frequency.
6. The LED lights used in the smart street lights are long lasting, power efficient and contain minimal losses for the following reasons.

   - LED light bulbs are much brighter than incandescent and also compact fluorescent lamps of the same wattage.

- Incandescent bulbs use approximately about five times as much as power to generate the same amount of light as LED bulbs.
- A typical LED bulb is pretty much long lasting and a 12 W bulb would last for almost around 25,000 h.
- LED bulbs actually do not produce any heat but produce minimal amount of heat occasionally.
- The electricity used by an incandescent for the entirety of its life would be 5-10 times its original price. Whereas, the LED bulb has a different dimension and is energy efficient.
- In comparison an LED bulb when turned for 5 h a day would only help my electricity bill go up by 1 INR, but an incandescent bulb for the same amount of time yields a power consumption at the price of 3 INR.
- For a week it is almost 14 INR of power that is used by the incandescent bulb when compared to the LED bulb.
- For a month the ratio will be around 30:90 in favor of LED and in a year it will be 365 INR and 1095 INR respectively.
- So if only a single LED can save around 2 INR when it is used for 5 h a day, 700 similar bulbs would almost save 1400 INR worth electricity every single day.

6 Conclusion

By this adaptation the technology lot of power can be saved and the park is fully automated. The Lights glow only in the presence of any Motion else they don't. Environmental parameters can be monitored and also amount of harmful gases like CO etc. present in the Ecosystem can also be observed through Environmental Sensor Board. Through this enterprise it is possible to optimize a vast area into a completely developed smart park. This in the near future would play a pivotal role in the set up for a smart grid or an entire smart city.

References

1. Caragliu, A., Del Bo, C., Nijkamp, P.: Smart cities in Europe. J. Urban Technol. **18**(2), 65–82 (2011)
2. IBM. (2010). Smarter thinking for a smarter planet
3. Su, K., Li, J., Fu, H.: Smart city and the applications. In: IEEE International Conference on Electronics, Communications and Control (ICECC), pp. 1028–1031. IEEE Xplore (2011)
4. Pardo, T.A., Nam, T.: Smart city as urban innovation: focusing on management, policy and context. In: Proceeding of the 5th International Conference on theory and Practice of Electronic Governance, pp. 185–194. ACM, New York (2011)
5. Chourabi, H., Nam, T., Walker, S., Gil-Garcia, J.R., Mellouli, S., Nahon, K., Pardo, T.A., Scholl, H.J.: Understanding smart cities: an integrative framework. In: 45th Hawaii International Conference on System Sciences, pp. 2289–2297. IEEE Xplore (2012)

6. Schuler, D.: Digital cities and digital citizens. In: Tanabe, M., van den Besselaar, P., Ishida, T. (eds.) Digital Cities II: Computational and Sociological Approaches. LNCS, vol. 2362, pp. 71–85. Springer, Berlin (2002)
7. Schuurman, D., Baccarne, B., De Marez, L., Mechant, P.: Smart ideas for smart cities: investigating crowdsourcing for generating and selecting ideas for ICT innovation in a city context. J. Theor. Appl. Electron. Commer. Res. **7**(3), 49–62 (2012) (Universidad de Talca, Chile)

Work-in-Progress: A Novel Approach to Detection and Avoid Sybil Attack in MANET

Anitha S. Sastry[1(✉)], Sadhana S. Chitlapalli[1], and S. Akhila[2]

[1] Department of ECE, Global Academy of Technology, Bengaluru, India
anitha.sastry@gat.ac.in, sadhanachitlapalli@gmail.com
[2] Department of ECE, BMSCE, Bengaluru, India
akhilas.ece@bmsce.ac.in

Abstract. Mobile ad hoc networks (MANETs) require a unique, distinct, and persistent identity per node in order for their security protocols to be viable, Sybil attacks pose a serious threat to such networks. Fully self-organized MANETs represent complex distributed systems that may also be part of a huge complex system, such as a complex system-of-systems used for crisis management operations. Due to the complex nature of MANETs and its resource constraint nodes, there has always been a need to develop security solutions. A Sybil attacker can either create more than one identity on a single physical device in order to launch a coordinated attack on the network or can switch identities in order to weaken the detection process, thereby promoting lack of accountability in the network. In this research, we propose a scheme to detect the new identities of Sybil attackers without using centralized trusted third party or any extra hardware, such as directional antennae or a geographical positioning system. Through the help of extensive simulations, we are able to demonstrate that our proposed scheme detects Sybil identities with 95% accuracy (true positive) and about 5% error rate (false positive) even in the presence of mobility.

Keywords: Ad hoc network · MANET · Mobility · Sybil attack

1 Introduction

Wireless Sensor Network compromise randomly deployed sensor nodes which monitors various environmental phenomena's, considering “temperature”, “sound”, “vibration”, “pressure” etc. at various locations. WSN compromises of n number of mobile nodes which gets randomly deployed and they might have a fixed location to look after the environment. WSNs are the development considering past years and indulge in scattering sensor nodes in the network. Usually sensor within network observes all environment changes and informs other sensors within the network over system architecture. Node deployment takes in hostile environment or even over geographical areas.

Each sensor nodes in a network has a functional unit such as sensing, processing, communication and storage unit. In the network since sensor nodes are randomly

M. E. Auer and K. Ram B. (Eds.): REV2019 2019, LNNS 80, pp. 429–441, 2020.
https://doi.org/10.1007/978-3-030-23162-0_39

deployed the position of sensors are not predetermined. The needs for the implementation for low-power consumption, low-cost utilization, multifunctional sensors have made WSN an important data gathering mechanism. These multifunctional sensors in the network underwent a rapid growth as a new information gathering mechanism taking into account various different examples of applications. In many such applications, sensors in the network are randomly positioned & dispersed entirely over the detecting (sensing) area and are unattended once deployment, hence leading to battery replacement.

Four basic components to consider in WSN: (i) an interconnecting wireless network (ii) group of randomly distributed sensor nodes. (iii) Sink node (iv) a group of computing devices at base station which helps in analyzing & interpreting the data obtained from other sensor nodes in a network.

As shown in Fig. 1, sensor nodes are outfitted with sensing unit", a wireless communication device (or radio transceiver), a micro controller and source of energy usually a battery. In network while relaying data from one sensor node to another, the sensor which is near to the sink node typically exhaust their energy much faster than other because of further relaying data traffic. Due to the depletion of energy by the sensors it may not guarantee the connectivity and coverage of network. Limiting to these, it is very tough to build (or design) an energy-efficient data collection scheme which aims in consuming uniform energy throughout the sensing area to accomplish extended lifetime of network. Some applications considers certain time specification for data gathering, hence effective, significant data gathering mechanism must be selected which aims at various network parameters such as high scalability, long network lifetime and low data latency.

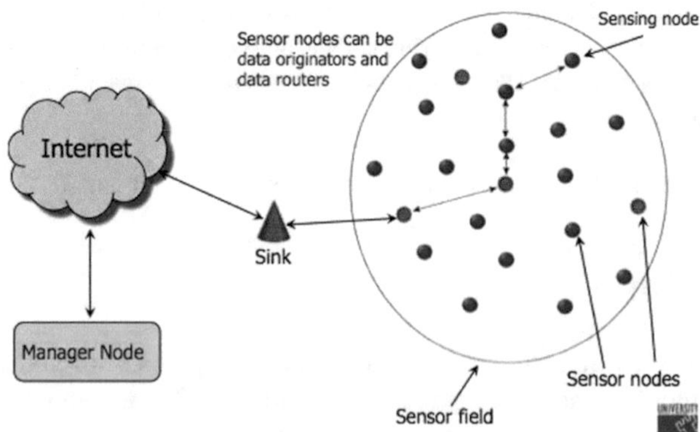

Fig. 1. WSN components

2 MANETs

A mobile ad hoc network consists of moving nodes which have the ability to configure themselves without the involvement of central surveillance. The nodes converse with other nodes within communication range through wireless medium. If the node is out of the communication range then it's not possible to communicate directly. In this case MANET relies on multi hop data transmission. Each node in a MANET is allowed to move self-rulingly in any direction which influences the network to various kinds of attacks because node can join and move the network at any point of time and will therefore the connectivity with other nodes will be evolving often. The key test in building a MANET is preparing every node to ceaselessly keep up the information needed to legitimately route the traffic.

When compared with wired networks where the nodes must have physical access to the network medium, mobile ad hoc networks have no obvious secure limit. The attackers can communicate with all nodes once they are in the transmission scope of any other device. Henceforth security is the significant concern with the MANET. Research works have concentrated on the security of MANETS. The vast majority of them deal with prevention and detection approaches to combat individual misbehaving nodes. With this respect, the viability of these methodologies gets to be feeble at the point when different misbehaving nodes conspire together to launch a collaborative attack, which may result to all the more destroying harms to the network.

3 Problem Definition

MANET has dynamic topology. A node can enter into the network and can leave the network at any instance. Because of this MANET is exposed to various attacks. The prominent attack which degrades the network performance is Sybil attack. In this regard the objectives of this project are:

- To find the path cost between the nodes in the network.
- To detect the nodes with larger packet drops and confirm the type of attack is a Sybil attack.
- To authenticate the attack by estimation of MAC address of the neighboring node.
- To find a path to successfully route a packet to the destination.
- To estimate the packet loss w.r.t the network parameters with throughput, overhead, packet delivery ratio and end to end delay.

3.1 Motivation

- Network security in Mobile Ad hoc Networks is a major issue.
- MANET efficiency and security is compromised to create an instability and disrupt the performance of the network.
- This issue of the Mobile Ad hoc Network creates an authentication and credibility issue in the network. This motivated to present a work that is a comprehensive study on the Sybil attack and its effects on the performance is presented in this work.
- The malicious node is either replaced or a new route is created to route the information from source to destination.

3.2 Design and Implementation

3.2.1 System Architecture

The framework is constantly disintegrated into sub- frameworks that give some related set of services. The beginning outline process of recognizing these sub-systems and setting up a framework for sub-frameworks control and correspondence is called Architecture design and the yield of this configuration procedure is a depiction of the software architecture. The structural outline procedure is concerned with building up a fundamental structure framework for a system. It includes recognizing the significant components of the system and correspondences between these. In the above architecture diagram shows the nodes deployment along with the source, destination and attackers node. While attacks, the attackers may launch the types of attack to compromise the authentic node.

3.3 Data Flow Diagram

Data Flow Diagram—Level 0:

The level 0 is beginning level Data Flow Diagram and it is generally called as the Context Level Diagram. A diagram giving a whole framework's data flow and handling with a solitary process is known as a context diagram. Every process may be further consumed into a set of inter-connected sub process. This strategy of extending a DFD is known as leveling (Fig. 2).

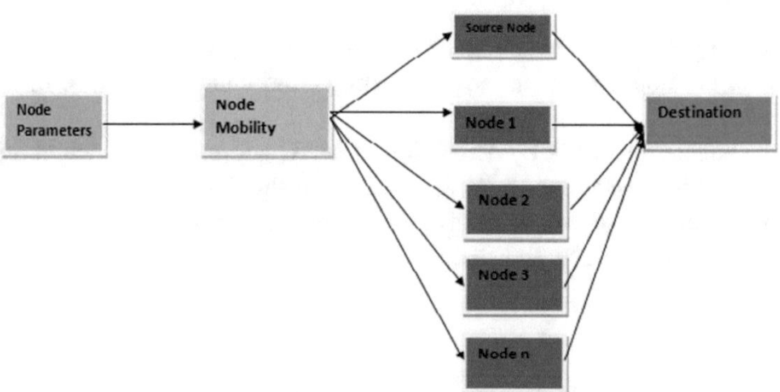

Fig. 2. Data flow diagram—level 0

Data Flow Diagram—Level 1:

In Level-1DFD, a fake address is given as input to the reverse tracing procedure. The reverse tracing procedure will identify the Sybil nodes present in the network. The output of the reverse tracing procedure, Sybil node lists, is fed into notification and route setup procedure (Fig. 3).

The WSN nodes are randomly deployed and communication is done between two nodes. While communicating the nodes are checked for its authentication in each

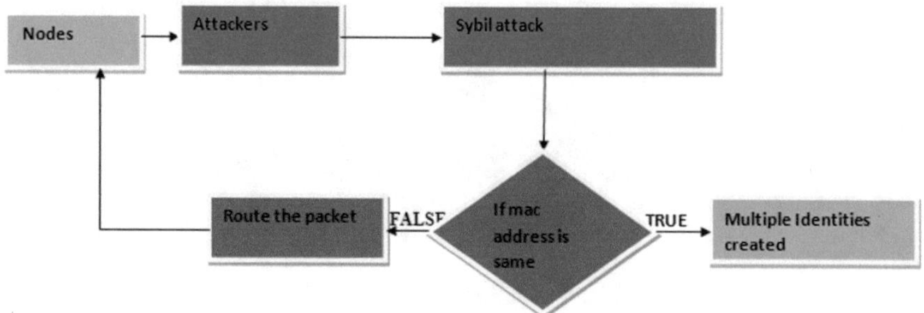

Fig. 3. Level 1: partition

hop. When the communicating node is authentic it does not do any misbehavior. If the nodes are not authentic it tries to drop packets or tries to alter the data. Nodes communication detects types of attack, if the node is passive attack it tries to components. Drop packet, and if node in active attacks it tries to tamper the data.

Implementation Modules

Topology module: Topology module can also be referred as network module. It describes script's functionality used for building network topology.

The steps involved in this module are as follows:

- **Wireless Network topology setup**: This considers setting up of node configuration, topology creation and environmental settings.
- **Transmission Parameters setup**: This includes the parameters that are being used such as node functions, its range and channels assigned are defined.
- **Bandwidth and threshold setup**: Always there will be certain bandwidth and threshold associated to each and every node in the network.
- **Finding the neighbour nodes**: Euclidian distance concept is used to know the neighbours of a particular node.
- **Stating source node, destination node and data**: The node from where data needs to be sent i.e. "source" and the node which must receive data i.e. "destination". Data is nothing but the quantity of information sent in specific time period.
- **Stating simulation start and end time**: In NS2, whole transmission gets over in short period of time and it can be seen from NAM window at any instant of time period. Hence considering simulation it is mandatory to specify the start and end time (Figs. 4 and 5).

The pseudocode for initialization of nodes randomly is as follows

a. for i = 0 to num_of_nodes
b. Initialize mobile nodes
c. Enable random motion
d. Nodes Energy
e. Define initial node position
f. Initialize agent

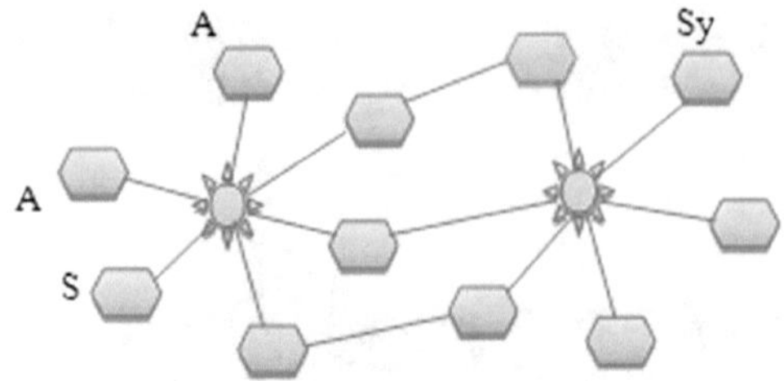

Fig. 4. Implementation module

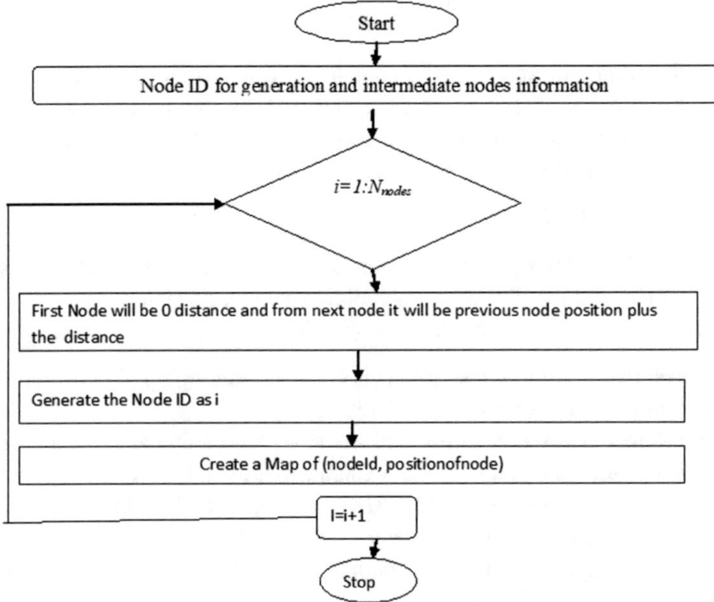

Fig. 5. Random node deployment algorithm

g. Attach agent to node.
h. End for

Algorithm:
The pseudo code for initialization of nodes randomly is as follows

a. Start the process.
b. Calculate the distance between the nodes and how much number of nodes should be present.
c. Initialize i = 1 & N number of nodes.
d. The distance of the first node will be zero, and for the next node we should take the preceding node adding with the distance.
e. I will be produced as the node id.

f. For the position of the node and the node id created produce the map.
g. Increment the value of i once and repeat from step c.
h. End the process.

4 Energy Module

Energy module is referred to as a node attribute when implemented. It specifies the level of energy a node has in the network. The node owns the initial energy value which is referred as the energy level the node has at the start of the simulation. This energy level is called initial energy. The node reaches certain energy level after transmitting and receiving the data, hence that energy level is referred as txPower and rxPower.

The calculation of energy of nodes is very important as the nodes lifetime has to be increased, for prolong network, many optimization techniques have be used to optimize and monitoring of the nodes.

In energy module-assigning the nodes initial energy, calculating the consumed energy and residual energy of individual nodes is carried out.

Residual energy is calculated by

$$\mathrm{RE} = \mathrm{IE}(\text{initial energy}) - \mathrm{CE}(\text{consumed energy}) \tag{1}$$

5 Node Energy Calculation Algorithm

This algorithm is used for energy calculation of node. Pseudo code for random nodes deployment with energy levels. All the nodes will be created at the same location in the topology. The following pseudo code is to create random node position (Fig. 6).

a. Start the process.
b. Calculate the nodes primary energy in joules.
c. Initialize i = 1 and n number of nodes.
d. Calculate the amount of energy taken by node.
e. The residual energy of the node is calculated.
f. Increment the value of i once and repeat from step c.
g. End for process.

6 Sybil Attack Detection and Avoidance

In this implementation, the attacks can be classified as Control and Data attacks in network. We propose a trusted route using AODV routing protocol and simulate the nodes dropping the packets which reduces the efficiency of the network. Blackhole attacks deals with data attacks, where the node intentionally drops the packet which it has received and does not forward to the next node, this creates a malicious activity in

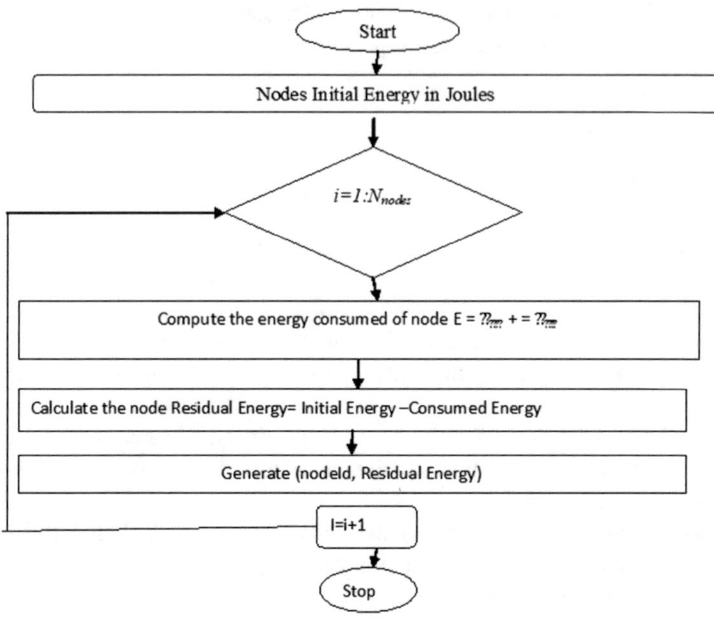

Fig. 6. Energy calculation algorithm

the network. Sybil node starts dropping packets, generates same MAC ID and misdirects the requester node to some other destination.

In the above figure shows the working of black hole, where the node collects route reply message (RREP) from multiple nodes and thus forms the redundant routes to the destination. In the routing table, the routing table has to be updated while the sender node sends the route request packet (RREQ) to the neighbor nodes which are within the transmission range, until it reaches to receiver node or destination. The receiver nodes send the route reply sequence number in the order which the request came from, intermediate checks for the sequence number, if any mismatches happen, this calls for a malicious activity in the network.

Sybil attack manifests itself by faking multiple identities by pretending to be consisting of multiple nodes in the network. So role of multiple nodes can be assumed by single node and can monitors or hampers multiple nodes at a time (Fig. 7).

7 Testing

Testing is the method of examining the system or its module and its main motive is to know whether it meets the necessity requirements. In short, testing is all about executing the system parameters to know if any error occurs in the system in contrast to actual system. The main aim of testing is to identify if any error (or bugs) occurred in the system.

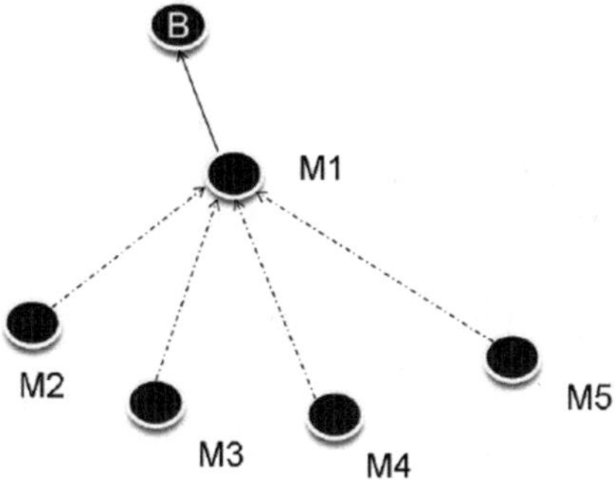

Fig. 7. Energy calculation algorithm

It is actually the mechanism to validate and verify a software program or any application running over the system. Identifying bugs in the program, verifying and validating are the three main purposes for testing. The actual intension of testing the application developed is to identify if any error occurred while executing the program. Testing is the last stage to know if any error occurred in the system. Various test cases for a particular application or program or system exists. Hence the test cases should be selected in such a way that it can be checked or examined for all combination or mixture of conditions. While selecting the test cases, we need to have expected results as well.

8 Purpose of Testing

The main purpose of testing is to know whether the application developed or the system is running as expected and without any bugs. The goals set have to be measurable and accomplishable.

For testing purpose, the goals can be set as per the expected output for software testing procedure. Testing is done to check whether the application is developed according to the specifications and yields the results as expected. The other reason for testing is to know if any error occurs in the application developed or the system.

During the testing, a set of test cases that needs to be accomplished with the plan that has to be tested and it needs to be checked whether the program was performing in same manner.

Testing aims in removing bugs, the more we get rid of the errors the best would be the system performance. The drive of testing could be the guaranteed quality, verification/and validation.

9 Levels of Testing

The vital objective of the testing is to complete functional requirement validation. The test will be severe when considered all the test cases, which are expected in real time environment and also which aren't in real time environment. Test team needs to be verify the entire test cases.

The project considers two level of testing:

- Initialization testing
- Functional testing

A. *Initialization testing*

In the initialization testing, the network elements are initialized that specify the system in the simulation needs to be carried out. Mobile nodes get deployed in the network initially so the area needs to be known where they are going to deploy. Upon selecting source and destination nodes, the packets are transmitted to the destination in the network.

B. *Functional testing*

Functional testing resembles testing of black box which locates its test cases on the specifications of software component under the test with giving the inputs, various functions are tested and out is checked (or examined).

10 Result and Analysis

The result step or phase of project is the last step where the system can be evaluated in terms of performance and the results are verified using the graphs if the goals in the project that are described in the starting are met or not. The performance is checked using the values obtained. The performance is evaluated using the graphs.

A. *Parameters of simulation*

Simulation Parameters required for Performance Evaluation (Table 1).

Table 1. Simulation parameters

Simulation area	500 * 500 m^2
No of nodes	50
Mac	802.11
Antenna	Omni antenna
Transmission range	550

Few parameters are considered in this project which yield in estimating the system performance and also which provides proof for network lifetime improvement.

Throughput: Number of packets sent and received per unit of time. It is expressed in terms of kbps.

Packet delivery ratio: It is the measure of ratio of number of packets transmitted by source and the number of packets acknowledge by destination.

End-to-End Delay: Delay can be calculated by sent time of packet by source—received time of packet by destination. It is expressed in milli seconds (ms).

Overhead: It is the number of routing packet processed. The Below Snapshots of Graphs shows the crystal clear difference between existing system and the proposed system.

The graphs are with respect to the Parameters are represented in the Figs. 8, 9, 10, and 11.

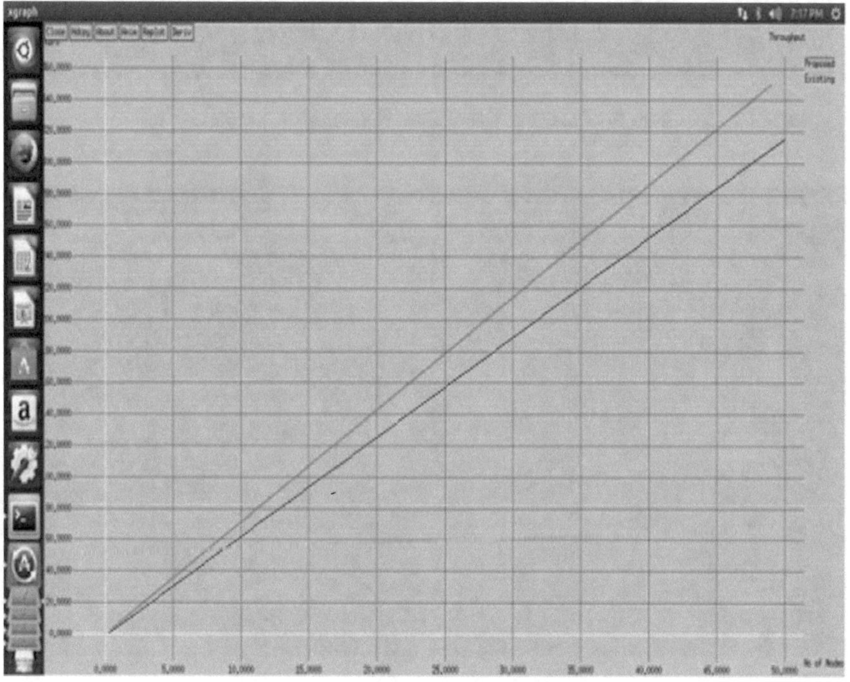

Fig. 8. Throughput (existing vs. proposed)

11 Conclusion

Here we proposed an RSS-based detection mechanism to safeguard the network against Sybil attacks. The scheme worked on the MAC layer using the 802.11 protocol without the need for any extra hardware. We demonstrated through various tests that a detection threshold exists for the distinction of legitimate new nodes and new malicious identities. We confirmed this distinction rationale through simulations. We also showed the various factors affecting the detection accuracy, such as network connections, packet

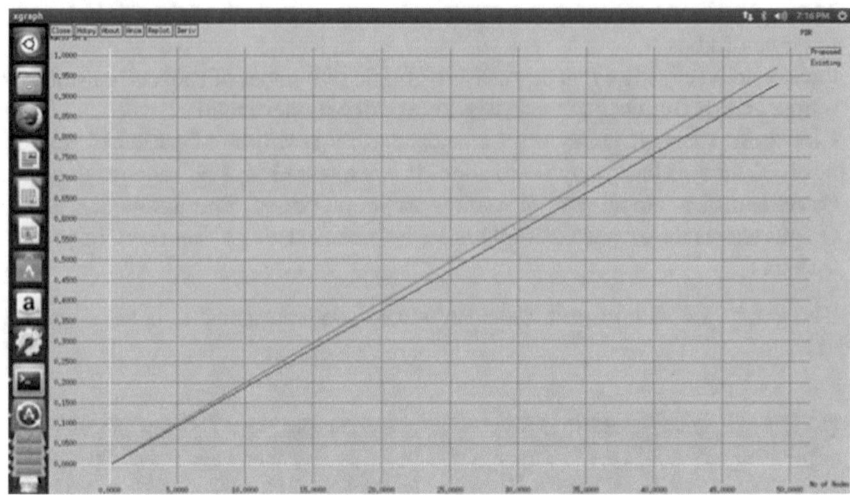

Fig. 9. Packet delivery ratio (existing vs. proposed)

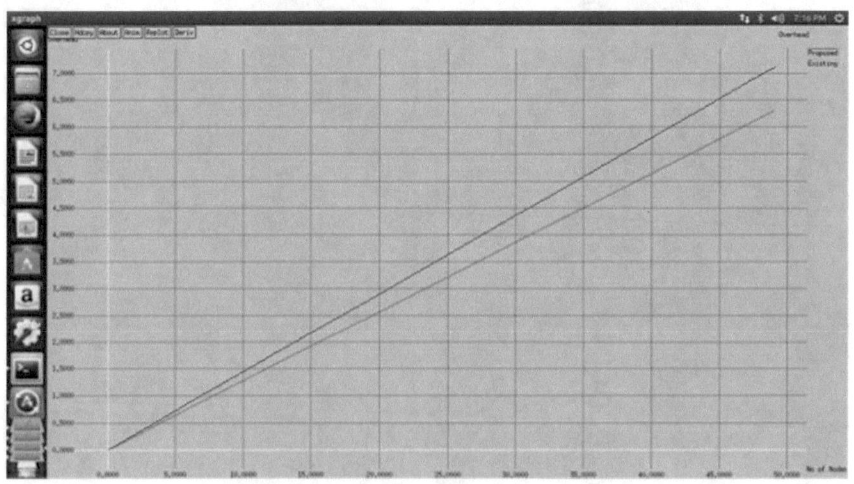

Fig. 10. Overhead (existing vs. proposed)

transmission rates and node density. The simulation results showed that our scheme works better even in mobile environments and can detect both join-and- leave and simultaneous Sybil attackers with 95% accuracy (true positive) and about 5% error rate (false positive) even in the presence of mobility.

Our future work includes increasing the accuracy tackling issues related to variable transmit powers, detection of masquerading attacks in the network and reduction of error rate.

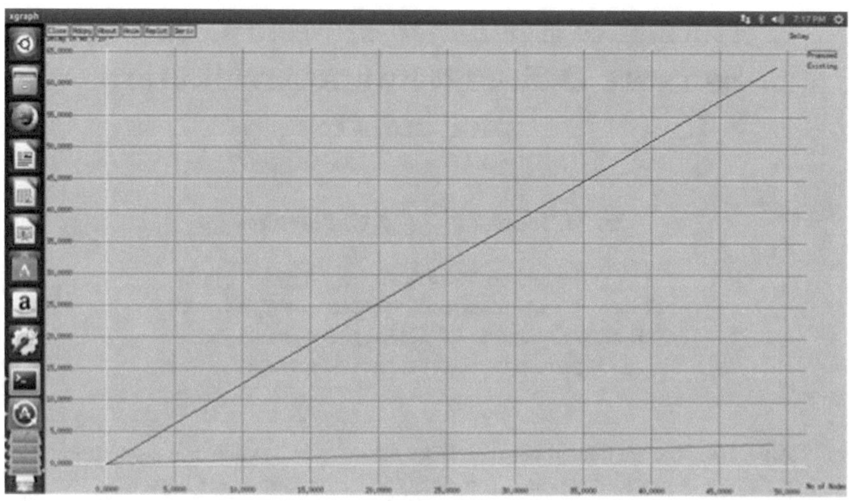

Fig. 11. End to end delay (existing vs. proposed)

References

1. Douceur, J.R.: The Sybil attack. In International Workshop On Peer-To- Peer Systems, Mar 2002
2. Johnson, D., Maltz, D.: Dynamic source routing in ad hoc wireless networks. In: Mobile Computing, vol. 353. Kluwer Academic Publishers (1996)
3. Hu, Y., Perrig, A., Johnson, D.: Ariadne: a secure on-demand routing protocol for ad hoc networks. In: Proceedings of International Conference on Mobile Computing and Networking, Sept 2002
4. Newsome, J., Shi, E., Song, D., Perrig, A.: The Sybil attack in sensor networks: analysis & defenses. In: Proceedings of International Symposium on Information Processing in Sensor Networks (2004)
5. Nadeem, A., Howarth, M.P.: A survey of MANET intrusion detection & prevention approaches for network layer attacks. IEEE Commun. Surv. Tutorials 1–19 (2012)
6. Newsome, J., Shi, E., Song, D., Perrig, A.: The Sybil attack in sensor networks: analysis & defenses. In: Proceedings of the third international symposium on Information processing in sensor networks. ACM, Berkeley, California, USA (2004)
7. Douceur, J.R.: The Sybil attack. Presented at the Revised Papers from the first International Workshop on Peer-to-Peer Systems, pp. 251–260 (2002)
8. Abbas, S., Merabti, M., Llewellyn Jones, D., Kifayat, K.: Lightweight Sybil attack detection in MANETs. IEEE Syst. J. **7**(2) (June 2013)
9. Kavitha, P., Keerthana, C., Niroja, V., Vivekanandhan, V.: Mobile-id based Sybil attack detection on the Mobile ADHOC Network. Int. Journal of Communication and Computer Technologies **2**(2) (Mar 2014)

Performance Evaluation of Multi Controller Software Defined Network Architecture on Mininet

B. M. Rashma[1(✉)] and G. Poornima[2]

[1] PESIT Bangalore South Campus, Bangalore, India
rashma.bm@gmail.com, rashmabm@pes.edu
[2] B.M.S. College of Engineering, Bangalore, India
gpoornima.ece@bmsce.ac.in

Abstract. The traditional network lacks flexibility when network needs to be scaled-up or scaled-down as it completely relies on hardware components like routers and switches. This can be solved by a paradigm technology SDN (Software Defined Networks). SDN facilitate network administrators to decouple control plane from data plane, and logically centralize control functions in a software controller. Proposed work in this paper is a rapid SDN prototyping of Single Controller (SC) and Multi Controller (MC) architecture on mininet, a programming flexible simulator. The emulation displays a Multi-Controller (MC) SDN architecture which improves efficiency and exhibits competence in handling scalable network on contrary to Single-Controller (SC) architecture. The proposed work demonstrates defining user defined controller on inbuilt Open Flow controller which also brings intra-cluster and inter-cluster communication in a hierarchical network. We empirically ratify the scalability of network by increasing number of host nodes.

Keywords: Mininet · Multi controller · OpenFlow · Ethernet · Round Trip Time (RTT)

1 Introduction

In current scenario data centre provide extensive services to run diverse range of applications and programs which are evident part of campus networks, enterprise networks and networks of even more higher order like World Wide Web. When such data centres alter their infrastructure for various reasons, network is scaled up or down, which involves a lot of hardware configurations which can be very expensive and time consuming. This problem can be addressed more effectively using SDN based network.

Primary difference between SDN and conventional data networks is the partitioning of control plane from the forwarding plane [1]. This decoupling promotes independent evolution of data and control plane components. With introduction of simple standard Application Programming Interface (API) between the two, strengthens network architecture. Scalability is the major advantage of having a network based on SDN. A controller is required to support control software which manages under lying network devices. OpenFlow is one of such renowned open and standardized API.

M. E. Auer and K. Ram B. (Eds.): REV2019 2019, LNNS 80, pp. 442–455, 2020.
https://doi.org/10.1007/978-3-030-23162-0_40

As SDN has centralized control paradigm it is being adopted in the diverse types of network such as data centre, mobile network, transport network, and enterprise networks. It has layered architecture with the decoupling of control and data layer, unlike traditional networking.SDN based networking separates data plane and control plane because of which SDN has provided a novel network technology that offers high interoperability in most cost-efficient way. SDN has enhanced control and network programmability in data centre networks. In SDN technology, dedicated controller centrally manages the network. These dedicated switches uses specific protocol by which it interacts with network switches.

In huge networks like telecommunications network congestion control is one of the major concerns in controlling the network traffic. Preventing the congestive collapse by trying to avoid the oversubscription brings down the popularity of the network provider. SDN is a ground-breaking network technology that offers innovative solution to this problem in most cost-efficient ways instead of restriction on subscriptions.

1.1 OpenFlow

SDN controller is an important aspect in any Software Defined Network which acts as interface between South Bound (SB) API's and North Bound (NB) API's i.e., switches/routers and applications/business logic respectively. In SDN environment, the SDN Controller directly interacts with the forwarding plane of network devices. This is achievable because it originally defines the communication protocol, OpenFlow (OF) is one of such protocol in Software Defined Networking (SDN) standards [2, 3].

1.2 Mininet

In the domain area of computer networks Software-Defined Networks (SDNs) is a novel approach, Not only in small networks data packets can be controlled and routed in large networks like the World Wide Web. Only drawback in current scenario is that practical implementation of SDN architecture in various network technologies is very expensive because network devices such as routers and switches with SDN functionality implemented are not available in abundance. Existing devices with such functionality are very expensive.

Using virtual network emulators is one of the best solutions to overcome huge cost. These emulators help the researchers to carry out research on new concepts and to assess features of the new paradigm in practice at a low monetary cost. The current work focuses on simulation and assessment of a SDN based architecture on SDN emulation tool called Mininet [4]. The performance fidelity amidst the real environment and simulated of course persists. But it is backed up with advantages such as rapid and simple prototype that ensures applicability, the likelihood of sharing outcome and tools at zero cost. These features help researchers to enhance their research ideas despite the inadequacy of physical devices.

2 Literature Survey

As the product manufacturers dislike to render the internal workings of their products and it is also laborious to write product-specific control software for each product. There is need for generic software or protocol that provides pragmatic compromise that permits researchers to run experiments on heterogeneous switches and routers in a uniform way hence OpenFlow was developed [1]. OpenFlow is a centralised approach and still in its premature stages.

When basic model of OpenFlow architecture is analysed with respect to processing speed of the OpenFlow controller the inference is that, blocking occurs in the forwarding queue of the controller. There is probability of reduced forwarding speed because of occurrence of blocking [3]. The result illustrate that the total turnaround time of the network mainly is based on the processing speed of the OpenFlow controller. If the probability of new flows arriving is higher at the switch, the coefficient of variation is lower but the longer is time.

Distinguishing feature of Mininet are firstly flexibility, that is, new topologies and new features can be set in software, using programming languages [4] the current proposed work use python as programming language and common operating systems used is Ubuntu 14.4 LTS. Hence network which provides scalability, with hundreds or thousands of switches on a single computer can be emulated on mininet. The prototype behaviour represents real time behaviour with a high degree of confidence. Created prototypes can be easily shared with other collaborators. Protocols stacks should be designed in such a way that it can be used without any code modification.

SDN default controllers have drawback i.e., one controller cannot administer a large number of devices efficiently, and as a result, the controller may pretence restricted access. The SDN controllers' safety and performance is also of concern as all the communication should go via central controller [5]. Carrier-grade SDN controller proposes a modularized approach that is designed to resolve the problem of controlling large-scale networks of carrier. A prominent characteristic of carrier-grade Software Defined Networks is the introduction of processing logic in communication mechanism of the controller. This approach makes sure that OpenFlow-based switches establish a good interaction with the OpenFlow specification.

SDN-enabled networks have two modes of operation i.e., proactive and reactive forwarding mode. If most of the traffic traversing the topology is assumed to be known then it's called as proactive mode [6] hence traffic flow rules can be pre-installed in the network. If network is highly dynamic traffic where no information on the traffic mix is previously known then we need to choose reactive mode. Each new arriving flow at the switch prompts a signalling request to the controller. In SDN, in general, multiple controllers per switch, with various controller architectures, are possible.

Periodic collection of the statistics of all flows in the network [7] is possible through flow monitoring schemes available in SDN controller. If multiple switches are involved in flow, the rate of a flow across can be collected from any switch along the path. But if single switch is involved there will be lonely flow in such case rate of flow need to be collected from that particular switch only. In order to collect information about flow rate with respect to switches we can use polling. Polling can be either poll-single or poll-all.

If the flow table information generated is superfluous, network bandwidth will be under used. To efficiently use network bandwidth can employ polling. The most efficient polling is Lonely Flow First (LFF) [8]. LFF is designed as follows; first consideration is given to switches that carry lonely flows with the minimum polling cost. Based on the actual bandwidth consumption polling mechanism can be determined such as poll single or poll-all. When compared with all existing greedy algorithms for polling, LFF has lower time complexity and consumes less network bandwidth.

Not only in normal network, SDN can be used in IoT. One of the important challenges of traditional IoT, is that the topology is fixed and constant. On break down of any gateway, the networks cannot continue to work by dynamically regulating themselves and reconfigure base on dynamic change that has occurred in the network environment [9].

Network Address Translation (NAT) is one of the potential schemes in networking. SDN based fully distributed NAT traversal scheme dispenses workload among devices and reduce transmission delay by packet switching instead of packet modification. Furthermore, SDN based IoT connectivity management architecture supports IoT global connectivity and enhanced real-time communication [10].

SDN model can be expanded to IoT. Node in SDN is combination of interfaces, programmable layer and controller. Controller has full access to the data-link layer switches [11]. SDN controller can improve the security and accessibility among the nodes. In peer-to-peer to communication each forwarding device snoops to events from its peers and makes self-directed decisions based on a local view of comprehensive state.

SDN's Decoupled architectures are closer to client-server architecture. Where controller plays role of server which can take independent decision provide service to all packet-forwarding devices which can be projected as client, which has limited decision-making capabilities and must implement control decisions made by controllers.

OpenFlow protocol can simulate both switch and controller. OpenFlow controller is capable of finding path from node to node and populates flow tables. OpenFlow switch has no local control-plane logic and relies entirely on the controller to populate its forwarding table [12]. SDN Control network is flexible to implement in any network topology including a star which will include single controller, a hierarchical network with set of controllers connected in a full mesh, which connects to forwarding nodes below, dynamic ring a set of controllers in a distributed hash table.

NOX-MT was the first controller which was designed over NOX to boost performance and motivated other controllers to improve. All controllers have different performance characteristics for the reason that SDN controllers can be optimized to be very fast [13].

In a campus network, the network administrator may delegate authority to several teams such as traffic engineering team, media services team for simultaneous casting, in case of bulk data such as scientific data. Systems natively can use a tree structure for accounting and delegation of tasks, example for such networks is cgroups and Cinder

[14]. Software Defined Networks (SDN) enable custom packet handling at each switch but cannot establish hierarchical policies directly on commodity hardware.

Link recovery in an OpenFlow network completely depends on the controller. CoD is one of such example. In CoD on link failure, the OpenFlow switch in a link detects link failure when it senses port failure in that link and sends a port failure notification to the controller. In response to the failure notification, the controller transmits flow modification messages based on this message current flow modifies and disrupted flows. In CoD design Constraint is that the network topology should be known to the controller in prior.

Network topology information is fed in prior to the controller. The controller finds and stores primary path and secondary (alternate) path for each link in the network in a data structure which is normally a nested list. The following information is stored in nested list by CoD primary port, alternate port, and alternate path. The stored alternate route consists of all intermediary switches between the source node and destination node attached to the link [15]. When the link fails alternative route flows from a failed link to its intended alternate path. There should be a port change at this point i.e., the previous path switch from outdated flow rule must change its action output port from the failed primary port to a working alternate port of the new alternative path.

3 Proposed Work

In this work, we propose the use of SDN as the framework for the networks. We improve efficiency and scalability of SDN. We accomplish this by incorporating multiple controllers on contrary to that of single controller in general SDN architecture.

The proposed solution makes use of the core concepts of Software Defined Network, with OpenFlow protocol. General SDN architecture is shown in Fig. 1. SDN communicates to application layer and physical layers by North bound and South bound API's. Figure 2 is a customized SDN architecture that depicts accessing various layers of general SDN architecture by means of programming.

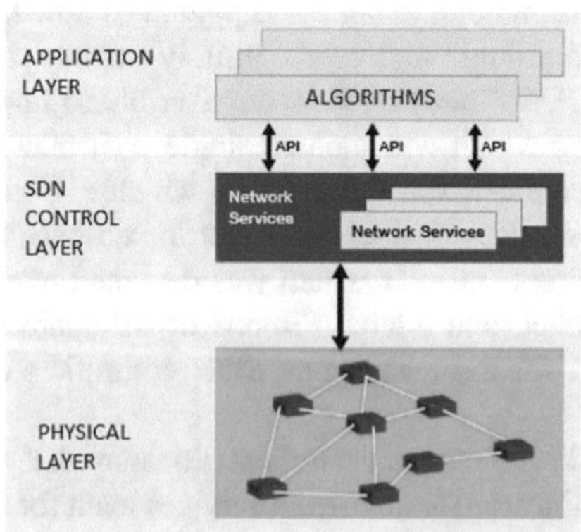

Fig. 1. General SDN architecture.

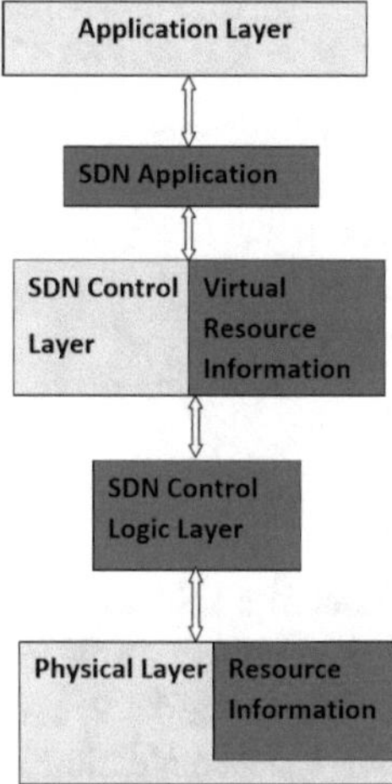

Fig. 2. Customized SDN architecture

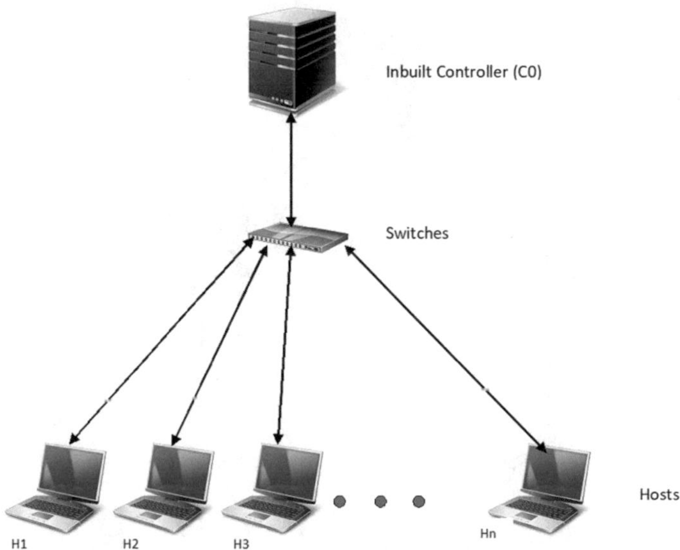

Fig. 3. Single Controller SDN architecture.

4 Implementation

Implemented in a simulation network environment. Simulation is done on Mininet and Openflow protocol. We have a POX controller installed which has a default controller. Under the default controller we define user defined controllers. Programming language used is python.

4.1 Assumptions

- Topology hierarchical 3 level tree.
- Network: Ethernet.
- Link type is EN10MB (Ethernet).
- Protocol at network layer is ICMP.
- Protocol at application layer is TCP.
- Network is static.

4.2 Clusters Communication

Since it is a multi-controller architecture this will emulates cluster/Sub-network communication. Cluster/sub-network is a collection of switches and host. This module is communicating between the two sub-networks 2 user defined controllers and a main controller which is a pox controller. Communication between two network devices that belongs to different clusters involves a lot of complexity. The simulated network set-up is shown in Fig. 4.

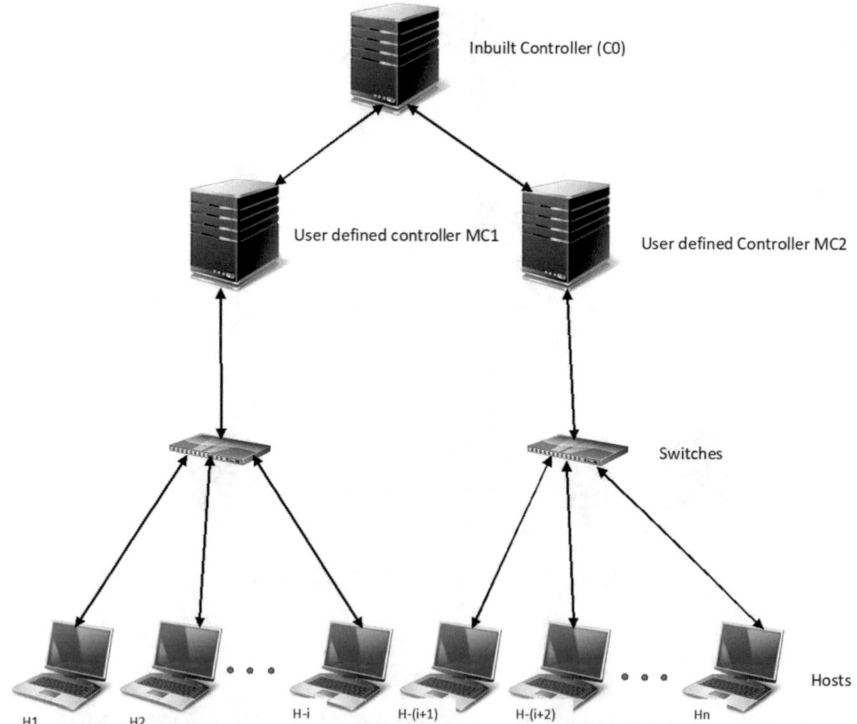

Fig. 4. Hierarchical network of 3 levels.

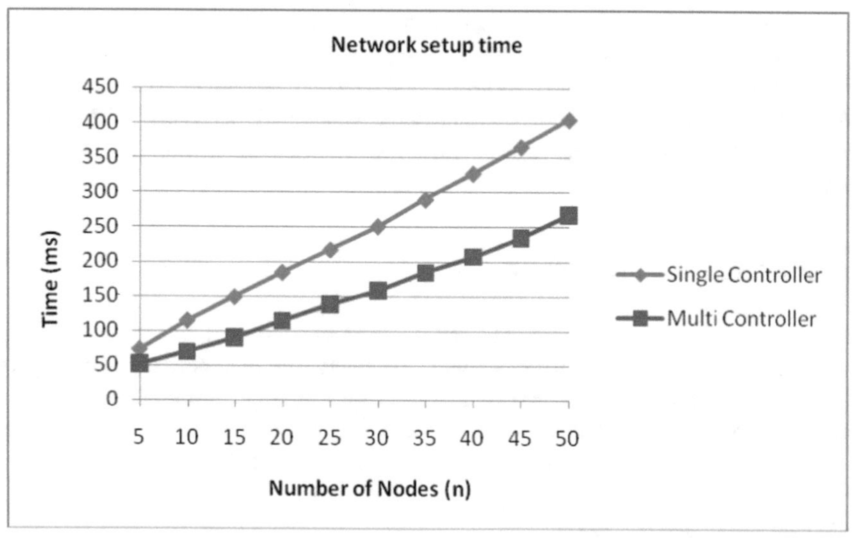

Fig. 5. Network set up time for MC and SC architecture.

4.3 Experimentation Scenario

Single Controller Architecture: A python simulation program is written to set up a static user defined network. A controller program is written to establish Ethernet connection with each host and OpenFlow switch. OpenFlow switches are in turn connected to user defined network. A ping message is sent from every host node to other host node. Figure 3 depicts single controller architecture.

Multi Controller Architecture: A python simulation program is written to set up a user defined network which has two user defined controllers Connected to build in OpenFlow controller which is a pox controller. We define OpenFlow switches connected to user defined controllers and host nodes. Since it is a multi-controller architecture this will emulate cluster/Sub-network communication. A controller program is written to set up Ethernet connection this work also includes a distance vector algorithm to find shortest path. In the simulated network once network is configured, each user defined controller is called remotely and functions in parallel. The simulated network set-up is shown in Fig. 4.

We configure each device of the network by assigning the addresses and bandwidth. We define user defined controller MC1 and MC2 and assign IP address 127.0.0.1 and 127.0.0.2 respectively, where MC1 is listening on port number 6633 and MC2 is listening on 6634. We connect both the sub-networks to default POX controller C0, C0 is assigned address 10.0.0.1. Using a links EN10MB we connect controllers, switches and hosts. A ping message is sent from every host node to other host node.

Working of MC architecture, number n of nodes in network, network is configured in such a way that nodes are distributed among two controllers MC1 and MC2. Where MC1 is assigned with handles hosts H − 1 to H − i where ‘i’ is an integer second controller is assigned with H − (i + 1) host to H-n. consider a scenario when a Host

H − i of cluster MC1 wants to communicate with Host H − (i + 2) of cluster MC2, the devices have to first communicate with the controller which is connected to both the sub networks that is default pox controller C0. If Host H − i and Host H − 2 which belongs to same cluster needs to communicate it is handled locally by MC1 it will not consult built in controller. Inter cluster and intra cluster Communication is tested using ping.

5 Results and Comparison

Simulation is started with execution of controller on one terminal, on successful start of controller. User defined network program is executed, in which network is configured. Ethernet connection is established and reflects on controller terminal. Simulation is done for maximum of 50 hosts with initial deployment of 5 hosts and subsequent increment of 5 hosts in each iteration. Ten such iterations are captured.

5.1 Observations Recorded

- Network set up: User defined network is set up before transfer of data starts. Static user defined network that is deployed in this work is shown in Figs. 3 and 4 for SC and MC architecture respectively. During setting of network, nodes are created, links are established and flow tables are generated implicitly by the simulator. Flow table in case of SC architecture is generated for main controller to all other nodes. IP address assigned to main controller is 10.0.0.1 flow table is generated from this controller to all other nodes to which address is assigned during runtime. Flow table in MC architecture is generated for main controller with IP address 10.0.0.1 and for user defined controllers MC1 and MC2 with IP address 127.0.0.1 and 127.0.0.2 respectively.
- Trace file: Trace file is generated during execution with has MAC address assigned to each node with other details such as number of packets transferred successfully, number of packets dropped.
- Total time taken to communicate: Time taken for transfer of data from source node to all other nodes is inferred where the source node is controller and destination nodes are all other nodes. Time taken for data transfer in SC architecture is higher than MC architecture. Consider number of nodes n = 45 time consumed by SC architecture is 400,000 ms and in MC architecture is 150,000 ms this is depicted in graph Fig. 6.
- Throughput of the both SC and MC architecture is analysed by considering Minimum, Maximum, Average and Moving standard deviation Round Trip Time (RTT). Consider n = 45 Minimum, Maximum, Average and Moving standard deviation RTT in ms is as follows in Table 1.

The above results are revealed in detail for various values of n in Figs. 7, 8, 9 and 10.

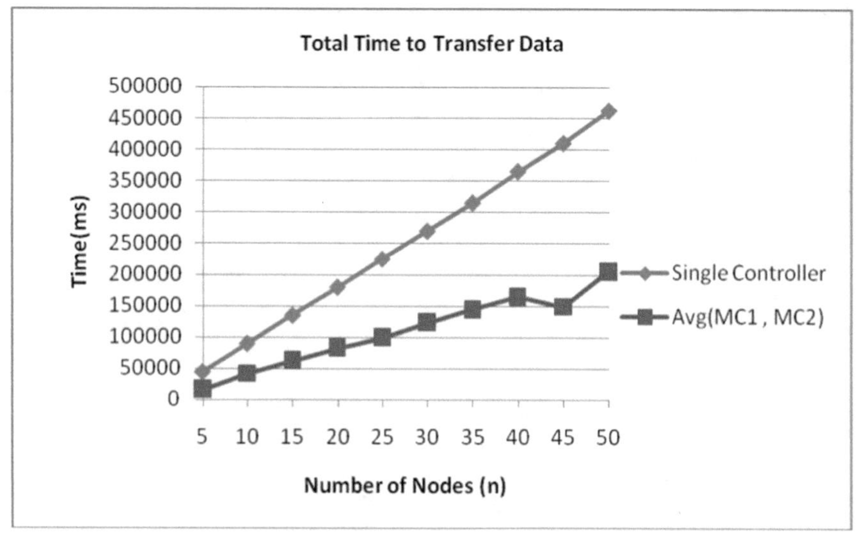

Fig. 6. Total data transfer time for MC is efficient than SC.

Table 1. Minimum, Maximum, Average and Moving standard deviation RTT values for n = 45.

Number of nodes 45		
Minimum	190.1	9.5
Maximum	425.2	1.8
Average	219.8	1.5
Moving standard deviation	210.5	4.2

5.2 Comparison

Simulation of Single-Controller (SC) versus Multi-Controller (MC) architecture: In the same simulation environment. Following are the inferences.

In MC controller architecture network set up process with respect to flow table generation is parallel, as the user defined controller MC1 and MC2 generates flowtables independently. Where as in SC architecture flow table generation is generated only for main controller C0 to all other nodes in the network. Considering n = 30 network set up time in SC architecture is 250 ms and MC architecture is 150 ms. Hence MC architecture network set up is faster than SC. This can be inferred for various values of n from Fig. 5.

The minimum RTT for MC architecture to that of SC architecture for 50 nodes are 0.054 and 200.05 ms respectively shown in Fig. 7. Similarly consider maximum RTT in Fig. 8 it is observed that maximum RTT is almost negligible in MC architecture where as for SC architecture it is 224 ms.

Study on Moving Standard Deviation indicates that higher the moving standard deviation RTT values, there will be speed issues with bulk data transfer. From Table 1

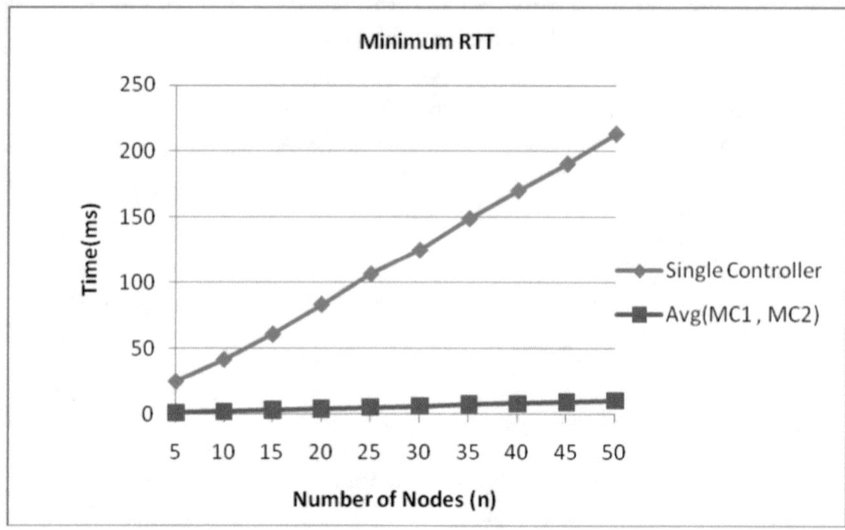

Fig. 7. Minimum RTT for MC is efficient than SC.

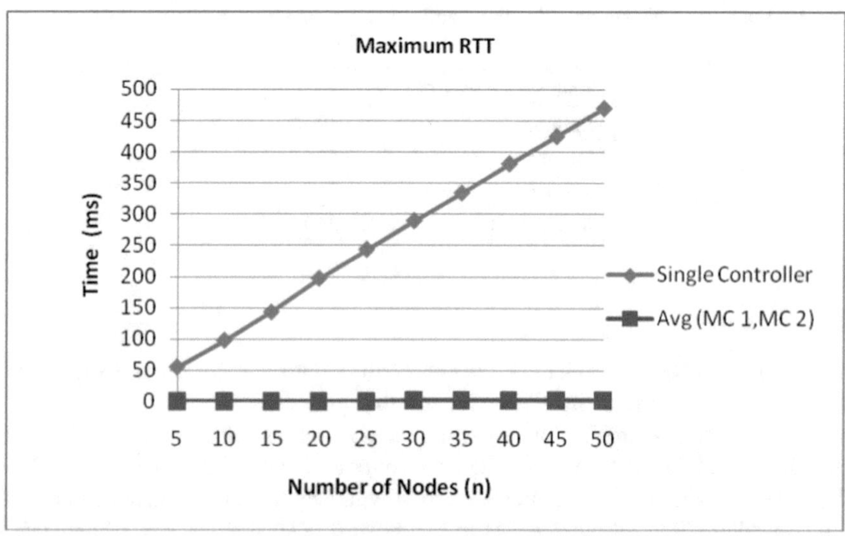

Fig. 8. Maximum RTT for MC is efficient than SC.

for number of nodes n = 45 Average and Moving standard deviation is 219.8 and 210.5 respectively for SC architecture. Average and Moving standard deviation for MC architecture is 210.5 and 4.2 respectively. For other values of n, Average and Moving standard deviation can be inferred from Figs. 9 and 10.

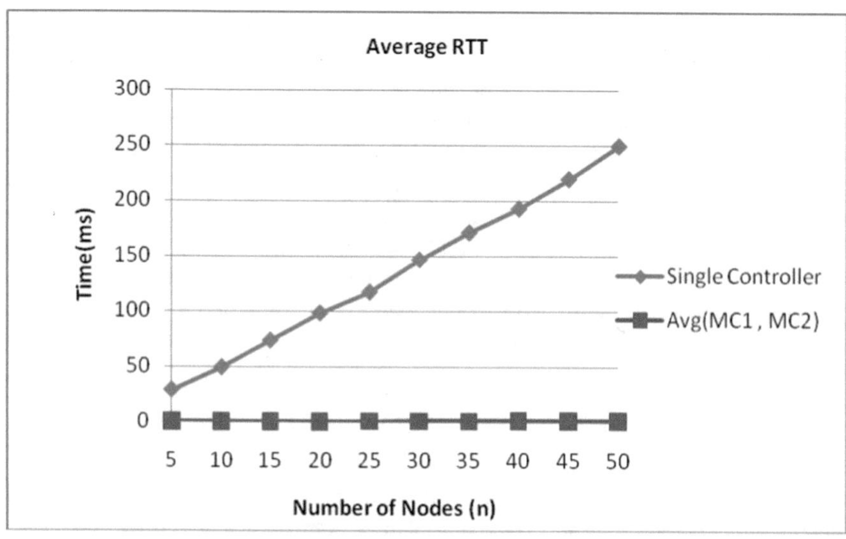

Fig. 9. Average RTT for MC is efficient than SC.

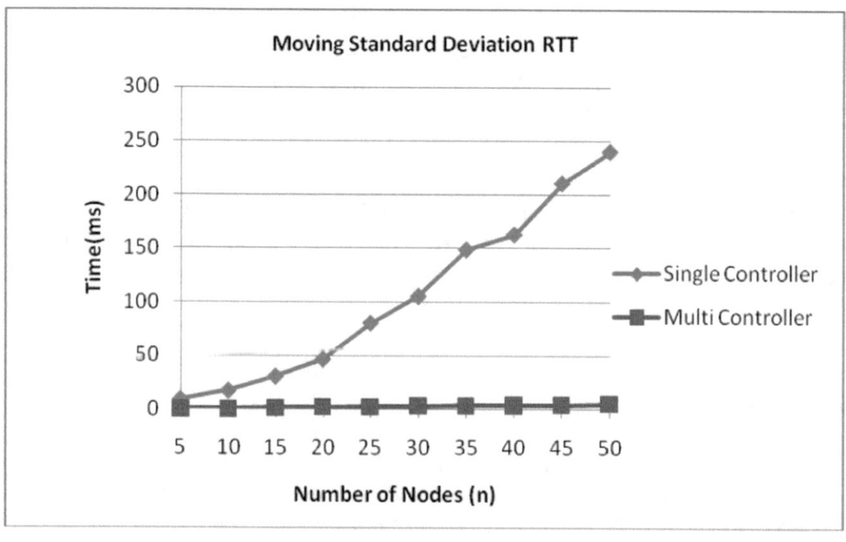

Fig. 10. Moving Standard Deviation RTT for MC and SC.

6 Conclusions

Considering comparison of Round Trip Time (RTT) between SC and MC architecture i.e., ping results Minimum, Maximum, Average and Moving standard deviation RTT values are much less in case of MC architecture but in SC architecture there is linear growth. This is depicted in Figs. 7, 8, 9 and 10.

It is observed that the efficiency of the network tends to decrease with an increase in the number of nodes. It is inferred that the RTT increases at a very high rate as the number of node increases in case of SC architecture. In case MC architecture the RTT does not display a credible increase with the increase in number of nodes. Hence the MC architecture can be scaled to a large extent retaining its efficiency.

This work has illustrated defining user defined controllers. On simulation of Multi Controllers in a network we have brought out clustering of networks which has enormously increased efficiency. Load on built in controller is reduced by balancing it among user defined controllers. The emulation results of the proposed mechanism significantly shows that the network is constantly monitored and also enables a faster and much more efficient way of communicating between two clusters of devices.

7 Future Enhancements

There is a lot of scope in field of SDN to explore, in a vision to virtualize network infrastructure. The above proposed system can be improvised or further enhanced by making the system enabled to communicate with wireless networks. This work can be extended to be used in dynamic reconfiguration of network in IoT infrastructure with huge number of devices. This can also provide platform for Edge/Cloud computing in scenario where huge number of device data needs to be handled.

Heterogeneity of IoT network can be handled by assigning each homogenous set of devises to a particular cluster. Data received by each cluster can then be handled efficiently at the root level. Compared to other methods like introducing a new protocol stack or introducing layer to a existing protocol stack, this approach simplifies handling heterogeneity.

Cloud applications implemented using single controller SDN may disturb functioning specially deterioration of security. Therefore, we can implement multi-layer access control system to mitigate the consequences and focus on securing both SDN's application layer as well as its control layer.

References

1. McKeown, N., Anderson, T., Balakrishnan, H., Parulkar, G., Peterson, L., Rexford, J., Shenker, S., Turner, J.: OpenFlow: enabling innovation in campus networks. In: ACM SIGCOMM Computer Communications Review, Apr 2008
2. Coto, J.: A holistic presentation and recommendation of OpenFlow its challenges and future research needed. Int. J. Sci. Res. Publ. **5**(10) (Oct 2015). ISSN 2250-3153
3. Jarschel, M., Oechsner, S., Schlosser, D., Pries, R., Goll, S., Tran-Gia, P.: Modeling and performance evaluation of an Openflow architecture. In: Proceedings of the 23rd International Teletraffic Congress (2011)
4. de Oliveira, R.L.S., Shinoda, A.A., Schweitzer, C.M., Prete, L.R.: Using mininet for emulation and prototyping software-defined networks. In: Communications and Computing (COLCOM), INSPEC Accession Number: 14468968, IEEE Xplore: 21 July 2014

5. Wang, F., Wang, H., Lei, B., Ma, W.: A research on carrier-grade SDN controller. In: 2014 International Conference on Cloud Computing and Big Data, 978-1-4799-6621-91/14 $31.00 ©2014 IEEE. https://doi.org/10.1109/ccbd.2014.41
6. Metter, C., Seufert, M., Wamser, F., Zinner, T., Tran-Gia, P.: Analytical model for SDN signaling traffic and flow table occupancy and its application for various types of traffic. IEEE (2016)
7. Yang, Z., Yeung, K.L.: An efficient flow monitoring scheme for SDN networks. 978-1-5090-5538-8/17/$31.00 ©2017 IEEE
8. Schwartz, C., Hossfeld, T., Lehrieder, F., Tran-Gia, P.: Angry apps: the impact of network timer selection on power consumption, signalling load, and web QoE. J. Comput. Netw. Commun. (2013)
9. Huang, H., Zhu, J., Zhang, L.: An SDN_based management framework for IoT devices. ISSC 2014/CIICT 2014, Limerick, June 26–27
10. Kim, G., Kim, J., Lee, S.: An SDN based fully distributed NAT traversal scheme for IoT global connectivity. ICTC 2015, 978-1-4673-7116-2/15/$31.00 ©2015 IEEE
11. Sahoo, K.S., Sahoo, B., Panda, A.: A secured SDN framework for IoT. In: 2015 International Conference on Man and Machine Interfacing (MAMI)
12. Heller, B., Sherwood, R., McKeown, N.: The controller placement problem. In: HotSDN'12, August 13, 2012, Helsinki, Finland. Copyright 2012 ACM 978-1-4503-1477-0/12/08
13. Tootoonchian, A., Gorbunov, S., Casado, M., Sherwhood, R.: On controller performance in software-defined networks. In: HotSDN'12, 13 Aug 2012, Helsinki, Finland. Copyright 2012 ACM 978-1-4503-1478-0/12/08
14. Ferguson, A.D., Guha, A., Liang, C., Fonseca, R., Krishnamurthi, S.: Hierarchical policies for software defined networks. In: HotSDN'12, 13 Aug 2012, Helsinki, Finland. Copyright 2012 ACM 978-1-4503-1477-0/12/08
15. Thorat, P., Raza, S.M., Kim, D.S., Choo, H.: Rapid recovery from link failures in software-defined networks. J. Commun. Netw. **19**(6), 648–665 (2017)

A Smart HEMS Architecture Using Zigbee

M. C. Srivas, T. Anushalalitha(✉), Sri Kumar, and K. Koushik

Department of Telecommunication, BMS College of Engineering, Bengaluru 560019, Karnataka, India
{sriv95,Srikum95,Koushik4u95}@gmail.com,
anusha.tce@bmsce.ac.in

Abstract. This project aims to implement Smart Home Energy Management System (HEMS). The simulation is implemented using arduino and MATLAB tool. Currently, the current energy crisis has required significant energy reduction in all areas. Energy saving and renewable energy sources are considered as methods of solving home energy problems. Two Arduinos are communicated using Xbee where different devices or home appliances are connected and the output is displayed on the monitor screen at a central home computer. The power consumption of home appliances and of renewable energies are collected from home server, which is used for analyzing the total energy estimation. The overall idea of a Smart Home Energy Management System monitors energy-consuming home appliances and renewable energy devices. This system will help in optimizing home energy use and result in home energy saving cost.

Keywords: Smart home energy management system · Power consumption · Total energy estimation

1 Introduction

The current has required significant energy reduction in all areas. The energy consumption in home areas has more home appliances installed. Energy saving and renewable energy sources are considered as methods of solving home energy problem. Both energy consumption and generation should be simultaneously considered to save the home energy cost. Several researches have proposed Home Energy Management System (HEMS). Optimization of home power consumption based on power line communication has been studied to provide easy-to-access to home energy consumption.

A smart HEMS architecture considers both energy consumption and generation based on zigbee and PLC based renewable energy gateway. The home servers gather data through zigbee and energy consumption data through renewable energy gateway (REG). This system enables ease of use and monitor power from a remote device located at a central place. A monitoring system and renewable energy devices will make a big difference in the energy usage patterns. Renewable energy is the main source of energy in future and it is to be fully utilized for the benefit of the society. Home Energy Management System will notify and determine the type of energy and power usage patterns thereby helping in saving energy when appliances are not in use.

M. E. Auer and K. Ram B. (Eds.): REV2019 2019, LNNS 80, pp. 456–462, 2020.
https://doi.org/10.1007/978-3-030-23162-0_41

2 Related Work

The idea of implementation of the project has varied facets and analysis, which thus gave improvisations in different stages. With the basic knowledge of Embedded Programming and also to implement the above mentioned objective, survey was conducted which had similar idea's but differed in the implementation. Han et al. [1] has implemented the real working model of the HEMS system using Zigbee, Power line communication and a Home server. This paper proposes a Smart HEMS Architecture that considers both energy consumption and generation simultaneously. Sigsbee based energy measurement modules are used to monitor the energy consumption of home appliances and lights. The home server gathers the energy consumption and generation data, analyzes them for energy estimation, and controls the home energy use schedule to minimize the energy cost. Sathish kumar and S. Jayamani et al. [2] had certain idea about the implementation and discusses about monitoring energy in home environment using HEMS, Zigbee Communication. Han et al. [3] proposes a System that unifies various home appliances using a zigbee network.

Kumar et al. [4] shows that Energy consumed is displayed on the LCD by using microcontroller and the transmitter section information is transmitted to receiver section and displayed on the PC through Wireless Communication by using Zigbee technology. Sethuraman and Jayanthi [5] wrote a paper where in the energy consumption of the renewable energy devices and home appliances are collected from home server to analyze the total energy and monitor the energy consumption in home to minimize the energy cost.

3 Aim and Methodology

Today's technology uses immense amount of electrical power all over the world. The monitoring system are never in the home environments and domestic areas. All the monitoring and control are present in the main substation. Considering today's energy crises, a system has to be developed to save energy use. When energy usage increases, the cost also increases. There is also energy wastage, because of negligence and misuse. A lot of problems are faced in the field of energy distribution and planning, leading to frequent power cuts and outages. There are more losses in distribution of energy and no machine achieves full efficiency. Considering these factors, energy is to be saved to the maximum extent possible. The main aim of the project is implementing a Smart Home Energy System using Arduino, Zigbee and MATLAB©. This system monitors the power consumption in all respects. Implementing effective power usage and displaying at the central computer is the main objective of the project. A transmitter block and receiver block is employed for the operation of the system. The transmitter section of the system uses the solar panel and battery as the sources. Two separate LEDs are connected to each of the sources and the voltage and current is read using Arduino and current sensors (Fig. 1).

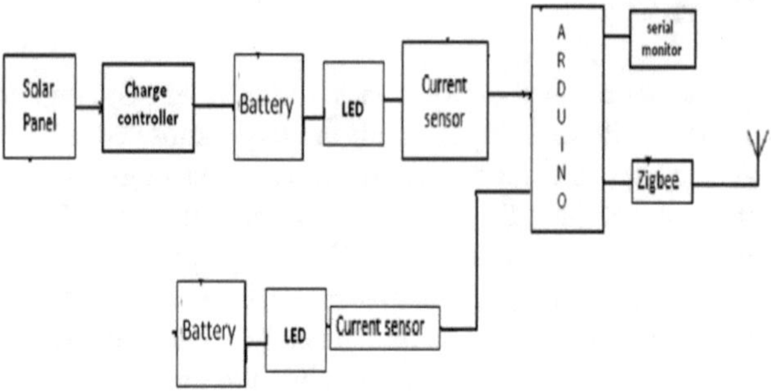

Fig. 1. Transmitter block

The Serial monitor displays the power utilized as the output for reference at the transmitter. Zigbee is used to transmit this data to the receiver block (Fig. 2).

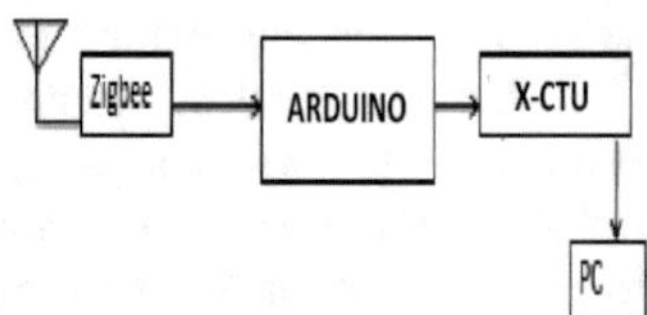

Fig. 2. Receiver block

The receiver block receives the incoming data at Zigbee using Xbee communication. It then sends it to the Arduino and then display the power utilization on the computer, where the user can record and analyse the data.

3.1 Implementation

The system is implemented to analyse the operation of Home Energy Management System. Considering a small scale implementation as opposed to a real structure, the system is implemented in the following way.

- It gets power from the solar and DC sources and powers the LED.
- Power measurement is done using current sensors and Arduino boards.
- Data is then transmitted to the zigbee which then communicates wirelessly to the receiver Xbee.
- Receiving Xbee sends information to the Arduino to display the obtained values.
- On the receiving Xbee X-CTU results can be seen or the system can make use of Arduino screen monitor.
- Matlab is used to plot the obtained outputs of power usage and analysed (Figs. 3 and 4).

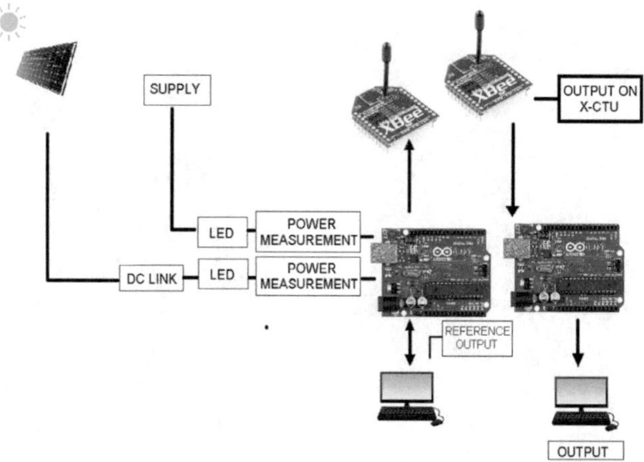

Fig. 3. Implementation Strategy

Fig. 4. System implementation

4 Results and Discussion

The results are seen on the X-CTU screen on the receiving side. The values that are calculated at the transmitting end are transferred wirelessly and displayed on the X-CTU screen. A user can easily find the power usage on this software than having additional devices (Fig. 5).

Also, the same data is seen using Arduino boards. Data is obtained by the hardware and software implementation. The measurements are taken for three trials. The following table is generated from the Arduino output (Table 1).

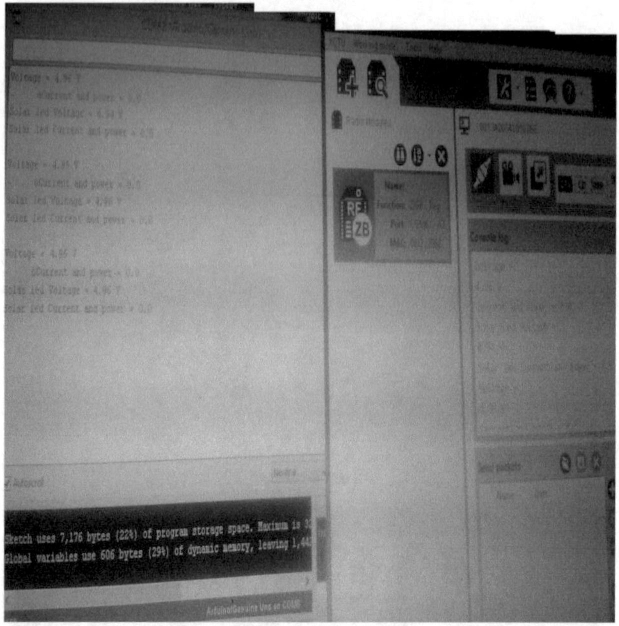

Fig. 5. Data obtained on X-CTU

Table 1. Recorded data for different trials

(Arduino/Genuino UNO)					
Voltage (V)	Current (A)	Power (W)	Solar led voltage (V)	Solar led current (A)	Solar led power (W)
1.46	0.0050	0.0073	5.00	0.0051	0.0075
1.48	0.0050	0.0076	5.00	0.0051	0.0076
1.52	0.0047	0.0081	5.00	0.0051	0.0081

These results as a secondary output option can be obtained on the Arduino monitor screen wherein data on the transmitting port and same data as received on receiving port are shown in Fig. 6.

The results obtained on different set of trials are then plotted using Matlab and corresponding total power usage, power distribution can be analysed.

Power distribution, Total Power usage and average power are obtained, considering a number of trials. Here three trials are considered (Fig. 7).

5 Future Work

This system has been considered in a small scale level and has a scope for future work. The work to be followed can be classified as follows.

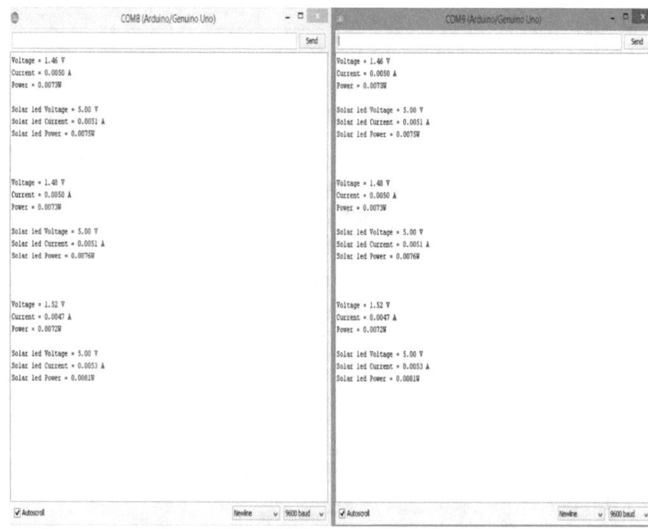

Fig. 6. Arduino Serial Monitors

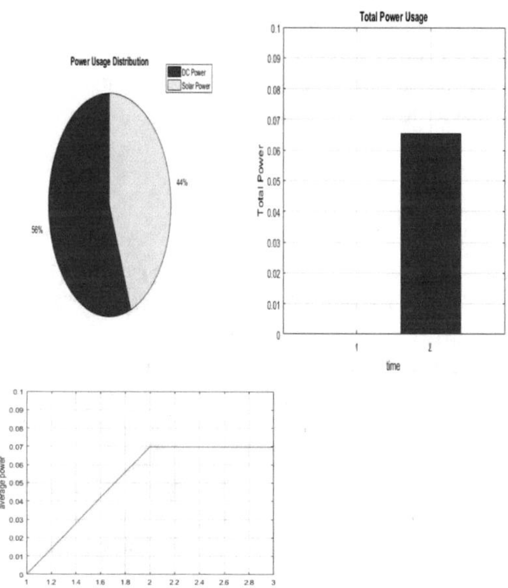

Fig. 7. Power distribution—DC, Solar, total power usage and Average Power

Use of AC Supply: Real world scenarios can be considered and Inverters can be used to get power reading. Complex appliances: Motors, Fan, Refrigerators can be tested. Analysis of each device: A report of power usage for each device can be done. User account login and control of power can be introduced. Enhancement of security features at different layers and protection of data can be done.

6 Conclusion

Renewable energy sources have to be installed in residential area to save the energy cost, for effective usage and for longevity of the appliances. In the home environment, this system is installed in outlets and lights to measure the energy and power usages of home appliances. The Zigbee is used to transfer the gathered data to the home server. The home server or the central computer figures out the home energy usage pattern. Power consumption details are successfully uploaded into the home server continuously. It also provides the comparison and analysis of each home energy usage. Home energy is monitored and utilised efficiently for cost savings and record of data. This system enables green home practices.

Although this system is found to be beneficial to save energy, it has certain faults on its own. The cost required to set up this system is quite costly and users may not feel the need.

Measurements may be not very accurate. Loose connections and minor damages can occur over time.

Acknowledgements. We are greatly indebted to our college, BMS College of Engineering, for providing us a healthy environment; conducive to learning that made us reason out, question and think out of the box. It has also strengthened our confidence and has enabled us to conceptualize, realize our dream project. We would like to thank our college for providing support and encouragement during this research study.

We would also like to thank Mrs. Rajeshwari Hegde for the guidance and encouragement extended to us that helped us in the successful completion of this paper.

References

1. Han, J., Choi, C.-S., Park, W.-K., Lee, I., Kim, S.-H.: Smart home energy management system including renewable energy based on ZigBee and PLC. IEEE Trans. Consumer Electron. **60**(2) (May 2014)
2. Sathish kumar, A., Jayamani, S.: Renewable energy management system in home appliance. In: 2015 International Conference on Circuit, Power and Computing Technologies [ICCPCT]
3. Han, D.-M., Lim, J.-H.: Smart home energy management system using IEEE 802.15.4 and zigbee. IEEE Trans. Consumer Electron. **56**(3) (Aug 2010)
4. Kumar, S.V., Prasannanjaneya Reddy, V.: Zigbee based home energy management system using renewable energy sources. IJMET **2**(1) (Jan 2015)
5. Sethuraman, M., Jayanthi, S.: Low cost and high efficiency Smart HEMS by using Zigbee with MPPT techniques. IJARC **4**(11) (Nov 2014)
6. Arduino-energy-meterhttp://www.instructables.com/id/ARDUINO-ENERGY-METER/step4/Current-Measurement/

Part V

Virtual and Remote Laboratories

Development of Remote Instrumentation and Control for Laboratory Experiments Using Smart Phone Application

N. P. Arun Kumar[1(✉)] and A. P. Jagadeesh Chandra[2]

[1] Honeywell Technology Solutions Lab, Bangalore, India
arunkumar.np@gmail.com, arunkumar_np@rediffmail.com
[2] Adichunchanagiri Institute of Technology, Chikmagalur, India
apjchandra@gmail.com

Abstract. The unprecedented growth in Internet technologies has created revolutionary changes in the use of collaborative learning tools with remote experimentation. These tools enhance the experiential learning aspects of engineering education. Laboratory experiments are integral part of science and engineering education. Automation is changing the nature of these laboratories, and the focus of the system designer is on the availability of various interfacing tools to access the laboratory hardware remotely with the integration of computer supported learning environment. Work on laboratory experiments and project works requires access to expensive hardware equipments. The high cost of these instruments along with time consuming development process of experimentation in the educational process creates a significant bottleneck. There is a need for the development of remote laboratory using which the users can access the laboratory instruments and the programmable devices remotely on their smart phones/tablets to perform the laboratory experiments. This implementation avails laboratory facility for complete twenty four hours a day and will increase the productivity of the laboratory hardware and measuring instruments. This paper presents the detailed architecture and the implementation details of remote laboratory by which user can perform laboratory experiments remotely. Develop mobile based remote laboratory where user can access remote laboratory on his smart phone or tablet to perform the experiments. Software application is developed on Android platform for the implementation.

Keywords: Remote laboratory · Virtual instrumentation · Remote access · Android Studio

1 Introduction

With the advent of wireless technology, mobile phones are being used for many other applications other than communication. Mobile applications are being developed for aiding students, teachers and universities in academics. Apps are developed for research application which consolidates the journals across the publishers and allows user to read as a single interface. The increased power and capabilities of the mobile operating systems and the smartphones, it is possible to access the desktop remotely

M. E. Auer and K. Ram B. (Eds.): REV2019 2019, LNNS 80, pp. 465–476, 2020.
https://doi.org/10.1007/978-3-030-23162-0_42

from a smart phone or the tablet. This provides mobility to the user to access the laboratory server from anywhere within the network operating range. Software applications are developed to provide remote access to the desktops where user can get complete access of the remote computer from a mobile phone or a tablet.

Remote laboratory allows users to access the laboratory instruments including the programmable devices remotely to perform their laboratory experiments. The existing remote laboratories on Digital Signal Processor hardware uses either the server machine to control the test instruments using GPIB interface or control is established through the DAQ cards [1–3]. This makes architecture more specific and cannot be reused for other laboratory implementations. This implementation avails laboratory facility for complete twenty four hours a day and will increase the productivity of the laboratory setups and measuring instruments. Remote instrumentation laboratory for DSP training uses client server methodology and connects multiple clients to the server using Virtual Instrument application [2]. Thin Client Server manages input and outputs between client and servers. Remote access tool [9] used for this type of laboratory implementation is selected based on real time access parameters like data speed, security protocol and ability to establish multiple user environment etc.

Proposed architecture is more generic and uses customized interface board to interface server machine with the test instruments and unit. By making minor modification on the interface card and adding test instruments, whole setup can be used for other laboratory implementations. This makes remote architecture more generic and capable of adopting new requirements with few minor modifications in the setup. It provides controlled server access to multiple clients using the mobile application developed using Android Studio.

Remote access facility is provided through the server machine using webserver has limitations on the usage of software and less control on the hardware [5, 6]. Proposed architecture uses remote access method which provide complete controlled access to the server machine where user can access all the softwares and data in the server machine. This tool also provides the log history which details out the client machine details with login and logout data.

This paper describes the design and implementation details of development of Remote Instrumentation and Control for Laboratory Experiments using Smart Phone Application. Paper is organized as follows. Section 2 describes Remote Laboratory architecture. Section 3 details out the remote access tool used for the implementation with client server access methods. Section 4 describes the LabVIEW user interface developed for the remote client access and control. Section 5 describes mobile application development to access laboratories remotely. Section 6 provides a case study for setting up and programming a microcontroller board using the same test setup. Conclusions are drawn at Sect. 7.

2 Generic Remote Laboratory Architecture

The generic remote laboratory is designed to provide remote access to test unit, test and measuring instruments and the required software at distant location. Detailed Interface architecture of remote laboratory is as shown in Fig. 1.

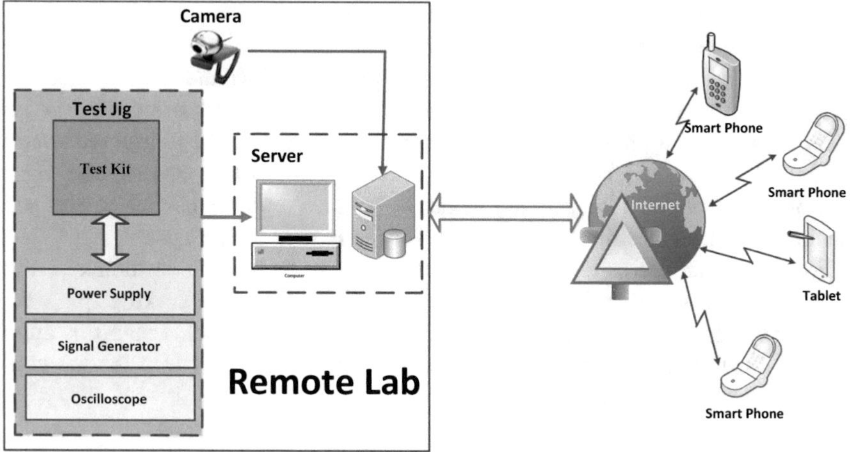

Fig. 1. Generic remote laboratory architecture

Remote clients are connected to the laboratory server via mobile application developed on the smart phone or the tablet. Microsoft remote desktop or Radmin tool is used for remote access implementation as it is best suited for real time applications. User interface is developed using LabVIEW on the server machine which communicates and controls the test/measuring instruments and interface hardware DSP kit and microcontroller kits. Customized hardware interface board is designed and developed which acts as interface between the Data Acquisition cards in server machine and the unit being tested.

Camera is connected to server machine which provide continuous video of the experimental set-up to the remote user logged onto server machine. Power and position of the camera is controlled by LabVIEW VI located on the server machine.

3 Remote Access Methods

Remote access technologies currently available have their own pros and cons. Selection of the tool is based on its end application. Available remote access tools and methodologies includes Team Viewer, Remote desktop access, Virtual network computing, Web server and Radmin. Each tool is investigated for its ability to meet the virtual laboratory implementation requirements. The tool selected should operate with maximum reliability and provide high security.

Based on the comparison and analysis, Radmin tool is one of the best tools available for Remote Laboratory applications. This supports simultaneous multiple user access and data transfer between the machines. It is suitable for real time applications [12].

Radmin server software is installed in the server machine and Radmin viewer on the client machine. Server and client software required is of lower cost. Client software can be installed in any number of systems. Multiple users can work and watch simultaneously on single server machine. This tool also provides the log history which details out the client machine with login and logout data.

Radmin server on the server machine is switched ON for remote access enable and admin can set access permissions for specific remote computers using server configuration window. On client machine, Radmin viewer application is triggered and the remote server machine is identified by its IP address. Server access user ID and password are entered into security window to get complete access of the server machine. This provides first level security for the remote access system.

For the remote access through the smart phones or the tablets, customized android application is developed using Android Studio. Microsoft remote desktop app is being called internally based on the remote laboratory application need. Using this app the connection can be established to the laboratory server PC to perform the lab experiments and to files and network resources.

4 Remote User Interface Using LabVIEW

User interface developed using LabVIEW can be used for remote accessing, data sharing and other desktop sharing operations between two remote computers. It supports most of the operating systems like Windows, Linux, Mac OS, Android, Windows RT, Windows Phone and BlackBerry systems. It is the freeware which can be used for non-commercial applications. It allows user to establish Virtual Private Network connection with partner. It facilitates data acquisition using DAQ cards to capture analog and digital data.

For remote laboratory implementation, user interface is developed using LabVIEW. This user interface developed is loaded on to the laboratory server machine to which the test kit and all the test/measuring instruments are connected. Many VIs are developed and interlinked to each other to provide secure laboratory connectivity to the remote user. User login VI checks for the user credentials to provide access to the main operating window as shown in Fig. 2.

It has additional tab for administrator login is provided. This allows administrator to add new users, change password and deletion of the existing users. Once the user login with valid user ID and password, the control is routed to main VI as shown in Fig. 3. This window provides access to all the hardware and softwares to conduct laboratory experiments with live visual images.

The main user interface developed has customized buttons to control below mentioned features

1. Power Supply Control

Power input to DSP kit is controlled using the power button on the user interface. This sends command to the power relay on the interface board which switches the power input to the DSP kit.

2. Code composer Studio

This button will launch code composer studio software. User needs to select the workspace location from user interface and configure the code composer studio using

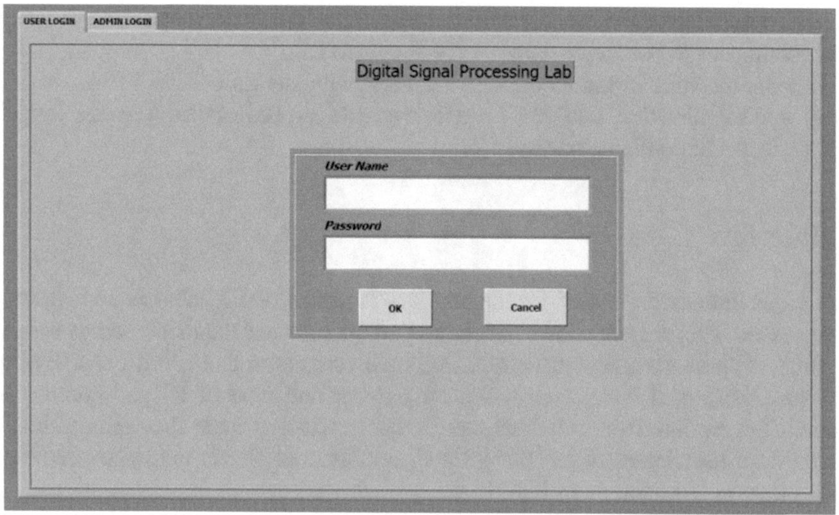

Fig. 2. User login window

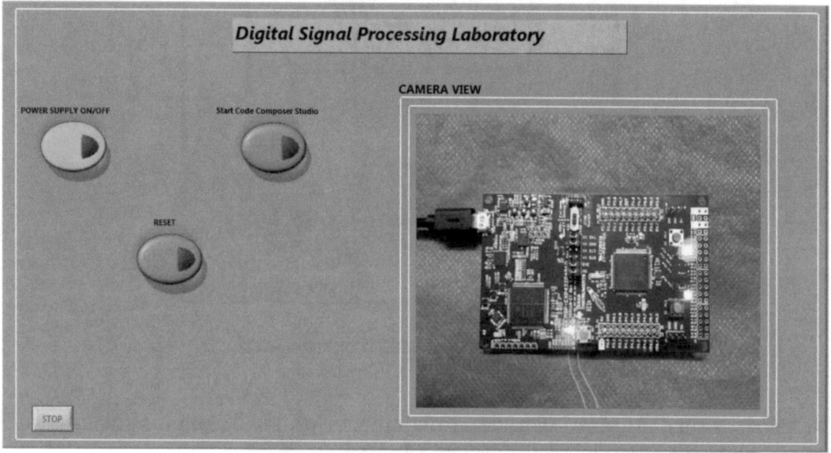

Fig. 3. User interface for remote DSP laboratory

target information. Target can be connected by selecting a proper connection and device type on the user menu. Code to be executed is written in new source file and further built, debugged and run on the target to perform the experiments.

3. Reset

As part of the code execution using code composer studio, a control to reset the DSP controller is provided which can be used while debugging the code. Remote reset on the hardware can be achieved by designing reset circuitry on the interface board. It is controlled remotely using the Reset button on the LabVIEW user interface. This button sends command to the Reset relay on the interface card.

Remote machine in the laboratory is loaded with the LabVIEW VI that is developed for this application. LabVIEW run time engine can be installed if we are installing only the executable file developed.

5 Interface Card

Data acquisition card required for the interface is placed on PXI chassis and connected to the server. This chassis allows user to add additional data acquisition cards based on the future requirements. All lines on DAQ cards are controlled using LabVIEW user interface. DAQ card is selected based on number and type of IO and control lines required for the interface. Interface card is customized to route the control and data signals from user interface to DSP kit and test instruments via relays. Interface card block diagram is as shown in Fig. 4.

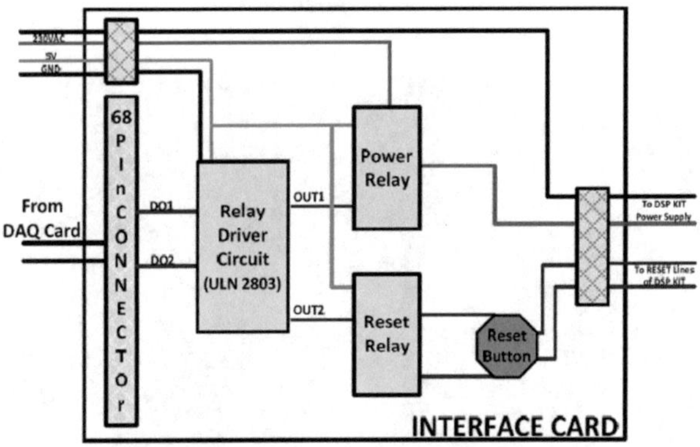

Fig. 4. Interface card block diagram

All lines form DAQ card is routed to interface card via 68 pin interface connector. Interface card comprises below mentioned blocks

1. Relay Driver

Darlington transistor arrays in ULN2803A are used to drive power and reset relays. Digital output signals from DAQ card are routed to relay driver circuit. OUT1 and

OUT2 signals from relay driver circuit is connected one of the coil on the relay. It provides the grounding path to the coil voltage when it is triggered. Digital signals at the input of relay driver circuit is controlled by the power supply and reset button on the user interface.

2. Power Relay

This relay controls the power input for the DSP kit. User commands the power input using power supply button on the LabVIEW user interface. Digital output line DO1 on DAQ card is configured for the power button. Signal from the DAQ card is driven using relay driver circuit and is fed to the coil of power relay. Enabling the power button will provide grounding path for the relay circuit and relay will be switched to link 230 VAC to the input of the DSP kit power adapter. Power input control relay circuitry developed on interface card is as shown in Fig. 5.

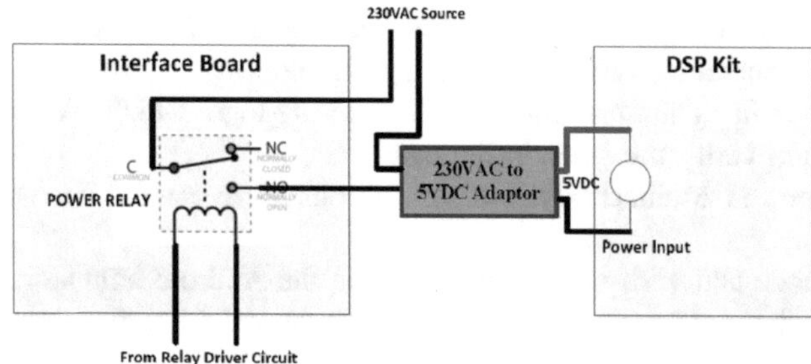

Fig. 5. Power input relay circuit

3. Reset relay

User need to reset the DSP kit while flashing the code as part of the experiment. Reset button on LabVIEW user interface is provided to offer this functionality. This button will trigger the digital output line DO2 on DAQ card and further, the signal is driven from the DAQ card using relay driver circuit and is fed to the coil of the reset relay. Trigger on OUT2 line from relay driver will provide grounding path to the relay coil voltage and triggers relay which shorts the two reset lines on the DSP kit. Additional reset button is also provided on the interface card to perform hard reset. Reset control relay circuitry developed on the interface card is as shown in Fig. 6.

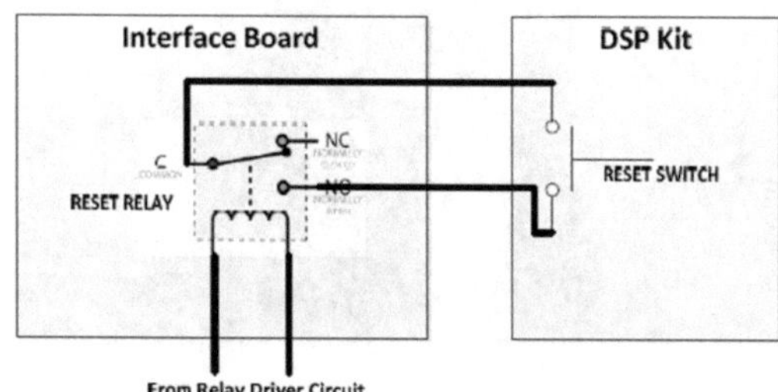

Fig. 6. Reset relay circuit

6 Mobile App Development

Mobile app is developed using Android studio to provide access to remote laboratory. Code is developed to create new user account and store the data in the database. For secured login, network link established to connect the remote laboratory PC using smart phone or the tablet.

Below mentioned algorithm used to develop mobile application for remote access

1. Create new Android project and open Android new application wizard.
2. Develop the desired layout of the app in activity_main.xml.
3. Declare all the required components in the manifest file. Develop display.xml and signup.xml file based on the application requirement. This file works as interface between the operating system and the app developed.
4. Develop code in signup.java file to create a secure login which takes user name and password and verify the same in the database.
5. Develop code in Mainactivity.java file to connect to remote access tool on successful login.
6. Build the code and verify its functionality on the Android emulator.
7. Connect to the mobile phone or tablet and download the app developed.

The app developed for remote access should be installed on to the smart phone. Remote laboratory login window is developed as security check-in for the user login. User willing to access the remote laboratory should have the valid login user id and password. New user can sign up with valid user data and create the user account as shown in Fig. 7.

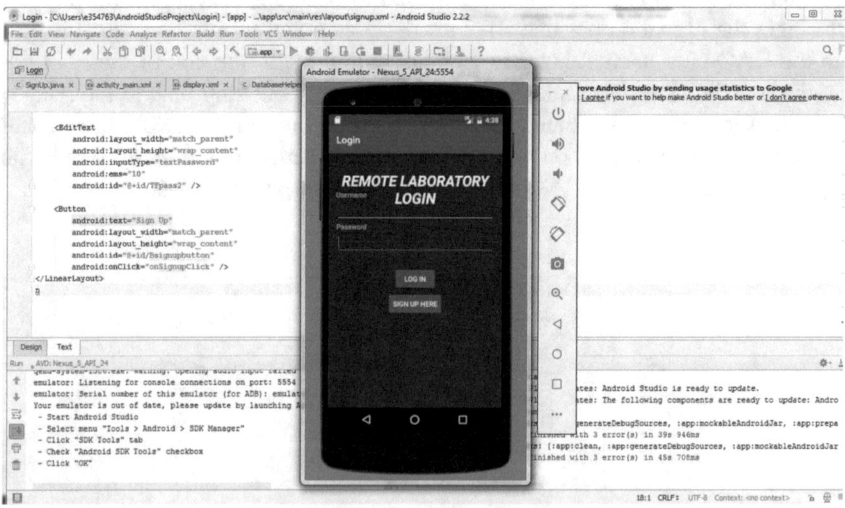

Fig. 7. User login window of the app

With valid user id and password, user is routed to next window where it will provide access to remote laboratory. User will be directed to laboratory server machine through this window.

App lists out all the available laboratory server machines on the network. User can select the specific laboratory machine based on the experiment being conducted. With valid login credentials of the laboratory machine user can have complete access of the laboratory machine. User interface developed using LabVIEW is installed on the laboratory server machine. Remote client can use this user interface to get control of the laboratory test setup along with test instruments and can conduct the experiments remotely.

7 Remote Access Experimentation

To demonstrate the generic application of the remote laboratory test setup, a microcontroller board is programmed remotely using the same test setup and verified for the functionality. Microcontroller board is connected to the server and two reset lines from microcontroller board is connected to interface card as shown in Fig. 8.

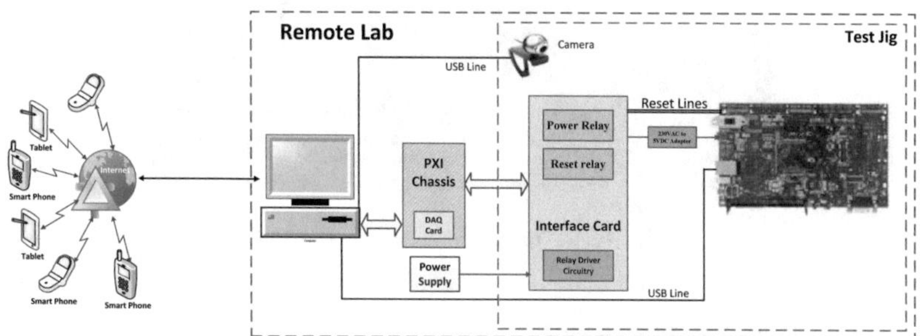

Fig. 8. Remote DSP laboratory architecture

Remote app installed on the smart phone is used to login into the laboratory machine. This is the customized app developed using Android studio to provide remote access to the laboratory server machine. Valid user credentials are entered in the App login window as shown in Fig. 9. This provides the first level security to avoid unauthorized access to the remote lab server. New user can create his account using valid user data using “Sign Up Here” button.

Client user is logged into server machine using mobile application. Server machine to be accessed is identified using its ip address. Server access user id and password is entered into security window to get complete access of the server machine.

LabVIEW user interface is opened on the server machine to get access to the DSP board. Live visual images of the controller board is displayed as part of the user interface. Controller board is powered ON using the control on the user interface. Further, code composer studio application v6.1 is opened to start the programming. TI resource explorer of code composer studio is opened and workspace is selected on the popup window as shown in Fig. 10.

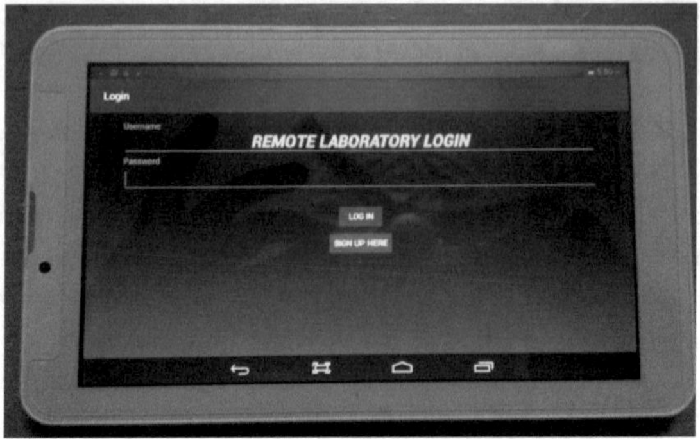

Fig. 9. App login window

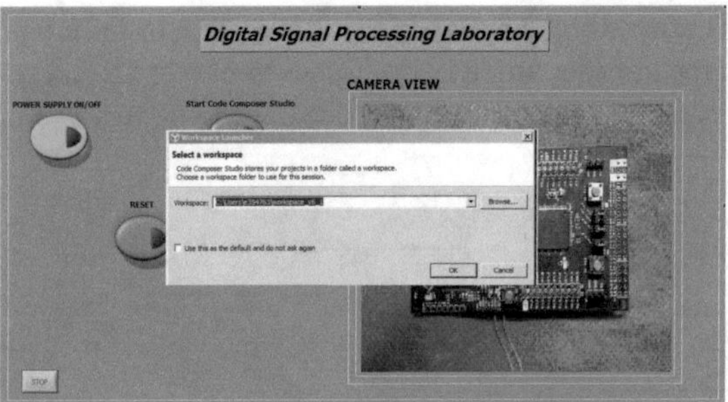

Fig. 10. Workspace selection window

Target is connected by selecting a proper connection and device type on the user menu. New project file is opened and code for blinking LED is written and further built, debugged and run on the target in the source code entry window as shown in Fig. 11. Program functionality is verified by observing the LED blinking on live image capture.

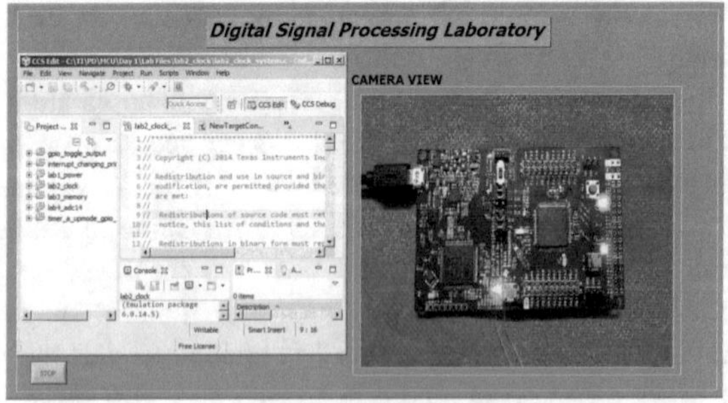

Fig. 11. Source code entry window

8 Conclusion

Design of virtual instrumentation of remote laboratory is presented with the detailed architecture. Laboratory experiments can be performed remotely by logging into the laboratory server machine using LabVIEW user interface. Power ON/OFF and Reset functions on test kit can be controlled remotely using DAQ cards and relays on interface board. This implementation facilitates the remote access of the DSP laboratory hardware and visualizes the real experiments on the user smartphone/tablets.

This implementation avails laboratory facility for students all over the day for entire year. This will increase the productivity of the laboratory setups and measuring instruments. As the laboratory requirements changes with the change in the syllabus, the laboratory developed is capable of adopting new requirements with few minor modifications in the setup. Further this facility can be extended to perform other laboratory experiments and project related activities. This architecture is independent of the operating system being loaded on the remote laboratory machines as we are accessing the complete remote machine instead of specific application. Smartphone/tablet application is developed in Android platform which is most common IOS used in all the mobile phones and tablets. As this mobile application is customized as per the requirement, further this app can be integrated with collaborative learning features and multi user access facilities.

Acknowledgements. We thank the department of Electronics and Communication engineering of Adichunchanagiri Institute of Technology, Chikmagalur, Karnataka for providing the required facilities.

References

1. Kalantzopoulos, A., Karageorgopoulos, D., Zigouris, E.: A LabVIEW based remote DSP laboratory. iJOE **4**(Special Issue 1): REV2008 (July 2008)
2. Gallardo, S., Barrero, F.J., Sergio, L.: Remote Instrumentation Laboratory for Digital Signal Processors Training. Toral University of Seville, Spain
3. Dvir, Z.: Web-based remote digital signal processing (DSP) laboratory using the Integrated Learning Methodology (ILM). In: 2006 International Conference on Information Technology: Research and Education
4. Shelke, S., Date, M., Patkar, S., Velmurugan, R., Rao, P.: A remote lab for real-time digital signal processing. In: Proceedings of the 5th European DSP Education and Research Conference (2012)
5. Hashemian, R., Riddley, J.: FPGA e-Lab, a technique to remote access a laboratory to design and test. In: Life Member IEEE, Northern Illinois University
6. Jagadeesh Chandra, A.P., Venugopal, C.R., Novel design solutions for remote access, acquire and control of laboratory experiments on DC machines. IEEE Trans. Instrum. Measur. **61**(2), 249–357 (Feb 2012)
7. Jagadeesh Chandra, A.P., Sudhaker Samuel, R.D.: Design of novel online access and control interface for remote experiment on DC drives. Int. J. Online Eng. (iJOE) **5**(2), 11–17 (May 2009)

8. Richardson, T., Stafford-Fraser, Q., Wood, K.R., Hopper, A. (1998). Virtual network computing (PDF)
9. "Radmin vs Team viewer" Radmin support center
10. "Radmin Remote Access Software—Key Features for Remote Computer" Control: 25
11. ArunKumar, N.P., Jagadeesh Chandra, A.P.: Development of remote access and control features for digital signal processing laboratory experimentations. Int. J. Online Eng. (IJOE) **12**(8) (2016)
12. Arun Kumar, N.P., Jagadeesh Chandra, A.P.: Internet protocol for Multimedia Communications. Int. J. Netw. Secur. **1**(7) (May 2013)

WebLabLib: New Approach for Creating Remote Laboratories

Pablo Orduña[1,2(✉)], Luis Rodriguez-Gil[1], Ignacio Angulo[2], Unai Hernandez[2], Aitor Villar[1], and Javier Garcia-Zubia[2]

[1] LabsLand, Bilbao, Spain
{pablo,luis,aitor}@labsland.com
[2] DeustoTech, Fundacion Deusto, Bilbao, Spain
{pablo.orduna,ignacio.angulo,unai.hernandez}@deusto.es

Abstract. Remote laboratories are hardware and software tools that enable students to access real equipment through the Internet. Remote Laboratory Management Systems (RLMS) are software tools developed for creating remote laboratories in an easier way, providing some of the transversal features common in most remote labs (such as authentication, authorization, scheduling platforms or administration tools), and some protocols or APIs (Application Programming Interfaces) for creating the laboratories. WebLab-Deusto is a popular open source RLMS used in different universities to create or administer their remote laboratories; and it offers two approaches for developing remote laboratories: managed (where all the communications go through WebLab-Deusto) and unmanaged (where the communications are managed by the remote lab developer). While originally the managed approach had a number of advantages over the unmanaged, nowadays, with web development technologies fastly changing and increasing productivity, it became important to provide a proper support for the unmanaged by creating a completely new framework called weblablib, developed by LabsLand and also Open Source. This article describes this framework, and the different trade-offs that remote lab developers have to deal with when implementing a remote laboratory.

1 Introduction

An Educational Remote Laboratory is a software and hardware solution that enables students to access real equipment located in their institution, as if they were in a hands-on-lab session, using an standard web-browser. The laboratories are typically hosted in universities or research centers.

A key factor of remote laboratories is that once they are available through the Internet their usage can be scaled up and used by students of other institutions. Thus, two or more institutions can share different equipment to reduce costs by requiring less duplicated equipment: it is typically only used in certain hours of the day and in certain days of the year. Furthermore, this empowers a sharing economy where multiple providers provide access to their laboratories to each other, freely or not.

M. E. Auer and K. Ram B. (Eds.): REV2019 2019, LNNS 80, pp. 477–488, 2020.
https://doi.org/10.1007/978-3-030-23162-0_43

In the literature there is a wide range of remote laboratories in many fields (e.g., robotics, electronics, physics, chemistry). Software frameworks have been developed to make the development of remote laboratories more affordable (Remote Laboratory Management Systems such as WebLab-Deusto[1] [19], iLab Shared Architecture,[2] RemLabNet[3] [23] or Labshare Sahara[4] [15]) and tools (e.g., gateway4labs[5] [20]) to provide integrations with educational tools (such as Moodle, Sakai or other LMS, both through ad hoc solutions and through standards such as IMS LTI) or repositories linking remote and virtual laboratories (such as Go-Lab [8,14], LiLa [21] or iLabCentral).

Most technologies are motivated by the idea of leveraging remote laboratories for shared growth: if two universities have 3 remote laboratories each, technically, it would become possible (through sharing them) to have 6 laboratories together. However, there are a set of organizational challenges associated to this that traditionally has made it more difficult, such as reliability or consumer trust. For example, how can the lecturers of one of the two universities know that the laboratories in the other university are going to be maintained for several years? In small environments (e.g., within a funded project or among universities that have extensive collaborations), this is possible. However, at a larger scale, these issues become a problem.

With the goal of scaling the field of remote laboratories, the WebLab-Deusto team created LabsLand[6] as a spin-off company of the University of Deusto, focused on providing the necessary reliability and business background that guarantees trust for the involved partners. There are multiple institutions in several countries who are already either providing resources to build new laboratories or consuming laboratories well supported from the LabsLand network (using different RLMS and not only WebLab-Deusto).

However, for LabsLand to be successful, many remote laboratories must be available in the network covering a wide range of subjects. With this in mind, LabsLand is also developing technologies to create remote laboratories in a faster and more reliable way. While some of these tools are proprietary, weblablib is an Open Source new framework for creating remote laboratories, that internally relies on WebLab-Deusto.

The focus of this article is to describe weblablib. To this end, this article covers the basics of RLMS, and of WebLab-Deusto in particular, and explains what are the trade-offs of the design of WebLab-Deusto before the development of weblablib. Then, the article focuses on weblablib itself, describing its features and what it is good for and what it is not.

---

1 http://weblab.deusto.es.
2 http://ilab.mit.edu.
3 http://www.remlabnet.eu.
4 https://remotelabs.eng.uts.edu.au.
5 http://gateway4labs.readthedocs.org.
6 https://labsland.com.

2 Current Solutions for Sharing Remote Laboratories

This section introduces the concepts of remote laboratories, Remote Laboratory Management Systems (RLMS), remote laboratory federations and portals for sharing remote laboratories.

2.1 Remote Laboratories

A remote laboratory is a software and hardware tool that allows students to remotely access real equipment located in the university. Users access this equipment as if they were in a traditional hands-on-lab session, but through the Internet. To show a clear example, Fig. 1 shows a mobile low cost robot laboratory described in [6]. Students learn to program a Microchip PIC microcontroller, and they write the code at home, compile it with the proper tools, and then submit the binary file to a real robot through the Internet. Then, students can see how the robot performs with their program through the Internet (e.g., if it follows the black line according to the submitted program, etc.) in a real environment.

In this line, there are many examples and classifications in the literature [9,10]. Indeed, remote laboratories were born nearly two decades ago [1,2,13], and since then they have been adopted in multiple fields: chemistry [3,4], physics [5,7], electronics [11,16], robotics, [22,24], acoustics [25], and even nuclear reactor [12].

2.2 Remote Laboratory Management Systems

Every remote laboratory manages at least a subset of the following features: authentication, authorization, scheduling users to ensure exclusive accesses - typically through a queue or calendar-based booking-, user tracking and administration tools. These features are common to most remote laboratories, and are

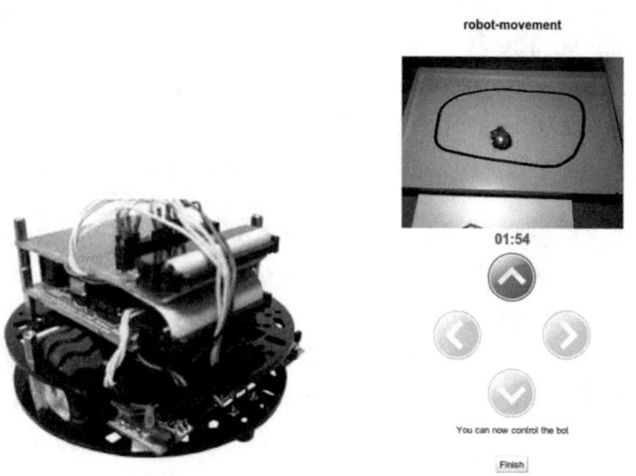

Fig. 1. Robot laboratory [6]. At the left, the mobile robot itself. At the right, the user interface once the program has been submitted

actually independent of the particular remote laboratory settings. For example, an authentication and queuing system is valid both for an electronics laboratory and for a chemistry laboratory.

For this reason, Remote Laboratory Management Systems (RLMSs) arose. These systems (e.g., MIT iLabs,[7] WebLab-Deusto[8] or Labshare Sahara[9]) provide development toolkits for developing new remote laboratories, as well as management tools and common services (authentication, authorization, scheduling mechanisms). The key idea is that by adding a feature to a RLMS (e.g., supporting LDAP, a Learning Analytics panels [18] or similar cross-laboratory features), all the laboratories which are managed with that RLMS will support this feature automatically.

2.3 Federating Remote Laboratories

As previously stated in the introduction, a key factor of remote laboratories is that once the laboratory is available on the Internet, it can also be shared with other institutions.

To do this, there are three general approaches:

- Leave the laboratories completely open, so whoever wants to use them can use them. This may reduce the chances of providing proper Learning Analytics or supporting proper accountability mechanisms, in addition to avoiding priorities among students coming from different institutions, leading to a tradeoff between accessibility and advanced features [20].
- Share accounts between the different RLMS: if *University A* want to use laboratories of *University B*, then someone in *University A* will provide a list of usernames to *University B* and students will go to this institution using credentials in *University B*. Ideally, some federated authentication could be used to avoid providing credentials in different domains (such as Shibboleth, OAuth or similar), but it is not typically the case.
- Federate laboratories: if a RLMS supports federation, then if installed in two different institutions (e.g., *University A* and *University B*), students of *University A* will go to the RLMS of *University A* and they will transparently use laboratories in *University B*, working in a institution-to-institution basis (so *University B* does not need to know the list of students of *University A* and simply rely on an existing agreement with that university).

From the items in this list, the most advanced mechanism is the federation of remote laboratories through proper protocols oriented to market-like situations. These federation protocols have been used for fostering interoperability between RLMS [19]. These interoperable bridges between different systems can be enhanced if properties such as transitivity or federated load balance are provided [17].

[7] http://ilab.mit.edu.
[8] http://weblab.deusto.es.
[9] http://github.com/saharalabs.

3 WebLab-Deusto Software

WebLab-Deusto is an Open Source Remote Laboratory Management System that allows remote laboratory developers to develop new laboratories without dealing with most of the pains. In particular, it provides:

- An advanced queueing mechanism that supports both local and federated load balancing (e.g., if there are two copies of the same laboratory, the queue will split among the different copies automatically).
- Administration tools and dashboards for learning analytics. Administrators and teachers can see who used what and what they did in the lab.
- Sharing laboratories with other WebLab-Deusto systems in other institutions without sharing credentials or duplicating users.
- Multiple authentication mechanisms. Also, through gateway4labs, support for LTI so it can be used from Learning Management Systems.
- Native support in the LabsLand network and in the Smart Gateway of the Go-Lab project [20].

WebLab-Deusto has been designed with the following notions:

- Remote laboratory developers have very different backgrounds and skills. Some are more comfortable with one technology while others with other. For this reason, WebLab-Deusto must support multiple technologies for its adoption (instead of a single technology).
- Some remote laboratory developers might not have heavy IT skills and should not need to deal with them. It is better if WebLab-Deusto provides some abstraction level for them. However, other remote laboratory developers might have these skills and might be willing to work with the latest technologies without limitations.

For these reasons, WebLab-Deusto was designed with two development approaches in the same system. The first approach, the *managed approach*, enables remote laboratory developers to avoid dealing with networks or security systems in detail. It provides a basic model where some code on the client side (e.g., in JavaScript) send commands or messages through an API (without dealing with any network), and WebLab guarantees that those commands or messages are sent all the way to the final experiment, which will be running other code with a simple API. The developer does not need to know how many instances of this remote laboratory there are, where are they, how is https configured, etc.: they just send messages and receive messages. This protocol was implemented in many programming languages, including Python, C, C++, LabVIEW (code), .NET, Java and Node.js. So remote laboratory developers familiar with any of these systems could start implementing a remote laboratory.

The second approach, the *unmanaged approach*, provided an HTTP interface for creating remote laboratories. WebLab-Deusto would call this interface, and the remote laboratory developer would need to manage all the deployment, addressing, security, etc. In general, for a simple laboratory, it would require

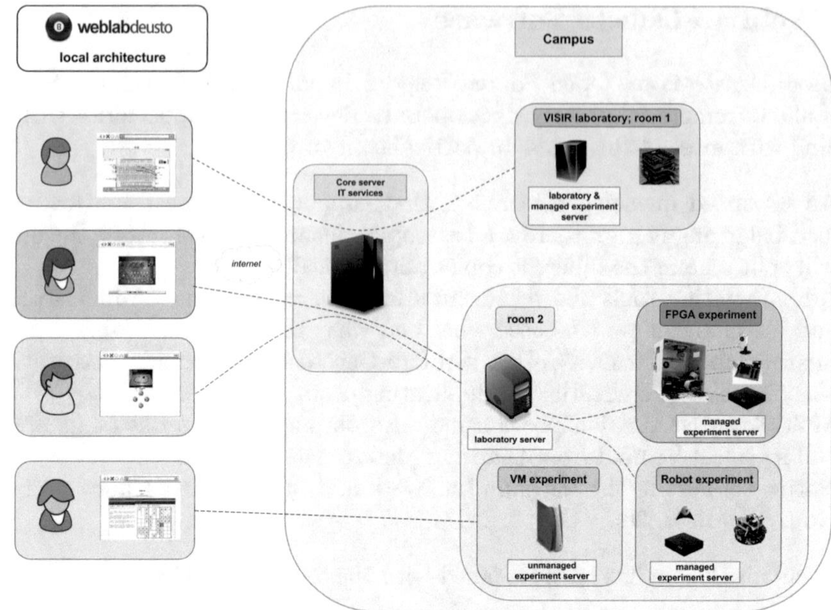

Fig. 2. WebLab-Deusto architecture

more work to use the *unmanaged approach*. However, the *unmanaged approach* provided advanced developers a customizable interface that they could extend in any web framework and use any web technology; and therefore it was intended to be used by more advanced developers.

In Fig. 2, the basic local architecture of WebLab-Deusto is displayed (local of a single university; as compared with the federated architecture that explains how laboratories can be shared). Different users will go through the same WebLab-Deusto interface to access different laboratories, that can be deployed in different locations of the university. This allows remote lab developers to have one or multiple WebLab-Deusto "core servers" on the campus (audited by IT services) and different remote laboratories that can be using unexpensive technologies (such as Raspberry Pi) in different locations of the university. In the managed laboratories, all the commands/messages are sent through these "core servers" and "laboratory servers", while in the case of the unmanaged laboratories, the reservation process is done through WebLab-Deusto, but then it forwards the user to the final laboratory directly.

4 WebLabLib

As explained in the previous section, WebLab-Deusto supports the *managed approach* and the *unmanaged approach*. Libraries for different languages were provided for the *managed approach*; however, this cannot be done in the *unmanaged approach* since the variety of potential web frameworks that a developer can choose are too many and are going to change too often. For this reason, the HTTP interface is public (so developers of other frameworks can implement it; and few small examples are provided for PHP and Python), but a full implementation with support for many libraries have been implemented and called weblablib. This full implementation -weblablib- relies on a concrete popular Python framework called Flask, and its ecosystem (database, networking, authentication mechanism, etc.). This way, weblablib is used for most new WebLab-Deusto laboratories, and it is oriented to make the development process faster.

4.1 Features

weblablib provides the following features:

- It is a pure Flask plug-in. Flask is a popular Open Source Python web microframework, highly extensible and easy to learn. weblablib integrates very easily and interacts with the rest of the components in an easy way. Developers can rely on the Flask ecosystem documentation to see details on how to deploy the system in production.
- Simplified model: the remote lab developer does not need to deal with credentials: the user is already authenticated in WebLab-Deusto (or Moodle or other Learning Management System through LabsLand or gateway4labs), and WebLab-Deusto tells weblablib the relevant user data. The same applies to scheduling or authorization: in production, WebLab-Deusto manages the groups, users or who has access to what copy of what laboratory. weblablib is only called whenever the user is valid.
- User information: weblablib integrates in Flask-Login so that whenever a user uses the laboratory, the application has easy access to the name of the user and a global unique identifier. It also integrates with Flask-SQLAlchemy to be able to store data in a local SQL database.
- WebSockets: weblablib integrates natively Flask-SocketIO, therefore enabling the remote lab developer to implement WebSockets in a secure way (relying on the authentication mechanism, etc.). WebSockets are a HTML5 modern protocol that enables the server to asynchronously push information to the client, which is very relevant in the remote laboratory context.
- Concurrency and task management: the system enables an easy interface for launching background tasks (such as programming a device) that have access to the rest of the features (user identification, etc.) but can be processed in batch.
- Internationalization (i18n): the system already understand what language (English, Spanish ...) WebLab-Deusto requests and supports on Flask-Babel to manage the internationalization of the remote lab.

The framework also supports its own debugging system, and developers do not need to start or configure a WebLab-Deusto system for development. This way, the development process is lightweight and fast, and when the laboratory is ready, it can be configured in a WebLab-Deusto in production.

```
from flask import Flask, url_for
from weblablib import WebLab, weblab_user, requires_active

app = Flask(__name__)

weblab = WebLab(app)

@weblab.on_start
def on_start(client_data, server_data):
    # ...
    print("Starting user")

@weblab.on_dispose
def on_dispose():
    # ...
    print("Ending user")

@weblab.initial_url
def initial_url():
    return url_for('index')

@app.route('/')
@requires_active
def index():
    return "Hello, {}".format(weblab_user.username)
```

Fig. 3. Weblablib sample lab

4.2 Examples

The full documentation[10] provides examples on how to use weblablib. Figure 3 contains a very simple code using weblablib. In this example code, the developer establishes what functions should be called when the user session starts (where tasks for preparing the session should be run) and when the user session ends (e.g., time finishes for the user and it needs to clean resources -such as stopping motors if running- or make automated checks so the next user will be in the initial state again). It also shows how it is possible to use standard Flask web methods (*@app.route('/')*), but supporting authentication mechanisms (*@requires_active*) that makes that if someone enters in that website it will fail unless the student has been redirected from WebLab-Deusto and WebLab-Deusto using the unmanaged HTTP protocol has sent information about the student. Also, there are global variables such as *@weblab_user.username* that provides data about the current user each time it is called in a simple way.

In Fig. 4, it is shown how tasks can be defined. By adding *@weblab.task* it becomes possible to define that a function can be called either as usual or in background, returning an object that defines if the task is still running, what is its identifier, status, or be able to wait for it if necessary. By default, if a task

[10] https://docs.labsland.com/weblablib.

```
@weblab.task()
def program_device(contents):
    """ Programs a device. Typically takes 5-10 seconds """
    if weblab_user.time_left < 10:
        raise Exception("Error: user does not have "
                        "enough time to run it!")

    arduino.program("my_file.bin") # In this case
    return len(contents)
```

```
# This other code runs it in a different
# process
task = program_device.delay(code)

# The following is a string that you can store in
# Flask session or in weblab_user.data
task.task_id

# a string 'submitted', 'running' or 'failed'/'done' if finished.
task.status

task.submitted  # bool: not yet started by a worker
task.running    # bool: started by a worker, not yet finished
task.done       # bool: finished successfully
task.failed     # bool: finished with error
task.finished   # task.failed or task.done

# These two attributes are None while 'submitted' or 'running'
task.result # the result of the function
task.error # the exception data, if an exception happened

# Join operations
task.join(timeout=5, error_on_timeout=False) # wait 5 seconds
```

Fig. 4. Weblablib tasks usage

is long, weblablib will wait for it to finish during the clean resources period. The example is very simple so it does not show the details about how it also supports communication with the task, so it is possible to request the task to stop, or exchange data about its status, and how inside the task, the developer has access to all the information about the current user (e.g., username, full name, language, etc.).

In Fig. 5, it is displayed how WebSockets can be used with weblablib. It shows that there are certain methods such as *@socket_requires_active* which have been adapted to guarantee that the WebSocket can only be opened by valid users with a valid scheduling slot; both on connection and when emitting any data. This is important so once the user session is over, weblablib already provides a security mechanism by default to avoid the client to push information to the hardware.

The code in Fig. 6 shows how the system also supports a native integration with databases using the Flask-SQLAlchemy library. The global *weblab_user* object supports the concept of calling its *user* object to obtain internally access to an object retrieved from the database, that could include information such as some folder with certain files or so.

```
from weblablib import socket_requires_active

@socketio.on('connect', namespace='/mylab')
@socket_requires_active
def connect_handler():
    emit('board-status', hardware_status(), namespace='/mylab')

@socketio.on('lights', namespace='/mylab')
@socket_requires_active
def lights_event(data):
    switch_light(data['number'] - 1, data['state'])
    emit('board-status', hardware_status(), namespace='/mylab')
```

Fig. 5. Weblablib WebSocket usage

```
# Using Flask-SQLAlchemy ( http://flask-sqlalchemy.pocoo.org/ )
from .models import LabUser

@weblab.user_loader
def load_user(username_unique):
    return LabUser.query.filter_by(username_unique=username_unique)

@app.route('/files')
@requires_active
def files():
    user_folder = weblab_user.user.folder
    return jsonify(files=os.listdir(user_folder))
```

Fig. 6. Weblablib database usage

The complete structure of this API is available in the weblablib documentation.

5 Conclusions and Future Work

Developing a remote laboratory is a complicated task, and it requires an important effort on the programming side. WebLab-Deusto is focused on delivering different approaches for different profiles of remote laboratory developers. One of them is the managed approach, with multiple APIs for different programming languages. The other is the unmanaged approach, that is more complicated but better for experienced developers. This one supports an HTTP interface, but so as to provide an easier to adopt toolkit, it provides weblablib as an open source framework for a particular technology (Flask for Python and its ecosystem), making the effort in that framework considerably simpler.

References

1. Aktan, B., Bohus, C., Crowl, L., Shor, M.: Distance learning applied to control engineering laboratories. IEEE Trans. Educ. **39**(3), 320–326 (1996)
2. Carisa, B., Burain, A., Molly H.S., Lawrence, C.: Running control engineering experiments over the internet (1995)

3. Cedazo, R., Sanchez, F., Sebastian, J., Martínez, A., Pinazo, A., Barros, B., Read, T.: Ciclope chemical: a remote laboratory to control a spectrograph. Adv. Control Educ. ACE **6** (2006)
4. Coble, A., Smallbone, A., Bhave, A., Watson, R., Braumann, A., Kraft, M.: Delivering authentic experiences for engineering students and professionals through e-labs. In: Education Engineering (EDUCON), 2010 IEEE, pp. 1085–1090. IEEE (2010)
5. Del Alamo, J., Brooks, L., McLean, C., Hardison, J., Mishuris, G., Chang, V., Hui, L.: The mit microelectronics weblab: a web-enabled remote laboratory for microelectronic device characterization. In: World Congress on Networked Learning in a Global Environment, Berlin, Germany (2002)
6. Dziabenko, O., García-Zubia, J., Angulo, I.: Time to play with a microcontroller managed mobile bot. In: Global Engineering Education Conference (EDUCON), 2012 IEEE, pp. 1–5. IEEE (2012)
7. Gillet, D., Latchman, H., Salzmann, C., Crisalle, O.: Hands-on laboratory experiments in flexible and distance learning. J. Eng. Educ. **90**(2), 187–191 (2001)
8. Gillet, D., de Jong, T., Sotirou, S., Salzmann, C.: Personalised learning spaces and federated online labs for stem education at school. In: Global Engineering Education Conference (EDUCON), 2013 IEEE, pp. 769–773. IEEE (2013)
9. Gomes, L., Bogosyan, S.: Current trends in remote laboratories. IEEE Trans. Ind. Electron. **56**(12), 4744–4756 (2009)
10. Gravier, C., Fayolle, J., Bayard, B., Ates, M., Lardon, J.: State of the art about remote laboratories paradigms-foundations of ongoing mutations. iJOE **4**(1) (2008)
11. Gustavsson, I., Zackrisson, J., Håkansson, L., Claesson, I., Lagö, T.: The visir project—an open source software initiative for distributed online laboratories. In: Proceedings of the REV 2007 Conference, Porto, Portugal (2007)
12. Hardison, J., DeLong, K., Bailey, P., Harward, V.: Deploying interactive remote labs using the ilab shared architecture. In: Frontiers in Education Conference, 2008. FIE 2008. 38th Annual, pp. S2A–1. IEEE (2008)
13. Henry, J.: Running laboratory experiments via the world wide web. In: ASEE Annual Conference (1996)
14. de Jong, T., Linn, M.C., Zacharia, Z.C.: Physical and virtual laboratories in science and engineering education. Science **340**(6130), 305–308 (2013)
15. Lowe, D., Machet, T., Kostulski, T.: Uts remote labs, labshare, and the sahara architecture. Using Remote Labs in Education: Two Little Ducks in Remote Experimentation, p. 403 (2012)
16. Nedic, Z., Machotka, J., Nafalski, A.: Remote laboratory netlab for effective interaction with real equipment over the internet. In: 2008 Conference on Human System Interactions, pp. 846–851. IEEE (2008)
17. Orduña, P.: Transitive and scalable federation model for remote laboratories. Ph.D. thesis, Universidad de Deusto, Bilbao, Spain (May 2013). https://morelab.deusto.es/people/members/pablo-orduna/phd_dissertation/
18. Orduña, P., Almeida, A., Ros, S., Lpez-de Ipiña, D., García-Zubia, J.: Leveraging non-explicit social communities for learning analytics in mobile remote laboratories. J. Univ. Comput. Sci. **20**(15), 2043–2053 (2014)
19. Orduña, P., Bailey, P., DeLong, K., López-de-Ipiña, D., García-Zubia, J.: Towards federated interoperable bridges for sharing educational remote laboratories. Comput. Hum. Behav. **30**, 389–395 (2014), http://www.sciencedirect.com/science/article/pii/S0747563213001416
20. Orduña, P., Garbi Zutin, D., Govaerts, S., Lequerica Zorrozua, I., Bailey, P.H., Sancristobal, E., Salzmann, C., Rodriguez-Gil, L., DeLong, K., Gillet, D., et al.:

An extensible architecture for the integration of remote and virtual laboratories in public learning tools. IEEE Rev. Iberoamericana Tecnologias Aprendizaje **10**(4), 223–233 (2015)

21. Richter, T., Boehringer, D., Jeschke, S.: Lila: A European project on networked experiments. In: Automation, Communication and Cybernetics in Science and Engineering 2009/2010, pp. 307–317 (2011)
22. Safaric, R., Truntič, M., Hercog, D., Pačnik, G.: Control and robotics remote laboratory for engineering education. Int. J. Online Eng. (iJOE) **1**(1) (2005)
23. Schauer, F., Krbecek, M., Beno, P., Gerza, M., Palka, L., Spilakov, P., Tkac, L.: Remlabnet iii—federated remote laboratory management system for university and secondary schools. In: 2016 13th International Conference on Remote Engineering and Virtual Instrumentation (REV), pp. 238–241. IEEE (2016)
24. Torres, F., Candelas, F., Puente, S., Pomares, J., Gil, P., Ortiz, F.: Experiences with virtual environment and remote laboratory for teaching and learning robotics at the university of alicante. Int. J. Eng. Educ. **22**(4), 766–776 (2006)
25. Zappatore, M., Longo, A., Bochicchio, M.A.: Enabling MOOL in acoustics by mobile crowd-sensing paradigm. In: 2016 IEEE Global Engineering Education Conference (EDUCON), pp. 733–740. IEEE (2016)

Self Optimizing Drip Irrigation System Using Data Acquisition and Virtual Instrumentation to Enhance the Usage of Irrigation Water

R. Raj Kumar, K. Sriram, and I. Surya Narayanan(✉)

Department of EIE, Kongu Engineering College, Erode, India
{rajkumarpublications, sriramkl46, Suryanarayanan40453}@gmail.com

Abstract. Water scarcity is one of the major issues that make the farmers to spend more for the efficient irrigation for their farmlands. The government needs to pay compensation cost in millions to the farmers during drought season. Utmost care should be taken for the effective utilization of the available water source in present situation. The effective way of irrigation that had been introduced in near decade is the Drip Irrigation System (DIS). The wastage of water to irrigate crops and plants in the agricultural fields are reduced to 60–70% at the maximum but, there is approximately 30–40% of water being not properly utilized. Further, an idea is needed to maximize the utilization and to reduce the wastage of water in drip irrigation. The Actual water flow through drip irrigation is 4 gph. The proposed idea consists of Earth Electrodes, NI-DAQ 6009, Relay Module, Solenoid valve and aims to reduce the water wastage which is approximately 2 gph while irrigating plants in existing drip irrigation system. So, the optimization made in this proposal idea will help the farmers in water conservation and effective utilization of water.

Keywords: Drip irrigation · Earth resistance · NI-DAQ 6009 · Water conservation

1 Introduction

The Infrastructure development to conserve the water resources has been a common agenda in many developing countries like India. India has both arid and semiarid tropical condition. The Water demand will increase 50% in 2025 in India according to the survey of International Water Management Institute (IWMI) done recently [1]. The average consumption of water per citizen in India is 100–150 litres per day [2]. This 'high demand low supply' state of water has to be considered seriously and effective water conservation techniques are needed to meet the water demand. The one among the major water consumptions is the irrigation of crops in agricultural land by farmers in India. The rising global challenge of water includes Irrigation for crops. The farmers need to spend more money to irrigate their crops despite of water demand [3]. The effective Irrigation Method introduced in the recent decade is Drip Irrigation System

M. E. Auer and K. Ram B. (Eds.): REV2019 2019, LNNS 80, pp. 489–495, 2020.
https://doi.org/10.1007/978-3-030-23162-0_44

(DIS) [4]. The DIS produces 4gph water flow to the Agricultural Fields [5]. The Water flow produced by the DIS is not utilized at maximum [6]. Based on the literature survey, DIS has a drawback that it is not utilizing the water for irrigation. The optimization is required in the existing DIS. Eventhough the DIS is more effective than other irrigation techniques, there is still a part of water is underutilized and therefore, water conservation is affected. As a result, Government has to pay compensation in millions at the time of water scarcity to farmers [7]. The other water Irrigation system which are practiced before the DIS and introduced after the DIS are not proved as better practices [8, 9]. The observations from *existing irrigation systems* are: (i) Reduction in wastage of water, (ii) Installation should be easy and (iii) proper water distribution. The *outcomes of the proposed self optimized drip irrigation system* are: (i) Reduction in wastage of water, (ii) Enhanced water utilization and (iii) optimized efficient irrigation system. Water conservation is agricultural field and the irrigation systems are needed to be upgraded.

2 System Description

The major functional blocks of the proposed system is shown in Fig. 1 which includes 6 V DC supply, earth electrodes, NI DAQ card 6009, a processing system, flow control valve setup and relay circuit. The voltage supply is given to the earth electrodes and the terminals of the electrodes acts as the terminals of a resistor (earth resistance). Therefore, when the voltage is given, there will be a voltage drop across electrodes due to the earth resistance. The voltage drop obtained is fed into the signal conditioning and controller circuits and according to the changes in the voltage drop due to the change in the state of wetness in the soil the flow valve control will be actuated by the control pulse generated through the LabVIEW powered NI-DAQ card 6009. According to the control signal the flow of water drops to the plants through the drip tube will be controlled by controlling the valve that lets the water to flow through the drip tube.

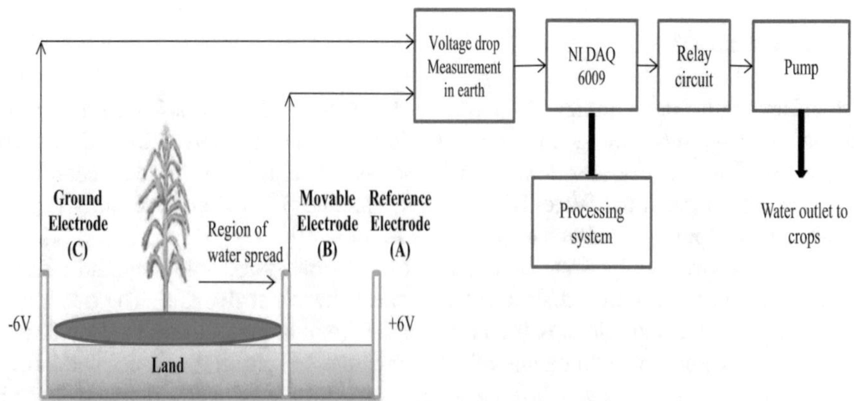

Fig. 1. Block diagram of proposed system

3 Design and Methodology of the Prototype

The prototype is designed as shown in Fig. 2. The 6 V DC supply is given to the Electrodes A and C. The electrodes are placed in such a manner that the crop or pit is located Electrode B1 and C as shown in Fig. 2. The movable electrode is placed at a particular distance from the specific plant at the maximum absorbance point of the crop. Every crop variety has different maximum water absorbance point (e.g. Turmeric plant 15 cm).The absorbance point can be calculated by field surveying the particular crop used for drip irrigation. The distance to place the movable electrode (B1) for different crops is found, based on the water absorbance point of the particular crop or pit. The voltage drop (D) is observed between the B1 and C in both wet and dry conditions of the soil. It is inferred that the soil exhibits conductivity properties. The conductivity varies according to the type of the soil and the various soil types contain different ion contents naturally. This prototype is tested in the Red soil and the voltage drop (D) readings are tabulated in Table 1 shows the conductivity in soil at wet and dry conditions. The Conductivity of the soil is decreased due to the resistance created by the water when soil is irrigated. The decrease in conductivity happens due to the resistance created for flow of ions in the soil when soil gets wet. This phenomenon is sensed in terms of change in voltage drop (D). To achieve our required outcome, conductivity phenomenon is used for controlling action.

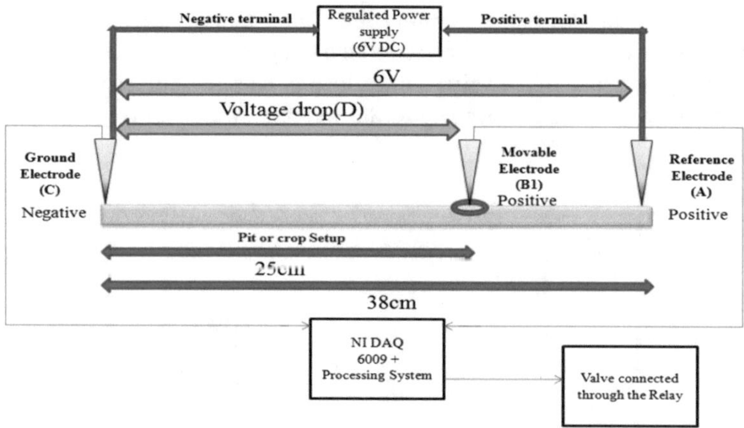

Fig. 2. Layout of the prototype

Table 1. Voltage drop in dry and wet soil

Distance between the reference and ground electrode	38 cm	
Distance between the movable and ground electrode	25 cm	
Supply given to the reference and ground electrode	6 V	
Voltage drop between the movable and ground electrode in normal (dry) condition	2.78 V	Difference in voltage drops = 0.53 V
Voltage drop between the movable and ground electrode in wet condition	2.25 V	

4 Signal Conditioning Circuit

In the signal conditioning process, the NI DAQ card 6009 is used to process the signals acquired from the Earth Electrodes which is powered by the NI LabVIEW software. The powered NI DAQ card 6009 is programmed to have one accessible analog input terminal which takes the voltage drop from the electrodes B1 and C in real time and one accessible digital output terminal which handles the state of the solenoid valve for the flow of water. As per Fig. 3 programming is done in such a way that until the acquired analog input is greater than the value 2.25 V (as per the observations in Table 1), the digital output from the DAQ card should be high (5 V) and when it drops below 2.25 V, the digital output from the DAQ card should go low (0 V). The real time function named 'DAQ Assist' is used in the programming for both acquiring the analog input from the electrodes and generating the digital output [10]. The digital output terminal from the DAQ card is connected to the relay and the solenoid valve circuitry. This will ensure the state of the valve (either open or closed) with respect to the voltage drop (D) in the electrodes accordingly to the wetness of the soil (Fig. 4).

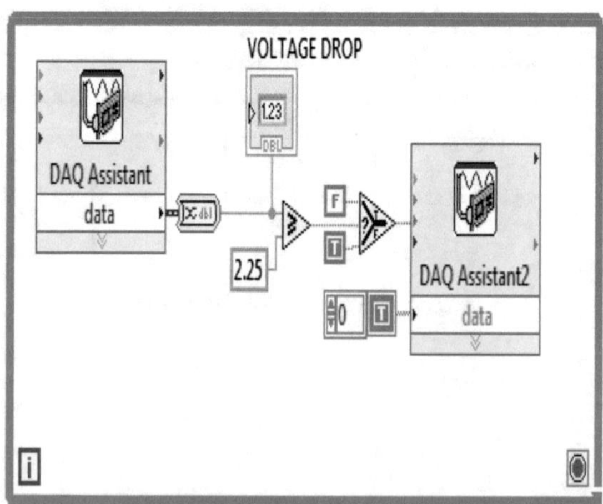

Fig. 3. LabVIEW block diagram

Fig. 4. Real time prototype model

5 Results and Discussion

It is inferred from Table 1 that the voltage drop (D) occurred during the change in state of the soil wetness can be used to control the flow valve of the Drip Irrigation System in the Agricultural Fields. The outlet flow rate of the drip tube is absorbed as 4 gph the valve shuts down when the water spreading reaches to maximum water absorbance point and the Irrigation is controlled. All these occurred within the time period of 90–120 min which shorter than the time period practiced by the farmers (150–180 min). As per the field survey conducted for the turmeric plant, it is observed that the maximum absorbance point turmeric root is 15 cm and any quantity of water irrigated beyond this distance will not be absorbed by it. This fact can be explained in Fig. 5. The analysis of the obtained results on the quantity of water and time saved is described in Fig. 6. It is observed from Fig. 6 that the actual time taken for the optimized irrigation is 90–120 min where as in the conventional irrigation practice it is 150–180 min. Therefore, the implementation of the proposed idea in the agricultural fields results in the conservation of 4 to 8 gallons of water per irrigation.

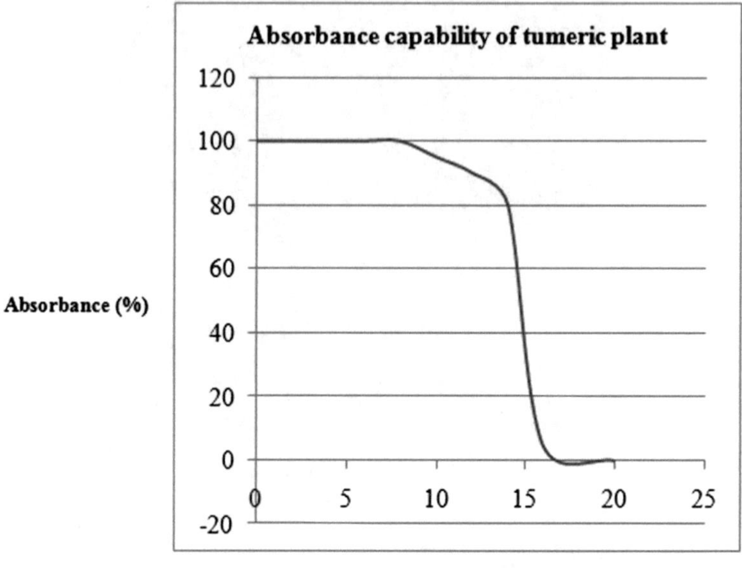

Fig. 5. Absorbance capability of turmeric plant

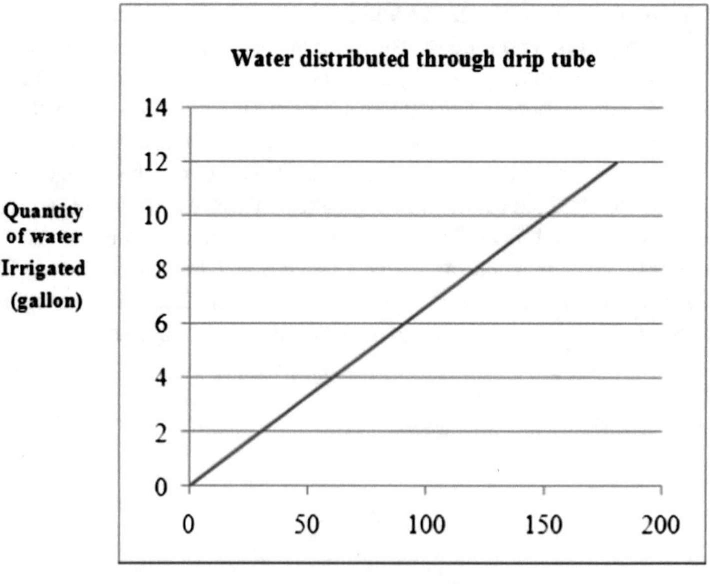

Fig. 6. Water distribution from drip tube to soil

6 Conclusion

To entrust the awareness about the water wastage in the drip irrigation among the farmers of this generation and the future generations is the sole purpose of this whole attempt. As from all that has been discussed in this real time proposed system, it is inferred that with a controller and the change in voltage from electrodes, the wastage of water in the drip irrigation system being reduced completely. In the upcoming days, the essentiality of water will grow humongous and every drop of water matters. Hence this prototype ensures that the water wastage will be drastically abridged and there is a high possibility that the unutilized 30–40% of irrigation water get retained completely.

References

1. Giordano, M., Turral, H., Scheierling, S.M., Tréguer, D.O., McCornick, P.G.: Beyond More Crop per Drop: Evolving Thinking on Agricultural Water Productivity. IWMI Research Report-169 (2016)
2. Viswanath, S.: How Much Water an Urban Need? The Hindu-Feb 15 (2013)
3. International Water Management Institute (IWMI). IWMI Annual Report (2017)
4. Suresh Kumar, D., Palanisamy, K.: Impact of Drip Irrigation system on Farming system: Evidence from South India. Management, Performance and Applications of Micro Irrigation Systems (Chap. 2) (2016)
5. Tamil Nadu Agricultural University. Drip irrigation methodology and statistical data on water consumption. TNAU http://www.tnau.ac.in/agritechportal
6. Information gathered from interviewing Mr. Iyappan N., a farmer in Mulanur and his subordinate farmers about “Drip Irrigation”
7. Tamil Nadu State Government, “TNAU drought survey 2016”, TNAU portal
8. Karagiannis, G., Tzouvelekas, V., Xepapadeas, A.: Measuring irrigation water efficiency with a stochastic production. Front. Environ. Resource Econ. **26**(1), 57–72 (2003)
9. Dawalibi, F.P., Ma, J., Southey, R.D.: Behaviour of grounding systems in multilayer soils: a parameter analysis. IEEE Trans. Power Delivery **9**(1), 334–342 (1994)
10. User Manual for DAQ 6008/6009 Series, 2015 issued by National Instruments

Development of a Remote Compression Test Lab for Engineering Education

Alessandro Selvaggio(✉), Siddharth Upadhya, Joshua Grodotzki, and A. Erman Tekkaya

Institute of Forming Technology and Lightweight Construction (IUL), TU Dortmund University, Dortmund, Germany
{alessandro.selvaggio, siddharth.upadhya, joshua.grodotzki, erman.tekkaya}@iul.tu-dortmund.de

Abstract. A fully automated compression test for material characterization in forming technology is presented. This formability test is performed in a tele-operative testing cell consisting of a universal testing machine Zwick Roell Z250, an industrial robot KUKA KR 30-3, and other necessary components for the automation and execution of experiments. First, a methodology is introduced explaining how the remote compression test is realized. Afterwards, the integration of the compression test into the existing tele-operative testing cell is presented. The practical application of theoretical concepts is realized through the analysis of the final results of the experimental data.

Keywords: Compression test · Robot control · Remote laboratory · Automation · Forming technology · Specimen handling

1 Introduction

In the field of forming technology, the mechanical properties of materials can be quantified by several material characterization processes. Common tests in this field are tensile tests, compression tests, cupping tests, and FLC tests. These experimental tests are also important for students of mechanical engineering since they will be able to connect theoretical models and reality much better by understanding material properties and the influence of test parameters through their own experiences in lab sessions. Generally, such experiments require highly specialized testing machines and equipment. Some of the main issues in carrying out meaningful experiments are the safety conditions and the knowledge about the use of these machines. Due to time personal constraints, the students may not have any or just restricted access to free experimentation through on actual hands-on laboratory. Under such circumstances, remote laboratories have a great potential for teaching [2–4]. In order to provide unhindered access to such kind of experiments, a remote laboratory was implemented at the Institute of Forming Technology and Lightweight Construction at TU Dortmund University [5]. For this laboratory, the iLab Shared Architecture (ISA) [6] is used. The laboratory allows the operation of real testing machines through the Internet, performing a test, independent of time and location, and observing the experiment as well as its results in real time with the help of cameras. Besides, the tests are fully automated

M. E. Auer and K. Ram B. (Eds.): REV2019 2019, LNNS 80, pp. 496–505, 2020.
https://doi.org/10.1007/978-3-030-23162-0_45

with the need for zero human intervention which directly leads to high safety conditions. The students can, hence, focus on the observation and the experimental outcomes rather than the carrying out of the lab as such.

2 Tele-operative Testing Cell

With the realized tele-operative testing cell for material characterization, fundamental common experiments (tensile, compression, cupping, and FLC tests) are automated and provided to students and teachers over the internet. For this purpose, several material testing machines and other necessary components were used.

In Fig. 1, the relevant components are presented:

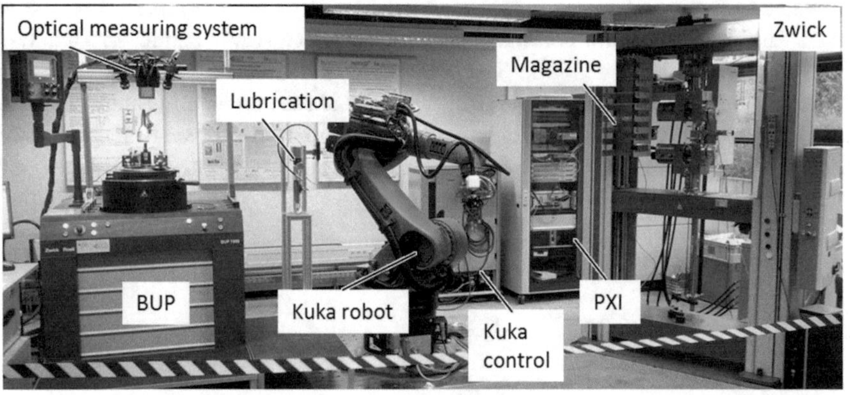

Fig. 1. Remote laboratory for tensile, compression and cupping tests

- Universal testing machine Zwick Z 250 to conduct tensile and compression tests with a force of up to 250 kN
- Sheet metal testing machine BUP 1000 for Cupping and FLC tests
- GOM ARAMIS 4M as an optical measuring system to measure the strain of the specimen during a sheet metal test
- An industrial robot with six axis of freedom for the handling
- Control unit for the KUKA Robot KR 30-3 and PXI-platform for real-time automation functions
- Multiple cameras in order to provide a comprehensive view of the experimental field to the user
- Magazines to store the specimens for the tests.

The robot allows the handling of the specimens for the Zwick Z250 as well as for the BUP1000. Due to the fact that there are several kinds of specimens with differing geometries, different grippers are needed. These grippers are, on the one hand, typical two finger parallel grippers, and on the other hand, pneumatic suction grippers. The two finger grippers are used to deal with the specimens for the tensile or the compression test. A second type of gripper was designed with a pneumatic suction system. This gripper is needed for the handling of specimens for the sheet metal testing machine. With a gripper change system, which itself is also based on pneumatic suction, it is possible to switch between the different grippers in the automatic mode of the tele-operative testing cell. Furthermore, an additional system is installed between the quick-change system and the mounting flange to protect the robot from collisions and overloads. If this system detects a collision, the automated movement is stopped. For the safety of the others working in the lab, Laser sensors are installed along the perimeter of the testing zone which can stop the movements of the system instantaneously in case of any trespassing.

The user can communicate with the tele-operated testing cell via internet. For this he can use his normal computer, since there are no special requirements for the hardware. Communication takes places in the first stage via a homepage on the web server. The website in turn communicates with a proxy server connecting the World Wide Web with a local network. The proxy server is used to prevent a direct communication between user and testing cell, whereby abuse can be prevented. In the next stage the proxy server communicates with the PXI-platform for real-time automation which is directly connected to the testing machines and the robot. An overview of the communication structure is shown in Fig. 2.

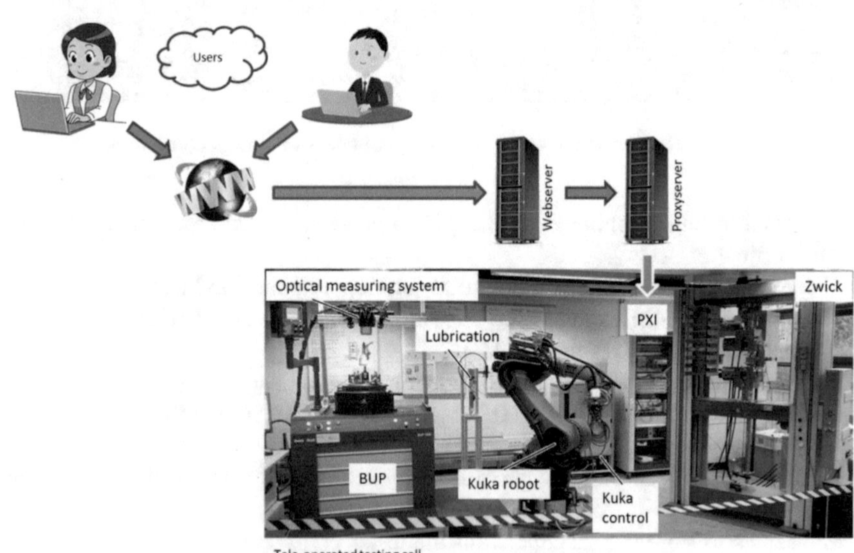

Fig. 2. Overview of the communication structure

3 Tele-operated Compression Test

3.1 Theoretical Background

Compression tests, as well as tensile and torsion tests, are different methods to determine flow curves in order to study the behavior of any material during plastic deformation. According to DIN 50106 [7], a compression test represents the deformation of a metallic specimen under room temperature under conditions of a homogeneous and uniaxial stress state. A constant loading with a compressive force is performed till the failure of the material or till a certain strain value is reached (at least 50% of the initial height h_0 should be compressed). During the experiment, the applied force and the change in height of the specimen are measured. This data is used to plot the force-displacement curve. Afterwards, the engineering stress-strain curve or the true stress-strain curve can be calculated. When true stress and true strain are calculated, it becomes possible to determine the flow curve in order to evaluate characteristics of metals such as formability or hardening effects. Flow curves can be determined as a function of true strain by plotting the true stress-true plastic strain diagram [8].

During compression of a sample, its height decreases while the diameter uniformly increases. In reality there is a difference of the diameter along the height of the sample which is known as barreling. Friction between the contact surface of the specimen and the compressive plates of universal testing machine leads to this inhomogeneous transverse deformation. In order to minimize the effect of barreling, proper lubrication of specimen's end-surfaces before the test is required.

A second defect that can occur is the so-called buckling, which can occur using high specimens. Buckling can be reduced by using cylindrical samples where the ratio of specimen height h_0 to diameter d_0 fulfills Eq. 1.

$$1 \leq \frac{h_0}{d_0} \leq 2 \tag{1}$$

For illustration, Fig. 3 compares both types of errors.

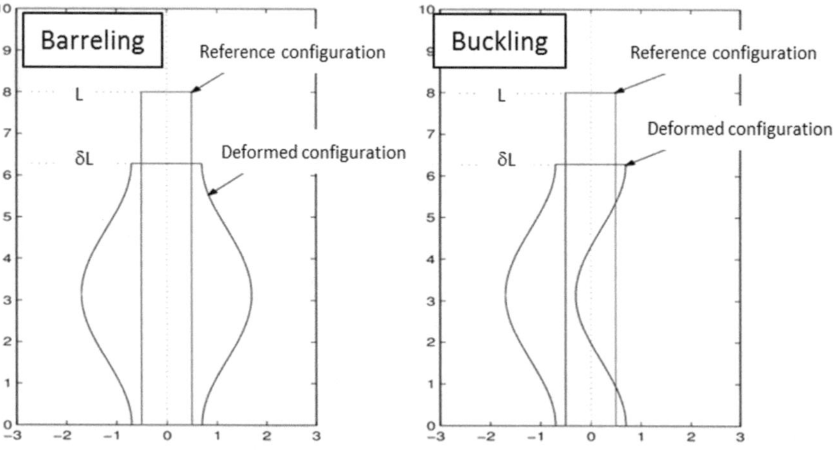

Fig. 3. Comparison of barreling and buckling during compression tests [9]

3.2 Implementation

For the tele-operated compression test, cylindrical specimens are used. The diameter of each specimen is 10 mm and the height is 15 mm. The sizes fulfill Eq. 1 to avoid buckling. In Fig. 4 the specimen before and after compression test is shown. After compression test, the diameter of the specimen is almost two times bigger while its height is three times smaller. Barreling can be clearly observed in the compressed specimen in the preliminary tests. For this reason, wd40 will be used for lubrication for the tele-operated experiments.

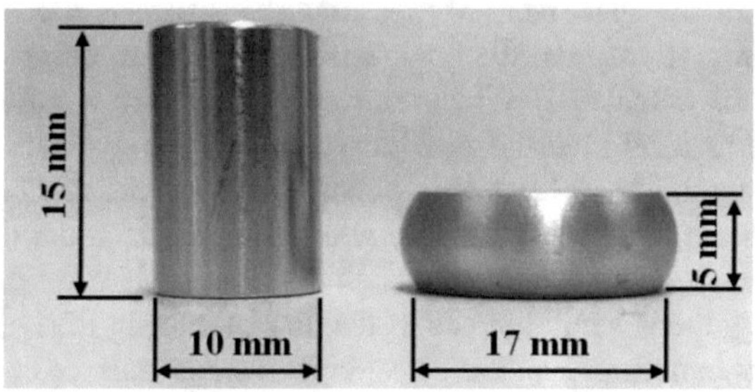

Fig. 4. A specimen before the compression test (left) and after the test (right)

In order to meet the requirements for gripping, handling, and releasing the specimen before and after the test, a new gripper is designed (Fig. 5). The distance between the fingers in the open position is 17.09 mm. The geometry of the gripper fingers allows gripping of specimens from an initial diameter $d_{0min} = 7.3$ mm onwards.

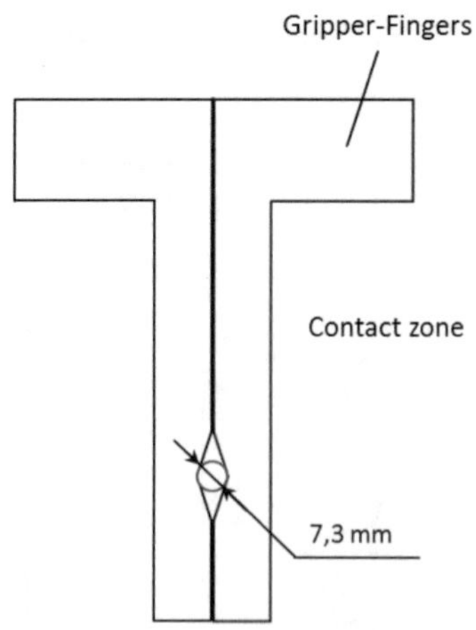

Fig. 5. Design of the new gripper

For the remote compression test, a magazine to hold up to 18 cylindrical specimens has been installed. The magazine is designed so as to allow the robot to grip the specimens one by one. In Fig. 6 the realized gripper and magazine is shown.

Fig. 6. Top view of the realized gripper and the magazine for the compression test

In the LabVIEW interface, once all controllers are started, it is possible to switch to the tab *Magazine* (Fig. 7). The user has the possibility to choose the magazine created especially for the compression test and to fill the virtual magazine mirroring the real model: 9 specimens in the first row and 9 specimens in the second row. Here the material, height and the diameter of the specimens need to be defined by the user.

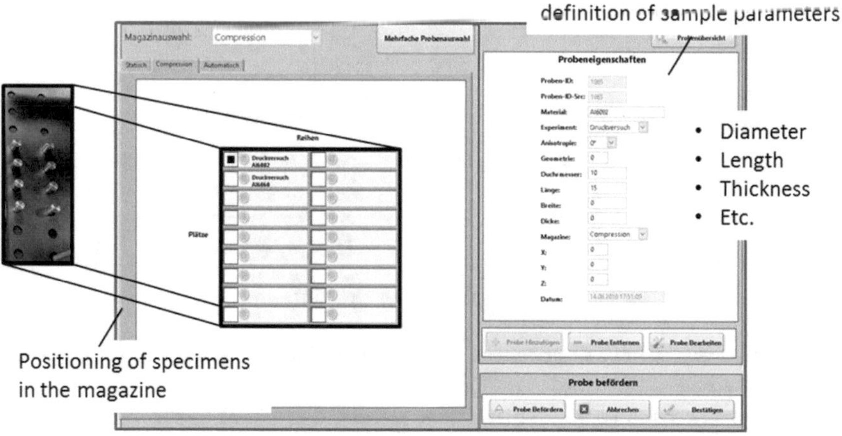

Fig. 7. LabVIEW tab “Magazine”

As a part of the educational process, students have the possibility to change some influencing parameters of the experiments from LabVIEW. In the compression test, it is possible to adjust the strain rate during the plastic deformation. As a default value, it is set at 0.0067 1/s and can be changed according to the requirements, within the range of pre-defined maximum and minimum values, which can be changed only by the supervisor.

In addition to the strain rate value the students have the possibility to choose the material of the specimen. To show the influence of the used material two different aluminum alloys were selected for the tele-operated test. Aluminum is light and recyclable, its formability, conductivity and corrosion resistance features are high which makes Aluminum a highly utilized material in many fields. Al–Mg–Si (Aluminium–Magnesium–Silicon) alloys are widely used in the automotive industry. Two different alloys of this family are used in the tele-operated compression test: Al6060 and Al6082. The chemical composition of each alloy is shown in Table 1.

Table 1. Chemical composition of Al alloys

Specimen material	Si	Fe	Cu	Mn	Mg	Cr	Zn	Ti
Al6060	0.3–0.6	0.1–0.3	0.1	0.1	0.35–0.6	0.05	0.15	0.1
Al6082	0.7–1.3	0.5	0.1	0.4–1.0	0.6–1.2	0.25	0.2	0.1

Al6060 and Al6082 alloys are heat-treatable ones, consequently, their strength parameters depend on the treatment methods. In general Al6082 is an alloy of higher strength as shown in Fig. 8.

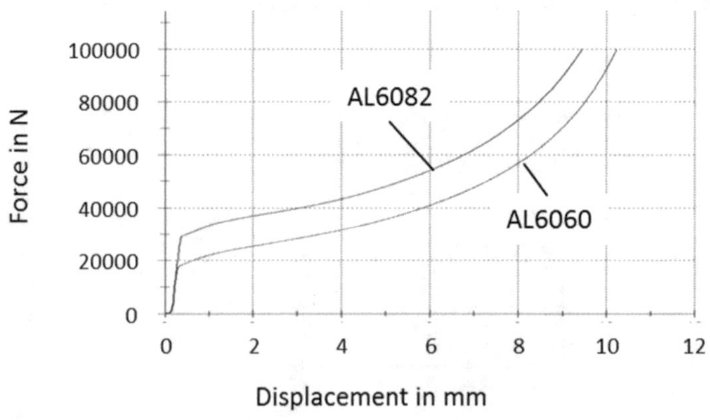

Fig. 8. Comparison of force-displacement curves (Al6082 and Al6060) generated in testXpert I

Each test has its own unique ID with information about experiment type, date, status and description being saved under this ID. All tests are saved in a database and can also be evaluated later. To start an experiment, the user has to confirm all inserted data. After that the robot is ready to perform its movements and the experiment can be started. To do this the robot moves to the magazine, gets the specimen from the magazine and puts the specimen into the universal testing machine (Fig. 9).

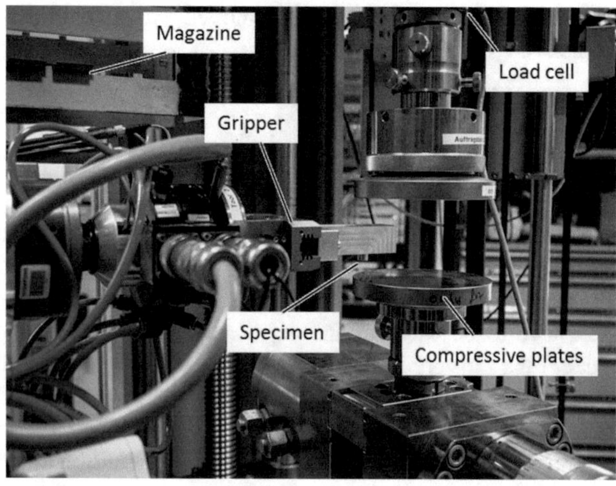

Fig. 9. Gripper loading the specimen into the universal testing machine

During the test a specimen is compressed between the compressive plates. Sensors for force and displacement measurements are placed in the crosshead of the machine so no additional extensometers attached to the specimens are needed. The movements of the Zwick are controlled by a drive unit.

The main window of the experiment is shown in Fig. 10. On left side of the window, the general parameters about the experiment are shown.

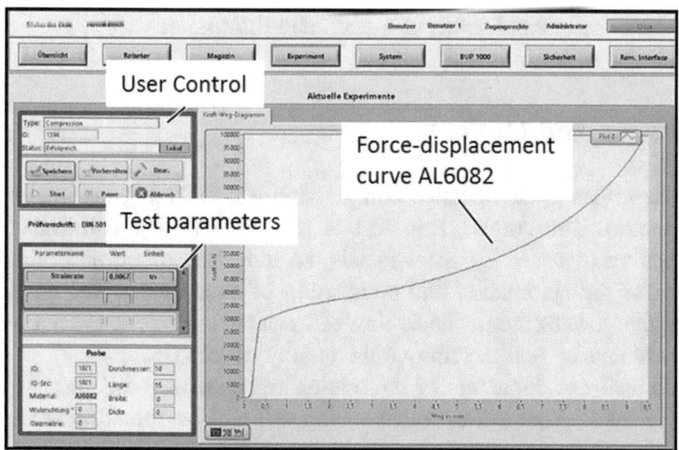

Fig. 10. Main window of the experiment

As soon as the robot puts the specimen into Zwick Z250 machine and returns back to the home position, the compressive plates close and the preload is applied. Once the preload value of 25 N is reached, the test is ready to be performed and can be started by the user. The force-displacement curve is shown on the right side of the window. When the test is ended, the compressed specimen is removed by the robot and the system is prepared for the next experiment.

4 Integration into a Learning Scenario

Since material characterization is an essential part in forming processes, students deal with this aspect during lectures [10], different labs [11] or seminars in various facets. The tele-operated testing cell was integrated into a fundamental lecture of forming technology in the form of a live experiment. During the lecture the students choose the parameters which have to be used for the experiment and afterwards the results are discussed. In general the observed effects can be derived theoretically and afterwards shown practically or vice versa. By integrating this concept into the lecture, many students can see and understand the process and the material behavior without being at the machine itself. The tele-operated compression test is an additional building block to improve the understanding of the behavior of materials during deformation.

The tele-operated compression test can also be integrated into labs, so that students can do experiments with the machine and investigate the different phenomenon from their homes.

A plan for the implementation of the lab could look like this:

1. Literature research of influencing parameters for the compression test
2. Create a meaningful experimental program
3. Realization of the experimental program and observation of the phenomena
4. Attempt to explain the phenomena on basis of the observation and the literature research
5. Creating a lab report
6. Present the experimental result as a team and discussion.

5 Conclusion and Outlook

This paper describes a tele-operated compression test including the integration into a learning scenario. The compression test is part of a testing cell consisting of an additional testing machine for sheet metals, an industrial robot, and other necessary components for the automation and conduction of experiments like control units or cameras for the visualization. The testing cell can be controlled via internet from all over the world and is used to improve the quality of lectures.

The described developments of the tele-operative testing cell present the current stage. In the next step the tele-operated testing cell will be offered to students for free experimentation once a month.

Acknowledgements. The developed remote laboratory is a part of the project "ELLI – Excellent Teaching and Learning in Engineering Education", which is funded by the Federal Ministry of Education and Research in Germany (project number: 01PL11082C). Three universities are involved in the project: RWTH Aachen University, Ruhr University Bochum, and TU Dortmund University.

References

1. Tekkaya, A.E.: Metal forming. In: Grote, K.-H., Antonsson, E.K. (eds.) Handbook of Mechanical Engineering (Chap. 7.2), pp. 554–606. Springer, Berlin (2009)
2. Gomes, L., Bogosyan, S.: Current trends in remote laboratories. IEEE Trans. Industr. Electron. **56**(12), 4744–4756 (2009)
3. Corter, J.E., Nickerson, J.V., Esche, S.K., Chassapis, C., Im, S., Ma, J.: Constructing reality: a study of remote, hands-on, and simulated laboratories. ACM Trans. Comput.-Human Interact. **14**(2), 1–27 (2007)
4. Ma, J., Nickerson, J.V.: Hands-on, simulated, and remote laboratories: a comparative literature review. ACM Comput. Surv. **38**(3), 1–24 (2006)
5. Ortelt, T.R., Sadiki, A., Pleul, C., Becker, C., Chatti, S., Tekkaya, A.E.: Development of a tele-operative testing cell as a remote lab for material characterization. In: Proceedings of 2014 in International Conference on Interactive Collaborative Learning (ICL), pp. 977–982 (2014)
6. Harward, V.J., Del Alamo, J.A., Lerman, S.R., Bailey, P.H., Carpenter, J., DeLong, K., et al.: The iLab shared architecture: a web services infrastructure to build communities of internet accessible laboratories. In: Proceedings of the IEEE, vol. 96, pp. 931–950 (2008)
7. DIN 50106:2016-11, Prüfung metallischer Werkstoffe – Druckversuch bei Raumtemperatur (DIN 50106:2016)
8. Chakrabarty, J.: Applied Plasticity (Mechanical Engineering Series), 2nd edn. Springer, New York Inc (2009)
9. Negron–Marrero, P.V., Montes–Pizarro, E.: The complementing condition and its role in a bifurcation theory applicable to nonlinear elasticity. New York J. Math. **17**, 1–21 (2011)
10. Sadiki, A., Ortelt, T.R., Pleul, C., Becker, C., Chatti, S., Tekkaya, A.E.: The challenge of specimen handling in remote laboratories for engineering education. In: 12th International Conference on Remote Engineering and Virtual Instrumentation (REV) (2015)
11. Pleul, C., Hermes, M., Becker, C., Tekkaya, A.E.: ProLab@Ing – Projekt-Labor in der modernen Ingenieurausbildung. TeachING-LearnING.EU innovations. In: von Petermann, M., Jeschke, S., Tekkaya, A.E., Müller, K., Schuster, K., May, D. (eds.) Flexible Fonds zur Förderung innovativer Lehre in den Ingenieurwissenschaften, pp. 16–21 (2012)

Immersive Communication with Augmented Reality Headset

Mohammed Misbah Uddin and Abul K. M. Azad(✉)

College of Engineering and Engineering Technology, Northern Illinois University, Dekalb, IL, USA
aazad@niu.edu, aazad2005@gmail.com

Abstract. The system aims to run an immersive communication environment by mixing synthetic and natural images in real-time using an Augmented Reality (AR) headset. Images captured by a camera are processed to extract the foreground object and then stream it over the web. The foreground object is then mixed with real-time natural images using augmented reality technology that can be viewed from an AR headset.

Keywords: Immersive communication · Augmented reality · Hololens · Video streaming · JSON script

1 Introduction

Over the years, technological developments in the field of communication have attempted to make the experience of communicating from a remote location as seamless as possible. The ability to make video calls is the closest to what we currently have for a natural communication medium, and inexpensive video conferencing applications like Skype, TinyChat, and Line have made this process freely available to their users. People have even started to use virtual environments like 'Second Life' to conduct meetings and conferences [1]. However, with the advent of Augmented Reality / Virtual Reality (AR/VR) headsets, video processing has been brought to new heights. Natural images can easily be augmented by mixing synthetic images and projecting them to a headset, which allows the user to experience an immersive environment for communication [2].

Microsoft Research developed a project called Holoportation, which was intended to develop an immersive method of communication using the Microsoft AR headset called MS Hololens. It used a network of expensive 3D cameras that were focused on an object/person and could continuously capture images and render a 3D model of the object/person. This 3D model was then transmitted to the remote AR headset, which was then able to augment it in the real world using the AR headset's augmented reality technology [3]. However, this process is not only expensive because of the extensive use of 3D cameras but also very computationally inefficient for making the system process images in real-time.

Figure 1 explains the overview of the system design for this reported work. Images are captured from location 1 by a camera and processed in real-time to extract the

M. E. Auer and K. Ram B. (Eds.): REV2019 2019, LNNS 80, pp. 506–514, 2020.
https://doi.org/10.1007/978-3-030-23162-0_46

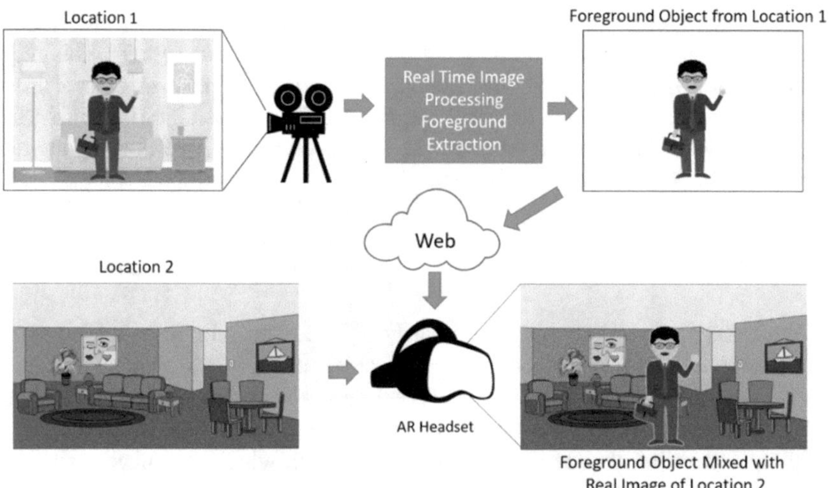

Fig. 1. System overview.

foreground objects. The foreground objects are then transmitted over the web, which is then accessed by the AR headset (present in location 2), which mixes the foreground image with the natural image of location 2. The user wearing the AR headset can visualize the foreground object in its current space and enjoy the experience of feeling the object to be actually present in his immediate vicinity. Since the system is using images from a simple web camera, it is very cost effective and also very minimal in respect to the computational aspects (image processing).

To design an optimal algorithm, a number of studies were conducted in the field of Background Removal/Segmentation. The most widely used algorithms of background removal involve foreground estimation using color information or depth information. However, both of these algorithms have their own challenges and shortcomings. By using color information alone, issues arise when the foreground and background objects are a similar color. Hence, it becomes difficult to differentiate and parts of foreground and background can become mixed. Using depth information alone causes problems extracting the exact bounding region of the foreground object, and also it becomes difficult when parts of the background and foreground are at the same depth. G. Gordon and his team at Interval Research Corporation proposed a system that utilizes a combination of both color and depth information and then computes the background and foreground [4]. While this technique provides good results, our system cannot handle the computation expenses of both these systems simply because they have to be processed in real-time.

Microsoft Research designed the Grab Cut algorithm, which is now part of Microsoft Office Suites [5]. Grab Cut is designed as an interactive foreground extracting tool, which requires defining a bounding box that may contain the object in question. Since our system is a real-time automated system that has to process the images on its own without any supervision, we could not implement this approach. Instead the researchers proposed a context-aware background subtraction system that uses information about a foreground object's position in the previous frame and assumes the object is somewhere in that region in the next frame [6, 7]. This reduces the computation time; however, this technique cannot be utilized in case of moving foreground objects. Microsoft has also proposed a Flash Cut algorithm that utilizes images taken with flash-on and flash-off [8]. However, this system is computationally expensive and is not feasible in real-time videos.

2 Our Approach

The developed system is composed primarily of two parts. The first part is the background removal or image processing part, which extracts foreground objects and uploads them on an online server. The second part is the projection part, which takes the extracted images from the web and projects them onto the AR headset.

Figure 2 describes the layout of the system through which the images captured by the camera are processed to extract the foreground object. The images are then uploaded onto an online web server that can be accessed by the AR headset (with internet connectivity) and the foreground image can be visualized in an augmented reality manner through the AR headset.

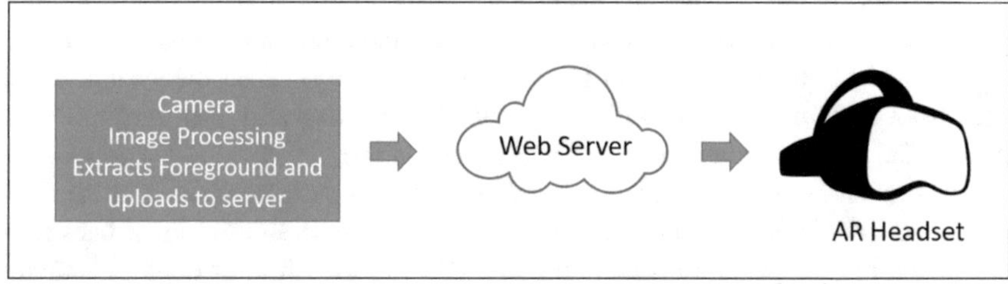

Fig. 2. System design layout.

This paper is based on the concept that the color black cannot be projected. Sir Isaac Newton's book Opticks: or, A Treatise of the Reflexions, Refractions, Inflexions and Colours of Light in 1704 demonstrated that black/darkness is merely an absence of light [9]. Modern day projectors are not capable of projecting pure black images. To project an immersive experience, the image can undergo the following processes: (a) Segmentation/Background removal (b) Replacing the background with black. Figure 3a shows the real image; Fig. 3b shows the background segmented image in which the background is replaced by a black background.

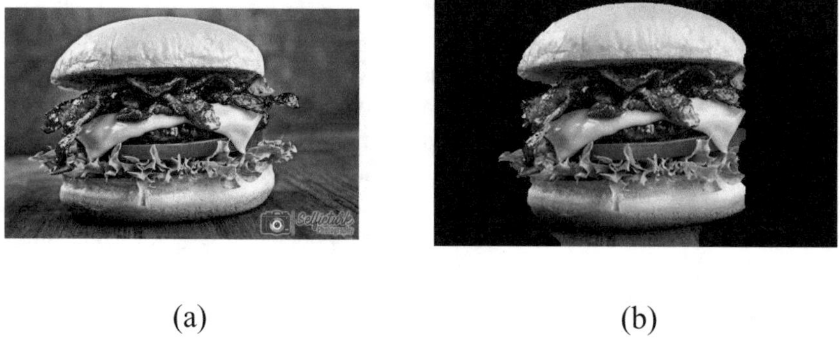

(a) (b)

Fig. 3. **a** Image of a Burger [10]; **b** Background segmented image, replacing background with a black background.

The background of an image is removed by subtraction of foreground image F(x,y) (without background) of a given image I(x,y). The image is then sent through morphological operations (erosion and dilation) to remove any noise in the image. One can read the book by Richard Szeliski Computer Vision for more information on erosion and dilation [11].

$$Background\ Subtracted\ Image = |I(x, y) - F(x, y)|$$

The background subtracted image when projected by any modern day projector will fail to project the black portion of the image. This technique can then be applied onto VR-AR devices like the Microsoft Hololens to create an immersive experience of the images being projected. This technique can be used into-and-fro communication to provide an immersive experience for the user in which the foreground object in an image can be projected onto the screen of a Hololens, making it augment the user's vision. Figure 4 shows the process flow-diagram.

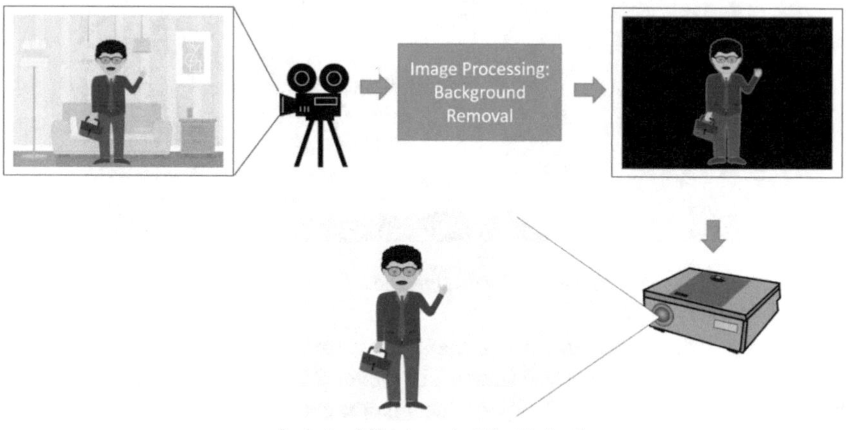

Fig. 4. Process flow diagram.

For communication purposes, this technique can be used to stream the live video feed onto a remote location, which can be processed and projected by a projector. This serves as an augmentation of the contemporary video call feature. For our experiment, we used the Microsoft Hololens. Video feed from a normal webcam (Logitech Web-Cam C270) is processed for the immersive communication effect. The challenge is to transmit the processed image with the background removed video feed over the web and project it onto the MS Hololens. We tried the following three methods: streaming images, server playing offline videos, and server updating images at regular intervals.

2.1 Streaming Images

The video feed is streamed on a web page that can be viewed from any Internet-enabled device. This is done by making a flask-based web server that can process the image from a webcam (for background removal) and then display it onto a web page. Flask is a micro web framework that along with its other uses can act as a server for hosting html pages [12]. In our case, we used Flask to host images processed with background removal. Figure 5 shows a background removed image of a hand that is hosted on a web page.

Fig. 5. Background image hosted on a web page.

The images displayed are in a moving jpeg format. The OpenCV was used to generate the background removed images. However, Microsoft Edge, which is the only supported web-browser in the MS Hololens, is not capable of displaying moving jpeg images. This made it difficult to project the images on an MS Hololens.

Flask is a Python-based micro web framework that allows developing web applications. It can act as a web server and run backend applications on web pages. Flask is used as a backend server that can retrieve the processed background removed images and display them on a web page, which can be later accessed by MS Hololens. Figure 6 describes the flow diagram of the process. The key element of this web server is the use of generator functions.

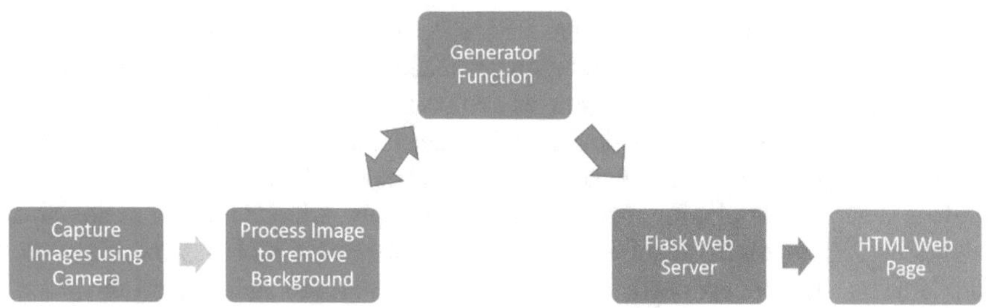

Fig. 6. Flask web server.

Generators are special functions that, unlike any other functions or sub routines, perform the regular tasks like initialize local variables, process and return values. However, they have the ability to store the current state of the subroutine, and every time it is called, it can retrieve and execute the next step of the same subroutine (without the need to re-initialize the local variables and other initial processing steps). This is normally employed when a function is tasked with processing and returning large data and do not hold the other processes in the main program. As an example, the processing of images in a continuous video stream. Here a generator function should be continuously capable of retrieving the processed images from the camera and also sending it to the web server for projection on the web page.

```
def gen(camera):
  while True:
      frame = camera.get_frame()
      yield(b '--frame\r\n'
          b 'Content_Type: im-
age/jpeg\r\n\r\n'+frame+b '\r\n')@ app.route('/video_feed')
def video_feed():
  return Response(gen(Camera()), mimetype = 'multipart/x-
mixed replace; boundary=frame')

if __name__ == '__main__':
  app.run(host = '0.0.0.0', debug = True)
```

In the code snippet above, the generator function here is tasked to get_frame() from the Camera class that continuously retrieves images and feeds them to the video_feed() of the web server. The web server places this newly grabbed image on a html web page. This image on the web page is continuously replaced by the newly retrieved image, giving it an illusion of a video file being played.

2.2 Server Playing Offline Videos

In this approach, an attempt was made for background removal on recorded videos of short duration and then the file was made available for video projection, where a camera captured the frames of the images and sent them to the background subtraction module. The background subtraction module then stored the processed and background subtracted images in a video format (avi or mp4, mp4). However, we found that MS Hololens is capable of playing mpeg video formats only, which OpenCV is not capable of exporting. This process required another step before we could play the video on the MS Hololens. The video needed to undergo a format conversion step and then be uploaded on the web page for the MS Hololens to play. This was not only time consuming and inefficient but also not real time.

Figure 7 shows the different offline steps carried out on the video before it could be played on the HTML page supported by MS Hololens. The code below shows how to redirect an offline video to be played on the Flask server.

```
@app.route('/done')
def done():
    return redirect('/static/formattedVideo.mpeg')
```

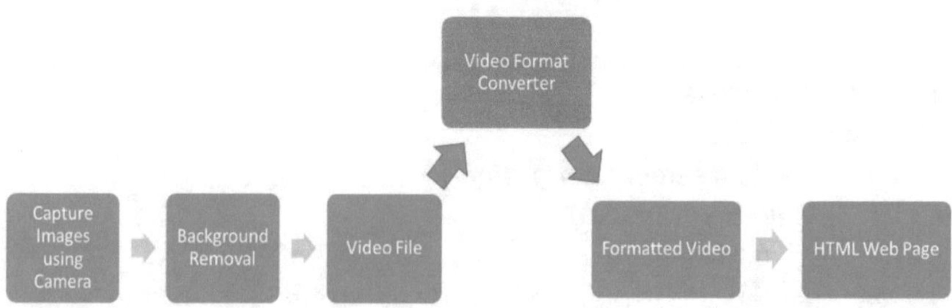

Fig. 7. Playing recorded videos on HTML page.

2.3 Server Updating Images at Regular Intervals

This technique relies on the fact that an image (.jpeg) can be viewed on a web page (supported by MS Hololens). Here the images that are processed for background subtraction are saved on the local drive. The Flask web server then uploads this image onto the web page, in the meantime replacing the existing image in the image box. This technique utilizes JSON scripts that update the images on the web page at constant intervals, which forces the web browser to refresh the web page every time a new image is uploaded. This process helps in creating a video playback effect on the web page. However, it is also affected by a slight delay, which is caused in the saving of the original image and uploading it to the web server. We noticed a lag of 0.2 s in consecutive frames.

```
$("#images").remove();
$("div").html("<img src='http://localhost/image_data.html’
class=’processed’>")
```

The code above is the JSON script, which is embedded in the Flask web server.

3 Conclusion

This paper describes a novel method to process images for developing an immersive communication experience. The process is cost effective and requires minimal computational processes to achieve the desired effects. This technology is still in its elementary stages, but with the progress being made in the VR/AR industry, this process can offer successful strides in changing the way we communicate. By developing an AR compatible browser or web server, this process can be improved and made more seamless.

References

1. Tutton, M.: Going to Virtual World. http://www.cnn/2009/business/11/05/second.life.virtual.co.llaboration/ (2009)
2. Rosenberg, L.B.: The Use of Virtual Fixtures as Perceptual Overlays to Enhance Operator Performance in Remote Environments, Center for Design Research, Stanford University, Stanford (1992)
3. Holoportation, Microsoft, Web Link. https://www.microsoft.com/en-us/research/project/holoportation-3/
4. Gordon, G., Darrell, T., Harville, M., Woodfill, J.: Background estimation and removal based on range and color. In: Proceedings of IEEE (1999)
5. Rother, C., Kolmogorov, V., Blake, A.: GrabCut: interactive foreground extraction using iterated graph cuts, ACM SIGGRAPH 2004 Papers, August 08–12, 2004, Los Angeles, California https://doi.org/10.1145/1186562.1015720
6. Lakshmi, S., Sankaranarayanan, V.: A robust background removal algorithms using fuzzy C-means clustering. Int. J. Netw. Secur. Appl. (2013)
7. Garcia, A., Bescós, J.: A Real-Time Video Foreground Extraction Based on Context-Aware Background Substraction, Technical Report TR-GTI-UAM-2007-02 (2007)

8. Sun, J., Kang, S.B., Xu, Z.-B., Tang, X., Shum, H.-Y.: Flash cut: foreground extraction with flash and no-flash image pairs. In: IEEE Conference on Computer Vision and Pattern Recognition, 2007. CVPR07, pp. 1–8. IEEE, New York (2007)
9. Isaac Newton. Opticks: or, A treatise of the reflexions, refractions, inflexions and colours of light. Also two treatises of the species and magnitude of curvilinear figures. London, Printed for S. Smith, and B. Walford, 1704. Pdf. Retrieved from the Library of Congress. www.loc.gov/item/46039060/
10. Selfieture Photography, Web Link. https://selfieturk.com/neler-cekiyoruz/urun-mekan-cekimi/
11. Szeliski, R.: Computer Vision: Algorithms and Applications. Springer, New York Inc (2010)
12. Flask, Web Link. http://flask.pocoo.org/

Hybrid Teaching Model to Persuade Different Dimensions of Felder-Silverman Learning Style Model

B. Hareesh[1], N. PrathviRaj[2(✉)], and S. Gururaja[1]

[1] Department of Computer Applications, SJEC, Mangaluru, India
{hareeshb,gururajs}@sjec.ac.in
[2] Department of Computer Science and Engineering, SJEC, Mangaluru, India
prathvirajn@sjec.ac.in

Abstract. The student learning and listening style always changes generation by generation. It is a need and challenge for the teacher to keep the class environment active for the entire session. In this busy and competitive era, to cope-up with the student's skill sets and mindset, the teaching-learning should happen efficiently within the time bound. The available infrastructure should be well synchronized along with the current generation, educational type, topic and situation etc. The observation in the class room as well in the as student evaluations strongly proven that when new technologies are adopted in teaching modalities like remote class room, virtual class room, google class room and even a power point presentation always some section of learners not actively participated in the learning process. The reaching out of all section of students if only traditional class room technique where used is also not good for all the learners. The goal of a teacher is to map the course design to the student skill set along with the cost and time efficient teaching methodologies. This could be done by adopting the technology bounded teaching skills. The Silverman Learning Style model gave us a room to think about the different levels and capacities of student learnings. It means that when different types of modalities adopted for the particular topic there will be a greater possibility of imbalance in the learning at the receiving end. The proposed hybrid model expects to address all the types of learners in the Silverman Learning Style Model. It will ensure a balanced traditional and technology learning happens within the available infrastructure efficiently managed with time. The efficiency of this model enables the students to perfectly undergo with the basic conceptual learning and feel the power of technology in the teaching.

Keywords: Hybrid · Teaching model · Silverman learning style · Student skill and mindset

1 Introduction

1.1 Technology in Teaching

Current engineering education needs to change significantly in order to prepare our graduates for a world of rapidly accelerating changes. The boundary barriers between the different engineering discipline and science disciplines are diminishing rapidly to

M. E. Auer and K. Ram B. (Eds.): REV2019 2019, LNNS 80, pp. 515–531, 2020.
https://doi.org/10.1007/978-3-030-23162-0_47

cope up with globalized automation challenges. We also need to prepare our graduates to meet the cross cultural and cross-national challenges. This can be done by redefining the engineering education teaching trough hybrid instructions. The biggest question here is the implementation of the proposed hybrid teaching, with a cost-effective and time efficient manner into to this generation particularly in our college class room. The majority of current generation student learning styles relinquished from the traditional class room leaning, more-over they likes to learn independently through advanced technology teaching like remote/virtual laboratories, flipped class room etc. These technologies considerably decrease the face-to face learning environment rather than increase the enhanced conceptual skilled learnings and benefit the reasoning skills among students. The following technology in teaching is adopted to meet the globalized and psychological challenges among the student learnings.

1.2 Remote/Virtual Laboratory

Laboratory based courses plays a major role in engineering education. The traditional Hands-On learning incorporates the experiential learning among the students and also it back-up the lacunae of the traditional Black-Board teaching up to certain extent. But the major drawback of the Hands-On learning is its close connotation between the installed computers at a particular location with instructor assistance. This overrides the infrastructure cost for the college as well as to the students. On the other hand the remote/virtual laboratories implemented by some of the universities or organizations helps the students to learn the content remotely without an instructor support, which greatly improves the cost and time effectiveness for both the teachers as well as learners in engineering education. These remote/virtual laboratories simulated with a clear instruction manual, which greatly helps the learners to follow the steps to solve their problem domain along with a visualized effects. This will boost the learning behavior of the students.

1.3 Flipped Class Room

This is one more effective technology teaching where the students benefit an active and interactive class room environment where they involved through online videos. A technical videos related to the content of the future class is facilitated prior to the class through some course design tools like google class room or canvas etc. The online videos are chosen such that the same contents were solved by different technical experts with different conceptual skills. The students are instructed to go through these videos, so that the students will get more than one problem solving skills. At the same time they will experience the technical expert's presentation of the technical content along with the instructor view point. After this the students are asked to discuss the content in the class. This majorly benefits all levels of learners which are discussed in the Chapter "Cyber-Physical Control and Virtual Instrumentation".

1.4 The Felder-Silverman Learning Style Model (FSLSM)

The result of this technology teaching in the learners end gives us diverse attributes of the learning success. It motivates us to measure the attributes for the diverse behavior. We found that the psychological behaviors of the students learning are particularly changes person by person and we found that the Felder-Man model effectively measured the different attributes of the learnings. So the proposed hybrid teaching model is sculpted with reference to the measurable learning capacities of the students psychological learning behaviors.

So the Personalized teaching learning process is very essential tool for enhancing the teaching learning process in the class. The requirement and expectation of all students should be addressed individually. Personalization in the process will bring all students in the class into an account. The facilitator in the class should manage to address the expectation of different category of students.

The Felder-Silverman learning style model (FSLSM) [5] is a unique model to bring personalization into traditional teaching learning process. This model brings technology enhanced learning environment to handle students from different dimensions of the learning styles. Understanding the learning style with FSLSM will be helpful for enhancing teaching style to reach out all category of students present in the class. The learning style is categorized into 4 dimensions by Felder-Silverman based on specific preference for each of these dimensions. The distinction between active and reflective learners based on how they process the information. Here active learners prefer to work in group by interacting and communicating with group members. On the other hand reflective learners prefer to revise individually on what information they have. Under the second category sensing and intuitive learners dimensions are covered. Sensing learners follow facts and concrete learning information with standard approach to solve the problem. In contrast to sensitive learners intuitive learners on other hand with the little abstract knowledge explore different possibility and try to relate to new concept. Third category is based on how learners will be able to remember i.e. visual—verbal dimensions. Here visual learners can remember the concept based on pictures, diagram, flowchart they saw, on the other hand verbal learners depend on textual information. The fourth-dimension deals with students understanding capability. Sequential learners on one side concentrate on step by step development of concepts, which is a linear progressive learning. Global learning on another side think about big picture and concentrate on entire process.

The learning environment in educational institution should be suitable for education. These environments might be virtual or physical like traditional classroom or e-learning classroom. The sessions or classes in educational institution will happen in environment like defined, describable and observable. The teaching process in university model is always course-steered and course structure and timing should be clearly followed. In this context providing personalization to different category of students is challenging job for faculties.

The prominently used teaching methods for course steered model are traditional face to face teaching and PowerPoint presentation. The face to face teaching provides real time direct interaction with the students. The faculty will get maximum benefit to restructure his course delivery in real time by looking at spontaneous feedback.

The power point presentation will be helpful in bringing faculties level of understanding to class with help of animation, images and case studies.

Online virtual learning opportunities on the other hand provide extensive support to learner's community. Technologies used in these online learning portals will fill the gap of faculty's absence and enhance the students learning capability with useful animation, video, images. The students can use this model to independent content walkthrough, content mastery, innovative collaboration and pair up with their peer member for extensive study.

The advantage provided in each model is unique and it is impossible to replicate one model in another along with added advantage. Each model is identified with its own drawbacks. For example, in traditional face to face model teaching is limited to available teaching aids, here it is difficult to provide big picture. Online virtual learning comes with absence of instructors physical presence, since no one monitors students, may easily distract the students.

One of the immense reason to incorporate hybrid model is not just the flexibility in terms of how time, volume of content for a course taught, rather it is more important how students are engaged with material, teaching learning method implemented in the class. Whereas with the traditional or online model one content delivery format is chosen and is used exclusively throughout the session this is not benefitting all section of students in the class. So hybrid learning model is implemented which offers the best of both in one unified experience.

1.5 Overview

The essence of technology teaching is a frontier requirement in the current engineering education teaching. The technology teaching sometime filters out specific type of learners from the learning environment which may affect the cumulative outcome of the learning. This definitely will impose a restrictive use of technology teaching especially in engineering education with typical university curriculum. To overcome from these restrictions a well-balanced mixture of teaching methodologies called hybrid model is proposed which satisfies almost all types of learners as sorted out in the Felder-Silverman model. The proposed model constituted with a combination of technology teaching and traditional backboned hand-on experiential learning for the engineering graduates. Here in this model the instructor can free to choose any teaching component of the Hybrid model to satisfy the learners learning attitude. To measure the effectiveness of this model the various academic literatures are taken into the study. Based on the reflections of this dynamic selection of traditional black-board teaching, Hand-on laboratories, remote laboratory and flipped class room techniques are incorporated in the hybrid model. The actual reflections of the components of the hybrid model help us to assess the effectiveness of each component of the model. The actual benefit seen from this model is the students engagement in the active learning which greatly benefit us to promote outcome based education in the current engineering education and also nowadays it is a global need to meet the innovative challenges in the global automotive field.

2 The Psychological Learning Styles

The study on psychological aspects of the students learning will vary person to person, course to course and as well as generation by generation. The generation crossover highly affects the student learning. In the global and competitive world it requires the outcome based activities in the teaching learning process. To meet these global requirements educational institutions initiated the outcome based education which clearly defines the outcomes for each learning contents among students.

To override the different basic and psychological learning differences among students and to bring a uniform teaching environment among students, several research studies presented their own theories.

The Felder-Silverman learning theory discussed in this chapter focuses on uniform learning and teaching theories required to address the psychological problems with the students learning styles and also to achieve a clear outcome specified by the competent authority on topic-by topic.

The Felder-Silverman learning theories are categorized with the following teaching methodologies and learning styles.

2.1 Blackboard Versus Hands-On Laboratory

The intuitive learners always tend towards exploring the different possibilities of the concepts and its applications. They are also tending towards memorizing, reflections and imaginations, which can be addressed through the blackboard teaching. Usually these types of learners are quick in imaginations and capable of finding immediate alternative solutions. This category of will be adjustable with any type of teaching methodologies. But the sensing learners learning style clearly focused to the context of observations and procedures. They learn only with well established procedures and in a detailed step and do not like unpredicted twists and complexities in the content delivery from the facilitator. This group of learners is more practical oriented and most of the times uncomfortable with the verbal mode of learning even symbols make them slow in understanding. So they require a hand-on and laboratory work. At the same time they require detailed observation and procedures to have an experiential learning.

2.2 Picture Presentation Versus Written and Spoken

The verbal learners remember the words written and spoken. Most of the courses taught verbally in the classroom. This category of students likes to write outlines of the content of the course in their own words. They will get more clarity if the facilitator repeats the explanation. Unlike reflective learners, this category of learners more comfortable with the group activities. So it is cleared that verbal learners also, most of the time more comfortable with the traditional lecture sessions to outline the contents delivered by the faculty. Even they may be happy if the session is conducted with the PowerPoint presentation. One of the practical problems with the PowerPoint presentation with the active, reflective and sometime with the verbal learners is the problem of noting down the points. If they miss one ppt, sometime they may miss the link. So it proves that the verbal learners also expects some good traditional lecturing sessions for the better understanding of the content.

When it comes to the visual learners learning attitude their expectation exactly reverse of the verbal and reflective learner attitude. They expect the content delivery more in the form of picture, diagram, charts, film presentation and demonstrations. Most of the people enjoy learning through visuals. Sometime in the classroom we see that a blackboard presentation followed with a visual presentation having more effects on both of the learners. So to benefit the visual learner a lecture ppt and a laboratory exercise can be adopted in the teaching methodology.

2.3 Group Discussion and Hands-On Versus Brief Thinking and Running Notes

The reflective learners process the information with a brief thinking with short running notes in hand. Usually a black board teaching is a best methodology to take the running notes. They usually want to work alone instead of groups. It means that this type of learners slightly uncomfortable with the group activities. They not only memorize, but also review what they have read, so ultimately a favorable teaching methodology for them is a lecture with summaries which ultimately requires a black board teaching.

But when it comes to the active learners learning attitude they have high tendency towards group discussion and they really interested in trying something with hands-on. They really cannot sit and take notes and think about the problem, they just want to try it out. They are most favorable towards the hand-on sessions, laboratories and discussions other than the classroom black board teaching.

2.4 Logical Steps Versus Jump and Act with Pictures

Sequential learner usually tends towards studying in logical steps to solve problems. They need a facilitator assistant to link in between the steps. But the global learners can learn without serialization of the contents. They can solve complex problems through pictures. Even they can understand the entire chapter with a series of picture explanations. It means that the global learners can set with any type of teaching mode. But the sequential learner requires a combination of teaching mode like lecturer/ppt and other procedural teaching methodology to understand the concept with the logical steps.

It is true that in the classroom we experience all the above discussed category of learners. As we told in the beginning it is our duty to create an active learning environment in the classroom to contribute towards competitive global world. When we have different mindsets in the classroom it is one of the biggest challenges to satisfy their learning commitments, and to achieve outcomes of the course. Here in the Felder-Silverman model author clearly specify the student's comfortableness towards content presentation. It looks like different levels of learners require different teaching methodologies as their comfort modality to understand the subject. This is another bottleneck scenario to the facilitator to select the appropriate teaching style to balance all levels of learners.

3 Hybrid Model

The learning methodology proposed by the Felder-Silverman Model is discussed in detail in the previous chapter. The model clearly explains the different learning capacities of a student, and best possible methods to balance the student learning capacities. The Felder-Silverman theory of learning proposes different teaching methodologies for each learner. It is proved that in each class we have different categories of students with different learning styles. The question here is how to satisfy all these learners in a single session extended with minimum one hour to three hours. It is proven that when you follow Felder-Silverman model it only suffices a minimum set of learners. This problem motivated to us to adopt a blend of teaching methodology and our intention is to address all level of learners who has different learning styles. A proper plan and schedule may bring lot of joy in this active learning process.

The planned execution of a particular session requires lot of commitment from the faculty. The course content design for the particular course should address the outcomes of the topic/module clearly. To achieve each topic learning outcome, it is our practice to clearly describing the learning methodology like chalkboard, PowerPoint presentation etc. Most of the time the teachers only limit on to these two methodologies because of various reasons like syllabus coverage, organization culture etc. When it comes to the laboratory exercises our focus and concern is to get only the experimental results. This will be a routine practice almost for all the teachers in engineering education. When we think from the student perspective, they will only experience the two teaching methodologies which is one other the same. In PowerPoint presentation, we normally copy the contents form the web resource or from the online reference text books present it for the student as a mode of advanced teaching methodology. We observed and experienced that this flat PowerPoint presentation one of the dry teaching methodology compare to the black-board teaching. The outcomes of these two methodologies are discussed in detail in the previous chapter with reference to the Felder- Silverman learning style. The black-board teaching majorly addresses the reflective and verbal learners and rest of the students will be idle throughout the session. The teaching only through power point presentation is seems to be the most ineffective methodologies because it excludes almost all types of learners except visual learners. So we are in a situation in the class where we can address only 10% of the total strength using black-board and power point presentation. But we accept an active involvement of the class in the session. When our teaching methodology itself is not suitable for all the learners learning style it is too challenging to the teacher to expect the active learning in the class.

Let us introduce the power of technology teaching like, remote lab or virtual lab, flipped class room and collaborative learning in the class room. It requires good infrastructure facility from the institution for the learners and a teacher, which is not a constraint nowadays but the learning discipline of a student, can affect the performance of this type of methodology. This methodology can address certain group of learners like active, visual, sequential and global learners. It means that when we adopt technology teaching it almost covers certain percentage of learners in the class. But only constrain here is the learning discipline of the students. When we use modern teaching modalities like virtual laboratory and flipped class room, we observed that students are not really undergone with the assignments because of its learning time flexibility. So it is a matter of challenge for the teachers to select certain teaching modalities which suits for the whole group of learners discussed in the previous chapter. Already we discussed that each learning style has its own limitation and it address only addresses one or two types of learners and rest will be omitted from the active learning environment.

Other than this we can have a collaborative learning environment. This learning methodology can be adopted easily with certain constraints, but when we look at the habit of collaborative learning among the students it seems to be discouraged from both the ends. Certain students feel uncomfortable with this type of methodology as it requires individual efforts in certain areas where the students may be weak. Certain students may not be good in the activity particularly they will have their own personal problems to collaborate in the groups. On the other hand when we form some groups somehow a reflective learner may be the part of the team and who cannot contribute towards group activity, because he or she always discomfort in group activities.

So when we tried all these different teaching modalities individually as part of active learning activity, we found that one or the other way at least one or two learning groups are missing inline to the content delivery.

So we tried to implement the proposed hybrid teaching model to address all level of learners. This model uses all the teaching methodologies proposed in the Felder-Silverman model which addresses the different types of learners learning styles. This hybrid model is composed of black-board teaching, virtual lab and a flipped class room.

Working Background

This hybrid model is applied for the topic class modelling in the subject object-oriented modelling which is taught for the final year MCA course. This subject also related with a laboratory subject software design laboratory. The implementation of the model is discussed in the below steps.

In the object oriented modeling a class model basically captures the real-world objects involved with the system. It is also used to deduce the attributes and operations of the class diagram. The class model also identifies the relationship among the different classes.

An online test is conducted for the students to view their tendency towards their learning style.

Hybrid Model Architecture

Figure 1 explores the work flow of the Hybrid Model of teaching. Hybrid model of teaching is closely interrelated with the Felder-Silverman Model learning. This proposed model constructed with different levels which includes a short span of blackboard teaching for minimum 20 min, continued with a hands-on experience with a virtual laboratory about to end within an hour and a flipped class room with an experiential learning through a tool at least requires one hour fifteen minutes. The inputs taken from the Felder-Silverman Model recursively applied with a teaching methodology which particularly addresses at least a typical level of learner, which inturn contribute towards the success of hybrid model. These different levels are explained below.

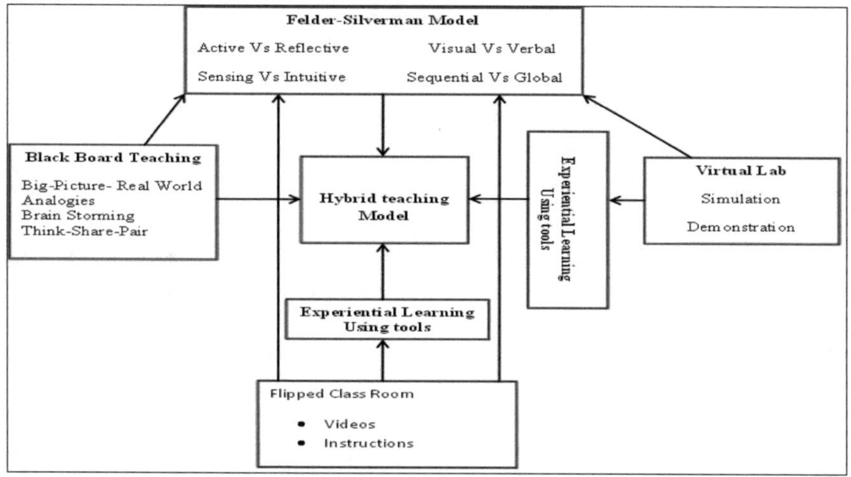

Fig. 1. Architecture of hybrid model

Hybrid Model Level 1

The object-oriented methodology and the basics of class modeling are clearly explained in the session at the beginning within 20 min. To explain this concept at the first step initially the black board teaching methodology is used for the first 20 min.

It is always good to bring real world scenarios to the class room. Most of the time, all level of learners will be attracted if we bring a case study or big picture of the topic. To explain the particular concept we brought the example of Banking System. The brainstorming session conducted to identify the objects, classes and its parameters for this case study addressed some level of learners. We observed that the reflective, global, verbal and reflective learners really participated well in the session. But the active, sequential, visual and sensing learners did not easily get into the topic because their learning behavior. This is measured with a think share pair activity for two minutes which is followed with the brainstorming activity.

For next 10 min the opinions from the students regarding the objects, attributes and operations for the classes are recorded in the black board in brief. This activity is once again very much helpful especially for the reflective and verbal learner who has the habit of taking running notes effectively in the class room. Sometime the intuitive learners also used to take procedures as they like to learn with proper serialized procedures.

For other 20 min the basics concepts of class, objects, attributes and different types of association are demonstrated with a power point presentation along with the points noted in the black board. Even if there is a visual aid for the above said topic, it is failed to interact the visual learner types because, they did not understood the problem from the basic. These types of learners cannot jump into the next topics easily. The PowerPoint point presentation recursively fails to address any levels of learners.

But point to be noted that the visual and active learners are not clearly understood, as they need some time to learn the things. As per the Felder-Silverman methodology they will understand the concept only after the practical hands-on and group activity. So in this almost 20% of the students able to note down the objects involved with the system and their relationships.

Hybrid Model Level-2

In this step a virtual lab link http://vlabs.iitkgp.ernet.in/se/5/ is shared in the google class room course page. Students were experience the power of virtual lab. The detailed manual also uploaded to the course page. A virtual lab or a remote lab is one of the very good and effective teaching modality for the visual, active, sensing and sequential learners.

But there may be some chances of failure in this method, as they may be loosely disciplined towards the online modalities, as it requires some time among the students to synchronize to any new teaching aids. At the same time students may not have enough resource requirements to run the virtual or simulation laboratory. So students are taken into the laboratory in the very next class session. The students are asked to go through the virtual lab link provided in the from the course page. The virtual lab program for the class modeling is demonstrated by the facilitator in the laboratory. Students are also experienced the virtual lab. Now one more quiz conducted to learn the understanding of the students it clarifies that more number of students is addressed with this virtual laboratory. As the virtual laboratory shows the simulation with a clear step by step procedure the active, visual, sequential and sensing learners are actively involved in the session. This methodology also helpful to the other types of learners because, they are already jumped into the problems by their learning nature and now by experiencing the simulation link they are more clarified and could clearly understood the concept.

Hybrid Model Level-3

Now almost all the students are understood the concept very well. They got the class diagram which is raised in the first session through black board teaching. But still there is a small learning gap is left behind the active, verbal, sequential and sensing learners as they naturally wanted to experience the learning through hands-on. In the laboratory we use IBM Rational Rose Software Architect Version 7 as a design tool.

Now the students are asked to draw the class model through this tool, which is constructed by some students in the first step, some students viewed and experienced in the step 2. This technique is very much helpful for all levels of learners because it will helpful to each level learner at their own context. The first set of learners like reflective, global and verbal drawn the model and analyzed it modified and came with some good results. It is shows that a experiential learning brings some innovative, challenging and research oriented attitude among students. The third step is finished within 1 h 15 min.

When it comes to the end of the session 95% of the students understood the concept with great practical approach. As this is one of the important model among 2 other model also it is a prerequisite to the other two models of the curriculum, it was fortunate to apply this Hybrid Model to this topic.

Hybrid Model Level-4

A flipped class room activity is also conducted inline to this topic for the greater understanding of the subject. This activity brings certain levels of critical concepts among the learning. It was easy to adopt the technology teaching when our student's mindset synchronized with the technology in teaching.

In this activity a video for a design pattern is shared among the students along with an introductory video of the facilitator. Students are asked to analyses the video before the next session. This advanced level of teaching modality really satisfies all levels of learners in the loop. The verbal and visual contents demonstrated in the video really work very well in all the categories of learners. The reflective, verbal, intuitive and global learners really feel a change in teaching mode, as they can view the online facilitator along with the verbal discussion in the white board. Wherever and whenever they want they can stop this video and can take a running note of the class model along with pattern. Another advantage is the schedule of the teaching. As it is shared through the course page, they can view and experience the class at anytime and anywhere. This flexibility is really helpful to the current generation students, because their physiological learning behavior changed because of various distractions of the global competitiveness. Even the other set of learners also enjoyable because the visual aids they can see in the video. The presenter also follows some constructive, sequential, well defined procedures to explain the class model implementation which really benefit the slow learners.

In the next session a group activity is conducted, where for each team 20 min of time is given to discuss the concepts studied in the video. The groups are formed in such a way that it includes different types of learners and care is taken such that there will be a leader who scored a considerable grade in the quiz conducted at the end of step-3. The students are asked to write a class model and a sample code for the classes by sitting in the groups. This activity really favors the active learners and their descendants who really work well in groups. The 6 member groups finally should present a best suitable model through Google classroom to the facilitator. This really worked very well as we got six best models as a solutions for the problem given in the first class. It is better to give an open ended problem to solve it in the Hybrid model.

4 Performance Analysis

The performance of the hybrid model executed in the class gives great benefits to the learner as well as teacher.

To measure the performances of the different categories of learners in the class as proposed by Felder-Silverman model, as a prerequisite study we are taken the first internal marks of the previous year final year students from the same course and topic. The questions paper contains a question to frame a class model, which was clearly taught in the class room using traditional teaching aids. The below average marks scored by the students in the particular question explore the students understanding capability of the topic is under motivated the teachers effort. It is true that the students are not too bad in their studies, still the marks scored in the particular subject shows the lack of learning happened at the receiver side. The below chart show the student performance in the particular topic said above in the year 2018 first internal test.

The below chart shows that student performance in the content "class modeling" was very poor. Most of the students could not understand the key components required to create a class diagram. If a student unable to identify the common attributes and operations, the key components in the problem statement, surely, they cannot identify the class names, further they cannot build a relevant class diagram. The below chart shows the marks scored for the students gives a clear picture of understanding of their learning. Most of the students poorly understood the attributes and operation which is very important to group the genuine classes, which results in an effective class diagram. It proves that the impact of the black board or power point presentation is very limited among the students as per the Felder-Silverman theory (Fig. 2).

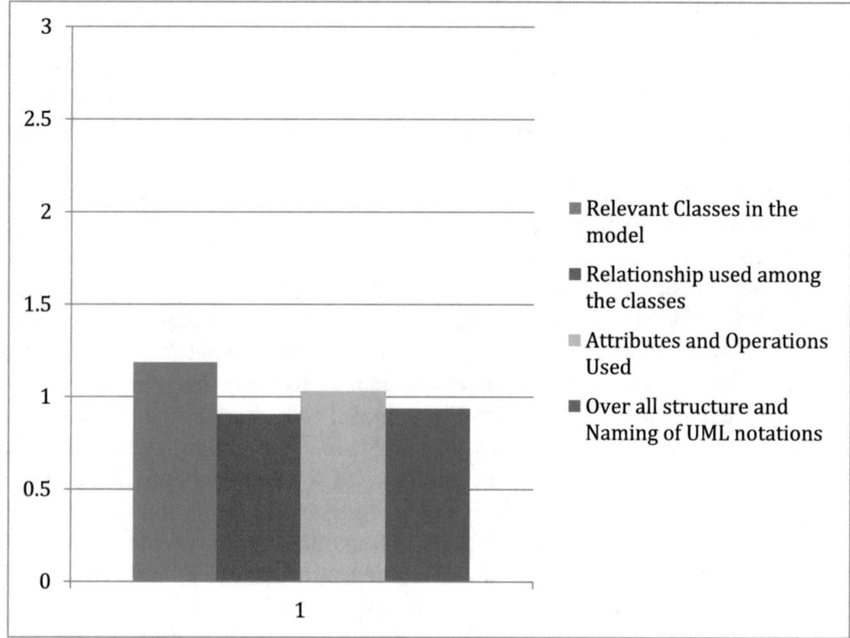

Fig. 2. Previous year marks chart scored by the students

Hybrid Model Level 1

The analysis of the previous year internal marks motivated us to use a hybrid model of teaching especially for the class modeling. The class model is clearly explained in the session using blackboard teaching. To know the understanding capability of the student, a brainstorming session is conducted in the class session. To evaluate the outcome of the brainstorming session think-share-pair active learning technique is conducted. A rubric is designed to evaluate their understanding level and the students itself asked to evaluate their model using the below rubrics. It is a self-activity to promote their ethical responsibility as the students are in the verge of completing their academics. Each student mark is recorded in a sheet.

The below rubrics designed clearly explains the level of understanding of the students. When we apply Folder-Silverman theory of learning against this chart, it clearly shows very few categories of students able to solve the problem. It proves that the black board teaching only addresses a minimum set of learners and major students are unaddressed.

The evaluation based on Table 1, clearly shows the understanding levels of the students in different aspects of the content. Figure 3 shows that student's capability to draw a complete class diagram for the given problem is not convincing and it is a clear proof of the students level of understanding of the concept.

Table 1. Rubrics for evaluating the student learning immediately after the black board teaching

Objects identified	Well framed (3 points)	Moderately framed (2 points)	Improvement required (1 point)	No objects (0 points)
Attributes identified	Well framed (3 points)	Moderately framed (2 points)	Improvement required (1 point)	No attributes (0 points)
Operations identifies	Well framed (3 points)	Moderately framed (2 points)	Improvement required (1 point)	No operations (0 points)
Class diagram	Well drawn (2 points)	Moderately drawn (1 points)	Improvement required (0.5 points)	No class diagram (0.5 points)

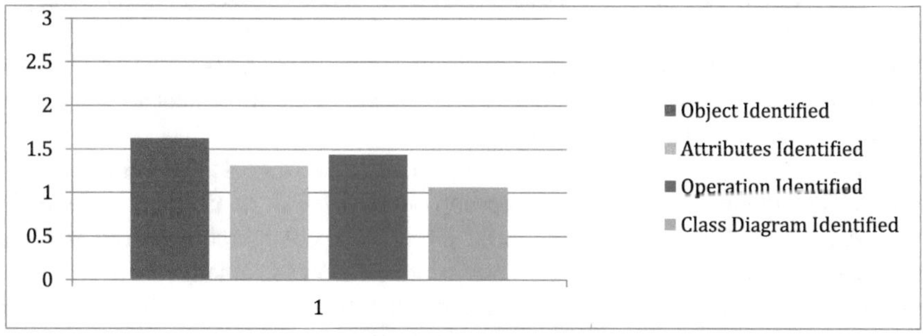

Fig. 3. Analysis of marks scored by the students, immediately after the black-board teaching aid

Hybrid Model Level 2

The major drawback of the phase one is, unable to reach towards the visual learners. It only addressed the verbal learners usually very minimum in the crowd. So, it is decided to use the virtual lab aid to reach out these levels of learners. The virtual laboratory link is shared among the students and directed them with a lab manual. To measure the student performance a quiz is conducted. The below chart shows the performance which is not so much satisfactory (Fig. 4).

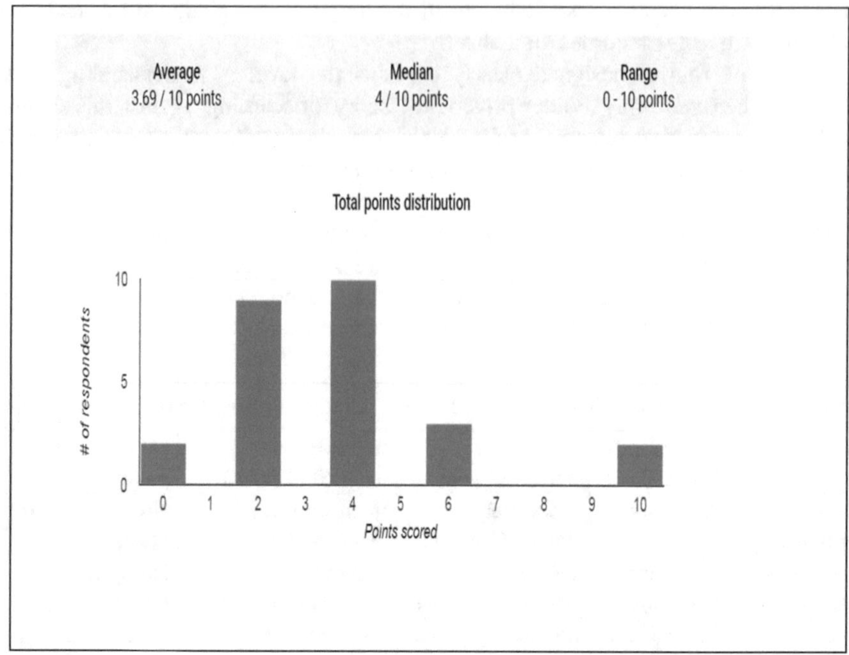

Fig. 4. Analysis of marks scored by the students immediately after the virtual lab session

We can see that the results are not satisfactory as it shows that average scoring point is 3.69. It is because the students are not disciplined to the online modalities because of various physiological behaviors and disruptions very few students accepted virtual lab link. But it clearly explains that there is a considerable improvement in the students learning. There are some high scores too.

This is may be because the visual impact reached towards the certain level of learners. They could see the step by step implementation of the designing of the class model. A problem statement is show in the simulation tool, which is solved as per the steps mentioned in the text book. These steps are difficult to some levels of learners as they are not good at concentration. But the simulation tool makes their attention with little time, with definite steps and with an interactive user interface, which is appreciated with most of the students. So it seems that this virtual laboratory tool is a good alternative to the traditional book reading which requires lot of patience.

But we also observed from the above figure that the there is no much improvements in the assessment part. The reasons for this also we discussed.

So, in step 3 we decided to demonstrate the tool among the students. It works well as most of the students used sit in the common laboratory and they experienced the virtual laboratory.

In the step 3 and 4 we found that significance numbers of students are addressed as per the Silverman-Felder learning model still it kept certain room for the improvement. The students without intuitive behaviors found difficulty in realizing the step-2 and 3. So in the step 4 as per the model a flipped class room model is imposed to the topic.

The assessment of this teaching aid is done through an experiential learning. It means that the flipped class room video suggest to do an experiment to construct a class model. So students are instructed to look into the videos shared through Google class room before coming to the session. They are asked to note down the important points reflected in the video. It is a good practice to assess these learning with an experiment. So using IBM Rational Rose Software Architect tool students are instructed to visualize the practical aspects of the topic class modeling. Here students can design the class model, as they are seen in the virtual lab link, discussed in class trough black-board and also visualized from some other authors in the flipped class room.

So, to explain a content if we effectively use all these or at-lest few teaching modalities it addresses all levels of learners with a greater practical impact on the student learning's. It is proved with our Hybrid Model of teaching also. The below chart proves that certain amount of learning is happened because of the proposed hybrid model.

Hybrid Model Level 4

We can see that this time students are able to figure out the attributes and operation, which resulted into an effective model. This effective model is sketched out in the tool and students are instructed generate the java code for the class. This java code is executed, and the expected results are analyzed and the students are asked to modify certain attributes, operations and relationships among the classed to build a best possible application for the given problem statements. The blow rubrics are used to measure the student's participation in this activity. The rubrics prepared giveclear and justifiable points to measure the student learning in this topic (Fig. 5; Table 2).

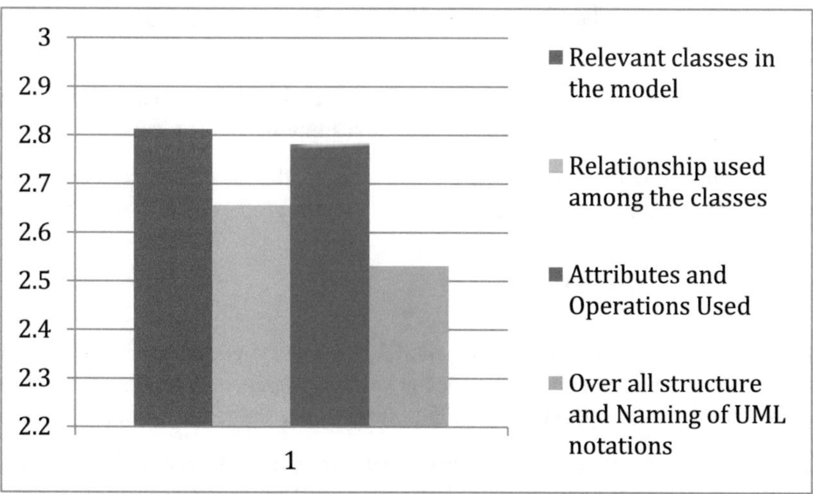

Fig. 5. Analysis of marks scored by the students after applying all levels of hybrid model

Table 2. Rubrics for evaluating the student learning immediately after the black board teaching

Relevant classes in the model	Well justified (3 points)	Moderately justified (2 points)	Improvement required (1 point)	No classes (0 points)
Relationship used among the classes	Relevant (3 points)	Partially relevant (2 points)	Improvement required (1 point)	Completely irrelevant (0 points)
Attributes and operations used	Relevant (2 points)	Partially relevant (1 points)	Improvement required (0.5 points)	No list of applications (0.5 points)
Over all structure and naming of Class UML notations	Very good (2 points)	Good (1 points)	Improvements required (0.5 points)	Discouraged (0.5 points)

Benefits of the Hybrid Model

As we are seen and experienced that this model brings a lot of joy in the learning and finally ensures that a great learning's is happened from various learning angles. It is worthy to apply this Hybrid Model for the important contents we are going to teach in the class. Some specific benefits are discussing below.

- It addresses all the different categories of learners as discussed in the Felder-Silverman Model.
- It requires maximum three hours for the execution of this model, so easily we can accommodate the time.
- It effectively covers the large content delivery compared to the traditional teaching mode.
- It also links the module wise contents which give a big picture to the global learners and really it initiates innovative research ideas among the students.
- The global challenges can be addresses in this model as students are experiencing technology teaching they can easily adjusted with the global challenges in the industry.

Disadvantages the Hybrid Model

- The infrastructure facility for this model, especially virtual lab requires a high bandwidth of internet connection for the smooth visuals may hinders the execution at the learners end, especially when they want to use the facility, out of the campus.
- Certain level of online learning discipline and ethical responsibility is required while undergoing with the Virtual Laboratory and Flipped Class Room activity.
- This model may not find relevance to certain topics for which computer oriented experiential learning may not available.

5 Conclusions

The Hybrid Teaching Model is coursed into the final year post graduation students to assess their learning outcomes in the specific contents. It was observed that the technology teaching is one of the important need of the century, but the literature explores it's inefficiency in addressing the different levels of learners as per the Felder-Silverman Learning style context. The hybrid teaching method is designed to meet the different levels of learners specifically for their own psychological learning capabilities. The content delivery by using this model shows a considerable improvement in their learning. Along with this the hybrid model brings a lot of joy and best practices in the teaching—learning process. The different components of the model synergistically address all the level of learners.

References

1. Graf, S., Viola, S.R., Kinshuk, Leo, T.: Representative Characteristics of Felder-Silverman Learning Styles: An Empirical Model, Austrian Federal Ministry for Education, Science, and Culture, and the European Social Fund (ESF) under grant 31.963/46-VII/9/2002
2. Graf, S., Viola, S.R., Leo, T., Kinshuk, R.: In-depth analysis of the Felder-Silverman learning style dimensions. J. Res. Technol. Educ. **40**(1), 79–93 (2007)
3. Felder, R.M.: Hoechst Celanese Professor of Chemical Engineering North Carolina State University, Learning Styles And Strategies, Abrahim H. Maslow (1908–1970), Humanistic Theory of Learning. http://www.lifecirclesinc.com/Learningtheories/humanist/maslow.html
4. Felder, R.M., Brent, R.: Active learning: an introduction. ASQ High. Educ. Brief **2**(4) (August 2009)
5. Felder, R.M.: Learning and teaching styles in engineering education. Eng. Educ. **78**(7), 674–681 (1988)
6. Wirz, D.: Students' learning styles vs. professors' teaching styles, from inquiry **9**(1) (Spring 2004)
7. Felder, R.M.: Reaching the second tier: learning and teaching styles in college science education. J. Coll. Sci. Teach. **23**(5), 286–290 (1993)

A Skill Enhancement Virtual Training Model for Additive Manufacturing Technologies

Arjun C. Chandrashekar(✉), Sreekanth Vasudev Nagar, and K. Guruprasad

Department of Mechanical Engineering, B.M.S College of Engineering, Bengaluru, India
{arjuncc.mech, snv.mech, kgp.mech}@bmsce.ac.in

Abstract. Additive Manufacturing (AM) is one of the advanced manufacturing technology which has shown tremendous promise through revolutionizing the manufacturing paradigm. AM has made inroads into various domains through its versatility and distinct advantages of manufacturing. However, access to AM machines for training of personnel has been challenging due to high costs involved in the process. This research is undertaken to provide a feasible solution by incorporating virtual training modules pragmatically. This work aims at exploring opportunities to potentially solve the aforesaid concerns using training modules which can be accessed remotely using mobile connectivity. Purpose of this research effort is to provide a structural framework of design to enhance the productivity and skill sets of youth population of age group of 18–35 years in the domain of Mechanical Engineering; thereby making them industry ready through effective virtual training. The research is carefully crafted keeping the high costs involved in the conventional training using Industrial grade AM machines. For developing economies which may not be able to afford the usage of AM machines for hands-on training, this approach may become a game changer. The authors feel the need to explore the grey area and find a common point wherein neither access to training of personnel nor machinability of the equipment is not compromised. The authors believe this model will bring a new dimension in the context of advanced skill enhancement in the manufacturing sector in general and AM Technology in particular.

Keywords: Additive · Manufacturing · Virtual training · Skill development · 3D printing · Simulation software

1 Introduction

The greatest advantage of additive manufacturing over conventional manufacturing technologies is its ability to produce parts with complicated geometries keeping the structural integrity under consideration. This attribute of AM offers a wide spectrum of operational capabilities. It is suitable in diverse fields like Military Technology, Food Industry, Aerospace, Mechanical Engineering etc. [1]. Along with this, it also provides a distinct advantage of producing parts with convenient ease from various geographical

M. E. Auer and K. Ram B. (Eds.): REV2019 2019, LNNS 80, pp. 532–543, 2020.
https://doi.org/10.1007/978-3-030-23162-0_48

locations remotely. In an ever changing dynamic manufacturing environment, where focus is on Cyber Physical Systems (CPS), a virtual supply chain model can play a vital role wherein: each component of a sub system can be remotely manufacturing and shipped to an assembly center through AM Technology. A blend of CPS and AM can bring about an enormous traction into the manufacturing space which can potentially disrupt conventional binding of regular supply chain issues and make way for disrupted supply chain systems. Apart from industrial applications, AM has made inroads to homes of enthusiasts where household articles, toys and toy parts are being created which was otherwise not possible earlier. Schools and colleges make use of 3D Printers to demonstrate concepts by printing a 3 dimensional object which is under discussion. This is visualized to give the students an intuitive understanding of the subject which otherwise may be abstract.

From the industry stand point, we can observe a sea change in the methodology of working in the last decade, where modern and higher efficient engineering systems are taking positions of the conventional methods and bringing in a welcome change with higher productivity. Figure 1 shows the trend of increase of procurement of Metal Additive Manufacturing Machines year after year starting from 2000 till 2017. According to Wohler's report of 2018, a steep rise in trend pertaining to the rise in metal additive manufacturing machines may be observed. In 2017, an estimate of 1768 machines are procured worldwide which is around 80% increase compared to 2016s sales of 983 machines. In the year 2017 alone, the AM industry is said to have grown by approximately 21% with an expanded share of about 1.25 billion $.

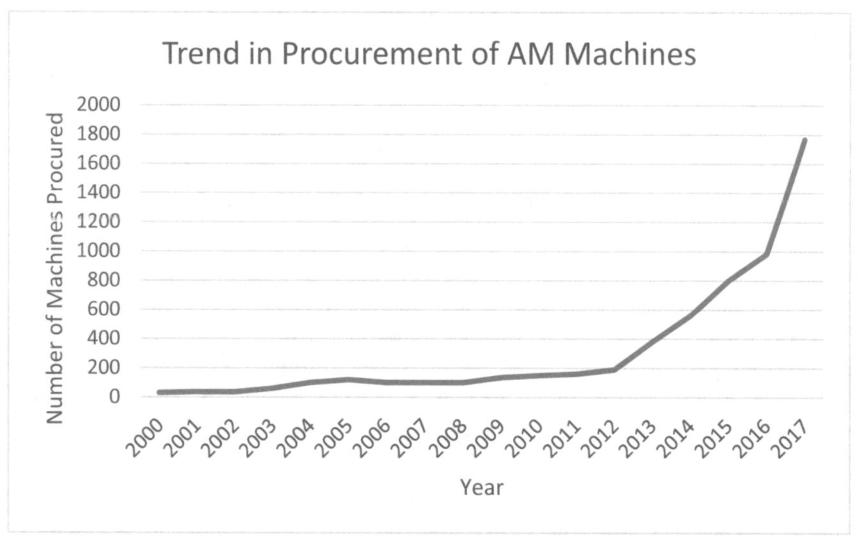

Fig. 1. *Source* Wohler's Report 2018 via www.plasticstoday.com [2]

With an upward surge in the demand for AM, various companies have stepped into production and service of AM machines. As evident as it is from the pattern in the graph, there is a growing demand and acceptance for AM related technology and companies across the world are using AM to realize their components. With this increase in demand, a definite need to provide the essential skillset plays a pivotal role in the entire framework of this manufacturing model. In the Indian scenario, according to the Government of India's Ministry of Heavy Industries and Public Enterprises policy of National Capital Goods 2016, emphasis is laid to Industry 4.0 technologies and recommendations to upgradation and provisions for skill based training for existing and emerging workforce [3]. Overhauling of technologies are leading to faster changes in the manufacturing paradigm; whose corner stone is the skillset required to undertake operations. With the demand for the new skilled force, Mckinsey's report on India's Labor Market: 2017 edition, suggests that the manufacturing industry in India may hire 2.5–3 million more workers by 2025 provided they acquire the skills suiting the changing needs of the industry [4].

2 Literature Survey

Nasscom report on "Skilling for Digital relevance" published in 2017 shows describes the direction which the industry is treading in the past few years. Automation is creating a pragmatic change in the industry. For example, in the US, 21 million jobs are lost due to technological and business related shifts every year but US also creates 23 million new jobs during the same time. Inspite of rapid changes in the dynamics of the market due to embracing technology, the unemployment levels in the UK and US is at its lowest. It has become imperative to the industry to find renewed ways to stay in the helm by reskilling its workforce. New shifts in skilling like virtual classrooms, e-learning platforms, self-learning, cloud based and app based platforms, MOOCs, hackathons, gamification and tech-talks are the newer ways firms are employing to engage their workforce in development of newer skills [5]. Songkram et al. [6] have developed an e-Learning module to enhance the cognitive skills of an individual (learner). Cognitive skills refer to the problem solving, analytical thinking, creative thinking and critical thinking capabilities of the learner. The authors have created subsystems in Blended Learning Environment and Virtual Learning Environment with 240 participants across varied disciplines. The study resulted in creation of a robust model which will evaluate the learner's aptitude on formative and summative assessment pertaining to the learner's cognitive skillset. Virtual Learning through Mobile Applications is gradually gaining traction in the last decade. A growth opportunity through mEducation (Mobile Education) is worth about 32 billion $ by 2020 [7]. Skillsets which are required through training can be systematically analyzed summative evaluation, formative evaluation, direct engagement, structured delivery modules with the aid of mEducation platforms. Multinational technological companies are employing newer virtual ways to engage their workforce in enhanced skill development. Video On-Demand, Webinars, eLearning modules, Virtual Instructor lead video training sessions, web enabled task tools for online learning and support are some verticals companies are engaging these days [8]. Along with virtual engagement, major global companies these days offer web based training resources and virtual classroom experiences to connected stake holders. The technical competency and skillset of a prospective learner can also be certified by virtual trainings these days and various levels of certifications viz., Professional, Expert and Master are offered. Depending on the level of certification achieved, professionals can thereby apply for jobs in the corresponding segments and explore employment opportunities [9]. With advancements in the field of Information Technology and Digital Connectivity, learning virtually has become very affordable and accessible. In the mobile internet booming world of 4G and 5G connectivity, virtual learning platforms can be a great opportunity for all the stake holders. The mobile based education segment is a path breaking shift in the learning space. Educational content is accessible anywhere even the remote locations of countries with 2G, 3G or 4G supported by WiFi enabled mobile devices. Already, in remote villages of India, primary schools train students in English using mobile applications [7]. This shows the enormous potential which is available through mEducation. Apart from the learning potential, the business potential available through accessible virtual learning is superior. Mobile operators can ride along this wave and create more revenue for themselves. Creation of Virtual ecosystem encompassing IT and Tech Support, Content, Hosting and Data Management services, mobile operators can tap into a revenue pool of a market size of 20 Billion USD. Another opportunity for the mobile operators is that, they can lead the technology as an end-to-end service provider. Here, the operator can become a one stop platform to provide all the solutions across the horizon pertaining to a specific virtual learning technology. Mobile network companies can partner with other players in the virtual education ecosystem to provide better offers and attractive discount plans. The virtual learning model primarily depends on the mobile connectivity, therefore, the mobile network companies can rely on connectivity services partnership and also extend to other services.

Other aspects which is critical to this entire framework is the allied services include cloud based storage and hosting, DBMS, Analytics, content management and delivery. Many mobile operators are engaging in partnerships with educational companies and institutions to provide comprehensive education supplemental learning solutions. With this emerging multi-billion potential market which is disruptively revolutionizing the education space, an offshoot of its projection can easily be panned towards the corporate learning or skill learning space. The mobile learning platform is visualized to dent the existing conventional learning methodology drastically in the years to come. As mobile connectivity is abundantly available in almost all parts of the populace in any habitable part of our world, integrating learning and skill enhancement services to this will provide a feasible alternative to the skill mapping potential of the society.

Mohammad Shorufuzzaman et al. in their work have shown a model how big data analytics in the cloud supports learning analytics for mobile learning. A tasking challenge in this quest of data handling is the limited processing capabilities of the mobile devices. This is overcome by offloading heavier computational parts to the cloud thereby providing enough space for smoother handling. The research provides cloud based mobile learning methodology which employs big data analytics technique in mobile learning interface platforms [10].

3 Analysis of Existing Model

3.1 Financial Aspects of Additive Manufacturing

Let us consider an example to understand the existing model in a manufacturing set up of a 3D Printed part. This model can explain how the new virtual training model can financially be a better model as compared to the existing set up. Considering an example of a 3D Printer whose cost is $700,000, at a rate of 20% depreciation each year, the ROI of the machine will be possible at the end of fifth year as shown in Fig. 2. However, there are many other parameters which go into the calculations of the actual cost model during manufacturing.

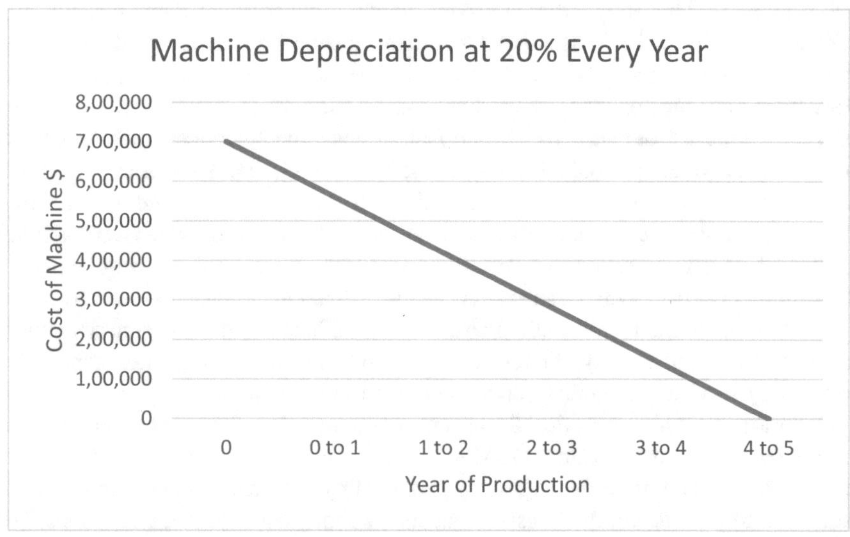

Fig. 2. Graph of straight line depreciation of actual machine cost over 5 year period

Additionally, there are other parameters of the 3D printing cycle which go into the budget as costs viz., designing of the component, production of the component, post processing and qualification. This being an iterative process, it will keep repeating over and over again. Each of these aspects can be understood further as explained in the example provided in Table 1.

Table 1. Example calculation considering various costing aspects in Additive Manufacturing

Parameter #1	Design of the 3D printed component
Design time	Approx. 1 month
Design engineer salary	$100,000/year
Software licensing cost	$800 per year
Labor cost per month/operator	$100,000/12 months = $8334
Software cost per month	$800/12 months = $67
Parameter #1 Total cost/month	$8334 labor + $800 software = $9134
Considering 20 components per month as production rate, per component cost would be:	
Per component cost	$9134/20 components = $457 for each component
Parameter #2	Material Cost @ 40% of per component cost = $183
Parameter #3	Post Processing @ 30% of per component cost = $137
Parameter #4	Overheads @ 30% of Per component cost = $137
Total parameter cost/month	Parameters #1 + #2 + #3 + #4 = $914

It can be noted from the example considered that, it requires approximately $900 to manufacture a component using a $700,000/- machine considering separate costs for design, software, materials, post processing and overheads. This cost of $914/- is exclusive of the training cost which goes into the component. With the present methodology of live training, the lead time involved in production will affect the ROI of the overall expenses as the machine is off from production during training of prospective employees. As these AM machines are highly expensive and also durable, they are generally used all days of the year in three circular shifts with very little idle times for periodic maintenance. However, if any given machine is stalled for training purposes, potentially, it will be using up hundreds of dollars which otherwise would be utilized during production. Presently, as there are no other alternative available, companies are forced to compromise when it comes to training: either they are required to slowdown the production by giving hands-on experience to trainees or give offline training through textbooks and other teaching methods which has a negative impact in production quality as the trainee does not have an intuitive experience of working on the machine. To resolve the two stated issues, a pragmatic virtual training module sits perfectly catering to both the points under consideration.

3.2 Drawbacks in Traditional Training Methodologies

Most of the companies in today's manufacturing scenario depend on traditional training methods like referring from manuals, chalk and talk teaching methods. Most of the content are picked from traditional sources like textbooks and handbooks which might be highly theoretical in nature. Traditional methods fail to address the critical thinking faculties of the learner. Process oriented learning techniques and dynamic interaction may not be available in the traditional methods of training. The text book style of training fails to give an immersive experience to the prospective learner which might not kindle his learning acumen.

3.3 Other Factors Influencing Training and Development

Not all the towns and cities are equipped with state of art training and technology enabled centers. For all the populace who are stationed in remote parts of the country, access to training and material will be not possible.

Another challenge for trainees to get an intuitive experience is the lack of availability of machines themselves; As the machines are highly expensive and always on production demand, they are either very few in number or not accessible due to high demand and high investment cost involved.

4 Parametric Variables Responsible for Development of Virtual Training Module

4.1 Mobile Network/Internet Connectivity

As the crux of the virtual training module is dependent on the internet, it is of paramount importance to create platforms which function seamlessly giving the user a hassle free experience while working on the training platform.

4.2 Language

In the Indian context, there are many regional languages apart from English and Hindi. To cater to a wide populace of diverse backgrounds, regional language interface must be an inbuilt feature in the virtual training module. Having an opportunity to understand a specific concept in the regional language will provide an enriching experience to the user and makes learning simpler.

4.3 Subject Matter Expertise

It is highly relevant that the concepts, illustrations and examples provided in the training module is accurate and wholesome. As the learner is expected to learn the content comprehensively only through a virtual training platform, the relevance and authenticity of the content is to critically evaluate for correctness and accuracy.

4.4 Catering to Variation in Technological Aspects

In the context of AM, there are multiple technologies namely Fused Filament Fabrication, Selective Laser Sintering, Digital Light Processing, Electron Beam Melting, Selective Laser Melting, Stereolithography etc. Each of these technology has a different methodology of working with diverse process parameters and machining configurations. A comprehensive understanding of each of these modules individually might become a costly affair to a prospective learner. However, a virtual training module can provide concurrent access to all the variants which are embedded into a single training module.

4.5 Devices for Accessing the Virtual Training Module

Holistically, the virtual training module is to be made available across all peripherals like mobile phones, desktop/laptop computers, tablets. The module needs to be configured for all generic operating platforms like Windows, ioS, Android etc.

4.6 Servers and Networking

Appropriate servers with enhanced specifications, network capabilities to maintain large data need to be taken account for. Multi mirrored servers need to be incorporated to ensure concurrent access to many users at the same time.

4.7 DBMS and Multi User Access

Effective systems to store and fetch data must be made available in the software, multi user concurrent access capabilities must be made available to the users.

4.8 Network Security

Robust network security mechanisms need to be in place to safeguard the data of the users.

4.9 Data Integrity and Responsive Website

Critical systems to ensure safeguard of the credentials and data of users need to be safeguarded. Responsive website which enables transition between various interfaces must be made available for enhanced user experience.

5 Development of a Generic Training Template

It is imperative that the virtual training module be as generic as possible encompassing various aspects of the technological horizon. Irrespective of the training one visualizes for developing a module, this generic model encompasses all the critical aspects required for a robust interface. As showcased in Fig. 3, a common framework across all OS is a mandatory requirement to cater to diverse populace of users. A generic process capability index comprising data for various machine types and models need to be integrated to give the user a working experience of various possibilities he/she may encounter in the real working environment. A user friendly GUI is essential for the ease of navigation. Immersive and intuitive course content which should be adaptable across all the output peripherals. Data compression through *lite* apps based and cloud based platforms will benefit the users multiple ways. All the content in the virtual training module to be discretized into frames; uncluttered and sequential streaming of content will enable the users to get an intuitive and engaging experience while learning the content.

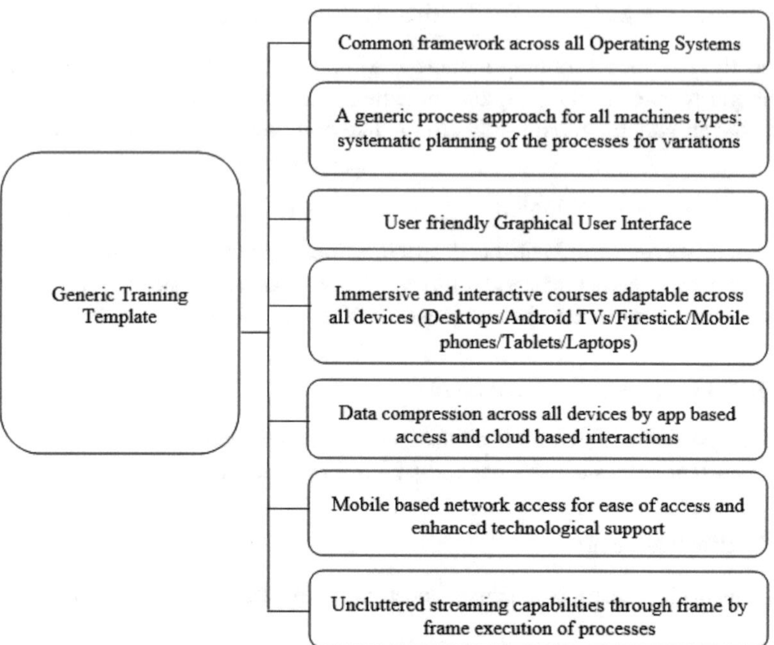

Fig. 3. Flow of parameters involved in the creation of a generic template for virtual training module development

6 Possible Impact Assessment in the Industry Through Virtual Learning Platforms

- Virtual learning through mobile and internet based platforms is visualized to create newer dimensions of learning across industrial spectrums.
- Its easy access to training on-the-go provides the user a great deal of comfort to learn during multiple intervals.
- A platform of this nature provides a network of like-minded people through forums leading to an enormous possibility of data exchange and knowledge bank creation.
- Virtual training methodology can have a positive impact on other developing economies in creation of need based modules in complex technologies where training of personnel is a tasking challenge.
- Education paradigm will greatly benefit from virtual learning as the teachers and students can interact with an intuitive experience considering a learning module during discussions.
- Illustrations, videos, animations, flipped classes etc. can be integrated to the learning modules to provide the user an experiential learning outcome.
- This module can be stepped up to understand various parameters of the machine by the means of simulations.
- New component and feature testing can be done virtually and assessment of the same can be obtained dynamically.

- Multiple users can interact concurrently to develop challenging scenarios to simulate a real world environment.
- This technology can be utilized to train a parallel skill force in tandem with the existing workforce; critical dependencies on skilled workers can be greatly minimized.
- Governments can understand the impact assessment on the magnitude of the technological intervention and develop corresponding policies to aid in mutually beneficial systems for effective governance and businesses.
- Online competitions and Hackathons can be organized to enhance the productivity of the technology, find niche domains for improvement and assessment of probable areas for future possibilities.

7 Proposed Revenue Generation Model Through Virtual Training Methodology

Although there are multiple technologies within the field of Additive Manufacturing, all the technologies follow the same process flow to get the finished component. The eight step process in AM is given in Table 2. The initial step is the design of the component through a CAD model. Next, the 3D model is converted to the standard tessellation file through .stl extension.

Table 2. Step by step process of a 3D printed part manufacture

Process No.	Description
1	CAD model
2	CAD to STL creation
3	Model preparation
4	Sliced model file along with G-Code
5	3D Printing process
6	Material removal from machine
7	Post processing of 3D printed parts
8	Application of 3D printed part

Then, the file is fed into the 3D printer, followed by the slicing of the model. Next, the model is prepared according to the build geometries and other design aspects. The model is further converted into G-codes for the machine to print in the prescribed coordinate system. Further, the actual printing takes place layer upon layer till the final build geometry according to the CAD model is achieved. After printing, the component is removed from the machine and undergoes the post processing before it is sent for the application after quality checks. All of these processes need to be embedded into a software and built for commercialization. This can be achieved by the following steps as showcased in Table 3.

Table 3. Steps involved in revenue generation through virtual training modules

Step 1	Development of 3D printer related courses for R&D centers, academic institutions and consultancy centers and creation of software for the same
Step 2	Tie up with companies like Protiq, Shapeways, 3D Hubs, Materialize etc. who offer virtual 3D Printing services.
Step 3	Establish connections between Need (Step 1) and Supply (Step 2) and set up an online market place and initiate an e-commerce venture for development and testing of 3D printed parts
Step 4	Initiate an affordable costing based subscription model for sustainability and periodic maintenance of training modules
Step 5	Dynamic updates and add ons after signing up with various machine manufacturers
Step 6	Establish an online pool of technological database for the benefit of all stake holders

8 Conclusion

This research is aimed at understanding the critical aspects required for training of personnel in advance manufacturing technology like Additive Manufacturing. The authors opine that internet based and mobile based connectivity potentially can solve the skill related lacuna in the manufacturing sector in emerging economies. The structured framework in this research work focusing on the fundamental aspects which go into the creation of virtual training modules. Using these concepts, a software module may be developed encompassing a host of training features and aid in bridging the skill level gap among the workforce in the manufacturing sector. Effective utilization of resources combined with optimized production planning with skilled personnel will be a winning combination to the manufacturing industry in today's rapidly changing world driven by cyber physical systems through Industry 4.0.

9 Scope for Future Work

In the future, a robust software module which can work across different OS is visualized. The software can comprise of all the different packages pertaining to AM Technology enabled with advanced features for the users to gain a intuitive experience of working on the actual machine. The software may be developed fully and shared across different domains across the society viz., academia, Research and Development Institutions and Industry. An open source database may be created to obtain feedback of the entire pool of knowledge which will be created with the use of the software over different intervals of time. The data which is available may be utilized by relevant professionals to create appropriate mechanisms in terms of policies to aid in better infrastructure and production facilities to benefit the society at large. Concurrent usage of the module will give rise to a uniform level of skill development across various regions of the manufacturing space from professionals from different nations. This data may be utilized to create advanced manufacturing processes which can contribute to solve problems which pose greater challenges to manufacturing and production.

Acknowledgements. The authors are grateful to the Department of Mechanical Engineering, Product Innovation laboratory supported by Dassault Systemes and Propel Lab—II (Additive Manufacturing Laboratory) at BMS College of Engineering, Bangalore for continually providing support and encouragement during the course of this research work.

References

1. Nieto, D.M., Lopez, V.C., Molina, S.I.: Large-format polymeric pellet-based additive manufacturing for the naval industry. Addit. Manufact. **23**, 79–85 (2018)
2. Wohlers Report: Dramatic Rise in Metal Additive Manufacturing (2018). https://www.plasticstoday.com/content/wohlers-report-2018-documents-dramatic-rise-metal-additive-manufacturing/193108598558520
3. National Capital Goods Policy: Building India of Tomorrow: Government of India, Ministry of Heavy Industries & Public Enterprises, Department of Heavy Industry (2016)
4. Woetzel, J., Madgavkar, A., Gupta, S.: McKinsey & Company Report, India's Labour Market-A new emphasis on gainful employment (2017)
5. Nasscom Report (2017). Skilling for Digital Relevance
6. Songkram, N., Khlaisang, J., Puthaseranee, B., Likhitdamrongkiat, M.: E-Learning system to enhance cognitive skills for learners in higher education. Proc. Soc. Behav. Sci. **174**, 667–673 (2015)
7. Transforming learning through mEducation: A Mckinsey & Company Report (2012)
8. Siemens Training Catalog, Siemens Power Academy. Energy Management Software www.siemens.com/learningcloud
9. Bosch Security and Safety Academy. https://www.boschsecurity.com/xc/en/support/training/
10. Shorfuzzaman, M., Shamim Hossain, M., Nazir, A., Muhammad, G., Alamri, A.: Harnessing the power of big data analytics in the cloud to support learning analytics in mobile learning environment. Computers in Human Behavior. In Press, Corrected Proof (2018)

Colour Histogram Segmentation for Object Tracking in Remote Laboratory Environments

Mark Smith, Ananda Maiti(✉), Andrew D. Maxwell, and Alexander A. Kist

Faculty of Health Engineering and Sciences, University of Southern Queensland, Toowoomba, Australia
{mark.smith, andrew.maxwell}@usq.edu.au, anandamaiti@live.com, kist@ieee.org

Abstract. Remote Laboratories are online learning environments where a major component of student's learning objectives is met though visual feedback. This is usually through a static webcam feedback at non-HD resolution. An effective method of enhancing the learning procedure is by tracking certain objects of learning interests in the video feedback. Detecting and tracking moving objects within a video sequence commonly employs varying segmentation methods such as background subtraction to isolate objects of interest. This paper presents two colour histograms models as a method to segment frames from a video sequence and an end-to-end tracking system. Six tests and their results are presented in this paper with varying frame rates and sequencing times.

Keywords: Computer vision · Image segmentation · E-learning · Remote laboratories · Cyber-physical systems

1 Introduction

Object detection and object tracking of live video is the focus of research within computer vision disciplines for its benefits within fields such as surveillance, robotics and recent autonomous navigation applications. Methods employed to extract meaningful data from the sequential two-dimensional digital data sets are complex, with varying levels of success. Image segmentation is a common method employed by image processing models to classify relevant from irrelevant data sets.

Detection and tracking of objects within a video stream has consistently been difficult to perform in real-time. Image segmentation classifies portions of a two-dimensional digital image into homogenous regions associated with various attributes such as colour/grey-level or distribution. Segmentation is both a detection method and an enabler of secondary detection processes. Most segmentation models are heavy on ITC resource usage, which is one of the primary obstacles to real-time video processing.

M. E. Auer and K. Ram B. (Eds.): REV2019 2019, LNNS 80, pp. 544–563, 2020.
https://doi.org/10.1007/978-3-030-23162-0_49

Real-time tracking becomes important for systems such as surveillance or within applications where synchronisation to time critical systems is important. Systems which require user interaction with the video stream demand real-time processing. Augmented Reality fails to be usable if computer enhanced imagery is unable to synchronise with the detected objects within the video stream. Repeatability and reliability is also critical for such systems.

Segmentation of the frame images has been effective at removing unnecessary data from the digital set. Background segmented data can generally be ignored, leaving the foreground or object of interest for further processing.

Colour Histogram indexing methods have previously been used to identify objects from a library of objects through the colour signature of the object. This paper uses knowledge of the key colour components of an object of interest and allows a two-dimensional video frame to be segmented based on the objects colour signature. The resultant data set highlights regions of the frame where the object of interest is located. Marking the attributes of the resultant object then allows further tracking. This tracking system is unaware of the object as a physical object; it is tracking the same data set between each frame.

In a Remote Laboratory (RL), experiments are viewed with a camera feedback on a webpage. The resolution of these videos is typically low. There is also a high degree of latency. For proper learning purposes, students must be able to view the objects properly and understand them. They must be able to keep track of them during the entire course of the experiment. This may be aided by the web-page and the Remote Laboratory management system (RLMS). The RLMS can keep track of the object and provide tracking features along tagging and highlighting of the objects with supplementary information. This is true for any cyber-physical systems that are observed though cameras only.

The assumption in this paper is the video feedback contains objects that perform a routine task. The number of such task could be very large, but not random. It also has a limited degree of freedom. In other words, there is no human or animals it the video as these entities generate a largely random behaviour.

This paper is structured as follows. Section 2 provides a brief overview of the current computer vision segmentation methods and colour histogram capabilities. Section 3 explains how the two methods of histogram segmentation occurs for digital frames and Sect. 4 details the tracking model, describing the histogram segmentation and interest feature point creation and tracking. Section 5 discusses the results of the newly developed tracking system while Sect. 6 concludes this paper.

2 Previous Work

Two dimensional digital images and digital frames from video streams may undergo segmentation for many reasons, using many techniques. Simple segmentation to achieve binary images was popularised through the Otsu technique [1] in which the Otsu threshold is calculated so that images can be segmented based on the overall grey-level. The Otsu method is still used for images where dark and light sections require clear discrimination. Binary markers such as used by ARToolKit, used in some Augmented Reality (AR) systems, use Otsu to filter noise before processing markers [2].

Binary segmentation does not address complex image attributes such as colour and composite objects. Multifaceted segmentation became necessary as the needs of the computer vision industry grew. Gaussian Mixture Models (GMM) determined foreground from background as a result of the probability density function. While popular, modelling heavily consumes ITC resources and the process was unable to function in real-time. This form of segmentation still provided reliable results with systems capable of tracking human faces [3], with cluttered backgrounds, and complex imagery [4].

Alternative segmentation methods, other than GMM involve clustering techniques. Popular clustering models such as DBSCAN [5], Fuzzy C-mean [6] and K-Means [7] are able to effectively segment [8] colour images. All clustering methods classify homogenous sections of the digital image based on model rules such as the spatial extent between pixels and colours. Fuzzy C-mean's clustering popularity for image segmentation is due to its simplicity yet effectiveness in classifying pixel membership. Clustering techniques, even so called *fast* clustering models, perform iterative and recursive processing, causing delays which are unacceptable for real-time operations.

Robust object identification through colour indexing is achievable utilising colour histogram techniques. Databases record and index images based on their histogram signature. Object identification occurs using standard histogram association methods such as histogram intersection [9] or Chi-squared comparisons [10]. Object identification through histogram matching is robust to resolution, camera pose and occlusion [11].

Previous histogram segmentation methods recursively split regions into subregions; based on threshold values calculated from the histogram of the features [12]. This method requires several steps in an iterative process where peak values in the RGB values define the thresholds for further region segmentation. Such methods do little to aide object detection.

3 Histogram Segmentation

The goal of 2D image segmentation is to filter the data set into two categories. The first category is of regions of interest, commonly classified as the foreground. The second category is the background, and usually discounted. Upon completion of a segmentation process, the image should consist of data which has been *isolated* from the clutter of the background.

It is assumed that a two dimensional digital image is defined as $I{:}[0, width-1] \times [0, height-1] \rightarrow [0, 255]^3$, where $I{:}[x] \times [y]$ is a colour point of the image consisting of three dimensional RGB colour bytes, e.g. pixel $p = (x, y)$ as shown in Fig. 1.

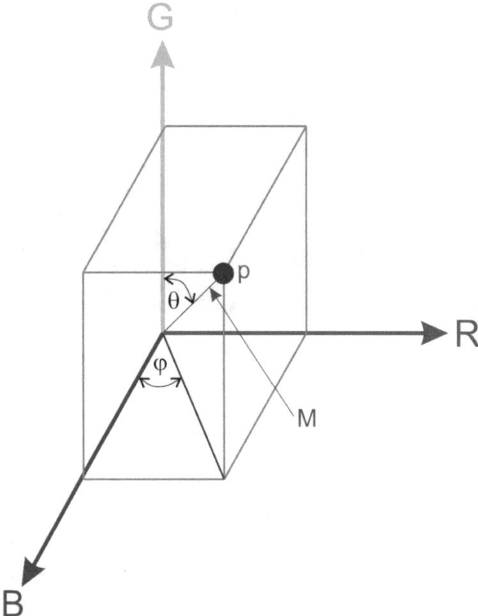

Fig. 1. Colour point within the RGB colour space (Values recorded as RGB values or polar coordinates)

3.1 Histogram Model

Within image processing, histograms analyse the pixel attributes and records the frequency of pixel colours. The frequency distribution of colours represents the occurrence of each colour within the image. A simple RGB colour number is not enough to show the relationship between colours, as a 24-bit colour scheme is simply a number. For example, there is no apparent correlation between colours 14,554,250 (pink) and 12,526,218 (purple), yet they share a common colour (14,554,250 is hexadecimal #DE148A and 12,526,218 is hexadecimal #BF228A which have the common blue colour 8A).

For a single pixel colour, the colour sits within the RGB colour space shown in Fig. 1. Storing the frequency distribution of a 24-bit full RGB colour image, colour histograms requires 16,777,216 separate data bins. Such fine precision causes problems for histogram comparison. Within unchanging consecutive video frames, factors such as noise from the image capture devices produces very different histogram distributions. Figure 2 shows two consecutive frames from a static video scene, yet the histograms are quite obviously different. Creating wider histogram frequency distribution bins, which accumulates pixel colour data for a colour range, improves the effectiveness. Histograms for colour image processing are multi-dimensional so as to record the individual RGB colour channels for each pixel.

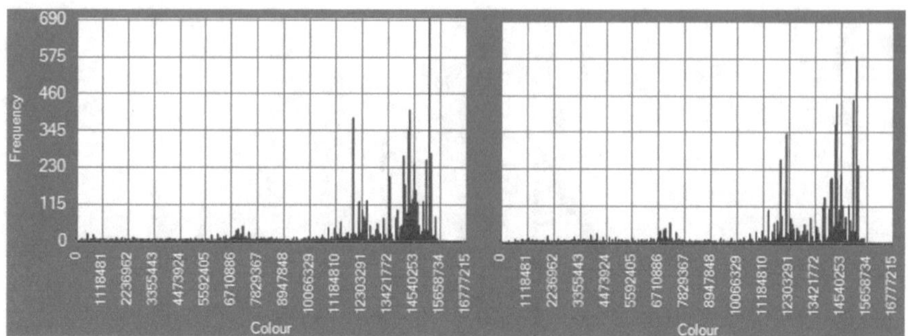

Fig. 2. Histogram of two consecutive frames of a video stream. Differences can be seen between the frames, even though the scene is static

3.2 Fast Histogram Model

Improving histogram performance balances with the needs of colour detail detection. Instead of standard hexahedron shaped accumulation bins within the RGB colour space, an alternative histogram model [13] utilises square pyramid shaped bins. The model shown in Fig. 3 requires only two dimensions to define the colour space, since the height (magnitude) of the square pyramid is not required. The two-dimensional histogram can then be defined as $H = \left\{ H_{\varphi\theta} = num\{p | I(p) \in B_{\varphi\theta}\}, (\varphi, \theta) \in [0, N-1]^2 \right\}$, where N is the bin count, and B the current bin for colour pyramid dimension (φ, θ).

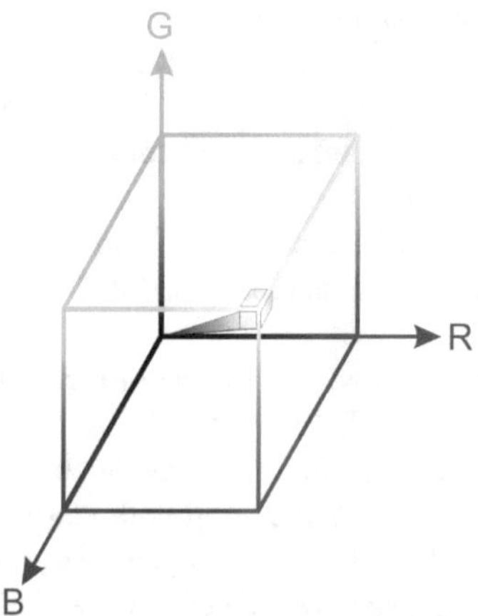

Fig. 3. Square pyramid histogram accumulation bin. Pyramid peak at 0,0,0 (black) to colour space border

Square pyramid histograms group similar digital defined colours, and can improve illumination stability [13]. Slicing the RGB colour space from the black (0,0,0) origin to the extremity of the colour space partially imitates a HSL or HSV colour model, in which illumination increases through the cylindrical model. Square pyramid histogram bin sizes are determined by the size of the pyramid base, and the accumulation bins are indexed by the polar angles phi (φ), theta (θ) as shown in Fig. 1.

Square pyramid histograms provide a faster method to capture the colour signature of an image. The model requires less dimensions and groups like RGB colours/hues together to create a robust symbolic representation for future referencing. Within the fast colour histogram model, some accumulation bins are ignored during data collection such as the colours that are located along the *grey-line*, which is the bin that centres on the line drawn from black (0,0,0) to white (1,1,1). Removing this zone filters large illumination variations, allowing other colours to be maximised.

3.3 Histogram Segmentation

Due to the colour histogram's ability to indicate the colour distribution of an object, this information can be used to segment a larger image in which the object appears. Most other data within the image data set can be filtered leaving data predominantly relevant to the object of interest. In Fig. 4, one of the blue gear pivot points has been chosen, and the objects histogram is then used to segment the full image.

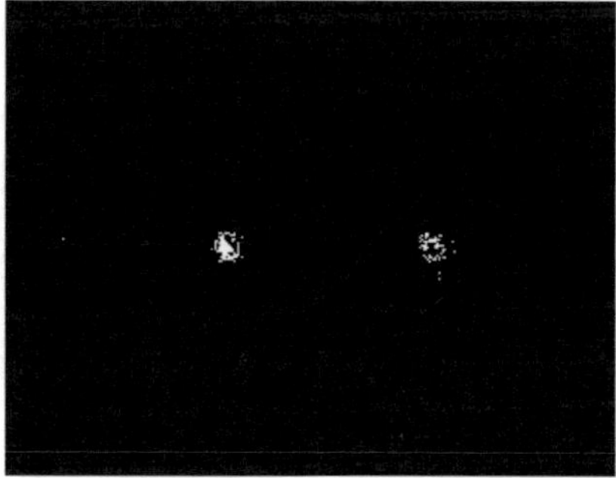

Fig. 4. Segmented Image. Blue gear fitting has been selected as the item of interest (top image), and then the image is segmented according to the histogram (lower image) resulting in a filtered image.

When an image is segmented, each pixel has its RGB colour converted to the colour histogram polar coordinate system and labelled with a histogram accumulation bin number. If the accumulation bin number does not match a bin number from the object of interest, the pixel is discarded and marked as background. Accumulation bin numbers that do match the object bin number are marked as foreground.

Choosing histogram parameters for segmentation requires only a decision about the level of precision to apply from the object of interest. Shown in Fig. 5 is the colour histogram of the blue pivot from Fig. 4. From the histogram, it can be seen that there are approximately ten accumulation bins that are significantly larger than the rest. It could be assumed that the less significant colour groups play little importance on the segmentation results. One estimate is that it only requires 200 discrete colours to identify a large number of objects [11]. Smaller values in bins, compared with the larger peaks, from a human vision point of view, may not be readily apparent.

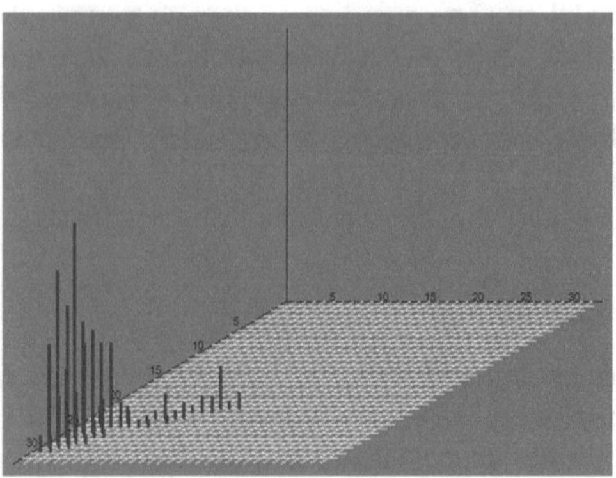

Fig. 5. 2D Colour histogram of the left blue gear from Fig. 4 (Pyramid base size 15 × 15)

Varying the number of peak accumulation bins to accept for segmentation purposes can result in largely different segmentation. Figure 6 shows the target image to locate within the displayed composite image. Table 1 shows the effect of adjusting the pyramid base size, and the number of accumulation bins to process for segmentation. The composite image consists of other regions with colours in a similar range. As can be seen in Table 1, the segmentation becomes cluttered with noise as the number of accumulation bins increases. Histogram precision also affects image segmentation, as shown in Table 1. Determining the base size for a square pyramid colour histogram balances processing needs, sensitivity to colour range, and the effectiveness to isolate or group like colours. Practice has shown that there is a large range acceptable for most applications.

Fig. 6. Target object for histogram segmentation (top) Composite image including target object (bottom)

Table 1. Histogram parameter variations (Pyramid base size and the number of larger peaks to include)

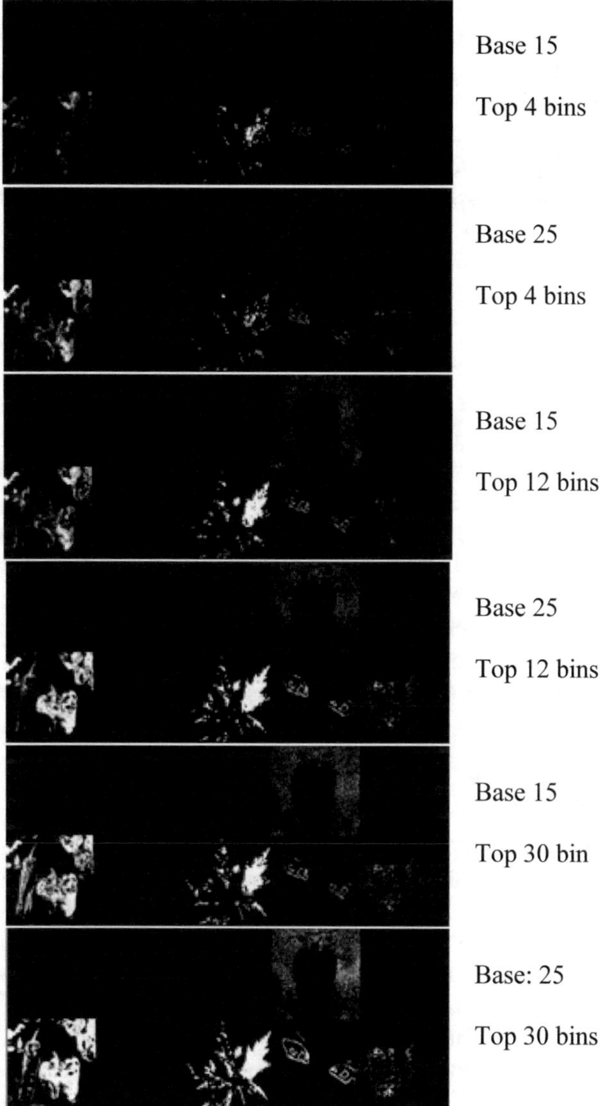

A faster segmentation technique takes advantage of a processing practice used when compressing data from image capture devices. Within the digital image, colour blending occurs as the objects natural colour varies; this is as a consequence of the image compression. shows a magnified portion of an image. When viewed at normal size, the colours appear solid, but in it is difficult to ascertain a specific colour due to the blending. Selecting a point somewhere on this image will select a pixel that is a different colour each time. Because most of the colour blending is of a similar hue, a point can be selected as the reference colour, and its applicable histogram bin as the primary colour bin. Adding the accumulation bins immediately adjacent to the central bin creates a virtual accumulation bin 3 × 3 the original base size.

The effect of using a single point colour and extrapolating the accumulation bin region is understandable when viewing Fig. 4. Fast segmentation is utilised by selecting a point on the blue gear pivot and expanding the accumulation bin influence. From this point, segmentation occurs in the same manner as the previous histogram segmentation method; classifying pixel colours which are not within the extrapolated accumulation bin region as background. Figure 12 shows the difference between the two histogram segmentation methods.

4 Tracking System

This tracking model applies colour histogram segmentation, and then discovers interest points from the resultant segmentation. The interest points become the object signature for tracking through out the video stream. Each phase of the tracking model is explained below. For all testing, a recorded video sequence of a gear assembly experiment was used. Three coloured pivot points were available to track and each was used for testing each model. The models are tested for repeatability, stability, loss of track and false positives. The key measure for these attributes is counting the number of frames that successfully locate the object selected for tracking as the video proceeds. Frames that do not find a target represent a loss of track, while an incorrect target is a false positive. A baseline tracking systems is also used to compare the effectiveness of the test models.

4.1 Histogram Segmentation

The two segmentation methods described in Sect. 3.3 represent histogram segmentation models that are clearly different with separate benefits. The histogram of the object of interest may segment an image with a large portion of colour range, based on the spread of accumulation bins. Segmentation from this model may be more precise if there are fewer occurrences of similar colours through the image. The opposite is also the case, where the segmentation result is very noisy because of similar colours which fall into the accumulation histogram bins. The second model provides faster segmentation due to its use of a single colour zone along the RGB colour lines.

For each tracking experiment, the square pyramid base size was set to 18 × 18. The objects of interest to track are the two blue and one red pivots shown in the gear experiment in Fig. 4. Applying colour histogram segmentation to the video stream frames quickly filters out the majority of data irrelevant to the tracking system.

(1) *Object Signature Segmentation*

Segmentation using the object signature technique involves selecting a square region around the object of interest or using a separate image of the object of interest. A colour histogram is then calculated, representative of the object. Figure 5 shows the fast histogram for the object of interest to track. The top four accumulation bins are selected from the histogram and maintained as the tracking bins during the tracking process (Fig. 7).

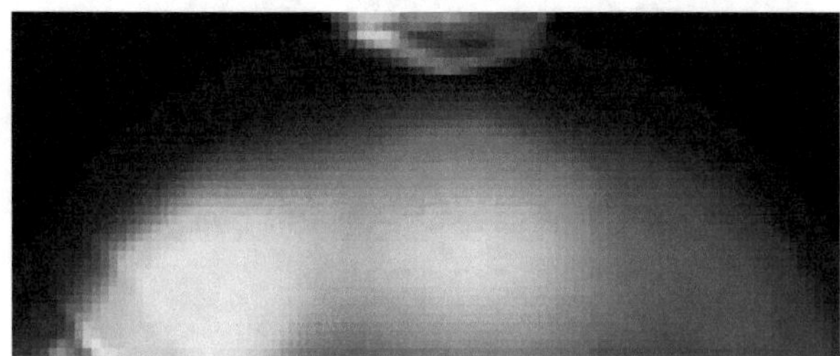

Fig. 7. Colour blending as a process of the image compression algorithm

As each video frame is received into the tracking system, it follows the steps listed in Fig. 8 where it is scanned and compared to the tracking bins. It is faster to convert the current pixel RGB colour to the polar coordinates of a histogram bin index number than it is to try to determine if the pixel colour falls within the range of the histogram bin. Each pixel is labelled with a histogram accumulation bin number. If the accumulation bin number does not match a tracking bin number, the pixel is discarded and marked as background. Accumulation bin numbers that do match the tracking bin number are marked as foreground.

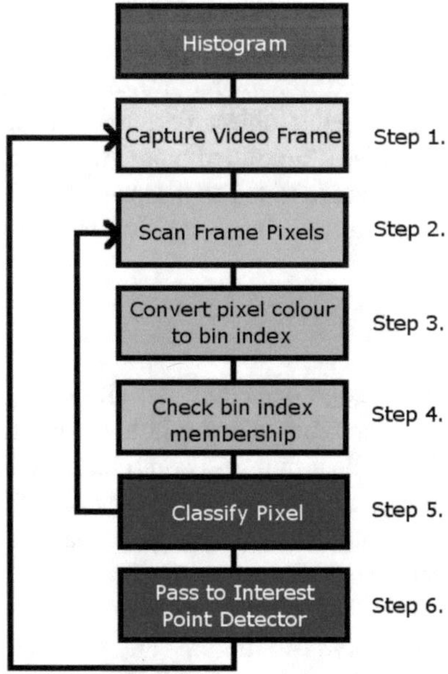

Fig. 8. Fast colour histogram segmentation steps

For each pixel of the frame, the question is: is this pixel colour within the range of the tracking histogram accumulation bin? If it is not, then it is flagged as background, else it becomes a foreground pixel.

$$pixel = \begin{cases} foreground, & if\ C \in T \\ background, & if\ C \notin T \end{cases}$$

where C is the colour of the pixel, and T = [t: t is the list of accumulation bins in the reference histogram].

At the completion of the object signature segmentation process for this frame, it is passed to the Interest Point process.

(2) *Colour Zone Segmentation*

Segmentation using the colour zone technique occurs by selecting any point on the object of interest, as shown in the digital image of the scene, which sets the reference colour. The reference colour is converted to the fast histogram accumulation bin index number. The eight surrounding histogram bin numbers are also recorded in the tracking histogram accumulation bin so that the effective tracking range covers nine times the area.

Segmentation occurs in the same manner as listed in steps 1 to 6 of Fig. 8 with the video frames captured, scanned and pixels classified. The difference is that only pixel colours that fall within the specific colour zone are marked as foreground instead of a variety of colours. At the completion of the colour zone segmentation process for this frame, it is passed to the Interest Point process.

4.2 Interest Points

Interest points may be discovered through computer vision techniques such as corner or edge detection. Binary segmented images require less processing from interest point detection methods. Another popular object detection method locates Binary Large Objects (BLOB's) and will utilise a simplified model when applied to the binary image.

Three interest point detectors are tested to discover the effectiveness of the tracking system. Each detector has its repeatability, stability, loss of track and false positives measured.

(1) *BLOB Detection*

BLOB detection is a form of edge detection and clustering techniques. During processing, if a pixel colour is greater then the threshold colour, the surrounding pixels are examined and labelled. The labels group pixels into like pixels, creating larger objects. In this model, the entire image undergoes BLOB detection. The BLOB located at the coordinates of the reference colour, or sub image is marked as the reference BLOB. This process is much faster after histogram segmentation, as the image is now only a binary image.

For BLOB detection, a number of attributes are available for each of the BLOB's discovered. The points contributing to the BLOB shape create a unique signature, and the area can be used to identify each BLOB. Each BLOB within the current video frame has its area compared to the reference BLOB, to a tolerance of ±10%. During video frame processing, each binary image undergoes BLOB detection. Once the reference BLOB is found, its coordinates are returned and processing of the current frame terminated.

(2) *Corner Detection*

A SUSAN corner detector [14] locates interest points among the edges and corners of the binary image. Interest points located within the perimeter of the selected segmented region are marked as the reference points. As the image is a binary image, corner detection is fast and relatively consistent as colour variations are already filtered. Selecting enough interest points as reference points creates a unique signature as each point remains in spatial relationship with neighbouring points. This allows tracking because as each point moves, the others move in coordination.

Video frames received from the histogram segmentation process undergo SUSAN corner detection. As each interest point is examined, it is compared to a reference point. The offset is calculated and then each reference point is checked on the image to look for a matching point. To compensate for noise variation in the source image, a minimum number of interest points must match the reference points. There is no optimum limit to the minimum number of required reference points to match. Tests have shown that variations from between 40% to 80% of reference points matching to the image interest points are sufficient for reliable object matches.

Once a match occurs and the reference found, the image coordinates are returned and processing of the current frame terminated.

(3) *Hotspot Tensor Detector*

Tracking systems require high standards to maintain cohesion to the tracked object. BLOB and corner detection are affected by noise, creating non-consistent data sets. As a result, a method of creating an object signature was necessary. A hotspot tensor was developed for this works, to discover a unique signature for the selected object. The tensor is derived from interest point detector models but does not rely on edge or corner detection.

The centre of gravity for the foreground object defines the first point of the hotspot tensor and grows the signature until a set ratio of foreground to background pixels for the current radius is reached. Radiating out from the centre of gravity, each pixel's classification is checked for foreground status. Through each iteration, as the radius increases, the numbers of foreground and background pixels are compared. While the ratio of foreground to background remains greater than 50%, the cycle continues. Once the discovery of the hotspot completes, the list of points is maintained as the reference. Each pre-processed binary video frame undergoes reference point detection the same as for the corner detection system.

Hotspot tensor works because it is selecting pixels spiralling out from the original pixel selection, which should closely follow the spread of the colour blending from the reference colour.

5 Experimental Results

To evaluate the effectiveness of the tracking system, a number of end-to-end tests are required. Six tests with combinations of segmentation and tracking models, as shown below in Table 2; evaluate the repeatability, stability, loss of track, and false positive object identification. The key measure for these attributes is counting the number of frames that successfully locate the object selected for tracking as the video proceeds. Frames that do not find a target represent a loss of track, while an incorrect target is a false positive. For each test listed in Table 2, three tests are performed where the object of interest is one of the three coloured pivots on the gear assembly.

Table 2. Test combinations

Test	Segmentation	Tracking
1	Object signature	BLOB detection
2	Object signature	Corner detection
3	Object signature	Hotspot tensor
4	Colour zone	BLOB detection
5	Colour zone	Corner detection
6	Colour zone	Hotspot tensor

The video feedback involves a gearbox experiment (Fig. 9). There are two moving parts with one degree of freedom. The gears rotate around their axis. The users give inputs to one gears and rotates it. They must measure the amount of rotational the other gear. The gears have different radii and the students have to figure the gear rations by counting rotations. To identify the rotations a small marker is used on both the gears one red and one blue. The tests are done to find out the feasibility of the discussed approach above in identify the markers in a video feedback. This could help the web-page to constantly highlight the objects int he video framework.

Fig. 9. Object Signature segmentation and BLOB Detection tracking (Test One). Red object successfully tracked as it rotates

For all testing, each video frame is converted from a bitmap into memory for fast processing. All image processing occurs on the memory array, and conversion back to bitmap format only occurs once all processing is complete. This method minimises processing costs and is far faster than operating directly on bitmap frames.

This section reports the results for each of the six tests, and comments on the model's suitability as a fast-real-time object tracking system. The summary tracking results are shown in Table 3, and discussed below. For tests One, Two and Three, the objects of interest have been previously cropped from the first frame of the video sequence and saved as separate files.

Table 3. Test results summary

Test	Blue left			Red			Blue right		
	Hit	None	Miss	Hit	None	Miss	Hit	None	Miss
1	96	404	0	145	331	24	58	404	38
2	14	472	14	58	420	22	74	400	26
3	372	24	104	398	0	102	223	0	277
4	269	230	1	187	244	69	281	175	44
5	218	162	120	277	106	117	150	237	113
6	494	4	2	500	0	0	421	38	41

5.1 Object Signature Segmentation

Segmentation of the sequential video frames using the histogram of the object of interest creates a binary image as shown in Fig. 4. Visually, it is clear that the major portion of background noise has been successfully removed, leaving the regions of similar colour available for the tracking models to interpret. Figure 4 demonstrates the possible problems associated with the histogram segmentation model. Within the test video are two gear pivot points of the same colour. Segmentation has resolved the image to two potential targets. The object of interest has a higher density of foreground pixels due to slight variations in the reflected colours back to the video camera of the two similar objects.

For the tracking models, the difference in foreground pixel density and foreground shape add to the overall signature of the tracked object. Applying the tracking models utilises the uniqueness of the foreground data set.

(1) *Test One*

The points contributing to the BLOB shape create a unique signature, but are also subject to noise and are therefore unstable for tracking purposes. The area encompassing the BLOB remains stable to a level of consistency that is sufficient to continue to validate the object within the frame.

As shown in Fig. 9, tracking can be improved by relaxing the tolerance of the area matching method. For instance, changing the tolerance to $\pm 15\%$ improves the tracking from 145 to 238 hits. False positives increased only moderately from 24 to 44.

Tracking was able to be performed for each target object, but with only marginal success. The best placed tracked object (red pivot) was only able to maintain tracking for less than 30% of the time. The large discrepancies between the three sub-tests show a lack of repeatability. False positive results for the blue objects were expected, but for the red pivot to also classify with incorrect regions of the image, was not expected. The signature of the shapes of the objects of interest was expected to add to a degree of uniqueness. Noise from segmentation produced unreliable foreground objects. While it is possible to isolate the objects of interest from the segmented image (see Figure), the object signature consists of too much variation for BLOB detection to function reliably.

(2) *Test Two*

Obtaining corner reference points from the histogram segmented image generates very unique signatures of each foreground object. The problems associated with Test One are also relevant to Test Two. This test was expected to improve on BLOB detection tracking, as only a small number of corner points will be necessary to create a unique signature of the object. All corner points are maintained for referencing, but only 50% are required to find a match.

As shown in Table 3, SUSAN corner detection tracking of the objects signature histogram segmented images, performed significantly worse than expected. From Fig. 10, corner detected points of the red pivot object; appear similar but not similar enough. The images within Fig. 10 are from consecutive video frames and consist of the SUSAN corner detection output. Decreasing the matching ratio of corner points from 50 to 40% drastically improves tracking hits. An increase from 58 to 233 hits was measured with only a slight increase in false positive rates, but in this instance, the false positives are not even a similar colour. This is because the random nature of noise and the small percentage of matching interest points overlap. Most interest points typically consist of 6–12 points. Matching only 40% of the points may mean only two points are required; too small for reliable tracking according to Lowe [15].

Fig. 10. Corner detection results. Red pivot from two consecutive video frames (circled)

(3) *Test Three*

The hotspot tensor tracking system uses the segmented foreground points to build a hotspot signature. The effects of noise from segmentation are minimised as a result of hotspot construction. The centre of gravity for the foreground object defines the first point of the hotspot tensor, and grows the signature until the ratio of foreground to background pixels for the current radius, is 50%. The signature is maintained for comparison to further frames.

This method has improved tracking over the previous two tests, as shown in Table 3. Both the tracking hits and false positive results are vastly superior to previous testing. Testing the right blue object of interest has higher false positive results purely because of the similarity of the two objects.

All interest points within the signature are validated against the foreground pixels of subsequent segmented frames. Tighter twinning used within the hotspot tensor model may either force a high percentage of tracking failures or bind to a greater degree. As Table 3 shows, the model has constricted tracking to the object of interest. A loss of track is unusual due to the tight object binding, but false positives are the new cause of concern. Histogram segmentation from a template object filters along the top accumulation bins. Within the test video stream, objects with similar colour signatures were selected to challenge the tracking models. While hotspot binding to relevant foreground objects is strong, it can become diverted by comparable secondary signatures. Secondary signatures that contain a subset of the object signature may redirect the tracking, creating false positive results (Fig. 11).

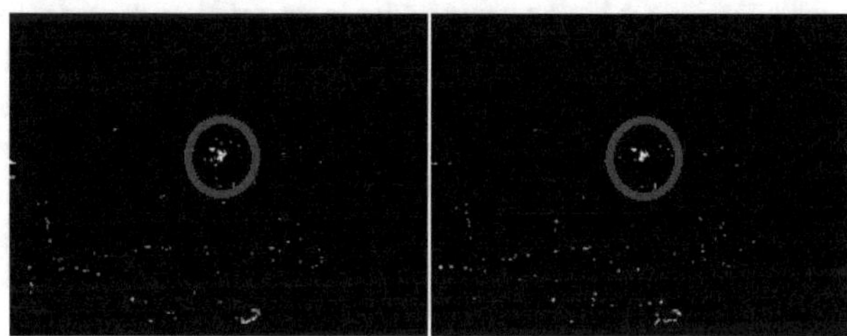

Fig. 11. Histogram segmentation. Red Pivot from two consecutive video frames (circled)

5.2 Colour Zone Segmentation

Using colour zone histogram segmentation of sequential video frames is effective in creating binary images. Comparing objects of interest and colour zone segmentation methods produces a visible difference. It can be seen from Fig. 12 that the foreground density is greater for colour zone segmentation, providing a stronger input to the tracking systems. Greater definition of the foreground object provides better interest points, which are also more reliable.

For each test, the selected reference bin is calculated from a random point on the object of interest, from the first frame of the video sequence. These tests validate the model's ability for repeatability, reliability, loss of track and false positives. The following three tests then utilise the enhanced segmentation data set.

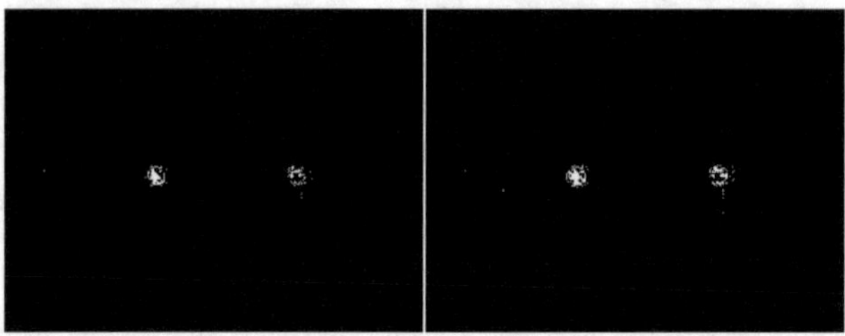

Fig. 12. Segmented image using object signature segmentation (left) and colour zone segmentation (right)

(1) Test Four

Higher intensity foreground objects provide data for larger BLOB's during the BLOB detection tracking phase. BLOB objects are more distinct using the colour zone histogram segmentation model. The results of object tracking for this model show a dramatic improvement over the results of Test One (see Table 3).

Some problems occurred as a result of the shadows changing while the object of interest moves. During tracking of the left blue object, the colour and shape of the object, to the eye, appears to change when at a particular location in its motion. In Fig. 13 it can be seen that the original BLOB reference shape is nothing like the shape from another frame where the object of interest has moved. Shadows and lighting have deformed the appearance, and both the histogram segmentation and the BLOB detection processes cannot compensate. Also, from Fig. 13, the clear definition of the reference BLOB provides clarity from other foreground objects, limiting false positive detection rates. This model is more likely to lose track than trigger a false positive result.

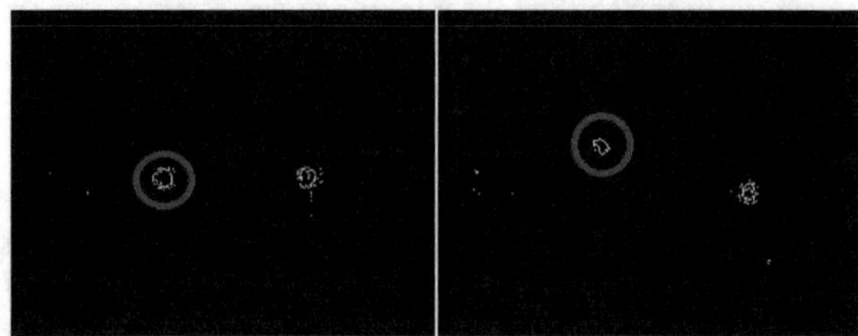

Fig. 13. BLOB shape change during tracking. Reference BLOB (Left), BLOB during motion (Right)

(2) Test Five

From the apparent improved foreground object density, shown in Fig. 12, from the colour zone segmentation, it was assumed that the SUSAN corner point detection tracking model would also increase its performance. The results appear to be a double-edge sword. Clearer segmented foreground objects have produced improved unique signatures for the object of interest. As shown in Fig. 14, the interest points in the right image are less cluttered and better representative of the shape and area of the object of interest. Alternatively, the model as it is within Test Two is sensitive to the noise from segmentation, causing a loss of tracking.

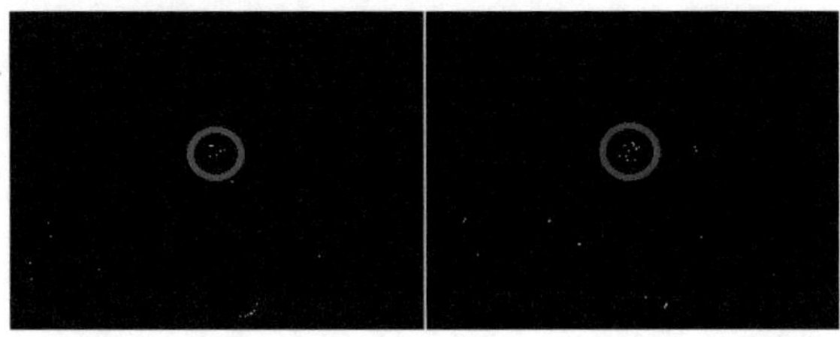

Fig. 14. Corner detection results. Red pivot from test two (Left) and test five (Right)

Segmentation using the colour zone histogram, improves the corner detection tracking when compared to the object histogram segmentation. This primarily is as a result of the superior segmentation results which provide better definition of the foreground objects. False positive detection rates are lower than Test Four because of the higher degree of uniqueness between reference signatures.

(3) Test Six

All models based on the colour zone segmentation system show an improvement in the tracking system due to the increased foreground object density. Hotspot tensor creation should enjoy larger and clearer foreground regions to exploit. Results for all tests utilising the sixth test model are significantly better, as shown in Table 3. The nature of creating the hotspot suits the higher density foreground object.

The uniqueness and strength of the hotspot tensor signature is as a result of the superior foreground density from segmentation. Both false positive and loss of track figures are vastly reduced due to the improved signature. Tests occurring with the red object of interest performed better than expected and demonstrated the combination colour zone segmentation and hotspot tensor as superior for this method of video system tracking.

5.3 Template Matching

Computer vision object tracking systems exist in many forms, and template matching is a current reliable system which is tested here for baseline purposes. For each test, the templates used are the separate files of the cropped gear objects. Template matching attempts to find the selected template item within the current video frame. The method is reliable but is not suitable for real-time operations.

Performing template matching with the three object of interest files used for test one, two and three produces an average result of: 426 hits, 74 lost track and zero false positives. The execution times for the model well exceeds the real-time needs. Processing times for 500 frames of the video stream are shown in Table 4. The *base* time reflects the template matching time and equates to an approximate frame rate of 6fps. Each colour histogram model achieves a frame rate that would be compatible with most real-time applications, where as the template matching model could not sustain real-time frame rates.

Table 4. Test execution times

	Tests						
	1	2	3	4	5	6	Base
Sequence time (s)	13.4	22.3	10.9	15.7	25.4	13.6	79.75
Frame Rate (fps)	37	22	45	31	19	36	6

The above results show that the methods are good enough for implementation in a RL environment.

6 Conclusion

Tracking objects within digital video streams requires consistent and reliable detection of the objects and the means to continually recognise the object as it progresses through the video scene. Many current tracking systems work well, but only in an offline mode where the extensive processing requirements are not impacting the output. For systems such as augmented reality, real-time object detection and tracking is necessary, or the system becomes ineffectual for the users. Common computer vision filtering processes attempt to isolate foreground and background regions of the incoming video frames. Isolation involves segmenting the images based on rules for the model. Colour histograms are capable of recognising objects', so segmenting video images according to the histogram colour frequency distribution seems plausible.

Applying colour histogram segmentation to classify foreground and background objects is effective. For tracking purposes, the histogram segmented data set requires a means to select the appropriate foreground object, and ensure it is able to track the object through the video sequence, in real-time. These works have created six end-to-end tracking models to ascertain the effectiveness of colour histogram tracking. While segmenting video frames with colour histograms of the selected object and capturing the foreground objects for tracking, the models are not reliable and efficient for tracking purposes. However, segmenting video sequences using the colour zone of a selected region of the object of interest and applying tracking techniques to the foreground objects does appear to be effective.

All tests with colour zone segmentation provided improved results compared to object histogram segmentation. All tests were significantly faster than the industry reliable template matching tracking system. Only colour zone segmentation along with hotspot tensor tracking combined all the best features, and was repeatable, stable and maintained good tracking. It also was fast enough for real-time processing and tracking.

The repeatability, reliability and speed allow the colour zone segmentation and hotspot tensor model to give simple object tracking capabilities to systems currently unable to process video in a real-time environment. Combined with colour histogram object identification, the system is able to recognise objects and easily track them.

Whether the colour zone histogram segmentation model is suitable for other computer vision tracking systems has not been addressed in this work. Other adjustments to the histogram model may be able to improve foreground detection and density levels, providing higher contrast for secondary tracking methods.

References

1. Otsu, N.: A threshold selection method from gray-level histograms. Automatica **11**, 23–27 (1979)
2. Ma, X., Shi, G., Tian, H.: Adaptive threshold algorithm for multi-marker augmented reality system. In: Presented at the VRCAI, Seoul, South Korea (2010)
3. Raja, Y., McKenna, S.J., Gong, S.: Segmentation and tracking using colour mixture models. In: Computer Vision—ACCV'98, pp. 607–614. Springer, Berlin (1998)
4. Wells III, W.M., Grimson, W.E.L., Kikinis, R., Jolesz, F.A.: Statistical intensity correction and segmentation of MRI data. Vis. Biomed. Comput. **1994**, 13–24 (1994)
5. Ester, M., Kriegel, H.-P., Sander, J., Xu, X.: A density-based algorithm for discovering clusters in large spatial databases with noise. In: Second Annual Conference on Knowledge Discovery and Data Mining (KDD-96), pp. 226–231 (1996)
6. Cai, W., Chen, S., Zhang, D.: Fast and robust fuzzy c-means clustering algorithms incorporating local information for image segmentation. Pattern Recogn. **40**, 825–838 (2007)
7. MacQueen, J.B.: Some methods for classification and analysis of multivariate observations. In: 5th Berkeley Symposium on Mathematical Statistics and Probability, pp. 281–297 (1967)
8. Zhang, C., Wang, P.: A new method of color image segmentation based on intensity and Hue clustering. In: 15th International Conference on Pattern Recognition, 2000. Proceedings, pp. 613–616 (2000)
9. Funt, B.V., Finlayson, G.D.: Color constant color indexing. IEEE Trans. Pattern Anal. Mach. Intell. **17**, 522–529 (1995)
10. DeVeaux, R.D., Velleman, P.F., Bock, D.E.: Intro Stats, 4th edn. Pearson Education, Boston (2014)
11. Swain, M.J., Ballard, D.H.: Color indexing. Int. J. Comput. Vision **7**, 11–32 (1991)
12. Ohlander, R., Price, K., Reddy, D.R.: Picture segmentation using a recursive region splitting method. Comput. Graphics Image Process. **8**, 313–333 (1978)
13. Smith, M.: Fast colour histogram matching. In: Presented at the Under Submission (2017)
14. Smith, S.M., Brady, J.M.: SUSAN—a new approach to low level image processing. Int. J. Comput. Vision **23**, 45–78 (1997)
15. Lowe, D.G.: Object recognition from local scale-invariant features. In: The Proceedings of the Seventh IEEE International Conference on Computer Vision, 1999, pp. 1150–1157 (1999)

Part VI
Remote Control and Measurement Technologies

Work-in-Progress: Prognostics in Arboriculture Using Computer Vision and Statistical Analysis

T. K. Sourabh[1], Veena N. Hegde[2(✉)], and Nishant Velugula[3]

[1] ModeliCon Infotech LLP, Bengaluru, India
sourabhtk96@gmail.com

[2] B.M.S. College of Engineering, Bengaluru, India
veenahegdebms.intn@bmsce.ac.in

[3] University of Illinois at Urbana-Champaign, Champaign, USA
nishant.velugula@gmail.com

Abstract. As engineers, we can analyze and mathematically relate the behaviors present in all life sciences. For example, a brain can be modelled as a neural network, or a heart can be modelled as a closed loop control system. Similarly, entities from the world of arboriculture or the plant kingdom as a whole can be modelled and fitted into mathematical models. With the increased use of high speed computers in recent years, process models have a high scope for getting transformed to their digital twins. The performance analysis of such digital twins can be carried out using various techniques and the future behaviour of these systems can be predicted. The non-linear and stochastic behaviour of inputs for these models can be weighed based on the history of their occurrences and the priority level can be assigned to obtain the performance evaluations. This concept has been applied here in this work, to predict the characteristics of trees in terms of their lifespan and strength. The methods of analysis proposed in this paper use widely available open source software platforms, which make the design dedicated, reliable, and accessible to all.

Keywords: Ree · Lifespan · Photogrammetry · Statistics

1 Introduction

Trees are the fundamental sources of life on this planet. They are responsible for maintaining the oxygen-carbon-dioxide balance which is crucial to ensure sustainability of all its species. Trees come in all shapes and sizes. They exist in almost all the regions of the world. Trees provide a multitude of benefits from reducing erosion and pollution run-off to cooling the surrounding areas. Many municipalities understand the many benefits that trees provide for their communities and have rules in place that specify the number, size, and type of trees used in any new development [1]. Since they are of significant importance in our daily lives, we must make an effort to understand their impact when they fall. Safeguarding people against the devastating effects of falling trees is a difficult task. The likelihood of them falling increases during the onset of strong winds, heavy rainfall and storms due to the accumulation of water on

M. E. Auer and K. Ram B. (Eds.): REV2019 2019, LNNS 80, pp. 567–578, 2020.
https://doi.org/10.1007/978-3-030-23162-0_50

branches and the uneven distribution of weight on these branches causing an overall imbalance. This could even result in the uprooting of the entire tree. Having presented an overall view and understanding of trees and their characteristics, it is now possible to easily relate with what many researchers have said in the past.

In one of the recent works, the estimation of the age and growth rate of trees were calculated based on tree ring analysis. The measurements made at base and breast height of the tree and equations predicting biomass showed the accurate age estimation. However, the method was not fully automated [2]. The digital close range photogrammetry (DCRP) approach has been used for tree's age estimation from its photographs with image analysis in which Age of a tree is determined using diameter [3]. One of the work proposes a binary coding method, a novel post classification change detection method that indexes multi-temporal satellite images into a single information layer [4]. The process of loading of trees by natural environmental conditions have been studied and various models have been developed to calculate stresses due to weight along tree branches. By evaluating the stress models, the stress developed inside the tree which can cause tree falling was estimated and their by tree life span was estimated [5]. High resolution, Image Processing techniques have been used to get the information about the internal decay of tree trunks, branches etc. due to moisture, and other environmental effects that can harm trees due to chemical reactions [6–8]. In all the research carried out, work focus is mainly on prediction of the Age of different species using imaging or other measuring techniques have tried to assess the Ageing of trees and arrive at inferences. But the solution not has been fully automated from the view point of providing data analytics and prediction based on classification. The paper here provides one such automated solution to know the Age and predict the remaining life of the trees, before it causes damage. The paper brings out automating the prediction process in the following subsections. Section 2 A represents the overall methodology with 3D imaging of trees with an innovative measuring methods of tree trunk diameter for redundancy. The feature extraction technique using photogrammetry and other statistical means are explained Sect. 2.2. Section 3 explains the 3D imaging and redundant methods. Section 4 gives validation of proposed technique. Statistical analysis is given in Sect. 4. The results of proposed integrated techniques of automation in deciding the tree Age is given in Sect. 5. The paper is concluded in Sect. 6. Now just place the cursor in the paragraph you would like to format and click on the corresponding style in the styles window (or ribbon).

2 Methodology

The research work was implemented in the following manner: Using hardware, including but not limited to cameras, and distance measurement sensors such as ultrasonic sensors, data was obtained using which 3D models of trees were computed. The purpose behind using both image capturing techniques, as well as standalone sensors is for redundancy, to ensure that error is minimized. Thus using this information, the diameter of the tree was effectively computed. Using this and the Mean Annual Increment (MAI) value, the age of the tree was computed.

All this information is stored in the database. This work can be extended by obtaining and storing in the database crucial information using Unmanned Aerial Vehicles (UAVs) and satellite imaging. Once all the necessary information was obtained, using the tree developed model, the loading effects, uprooting effects and other effects (which can be anything from the geographical location to any unique attribute of that particular tree) can be analyzed. Thus, all this analyses can be fit into the statistical model to predict the remaining lifespan of the tree (Fig. 1).

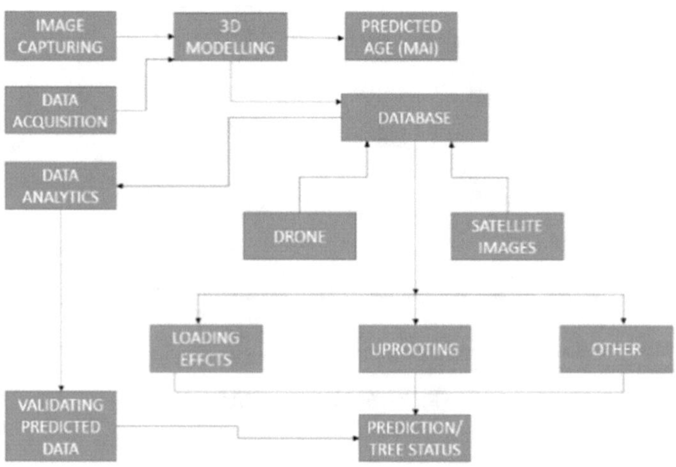

Fig. 1. Block diagram

2.1 Concepts Assessed Using Photogrammetry

(i) Diameter at breast height—is a method of measuring the diameter of the trunk of a tree. It is a highly frequently used mode of measurement in the field of dendrometry. The convention is to measure the diameter(s) of the tree at 1.3 meters from ground level [2]. The advantage of using a dbh measurement is that it is allows for the determination of many different traits pertaining to a particular species of trees. Conventionally, the dbh measurement can be performed with callipers if the trunk is small enough, but in this novel work, specialised hardware and the use of photogrammetric techniques obtain accurate measurements with redundancy.

(ii) The mean annual increment (MAI) or mean annual growth—refers to the average growth per year a tree or stand of trees has exhibited/experienced to a specified age. For example, a 20-year-old tree that has a diameter at breast height (dbh) of 10.0 inches has an MAI of 0.5 inches/year. MAI is calculated as MAI = Y(t)/t where Y(t) = yield at time t. Because the typical growth patterns of most trees is sigmoidal, the MAI starts out small, increases to a maximum value as the tree matures, then declines slowly over the remainder of the tree's life. Throughout this, the MAI always remains positive. MAI differs from periodic annual increment (PAI) because the PAI is simply the growth for one specific year or any other specified length of time. The point where the MAI and PAI meet is typically referred to as the biological rotation age. This is the age at which the tree or stand would be harvested if the management objective is to maximize long-term yield [3]. The differences between the two is elucidated in the following graph. This graph is of great significance since in this research MAI was used in the model (Figs. 2, 3, 4 and 5).

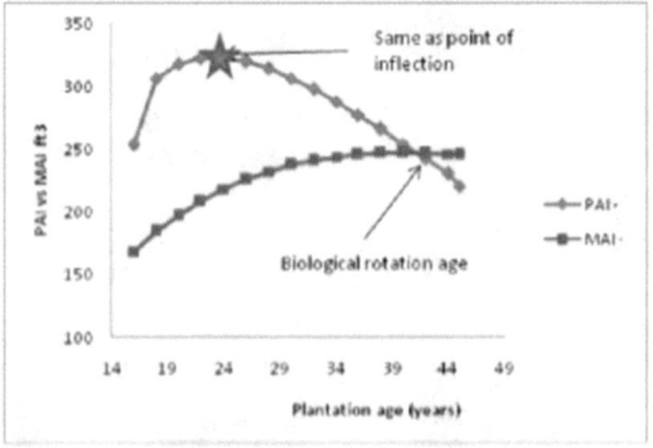

Fig. 2. MAI versus PAI

Fig. 3. Chosen scan object

2.2 Feature Detection Using Photogrammetry

Photogrammetry is the science of making measurements from photographs, especially for recovering the exact positions of surface points. Photogrammetric analysis may be applied to one photograph, or may use high-speed photography and remote sensing to detect, measure and record complex 2-D and 3-D motion fields by feeding measurements and imagery analysis into computational models in an attempt to successively estimate, with increasing accuracy, the actual, 3-D relative motions.

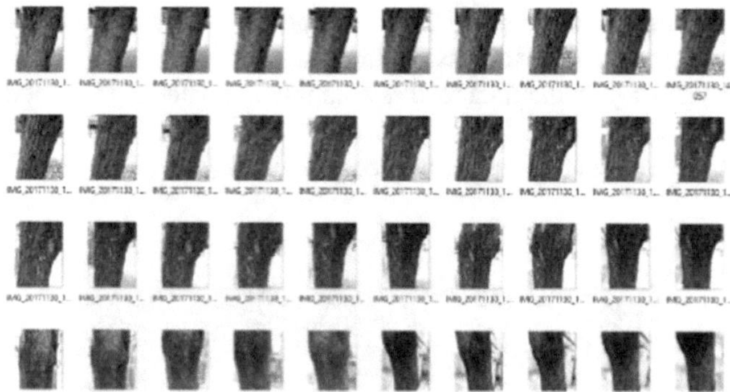

Fig. 4. Portion of image sequence

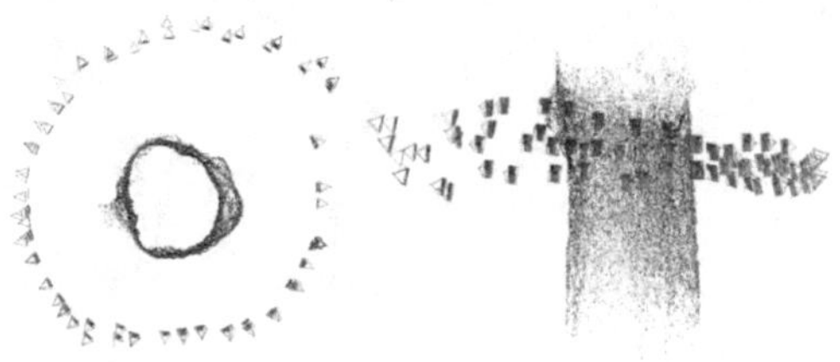

Fig. 5. Top and lateral cross section models

The following were the steps in implementing the first stages of photogrammetric techniques in this paper:

a. Importing Images
b. Missing Match Computation/ Pairwise Sequence Matching
c. 3D Reconstruction Computation.

These steps generated a sparse 3D data output in the form of point clouds and the availability of camera positions allow the user to see the exact sequencing of the missing match computation/ pairwise sequence matching, and the 3D reconstruction computation. The number of pictures needed to compute these images were around 70. The following sets of images represent the processing involved in getting the first renders of 3D data and allowed preparation for the complete 3D modelling. These images will be presented sequentially in the order that they were performed/computed.

The following steps were performed to obtain the final 3D Model.

a. Once the sparse point sample is saved, a CMVS dense reconstruction is run on the sample.
b. Once that is complete the model needs to be prepared for the texture computation.
c. All artefacts and unwanted surface structure are removed in the model clean-up stage.

d. To fill in the gaps and complete the model before the final texture computation a Screened Poisson Surface Reconstruction was used.
e. Following this stage is the final texture computation wherein the model is analyzed and the texture is computed.
f. Thus the final tree sample model was completed (Figs. 6, 7, 8 and 9).

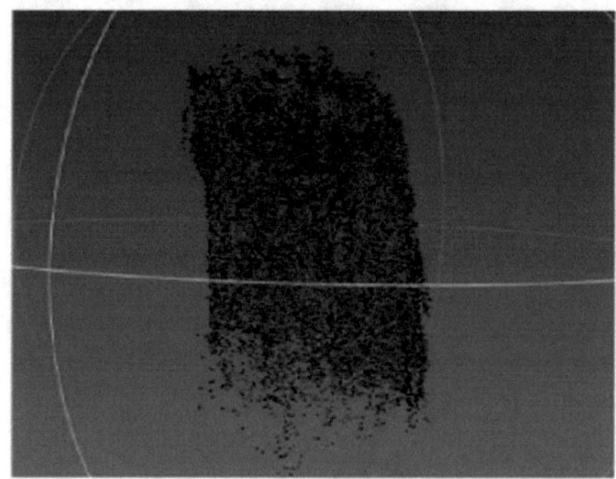

Fig. 6. Post CMVS dense reconstruction

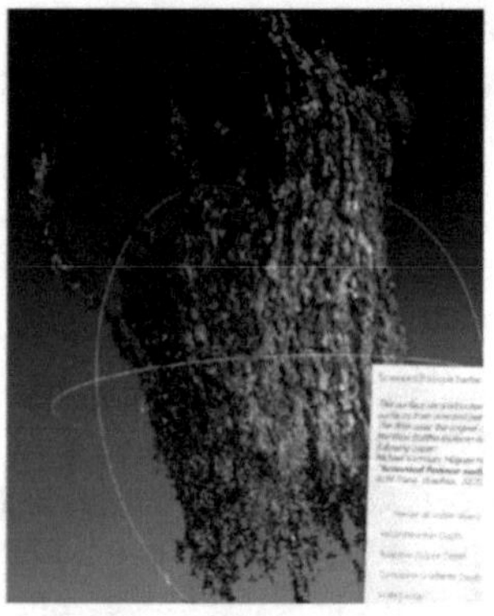

Fig. 7. Model cleanup

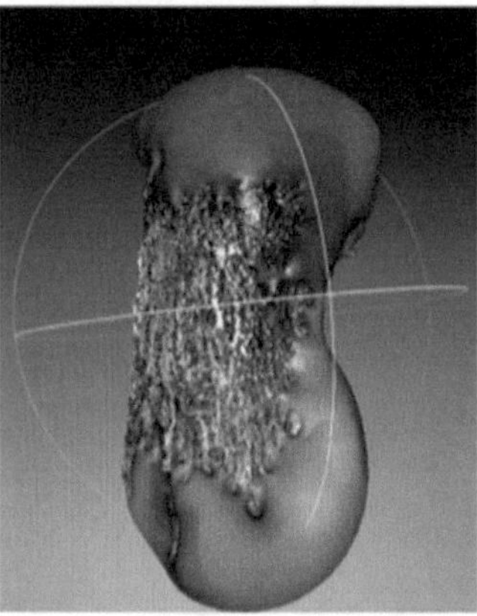

Fig. 8. Screened poisson surface reconstruction

Fig. 9. Reconstruction using photogrammetry

Poisson's equation is given by: $\Delta\phi = f$, where Δ is the Laplace operator, and f is a real or complex-valued function. In three-dimensional coordinates, is

$$\left(\frac{\partial^2}{\partial x^2} + \frac{\partial^2}{\partial y^2} + \frac{\partial^2}{\partial z^2}\right)\varphi(x, y, z) = f(x, y, z) \tag{1}$$

Thus this equation is applied throughout the 3D model to fill in the gaps generated by any missing information or the presence of any inconsistencies.

3 Photogrammetry Validation

To increase the accuracy of the tree diameter obtained, and for the purposes of verification, a mode of redundancy facilitated by diameter measurements using a custom-made hardware data capture device was used. This will be expounded in the following subsections.

The first stage of hardware implementation involved a removable and moving arm which wrapped around the tree trunk. On one end of the contraption is an ultrasonic sensor connected to an Arduino which is programmed to measure the distance between the sensor on the fixed arm and the wooden plate on the moving arm (Fig. 10). The final result builds on the fact that the measurement can be repeated multiple times and averaged out.

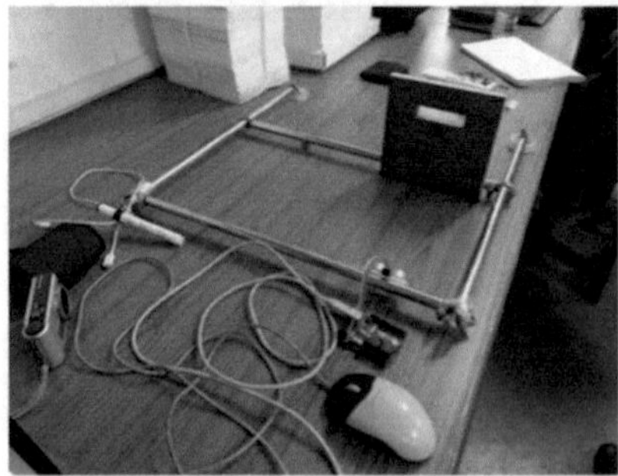

Fig. 10. V1 hardware implementation

The second version (Fig. 11) builds on the first and remains primarily same in terms of its functioning. There are two major improvements in the second version:

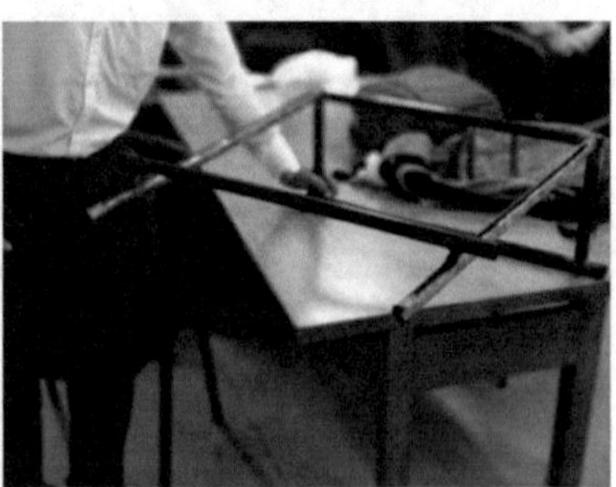

Fig. 11. V2 hardware implementation

1. The process of attaching the movable arm was improved, adding to the ergonomics of the contraption.

2. An additional support structure has been added to the frame. The purpose of this support structure is to ensure perpendicularity of the device during measurement. This will result in an increase in the consistency of measurements, reducing the error and increasing the accuracy of the overall tree trunk width.

4 Statistical Analysis and Its Challenges

To allow for the understanding and analyses of tree lifespan determination, difference parameters based on the user's choosing are considered to facilitate prediction of the remaining lifespan.

To train the machine a large dataset was required with multiple parameters to give the user multiple input features to strengthen the model. Each row is one training feature. The columns signified different parameters that the user can choose to implement in the machine learning algorithm namely—name of tree, species, average age, degree of root rot, degree of suffocation due to construction, rainfall, number of trees fallen, number of trees planted, height etc. The following tables summarize the input features and presents the dataset that was obtained. The information was mainly obtained from the IISc—Bangalore database.

The analysis was carried out in Python/R, using Pandas for data preparation. The main x axis input feature that was used is the species and the y axis target was the average age. To understand the input feature, all the categorical values were transformed into labels and then one hot encoded.

To validate the categorical values, Kendall's Tau method was employed. Kendall's Tau is a measure of non-parametric rank correlation. What it does is that it assesses the statistical associations based on the ranks of the data. Once that is complete, the data ranking is performed on the variable that are put in order and numbered separately. Correlation analyses can be used to test for associations in hypothesis testing. The null hypothesis is that there is no association between the variables under study. Thus, the purpose is to investigate the possible association in the underlying variables [4].

There are several advantages of using Kendall's Tau method for statistical analysis and they are as follows:

1. The Kendall's Tau distribution has good statistical properties.
2. The Kendall's Tau probability estimation of observing concordant and discordant pairs is direct.

The main challenge in trying to implement machine learning was the lack of available data. To train any model effectively, a substantial amount of data with a high correlation value between input and target output is required. With our limited dataset we managed to achieve RMSE values of 0.5 using XGBoost.

In addition to this, some data was not highly correlated with the average age of the trees that was used in the training set. Presently, the focus is on trying to procure good quality data with high correlation values so that we can further our research.

5 Results and Observations

The photogrammetric computations resulted in highly accurate renders of 3D tree models and the resulting measurements were accurate as well. As mentioned in Sect. 4 of this paper, the hardware rig allowed for verification of these tree measurements. The information from the curated dataset is displayed statistically as a box plot and a histogram to enable the user to observe the statistical representation of the information needed for the computation. The following images show the two sets of graphs. The box plot of all the three graphs i.e. number of trees planted, number of trees fallen and the average age (years) show the minimum, maximum values and the major 'box' region of all the values that are present in the input dataset. The red line in each box demonstrates the average or mean value. The histogram plot of all the three graphs i.e. number of trees planted, number of trees fallen and the average age (years) show the distribution of each input feature and presents to the user the 'trend' of information (Figs. 12 and 13).

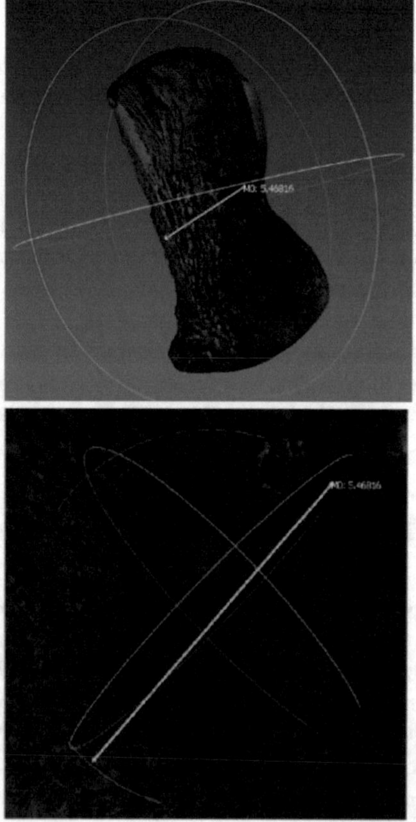

Fig. 12. Photogrammetric measurement extraction

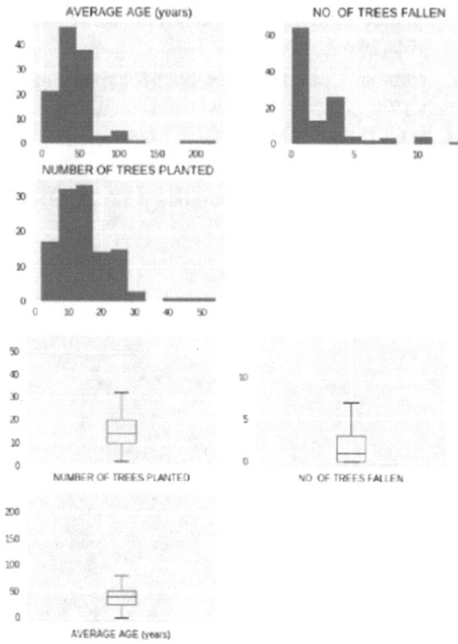

Fig. 13. Box and histogram plots of dataset

6 Conclusion

The algorithm understood the correlation between the species and age using all the input features mentioned in Sect. 5.

Thus, using the present age procured by the photogrammetric analysis, and by feeding that to the dataset, the remaining lifespan of the trees were statistically predicted.

References

1. Borges Silva, L., Teixeira, A., Alves, M., Elias, R.B., Silva, L.: Tree age determination in the widespread woody plant invader *Pittosporum undulatum*. **400**, 457–467 (15 September 2017) (Elsevier)
2. Chek Mat, M.S., Mohd Nor, M.A., Diah, J.M., Mohd Din, M.A., Hashim, K.A., Manan Samad, A.: Tree age estimation by tree diameter measurement using digital close range photogrammetry (DCPR). In: IEEE Transaction on Control System, Computing and Engineering (ICCSCE), 2014 IEEE International Conference, 28–30 November 2014
3. Lee, H.: Mapping forestation and age of evergreen trees by applying binary coding method to time–series landsat november images. IEEE Trans. Geosci. Remote Sens. **46**(11) (November 2008)
4. Shahbazi, Z., Kaminski, A., Evans, L.: Mechanical stress analysis of tree branches. Am. J. Mech. Eng. **3**(2), 32–40 (2015)

5. al Hagrey, S.A.: Geophysical imaging of root-zone, trunk and moisture heterogeneity. J. Exp. Bot. **58**(4), 839–854 (March 2007)
6. Brack, C.: Ph.D. (UBC) Standard point on tree bole for measurement. Forest Measurement and Modelling. Retrieved 18 Apr 2009
7. Mai Husch, B., Miller, C.I., Beers, T.W.: Forest Mensuration, p. 402. Wiley, New York (1982)
8. https://www.statisticssolutions.com/kendalls-tau-and-spearmans-rank-correlation-coefficient/
9. https://en.m.wikipedia.org/wiki/Tree

Remote Diagnostic Assessment Tool for Engines

Gautama Bharadwaj[1(✉)] and Anand M. Shivapuji[2]

[1] Department of EEE, BMS Institute of Technology, Bangalore 560064, India
gautama98@gmail.com
[2] Center for Sustainable Technologies, Indian Institute of Science, Bangalore 560012, India
anandms@iisc.ac.in

Abstract. This paper presents the evolution of a remote online diagnostic tool to evaluate the plausibility of the operation of any system having a signature sound. As a case in point example, a web enabled Fast Fourier Transformation tool has been developed for real time evaluation of pressure-crank angle trace(s) of a spark ignited engine to detect frequencies in the 5–25 kHz regime, typical of engine abnormal combustion. The goal of this work is to develop an interface for accessing real time temporal data representative of the system operation for typical electro-mechanical system, either from the on-board diagnostic system or from a central repository system in the cloud and process the same through appropriate DSP techniques to identify outlier frequencies and diagnose the potential cause of the same.

Keywords: Knock detection · FFT · Online diagnostic tool · Abnormal combustion

1 Introduction

The combustion in a spark ignition engine occurs due to the triggering of the spark by the spark plug, which ignites the air and fuel mixture compressed in the combustion chamber [1]. This gives rise to the propagation of a flame front. Engine knock occurs when the temperature or pressure in the unburned air and fuel mixture (end gases) exceeds a critical level, causing auto-ignition of the end gases [2]. Knocking consists of a sudden release of energy when combustion of the mixture in the cylinder does not result from the propagation of the flame front [3]. This produces a shock wave that generates a rapid increase in cylinder pressure [4]. Knock limits the spark advance, thereby limiting the maximum torque that could be achieved. Continued operation under knocking would lead to substantial power loss and can constrain the engine performance and thermal efficiency [5, 6]. Thus, detection of knock becomes extremely critical in order to maintain normal combustion in the engine.

For the detection of knock, first the operating conditions of the spark ignition engine, specifically, the pressure and the crank angle are extracted. Then, with the extracted data, the range of crank angle within which knock could occur is identified, thus predicting the occurrence of knock. Although there are several methods for knock

M. E. Auer and K. Ram B. (Eds.): REV2019 2019, LNNS 80, pp. 579–586, 2020.
https://doi.org/10.1007/978-3-030-23162-0_51

detection, the one based on the acquired in-cylinder pressure is possibly the most accurate of the available methods considering that the auto-ignition triggered acoustic signals are directly picked up by the pressure sensor [5–8]. In-cylinder pressure analysis has been a key tool for offline combustion diagnosis. Online applications for real-time analysis of combustion management and monitoring have become the need of the hour.

With this background, the main contributions of this work can be summarized as follows:

- We develop a real time web enabled remote diagnostic tool for combustion analysis based on the acquired in-cylinder pressure data.
- We also develop a system for indicating or forewarning the occurrence of knock with real-time updating of engine combustion information in the web enabled tool.

The rest of the paper is organized as follows. In Sect. 2, we present the considered experimental configuration. This is followed by the discussion of the methodology of the analysis in Sect. 3. The experimental data and results are presented in Sects. 4 and 5, respectively. Finally, conclusions are presented in Sect. 6.

2 Experimental Configuration

Experiments were conducted on a 6-cylinder engine having a bore of 120 mm, stroke length of 120 mm, and a compression ratio of 16:5:1. The engine speed was fixed at 1500 rpm. Data was collected at the rate of 90,300 samples per second. The in-cylinder pressure measurement requires access to the combustion chamber. In the current investigation, it is achieved using a spark plug adapter pressure sensor, thereby eliminating any need for engine cylinder head modification. The sensor used is an uncooled Gallium Orthophosphate ($GaPO_4$) differential pressure transducer mounted flush with the spark plug electrode end. The cylinder pressure is acquired in reference to the crank angle for which an optical encoder is used. The optical function is based on a slot mark disk and adopts the reflection light principle. The pressure crank angle data is acquired through a high-speed data acquisition unit with built-in plausibility check for safe operation.

3 Methodology

A web-enabled cross-browser interface is fabricated for the diagnostic assessment of the operation of the spark ignition engine. The temporal data from the engine is collected from the data acquisition unit. Figure 1 shows the architecture of offline diagnostic assessment. In offline diagnostic assessment, the data acquisition unit is connected to a computer and the data is processed by the diagnostic tool.

Figure 2 shows the architecture of the online diagnostic assessment. In online diagnostic assessment, the temporal data from the data acquisition unit is uploaded to a central cloud repository using Internet of Things protocols. The remote diagnostic tool collects the data from the central cloud repository. The server data thus obtained is

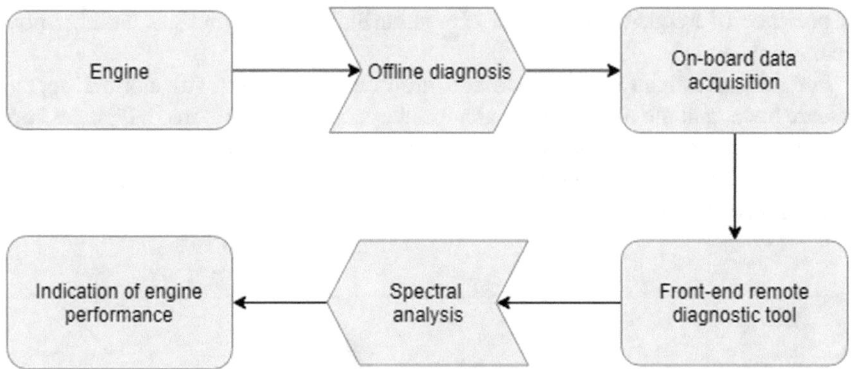

Fig. 1. Architecture of the offline diagnostic assessment

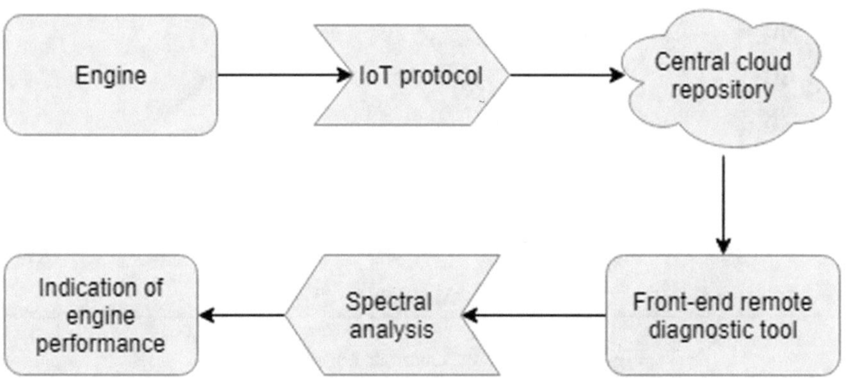

Fig. 2. Architecture of the online diagnostic assessment

processed on the front-end using techno business logic. By proceeding to analyze the engine parameters in the front-end, we reduce the stress on the server thus preventing any delay that would occur while the server delays to acquire sufficient data for analysis.

In the present investigation, the specific data collected includes in-cylinder pressure and crank angle information (updated every 0.1°). The in-cylinder pressure values are collected as absolute pressure and the crank angle is in degrees.

4 Experimental Data

For typical automotive sized engine, the knock free operation baseline pressure trace has sub 4 kHz frequencies with dominant frequency in the 2.5–3.0 kHz range [9]. The frequencies of pressure oscillations triggered by end-gas auto ignition (knocking) is reported to be between 5 and 25 kHz range. Further, the fundamental frequency in this case is typically found to be in the 5–7 kHz range [1, 10]. From this, it is evident that the presence of frequencies above 5 kHz is an indication of end gas auto-ignition or knock.

For this investigation, two data sets are considered, one with normal operating pressure trace, and the other with an abnormal operating pressure trace. Figures 3 and 4 represent the normal pressure trace and a typical knocking pressure trace. From the figures, the difference in the pressure-crank angle profiles between the normal operating condition and the presence of knocking is clearly evident.

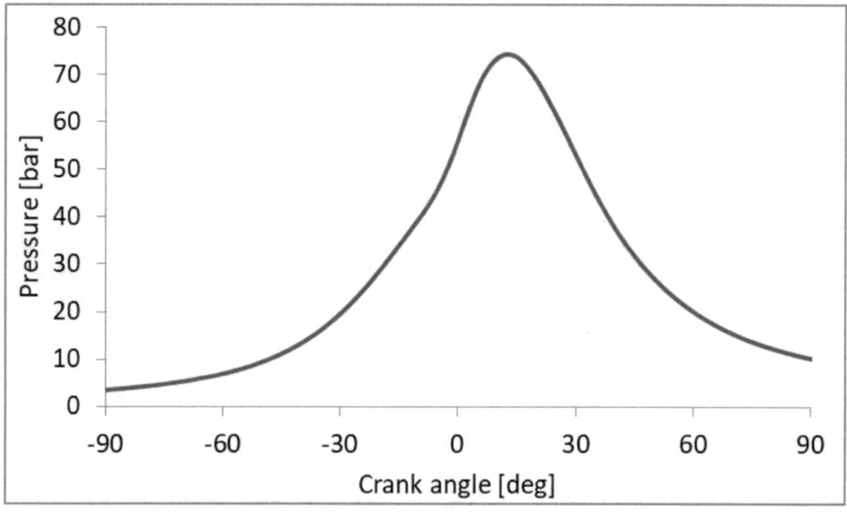

Fig. 3. Combustion chamber normal P-θ traces

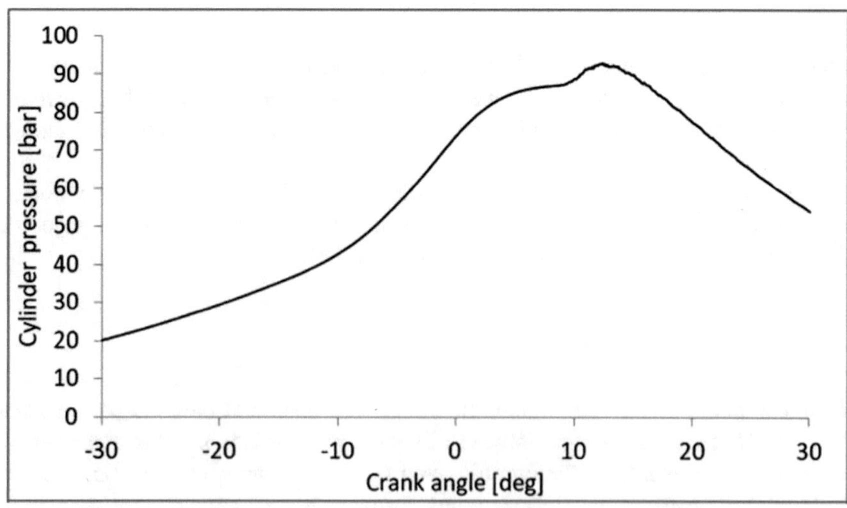

Fig. 4. Combustion chamber abnormal P- θ traces [Knocking occurrence]

Next, we carry out spectral analysis of the data to gain more insight about the presence of knocking.

5 Results

In order to perform spectral analysis, we need to analyze the difference of knocking pressure trace and normal pressure trace. Figure 5 shows the filtered pressure trace obtained by subtracting the knocking pressure trace with the typical normal pressure trace.

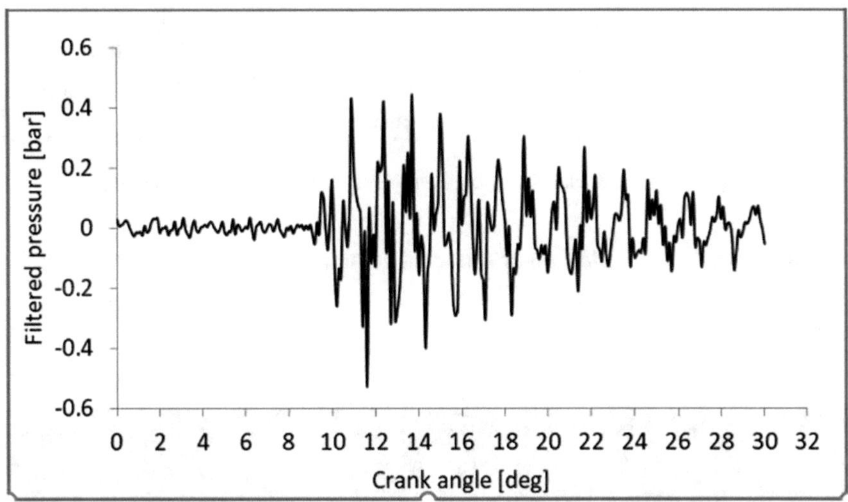

Fig. 5. High frequency components of filtered knocking pressure trace

The filtered signal shown in Fig. 5 is isolated by the remote diagnostic tool. The interface then processes the data, by computing its Fast Fourier Transform (FFT) in order to identify the frequencies in the region of typical knock. The spectral analysis occurs in real time, thus enabling the performance evaluation of the engine.

Figure 6 presents the results of spectral analysis of the filtered pressure trace seen in Fig. 5. It can be observed that there is a frequency spike around 5–7 kHz. This indicates the occurrence of knock in the engine. Thus, this tool forewarns of a break down when detecting frequencies in the specified range.

Identification of the engine operation is integrated with the web-enabled tool, and it enables real-time updating of engine combustion information. A color-coded mechanism is integrated for easier identification of engine operation as shown below:

- Red—Crossing of threshold towards abnormal combustion
- Yellow—Normal operation with small disturbances in base acoustic response
- Green—Smooth operation

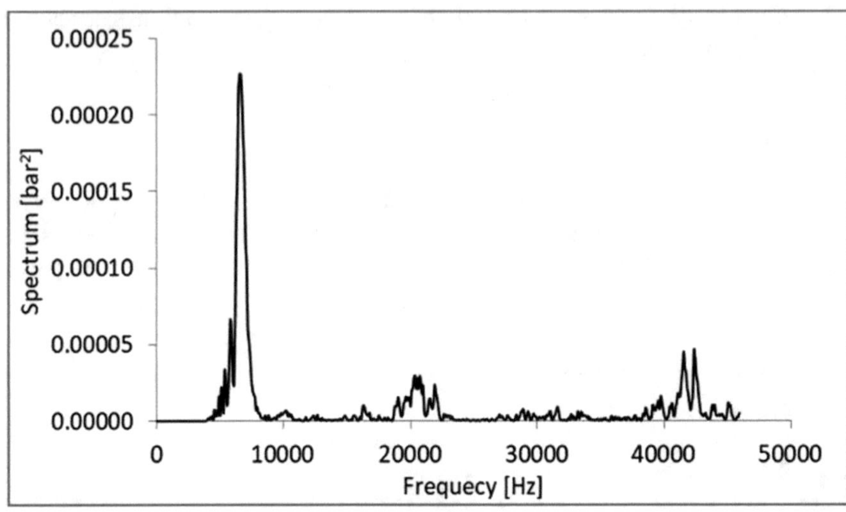

Fig. 6. Spectral analysis of cylinder pressure under normal operation

As an example, the indication of abnormal combustion in the remote diagnostic tool is shown in Fig. 7.

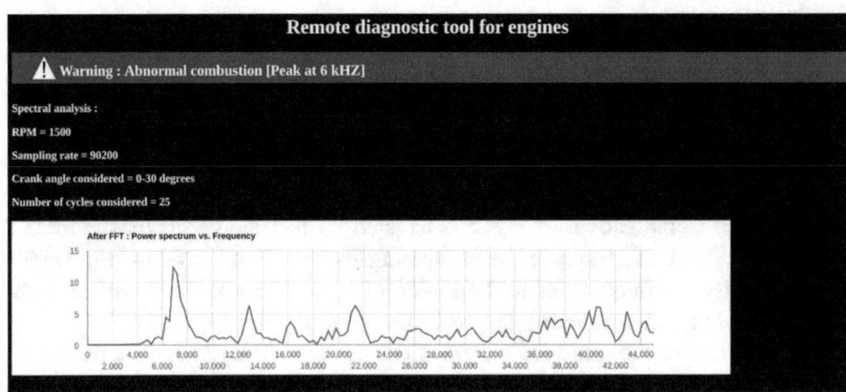

Fig. 7. Indication of abnormal combustion in the web-enabled tool

6 Conclusions and Future Work

This paper demonstrates a web-enabled tool that performs spectral analysis of in-cylinder pressure and crank angle traces of a spark ignition engine thereby determining and indicating the performance and operation of the engine. This enables analysis of the operation and forewarns a potential breakdown. The tool is expected to act in real time providing reliable supervision and enabling maintenance when essential.

Future prospects of such a tool can be diverged into multiple applications. For instance, on-board diagnostic tool can be used in individual vehicles. This would forewarn any potential breakdowns, and assist in acquiring maintenance when necessary.

By linking the performance analysis of every engine to a central cloud repository, a batch of engine models could be observed to determine the presence of systematic errors, if any. Thus, intelligent systems could be designed in order to diagnose and prevent problems. As shown in Fig. 8, spectral analysis of each unit under a certain engine model is performed. This enables us to recognize a pattern in the performance of a model, thereby preventing abnormal combustion in the models observed, or to improve the future models by considering the current diagnosis.

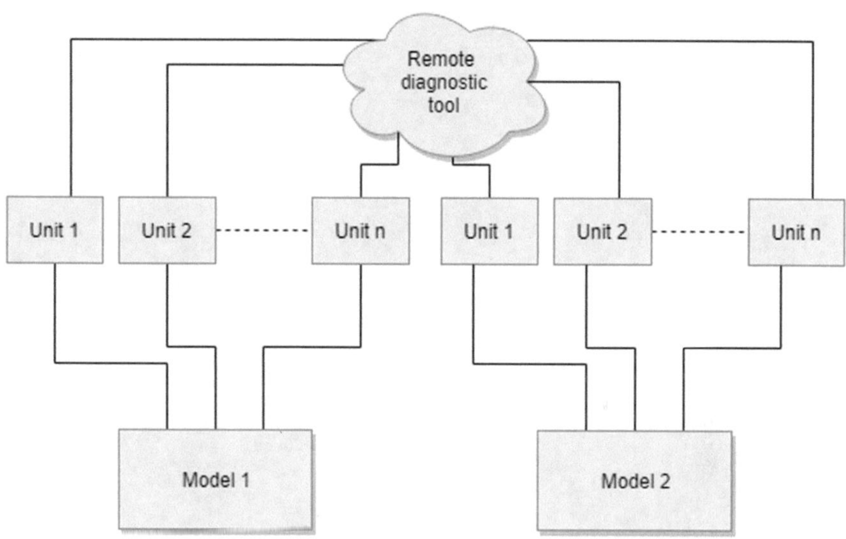

Fig. 8. Flowchart for monitoring a batch of engines

References

1. Heywood, J.: Internal Combustion Engine Fundamentals. McGraw-Hill Education, New York (1988)
2. Konig, G., Sheppard, C.G.W.: End gas auto ignition and knock in a spark ignition engine. SAE Trans. **99**, 820–839 (1990)
3. McKenzie, J., Cheng, W.K.: The Anatomy of Knock. SAE Technical Paper, SAE International (2016)
4. Zhi, W., Hui, L., Reitz, R.D.: Knock combustion in spark-ignition engines. Prog. Energy Combust. Sci. **61**, 78–112 (2017)

5. Shivapuji, A.M.: In-cylinder experimental and modeling studies on producer gas fuelled operation of spark ignited gas engines, PhD thesis. http://cgpl.iisc.ernet.in/dasappa/img/pdf/thesis/anandMS_PhD_Thesis.pdf
6. Zhen, X., Wang, Y., Xu, S., Zhu, Y., Tao, C., Xu, T., Song, M.: The engine knock analysis-an overview. Appl. Energy **92**, 628–636 (2012)
7. Schmillen, K.P., Rechs, M.: Different methods of knock detection and knock control. Technical report, SAE Technical Paper No. 910858 (1991)
8. Millo, F., Ferraro, C.V.: Knock in si engines: a comparison between different techniques for detection and control. Technical report, SAE Technical Paper No. 982477 (1998)
9. Fathi, M., Holland, A., Ansari, F., Weber, C.: Integrated Systems, Design and Technology 2010: Knowledge Transfer in New Technologies. Springer, Berlin (2011)
10. Shivapuji, A.M., Dasappa, S.: Knock and its prediction in producer gas fueled SI engines. In: Conference: Proceedings of the International Conference on Polygeneration Strategies, pp. 227–236

Development of Report Evaluation Portal for Remote Lab

Ankit Sharma(✉), K. N. Spurthy, K. Annapoorneshwari, Shorya Shubham, and K. C. Narasimhamurthy

Siddaganga Institute of Technology, Tumkur, India
{ankit34567, spukn98, anukrishnakanth, shorya007shubham}
@gmail.com, kcnmurthy@sit.ac.in

Abstract. Online engineering is being used as one of the efficient ways of learning platform by students. Remote lab is also being used extensively for conducting experiments and it can also be an alternative to conventional lab as it can be accessed by students from anywhere, as many times required. The learning through remote lab will complete only when the user effectively conducts the experiments and submits a detailed report to the instructor/Professor for evaluation. This paper gives the overview of development of Report Evaluation Portal for Remote Lab useful for both user and the lab instructor.

Keywords: Remote lab · Evaluation · Report · Grade · Analog discovery

1 Introduction

The Engineering education is adopting newer ways of learning. Online education is one of the major breakthroughs in that approach [1–8]. Students can take up online courses offered by reputed universities without being registered that university formally! The evolution process for these online classes has been well established that enables to provide grades and certification to students. On the same lines, Remote and virtual labs are also becoming popular. User can conduct experiments remotely from anywhere and anytime. This will help the user to solidify the concepts learnt in theory classes. These types of labs are very essential for Electronic courses which are based on circuit analysis. These courses can be better understood by conducting experiments. Even though Remote and virtual labs may help the user to conduct experiments, learning process will be incomplete if they are not evaluated. In this paper, development of Report evolution portal for Remote lab is discussed. This paper focuses on the development of the Report Evaluation Portal for an existing Remote lab facility of 25+ analog electronic experiments with 125+ variations. The electronic circuits available for remote control are Rectifiers, Regulators, Op-amp applications, 555 Timer applications, Filters, Amplifiers, wave shaping circuits etc. The report submission portal will provide the report template for all the experiments. It works as a bridge between the user and the in-charge/Professor of the remote lab. This will help the user to effectively utilize the remote lab facility by submitting the reports of the experiments conducted

M. E. Auer and K. Ram B. (Eds.): REV2019 2019, LNNS 80, pp. 587–594, 2020.
https://doi.org/10.1007/978-3-030-23162-0_52

and in-charge can grade the users as per their ability to understand the concepts during the experiment conduction.

2 Development of Remote Lab System for Analog Filters

In this section, details of the existing Remote lab system are presented. Figure 1 shows the block diagram of the Remote lab system that is built for conducting Analog Electronic Circuits. The selection of experiments for conduction and measurement of readings for circuit analysis is done using Arduino Board and National Instrument's product Analog Discovery Kit respectively. As remote lab system consists of verity of circuits, there is option to perform time response, frequency response and spectrum analysis. The remote lab system can be accessed the user using computer or smart phone. Experiments of the remote lab system can be accessed using laptop or smart phone and doesn't require any specific software at the users end. The advantages of using Remote lab are user can conduct experiments at anytime from anywhere. User can conduct the experiments at desired pace and repeat the experiments till the concepts are understood. User can conduct the experiments along with peers for better learning.

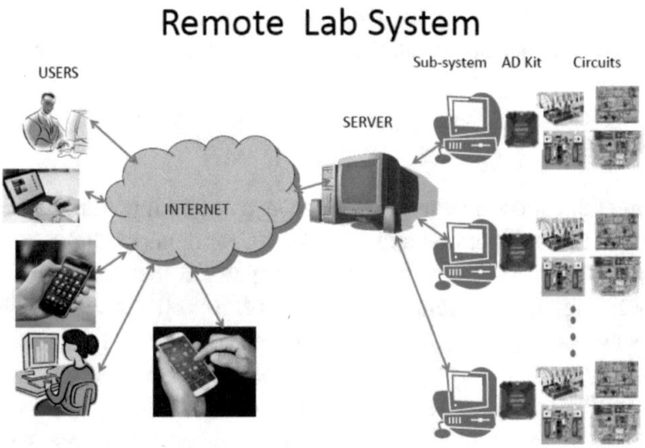

Fig. 1. Block diagram representation of remote lab system

The user can access the web based remote lab. User need to sign up to the system by providing necessary credentials. Since this is a evaluation portal for the Remote lab, both students and instructor have to sign up. Figure 2 shows the Sign up and login page of the portal.

User upon login to the remote lab system can access the any experiment available in the remote lab system. Figure 3 shows the sample of flow for selecting a filter experiment. As shown Fig. 3 all the experiments are grouped under nine category. After the selecting the category user has to narrow down on the experiment then further select any one circuit if that particular experiment has different variance. Then click on the conduction button.

Fig. 2. Evaluation portal front panel. **a** Sign up page, **b** Login page

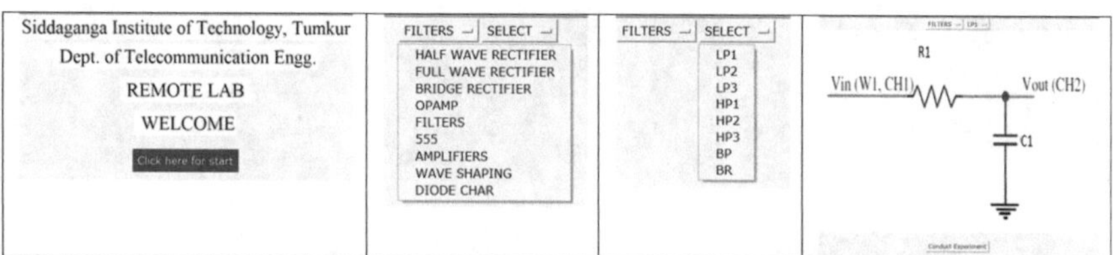

Fig. 3. Front panel views to conduct filter experiments in the remote lab system

User will conduct desired experiments by performing time and frequency responses. Figure 4 shows the screenshots taken during the conduction of various experiments. It may be observed that in each case user is performing desired analysis like, time response, Bode plot and so on. User can also measure the desired timing, voltage or frequency parameter.

During the process of conduction of experiment, user can take screenshots at various stages for future use. User has the option of down loading the template of the Report to be submitted for getting the grade from the lab instructor. Figure 5 shows the option to download the template. The webpage also has the option for user to check previous readings for the latest 10 trials.

Usually template of the experiment consists of circuit diagram, Design equations, theoretical values and a tabular column to edit the data obtained during the conduction. Figure 6 shows the report format of Half wave rectifier.

Figure 7 shows the sample report template of more complex experiment CE amplifier. To fill this template user has to perform both time and frequency response on the amplifier.

The report template will help student what all parameters to measure and which all analysis to perform for various options of the selected circuit. As in the sample template for CE amplifier, user gets a clue to perform both time and frequency responses. User also gets information about parameters to measure that would help in circuit analysis. For example in case of CE amplifier, it's necessary to measure peak to peak of input and output signal to estimate the gain and average voltage (V_C) of output signal give voltage at collector terminal, which decides the maximum swing of the output signal. User need to obtain the Bode plot to find the mid band gain, cut-off frequencies and hence the bandwidth and hence gain bandwidth product. This way the report template will guide the user to perform relevant analysis for better understanding of the

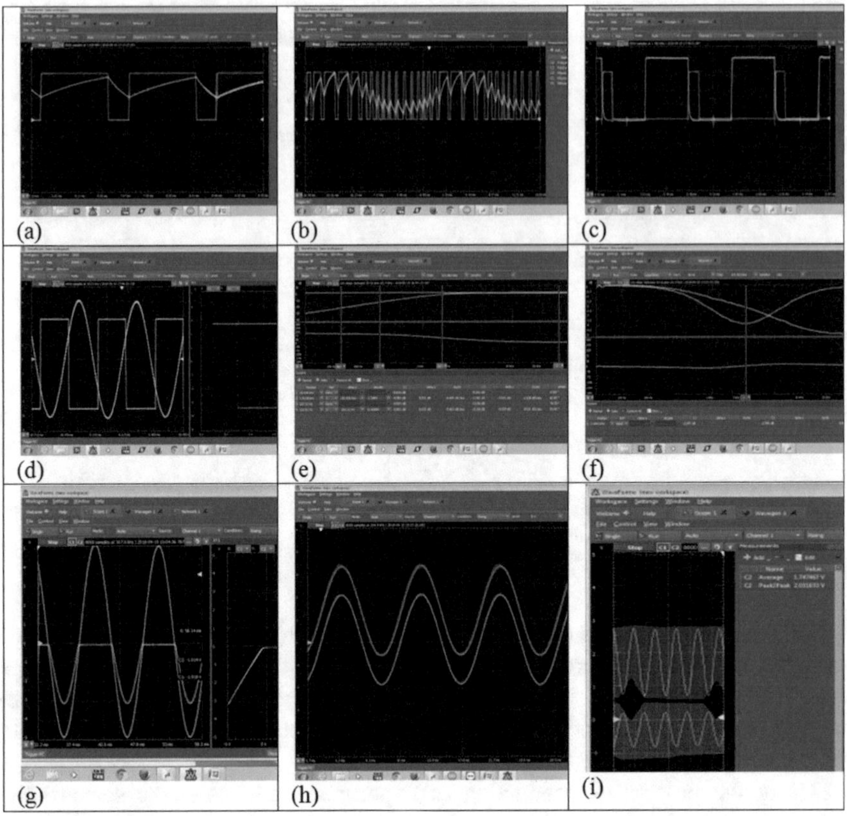

Fig. 4. Screenshots of various experiments. **a** 555 timer astable Multivibrator, **b** PWM generation using 555 timer, **c** Mono stable multivibrator using 555 timer, **d** Op amp based Schmitt trigger, **e** High pass filter, **f** Band stop filter, **g** Clipping circuit, **h** Clamping circuit, **i** Common emitter (CE) amplifier

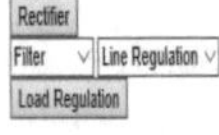

Fig. 5. Download template in student portal

concepts. The template will also bring uniformity in reports submitted by the users. This makes the instructor job easier to evaluate the student's analysis capability of the circuit.

Once the user is done with the report, it can be uploaded on the webpage. Figure 8 shows the web page option to upload the report after conducting Rectifier experiment. It may be noted user can upload the report of a specific experiment.

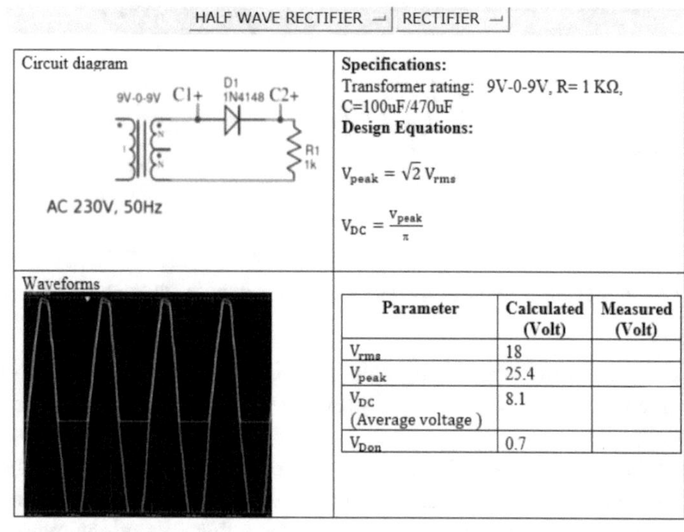

HALF WAVE RECTIFIER | RECTIFIER

Circuit diagram

Specifications:
Transformer rating: 9V-0-9V, R= 1 KΩ, C=100uF/470uF
Design Equations:

$$V_{peak} = \sqrt{2}\, V_{rms}$$

$$V_{DC} = \frac{V_{peak}}{\pi}$$

Waveforms

Parameter	Calculated (Volt)	Measured (Volt)
V_{rms}	18	
V_{peak}	25.4	
V_{DC} (Average voltage)	8.1	
V_{Don}	0.7	

Fig. 6. Report template of half wave rectifier

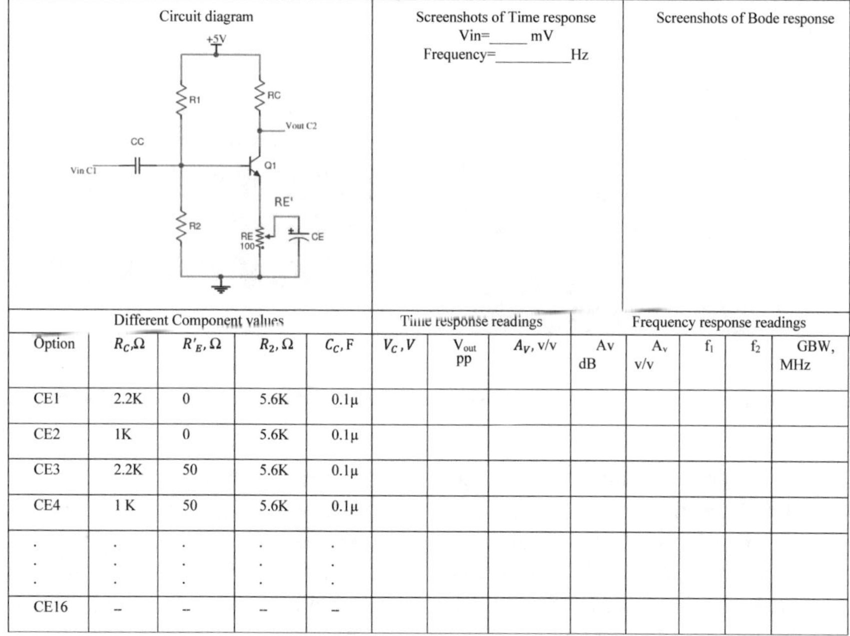

Circuit diagram

Screenshots of Time response
Vin=_____ mV
Frequency=__________Hz

Screenshots of Bode response

Different Component values					Time response readings			Frequency response readings				
Option	R_C,Ω	R'_E, Ω	R_2, Ω	C_C, F	V_C, V	V_{out} pp	A_V, v/v	Av dB	A_v v/v	f_1	f_2	GBW, MHz
CE1	2.2K	0	5.6K	0.1μ								
CE2	1K	0	5.6K	0.1μ								
CE3	2.2K	50	5.6K	0.1μ								
CE4	1 K	50	5.6K	0.1μ								
. . .	. . .	. . .	. . .	. . .								
CE16	–	–	–	–								

Fig. 7. CE amplifier report template.

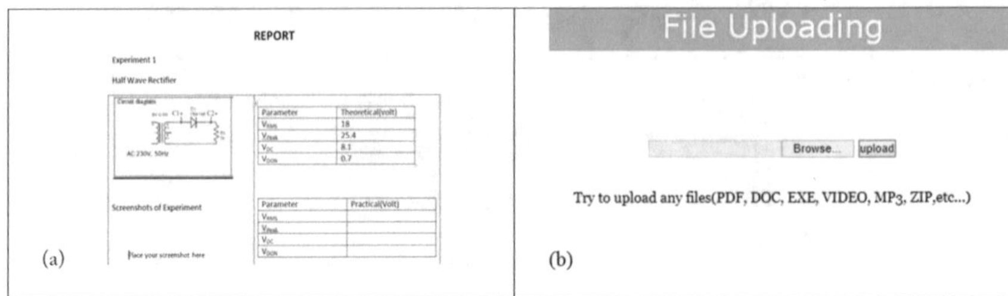

Fig. 8. File uploading feature. **a** Rectifier report template, **b** File uploading web page

User upon successful completion of the report, the document can be uploaded by clicking 'Upload Your Report' so that Professor/lab in-charge can evaluate the report and grade it.

Coming to instructor's login, lab in-charge/instructor will be given unique features. After login, the instructor can select any of the experiment from the list and then the details under that experiment like the list of total number of students who have conducted the experiments with their reports. Other than this it's possible to know duration of usage, repeated usage of experiments by the users. After evaluating the reports, instructor can assign the grade for the students. The web page view is shown in Fig. 9. The teachers are given access to reports of all students, but the student can see only their history of reports submission.

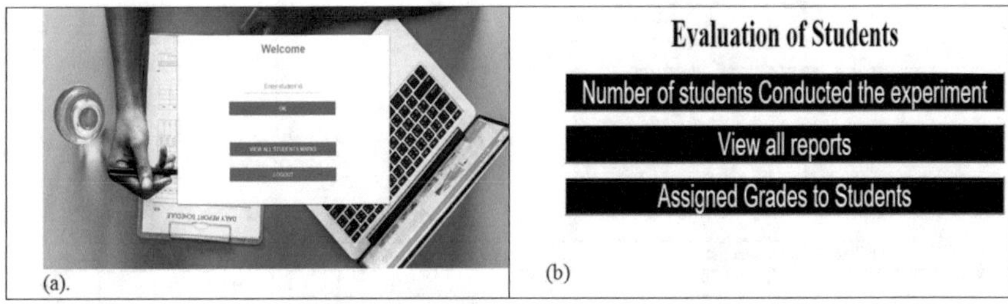

Fig. 9. Web page for instructor. **a** Welcome page, **b** Contents of the evaluation page

The front end designing is done using HTML, PHP, Bootstrap and CSS and the back end includes databases (MYSQL), tables. The database is most powerful and most necessary to store the data of students and teachers. Whenever a user signs up, that should be stored in database and whenever user log in, the username and password will be matched with the stored data and the access to the remote lab is given. Figure 10 shows the structure for the student database system.

To ensure effective use of the Remote lab system, all users' access details is stored in the data base and users who don't submit the report will be black listed. The instructor is given the authority to deny the access of remote lab system to such users.

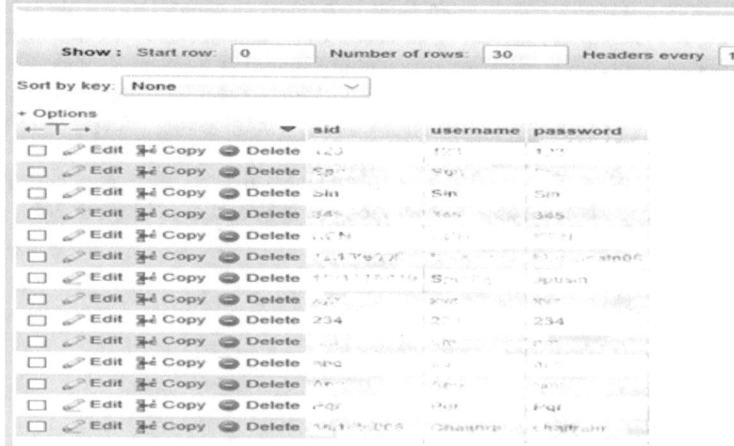

Fig. 10. The student database

3 Conclusions

Using this Report Evaluation Portal, the Remote lab system can be effectively utilized and students can get to know their grades quickly. Dummy users can be denied the permission of access. This will help the students to conduct experiments and analyze them and prepare a report using the template. Students can analyze their experimental data at anytime, from anywhere and send the same data to many teachers to get proper feedback regarding the experiment from them.

Acknowledgements. Authors thank the Management, Director and Principal of Siddaganga Institute of Technology (SIT), Tumakuru for their support in making this project. We also extended our gratitude to faculty members and students of Department of Telecommunication Engineering, SIT, Tumakuru for their support during the development of this innovative methodology.

References

1. http://vlab.co.in/
2. http://ohm.ieec.uned.es/portal/?page_id=76
3. http://www.uml.edu/IT/Services/vLabs/
4. Balamuralithara, B., Woods, P.C.: Virtual Laboratories in Engineering Education: The Simulation Lab and Remote Lab (2007)
5. Chen, X., Song G., Zhang, Y. :Virtual and remote laboratory development: a review. In: Proceedings of Earth and Space 2010, pp. 3843–3852, Honolulu (2010)
6. Chen, X., Lawrence O. Kehinde, P.E., Zhang, Y., Darayan, S., Texas Southern. David O. Olowokere, Mr. Daniel Osakue, USING Virtual And Remote Laboratory To Enhance Engineering Technology Education

7. Daniel Kruse, Sulamith Frerich from Ruhr-Universität Bochum, Project ELLI - Excellent Teaching & Learning in Engineering Sciences, Universitätsstr. 150 | 44801 | Germany and Marcus Petermann, Andreas Kilzer from Ruhr-Universität Bochum, Particle Technology, Universitätsstr. 150 | 44801 | Germany titled "Virtual Labs And Remote Labs: Practical Experience For Everyone", 2014 IEEE Global Engineering Education Conference (EDUCON)
8. Nedic, Z., Machotka, J., Nafalski, A. titled "Remote laboratories versus virtual and real laboratories", published. In: 33rd Annual Frontiers in Education, 2003. FIE 2003

Analysis of Operational-Amplifier Inverting and Non-inverting Amplifiers in Remote Lab

K. C. Narasimhamurthy[1(✉)], Bharat Malaviya[2], Kondamarri Reddy Pranesh[2], Alladi Jayashree[2], and Harikrishna Kamatham[2]

[1] Siddaganga Institute of Technology, Tumkur, India
kcnmurthy@sit.ac.in, kcnmurthy@gmail.com
[2] VJIT, Hyderabad, India
{bharatmalaviya7,rk.pranesh156,ajayashree2}@gmail.com, ecehod@vjit.ac.in

Abstract. In this paper, performance analysis of Opamp inverting and non-inverting amplifiers using Remote Lab is presented. Remote lab system has the feature of conducting experiments on Opamp Amplifiers with different voltage gain. Remote lab system is built using Arduino ATMEGA-2560, National Instrument's (NI) Analog Discovery kit and other circuitry. User has the freedom to apply input signal of desired amplitude and frequency. Opamp parameters like slew rate and unity gain bandwidth, can be evaluated. Paper describes the process of determining Opamp amplifiers like closed loop gain, 3-dB bandwidth, and phase and gain margin. Paper also presents analysis report of Opamp amplifiers conducted remotely.

Keywords: Op-amp amplifiers · Remote lab · Voltage gain · Bode-plot

1 Introduction

In the present Electrical Engineering education courses are taught by one faculty and the relevant lab courses will be conducted by students under the supervision of another faculty or teaching assistants usually after many days. Engineering education will be better understood employing "Experiential learning" concept. Electrical Engineering courses will better conveyed by conducting experiments relevant to the electronic circuits while discussing the theoretical concepts in the classroom. It's also necessary that students get ample opportunity to conduct these experiments to solidify their understanding and clarifying doubts beyond class hours. There are online virtual lab portals are available [1–3]. All these simulation based learning, in which user will not be able to perform real time circuit analysis. Even though experiments are conducted remotely without the assistance of instructor, user should feel comfortable. The methodology of experiment conduction should be similar to that of the conventional lab so that user can operate with ease. The remote lab system developed has necessary theoretical explanation, specifications, design equations, component values and ideal and typical values of the expected readings. It also provides report template so that students can easily submit the report. The remote lab for Opamp amplifiers can be

M. E. Auer and K. Ram B. (Eds.): REV2019 2019, LNNS 80, pp. 595–604, 2020.
https://doi.org/10.1007/978-3-030-23162-0_53

accessed easily on by providing the necessary credentials. The paper presents detailed time and Bode analysis of inverting and non-inverting Opamp amplifiers of various gain variations.

2 Development of Remote Lab System

Remote lab system is developed to provide access to faculty and students to encourage "Experiential learning". The block diagram of the Remote lab system is shown Fig. 1.

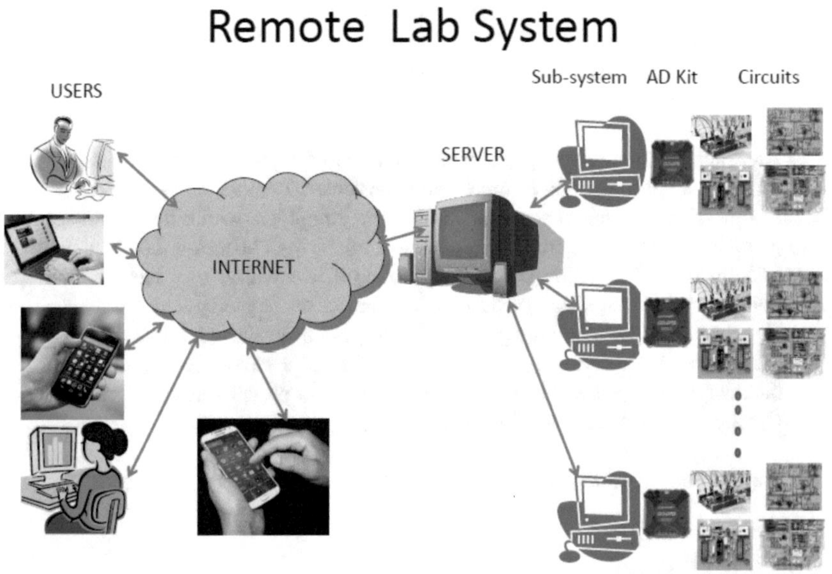

Fig. 1. Block diagram representation of Remote lab system

Figure 1 shows the conceptual Remote lab system implemented. Experiments of the remote lab system can be accessed using laptop or smart phone and doesn't require any specific software at the users end. The advantages of using Remote lab are that, user can conduct experiments at anytime from anywhere [4–8]. User can conduct the experiments at desired pace and repeat the experiments till the concepts are understood. User can conduct the experiments along with peers for better learning. User can conduct experiments by varying key circuit components. User can perform time response as well as Bode analysis. The circuits are analysed by measuring desired parameters of the relevant waveform or plots. Obtained readings along with the screenshots of the waveforms are used to generate report and summarize the readings with outcomes.

3 Opamp Amplifier Design and Analysis

Opamp inverting and non-inverting amplifier circuits are shown in Fig. 2. The design equations of both the amplifiers are shown in the Table 1. Voltage gain variations in the Opamp amplifiers are accomplished using CMOS switches for connecting different R_{IN} and R_F resistors. The actual Opamp amplifier circuits used in the remote lab are shown in Fig. 3. IC CD4016 is used for switching the desired resistor to the circuit. Using different switching combinations for the circuit in Fig. 3, it's possible to have 9 different voltage gains for each amplifier.

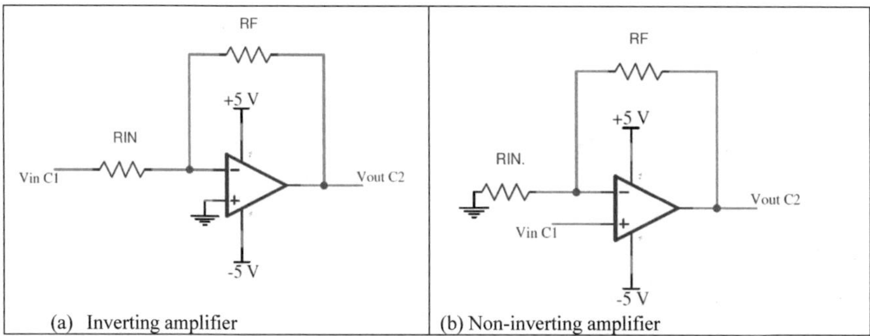

Fig. 2. Op-amp amplifiers. **a** Inverting. **b** Non-inverting

Table 1. Voltage gain expressions of Opamp amplifiers for ideal and typical Opamp

Voltage gain expression	Inverting amplifier	Non-inverting amplifier
With A_0 finite (Typical Opamp)	$A_V = -\frac{R_F}{R_{IN}}\left(1 - \frac{1}{A_0}\left(1 + \frac{R_F}{R_{IN}}\right)\right)$	$A_V = 1 + \frac{R_F}{R_{IN}}\left(1 - \frac{1}{A_0}\left(1 + \frac{R_F}{R_{IN}}\right)\right)$
With A_0 finite (Ideal Opamp)	$A_V = -\frac{R_F}{R_{IN}}$	$A_V = 1 + \frac{R_F}{R_{IN}}$

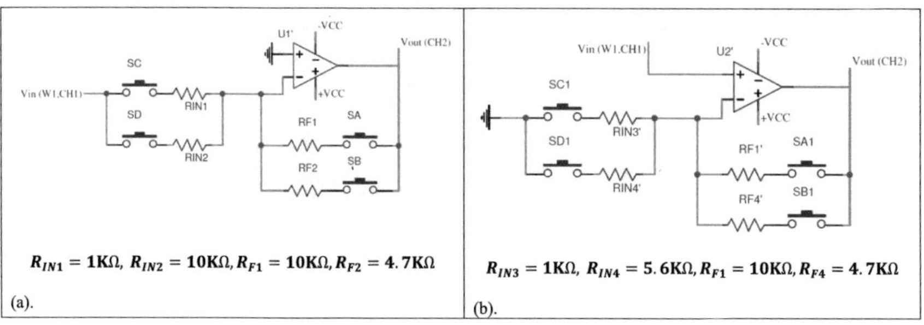

Fig. 3. Opamp amplifiers implemented in remote lab system. **a** Inverting amplifier. **b** Non-inverting amplifiers

Tables 2 and 3 shows all possible amplifier combinations for various resistor selections and the expected ideal voltage gain of inverting and non-inverting amplifiers respectively. The inverting amplifier voltage gain ranges from −0.319 v/v to −11 v/v and non-inverting amplifier gain ranges from 2.22 v/v to 23 v/v circuits, user has the option of varying the gain of the circuit. Python code along with Arduino is used to select desired voltage gain for inverting and non-inverting amplifiers and **National Instrument's Analog Discovery** kit serves the purpose of function generator, oscilloscope and network analyser. The software used for the implementation of Remote lab system are Arduino IDE, Python, and Waveform 2015.

Table 2. Inverting amplifier implemented in remote lab with various voltage gain

Option	Switch condition				R values		Voltage gain
	S_A	S_B	S_C	S_D	R_{IN}	R_F	$A_V = -\frac{R_F}{R_{IN}}$
Inv-1	ON	OFF	ON	OFF	1 KΩ	10 KΩ	−10
Inv-2	OFF	ON	ON	OFF	10 KΩ	10 KΩ	−1
Inv-3	ON	OFF	OFF	ON	1 KΩ	4.7 KΩ	−4.7
Inv-4	OFF	ON	OFF	ON	10 KΩ	4.7 KΩ	−0.47
Inv-5	ON	ON	ON	OFF	1 KΩ||10 KΩ	10 KΩ	−11
Inv-6	ON	ON	OFF	ON	1 KΩ||10 KΩ	4.7 KΩ	−5.17
Inv-7	ON	ON	ON	ON	1 KΩ||10 KΩ	10 KΩ||4.7 KΩ	−3.517
Inv-8	ON	OFF	ON	ON	1 KΩ	10 KΩ||4.7 KΩ	−3.197
Inv-9	OFF	ON	ON	ON	10 KΩ	10 KΩ||4.7 KΩ	−0.319

Table 3. Non-inverting amplifier implemented in remote lab with various voltage gain

Option	Switch condition				R values		Voltage gain
	S_{A1}	S_{B1}	S_{C1}	S_{D1}	R_{IN}	R_F	$A_V = 1 + \frac{R_F}{R_{IN}}$
Non-inv-1	ON	OFF	ON	OFF	1 KΩ	10 KΩ	11
Non-inv-2	OFF	ON	ON	OFF	5.6 KΩ	10 KΩ	2.785
Non-inv-3	ON	OFF	OFF	ON	1 KΩ	22 KΩ	23
Non-inv-4	OFF	ON	OFF	ON	5.6 KΩ	22 KΩ	4.928
Non-inv-5	ON	ON	ON	OFF	1 KΩ||5.6 KΩ	10 KΩ	12.785
Non-inv-6	ON	ON	OFF	ON	1 KΩ||5.6 KΩ	22 KΩ	26.928
Non-inv-7	ON	ON	ON	ON	1 KΩ||5.6 KΩ	10 KΩ||22 KΩ	9.102
Non-inv-8	ON	OFF	ON	ON	1 KΩ	10 KΩ||22 KΩ	7.875
Non-inv-9	OFF	ON	ON	ON	5.6 KΩ	10 KΩ||22 KΩ	2.227

4 Opamp Amplifiers Results and Discussion

In this section, performance of Opamp inverting and Non-inverting amplifiers implemented in Remote lab system will be discussed. The readings are obtained by conducting the experiments remotely. Both obtain time and frequency responses are performed to analyse various parameters of Opamp amplifier. During the conduction of both responses of experiments, it's necessary to know the voltage gain of the amplifier, because input sine wave amplitude has to be suitably adjusted such that the output signal is undistorted (non-clipped sine wave). Otherwise, input sine wave signal of 100 mV amplitude can be applied for all options, as this amplitude will give undistorted output signal even for amplifier with maximum available voltage gain is about 25 v/v. Input signal of lower frequency of about 100 Hz is preferred for signal measurement even at higher voltage gains to avoid slew rate effect of the Opamp.

Screenshots in Figs. 4 and 5 shows the Opamp inverting amplifier time and frequency response and Figs. 6 and 7 shows the same for non-inverting amplifier. Time response of amplifiers in Fig. 4 shows an expected 180° phase shift between input and output signal for inverting amplifier, however Fig. 6 shows output signal in-phase with input signal for non-inverting amplifier. The ratio of output to input sine wave will result in voltage gain (v/v). Bode plots (Both magnitude and phase plot) are obtained by setting frequency sweep as start frequency 50 Hz, end frequency 1 MHz and amplitude in the range of 100 mV. Using the magnitude plot, voltage gain of the amplifier, −3 dB bandwidth (corresponding phases) can be measured. Figures 5 and 7 shows the bode plots of inverting and non-inverting amplifiers respectively.

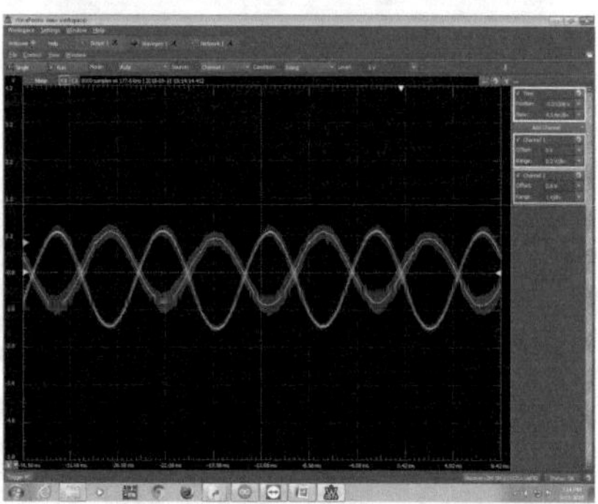

Fig. 4. Time response of inverting amplifier

Tables 4 and 5 shows the readings and calculations of inverting and non-inverting amplifiers all 18 possible variations. The table has voltage gain calculated from time response as well as from Bode plot. The −3 dB bandwidth along with gain bandwidth product for all cases are shown. These readings will help the user to analyse the circuits.

Tables 4 and 5 shows for all options the voltage gain Av (v/v) measured in time response is matching with that of Bode plot. It is interesting to notice that, the measured

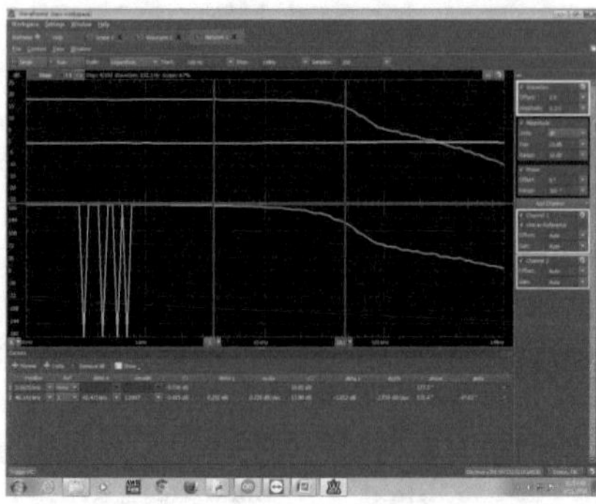

Fig. 5. Frequency response of inverting amplifier

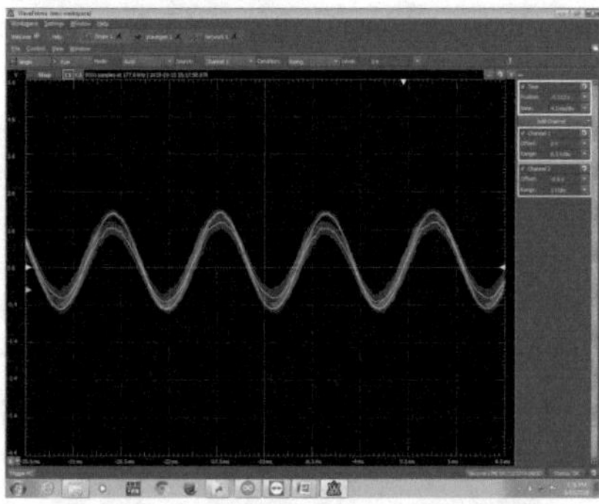

Fig. 6. Time response of non-inverting amplifier

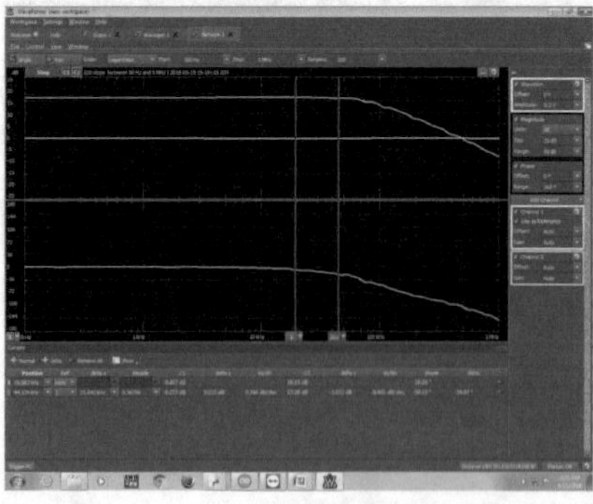

Fig. 7. Frequency response of non-inverting amplifier

voltage gain deviates from the ideal voltage gain for circuits with larger gain. This larger deviation for greater gain is expected as shown in the voltage gain expression in Table 1. It's also noticeable that as voltage gain varies, the −3 dB bandwidth also varies inversely to keep the gain bandwidth product (GBW) constant. Last column of

Table 4. Readings and calculations of Opamp inverting amplifier of remote lab system

Option	R_{IN}, Ω	R_F, Ω	V_{in}	V_{out}	A_V, v/v	$A_V = -\frac{R_F}{R_{IN}}$	Av dB	A_v v/v	Bandwidth KHz	GBW, KHz
Inv-1	1 KΩ	10 KΩ	0.962	6.7	6.96	−10	16.95	7.03	28.93	203
Inv-2	10 KΩ	10 KΩ	1.00	0.971	1.02	−1.00	−0.18	0.987	106.87	109
Inv-3	1 KΩ	4.7 KΩ	0.965	3.234	3.35	−4.7	10.67	3.415	41.686	142
Inv-4	10 KΩ	4.7 KΩ	1 009	0.489	0.486	−0.47	−6.4	0.47	319.48	151
Inv-5	1 KΩ||10 KΩ	10 KΩ	0.962	7.39	7.68	−11.00	17.84	7.79	27.055	210
Inv-6	1 KΩ||10 KΩ	4.7 KΩ	0.962	3.554	3.69	−5.17	11.53	3.77	40.114	151
Inv-7	1 KΩ||10 KΩ	10 KΩ||4.7 KΩ	0.962	2.39	2.48	−3.517	8.118	2.54	46.33	117
Inv-8	1 KΩ	10 KΩ||4.7 KΩ	0.962	6.767	7.03	−3.197	16.96	7.04	28.339	199
Inv-9	10 KΩ	10 KΩ||4.7 KΩ	1.00	0.316	0.316	−0.319	−9.99	0.316	558.81	176

Table 5. Readings and calculations of Opamp non-inverting amplifier of remote lab system

Amplifier selection			Time response			Ideal Voltage gain	Frequency response			
Option	R_{IN}, Ω	R_F, Ω	V_{in}	V_{out}	A_V, v/v	$A_V = 1 + \frac{R_F}{R_{IN}}$	Av dB	A_v v/v	Bandwidth KHz	GBW, KHz
Inv-1	1 KΩ	10 KΩ	0.426	3.63	8.50	11	18.4	8.40	47.27	390
Inv-2	5.6 KΩ	10 KΩ	0.417	1.12	2.69	2.785	9.16	2.87	112	320
Inv-3	1 KΩ	22 KΩ	0.226	3.8	16.8	23	23.3	14.62	30	440
Inv-4	5.6 KΩ	22 KΩ	0.225	0.98	4.38	4.928	13.88	4.94	64	310
Inv-5	1 KΩ||5.6 KΩ	10 KΩ	0.222	2.19	9.85	12.785	20.24	10.28	38.46	390
Inv-6	1 KΩ||5.6 KΩ	22 KΩ	0.221	4.45	20.07	26.928	25.8	19.5	15.84	380
Inv-7	1 KΩ||5.6 KΩ	10 KΩ||22 KΩ	0.224	1.56	6.96	9.1027	17.82	7.78	54.06	421
Inv-8	1 KΩ	10 KΩ||22 KΩ	0.225	1.33	5.9	7.875	16.47	6.66	44.9	260
Inv-9	5.6 KΩ	10 KΩ||22 KΩ	0.226	0.457	2.02	2.227	7.05	2.25	129.93	290

Tables 4 and 5 shows the GBW for all amplifiers is almost constant. Figure 8 shows the Bode plots of Inverting and non-inverting amplifiers obtained to estimate Unity gain bandwidth and gain and phase margin. The measured values indicate unity gain bandwidth of 87, 136 kHz and Phase margin of 40° and 77° for inverting and non-inverting amplifiers respectively.

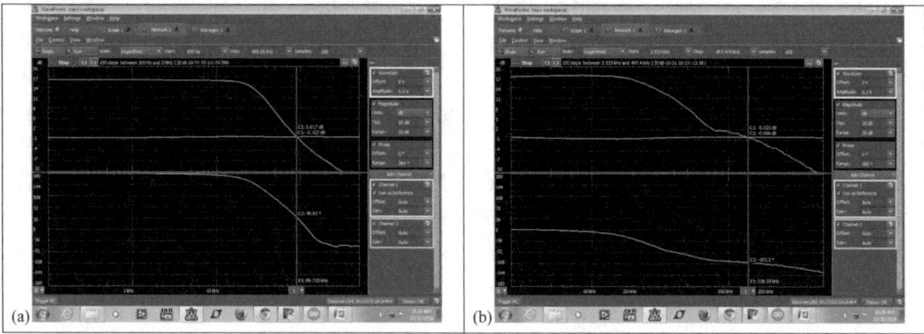

Fig. 8. Bode plots for Unity gain bandwidth and phase margin. **a** Inverting amplifier. **b** Non-inverting

It's also possible to measure one more important Opamp performance parameter slew rate. By applying square wave of 2 kHz with 1 V amplitude for Inv-2 option to conduct inverting amplifier with −1 v/v gain. It's possible to estimate the slew rate by measuring time required for the output to swing from +ve peak to −ve peak. Figure 9 shows the input and output waveforms of the amplifier used to measure slew rate, Fig. 9b show a slew rate of 0.362 V/μs of the Opamp. User can perform all these exhaustive analysis on the Opamp amplifiers remotely at the desired pace as many times and take relevant readings. By conducting experiments with many variations and options to measure many parameters will give confidence to the user in analysing the circuits for thorough understanding of the concepts.

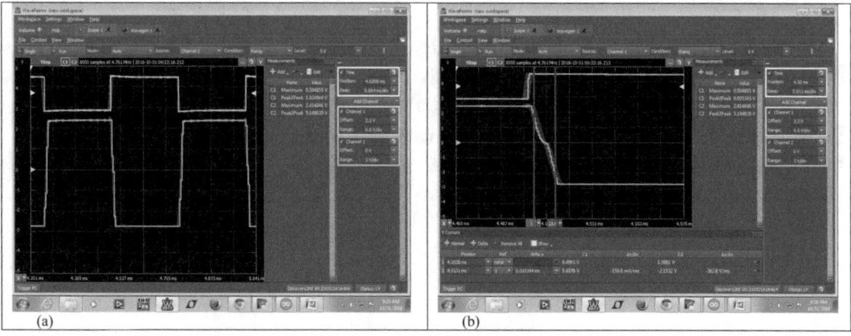

Fig. 9. Opamp time response. **a** Input and output waveforms of inverting amplifier. **b** Zoomed waveforms to measure slew rate

5 Conclusions

In this paper, the development of remotely controllable Operational amplifiers is discussed. Remote user can vary both resistors of the amplifier circuit, hence total of 18 different variations are available. Both time and frequency response can be performed for all the amplifiers. The measured voltage gain varies from −0.319 v/v to +23 v/v across inverting and non-inverting amplifiers. Analysis of these readings will help to understand gain bandwidth product, Unity gain bandwidth and phase margin of the amplifiers. There are many advantages of deploying a remote lab in an educational environment which provides hands-on-experience for engineering students in an efficient, flexible and in a cost-effective manner. User can do analysis on real-experimental data and learn more by repeatedly conducting experiments. Students have the liberty of accessing the lab at anytime and anyplace and perform at their own pace with their peer. This saves time, provides more understanding and students can spend more time with the circuit which will enhance their knowledge.

Acknowledgements. Authors thank the Management, Director and Principal of Siddaganga Institute of Technology (SIT), Tumakuru for their support in making this project. We also extended our gratitude to faculty members and students of Dept. Of Telecommunication Engineering, SIT, Tumakuru for their support during the development of this innovative methodology.

References

1. http://vlab.co.in/
2. http://ohm.ieec.uned.es/portal/?page_id=76
3. http://www.uml.edu/IT/Services/vLabs/
4. Balamuralithara, B., Woods, P.C.: Virtual Laboratories in Engineering Education: The Simulation Lab and Remote Lab (2007)
5. Chen, X., Song, G., Zhang, Y.: Virtual and remote laboratory development: a review. In: Proceedings of Earth and Space 2010, pp. 3843–3852, Honolulu (2010)
6. Chen, X., Lawrence O Kehinde, P.E., Zhang, Y., Darayan, S., Texas Southern. Olowokere, D.O., Osakue, D.: Using virtual and remote laboratory to enhance engineering technology education
7. Kruse, D., Frerich, S., from Ruhr-Universität Bochum, Project ELLI—Excellent Teaching & Learning in Engineering Sciences, Universitätsstr. 150 | 44801 | Germany and Marcus Petermann, Andreas Kilzer from Ruhr-Universität Bochum, Particle Technology, Universitätsstr. 150 | 44801 | Germany titled "Virtual Labs And Remote Labs: Practical Experience For Everyone", 2014 IEEE Global Engineering Education Conference (EDUCON)
8. Nedic, Z., Machotka, J., Nafalski, A.: Remote laboratories versus virtual and real laboratories. Published in: 33rd Annual Frontiers in Education, 2003. FIE (2003)

Analysis of Wave Shaping Circuits in Remote Lab

K. C. Narasimhamurthy[1(✉)], E. Sai Kumar[2], K. Likhitha[2], N. Navya Chowdary[2], B. Shilpa[2], N. Laxmi Sowmya[2], and Ajay Shiva[1]

[1] Siddaganga Institute of Technology, Tumkur, India
kcnmurthy@sit.ac.in, ajayshiva0077@gmail.com
[2] MLRIT, Hyderabad, India
{ellinthalasaikumar,kambampatilikhitha,navya.nelluri98,b.shilpa2397,sonu.nadimpally}@gmail.com

Abstract. In this paper, performance analysis of various wave shaping circuits using Remote Lab is presented. Remote lab system has ability to conduct clipping and clamping circuits with different wave shaping capability. User can remotely conduct the experiment by just click of a button. It's possible to apply any type of AC signal and any value of DC reference voltage. It presents series and shunt clipping circuits along with +ve and −ve voltage clampers. Along with input and output waveforms, user can also obtain transfer characteristics of all clipping circuits using this user can know the working of many wave shaping circuits. Remote lab system for filters is built using Arduino ATMEGA-2560, National Instrument's (NI) Analog Discovery kit and dedicated PCB. User has the freedom to apply input signal of desired amplitude and frequency.

Keywords: Remote lab · Wave shaping circuits · Transfer characteristics · Reference voltage · Analog discovery

1 Introduction

In the present Electrical Engineering education courses are taught by one faculty and the relevant lab courses will be conducted by students under the supervision of another faculty or teaching assistants usually after many days. Engineering education will be better understood employing "Experiential learning" concept. Electrical Engineering courses will better conveyed by conducting experiments relevant to the electronic circuits while discussing the theoretical concepts in the classroom. It's also necessary that students get ample opportunity to conduct these experiments to solidify their understanding and clarifying doubts beyond class hours. There are online virtual lab portals are available [1–3]. All these simulation based learning, in which user will not be able to perform real time circuit analysis. Even though experiments are conducted remotely without the assistance of instructor, user should feel comfortable. The methodology of experiment conduction should be similar to that of the conventional lab so that user can operate with ease. The remote lab system developed has necessary theoretical explanation, specifications, design equations, component values and ideal & typical values of the expected readings. Wave shaping circuits are one of the basic

M. E. Auer and K. Ram B. (Eds.): REV2019 2019, LNNS 80, pp. 605–611, 2020.
https://doi.org/10.1007/978-3-030-23162-0_54

analog electronic circuits being taught at the beginning of undergraduate courses. Students need to experiment on varieties of circuits to understand the concepts. Remote lab is very handy in such cases as user can tryout many variations of the circuit [4–8].

2 Development of Remote Lab System for Wave Shaping Circuits

Remote lab system is developed to provide access to faculty and students to encourage "Experiential learning". The block diagram of the Remote lab system is shown Fig. 1. Conceptual Remote lab system is implemented as shown in Fig. 1. Experiments of the remote lab system can be accessed using laptop or smart phone and doesn't require any specific software at the users end. The advantages of using Remote lab are user can conduct experiments at anytime from anywhere. User can conduct the experiments at desired pace and repeat the experiments till the concepts are understood. User can conduct the experiments along with peers for better learning. User can analyse experiments by varying key circuit components. User can perform time response as well as Bode analysis. The circuits are analysed by measuring desired parameters of the relevant waveform or plots. Obtained readings along with the screenshots of the waveforms are used to generate report and summarize the readings with outcomes.

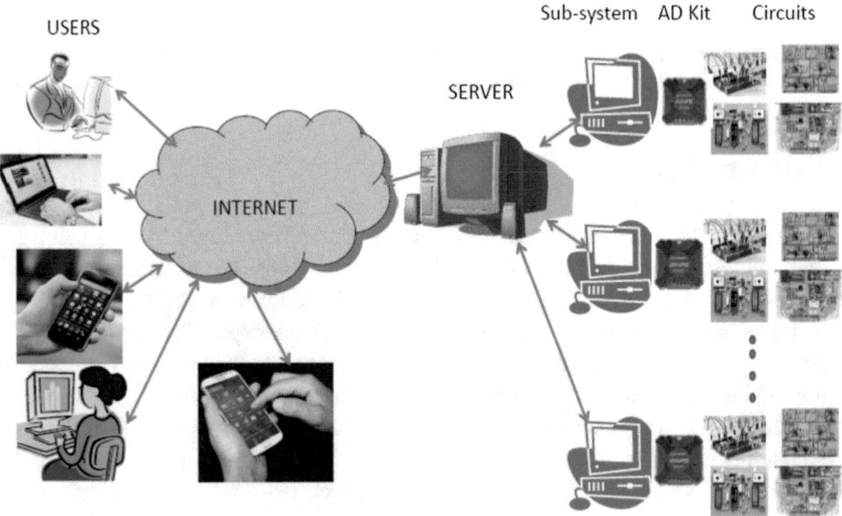

Fig. 1. Block diagram representation of remote lab system

3 Wave Shaping Circuits and Their Analysis

This section describes the development of wave shaping circuits namely clipping and clapping circuits for remote lab system. Remote lab will help the user to analyse the effect of modifications in the circuit and in reference voltages on the shape of the output signal. Figure 2 shows the possible shunt clipping circuits with three possible variances. Using these three circuits user can conduct experiments by varying input signal amplitude, frequency and reference voltage levels. Using this simple experiments user can understand the concept of Reverse recovery time and built in potential of the diode, in fact it's possible to measure them.

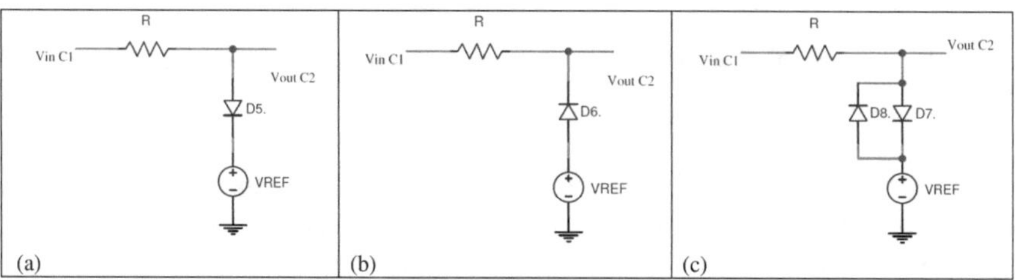

Fig. 2. Shunt clipping circuits **a** variance 1 **b** variance 2 **c** variance 3

Figure 3 shows the possible series clipping circuits with three possible variances. As in case of shunt clipping circuits it's possible to obtain waveforms of different clipping level.

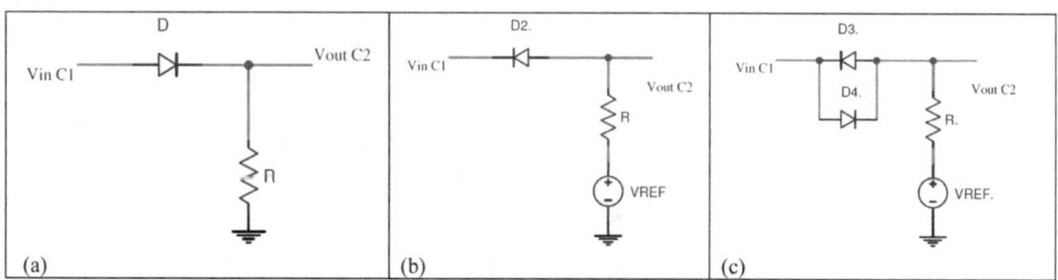

Fig. 3. Series clipping circuits **a** variance 1 **b** variance 2 **c** variance 3

As can be seen in Figs. 2 and 3 it is required to rig-up six different circuits to perform clipping circuits. However, using two innovative circuits it's possible to conduct all six variations of the clipping circuits. The circuits shown in Fig. 4 will enable user to conduct desired clipping circuits remotely.

Switches S1 and S2 are realized using CMOS switches (IC CD 4016) and can be controlled using Arduino board and python script.

Figure 5 shows the front panel screenshots of Remote lab system during the conduction of wave shaping experiments. As seen in the screenshots Remote lab system, it has many analog electronic experiments. To perform clipping experiments,

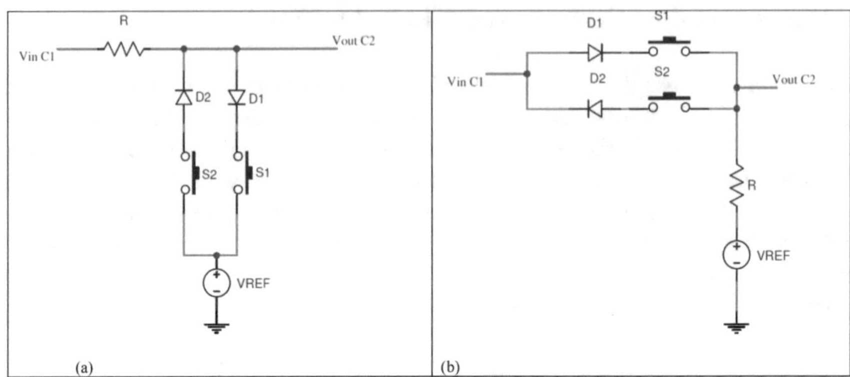

Fig. 4. Implementation of clipping circuits in remote lab system **a** Shunt clippers **b** Series clippers

User has to select the Wave shaping tab and then in the list clipping experiments any one experiment is selected for conduction. Using Analog Discovery kit through Waveform-15 tool user can perform timing analysis.

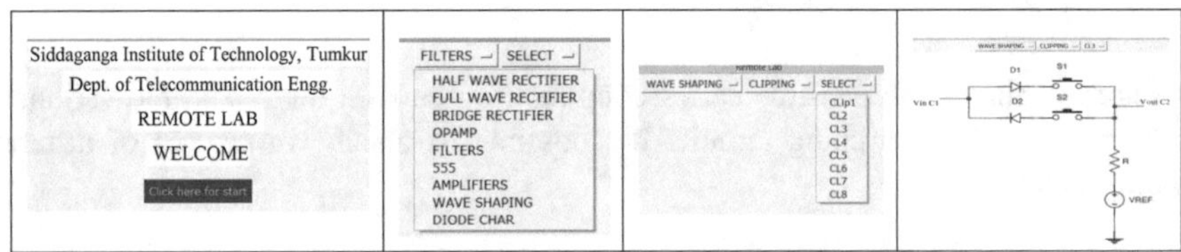

Fig. 5. Front panel views to conduct wave shaping experiments in the remote lab system

One more category of wave shaping circuits i.e. clapping circuits shown in Fig. 6 is considered for implementation in remote lab system. Figure 6a, b shows clamping circuit for inserting +ve and −ve DC voltages. Both the circuits are implemented in Remote lab system using the circuit in Fig. 6c. The switches shown in the circuit of Fig. 6c. are realized using CMOS IC CD 4016. User will have the opportunity to conduct clamping circuits with required DC voltage insertion.

4 Wave Shaping Results and Discussion

In this section, performance of clipping and clamping circuits are presented. Time response of both wave shaping circuits are discussed. During the conduction of time response of experiments, sine wave input signal of 5 V amplitude at 100 Hz is applied using Wavegen channel 1 of the Analog Discovery Kit.

Figure 7 shows the time response of various clipping circuits. Figure 7a is the response of the circuit shown in Figs. 2b and 3a with reference voltage as 0 V.

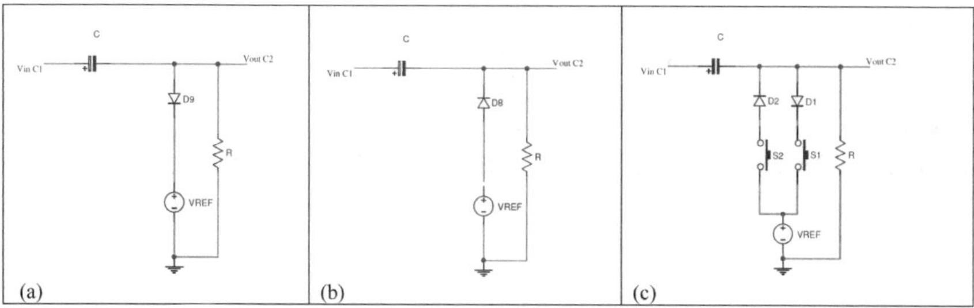

Fig. 6. Clamping circuits **a** +ve DC voltage insertion **b** −ve DC voltage insertion **c** Remote lab circuit

Similarly, Fig. 7b is of circuit in Fig. 2b with +ve DC voltage, waveform in Fig. 7c is due to circuit in Fig. 2b but with −ve DC voltage. Waveform in Fig. 7d is the time response of circuit in Fig. 3b. It may be noted that all the scope screenshots consists of transfer characteristics of the corresponding circuits. All these screenshots are obtained by conducting the experiments remotely.

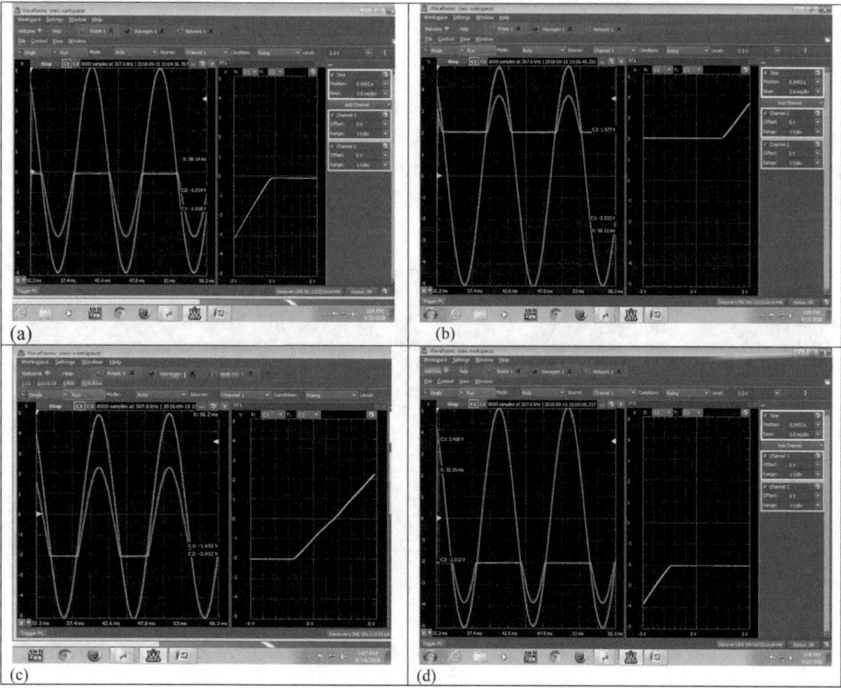

Fig. 7. Time response of various clipping circuits with different DC voltages **a** 0 V **b** +ve voltage, **c**, **d** −ve voltage

As mentioned earlier it's possible to measure diode drop and reverse recovery time of the diode. Using Fig. 8a voltage drop across the diode when it is ON can be obtained by calculating the difference in the maximum voltages of input and output waveform. The build in potential of the diode is 0.38 V. Similarly the reverse recovery time can be measured by applying sine wave of large frequency (10 kHz). As seen in the Fig. 8b, the reverse recovery time is 11.7 μs. In this way using Remote lab system user can understand the clipping circuits easily.

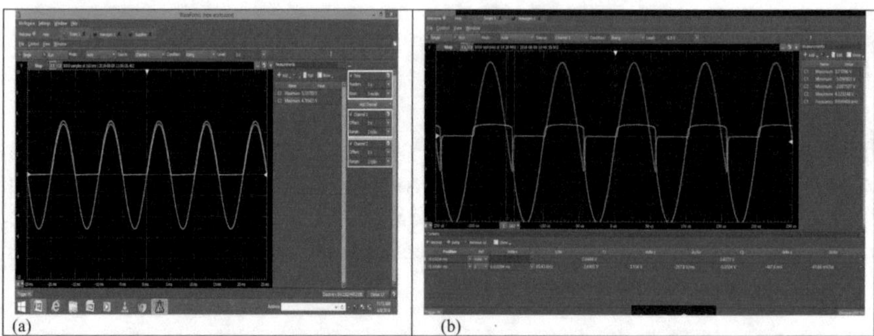

Fig. 8. Time response of clipping circuits **a** Built-in potential **b** Reverse recovery time

Waveforms of clamping circuits are shown Fig. 9. Using the clamping in the remote lab it is possible to clamp the sine wave to +ve or −ve DC voltage. The output waveform (blue) has been shifted down by inserting −ve DC voltage as shown in Fig. 9a, b shows sine wave with +ve DC voltage inserted.

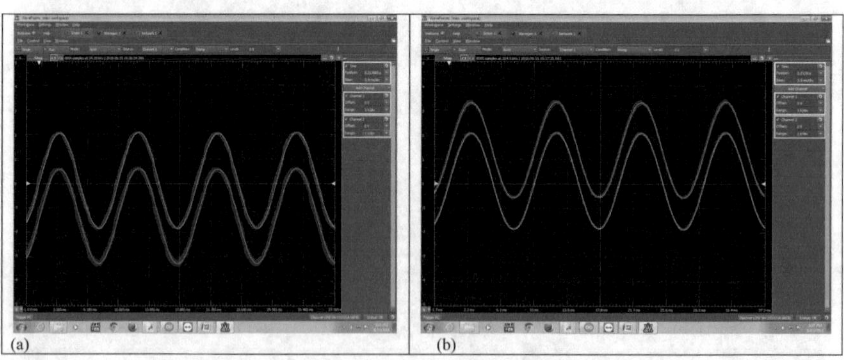

Fig. 9. Time response of clamping circuits **a** −ve DC insertion **b** +ve DC insertion

5 Conclusions

In this paper, the developments of remotely controllable wave shaping circuits are presented. Analog wave shaping circuits are very basic circuits to be understood in the initial stage of electronics engineering. Remote lab system will help user to conduct different varieties of clipping and clapping circuits. It also help user to vary input and reference parameters to analyse the circuit behaviour. By performing wave shaping experiments user can understand reverse recovery time and built in potential concepts of the diode. The paper also presents the performance of clamping circuit with different DC voltage insertions.

Acknowledgements. Authors thank the Management, Director and Principal of Siddaganga Institute of Technology (SIT), Tumakuru for their support in making this project. We also extended our gratitude to faculty members and students of Department of Telecommunication Engineering, SIT, Tumakuru for their support during the development of this innovative methodology.

References

1. http://vlab.co.in/
2. http://ohm.ieec.uned.es/portal/?page_id=76
3. http://www.uml.edu/IT/Services/vLabs/
4. Balamuralithara, B., Woods, P.C.: Virtual Laboratories in Engineering Education: The Simulation Lab and Remote Lab (2007)
5. Chen, X., Song G., Zhang, Y.: Virtual and remote laboratory development: a review. In: Proceedings of Earth and Space 2010, pp. 3843–3852, Honolulu (2010)
6. Chen, X., Lawrence O Kehinde. P.E., Zhang, Y., Darayan, S., Texas Southern. David O. Olowokere, Daniel Osakue, USING Virtual And Remote Laboratory To Enhance Engineering Technology Education
7. Daniel Kruse, Sulamith Frerich from Ruhr-Universität Bochum, Project ELLI - Excellent Teaching & Learning in Engineering Sciences, Universitätsstr. 150 | 44801 | Germany and Marcus Petermann, Andreas Kilzer from Ruhr-Universität Bochum, Particle Technology, Universitätsstr. 150 | 44801 | Germany titled "Virtual Labs And Remote Labs: Practical Experience For Everyone", 2014 IEEE Global Engineering Education Conference (EDUCON)
8. Nedic, Z., Machotka, J., Nafalski, A. titled Remote laboratories versus virtual and real laboratories. In: 33rd Annual Frontiers in Education, 2003. FIE 2003

Analysis of Filter Circuits in Remote Lab

K. C. Narasimhamurthy[1], K. C. Thanmayi[1], T. N. Bhuvana[1], H. R. Chaitra[1], K. N. Spurthy[1(✉)], and T. M. Raghavendra Kashyap[2]

[1] Siddaganga Institute of Technology, Tumkur, India
kcnmurthy@sit.ac.in, {thanmayikc,bhuvana.tn06, chaitraprabhu71,spukn98}@gmail.com
[2] REVA University, Bangalore, India
tmrkashyap@gmail.com

Abstract. In this paper, performance analysis of analog filters using Remote Lab is presented. Remote lab system has the option of changing the transfer function of the filter to analyze its performance. Eight analog filters are considered for analysis. Both time and frequency response are obtained to determine pass band gain, pole and zero locations. Using this user can know the effect of adding resistor to the existing filter circuit. Remote lab system for filters is built using Arduino ATMEGA-2560, National Instrument's (NI) Analog Discovery kit and dedicated PCB. User has the freedom to apply input signal of desired amplitude and frequency. Paper also presents typical analysis report of filters conducted remotely.

Keywords: Remote lab · Transfer function · Analog filters · Poles and zeros · Analog discovery

1 Introduction

In the present Electrical Engineering education courses are taught by one faculty and the relevant lab courses will be conducted by students under the supervision of another faculty or teaching assistants usually after many days. Engineering education will be better understood employing "Experiential learning" concept. Electrical Engineering courses will better conveyed by conducting experiments relevant to the electronic circuits while discussing the theoretical concepts in the classroom. It's also necessary that students get ample opportunity to conduct these experiments to solidify their understanding and clarifying doubts beyond class hours. There are online virtual labs portals are available [1–3]. All these simulation based learning, in which user will not be able to perform real time circuit analysis. Even though experiments are conducted remotely without the assistance of instructor, user should feel comfortable. The methodology of experiment conduction should be similar to that of the conventional lab so that user can operate with ease. The remote lab system developed has necessary theoretical explanation, specifications, design equations, component values and ideal and typical values of the expected readings. It also provides report template so that students can easily submit the report. The remote lab for Opamp amplifiers can be accessed easily on by providing the necessary credentials. The paper presents detailed

M. E. Auer and K. Ram B. (Eds.): REV2019 2019, LNNS 80, pp. 612–622, 2020.
https://doi.org/10.1007/978-3-030-23162-0_55

time and Bode analysis of inverting and non-inverting Opamp amplifiers of various gain variations.

2 Development of Remote Lab System for Analog Filters

Remote lab system is developed to provide access to faculty and students to encourage "Experiential learning". Remote lab system is implemented as shown in Fig. 1. Experiments of the remote lab system can be accessed using laptop or smart phone through internet [4–7] and doesn't require any specific software at the users end. The advantages of using Remote lab are user can conduct experiments at anytime from anywhere and get the output of the circuit and analyse the performance. User can conduct the experiments at desired pace and repeat the experiments till the concepts are understood. User can conduct the experiments along with peers for better learning by varying key circuit components. User can perform time response as well as Bode analysis. The circuits are analysed by measuring desired parameters of the relevant waveform or plots. Obtained readings along with the screenshots of the waveforms are used to generate report and summarize the readings with outcomes.

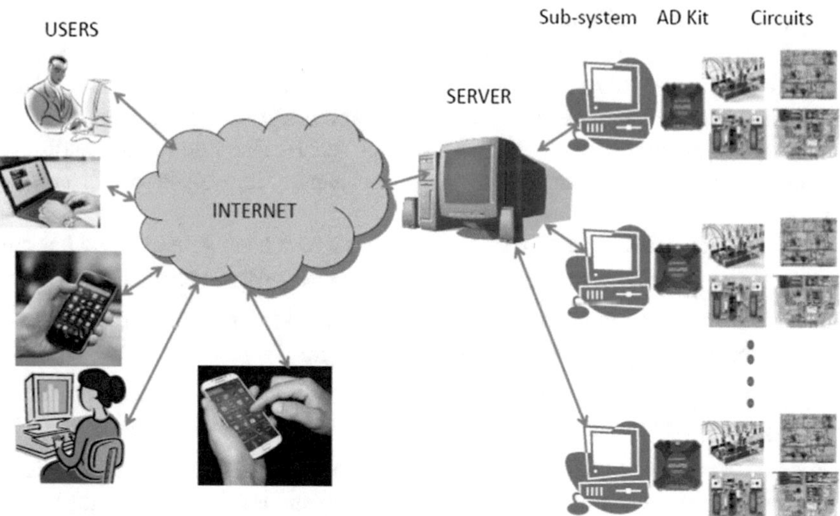

Fig. 1. Block diagram representation of Remote lab system

3 Analog Filter Design and Analysis

This section describes the development of Filter circuits for remote lab system, to analyse the effect of adding resistor to an existing filter circuit. Figure 2 shows typical first order Low pass filter (LPF) and realization two more circuits after adding resistor

R_2 in series with C_1 and in parallel with C_1. The effect of resistor R_2 insertion to LPF is discussed in Table 1, consisting of transfer function, pass band gain, pole-zero plot and Bode plot.

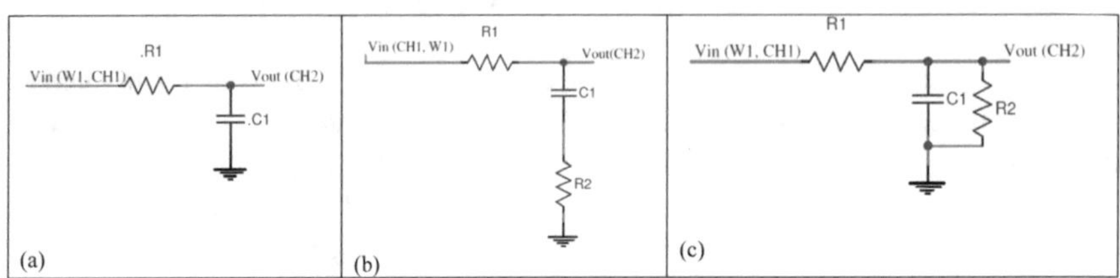

Fig. 2. Low pass filter and its variance. **a** LPF, **b** phase lag network, **c** LPF1

Table 1, shows that for the first order LPF has a pass band gain of 1 and has a pole at frequency $\frac{1}{2\pi R_1 C_1}$ and zero at infinity. Addition of resistor R_2 in series with capacitor results in major change in the pass band gain and the slight shift in the pole frequency. However, the zero will be shifted to $\frac{1}{2\pi R_2 C_1}$ from infinity. Table 1 reveals that addition of a resistor in parallel to capacitor will not modify the location of zero but the pole frequency is slightly modified to $\frac{1}{2\pi (R_1 || R_2) C_1}$. Table 1 shows the effect of R_2 on LPF through the movement of poles and zero positions both in pole-zero plot and Bode plot. All these variations on LPF can be performed by click of button in Remote lab system.

Similar type of analysis is done for first order high pass filter (HPF). Figure 3 shows first order HPF and realization two more circuits after adding resistor R_2 in series with C_1 and in parallel with C_1. The effect of resistor R_2 insertion to HPF is discussed in Table 2, consisting of transfer function, pass band gain, pole-zero plot and Bode plot.

Table 2, shows that for the first order HPF has a pass band gain of 1 and has a pole at frequency $\frac{1}{2\pi R_1 C_1}$ and zero at origin. Addition of resistor R_2 in parallel with capacitor results in major change in the pass band gain and the slight shift in the pole frequency. However, the zero will be shifted to $\frac{1}{2\pi R_2 C_1}$ from origin. Table 1 reveals that addition of a resistor in series with capacitor will not modify the location of zero, but the pole frequency is slightly modified to $\frac{1}{2\pi (R_1 + R_2) C_1}$. Table 2 shows the effect of R_2 on HPF through the movement of poles and zero positions both in pole-zero plot and Bode plot.

To visualize the effect of adding a resistor to LPF and HPF, user needs to rig-up the circuit for each case and perform the analysis. In conventional lab this process takes more time. Low pass filter and its variance and that of HPF shown in Figs. 2 and 3 are realized in the Remote lab system using the specially designed circuits as shown in Fig. 4. The switches are turned ON and OFF as per the filter circuit requited. Tables 3 and 4 shows the switch conditions for the required filter circuit for low pass filter and high pass filter category. These switches are realised using CMOS switches CD4016 and are controlled by Arduino. Control of these switches is given to the user through Python script. After selecting the desired filter circuit user can apply the sine wave

Table 1. Low pass filter and its variance realized in Remote lab system

Filter type	Pass band gain	Transfer function	Pole-zero plot	Bode plot
LPF	1	$\frac{1}{1+sR_1C_1}$	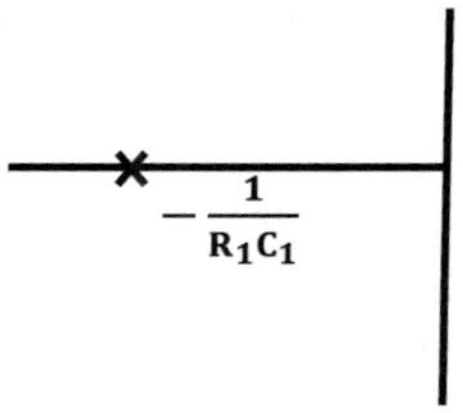 $-\frac{1}{R_1C_1}$	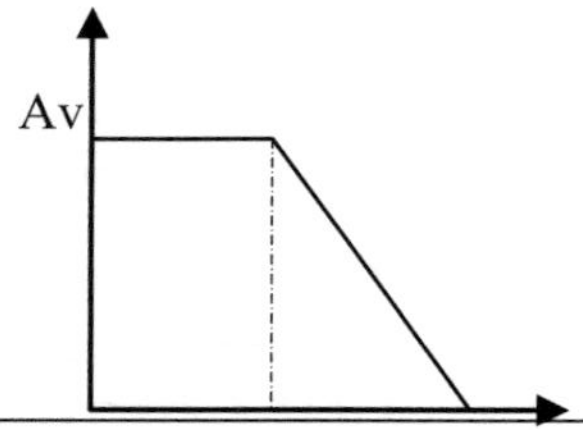 Av, f
Phase lag network	$\frac{R_2}{R_1+R_2}$	$\frac{1+sR_2C_1}{1-s(R_1+R_2)\,C_1}$	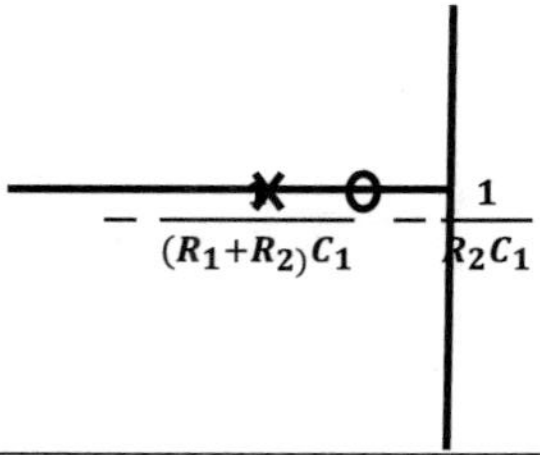 $-\frac{1}{(R_1+R_2)C_1}$, $-\frac{1}{R_2C_1}$	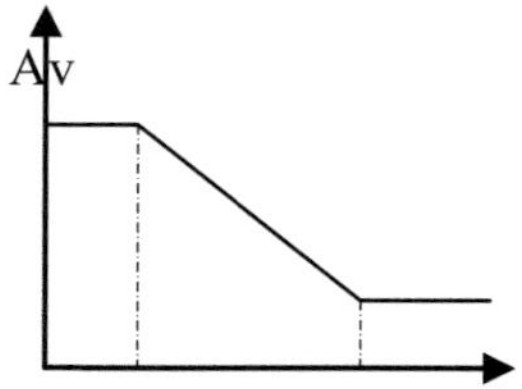 Av, f
LPF1	1	$\frac{1}{1+s(R_1\|\|R_2)\,C_1}$	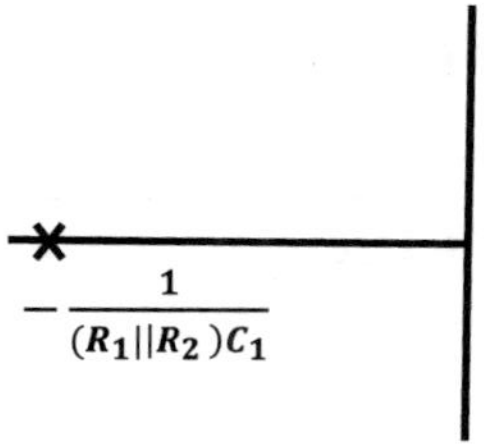 $-\frac{1}{(R_1\|\|R_2)C_1}$	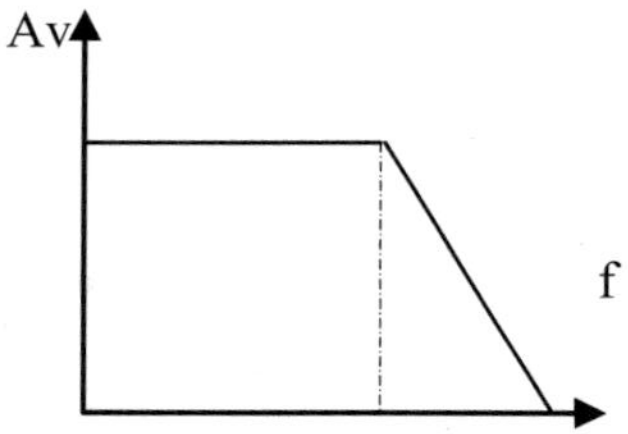 Av, f

Table 2. High pass filter and its variance realized in Remote lab system

Filter type	Pass band gain	Transfer function	Pole-zero plot	Bode plot
HPF	1	$\frac{sR_1C_1}{1+sR_1\,C_1}$	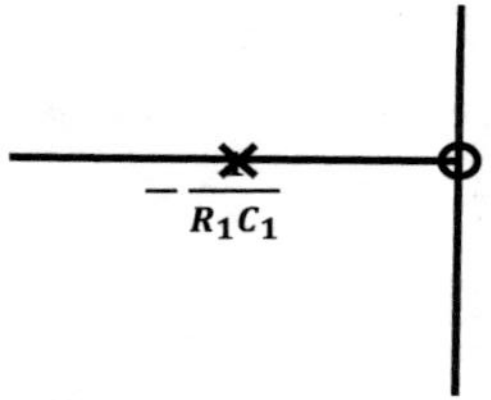 $-\frac{1}{R_1C_1}$	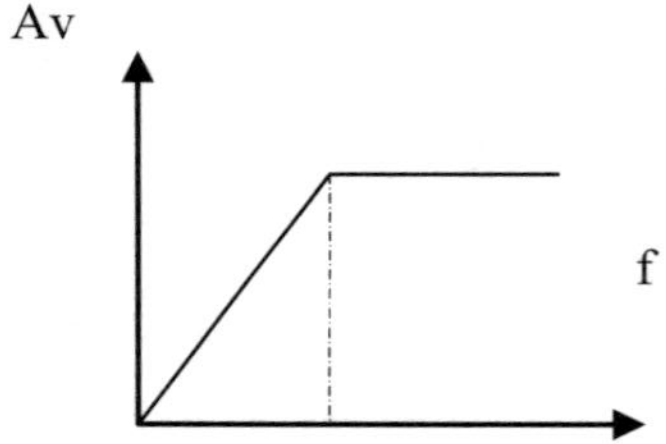 Av, f
Phase lead network	$\frac{R_1}{R_1+R_2}$	$\frac{1+sR_2C_1}{1+s(R_1\|\|R_2)\,C_1}$	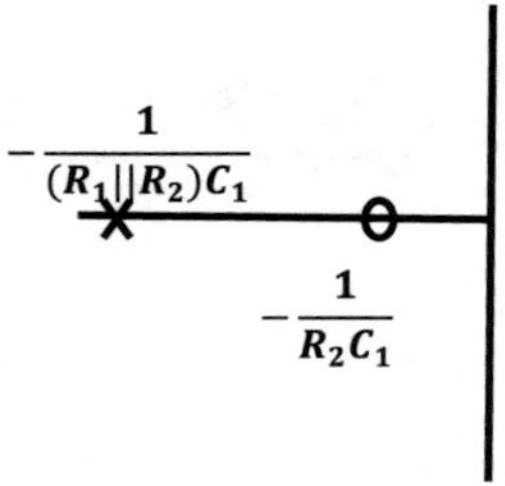 $-\frac{1}{(R_1\|\|R_2)C_1}$, $-\frac{1}{R_2C_1}$	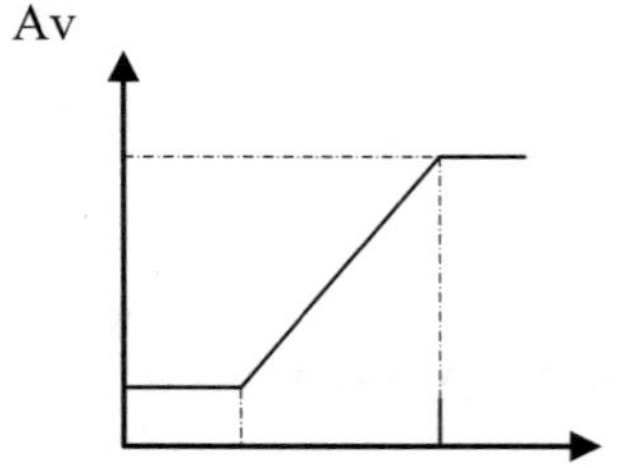 Av, f
HPF1	1	$\frac{sR_1C_1}{1+s(R_1+R_2)\,C_1}$	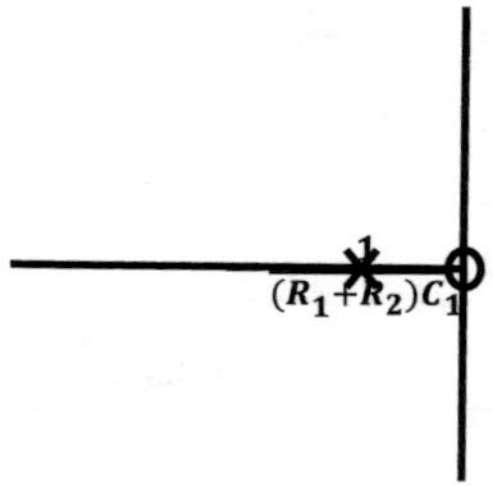 $\frac{1}{(R_1+R_2)C_1}$	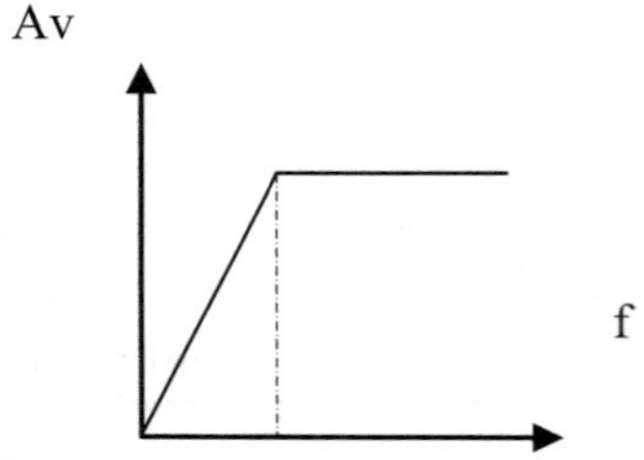 Av, f

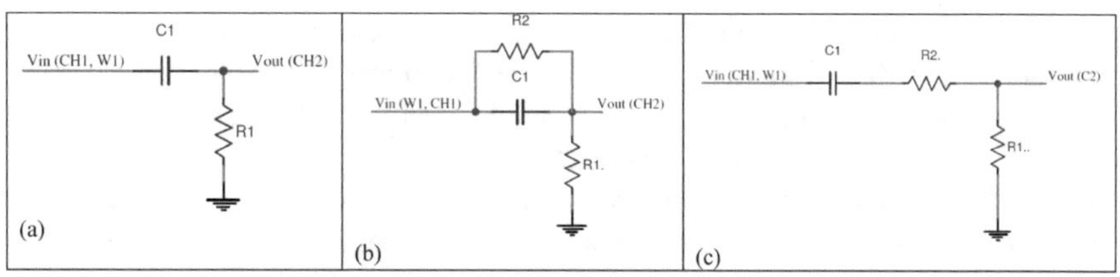

Fig. 3. High pass filter and its variance. **a** HPF, **b** phase lead network, **c** HPF1

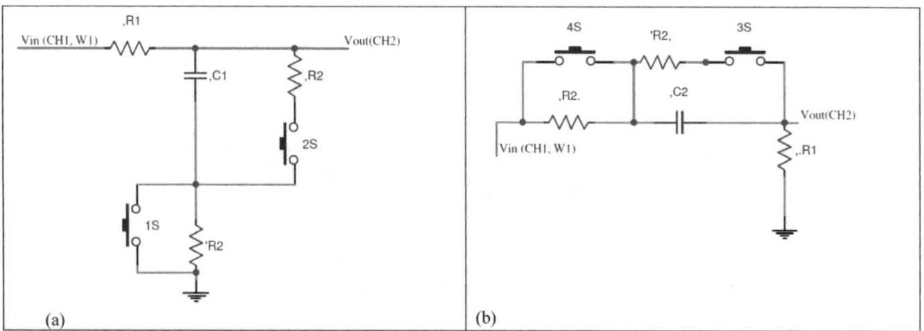

Fig. 4. Implementation of basic filter circuits in Remote lab system. **a** LPFs and phase lag network, **b** HPFs and phase lead filters in Remote lab system

Table 3. Switch conditions for LPF and phase lag network implementation.

Filter	Switch condition	
	1S	2S
LPF	ON	OFF
Phase lag	ON	OFF
LPF1	OFF	ON

Table 4. Switch conditions for HPF and phase lead network implementation.

Filter	Switch condition	
	3S	4S
HPF	OFF	ON
Phase lead	ON	ON
HPF1	ON	OFF

input signal of suitable amplitude and frequency and perform the timing analysis and Bode analysis. **National Instruments Analog Discovery** kit serves the purpose of function generator, oscilloscope and network analyser. The software used are Arduino IDE, Python, and Waveform 2015.

Figure 5 shows the front panel screenshots of Remote lab system during the conduction of filter experiments. Screenshots of the Remote lab system shows available analog electronic experiments for conduction. To perform filter experiments, user has to select the **Filter** tab and then in the list filter experiments any one experiment is selected for conduction. Using Analog Discovery kit through Waveform-15 tool user can perform time and frequency analysis.

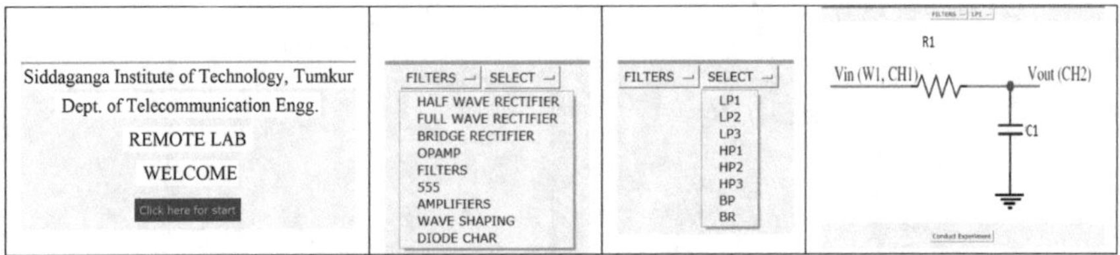

Fig. 5. Front panel views to conduct filter experiments in the Remote lab system

4 Analog Filter Results and Discussion

In this section, performance of eight analog filters will be discussed. The readings are obtained by conducting the experiments remotely. Both time and frequency responses are obtained to analyse transfer function of the filters. During the conduction of time response of experiments, sine wave input signal of 5 V amplitude at desired frequency is applied using Wavegen channel 1 of the Analog Discovery Kit. Large amplitude is chosen, so that even in stop band the output signal amplitude can be measured. As the maximum pass band gain of all the filters is 1 v/v, this large input amplitude signal will not distort the output signal. Bode analysis is very essential to analyse the performance of filters. Bode plots (Both magnitude and phase plot) are obtained by setting frequency sweep with, start frequency 50 Hz, end frequency 100 kHz and amplitude of 1 V (large amplitude can also be applied). Using the magnitude plot, pass band gain of the filter, pole and zero frequencies (corresponding phases) can be measured.

The analysis of Low pass filter and its variance are done using the Bode plots shown in Figs. 6 and 7 shows the bode plots of Low pass filter and its variance, that is due to the addition of R_2.

Analysis of Bode plots confirms the effect of additional resistor to LPF (in series and parallel to capacitor) will modify the pole zero position and the pass band gain. Table 5 shows the theoretical and the actual values of the same.

Table 5 indicates that the LPF has exhibited the desired characteristics after the addition of resistor. Theoretical and measured Pole frequency and pass band gains of all variance are matching whereas measured zero-frequency is finite (instead of infinite) due to parasitic of the measuring equipment.

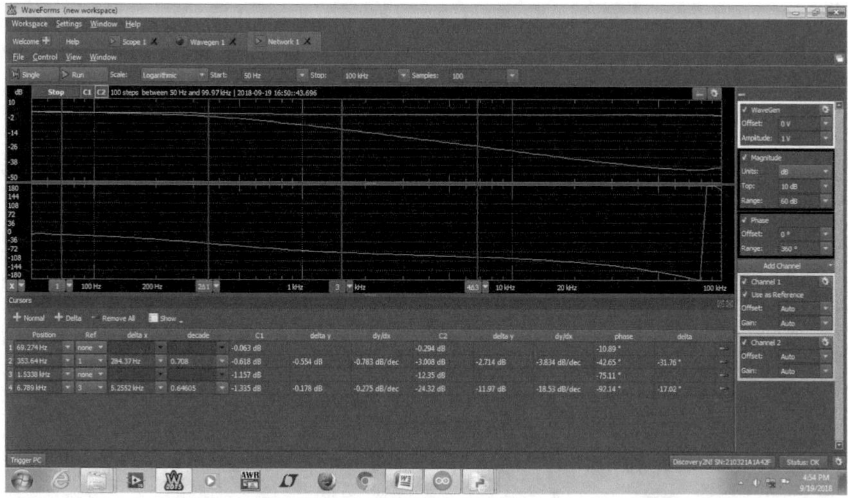

Fig. 6. Screenshot of Bode plot of LPF conduction on Remote lab system

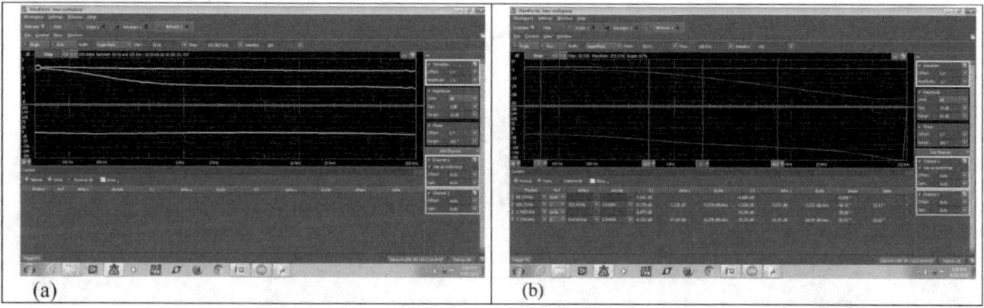

Fig. 7. Bode plots LPF variance. **a** Phase lag network, **b** LPF1

Table 5. LPF and its variance performance analysis

R_1 = 5.6 KΩ, C_1 = 0.1 μF, R_2 = 10 KΩ						
Filter type	Pass band gain (v/v)		Pole frequency (Hz)		Zero frequency (Hz)	
	Expected	Measured	Expected	Measured	Expected	Measured
LFP	1	0.96	284	353	Infinity	100 K
Phase lag network	0.64	0.62	102	148	159	285
LPF1	1	0.94	443	604	Infinity	120 K

Similar type of analysis is done for HPF and its variance. Figures 8 and 9 shows the Bode plots of HPF and its associated filters after adding the extra resistor. Table 6 shows the performance parameters of the same set of filters.

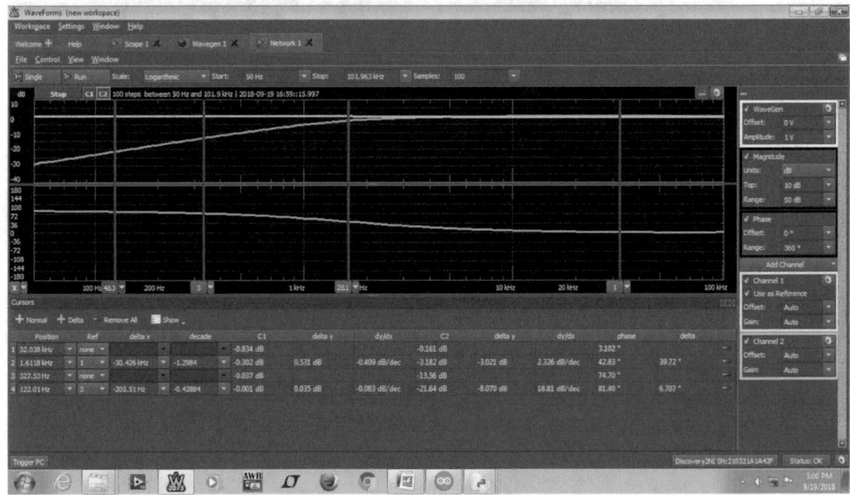

Fig. 8. Screenshot of Bode plot of HPF conduction on Remote lab system

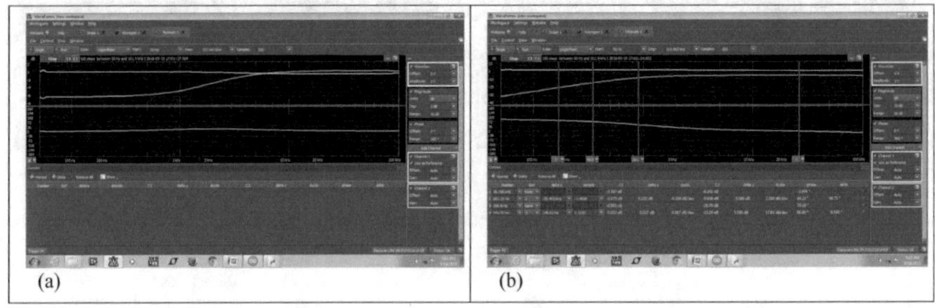

Fig. 9. Bode plots HPF variance. **a** Phase lead network, **b** HPF1

Table 6. HPF and its variance performance analysis

R_1 = 10 KΩ, C_1 = 0.01 μF, R_2 = 10 KΩ						
Filter type	Pass band gain (v/v)		Pole frequency (Hz)		Zero frequency (Hz)	
	Calculated	Measured	Calculated	Measured	Calculated	Measured
HFP	1	0.98	1.59 K	1.6 K	0	0
Phase lead network	0.5	0.48	3.18 K	3.5 K	1.59	1.7 K
HPF1	1	0.92	795	891	0	0

Table 6 indicates that the HPF has exhibited the desired characteristics after the addition of resistor. Theoretical and measured Pole frequency, zero-frequency and pass band gains of all variance are matching. Along with LPF, HPF and their variance filters, Remote lab system has two more basic filters band pass and band stop filters. Figure 10 shows the circuit diagram of these filters. Only basic filters are implemented in this category without any option to add any component to the circuit.

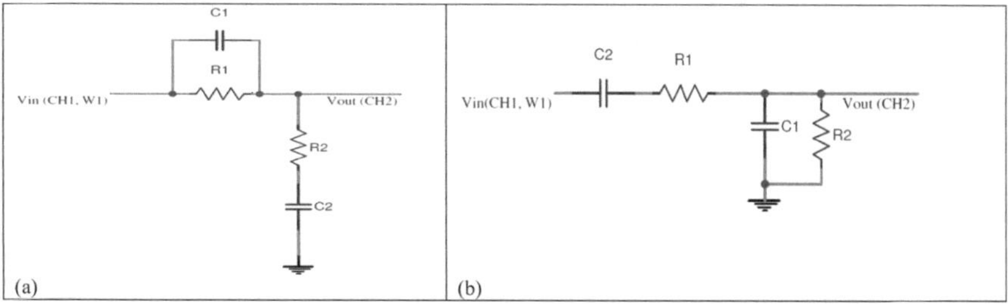

Fig. 10. Circuit diagram of **a** band pass filter, **b** band stop filter.

Observation of Bode plots in Fig. 11 shows that both filters have exhibited wide bandwidth characteristics. Band pass filter has a centre pass band frequency of 3.43 kHz whereas band stop filter has 2.34 kHz. This way user can conduct 8 different filter experiments remotely and perform with an exhaustive analysis to understand the concepts of filters.

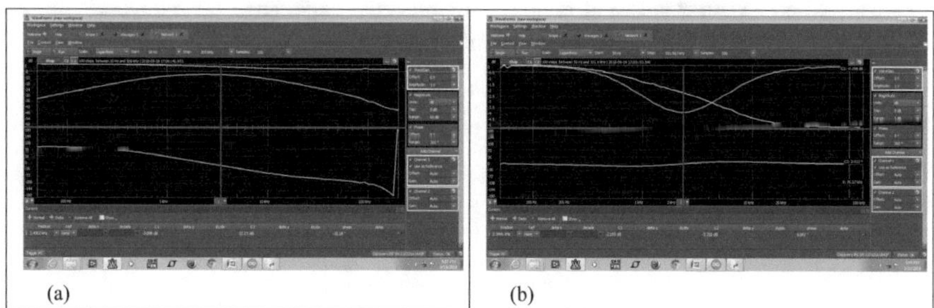

Fig. 11. Bode plots of **a** band pass and **b** band stop filters

5 Conclusions

In this paper, the development of remotely controllable Analog passive first order Low pass filter, High pass filters and their variance along with Band pass and stop filters have been discussed. A transfer function level of filter analysis is explained in detail. The user can understand the effect of addition of resistor on to filter circuit by observing the Bode response. Paper also demonstrates the movement of pole and zero position and also the change in pass band gain of the filters. It also gives the frequency response of Band pass and stop filters. User can perform these eight basic filters remotely, easily, repeatedly for better understanding of the filter concepts. This remote lab system with filters give confidence to the user the analysis of transfer function of RC networks that are very essential in electronic circuit design.

Acknowledgements. Authors thank the Management, Director and Principal of Siddaganga Institute of Technology (SIT), Tumakuru for their support in making this project. We also extended our gratitude to faculty members and students of Department of Telecommunication Engineering, SIT, Tumakuru for their support during the development of this innovative methodology.

References

1. http://vlab.co.in/
2. http://ohm.ieec.uned.es/portal/?page_id=76
3. http://www.uml.edu/IT/Services/vLabs/
4. Balamuralithara, B., Woods, P.C.: Virtual laboratories in engineering education: the simulation lab and remote lab (2007)
5. Chen, X., Song, G., Zhang, Y.: Virtual and remote laboratory development: a review. In: Proceedings of Earth and Space 2010, pp. 3843–3852. Honolulu (2010)
6. Chen, X., Kehinde, P.E.L.O., Zhang, Y., Darayan, S., Olowokere, D.O., Osakue, D.: Using virtual and remote laboratory to enhance engineering technology education
7. Kruse, D., Frerich, S., from Ruhr-Universität Bochum, Project ELLI—Excellent Teaching & Learning in Engineering Sciences, Universitätsstr. 150, 44801, Germany and Petermann, M., Kilzer, A., from Ruhr-Universität Bochum, Particle Technology, Universitätsstr. 150, 44801, Germany titled “Virtual labs and remote labs: practical experience for everyone”. In: 2014 IEEE Global Engineering Education Conference (EDUCON)

Exploration of Common Emitter Amplifier in Remote Lab

K. C. Narasimhamurthy(✉), T. S. Bindhu, Susheen Natraj, G. C. Bharath, Vismithata, Ankit Sharma, and Ajay Shiva

Siddaganga Institute of Technology, Tumkur, India
kcnmurthy@sit.ac.in, {binduts55, susheennatraj, bharathgc8, vismithatal, ankit34567, ajayshiva0077}@gmail.com

Abstract. Analysis of Common Emitter (CE) amplifier is essential to know many important concepts like voltage gain, bandwidth, constant gain bandwidth product, output voltage swing, etc. Realizing all these features of an amplifier is time consuming in the conventional lab. As it involves many component variations and there is a risk in getting the output after every variations. Hence, in depth analysis of amplifiers is usually not done in the conventional laboratory classes. Using Remote lab the analysis of amplifier is made easy because the output is guaranteed for every circuit parameter variations. The user need not have to worry about getting the output; instead can concentrate on the circuit performance analysis. In this paper performance of CE amplifier is explored by varying different parameters by performing time and frequency responses. The parameters analysed are voltage gain variation, Gain bandwidth product and phase and gain margins of the amplifier. Paper also presents details of readings obtained by remotely accessing CE amplifier.

Keywords: Remote lab · CE amplifier · Gain bandwidth product · Analog discovery

1 Introduction

Engineering education is all about learning by doing. In undergraduate Electronics engineering most of the courses are design oriented. Many courses deal with Analog electronic circuits. Concepts of these courses will be better understood by visualizing the concepts. Most of such courses will have associated laboratory course and nowadays SPICE simulations are also used to solidify the theoretical concepts. However, many times it's not possible to synchronize the lab experiments with concepts discussed in classroom. In many instances the instructors for theory class and lab class would be different. Therefore, in this paper an innovative platform of Remote lab for experiential learning is discussed. The paper describes how the Remote lab system can be used to analyze the working of CE amplifier. The remote lab helps the user to vary any desired circuit component and observe the effect of each modification on the amplifier response both in time and frequency domain. Remote lab enables users to solidify their theoretical concepts by performing experiments with peers.

M. E. Auer and K. Ram B. (Eds.): REV2019 2019, LNNS 80, pp. 623–633, 2020.
https://doi.org/10.1007/978-3-030-23162-0_56

2 Common Emitter Amplifier Circuit and Variations

The block diagram of Remote lab system is shown in Fig. 1. Any user can login remotely and perform experiments on the circuits at anytime, from anywhere [1, 2]. The remote lab can be accessed using Laptop or Smart phone. As there are many remote lab modules are active, many users can access remote lab simultaneously [3–8]. More users can use remote lab systems as they are available 24 × 7.

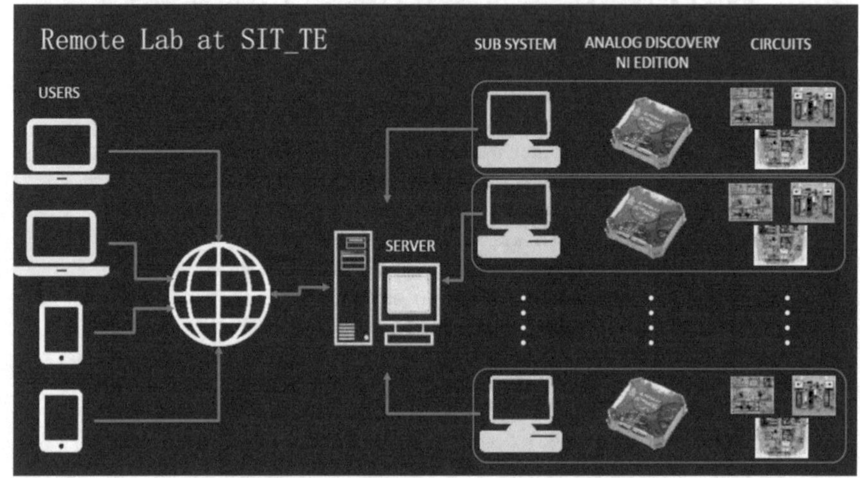

Fig. 1. Block diagram of Remote lab system

In this paper CE amplifier circuit with emitter degeneration is considered for the analysis. Verity of combinations in component selection is realized using CMOS switches that are controlled by Arduino board. Circuit selection options are created using Python programming.

In Remote lab system, to facilitate user to analyze the performance of CE amplifiers, four circuit components are identified and are switched accordingly using the switches as shown in Fig. 2b. The circuit has 4 components selected for variations are Cc, R_2, R'_E and R_C. The emitter de-generation resistor R_E of 100 Ω is realized using two 50 Ω resistors. In that, R'_E is defined as un-bypassed emitter resistance.

Following equations show the dependency of amplifier performance on the circuit components.

$$\text{Voltage gain} \quad A_v = -\frac{g_m R_C}{1 + g_m R'_E} \tag{1}$$

$$\text{Trans conductance} \quad g_m = \frac{I_C}{V_T} \tag{2}$$

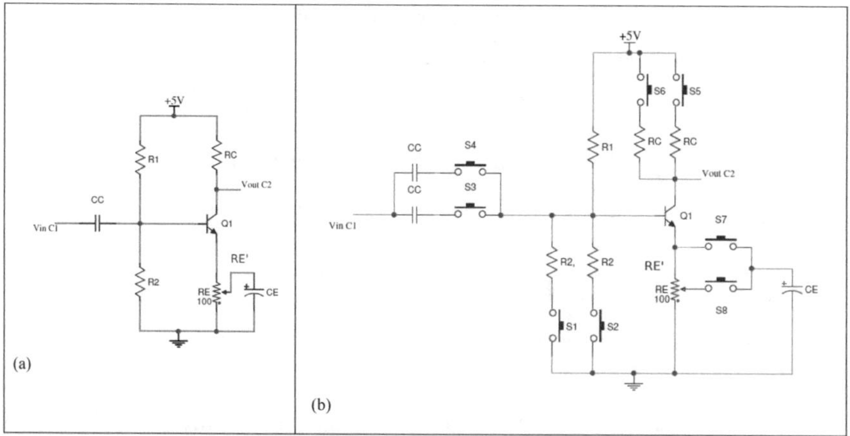

Fig. 2. Amplifier circuits. **a** CE amplifier, **b** CE amplifier implemented in Remote lab system with switches

$$\text{Collector current}\quad \mathrm{I_c} = \frac{V_{CC} - V_C}{R_C}$$

$$\mathrm{I_c} = \mathrm{I_{CBO}e^{-V_{BE}/V_T}} \tag{3}$$

$$\mathrm{V_{BE}} = \frac{\mathrm{V_{CC}R_2}}{\mathrm{R_1 + R_2}} - \mathrm{V_E},\quad \mathrm{V_E} \cong \mathrm{I_C R_E} \tag{4}$$

$$\mathrm{f_{c1}} = \frac{1}{2\pi R_{in} C_c} \tag{5}$$

$$\mathrm{R_{in}} \cong (\mathrm{R_1}||\mathrm{R_2}) + \mathrm{h_{fe}R'_E} \tag{6}$$

$$\mathrm{BW} \cong \mathrm{f_{c2}} = \frac{1}{2\pi R_C C_{out}} \tag{7}$$

$$C_{out} = C_{\mu}\left(1 + \frac{1}{A_V}\right) \tag{8}$$

$$\mathrm{GBW} = \mathrm{A_V} \times BW \tag{9}$$

The design equations are analyzed, and effect of circuit components of the amplifier performance are listed in Table 1. It has list of variable components, their values and corresponding effect on the circuit parameter and the relevant equation. Variation in coupling capacitor Cc will change lower cut-off frequency of the amplifier. R_2 will modify the bias voltage, which will modify the I_C, hence g_m. This change in g_m will slightly affect the voltage gain. Change in the value of un by-passed emitter resistor $R_{E'}$ from 0 to 50 Ω will affect the voltage gain inversely. However, voltage gain is varied in proportion to R_C as it capable of switching between 1 and 2.2 KΩ.

Table 1. Variable component with values and affected circuit parameters

Component	Values	Parameters affected	Equation Nos.
C_C	0.1 μF	Lower cut-off frequency (f_1)	(5)
	0.01 μF		
R_2	5.6 KΩ	I_C, hence A_V (but, not significantly)	(1)–(4)
	5.8 KΩ		
$R_{E'}$	50 Ω	A_V	(1)
	0 Ω		
R_C	2.2 KΩ	A_V and bandwidth (f_2)	(1), (7)
	1 KΩ		

The switches mentioned in Fig. 2b are implemented using CMOS switches and will be turned ON or OFF depending on the user requirement. Using 8 switches, Remote lab system has 16 different possible combinations of CE amplifier circuits as shown in Table 2. In remote lab system, use of **Analog Discovery kit** enables user to perform time and frequency response for the selected CE amplifier circuit. The main advantage of the Remote lab is that, user can perform any of the 16 variations by clicking the relevant button any number of times till the concepts are understood. User can apply sine wave input signal of any amplitude and frequency to observe the amplified signal and measure the required voltages to evaluate the voltage gain. User can perform the frequency response using Network option of Analog Discovery and evaluate mid-band gain and −3 dB bandwidth.

Table 2. Various circuit options of CE amplifier

Option	R_c (Ω)	R'_E (Ω)	R_2 (Ω)	C_c (F)
CE1	2.2 K	0	5.6 K	0.1 μ
CE2	1 K	0	5.6 K	0.1 μ
CE3	2.2 K	50	5.6 K	0.1 μ
CE4	1 K	50	5.6 K	0.1 μ
CE5	2.2 K	0	5.6 K	0.01 μ
CE6	1 K	0	5.6 K	0.01 μ
CE7	2.2 K	50	5.6 K	0.01 μ
CE8	1 K	50	5.6 K	0.01 μ
CE9	2.2 K	0	5.8 K	0.1 μ
CE10	1 K	0	5.8 K	0.1 μ
CE11	2.2 K	50	5.8 K	0.1 μ
CE12	1 K	50	5.8 K	0.1 μ
CE13	2.2 K	0	5.8 K	0.01 μ
CE14	1 K	0	5.8 K	0.01 μ
CE15	2.2 K	50	5.8 K	0.01 μ
CE16	1 K	50	5.8 K	0.01 μ

3 Results and Discussions

In this section, the reading obtained on the remote lab system by performing CE amplifier experiments has been presented. All 16 variations of the CE amplifier are conducted and for each option both time and frequency responses have been obtained for performance analysis. As a part of CE amplifier report, Oscilloscope and Bode plots screenshots of each circuit are taken with relevant measurement to analyse the circuit performance. Screenshots of time response of CE amplifier with 3 different component selections is shown in Figs. 3, 4 and 5. It can be observed that output signal is 180 degrees out-off phase with input waveform as mentioned in Eq. (1). It should be noted that, to get undistorted output sine wave, sine wave input amplitude of proper amplitude with mid-band frequency (10–200 kHz) should be applied using Wavegen feature of Analog Discovery.

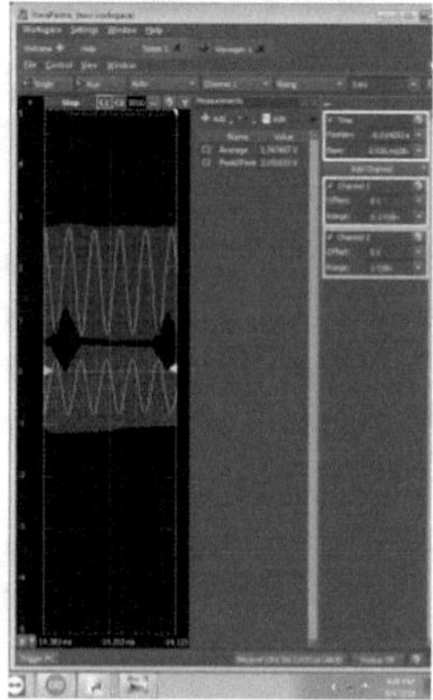

Fig. 3. CE amplifier time response $R_C = 2.2\ K\Omega$, $R'_E = 0$

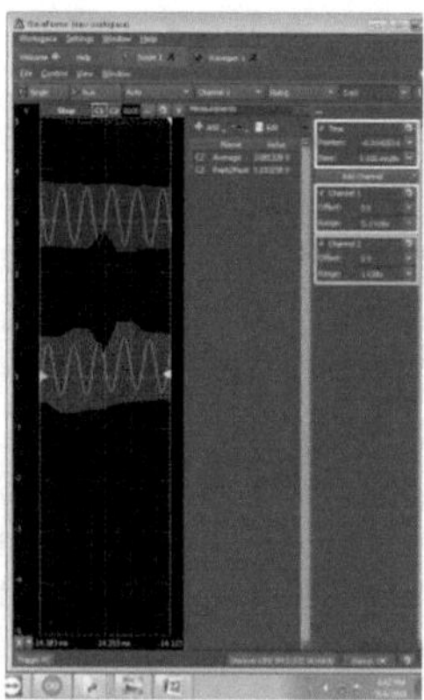

Fig. 4. CE amplifier time response $R_C = 1\ K\Omega$, $R'_E = 0$

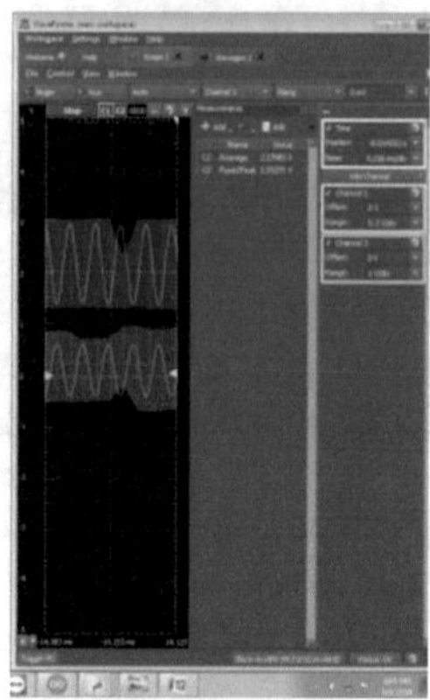

Fig. 5. CE amplifier time response R_C = 2.2 KΩ, R'_E = 50 Ω

Observation of time response in Figs. 3 and 4 revels that as RC decreases from 2.2 to 1 KΩ the DC voltage (Average voltage of channel 2) at the collector increases and voltage gain reduces, these variations are in consistent to design equations. The effect of variations in R'_E can be observed in Figs. 3 and 5, as R'_E increases from 0 to 50 Ω the voltage gain reduces as per the A_V expression in Eq. (1). Similar type of time responses can be obtained by varying R_2 and CC by selecting other CE amplifier options in Remote lab system. Results of which will be discussed later in this section. Using Network feature of Analog Discovery, it is possible to obtain the frequency response (Bode plots: both magnitude and phase plots). Using the Bode plots important frequency related performance parameters like mid-band voltage gain, pole positions (−3 dB frequencies), and phase margin. The user has the freedom to select the input signal amplitude and set the frequency swing. However, care should be taken to apply proper input amplitude otherwise it results in erroneous readings as the output would be clipped due to large input signal.

Figures 6, 7 and 8 is the Bode plots of CE amplifier for different RC and R'_E values. Figures 6 and 7 reveals that as R_C decreases from 2.2 to 1 KΩ, the mid-band gain almost reduces by a factor of 2 as per in Eq. (1). However, the cut-off frequency f_2 (amplifier bandwidth) increases by a factor of 2 as per Eq. (7). This will help to verify the concept of constant "Gain Bandwidth Product" of amplifier. By observing Bode plots in Figs. 6 and 8 of circuit with same value of RC but with different values of R'_E (0 and 50 Ω) shows that the amplifier voltage gain reduces. However, in contrary to the earlier case, this reduction in voltage gain will not increase bandwidth f_2, because the variation in AV won't vary COUT significantly hence dominant pole position f_2 is not affected. In summary, amplifier bandwidth (dominant pole f_2) depends heavily of Rc and weakly depends on R'_E. In conventional labs, these types of detailed analysis are time consuming, however in Remote lab it can done at click button.

User can do all these analysis at any time, at desired pace and from any place at free of cost by getting the permission to access the remote lab. Similar type of screenshots for all 16 options can be obtained. Data obtained from these screenshots will help user to strengthen theoretical concepts of CE amplifier. During the conduction of

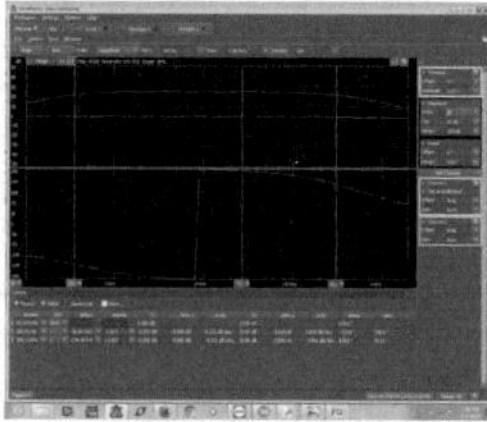

Fig. 6. Bode plots R_C = 2.2 KΩ, R'_E = 0

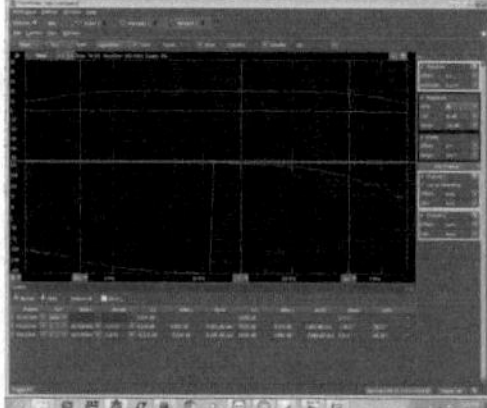

Fig. 7. Bode plots R_C = 1 KΩ, R'_E = 0

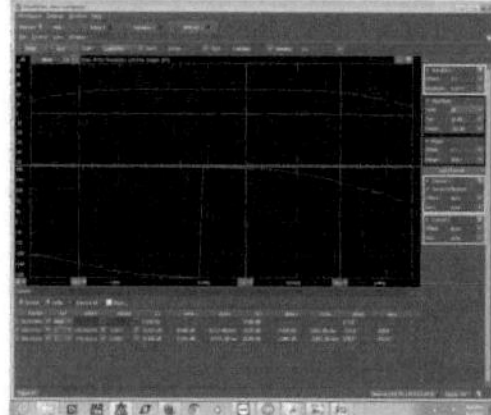

Fig. 8. Bode plots R_C = 2.2 KΩ, R'_E = 50 Ω

experiments, user need not have to worry about getting the output as the Remote lab system is made bug-free. Using remote lab system user can concentrate on circuit analysis with large data base. In conventional lab because of time constraint students will not try these types of explorations hence may not be able to perform such analysis.

The advantage of Remote lab is that user has options to vary the circuit components for circuit analysis. As discussed earlier, CE amplifier has 16 different combinations of components to vary. Table 3 gives the summary of readings obtained by conducting CE amplifier with all 16 variance. Table 3 includes DC voltage measured at collector terminal (V_C) of BJT (SL 100), peak to peak amplitude of output sine wave (V_{OUT}) for an applied 100 mV input sine wave of 20 kHz. A_V v/v next to V_{OUT} is the ratio of output to input voltages. A_V dB is the mid-band gain of the Bode plot, next to it is A_V v/v estimated from A_V dB(A_V (v/v) = 10 (A_V dB/20). The 3 dB cut-off frequencies f_1 and f_2 are also measured. GBW is the product of A_V v/v and bandwidth ($\cong f_2$).

Analysis of CE amplifier can be done with readings obtained by conducting 16 different combinations available in Remote lab system. The summary of readings of

Table 3. Summary of time and frequency response readings of all 16 combinations of CE amplifier

Different component values					Time response readings			Frequency response readings				
Option	R_C (Ω)	R'_E (Ω)	R_2 (Ω)	C_C (F)	V_C (V)	V_{out} pp (V)	A_V (v/v)	Av (dB)	A_v (v/v)	f_1	f_2	GBW (MHz)
CE1	2.2 K	0	5.6 K	0.1 μ	1.74	2.06	20.6	25.72	19.3	376	300 K	5.79 M
CE2	1 K	0	5.6 K	0.1 μ	3.08	1.1	11	20.85	11.02	461	527 K	5.8 M
CE3	2.2 K	50	5.6 K	0.1 μ	2.15	1.62	16.2	23.98	15.8	352	306 K	4.83 M
CE4	1 K	50	5.6 K	0.1 μ	3.29	0.95	9.5	19.06	8.97	415	434 K	3.89 M
CE5	2.2 K	0	5.6 K	0.01 μ	2.17	1.55	15.5	23.62	15.1	3.16 K	325 K	4.9 M
CE6	1 K	0	5.6 K	0.01 μ	3.07	1.05	10.5	20.52	10.61	3.43 K	572 K	6.06 M
CE7	2.2 K	50	5.6 K	0.01 μ	1.7	1.92	19.2	25.36	18.53	3.16 K	303 K	5.6 M
CE8	1 K	50	5.6 K	0.01 μ	3.26	0.89	8.9	18.96	8.87	3.3 K	572 K	5.0 M
CE9	2.2 K	0	5.8 K	0.1 μ	2.46	1.88	18.8	25.15	18.11	379	294 K	5.32 M
CE10	1 K	0	5.8 K	0.1 μ	3.36	1.08	10.8	20.61	10.72	451	527 K	5.64 M
CE11	2.2 K	50	5.8 K	0.1 μ	2.64	1.57	15.7	23.71	15.32	446	312 K	4.77 M
CE12	1 K	50	5.8 K	0.1 μ	3.51	0.89	8.9	18.9	8.81	446	538 K	4.73 M
CE13	2.2 K	0	5.8 K	0.01 μ	2.64	1.49	14.9	23.28	14.5	3.26 K	332 K	4.8 M
CE14	1 K	0	5.8 K	0.01 μ	3.42	1.02	10.2	19.91	9.89	3.26 K	590 K	5.8 M
CE15	2.2 K	50	5.8 K	0.01 μ	2.36	1.82	18.2	24.98	17.72	3.3 K	312 K	5.53 M
CE16	1 K	50	5.8 K	0.01 μ	3.54	0.89	8.9	18.39	8.3	3.4 K	615 K	5.1 M

both time and frequency response of CE amplifier of all combinations is shown in Table 3.

Table 3 shows voltage gain AV (v/v) evaluated by time response as well as A_V obtained by frequency response is almost identical across all options (compare both AV columns). Observations of CE1 and CE2 readings (R_C 2.2 and 1 K Ω and all other components values same) reveals that Av decrease as Rc reduces, but f_2 increases as per the design Eqs. (1) and (7). Hence the concept of constant GBW is verified. It may be noted that f1 almost remains constant as Rc will not affect f1 as per Eq. (5). Observation of set of readings with Cc of 0.1 μF and 0.01 μF indicate the f_1 shows that it almost increases 10 times for the later values of Cc. Few more performance analysis will be done using the bar charts in Fig. 9. Voltage gain variations measured against variations of different circuit components are shown in Fig. 9a. These are average of voltage gain obtained for the corresponding component values. The variation of A_V for two different values R_C has shown consistently expected values across all options. However, due to switch resistance the variations in the A_V for two R'_E are less than expected one. Mean while the voltage gain variation for other to components R_2 and C_C is negligible and is as expected across all options.

Similar types of analysis are done for bandwidth of the amplifier and are shown in Fig. 9b. The bandwidth of the amplifier almost decreases by a factor 2 for variation in R_C from 1 to 2.2 KΩ indicating the movement of dominant pole f_2. However for all other component variations the position of the dominant pole is not affected, hence no variation in the amplifier bandwidth. These readings are in consistent with the theoretical concepts. Readings to analyze all these can be obtained easily in remote lab very quickly compared to conventional lab. This way conduction of experiments on Remote lab system will help user to do more performance analysis.

One more performance parameter of amplifier, Phase margin is also measured using the Bode plot shown in Fig. 10. This particular option of CE amplifier has phase margin of 25.42° and an infinite gain margin. The circuit has exhibited poor phase margin, which results in more settling time.

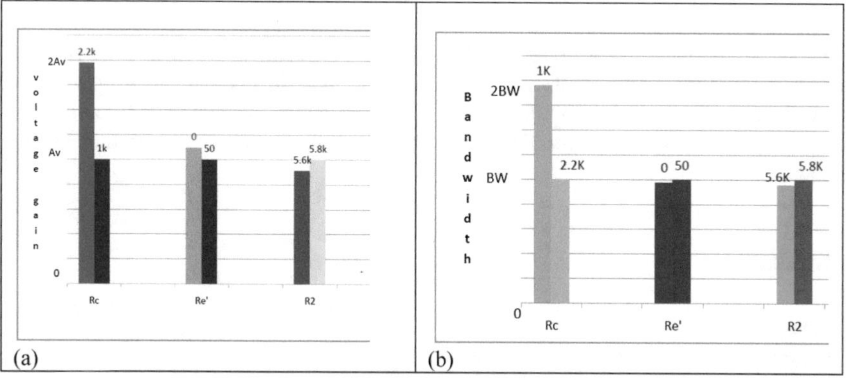

Fig. 9. CE amplifier performance analysis for various component variations. **a** Voltage gain, **b** bandwidth

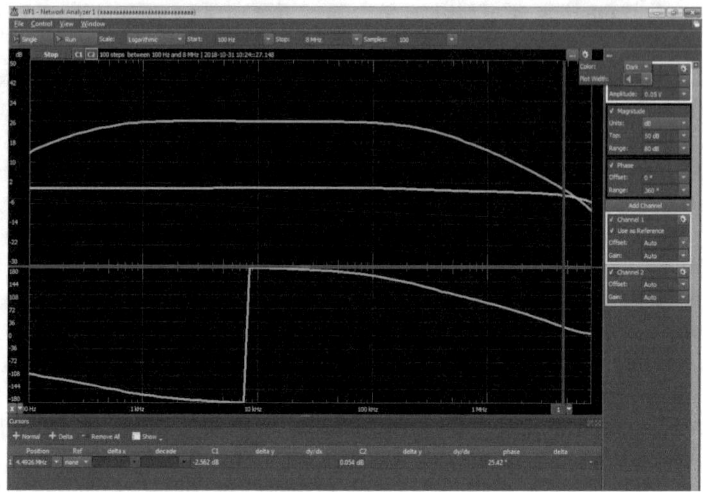

Fig. 10. Bode plots for phase and gain margin analysis

4 Conclusions

In this paper CE amplifier is explored using Remote lab system and discussion of the results obtained for various options are presented. User can conduct CE amplifier experiment with 16 different variations by click of a button. User has to take exhaustive readings to enable detailed analysis of amplifier circuit. Remote lab can also be used as platform for Experiential Learning by faculty members as they can show the live demonstration of the circuits in the classroom along with theoretical concepts. The analog electronic circuit remote lab system designed in this paper has the following advantages: web page accessing, convenient operation, no need to install the complicated client software. Time and frequency responses of CE amplifier are presented. Voltage gain, bandwidth, gain bandwidth product are presented for all combinations of CE amplifier.

Acknowledgements. Authors thank the Management, Director and Principal of Siddaganga Institute of Technology (SIT), Tumakuru for their support in making this project. We also extended our gratitude to faculty members and students of Department of Telecommunication Engineering, SIT, Tumakuru for their support during the development of this innovative methodology.

References

1. http://vlab.co.in/
2. http://ohm.ieec.uned.es/portal/?page_id=76
3. http://www.uml.edu/IT/Services/vLabs/
4. Balamuralithara, B., Woods, P.C.: Virtual laboratories in engineering education: the simulation lab and remote lab (2007)

5. Chen, X., Song G., Zhang, Y.: Virtual and remote laboratory development: a review. In: Proceedings of Earth and Space 2010, pp. 3843–3852. Honolulu (2010)
6. Chen, X., Kehinde, P.E.L.O., Zhang, Y., Darayan, S., Olowokere, D.O., Osakue, D.: Using virtual and remote laboratory to enhance engineering technology education
7. Kruse, D., Frerich, S., from Ruhr-Universität Bochum, Project ELLI—Excellent Teaching & Learning in Engineering Sciences, Universitätsstr. 150, 44801, Germany and Petermann, M., Kilzer, A., from Ruhr-Universität Bochum, Particle Technology, Universitätsstr. 150, 44801, Germany titled "Virtual labs and remote labs: practical experience for everyone". In: 2014 IEEE Global Engineering Education Conference (EDUCON)
8. Nedic, Z., Machotka, J., Nafalski, A.: Remote laboratories versus virtual and real laboratories. In: 33rd Annual Frontiers in Education, FIE 2003 (2003)

Power Intelligence and Asset Monitoring (PIAM)

Preeti Biradar[1(✉)], Shubham Mohapatra[1,2], N. Elangovan[1,2], Kalyan Ram B.[1,3], S. Arun Kumar[1], Panchaksharayya S. Hiremath[1], M. S. Prajval[1], Mallikarjuna Sarma[1], Nitin Sharma[3], Gautam G. Bacher[3], and Mounesh Pattar[1]

[1] Electronosolutions Pvt Ltd, Bengaluru, India
preeti1905@gmail.com
[2] Cymbeline Innovations Pvt Ltd, Bengaluru, India
[3] BITS-Pilani KK Birla Goa Campus, Sancoale 403726, Goa, India

Abstract. In a country such as India, wherein the annual budget is majorly occupied or predominantly dominated by the power sector, there exists a huge requirement for the effective utilization of power and simultaneously efficient monitoring of the same. Along with physical monitoring, if the same system or incoming power could be monitored remotely this would be an added advantage. Remote monitoring would indulge accessing of the system from any corner of the world leading to the concept of Remote engineering. Monitoring involves the accurate capture of live data on a timely basis and at the same time storing the obtained data in the respective database. The system mentioned above is a transformer being monitored collectively for its current and voltage values, which in turn are classified into 42 parameters. So after the successful comparison of values pertaining to the digital meter connected to the transformer and the values obtained from remote monitoring the exact loss incurred can be calculated and optimized.

Keywords: Power sector · Remote monitoring · Remote engineering

1 Introduction

Power is of utmost importance and it is turning into or is already a basic requirement for people residing all over the world. Any small power outage would result in significance delay or monetary loss. Time and money are the faces of a single coin in this modern day world. Both are of pretty high importance and priority and neither of the two can be afforded to be lost. There is a saying which goes as "money cannot buy you everything" and it is so true and evident in this scenario that Power cannot be bought at any price, hence only effective management of power is pivotal for its proper and adequate usage. Effective management of power can only be achieved only when it is monitored accurately throughout all its phases. Power has to be managed well and diligently as India is the third largest producer and fourth largest consumer of electricity the world as per statistics revealed in August 2018. The same stats also indicated that, in India the total installed capacity of power stations were at a likely of 344.69

M. E. Auer and K. Ram B. (Eds.): REV2019 2019, LNNS 80, pp. 634–640, 2020.
https://doi.org/10.1007/978-3-030-23162-0_57

Gigawatt (GW). The increasing demand for power across the nation has led to numerous initiatives and programmes that have been taken up by our Government. All of these have had a positive impact on this sector and the sector is now recognizing and undergoing a gradual notable rise. The power sector is highly prioritized when a decision is to be made about the Nation's annual budget. As per surveys in recent times, approximately 40% of the Budget is dedicated to our Power sector. So this takes us to a position wherein even if around 10–15% of the power being generated is lost or cannot be recovered, it directly hampers and has an adverse effect on the Budget. The approximate monetary loss calculation would result nothing less than a few hundred crores. Even if a small percentage of the 10–15% can be recovered and utilized efficiently then not only power and time will be saved but also the money involved can also be saved. Initially in order to save power, it has to be monitored perfectly such that not even a small amount of the power would go scathed. On receiving the entire power that was generated, the margin for error would obviously go down. Hence in order to do so, eventually a monitoring of a system on a small scale is ideal. The main objective of this paper is to discuss about the 24/7 monitoring of a transformer for its voltage, current and other 42 parameters. Thus to get started, a small scale monitoring involves a single transformer assigned to a specific geographical area and in turn this can grow into a larger scale finally giving rise to smart grids being the long term goal. The device to serve this purpose is termed as "Power intelligence and asset monitoring".

2 Approach

The fact that any substantial loss in power would not only affect individuals but also would result in monetary losses which are highly significant, actually led to this approach. In order to utilize power efficiently and to reduce power losses, firstly faults occurring at the base must be eradicated. The base in this context refers to the power that has arrived at the distribution stage after its usual generation and efficient transmission. Distribution is pivotal in any power transfer as it is this station which supplies electricity to the local homes. So monitoring the utility transformers which are responsible for dispatching power to individual homes is the most function to be carried out. The device must be mounted close to the local utility transformer such that PIAM is felicitated with three phase supply from transformer as its input for physical monitoring. The physical monitoring was verified with reference to the digital meter installed to observe the readings of the utility transformer. The idea was to implement such a physical device through which the live data of the utility transformer would obtained not only physically but also remotely after establishment of the graphical user interface. On the use of a data acquisition device within the PIAM, the incoming data can be continuously obtained and monitored remotely also. Similarly while monitoring, suppose if there was any outage or leakage of power, the device was able to shift out the exact reason and also which one among the three phases was faulty.

Figure 1 depicts the entire system in three main blocks. Firstly the transformer indicated is a utility transformer located which supplies power to the locals. PIAM is in contact with the local transformer and once all the connections are verified the physical monitoring of the system is done carried unless the remote connection is established.

The PIAM device includes an Intel NUC or Intel stick within it to carry out a successful remote connection. The Graphical user interface can be obtained once the user has the wright remote desk ID and password. Users can view the live data on a fruitful remote connection from their individual computers or mobiles.

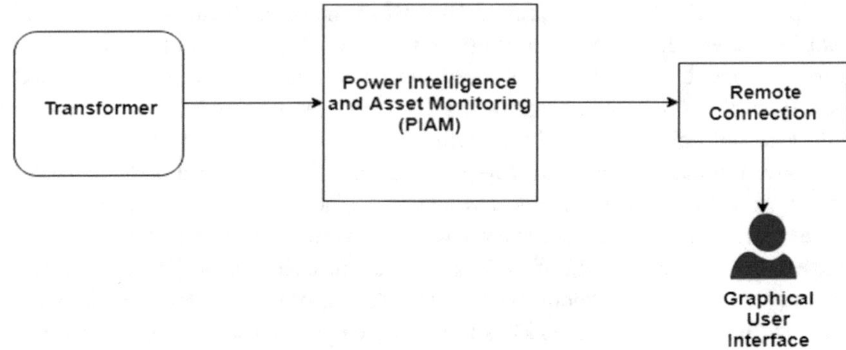

Fig. 1. Basic block diagram and overview of the Power Intelligence and Asset Monitoring system.

3 Implementation

The implementation predominantly holds all the cards for the full phase functionality of the proposed system. The planning was done with utmost importance keeping in mind of all the complexities that would arise during the time of implementation. All of these would have gone vain if the implementation did not go according to plan. Thus this plan was broadly classified into three broad parts. They are as follows:

Hardware description
Software involved.

3.1 Hardware Description

The hardware components used are mentioned in Table 1. Miniature circuit breaker (MCB) is used as a safety measure for the incoming three phase voltage. As a controller and to establish graphical user interface an OEM board which a chassis of DAQ is used. An Intel stick is used to run the code for the system and also stores the data being captured. The UPS provides back up power to the system. SMPS is required for triggering the relay board. The relay board in turn provides control. Current and voltage sensors are used for accurate measurements of current and voltage respectively. A temperature sensor (LM35) is used to monitor the inhouse temperature of the system.

Table 1. Overview of hardware and their specifications with price and utility

Hardware	Specifications	Price (INR)	Utility
Controller	OEM board 6001	11,800	Establishes control between system and software
Intel stick	Atom processor 2 GB RAM	15,000	A mini CPU to power the DAQ
Current sensors	CT 1270 5 A	350	To measure the current drawn across the system
Voltage sensors	Electrohms AC/440-4 V	250	Measures the incoming voltage to the system
Schneider MCB	415 V AC	1200	Means of protection
Relay board	Single channel 230 V AC	200	To establish control
4G modem	Jiofi	2000	Internet connectivity
SMPS	12 V, 5 V, 1.5 A	1500	To provide power supply
GM spike buster	1 socket 6 Amps	200	Supply for the UPS
UPS	600 VA	3000	To provide power backup
Exhaust fan	230 V AC	400	For ventilation and removal of excess heat

3.2 Software Involved

The code was developed by a software tool called LabVIEW. It stands for laboratory Virtual Instrument Engineering Workbench which is developed by National Instruments. It provides a graphical programming approach that helps visualization of the project including hardware configuration.

It was chosen for the following reasons:

Customized user interfaces
Data analysis algorithms
Accurate logic in a diagram
Precision measurement
Reading highly customizable.

Figure 2 shows the live data being obtained for all 42 parameters through Remote Desktop software.

SQL database server is used to store the obtained data in the database. This is a backup for the same data being stored in the Intel stick.

For remote connection, a software called ANYDESK is installed in the Intel stick of PIAM through which the system can be remotely accessed 24/7.

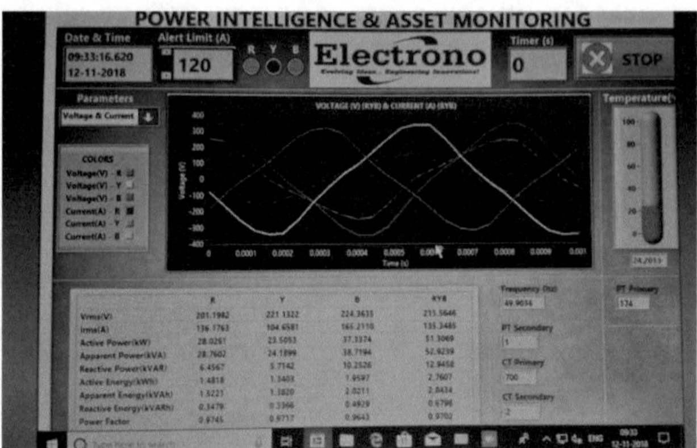

Fig. 2. Graphical user interface for live data after Remote connection.

4 Working

PIAM device was majorly implemented for the continuous monitoring of current, voltage, power factor, active, reactive and apparent power and energies summing up to 42 parameters. The transformer for which the device was decided to be implemented on had a delta connection on its input and comprised of a star connection on the opposite end. This star connection from the transformer output was is in contact with the digital meter through the current sensors with a rating of 250/5 turns ratio located above the digital meter. The digital meter was also configured for all its phase's current, voltage, and overall power factor.

Input to the PIAM device was tapped in from the digital meter which involved the three phases along with the neutral. This was connected to the 4-pole MCB present within the device. Two phases from the MCB act as the auxiliary 230 V supply for both the spike buster as well as SMPS (switched mode power supply). Though there is power all the time from the transformer, only when the MCB within the PIAM system is turned, the device will be active. Spike busters is made use of for the UPS for backup power. Similarly the Intel stick mentioned as NUC in the below diagram also needs an auxiliary supply. The LabVIEW code developed is added in the Intel stick and for remote connection ANYDESK software with a configured ID and password is set. The code is divided into smaller codes which comprises of codes for voltage and current sensors, the code for temperature sensor, and all the formulae for active, reactive and apparent power and energy for all the three individual and combined phases, power factor and frequency for all the three phases and combined.

The Intel stick is connected to the data acquisition device (OEM board) to turn it on by means of a USB cable. OEM board which does not have terminals, in turn is connected to a 34 pin FRC board with its terminals similar to USB 6001 through a connector. For acquisition of voltage and current data, the sensors were already configured and tested with a code. The voltage sensor board also required single phase supply for each of its three phases which was provided through the MCB again. The output of voltage sensors was connected to FRC board's analogous input pins by means of RMC connector. On similar terms the current sensors were also in contact with the analogous pins of the FRC board. A temperature sensor was placed within the system to monitor the temperature of the system. The temperature can be observed on the right hand side in Fig. 2. An exhaust fan which needed a 230 V supply was also mounted to clear the dissipated heat. An alarm was placed externally for indication of any fault or leakages.

Figure 3 depicts the exact working flow as mentioned in the working column.

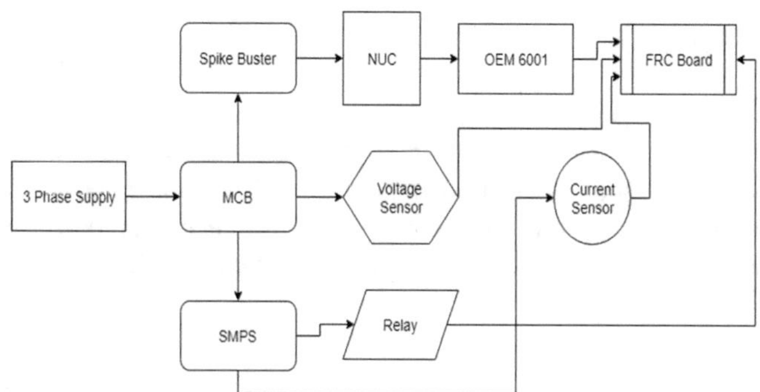

Fig. 3. An overview of the working flow of PIAM system

Figure 4 shows the entire real time PIAM working device located at the Utility transformer.

Fig. 4. Overall device installed near the transformer

5 Actual Outcomes

The major aim of the setup is to continuously power all day all night. Power outages can be detected through the software remotely as one of the phases will almost yield a nullified current reading. Power of both the individual phases as well as all the three phases combined can be restored. Along with continuous power monitoring, there are 42 other parameters which can be stored in the database too.

6 Conclusions

Power plays a very important role in all life forms. It is therefore considered as a basic need. In this era almost every minor stuff requires electric power and the adaptation of electric power for transportation and communication just signifies how priceless power is. So in this scenario, where there exists a huge demand for power, it has to be monitored and then utilized to the maximum possible extent. On the other hand it will help reduce power outages too.

References

1. Catterson, V.M., Rudd, S.E., McArthur, S.D.J., Moss, G.: On-line transformer condition monitoring through diagnostics and anomaly detection. In: 15th International Conference on Intelligent System Applications to Power Systems 2009, ISAP '09, pp. 1–6 (2009)
2. Giacomelli, F., Inoue, T.B., Morais, D.R., Rolim, J.G.: An ontology model for intelligent tools applied to transformer condition evaluation. In: 2011 16th International Conference on Intelligent System Application to Power Systems (ISAP), pp. 1–6 (2011)
3. Ahmed, E.B., Abu-Elanien, M., Salama, M.A.: Survey on the transformer condition monitoring. In: 2007 Large Engineering Systems Conference on Power Engineering, pp. 187–191 (2007)
4. Singh, J., Sood, Y.R., Jarial, R.K.: Condition monitoring of power transformers—bibliography survey. Electr. Insul. Mag. IEEE **24**(3), 11–25 (2008)
5. Morais, D.R., Rolim, J.G., Vale, Z.A.: DITRANS—a multi-agent system for integrated diagnosis of power transformers. In: 2007 IEEE Lausanne Power Tech, pp. 1917–1922 (2007)
6. McArthur, S.D.J., Catterson, V.M.: Multi-agent systems for condition monitoring. In: Power Engineering Society General Meeting 2005, vol. 2. IEEE, pp. 1044–1047 (2005)
7. Li, J., Wei, H., Xia, X.: The multi-agent model and method for energy-saving generation dispatching system. In: 2010 International Conference on Power System Technology (POWERCON), pp. 1–8 (2010)
8. Nagata, T., Ueda, Y., Utatani, M.: A multi-agent approach to smart grid operations. In: 2012 IEEE International Conference on Power System Technology (POWERCON), pp. 1–5 (2012)
9. Catterson, V.M., McArthur, S.D.J.: The practical implications of bringing a multi-agent transformer condition monitoring system online. In: Power Systems Conference and Exposition 2004. IEEE PES, pp. 12–16, vol. 1 (2004)

Remote Enabled Engine Vibration Measurement System

Preeti Biradar[1(✉)], Pranav Basavraj[1,2], Naveen Murugan[1], Kalyan Ram B.[1,3], Nitin Sharma[3], and Gautam G. Bacher[3]

[1] Electronosolutions Pvt Ltd, Bengaluru, India
preeti@electronosolutions.com, pranav@cymbelinein.com
[2] Cymbeline Innovations Pvt Ltd, Bengaluru, India
[3] BITS-Pilani KK Birla Goa Campus, Sancoale 403726, Goa, India

Abstract. Due to Higher Productivity and High Speed of Machinery, requirement of Resonant conditions are necessary to ensure safety margins of Machines. Hence here we proposed a vibration measurement kit which is used to monitor the health of vehicle engines through remotely. Every engine should have its own withstanding limit but how reliable it is, is the matter. When using vibration to observe engine health, the objective is to correlate observable vibration with typical wear-out mechanisms, such as bearings, gears, chains, belts, shafts etc. From this vibration measurement kit will get more information through remotely about the engine vibration to take action. The MEMS accelerometer based vibration sensor is used for vibration measurement. This sensor is used to collect the vibration data from the engine. The data can be acquired through 12 bit data acquisition device with the sampling rate of 8 kHz. Then Fast Fourier Transform (FFT) takes place to convert time domain data (acquired input data) to frequency spectral data. From the baseline of the signal the set point is fixed with tolerance of (+5, −5)%, whenever the engine vibration crosses the high and low level vibration set point it will generate the Time Transfer Logic (TTL) output pulse for second. The TTL output is used to count the number of times the engine vibration crosses its limit, depends on that the reliability of engine is being calculated.

Keywords: Machine health · MEMS accelerometer · FFT · TTL

1 Introduction

Vibration can be defined as a periodic motion of the particles of an elastic body or medium in alternately opposite directions from the position of equilibrium once that equilibrium has been disturbed.

Vibration level of machines will be low if it is designed properly, When Machine undergo operation it is subjected to fatigue, wear, deformation, and foundation settlement. These effects cause an increase in the clearances between mating parts, misalignments in shafts, initiation of cracks in parts, and unbalances in rotors all leading to an increase in the level of vibration. If Vibration Increases keep on it will results in failure or breakdown of the machine. The common types of faults or

M. E. Auer and K. Ram B. (Eds.): REV2019 2019, LNNS 80, pp. 641–652, 2020.
https://doi.org/10.1007/978-3-030-23162-0_58

operating conditions that lead to increased levels of vibration in machines include bent shafts, eccentric shafts, misaligned components, unbalanced components and loose mechanical parts.

Vibration Measurement system is a characteristic of all types of machinery specially needed for rotating machinery. Vibration Measurement is very important because it indicates state and health of the machinery. State of the machinery play a vital role since repair and maintenance decisions are based on it. Vibration measurement can be done by measuring displacement, velocity, acceleration, acoustic, magnetic, optical etc. The nature and location of the fault can be detected by comparing the frequency spectrum of the machine in damaged condition with machine in good condition. Another important characteristic of a spectrum is that each rotating element in a machine generates identifiable frequency, thus the changes in the spectrum at a given frequency can be attributed directly to the corresponding machine component.

The key point of our Vibration Measurement Kit is a sensor that measures acceleration of a specific object in terms of analog voltage signal. Once the Acceleration in a specific dimension is found, the velocity and displacement can easily be figured out by doing simple integration and double integration of acceleration with respect to time. An accelerometer is a sensor that measures the acceleration of a physical device as a voltage. Accelerometers can be mounted directly to bearings, gearboxes, or spinning blades. It provides acceleration data of an object it is attached to.

Though there are many vibration analysis systems are available, this paper work is intended to derive the process of building a low cost Vibration Measurement System using MEMS accelerometer, NI DAQ, Intel Stick, DC/DC converter and Relay Board.

Measuring the vibration of any particular machinery includes to mount an accelerometer on it and measure the accelerations produced by the vibration. The sensor provides an analog voltage signal output relative to the instantaneous acceleration. The measured value is then transmitted to the software installed in a personal computer which provides the overall information about the vibration phenomenon with a real time graphical representation.

This process is done by writing a Virtual Instrument (VI) script in the LABVIEW program. This includes configuring the serial port to communicate with the sensor, acquiring the data from the sensor, processing the data and finally analyzing the data. Once the data is in the computer, signal processing tools can be applied to examine the acceleration data. Using this data we can minimize or eliminate the vibration.

2 Approach and Working

The MEMS accelerometer based vibration sensor is used for vibration measurement. Typically the sensor is used to mount vertically on the surface of the engine head which contains more information. The data can be acquired through 12 bit data acquisition device with the sampling rate of 8 kHz. The FFT spectrum samples the input signal, computes the magnitude of the sine or cosine components and displays the spectrum of the measured frequency components. One of the key functions associated with process variable is storing FFT records of engine lifetime enables analysis of a variety of behaviours that may lead to a wear-out curve that contributes to maintenance and safety planning.

In order to evaluate the vibration data we need to compare the measured signal during operation with a good baseline signal which is obtained when machinery is operating properly. whenever the engine vibration crosses the high and low level vibration set point it will generate the Time Transfer Logic (TTL) output pulse for second (Fig. 1).

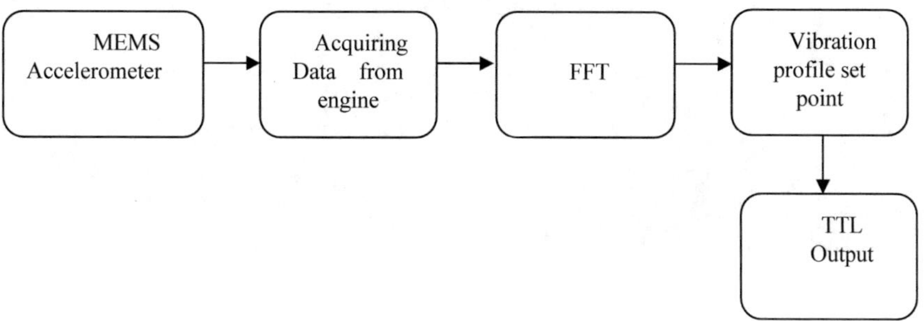

Fig. 1. Block diagram

2.1 MEMS Accelerometer Sensor

An accelerometer is an instrument which is used measure the acceleration of a vibrating body. Due to small size and high sensitivity, Accelerometers are widely used for vibration measurements.

A MEMS accelerometer measures vibration through the use of micro-fingers. When the accelerometer is subjected to vibrations the fingers are deflected and the distance between them and the electrodes changes. This causes a change in the capacitance between the anchor and the electrodes. This change in capacitance is converted to a change in the output signal which is used to produce analog signal to continuously indicate the acceleration of the sensor. If there is no acceleration, that is the sensor is at a constant velocity, the signal will be constant.

The acceleration-time history of a machine frame that is subjected to Vibration cannot be used to identify the cause of vibration. If the acceleration-time history is transformed to the frequency domain, the resulting frequency can easily be related (Fig. 2).

In our Vibration Measurement System, we are using two 3-axes ADXL1001 ($\pm$100 g) MEMS (Micro Electro Mechanical Systems) accelerometer. The ADXL1001 are high frequency, low noise single axis accelerometers that provide an analog output that is proportional to mechanical vibration.

The ADXL1001 have high ‘g’ (Acceleration due to gravity) ranges of 100 g and are suitable for vibration measurements in high bandwidth applications such as vibration analysis systems that monitor and diagnose machine or system health. The high ‘g’ range provides the dynamic range to be used in high vibration environments such as heating, ventilation, air conditioning (HVAC) and heavy machine equipment (Fig. 3).

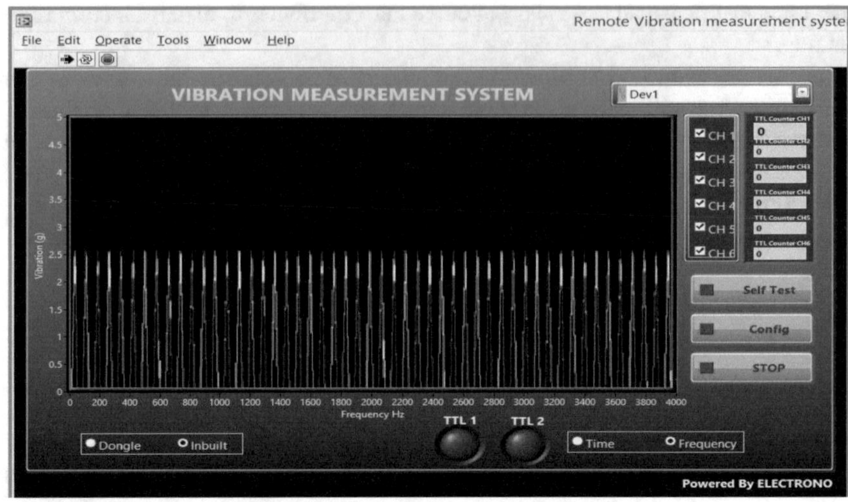

Fig. 2. Frequency domain measurement

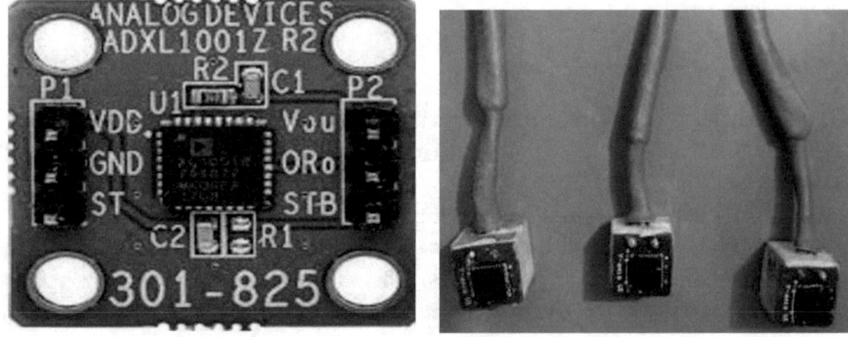

Fig. 3. ADX L1001 sensor board and wired ADXL1001 sensor for mounting

2.2 Data Acquisition

We are Using Intel Compute Stick as a Microcontroller to acquire the data from accelerometer sensor. In Intel Stick a Labview program Exe will be run and with the help of National Instrument's DAQ 6002, data is acquired at a Sampling Rate of 8 kHz. The Acquired data will be in terms of 'g', Acceleration due to gravity.

Labview is chosen for Excellent design in form of front panel and block diagram, Built in libraries and tools and High Precision measurement reading. A DC/DC Converter is used to provide input to Intel Compute Stick and LEDs. Data is stored in SD card in Intel Compute Stick and it is also possible to transfer data through internet using Dongles.

- In Fig. 4, X1, Y1 and Z1 represents the directions of First 3-axes Accelerometer Sensor whereas X2, Y2 and Z2 represents the directions of Second 3-axes Sensor.

	A	B	C	D	E	F	G	H	I	J	K
1	Date	Time	TTL X1 to Z	Freq Hz	X1	Y1	Z1	X2	Y2	Z2	
2	14-03-2018	18:40:00	0	0	183.9782	176.0924	168.331	160.7721	153.4029	146.1929	
3			1	1	181.1171	173.1113	164.576	155.9158	146.9781	137.6012	
4			1	2	63.01675	63.08746	62.95473	63.04191	62.98157	63.08674	
5			1	3	0.056765	0.094537	0.059257	0.050851	0.0888	0.129552	
6			1	4	0.087789	0.045965	0.079622	0.028513	0.103364	0.032426	
7			1	5	0.11808	0.010778	0.11321	0.088156	0.080238	0.125842	
8				6	0.160002	0.056348	0.05118	0.149433	0.045819	0.108866	
9				7	0.097039	0.085842	0.080787	0.118725	0.082319	0.028507	
10				8	0.109214	0.03107	0.077625	0.143085	0.104441	0.08167	
11				9	0.056907	0.033872	0.036644	0.153526	0.078214	0.109434	
12				10	0.042017	0.08516	0.084767	0.10201	0.034884	0.076746	
13				11	0.040831	0.050257	0.095276	0.05	0.030131	0.09718	
14				12	0.072301	0.07345	0.095399	0.073026	0.073932	0.029814	
15				13	0.08279	0.119433	0.079953	0.027487	0.133927	0.04733	
16				14	0.080565	0.126881	0.066875	0.082881	0.12174	0.062127	
17				15	0.05218	0.082232	0.048327	0.072211	0.048593	0.070849	
18				16	0.045795	0.107106	0.088552	0.085912	0.09051	0.035484	
19				17	0.030908	0.138541	0.123057	0.13185	0.042569	0.045991	
20				18	0.131616	0.135054	0.091133	0.082392	0.021978	0.094261	
21				19	0.104449	0.119384	0.105057	0.116491	0.053243	0.11764	
22				20	0.058099	0.092438	0.01634	0.065467	0.044867	0.082027	
23				21	0.053388	0.101346	0.08967	0.069631	0.076595	0.085648	
24				22	0.079926	0.057363	0.040538	0.099374	0.07612	0.043149	
25				23	0.06367	0.047129	0.056008	0.065727	0.118667	0.034343	
26				24	0.056729	0.107392	0.079993	0.076431	0.115162	0.091844	

40 Minute FFT data

Fig. 4. Output data

2.3 Fast Fourier Transform

In a real-time frequency analysis, the signal is continuously analyzed over all the frequency bands. Thus the calculation process must not take more time than the time taken to collect the signal data. Real-time analyzers are very useful since a change in the noise or vibration spectrum in the machine can be observed at the same time. There are two types of real-time analysis procedures: the digital filtering methodology and the fast Fourier transform (FFT) methodology. We are using FFT method in our VMS kit.

The Response signal is sent to FFT analyzer for signal processing. It computes the discrete frequency spectra of individual signals moreover as cross-spectra between the input and the different output signals. The analyzer converts the time-domain signals into frequency-domain data using Fourier series relations to facilitate digital computation and computes the spectral coefficients of these signals in the frequency domain.

The Frequency spectrum is a plot of the amplitude of vibration response versus the frequency. The frequency spectrum provides valuable information concerning the condition of a machine. This vibration analyzer may be used for fault detection by recording and storing vibration spectra from every measurement points. Any important increase within the amplitudes in the new spectrum indicates a fault that needs further investigation.

2.4 Profile Generation

Taking the average over several measurements a good baseline signal is produced and using this signal a profile is generated through Labview Exe. Then a Multiplication factor is provided to generate another profile of either a Positive Tolerance Profile (Multiplication factor will be Positive) or a Negative Tolerance Profile (Multiplication factor will be Positive). Another Profile will be Generated by Live Data of Sensors.

Hence Totally there will be a generation of 3 Profiles. Each Sensor is able to Set Different Profiles (Figs. 5 and 6).

Fig. 5. Profile generation front panel

Fig. 6. Remote vibration (live data of sensors) front panel

2.5 Time Transfer Logic (TTL)

Profile Generated by Live Data of Sensors is Compared with Profile Generated by Baseline Signal and Profile generated with Tolerance. The Live data Profile must be within the range of Baseline Profile and Tolerance Profile. Whenever the engine vibration crosses this vibration set point Range, It will generate the Time Transfer Logic (TTL) output pulse for second. The TTL data is Logged Separately to Each Sensor Channel. The TTL output is used to count the number of times the engine vibration crosses its limit, depends on that the reliability of engine is being calculated.

SET POINT: If TTL output crosses this value, Relay is Triggered. The Testing Will Stop if relay is Triggered. For example, if Set Point is 2000 and TTL value crosses 2000, the Relay is Triggered.

TARGET TIME: For Example, If a Testing Cycle is of 10 min and If we give Target time as 5 min then After 5 min Of Cycle Start the TTL Count will be reinitialised to 0.

2.6 Additional Features

1. To avoid Setting Profile Every time, one Button is given to retrieve the Previous Profile (Fig. 7).

Fig. 7. Saved data

2. Since we are using two 3-axes (6 Independent) Sensors, a Self test option is provided in the software to identify failure or misbehaviour of Sensor. If failure happens it will give warning and also indicates which axis sensor is failure (Fig. 8).
3. Data transfer is provided in such a way that whenever Dongle is connected to the system it will automatically copy the logged data to SD card in the Dongle (Fig. 9).

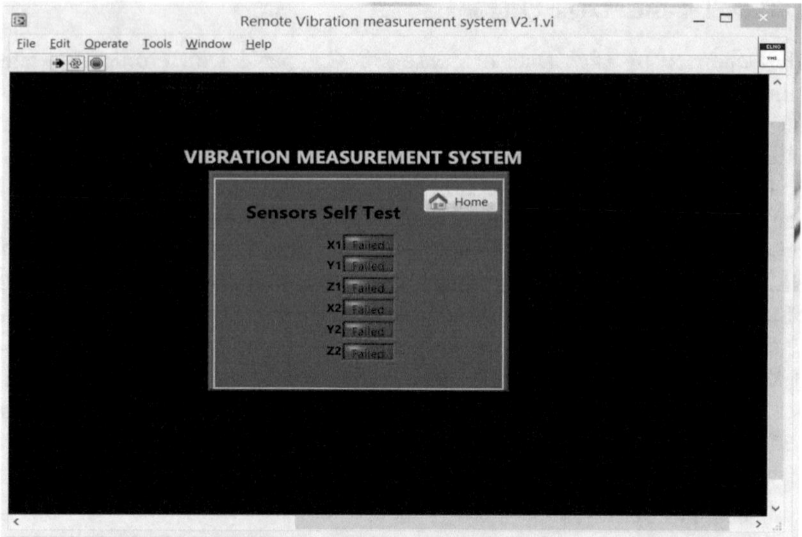

Fig. 8. Sensor self test window

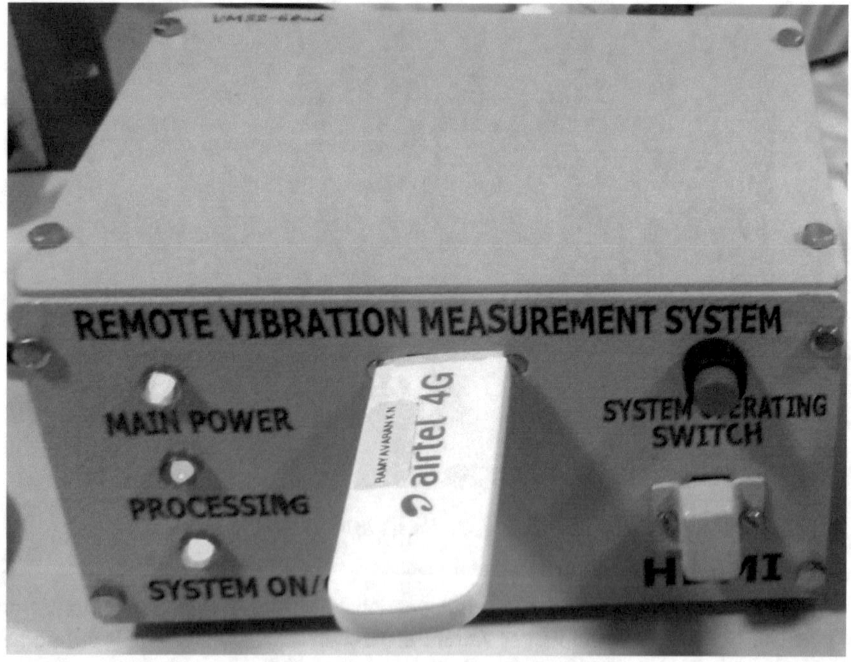

Fig. 9. Data transferring through Dongle

3 Actual Outcomes

The vibration measurement system is tested on field over a period of time it gives the good performance for evaluating the engine health monitoring using vibration signal. One of the most observations in vibration signal analysis is to remove the noise and mitigate the unwanted vibration signal (noise) originating from the sensor or the engine. The tests are conducted with different vibration levels for the bandwidth of 2 kHz. The designed system is cost effective solution for engine health monitoring system vibration analysis of up to 2 kHz (Figs. 10, 11, 12 and 13).

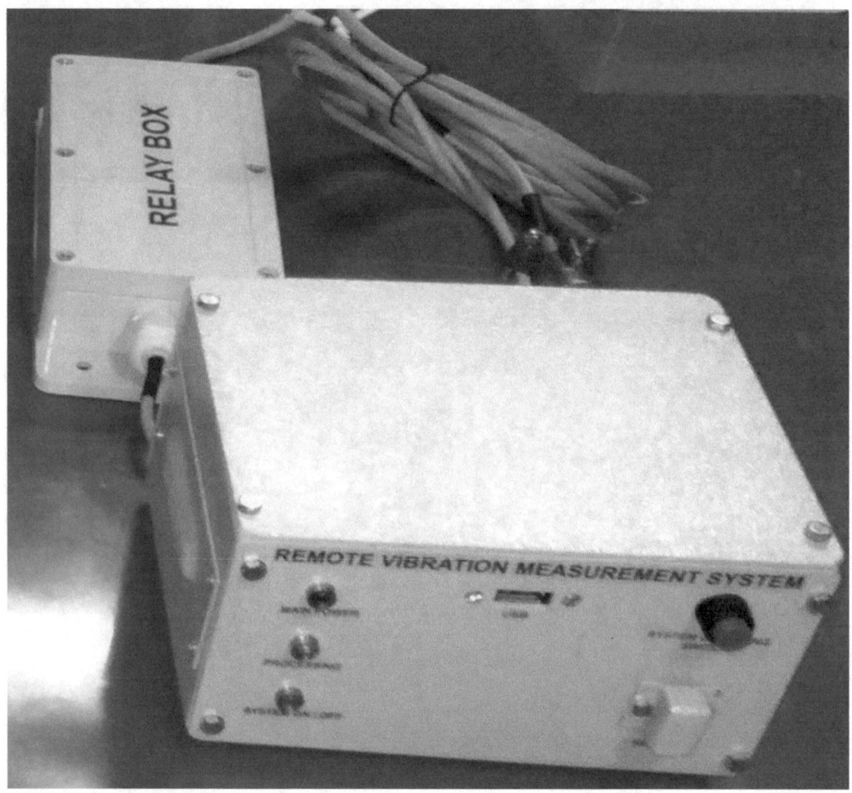

Fig. 10. Vibration measurement kit

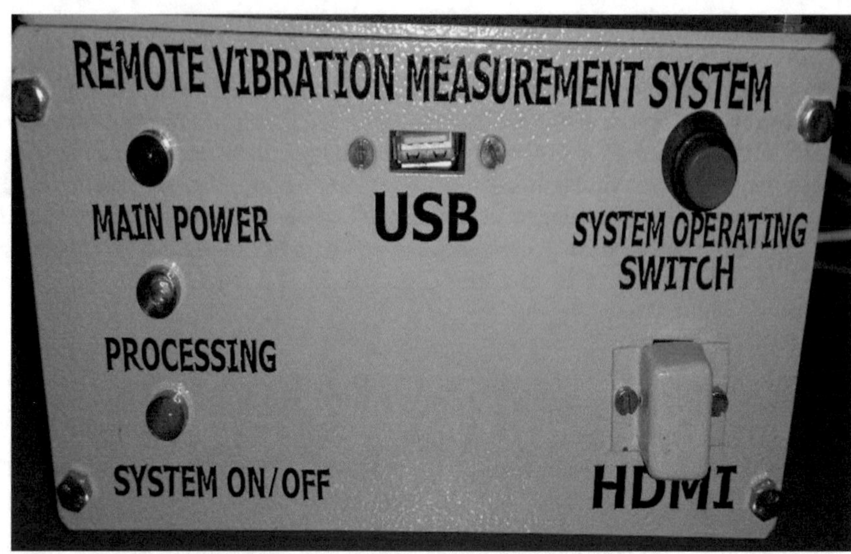

Fig. 11. VMS kit front view

Fig. 12. VMS kit back view

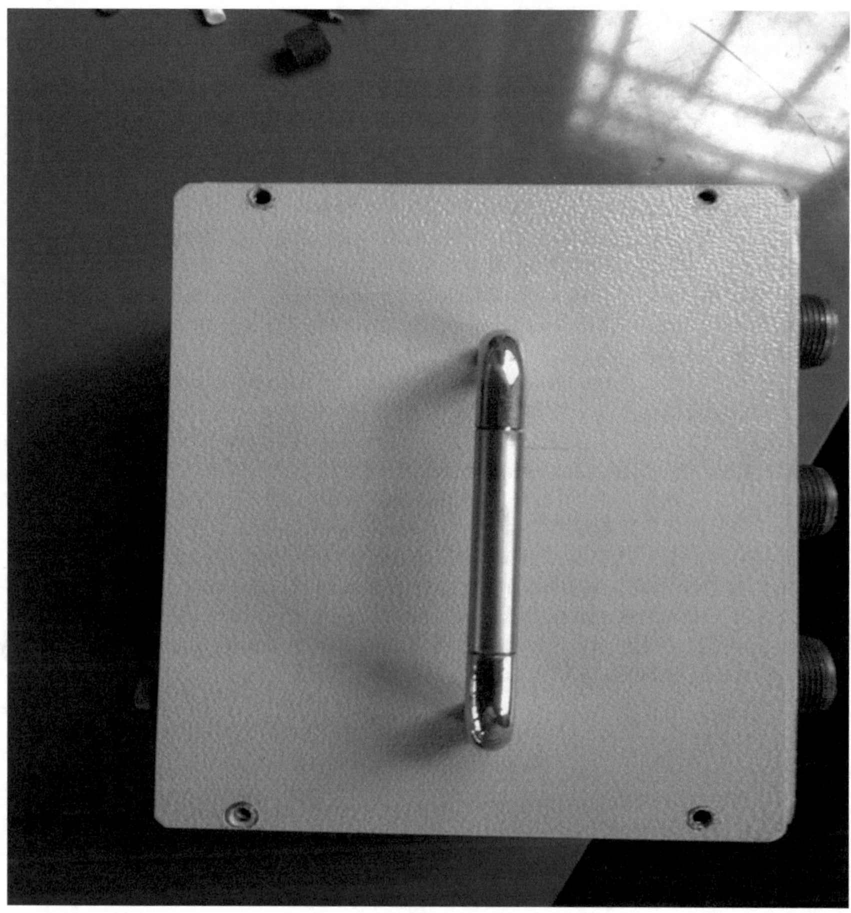

Fig. 13. VMS kit top view

4 Conclusions

From this kit through remotely, the data can be analysed from vehicle engine vibration carried out is observed that the vibration of engine is more on roads by compared with the nominal withstanding limits. This indicates that the vibration is increased by increasing the speed of the vehicle. Use of this kit will help us to find the failure rate of engine. Vibration Measurement System is provided for capturing 3 dimensional information. This paper has dealt with the design and the developmental aspects related to a low cost vibration measurement. This setup would be ideal for monitoring high value equipment using low cost.

References

1. Sinha, J.K., Elbhbah, K.: A future possibility of vibration based condition monitoring of rotating machines. Mech. Syst. Sig. Process. **34**, 231–240 (2013)
2. Elnady, M.E., Sinha, J.K., Oyadiji, S.O.: Condition monitoring of rotating machines using on-shaft vibration measurement. In: Proceedings of the IMechE, 10th International Conference on Vibrations in Rotating Machinery, London, UK, 11–13 Sept 2012
3. Jiang, F., Li, W., Wang, Z., Zhu, Z.: Fault severity estimation of rotating machinery based on residual signals. Adv. Mech. Eng. **4**, 1–8 (2012)
4. Divya, M., Ramgowri, S., Mangayarkarasi, T., Juliet, A.V.: A novel method of vibration measurement using MEMS accelerometer. In: International Conference on Computing and Control Engineering (ICCCE 2012), 12–13 Apr 2012
5. Perrone, G., Vallan, A.: A low-cost optical sensor for noncontact vibration measurements. IEEE Trans. Instrum. Measur. **58**(5), 1650–1656 (2009)
6. Sebastia, J.P., Lluch, J.A., Vizcaino, J.R., Bellon, J.S.: Vibration detector based on GMR sensors. IEEE Trans. Instrum. Measur. **58**(3), 707–712 (2009)
7. Stein, G., Chmúrny, R., Rosík, V.: Measurement and analysis of low frequency vibration. Measur. Sci. Rev. **7**(3, 4) (2007)
8. Arraigada, M., Partl, M.: Calculation of displacements of measured accelerations, analysis of two accelerometers and application in road engineering. In: Conference Paper STRC (2006)
9. Sinha, J.: On standardisation of accelerometers. J. Sound Vib. **286**, 417–427 (2005)
10. Mohn-Yasin, F., Korman, C.E., Nagel, D.J.: Measurement of noise characteristics of MEMS accelerometers. Solid-State Electron. **47**, 357–360 (2003)

Specialized Solar Panel Hinge Characterisation Test System

Panchaksharayya S. Hiremath[1], Kalyan Ram B.[1], Preeti Biradar[1(✉)], G. Harshita[1], and Ajay Kumar[2]

[1] Electrono Solutions Pvt. Ltd., Bangalore, India
{ps_hiremath, kalyan, preeti, harshita}@electronosolutions.com
[2] Ajay Sensors India, Bangalore, India
ajaysensors@gmail.com

Abstract. The upcoming spacecrafts are to be built such that it requires lesser space, lower cost and highly reliable, for implementing solar panels on the wings of the shuttles; in order to achieve these qualities the spacecrafts are built in such a way that the wings of these shuttles can be folded and unfolded during the utilization process. For obtaining this requirement the Spacecrafts are built by making use of various components, which also includes hinges placed at solar panel wings on the rockets and many other space shuttles, wings are built with hinges which helps the process of folding and unfolding of the solar panel wings on the shuttles during its energy generation operation. These hinges placed are of different types, the main purpose is to test its torque rotational and characteristics and to generate and provide the test report for all the combinations of hinges, before it has been deployed into the actual setup. The existing method was to manually test the hinges for its accuracy and proper working conditions, whereas this testing method is automated and is done on automated test equipment (ATE), with the help of sensors and actuators which is done to reduce the failure rate during the process by placing the hinge on the specialized solar hinge test bench in prior before deployment.

Keywords: Hinge · Torque · Pawls · Inch · Tare · PLC · ATE · Solar panel wings

1 Introduction

The upcoming spacecrafts are to be built such that it requires lesser space, lower cost and highly reliable, for implementing solar panels on the wings of the shuttles; in order to achieve these qualities the spacecrafts are built in such a way that the wings of these shuttles can be folded and unfolded during the utilization process. For obtaining this requirement the Spacecrafts are built by making use of various components, which also includes hinges placed at solar panel wings on the rockets and many other space shuttles, wings are built with hinges which helps the process of folding and unfolding of the solar panel wings on the shuttles during its energy generation operation. Survey

M. E. Auer and K. Ram B. (Eds.): REV2019 2019, LNNS 80, pp. 653–660, 2020.
https://doi.org/10.1007/978-3-030-23162-0_59

on provides the testing of the hinges done manually, this is an approach through which the testing is done by putting the hinges on ATE.

The hinges are tested by placing it on the automated test equipment (ATE) also known as the specialized solar hinge test bench. There are two steps involved namely Test sample and Analyse Data.

I. Test sample: This procedure includes testing the solar hinge by placing it on the vertical setup on the ATE, the arm on which the hinges are placed will rotate with specified angle of 30°, 60°, 90°, 120° and up to 180°; which in turn produces corresponding Torque value.

Furthermore, this test can be carried out either on sample test or on cycle test procedures.

a. Cycle test is carried out for different hinge types for understanding the abnormal conditions of the basic test of arm.
b. Whereas, the sample test procedure includes testing either the SADM types of hinges or on RDM types of hinges.

II. Analyse data: When the hinges are tested initially in test sample procedure the outcome or the data of the hinges will be saved, in order to analyse previously stored data this analysis of data is done by selecting the particular tested hinge file.

2 Architecture

- Software Flow

The flow of operation for testing different hinges by placing it on ATE is show below (Fig. 1).

- Hardware Architecture

The above block diagram shows the design of the ATE setup on which the hinges are placed and tested (Fig. 2).

3 Developmental Activity

- Hardware Setup

This setup includes the torque sensor and the angle sensor which is used for sensing the torque and angle values due to the movement in the arm where the hinges are placed for testing (Fig. 3).

- Software panel operations

1. Test sample: The test hinge is placed on the ATE; there are two tests that can be done on the hinges, namely Sample test and Cycle test.

(a) Sample test: This testing procedure is carried out for system working conditions, it has two types: SADM and RDM.

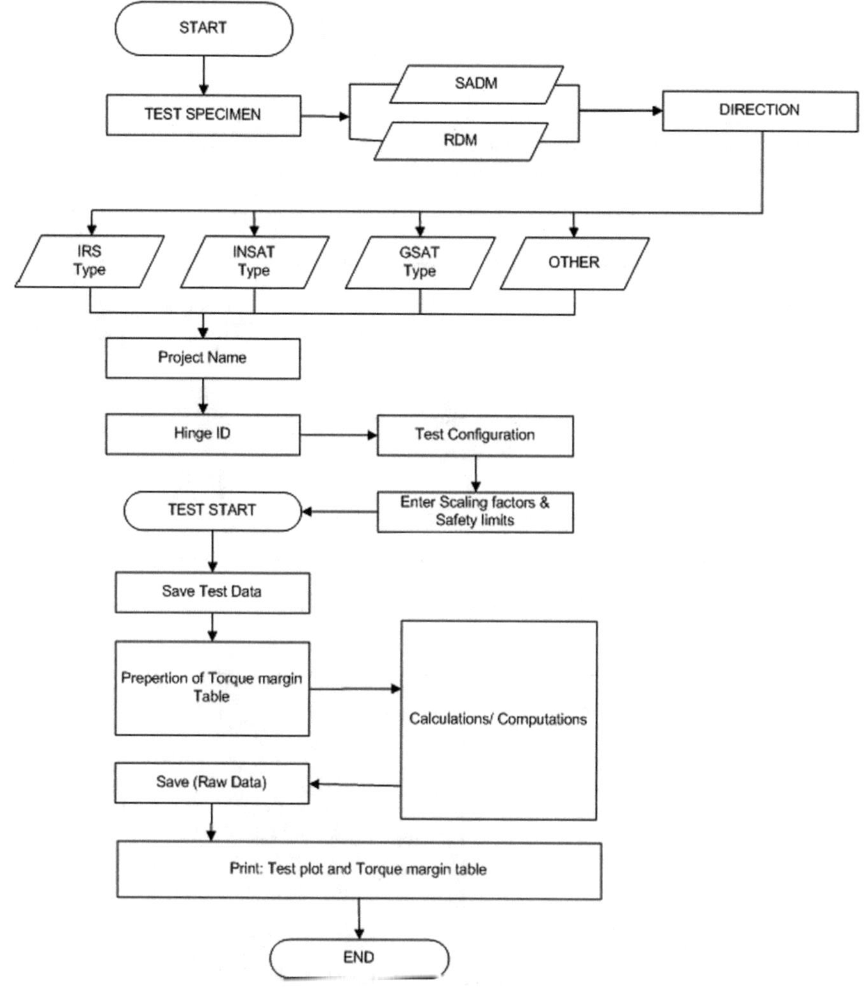

Fig. 1. Flow chart of operation for integrated hinge characterisation test system

i. **SADM**: In this test, type of hinge under test, the direction, test angle ranging from 0 to 180 can be specified based on the type of hinge and test configuration selected, Speed and delay must be specified, and the safety limits to the torque sensor must also be specified (Fig. 4).
ii. **RDM**: Here, the type of hinge and its sub-type along with the direction in which it is tested, RDM test configuration and the safety limits are to be entered. After specifying all the inputs the testing can be done (Fig. 5).

Figure 6 shows SADM/RDM test window, when all the input specifications are given and once the start test is selected, the test begins and the Torque versus Angle graph will start plotting the angle (deg) and torque (kg/cm).

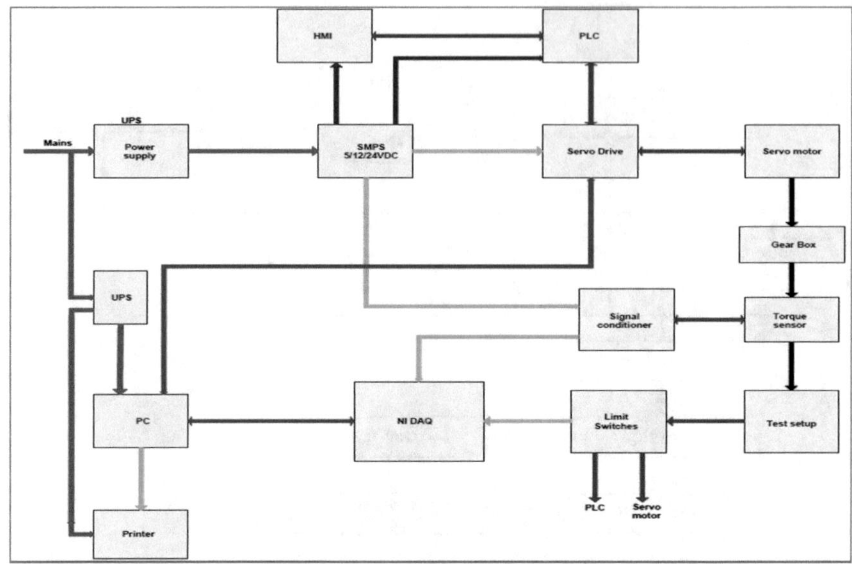

Fig. 2. Block diagram for ATE

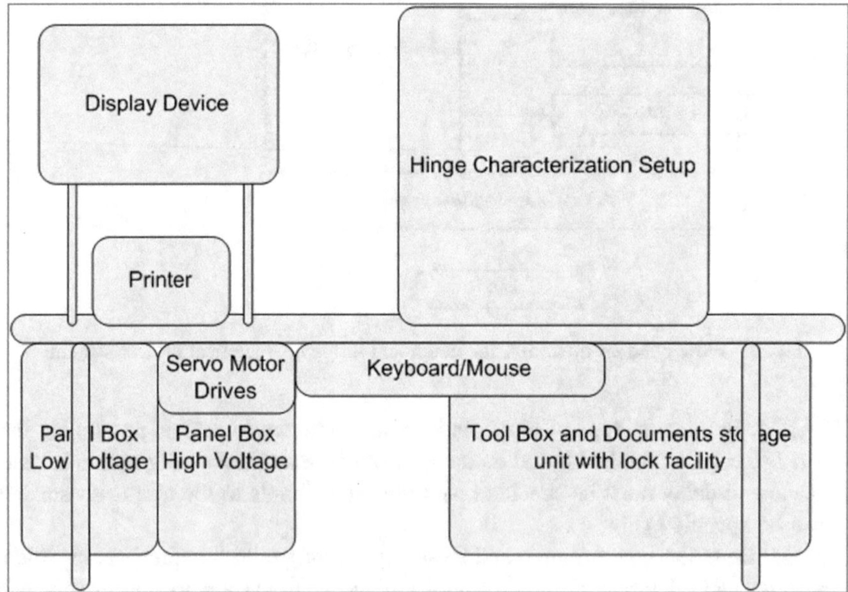

Fig. 3. Hardware setup

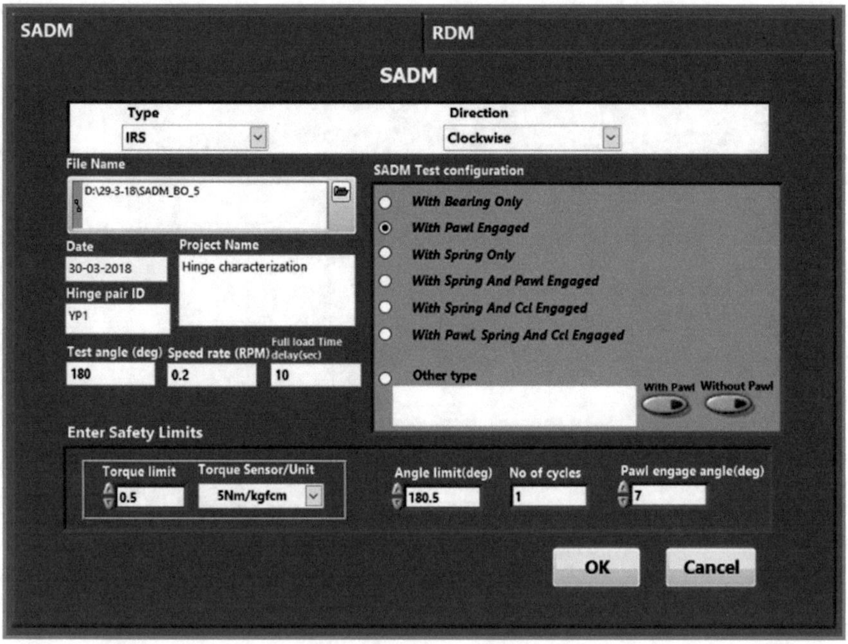

Fig. 4. SADM window

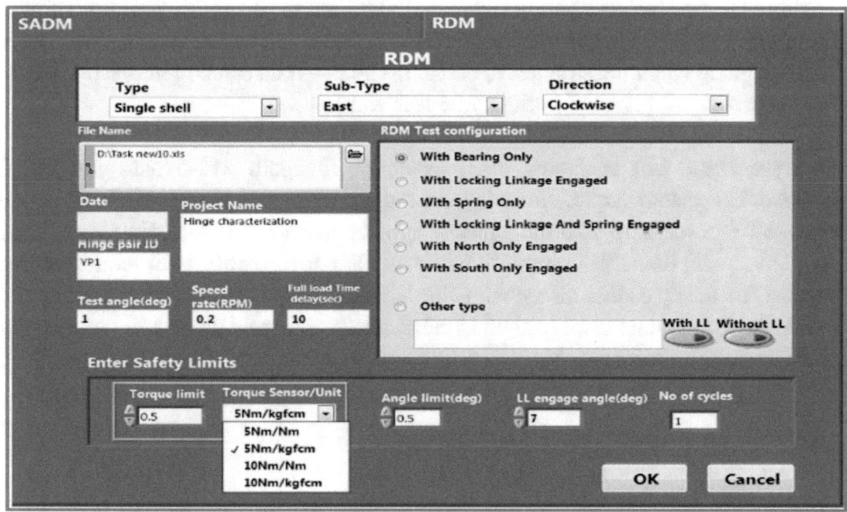

Fig. 5. RDM window

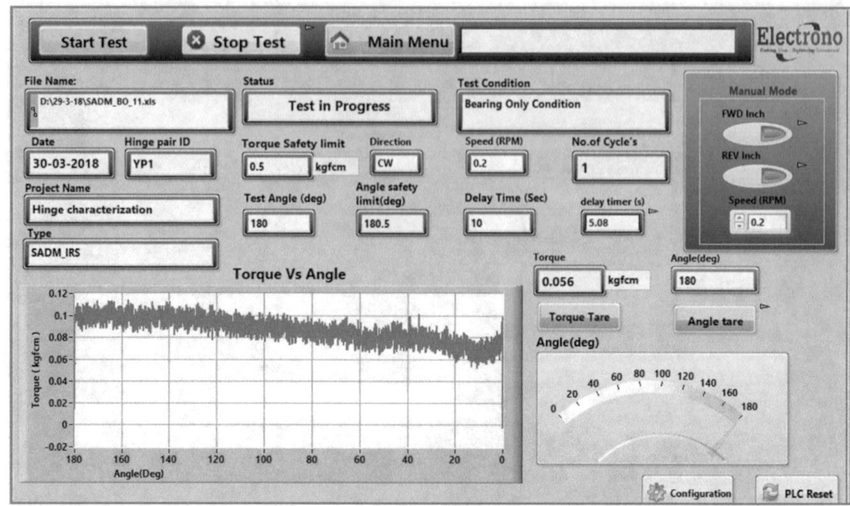

Fig. 6. SADM/RDM test window

(b) **Cycle test**: This testing procedure is done for different types of hinges to understand the abnormal conditions which occur during the basic testing of the arm.

- All the necessary inputs are given for performing the cycle test, and testing will begin once the start test button is pressed. The **angle** (degrees), **number of cycles completed** and the **torque values** (kg/cm) are indicated on the output side. The test will complete when the stop test is clicked. The raw data can be fetched by selecting the acquire data. Figure 7 is the cycle test window.

2. **Analyse Data**: For analysing data, select the file path which data you want to analyse. Enter start angle, end angle and angle interval and then click on **process data**, after processing data the window appears as shown below. Enter the harness torque value in the table column-5. Click on the **process data with harness torque button**, to retrieve data along with the harness torque. By clicking on the **show excel**, the data can be displayed on excel sheet. To get the data on MS word click on **print report** button (Fig. 8).

4 Actual or Anticipated Outcomes

The reports are generated for all the types of hinges with the different types of test configuration, the output graph of Torque versus Angle gives the characteristics of hinge depending on the particular combination of the solar hinge that is under test. The output torque value ranges from 0 to 10 Nm and the angle varies from 30°, 60°, 90°, 120° and up to 180° based on the type of hinge selected (Fig. 9).

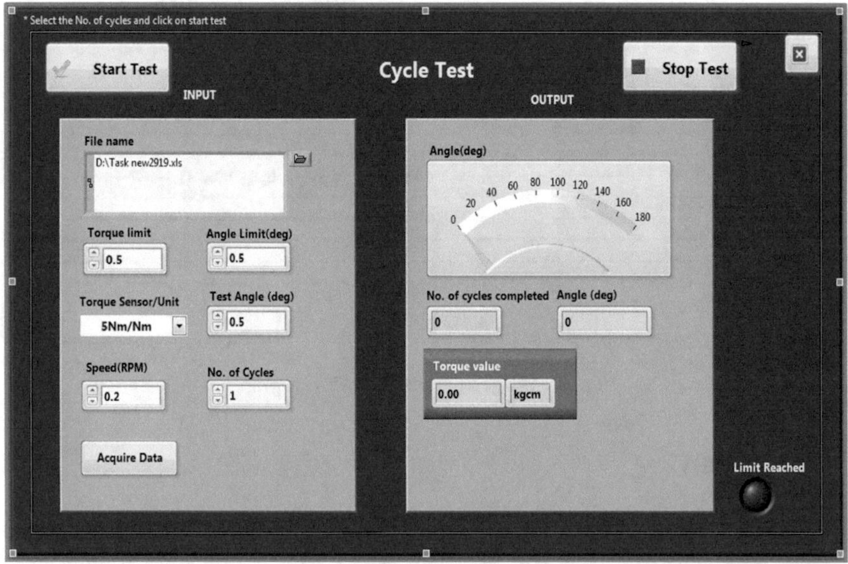

Fig. 7. Cycle test window

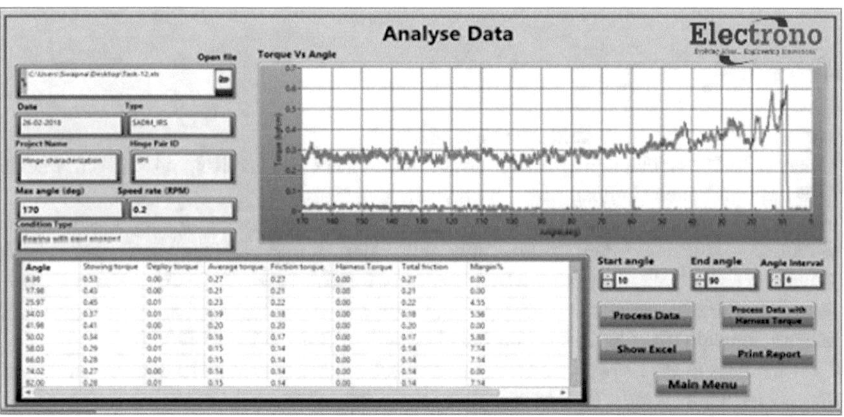

Fig. 8. Analyse data window

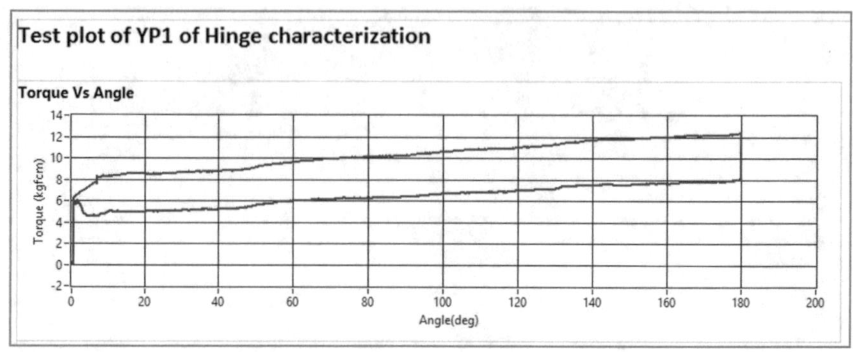

Fig. 9. Hinge characterization test graph

5 Conclusion

The hinges are tested on the vertical setup of the solar hinge test bench and its reports are generated along with the Torque versus Angle graph. Values of torque and the angle in degrees provide the overall details about the hinge. Since there are various types of hinges along with different types of test configuration, range of torque measured by hinges varies from 0 to 10 Nm and the angle in degrees varies from 0 to up to 180° but not necessarily to the same angle during all the tests; the value and the torque versus angle graph varies based on the hinge type selected.

References

1. http://www.dtic.mil/dtic/tr/fulltext/u2/a452620.pdf
2. Calassa, M.C., Kackley, R.: Solar Array Deployment Mechanism, pp. 80–85. Lockheed Missiles and Space Company Inc., Sunnyvale, CA (1990)
3. Francis, W., et al.: Development and testing of a hinge incorporating elastic memory composites. In: Presented at the 44th Structures, Structural Dynamics, and Materials Conference, Norfolk, Virginia, AIAA Paper No. 2003-1496, 7–10 Apr 2003; Clark, C., Wood, J., Zuckermandel, B.: Self-deploying, thin-film PV solar array structure. In: Presented at 16th Annual/USU Conference on Small Satellites, SSC02-VIII-5 (2002)
4. Beavers, F.L., et al.: Design and testing of an elastic memory composite deployment hinge for spacecraft applications. In: Presented at the 43rd AIAA/ASME/ASCE/AHS/ASC SDM Conference, Denver, CO, AIAA Paper No. 2002-1452, 22–25 Apr 2002
5. Lake, M.S., Munshi, N.A., Tupper, M.L.: Application of elastic memory composite materials to deployable space structures. In: Presented at the AIAA 2001 Conference and Exposition, Albuquerque NM, AIAA Paper No. 2001-4602, 28–30 Aug 2001

Biotelemetry Over TCP/IP in LabVIEW

S. Kumuda and M. N. Mamatha(✉)

Electronics and Instrumentation Engineering Department,
BMS College of Engineering, Basavanagudi, Bangalore 560019, India
{kumudamohan.intn, mamathamn.bms.intn}@bmsce.ac.in

Abstract. We aim at providing the necessary and sufficient information with respect to medical specification and condition of patient (may be an admitted patient or an outpatient of a hospital) to the concerned doctor over specific communication protocol like TCP/IP, here we set up server which necessarily in hospital is CNS (central nurse station) and concerned doctor on the hospital network would be a client to CNS, a nurse who thinks a patient requires certain doctors attention or the other way round based on patient ID the doctor can fetch various information analyze and send his/her opinion about the patient to the CNS.

1 Introduction

The Biotelemetry (or Medical Telemetry) involves the application of telemetry in the medical field to remotely monitor various vital signs of patients.

Biotelemetry is a method of measuring biological parameters from a distance. It is in fact modification of existing methods of measuring physiological variables to a method of transmission of resulting data. The transmission of data from the point of generation to the point of reception can be done in various ways. The stethoscope is the simplest device which uses this principle of biotelemetry. The device amplifies acoustically the heartbeats and transmits their sound to the ears of a doctor through a hollow tube system. Certain applications of biotelemetry use telephone lines for transmission. However biotelemetry mainly uses radio transmission by suitably modifying the biological data. Earlier times, the telemetry could be applied to measure.

(1) Body temperature.
(2) Electrocardiograms.
(3) Indirect blood pressure.
(4) Respiration.
(5) Pulse rate.

2 LabVIEW

LabVIEW [1] (short for Laboratory Virtual Instrumentation Engineering Workbench) is a system design platform and development environment for a visual programming language from National Instruments.

M. E. Auer and K. Ram B. (Eds.): REV2019 2019, LNNS 80, pp. 661–673, 2020.
https://doi.org/10.1007/978-3-030-23162-0_60

LabVIEW [2] (Laboratory Virtual Instrument Engineering Workbench) is a graphical programming environment which has become prevalent throughout research labs, academia and industry. It is a powerful and versatile analysis and instrumentation software system for measurement and automation. Its graphical programming language called *G programming* is performed using a graphical block diagram that compiles into machine code and eliminates a lot of the syntactical details. LabVIEW offers more flexibility than standard laboratory instruments because it is software based. Using LabVIEW, the user can originate exactly the type of virtual instrument needed and programmers can easily view and modify data or control inputs.

LabVIEW [2, 3] can communicate with hardware such as data acquisition, vision, and motion control devices, and GPIB, PXI, VXI, RS-232, and RS-485 devices. LabVIEW also has built-in features for connecting your application to the Web using the LabVIEW Web Server and software standards such as TCP/IP networking and ActiveX. Using LabVIEW, you can create test and measurement, data acquisitions, instrument control, datalogging, measurement analysis, and report generation applications. You also can create stand-alone executables and shared libraries, like DLLs, because LabVIEW is a true 32-bit compiler.

3 TCP/IP Communication

The Transmission Control Protocol (TCP) (Fig. 1) [4] is one of the two original core protocols of the Internet protocol suite (IP), and is so common that the entire suite is often called TCP/IP. TCP provides reliable, ordered, error-checked delivery of a stream of octets between programs running on computers connected to an intranet or the public Internet.

An application for transferring files with TCP, for instance, performs the following operations to send the file contents:

1. The Application layer passes a stream of bytes to the Transport layer on the source computer.
2. The Transport layer divides the stream into TCP segments, adds a header with a sequence number for that segment, and passes the segment to the Internet (IP) layer. A checksum is computed over the TCP header and data.
3. The IP layer creates a packet with a data portion containing the TCP segment. The IP layer adds a packet header containing source and destination IP addresses.
4. The IP layer also determines the physical address of the destination computer or intermediate computer on the way to the destination host. It passes the packet and the physical address to the Data-Link layer. A checksum is computed on the IP header.
5. The Data-Link layer transmits the IP packet in the data portion of a data-link frame to the destination computer or an intermediate computer. If the packet is sent to an intermediate computer, steps 4 through 7 are repeated until the destination computer is reached.
6. At the destination computer, the Data-Link layer discards the data-link header and passes the IP packet to the IP layer.

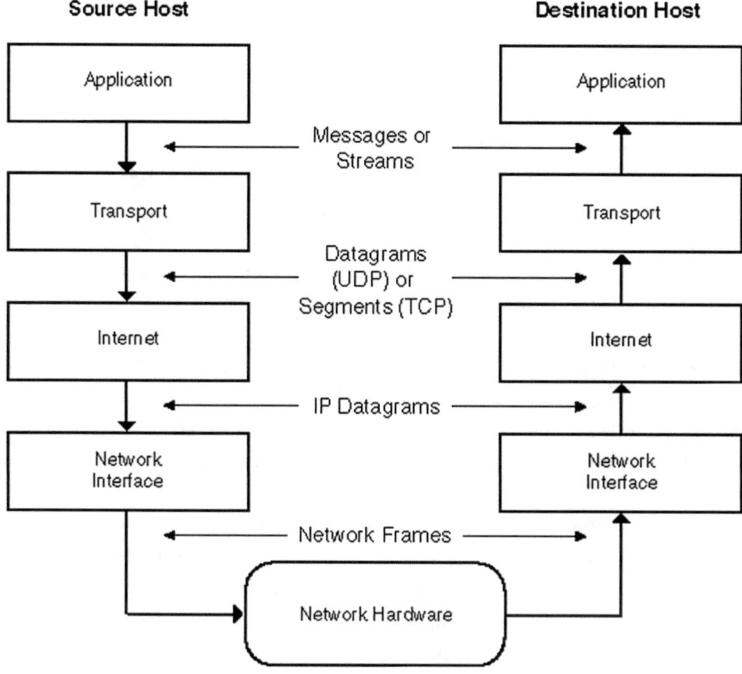

Fig. 1. TCP/IP model

7. The IP layer checks the IP packet header. If the checksum contained in the header does not match the checksum computed by the IP layer, it discards the packet.
8. If the checksums match, the IP layer passes the TCP segment to the TCP layer.
9. The TCP layer computes a checksum for the TCP header and data. If the computed checksum does not match the checksum transmitted in the header, the TCP layer discards the segment. If the checksum is correct and the segment is in the correct sequence, the TCP layer sends an acknowledgment to the source computer and passes the data to the application.
10. The application on the destination computer receives a stream of bytes, just as if it were directly connected to the application on the source computer.

The protocol corresponds to the transport layer of TCP/IP suite. TCP provides a communication service at an intermediate level between an application program and the Internet Protocol (IP). That is, when an application program desires to send a large chunk of data across the Internet using IP, instead of breaking the data into IP-sized pieces and issuing a series of IP requests, the software can issue a single request to TCP and let TCP handle the IP details.

IP works by exchanging pieces of information called packets. A packet is a sequence of octets and consists of a header followed by a body. The header describes the packet's destination and, optionally, the routers to use for forwarding until it arrives at its destination. The body contains the data IP is transmitting.

The Transport layer of the TCP/IP protocol [4] suite consists of two protocols, UDP and TCP. UDP provides an unreliable connectionless delivery service to send and receive messages. TCP adds reliable byte stream-delivery services on top of the IP datagram delivery service.

The ports numbered between 1 and 1023 are well-known port numbers. For dynamically bound ports, an application requests that UDP assign a port to identify which port the process uses. The port must be in the range of 1024 to 65,535.

UDP identifies applications through *ports*. The protocol defines two types of protocol ports: well-known port assignments and dynamically bound ports. For well-known port assignments, certain UDP port numbers are reserved for particular applications. Then the application can direct UDP datagrams to that port.

UDP enables multiple clients to use the same port number and different IP addresses. The arriving UDP datagrams are delivered to the client that matches both the destination port number and address. (A socket consists of an IP address and the port number.) If there is no matching client or if the ICMP destination is unreachable then a port unreachable message is sent and the packet is dropped.

For applications that must send or receive large volumes of data, unreliable datagram delivery can become burdensome. Application programmers might have to develop extensive error handling and status information modules to track the progress and state of data transfer for every application. The TCP/IP suite of protocols avoids this problem by using TCP, a *reliable byte-stream delivery protocol*. TCP establishes a connection between two applications and sends a stream of bytes to a destination in exactly the same order that they left the source. Before transmission begins, the applications at both ends of transmission obtain a TCP *port* from their respective operating systems. These are analogous to the ports used by UDP. The application initiating the transfer, known as the client side, generally obtains a port dynamically. The application responding to the transfer request, known as the server side, generally uses a well-known TCP port. The client side is typically the active side and initiates the connection to the passive server side.

4 Client–Server Architecture

The client–server model (Fig. 2) is a distributed application structure in computing that partitions tasks or workloads between the providers of a resource or service, called servers, and service requests called clients. [5] Often clients and servers communicate over a computer network on separate hardware, but both client and server may reside in the same system. A server is a host that is running one or more server programs which share their resources with clients. A client does not share any of its resources, but requests a server's content or service function. Clients therefore initiate communication sessions with servers which await incoming requests.

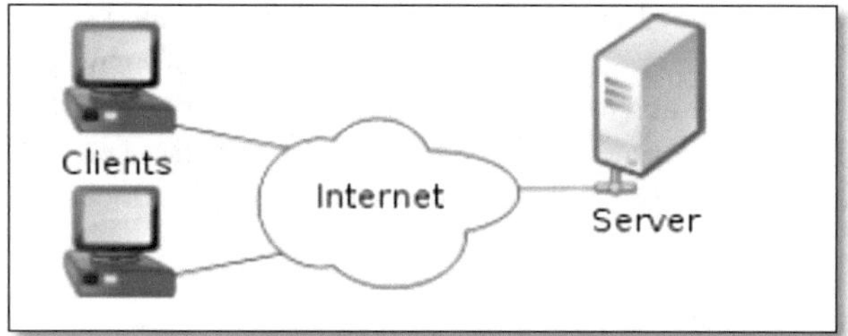

Fig. 2. Client–server model

5 Central Nurse Station (CNS)

Central Nurse station (Fig. 3) is usually located in the hospital, clinics or call centre and allows the doctor or other healthcare providers' direct access to the patient through the interactive session. An area in a clinic, unit, or ward in a healthcare facility that serves as the administrative centre for nursing care for a particular group of patients.

It is usually centrally located and may be staffed by a unit secretary or clerk who assists with paperwork, telephone, and other communication. Before going on duty, nurses usually meet there to receive daily assignments, review the patients' charts, and

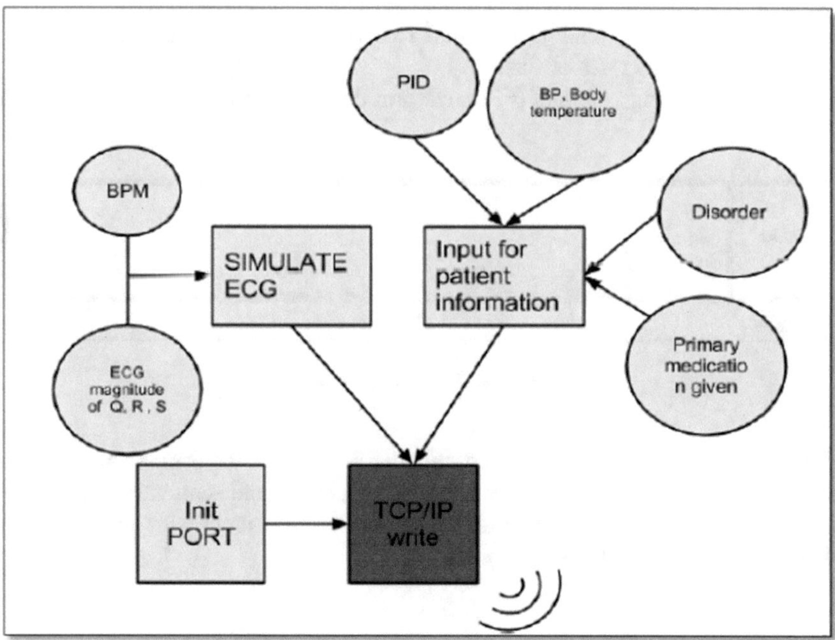

Fig. 3. Block diagram of Central Nurse station

update the files. In a critical care unit, the nurses' station may also contain panels of visual display terminals that allow centralized monitoring of many patients and computer terminals that allow access to information in the patients' records or to a databank of clinical information. In other parts of a hospital, the nurses' station is equipped in any of various ways appropriate to the care of the patients in that area or unit.

6 Proposed Idea of CNS

First an ECG wave of patient is simulated by configuring the VI's for ecg simulation and various details of the patient such as patient ID, age, his blood pressure, body temperature, Pulse rate are entered on page (Front panel of the VI), there is window where the recommendation from the Doctor is also received after seeing the patient's condition, a remote host is established at the CNS on a specific PORT number which essentially acts as a channel.

The server is established by adding a TCP/IP listen VI which creates a listener and waits for an accepted TCP network connection at the specified **port**.

A registration form is incorporated to facilitate storage of patient details and its retrieval, essentially creating the database. Each patient has a different Patient ID and an associated encrypted database file each file contains patients' medical details like blood pressure, Temperature, medical history, disorder, heart rate etc. And each connected client (doctor) has access to all information in the database from anywhere on the globe.

A separate loop for sending data and receiving data is incorporated in order to facilitate simultaneous rx/tx of data.

Format for sending ecg data [6] (waveform data) is as shown in Fig. 4.

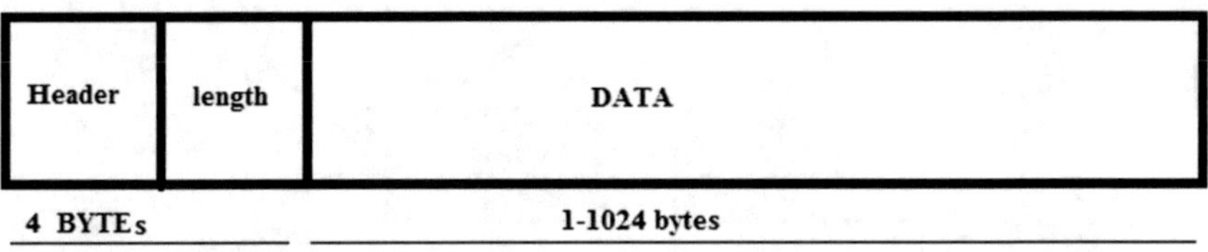

Fig. 4. ECG data format

Each value of the waveform, string, number is casted to the format above and sent over the internet, i.e. every data type is casted to a string and sent and on the receiver end it is decoded. By default 4 bytes of data is sent wherein it contains a header and the payload length and then the data is sent.

7 Implementation of CNS in LabVIEW

Basically the proposed system is based on client-server architecture.

The front panel of CNS designed in LabVIEW which act as user interface is as shown in Figs. 5 and 6.

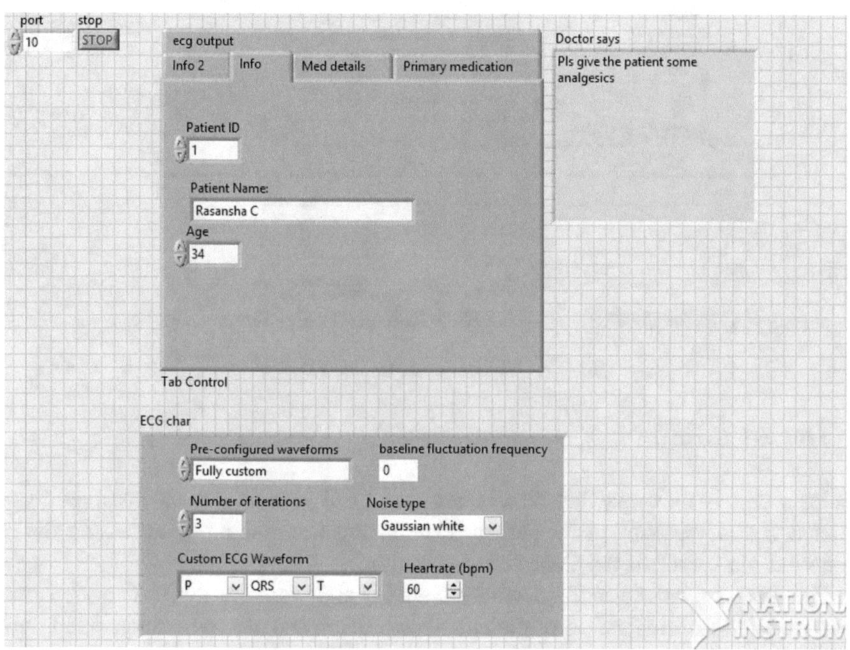

Fig. 5. Front panel of CNS designed in LabVIEW™

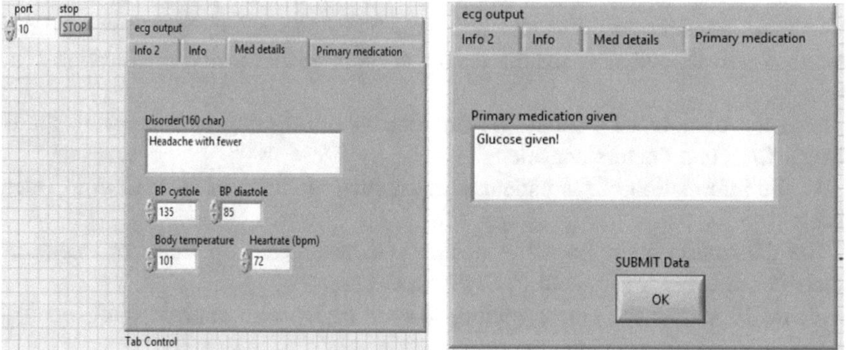

Fig. 6. Database development at CNS using LabVIEW™

It is designed to collect the complete details of the patients like ECG parameters, primary medication given, previous prescription etc.

The corresponding block diagram or G-code of CNS is as shown in Fig. 7.

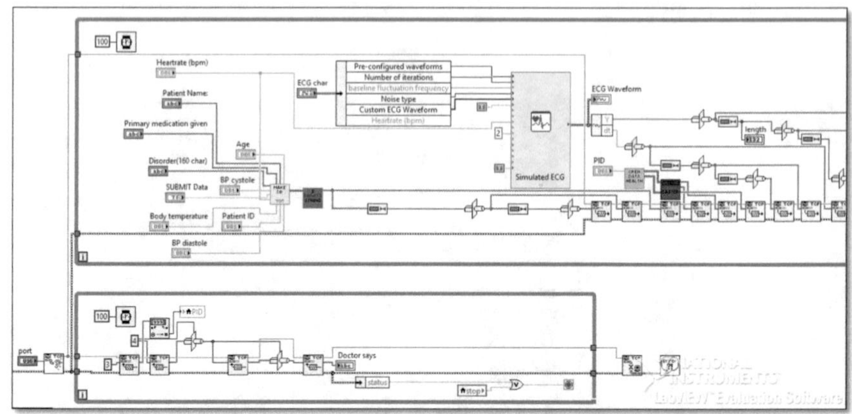

Fig. 7. CNS G-code in LabVIEW™

8 Doctor's End

Doctor's end is main part of biotelemetry of ECG signals, which involves recommendation and prescription for patient. Communication is done by accessing the port connected via server in the Central nurse station.

The data is received at CNS with patient ID and what are the further steps to be taken required for medication of patient. This process can also be implemented during emergency medevac situations in ambulance, doctor can analyse the condition of the patient and provide remedial measures (Fig. 8).

The front panel of doctors console (Figs. 9, 10, 11 and 12) is designed to obtain the information about the patient from CNS console remotely.

In the doctors console details like patient history, medical details like BPM, body temperature, systolic and diastolic pressure, ECG parameters, previous medication prescribed will be available for the doctor to refer so that further medication can be given.

Both the console VI's should be running for transferring and receiving the data between CNS and doctors console.

All the information of the patient is transmitted via the port from CNS to Doctors console.

The data/prescription from the doctor will be entered on the user interface of doctors console and transmitted to CNS via the port.

Figure 13 shows the corresponding G-code of Doctors console implemented in LabVIEW ™.

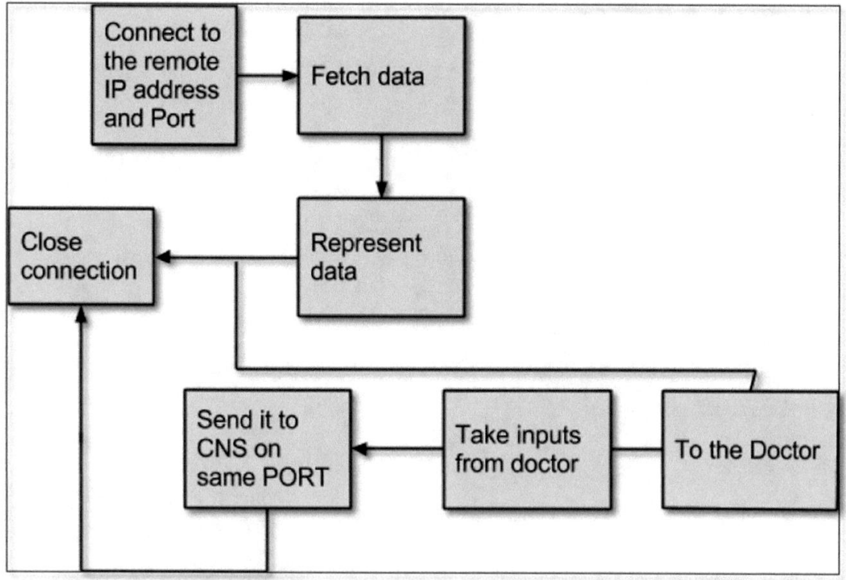

Fig. 8. Doctors console block diagram

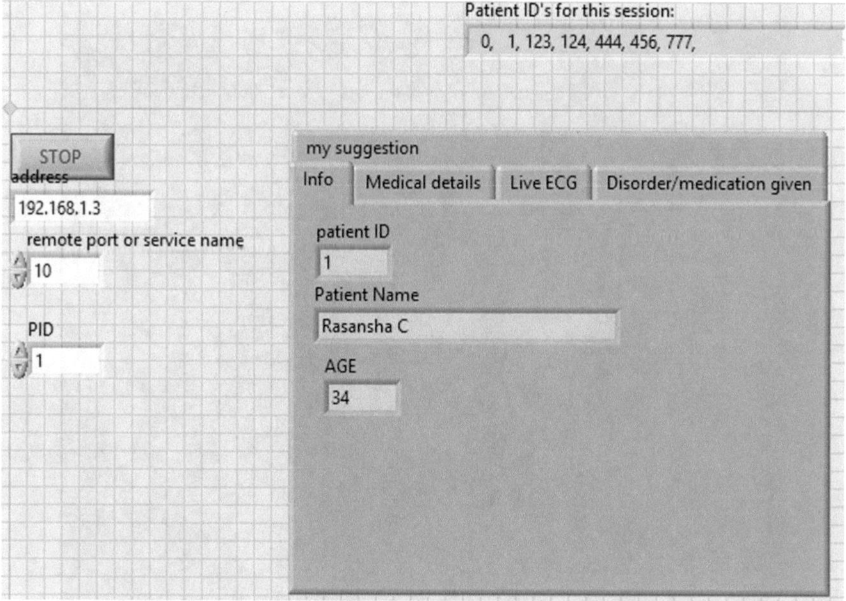

Fig. 9. Doctors console implementation in LabVIEW™

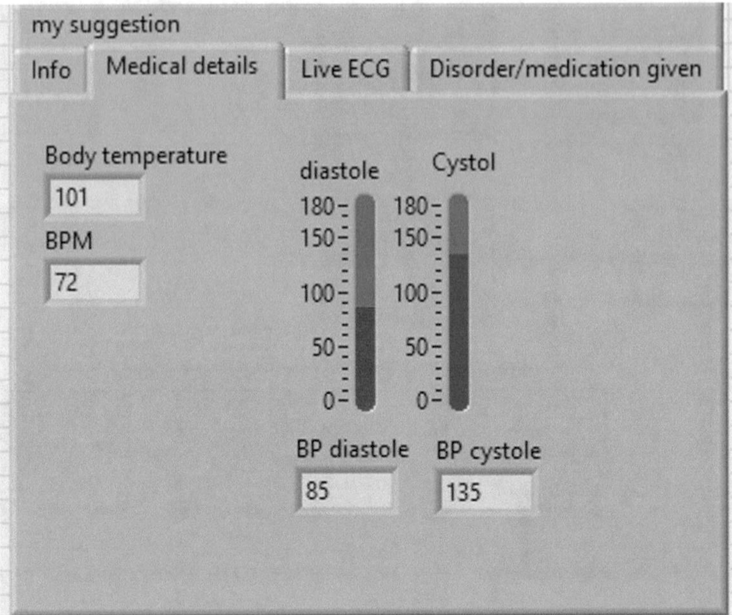

Fig. 10. Medical details of patient displayed on doctors console

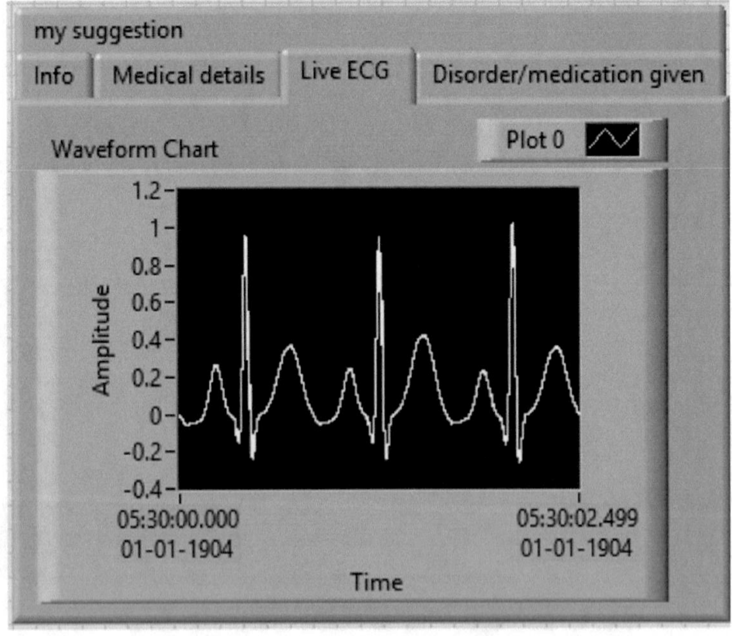

Fig. 11. ECG waveform of patient displayed on doctors console

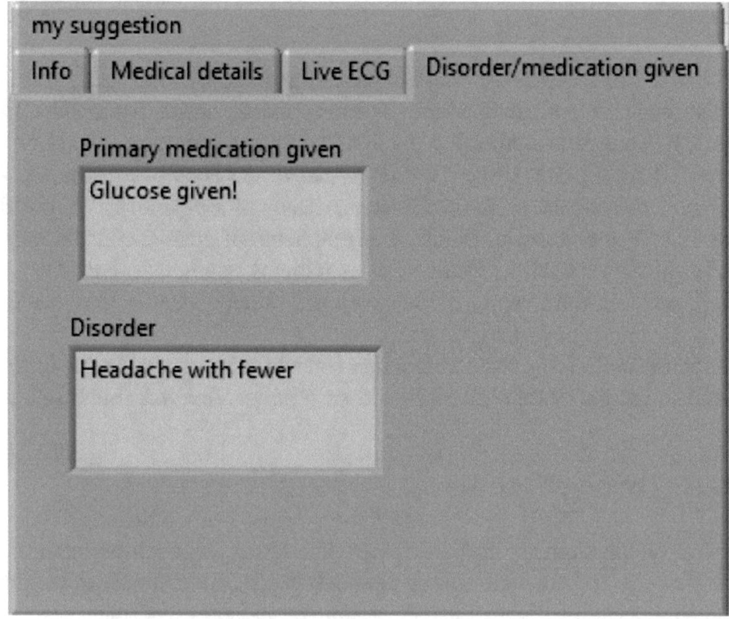

Fig. 12. Patients medication displayed on doctors console

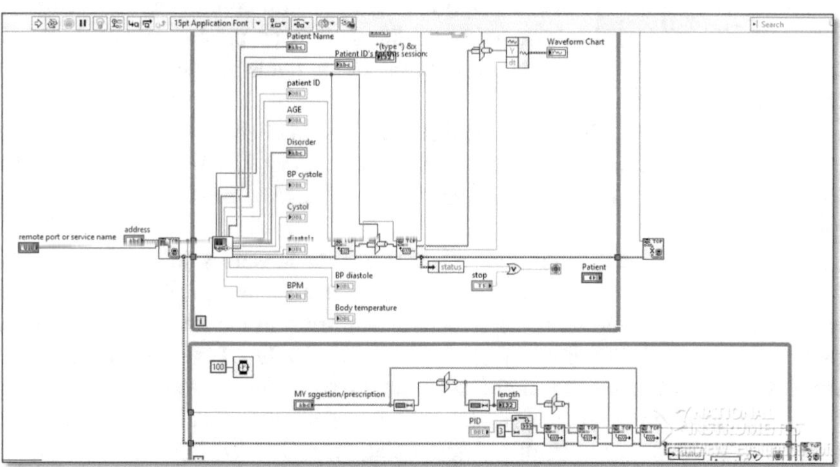

Fig. 13. G-code of Doctor's console in LabVIEW™

9 Conclusion

The CNS and doctors console model was developed successfully using LabVIEW.

The data such as patient database like age, name, health parameters like body temperature, BP, previous medication given to the patient is entered at the CNS console and transmitted via the port using TCP/IP format to the doctors console successfully.

At the doctors console, all the parameters, details of the patient is received via the port using TCP/IP successfully. Hence, the front panel of doctors console was updated successfully through TCP/IP. Details such as patients name, age, BP, body temperature, systolic and diastolic value, ECG waveform, previous medication was displayed successfully on the front panel of doctors console.

The implementation of CNS and Doctors console carrying the details of the patient was realized to the best extent possible and its functionality was found satisfactory.

10 Future Scope of the Work

Incorporation of biotelemetry has a number of advantages, such as reduction of the impediment of the information source (patient, subject or animal) reduction of the psychological effects on the information source, reduction of measuring artifacts, reduction of the risk for electroshock, reduction of the complexity of monitoring of physiological variables, as well as a potential reduction of the total cost of patient care. Therefore, biotelemetry may be incorporated advantageously not only in areas where this technique is absolutely needed (sport, work, function control during realistic circumstances of activity), but also in other areas such as patient monitoring during intensive care and in the operating room. Some of the fields in which biotelemetry can be useful are listed below.

Battlefield medicine, also called field surgery and later combat casualty care, is the treatment of wounded soldiers in or near an area of combat. Civilian medicine has been greatly advanced by procedures that were first developed to treat the wounds inflicted during combat.

Remote physiological monitoring: Technological advances in electronics, smart fabrics and wireless communications have led to increasing developments in remote physiological sensing hardware. These sensor platforms all measure raw physiological signal data from the body, however, for this to be of value it is necessary to convert these waveforms into meaningful and contextual information. Several of these devices provide real-time, on-board derivations of basic parameters such as heart rate and respiration rate. However, a wealth of additional, more detailed information is possible when measuring multiple, time synchronized data channels during non-laboratory, daily living conditions.

Remote patient monitoring (RPM) is a technology to enable monitoring of patients outside of conventional clinical settings (e.g. in the home), which may increase access to care and decrease healthcare delivery costs.

Incorporating RPM in chronic disease management can significantly improve an individual's quality of life. It allows patients to maintain independence, prevent complications, and minimize personal costs. RPM facilitates these goals by delivering care right to the home. In addition, patients and their family members feel comfort knowing that they are being monitored and will be supported if a problem arises

References

1. Basic TCP/IP communication in LabVIEW: http://www.ni.com/white-paper/2710/en/
2. Jerome, J.: Virtual instrumentation using Labview, vol 1
3. LabVIEW: www.ni.com/trylabview/
4. https://www.novell.com/documentation/nw6p/?page=/documentation/nw6p/tcpipenu/data/hbnuubtt.html
5. Client-server model: http://en.wikipedia.org/wiki/Client%E2%80%93server_model
6. ECG-simulation in LabVIEW: http://zone.ni.com/devzone/cda/epd/p/id/6189

Vegetation Index Estimation in Precision Farming Using Custom Multispectral Camera Mounted on Unmanned Aerial Vehicle

Sebastian Pop, Luciana Cristea, Marius Cristian Luculescu, Sorin Constantin Zamfira(✉), and Attila Laszlo Boer

Transilvania University of Brasov, Brașov, Romania
aerodrone.uav@gmail.com, {lcristea, lucmar, zamfira, boera}@unitbv.ro

Abstract. In the last decades several vegetation indices were introduced in order to quantify the spectral properties of green plants. This paper presents an integrated solution (hardware and software) affordable for acquiring and processing spectral data in order to estimate the vegetation status. A brief description of what are the vegetation indices and why are they important is followed by the hardware description, software setup and experimental workflow. The entire hardware system consisting of a modular multispectral camera and the Small Board Computer that controls the acquisition process, was developed within a research project taking into account certain conditions as modularity, low cost, small size, low weight, remote data access. The multispectral camera is carried by an Unmanned Aerial Vehicle (UAV) in order to take images of the investigated area. Regarding the software, the developed applications are mainly based on open-source solutions: a web-based interface designed for the acquisition control and one special program developed for vegetation indices computation. Other applications are used for data processing. Georeferenced orthophotomaps are generated using acquired images, each one for one wavelength of the multispectral camera. Using these data, different types of vegetation indices are computed and corresponding maps are generated. The key outcomes of the study are related to the multispectral camera that has a modular design, offering flexibility in changing the desired wavelengths and also to the data processing solutions.

Keywords: Vegetation index · Precision farming · Multispectral camera · Unmanned aerial vehicle

1 Introduction

Population in the world is currently growing at a rate of approximately 1.09% per year. This population growth will push up global demand for agricultural products. Consequently, new technologies must be incorporated in precision farming in order to face this challenge.

The reflection or absorption of electromagnetic radiation is unique for each type of vegetation and can be used to indicate its health and development.

M. E. Auer and K. Ram B. (Eds.): REV2019 2019, LNNS 80, pp. 674–685, 2020.
https://doi.org/10.1007/978-3-030-23162-0_61

The spectral response of the vegetation to the incident solar radiation can be measured and obtained data can be used to compute different parameters correlated with the plants health status. These parameters that are called vegetation indices were introduced in the last decades in order to quantify the spectral properties of green plants.

Usually these indices combine visible light radiation and non-visible spectra [1].

One of the first vegetation indices was proposed in 1969 by Jordan [2] and it is called Ratio Vegetation Index (RVI).

$$\mathrm{RVI} = \frac{\mathrm{Red}}{\mathrm{NIR}} \tag{1}$$

where *Red* is the reflectance for red light, whereas *NIR* is the reflectance in the near infrared domain. Its values are larger as the amount of green vegetation increases. For bare soil, RVI value can be a little more than 1, while for dense vegetation the value can increase to 20.

The Difference Vegetation Index (DVI) which values are between 0 and 1, was proposed later [3].

$$\mathrm{DVI} = \mathrm{NIR} - \mathrm{Red} \tag{2}$$

One of the most used indices calculated from multispectral or hyperspectral information is the Normalized Difference Vegetation Index (NDVI) [4] having a typical range of values between 0.1 for bare soil and 0.9 for dense vegetation.

$$\mathrm{NDVI} = \frac{(\mathrm{NIR} - \mathrm{Red})}{(\mathrm{NIR} + \mathrm{Red})} \tag{3}$$

Comparing NDVI and RVI, NDVI is more sensitive at low levels of vegetative cover, whereas the RVI emphasize variations in dense canopies. NDVI measurements can be perturbed by soil effects. In order to address these limitations Huete [5] introduced the Soil-Adjusted Vegetation Index (SAVI) which can be expressed as follows:

$$\mathrm{SAVI} = \frac{(1 + \mathrm{L})(\mathrm{NIR} - \mathrm{Red})}{\mathrm{NIR} + \mathrm{Red} + \mathrm{L}} \tag{4}$$

where L is the soil conditioning index, which ranges from 0 to 1 depending on the level of vegetation cover.

In order to minimize the influence of soil brightness, Optimized Soil-Adjusted Vegetation Index (OSAVI) was defined [6]. OSAVI is recommended to analyze crops in early to mid-growth stages.

$$OSAVI = \frac{(NIR - Red)}{(NIR + Red + 0.16)}(1 + 0.16) \tag{5}$$

Vegetation indices can be averaged over time in order to establish growing conditions for a given time of the year. By studying the time dependence of vegetation indices we can reveal vegetation stress as well as the influence of some human activities.

There are different solutions for acquiring these type of information: spectrometers, multi-spectral sensors with photodiodes; monospectral/multispectral cameras with single filter application on the image sensor; cameras with multispectral image sensors; cameras with hyperspectral image sensors.

In this paper is presented a solution for vegetation index estimation in precision farming using custom multispectral camera mounted on Unmanned Aerial Vehicle (UAV).

2 Data Acquisition System with Custom Multispectral Camera

2.1 Hardware Setup

The structure of the data acquisition system is presented in Fig. 1. It consists of a multispectral camera with four wavelengths and a Small Board Computer (SBC) that controls the acquisition process. An UAV is used for carrying them over the investigated area. On Ground Control Station (GCS) flight planning is carried out. This makes the UAV to have a steady height and a parallel path.

The flight plan must take into account the side lap and front lap to get more matches features in the imagery.

The superposition of the two photos is called end lap. Forward lap, also called end lap, is a term used in photogrammetry to describe the overlapping photos taken along the flight path.

Figure 2 illustrates how the two photographs are superimposed. The center of the two photos is separated by the distance B, which is also referred to as the "air base". Each photo covers a surface according to the height flight and sensor characteristics.

Side lap is a term used in photogrammetry describing the overlay of photographs taken by adjacent flight lines (W) (Fig. 3).

The completion of the flight pattern requires a minimum of 70% overlay (side lap and front lap).

The condition is mandatory for post-processing of photographs to obtain georeferenced maps.

During flight time, PIXHAWK autopilot of the UAV transmits to Raspberry Pi 3 SBC, at preset time interval, a trigger pulse signal taking into account the position received from the GPS. This signal is used for triggering the multispectral camera.

Raspberry Pi 3 writes simultaneously in a text file the GPS coordinates with a sampling rate of 10 Hz.

During the flight, images are stored on an SD card. After the flight is finished, the acquired photos are georeferenced on the SBC using the GPS data and downloaded to the GCS using a WiFi connection. Each camera has 1 Mpx resolution and is equipped with a proper interferential filter. Figure 4 presents the transmission factor for different used filters.

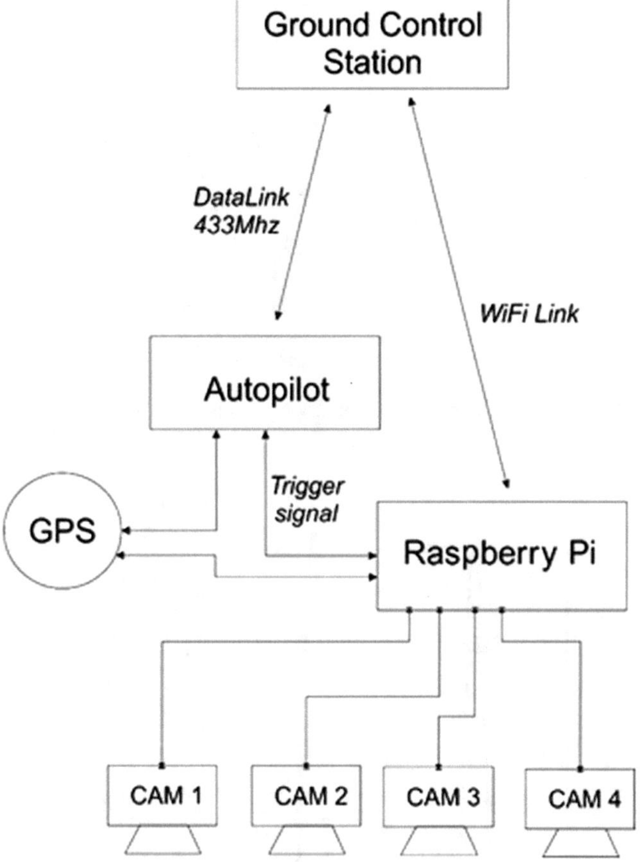

Fig. 1. The structure of the data acquisition system

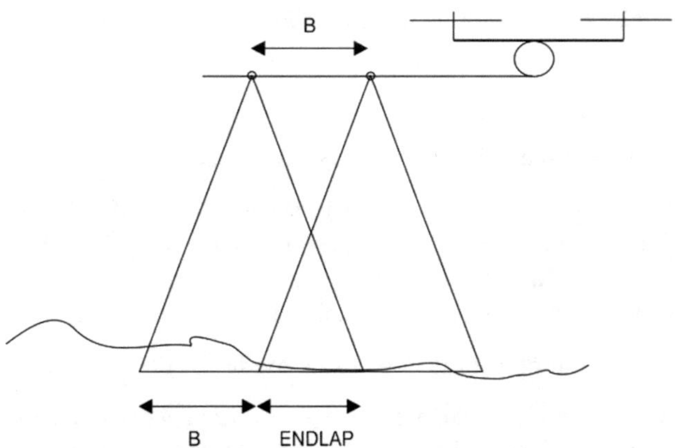

Fig. 2. End lap obtained by overlapping photos taken along the flight path

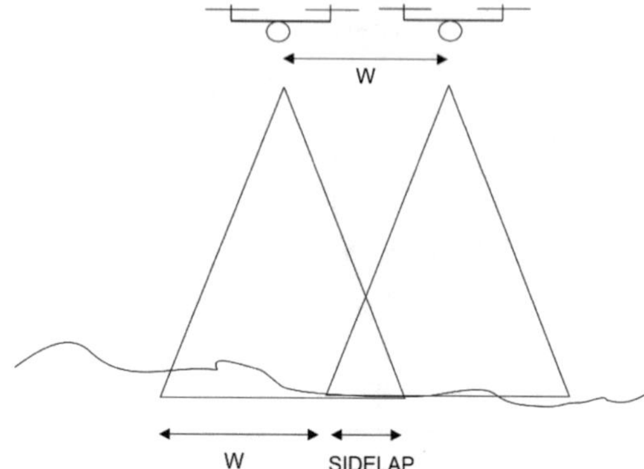

Fig. 3. Side lap—the overlay of photographs taken by adjacent flight lines

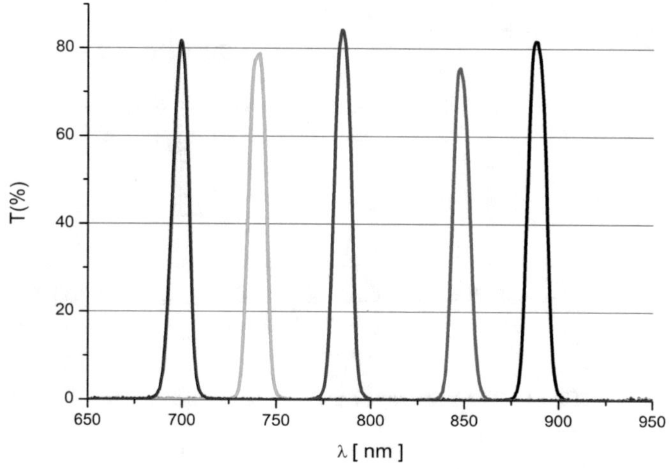

Fig. 4. Transmission factor for different interferential filters

2.2 Software Setup

The software setup consists of the following components:

- Linux operating system running on the Raspberry Pi 3 SBC (Raspbian);
- Custom applications for the image acquisition phase using hardware trigger;
- OpenDroneMap for the generation of the ortophotoplan;
- Grass GIS for image processing and analysis.

The data acquisition process is done using the following software components:

- Programs written in C++ which use the Pylon framework (camera identification, automatic exposure time settings for each camera, main acquisition sequence) [7];
- Ruby scripts which launch the above applications [8];

- Scripts written in Bash and Ruby for filter wavelength settings, GPS synchronization and Geotag) [8, 9].

The main applications are written in C++, based on the Pylon framework [7].

We must associate the filter's wavelength with the camera's serial number, because the device index of the camera may change when we reboot the Raspberry Pi 3. The camera's serial number can be queried with the following function:

```
cameras[i].GetDeviceInfo().GetSerialNumber()
```

In order to establish the optimal value of the exposure time we put the camera in continuous exposure adjustment mode and we read the obtained values for each camera after 10 s.

```
  cameras[i].RegisterConfiguration
(newCAcquireContinuousConfiguration,
RegistrationMode_ReplaceAll, Cleanup_Delete);
  cameras[i].ExposureAuto.SetValue(ExposureAuto
Continuous);
  cameras[i].AutoExposureTimeLowerLimit.SetValue(10);
  cameras[i].AutoExposureTimeUpperLimit.SetValue(500000);
  cameras[i].ExposureTime.GetValue()
```

After we obtain the optimal values of the exposure time we keep them fixed in the main acquisition process. The monochrome cameras have a rather slow adjustment of the exposure time, so for the capturing process we cannot set them in continuous adjustment mode.

```
cameras[i].ExposureAuto.SetValue(ExposureAuto_Off);
cameras[i].ExposureTime.SetValue(t_exp[i]);
```

The synchronization of the capturing sequence is done using software trigger.

```
  cameras[i].RegisterConfiguration (new
CSoftwareTriggerConfiguration,
RegistrationMode_ReplaceAll, Cleanup_Delete);
  cameras[i].ExecuteSoftwareTrigger();
```

The filename for the captured images was encoded as follows:

flight-number_wavelength_index_date(YYMMDD).jpg

When the Raspberry Pi 3 boots up we automatically start the gpspipe program which will write the GPS track file.

```
#!/bin/bash
/usr/bin/gpspipe -d -r -o /home/pi/camera-data/gps-
track.txt
```

After the image acquisition is done we will write the GPS information in the image file's EXIF data using a Geotag script.

```
#!/bin/bash
/usr/bin/gpsbabel -i nmea -f /home/pi/camera-data/gps-
track.txt -o gpx -F /home/pi/camera-data/gps.gpx &&
/usr/bin/exiftool '-datetimeoriginal<filemodifydate' -
if '(not $datetimeoriginal or ($datetimeoriginal eq
"0000:00:00 00:00:00")) and ($filetype eq "JPEG")' -
overwrite_original /home/pi/camera-data &&
/usr/bin/gpscorrelate --gps /home/pi/camera-
data/gps.gpx -z -f /home/pi/camera-data/*.jpg
```

The above script launches in order the following command line programs: *gps-babel*, *exiftool* and *gpscorrelate* [10].

The images can be transferred for further analysis and processing via SFTP (by using dedicated applications or the command line on UNIX-like systems).

In order to facilitate the acquisition control, we developed a web-based interface (Fig. 5).

The image capturing process involves the following steps:

- First of all, we must verify if date and time were correctly read from the GPS module which is connected via serial interface. Otherwise we must run the GPS synchronization script;
- We verify if all the camera modules are working correctly by running the camera identification program;
- Next we set the values for the filter's wavelength and flight number;
- At this moment we can launch the main acquisition process. The triggering for the image capture is controlled by UAV's autopilot;
- After we stop the image capturing process we may run the Geotag script (a backup of the GPS track will be created for each flight number which allows further Geotag processing).

Using OpenDroneMap for the acquired images, georeferenced orthophotomaps corresponding to the multispectral camera wavelength are generated.

Verificare data

Thu Sep 27 10:53:02 UTC 2018

Sincronizare GPS

Identificare camere

Camera 0: 22801928
Camera 1: 22801929
Camera 2: 22801932
Camera 3: 22801935

Lambda0 (nm): 500
Submit

Lambda1 (nm): 650
Submit

Lambda2 (nm): 800
Submit

Lambda3 (nm): 900
Submit

Nr zbor: 45
Submit

Completed correlation process.
Matched: 12 (12 Exact, 0 Interpolated, 0 Rounded).
Failed: 0 (0 Not matched, 0 Write failure, 0 Too Far,
0 No Date, 0 GPS Already Present.)

Calibrare

Porneste achizitia

Opreste achizitia

Geotag

Shutdown

Fig. 5. Web interface for data acquisition control

Starting from the obtained orthophotomaps corresponding to the optical interference filters of the multispectral camera, vegetation indices are computed and maps are generated.

3 Experimental Workflow

To accomplish this experiment, we started using the Mission Planner flight control interface (Fig. 6).

Fig. 6. Mission plan designed using Mission Planner flight control interface

In order to achieve a correct georeference of orthophotomaps, Ground Control Points (GCP) with known coordinates, have been used (Figs. 7 and 8).

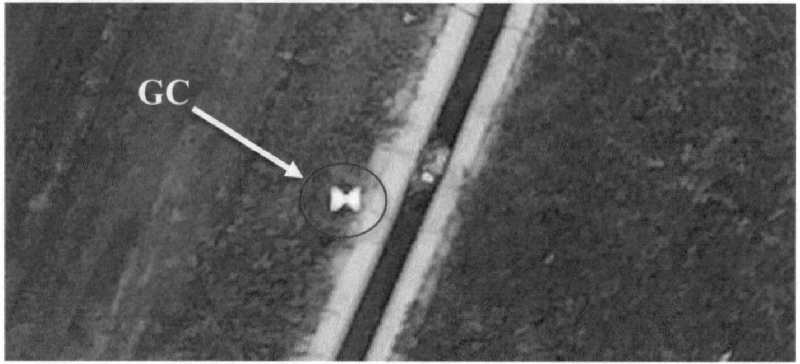

Fig. 7. Ground Control Point with known coordinates

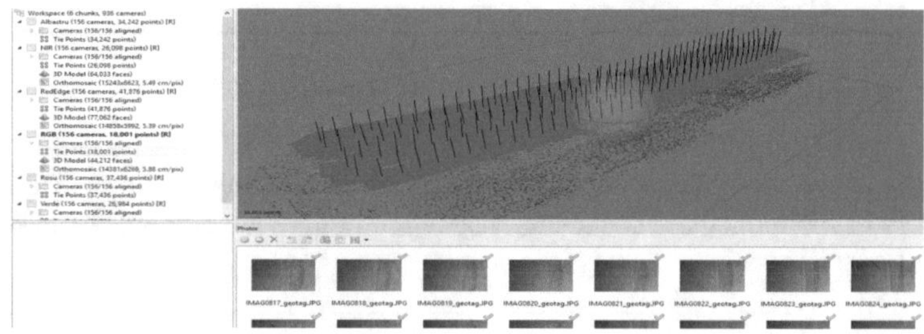

Fig. 8. A number of 156 georeferenced photos for each spectral band were obtained

Using OpenDroneMap for the acquired images, georeferenced orthophotomaps corresponding to the multispectral camera wavelength (NIR, Red Edge, Red + Green + Blue bands) were generated (Fig. 9).

Different types of vegetation indices can be computed and analyzed: OSAVI (Optimized Soil-Adjusted Vegetation Index), NDVI (Normalized Difference Vegetation Index), NDWI (Normalized Difference Water Index), CVI (Chlorophyll Vegetation Index), CCCI (Canopy Chlorophyll Content Index) and TCARI (Transformed Chlorophyll Absorption Reflectance Index).

Starting from the obtained orthophotomaps corresponding to the 4 optical interference filters of the multispectral camera, vegetation index maps are generated (Fig. 10).

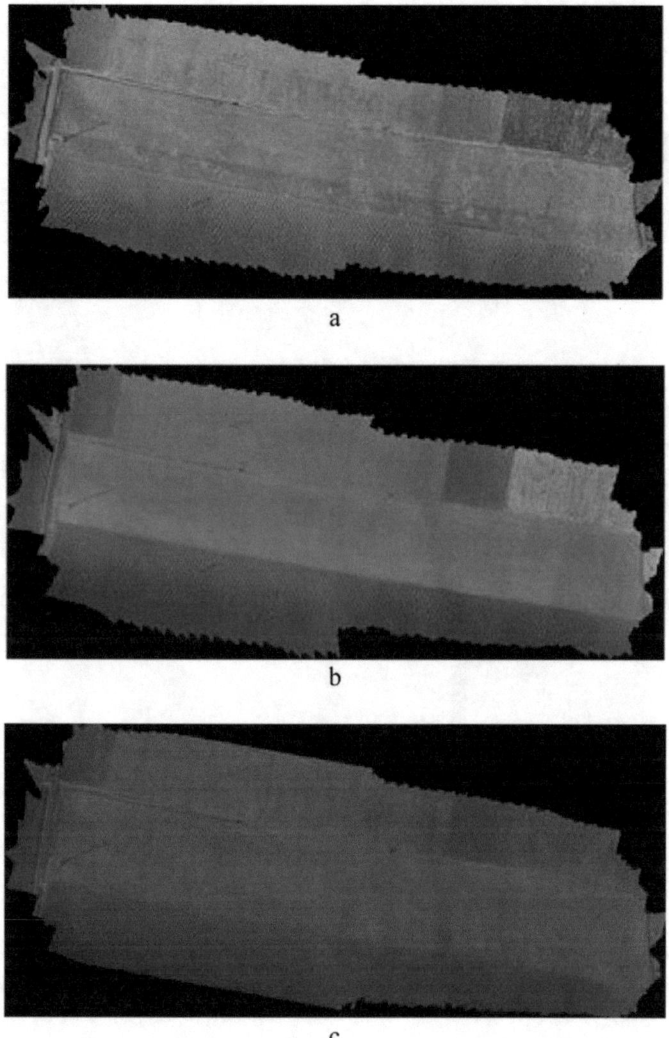

Fig. 9. Georeferenced orthophotomaps corresponding to the multispectral camera wavelengths: **a** Red + Green + Blue; **b** NIR; **c** Red Edge

4 Conclusions

The key outcomes of our study are related to the multispectral camera that has a modular design, offering flexibility in changing the desired wavelengths and also to the data processing solutions.

The proposed multispectral camera system allows an easy evaluation of the vegetation indices used in precision farming. The main advantages take into account the low cost, small size, remote data access and easy further development and/or

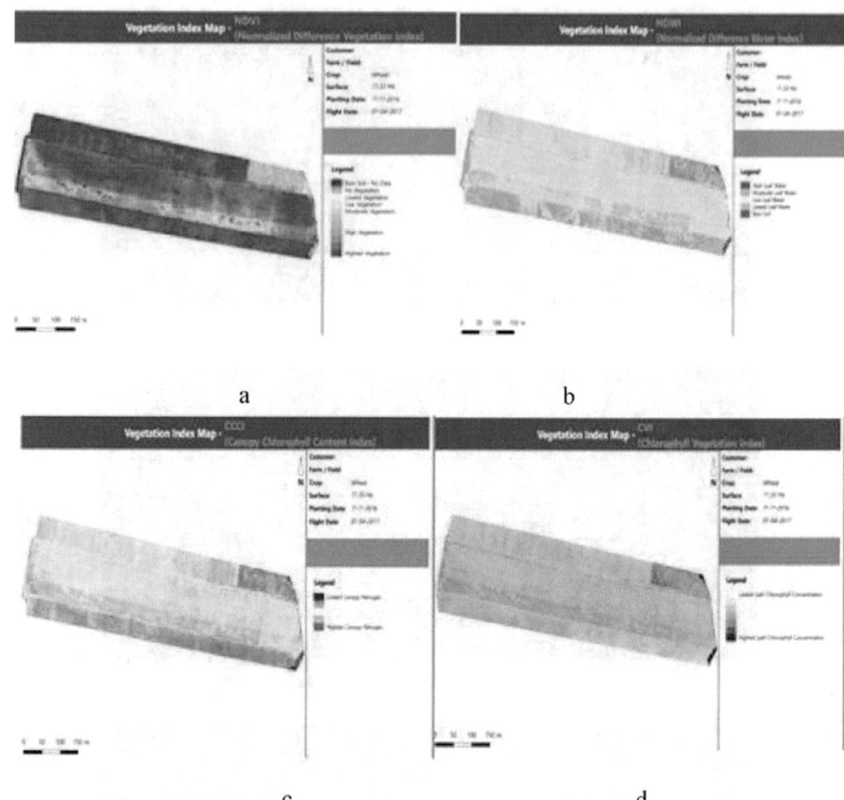

Fig. 10. Different vegetation index maps obtained after data processing: **a** NDVI; **b** NDWI; **c** CCCI; **d** CVI

modifications based on using open-source software. The obtained data can be useful also for specialists in order to monitor and optimize the vegetation process.

Acknowledgements. This paper was supported by a grant of the Romanian National Authority for Scientific Research and Innovation, CNCS/CCCDI—UEFISCDI, project number PN-III-P2-2.1-BG-2016-0132, within PNCDI III.

References

1. Xue, J., Su, B.: Significant remote sensing vegetation indices: a review of developments and applications. J. Sens. (2017)
2. Jordan, C.F.: Derivation of leaf-area index from quality of light on the forest floor. Ecology **50**(4), 663–666 (1969)
3. Richardson, A.J., Wiegand, C.L.: Distinguishing vegetation from soil background information. Photogram. Eng. Remote Sens. **43**(12), 1541–1552 (1977)

4. Rouse, J., Haas, R., Schell, J., Deering, D.: Monitoring vegetation systems in the great plains with ERTS. In: Third ERTS Symposium, NASA, pp. 309–317 (1973)
5. Huete, A.R.: A soil-adjusted vegetation index (SAVI). Remote Sens. Environ. **25**(3), 295–309 (1988)
6. Rondeaux, G., Steven, M., Baret, F.: Optimization of Soil-Adjusted Vegetation Indices. Remote Sens. Environ. **55**, 95–107 (1996)
7. Basler Pylon Camera Software Suite: https://www.baslerweb.com/en/products/software/basler-pylon-camera-software-suite
8. Thomas, D., Hunt, A., Fowler, C.: Programming Ruby: The Pragmatic Programmers' Guide. Pragmatic Bookshelf, Raleigh, NC (2005)
9. Newham, C., Rosenblatt, B.: Learning the bash shell: Unix shell programming. O'Reilly Media Inc, California (2005)
10. GPS Correlate application: https://github.com/freefoote/gpscorrelate

MTLinki Integration and ZDT

S. Santosh Kumar(✉), Santrupti M. Bagali, Nitin Sharma, Mallik Arjuna, V. Sabarish, S. P. Subramanya Swaroop, and K. Narmadha

B.M.S. Institute of Technology and Management, Bengaluru, KA 560064, India
{santoshks714, santruptibagali, spswaroop19, narmadha12400}@gmail.com, nitinn@goa.bits-pilani.ac.in, mallikarjuna@electronosolutions.com, info@electronsolutions.com

Abstract. MTLinki integration server are suites of operational management software. The suites enable to monitor multiple machines and collect their operational production results easily, supporting you to manage the factory effectively. The software creates a web screen and a URL address, persons with authority and skill can view these screens on their PC or tablet. MTLinki can make routine operations in industries faster. Monitoring operations can be done remotely. It can also analyse operational results. MTLinki can group multiple pieces of equipment such as CNC machines and robot controllers into unites, like process steps or line, and represent a wide range of data visually.

1 Introduction

The objective is to support the concept of Industry 4.0 and make industry operations smarter.

This can be implemented by integrating MTLinki and CNC machines. There should be minimum difference in time between the machine stopping and manufacturing process. Today there is a need to make all industry operations to be monitored and controlled remotely. These reduce cost for the companies as well as time and prevents any damages to the machines.

The concept of ZDT(Zero Downtime) can be approached.

2 Approach

With plants operating at high levels of capacity, unexpected downtime is expensive. The FANUC and Cisco developed ZDT (Zero Downtime) offers the ability to proactively detect and inform customers of potential equipment issues before unexpected

M. E. Auer and K. Ram B. (Eds.): REV2019 2019, LNNS 80, pp. 686–697, 2020.
https://doi.org/10.1007/978-3-030-23162-0_62

downtime occurs. It provides advanced analytics and reporting to help optimise equipment utilization in areas such as the sending of smart maintenance notifications to extend equipment life and optimise maintenance costs. Overall, this offer allows users to enhance their technical support services to increase productivity and overall customer satisfaction (Fig. 1).

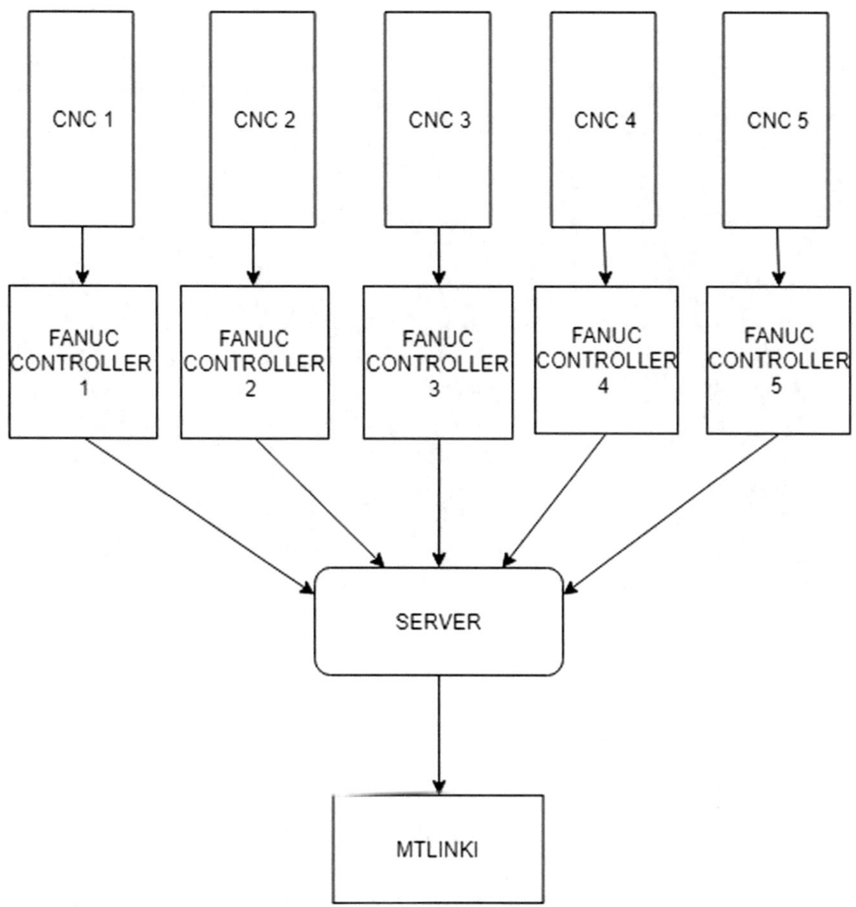

Fig. 1. The system: MTLinki integration

3 Working

The whole purpose of this is to allow users to monitor devices within a factory (to look at statuses, data history, log data, etc.), to be able to connect, collect, communicate and monitor each device individually for diagnosis and analysis. Not only FANUC devices can be connected, but also third-party devices that support OPC UA (Unified Architecture). This also allows more flexibility to connect to open management software such as MES (Manufacturing Efficiency System) or ERP systems. The target of the FIELD system is connecting each device within a factory, but also allowing flexibility to connect to upper host systems such as ERP (Enterprise Resource Planning), SCM (Supply Chain Management) and MES (Manufacturing Execution Systems).

Basic idea is to make smart machines smarter and interact with other smart machines.

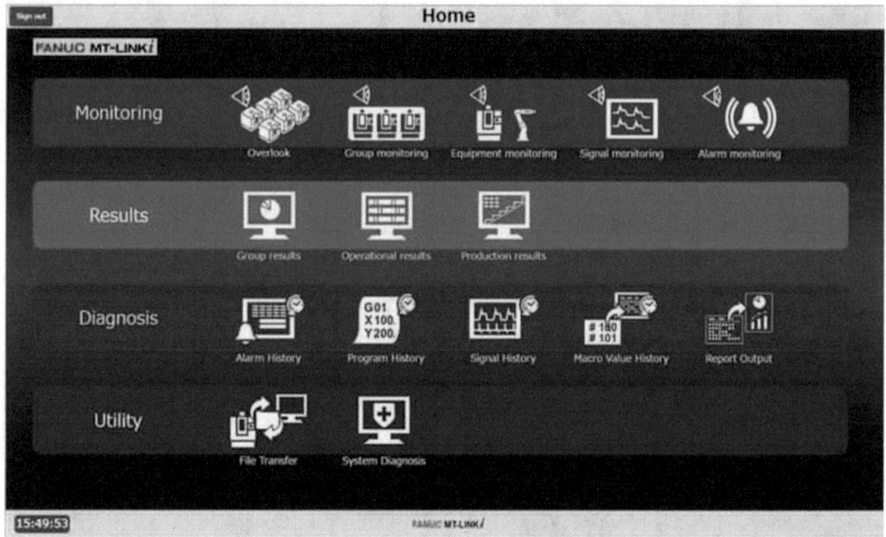

4 Monitoring

4.1 Overlook Screen

The overlook screen, enables us to monitor the state of the whole factory in real time so that you can recognise the states of the machines and promptly start the recovery work when abnormal events occur. When warnings occurs on the machines in the factory, this function alters you of the warning events by blinking the machine icons on the overlook screen. The screen also has the information of the machines on the current layout is displayed and listing of number of machines by each state and pie graph shows the proportion of each state is displayed. The bottom part also shows the status of total registered machines is displayed.

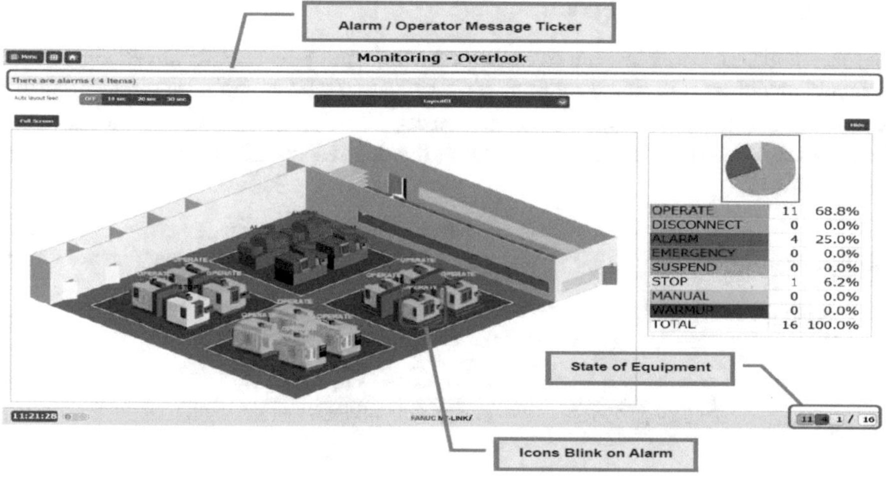

4.2 Equipment Monitoring Screen

The equipment monitoring screen displayed the detailed information of the machines such as current state of the machine, operational results, production results and alarm history. An operator message ticker appear on top of the page when alarm of operator message are issued. Clicking on the alarm or operation ticker moves you to the real time alarm monitor screen that displays detailed information on the current alarms or operator messages.

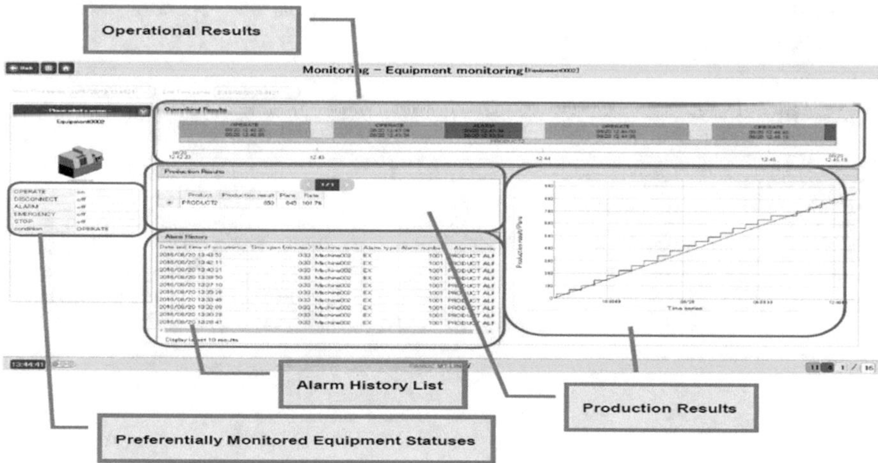

4.3 Signal Monitoring Screen

Signal monitoring enables you to monitor in real time how signals are changing on a selected machines. On signal monitor screen you can display the status transitions in 5 min and maximum of 3 signals on the screen, allowing you to view simultaneously how correlated multiple signals ware. All of the defined signals and machine status can be monitored.

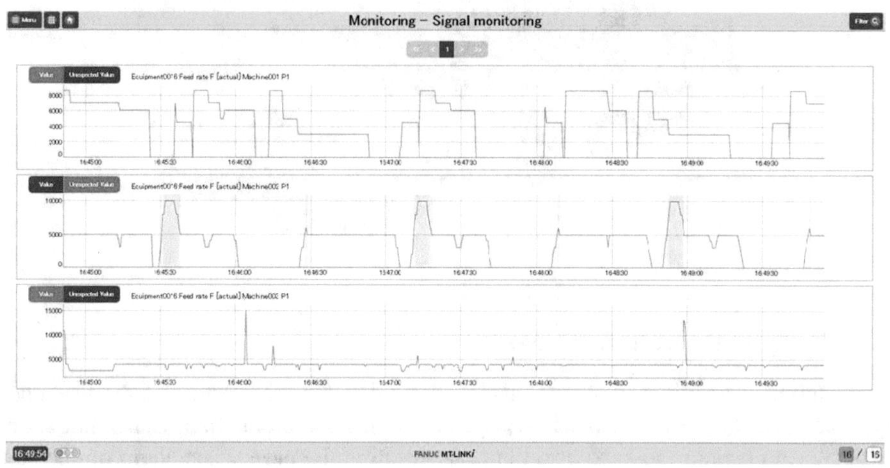

4.4 Alarm Monitoring Screen

Alarm monitoring screen enables you to monitor alarms and operator information occurring on the machines in real time.

Menu Monitoring - Alarm monitoring

Date and time of occurrence	Equipment name	Machine name	Alarm type	Alarm number	Alarm message
2017/08/10 13:54:13	OP-30A	OP-30A	EX	1003	EMERGENCY STOP.
2017/08/10 13:50:30	a-T14iFa Robodrill	a-T14iFa_Robodrill	EX	1003	EMERGENCY STOP.
2017/08/10 09:50:13	OP-30A	OP-30A	OPR	2037	PERIODICAL MAINTENANCE TIME.
2017/08/10 09:50:13	a-T14iFa Robodrill	a-T14iFa_Robodrill	OPR	2037	PERIODICAL MAINTENANCE TIME.
2017/08/09 20:15:16	FS32i	FS32i	OPR	2010	CABINET TEMPERATURE HIGH (X17.0)
2017/08/09 20:15:16	FS32i	FS32i	SV	1067	FSSB:CONFIGURATION ERROR(SOFT)
2017/08/09 20:15:16	OP-20A	OP-20A	OPR	2010	CABINET TEMPERATURE HIGH (X17.0)
2017/08/09 20:15:16	OP-20A	OP-20A	SV	1067	FSSB:CONFIGURATION ERROR(SOFT)

5 Results on Monitor

5.1 Operational Result Screen

Operational results screen provides you with a graph per machine which displays the operational results of different states such as operate, stop, suspend and alarm. You can also collect product name and store complementary information along with those operational results by manually inputting work comments related to the machines.

5.2 Production Results

Production results screen gives the information about productional results and the product plan of the machines on a graph and table. You can easily access the operational results per machine.

6 Diagnosis

6.1 Alarm History

Alarm history screen gives us the history of alarms and when they have occurred. Machine side alarm and operator messages, FANUC C&C alarms, custom marco alarms and messages are captured.

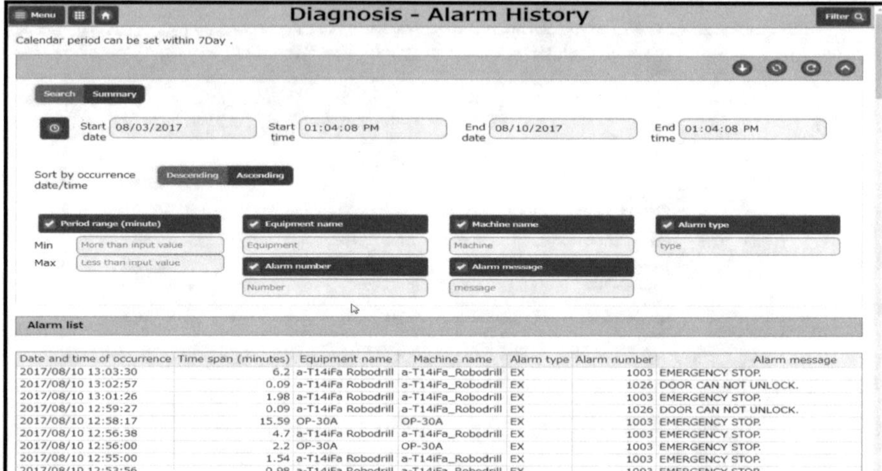

6.2 Program History

Program history function helps us to view the history of when main running programs have started and ended and how long cycle time was. This information can also be put out as a CSV file.

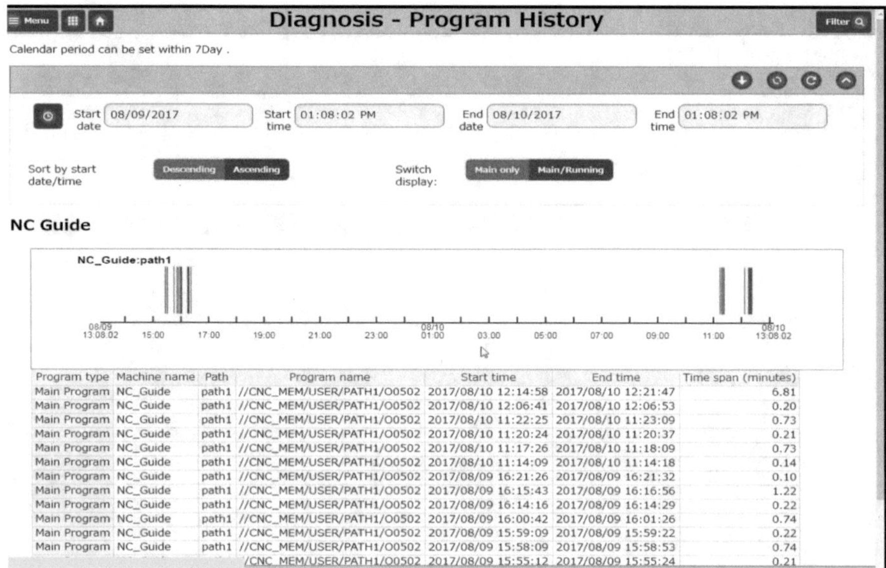

6.3 Macro Variable History

This function monitors given marco variable values and stores our values. Also for macro variables that preserve the measured data of works and tools, this history function can be used to collect and store those data. The names of the macro variables can be displayed with optional name, this information can be output as a CSV file.

Machine status and production information signals, alarms and macro variable history are stored into a Mongo database which has an open architecture and users are encouraged to develop their own custom applications. There is also a file transfer utility. This utility enables you to upload and download part programs. CNC parameters work and tool offsets marco variables and operator history data.

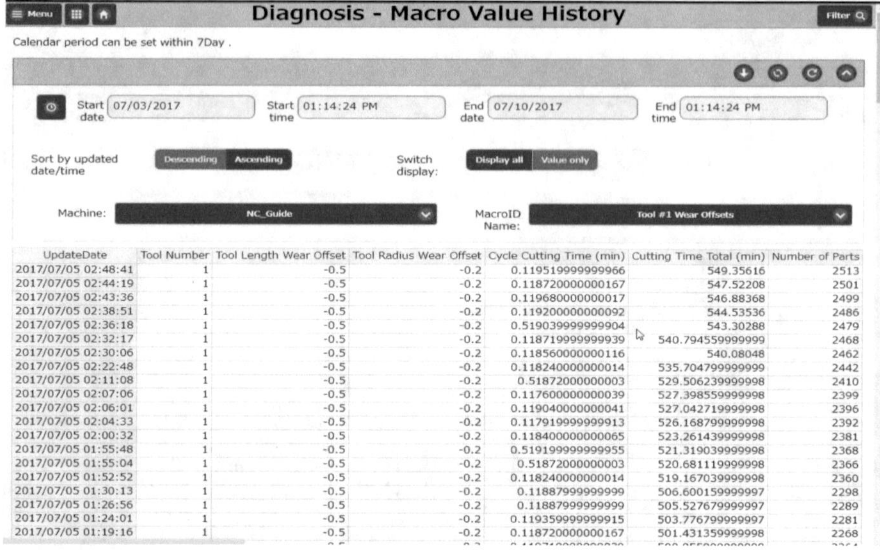

UpdateDate	Tool Number	Tool Length Wear Offset	Tool Radius Wear Offset	Cycle Cutting Time (min)	Cutting Time Total (min)	Number of Parts
2017/07/05 02:48:41	1	-0.5	-0.2	0.119519999999966	549.35616	2513
2017/07/05 02:44:19	1	-0.5	-0.2	0.118720000000167	547.52208	2501
2017/07/05 02:43:36	1	-0.5	-0.2	0.119680000000017	546.88368	2499
2017/07/05 02:38:51	1	-0.5	-0.2	0.119200000000092	544.53536	2486
2017/07/05 02:36:18	1	-0.5	-0.2	0.519039999999904	543.30288	2479
2017/07/05 02:32:17	1	-0.5	-0.2	0.118719999999939	540.794559999999	2468
2017/07/05 02:30:06	1	-0.5	-0.2	0.118560000000116	540.08048	2462
2017/07/05 02:22:48	1	-0.5	-0.2	0.118240000000014	535.704799999999	2442
2017/07/05 02:11:08	1	-0.5	-0.2	0.51872000000003	529.506239999998	2410
2017/07/05 02:07:06	1	-0.5	-0.2	0.117600000000039	527.398559999998	2399
2017/07/05 02:06:01	1	-0.5	-0.2	0.119040000000041	527.042719999998	2396
2017/07/05 02:04:33	1	-0.5	-0.2	0.117919999999913	526.168799999998	2392
2017/07/05 02:00:32	1	-0.5	-0.2	0.118400000000065	523.261439999998	2381
2017/07/05 01:55:48	1	-0.5	-0.2	0.519199999999955	521.319039999998	2368
2017/07/05 01:55:04	1	-0.5	-0.2	0.51872000000003	520.681119999998	2366
2017/07/05 01:52:52	1	-0.5	-0.2	0.118240000000014	519.167039999998	2360
2017/07/05 01:30:13	1	-0.5	-0.2	0.11887999999999	506.600159999997	2298
2017/07/05 01:26:56	1	-0.5	-0.2	0.11887999999999	505.527679999997	2289
2017/07/05 01:24:01	1	-0.5	-0.2	0.119359999999915	503.776799999997	2281
2017/07/05 01:19:16	1	-0.5	-0.2	0.118720000000167	501.431359999998	2268

6.4 Signal History

Signal history function helps us to view the status of the machine signals, up to 5 signals can be displayed in the histories of status transition of the relative signals can be checked simultaneously.

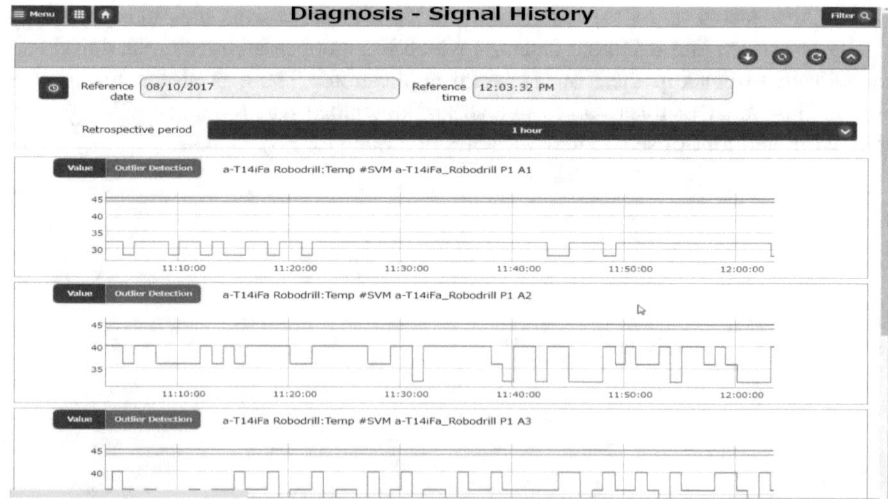

7 Embedded Ethernet Port

The embedded Ethernet function can be used by selecting the Embedded Ethernet port.

IP ADDRESS: Specifies the IP address of the Embedded Ethernet.
SUBNET MASK: Specifies a mask address for the IP address of the network.
ROUTER IP ADDRESS: Specifies the IP address of the router.
MAC ADDRESS: Specifies Embedded Ethernet MAC address.
AVALIABLE DEVICES: Embedded devices of the Embedded Ethernet.

There is also a basic Internet program called **PING** that allows the user to verify if the device (CNC's) is connected to the FANUC controller by ethernet.

We can also connect more than one CNC's to the FANUC controller.

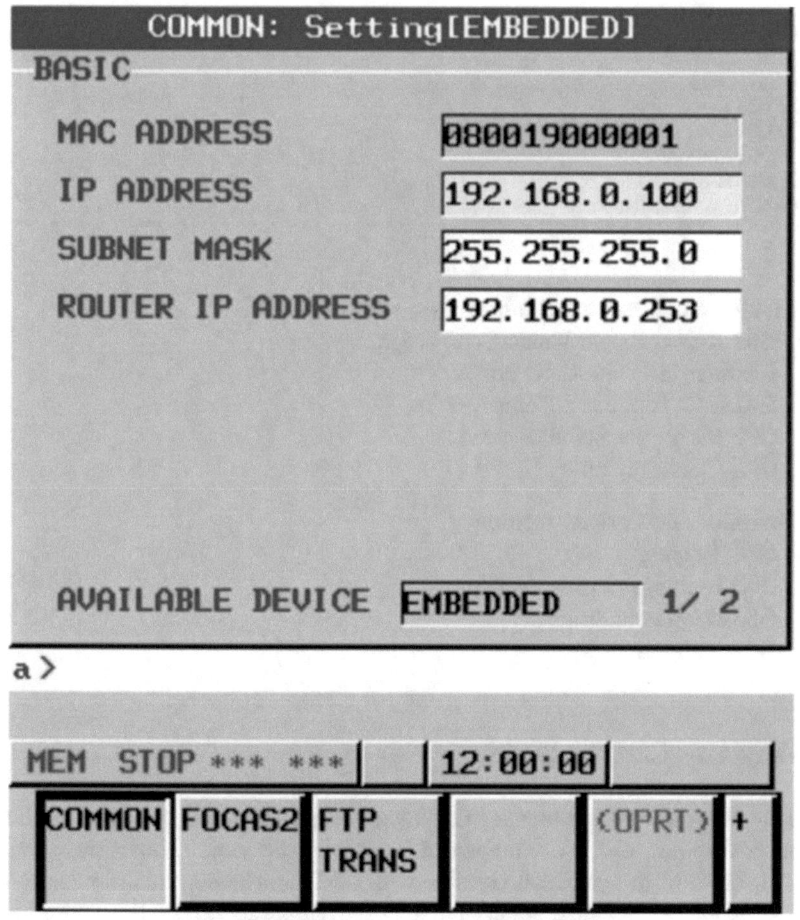

COMMON screen (BASIC)

7.1 Fanuc CNC Program [1, 2]

CNC Machining is a process used in the manufacturing sector that involves the use of computers to control machine tools. Tools that can be controlled in this manner include lathes, mills, routers and grinders. The CNC stands for Computer Numerical Control.

Example:

This is a how a typical Fanuc CNC program looks like:

```
N10 T2
N20 G92 S1200 M42
N30 G96 S150 M04
N40 G00 X-1 Z5 M08
N50 G01 Z0 G42 F0.2
N60 G01 X24 C2
N70 G01 Z-28
N80 G01 X32 Z-55
N90 G01 Z-56
N100 G02 X40 Z-60 R4
N110 G01 Z-75
N120 G01 X60 G45
N130 G00 X150 Z100
N140 M30
```

The above code uses the following:

- G02 Arc/Radius in Fanuc CNC program
- Chamfer in Fanuc CNC program
- G42/G40 Tool Nose Compensation
- G92 Maximum Spindle Speed
- G96 Constant Cutting Speed.

Some more operations include:

- G84 Tapping
- G73 High speed peck drilling
- G83 Deep hole drilling
- G81 Drilling.

8 Why Use CNC's? [1, 2]

There are many advantages to using CNC Machining. The process is more precise than manual machining, and can be repeated in exactly the same manner over and over again. Because of the precision possible with CNC Machining, this process can produce complex shapes that would be almost impossible to achieve with manual machining. CNC Machining is used in the production of many complex three-dimensional shapes. It is because of these qualities that CNC Machining is used in jobs that need a high level of precision or very repetitive tasks.

9 Outcomes

The live data of the current operations of an industry can be viewed, data for years together can be collected, this can be used for predictive analysis and prevent any accidents or damages, machine wear and tear can also be monitored, machine life can be increased, cost of manufacturing for the company can be largely reduced with the use of these Fanuc controllers with MTLinki.

Also, the physical type of monitoring the machines in an industry will not be needed.

All these can be done and the work is in progress

10 Conclusion

This MTLinki integration with CNC machines has provided a unique combination of solutions to many industries and the work is in progress for various operations and management of this system.

This can be a major part in the network of Industry 4.0.

The overall manufacturing cost can be reduced when implemented in real time.

References

1. Zuperl, U., Cus, F., Milfelner, M.: Fuzzy control strategy for an adaptive force control in end-milling. J. Mater. Process. Technol. **164**, 1472–1478 (2005)
2. Benardos, P., Vosniakos, G.-C.: Offline flexible optimisation of feed and speed in computer numerical control machining of sculptured surfaces exploiting dedicated cutting force metamodels. Proc. Inst. Mech. Eng. Part B J. Eng. Manuf. **228**, 878–892 (2014)

Work-in-Progress: Contemporary Barriers Faced by Precision Agriculture, New Paradigms and Proposals for Future Advance

Alin Cosma, Luciana Cristea, Constantin Sorin Zamfira(✉), and Marius Cristian Luculescu

Transilvania University of Brașov, Eroilor 29, CP 500036, Brașov, Romania
{alin.cosma, lcristea, zamfira, lucmar}@unitbv.ro

Abstract. Heightening concerns are surrounding world food capacity, especially in the context of modern demands in large-scale human activities. Precision agriculture has a prime role in optimizing the management of crops, but is currently blocked by a series of technical and economical obstacles. This paper endeavors to summarize and critically analyze the state-of-the-art in precision agriculture, while at the same time highlighting the current limitations and unknowns the research community is facing. It also contains suggestions for the best development directions for the future. The findings conclude that genuine progress will be undoubtedly based on large-scale multi-disciplinary research endeavors rather than individual effort, standardization, and development of agriculture-specific methods and equipment, as opposed to reusing existing technology. The most potent and promising approaches are identified and original contribution in the form of novel ideas is presented.

Keywords: Precision agriculture · Remote sensing · Vegetation index

1 Introduction

Lately, heightening concerns are placing the food problem at the top of global efforts. The challenge of alimentation is more and more pregnant in the modern world, especially in the context of other vital concerns. Agriculture seems pressed from all sides, and is required to increase its output even with decreasing inputs. Agriculture's inherent natural governing laws (the needed space, vegetation cycles, dependency on other natural resources, some uncontrollable: light, water, warmth etc.), together with modern concern for environmental preservation, leave small room for improvement.

A modern, intelligent agriculture is an information-based agriculture. A controlled management system requires a rapid and trustworthy method for acquiring information. This has been elusive due to the complexity of the environment that must be modeled and due to the limitations of existing technologies. This would constitute a prime step for an integrated management system, which is at the core of precision agriculture, a key driver of the agriculture of the future.

M. E. Auer and K. Ram B. (Eds.): REV2019 2019, LNNS 80, pp. 698–707, 2020.
https://doi.org/10.1007/978-3-030-23162-0_63

2 Current Trends in Precision Agriculture

Precision agriculture is an information-based approach of production. Its fundamental purpose is optimizing the entire agricultural process, especially regarding the use of soil, water and chemical resources, based on determined local specifics [1]. In other words, it targets localized management, that is, allocation of inputs exactly where, when and inasmuch as needed, thus realizing economic, environmental-friendly and sustainable agriculture.

Broadly, some stages in its process can be distinguished: data acquisition, data processing for spatial variability assessment of parameters, decision-making, and finally in-field application. At present, variable-rate technology seems to be the most advanced part of precision agriculture [2]. However, its implementation is dependent on availability of precise maps that would allow decision taking. Since it is unanimously recognized that the data capture phase is the most limiting at present, a large body of research is dedicated to treating its various aspects. Research is aimed at sensing technology (sensors), at mobility (vehicles), or data processing (software). A large proportion of research is done towards reflectance measurements [3]. Remote sensing offers rapidity, non-invasion, reproducibility and objectivity [4]. It also lends itself to easier automation.

An outlook on emerging technologies is useful together with an analysis of their potential as long-term solutions. Figure 1 outlines the technologies that show the most potential in advancing precision agriculture.

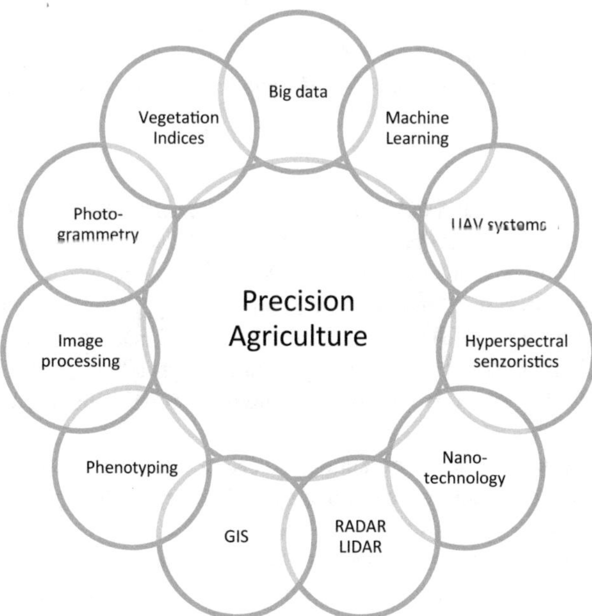

Fig. 1. Enabling and promising technologies for precision agriculture

2.1 Vegetation Indices

These are variables that comport information about the vegetation state of the plants, based on spectral reflectance. Du et al. [5] enumerate some of the correlations found by researchers. The obvious advantage is that this information can be gathered remotely, even from satellites, and the degree of correlation is shown to be very satisfactory. The last half of century has witnessed many such indices being developed, the most known being NDVI (Normalized Difference Vegetation Index), RVI (Ratio Vegetation Index) and DVI (Difference Vegetation Index) [3]. A plethora of studies show promising results in estimating LAI (Leaf area index), biomass, health, hydric stress, thermal stress, turgor, pest influence, vegetation dynamic, photosynthetic capacity, conductance, nutrition and maturity levels, species discrimination etc.

Lately, new sensor technology has opened the door for more advanced vegetation indices, based on narrow-band and hyperspectral or ultraspectral sensors. Hyperspectral data can be used to avoid inter-correlated bands, or devising multi-band indices. Thorp and Tian [3] describe a study that compares 119,805 spectral indices in order to determine the most relevant narrow bands.

Another way to use the cvasi-continuous nature of hyperspectral data is derivative reflectance indices. These employ the first or second-degree derivatives to analyze the slope and curvature of the spectral response. These can be successfully used to eliminate background noise and dependency on absolute levels.

Lussem et al. [6] is focusing on bands in the visible domain, in order to prove the capacity of ordinary RGB cameras to provide useful information for precision agriculture. The results are encouraging and can supplement readings from other sensors.

Thorp and Tian [3] mention an index especially developed for compensating for the influence of soil on readings. This is ever more so a problem where soil coverage is very reduced, however the adjusted NDVI equation shows good results.

2.2 Sensors

From all remote sensors applicable to agriculture (thermographic, chlorophyll, RGB, hyperspectral, LIDAR—light detection and ranging, RADAR etc.), spectral ones have the best potential and offer the possibility of identifying the pathogen source. Evolution has taken them from panchromatic to multispectral, hyperspectral and ultraspectral, according to the number of distinct spectral bands. Von Bueren et al. [7] compare all the sensor types and conclude that hyperspectral sensors, although the most expensive and complex in use (require calibration, intensive processing etc.), are the most performant.

Regarding the measurement principle, we can distinguish a few types of sensors, according to [4]: non-imaging, and imaging, which comprises push-broom, whisk-broom, filter and snapshot sensors.

Plant morphology can also provide valuable information about their state. Presently LIDAR and Structure-from-Motion (SfM) are used to estimate the density, the internal structure, the height, the population, biomass etc. [8].

de Paul Odabe et al. [9] make another differentiation between passive and active sensors. Most of the existing solutions are passive, i.e. they record existing radiation from the environment. Active sensors also contain a controlled radiation source. The study only mentions LIDAR and RADAR sensors as active sensors.

2.3 Vehicles

A major problem that is addressed is the way in which acquisition is done, namely the method or vehicle that carries the sensor. Since we are talking about large-scale applicability, the measurements are by nature dynamic, and as such, their quality is inextricably linked to that of the carrier equipment. The main vehicles used are manual systems, terrestrial systems, airborne systems, and spaceborne systems [4]. The winner seem to be UAV-s, especially drones.

Recently, UAV-based imagistic systems (UAS—Unmanned aerial system) are gathering more and more attention [10]. The technology seems to be mature enough to provide the default vehicle for smart agriculture. Nasi et al. [11] compare the performances of diverse vehicle types. Results suggest that amongst satellite, drone and aircraft, the satellite has the weakest results in estimating biomass. The aircraft comes just below the drone, which attains results that are above most similar systems.

2.4 Software

Another major side of an integrated agricultural process is the software component. It and influences all the steps of precision agriculture. A good software approach can result in compensating the drawbacks of other elements, or enhance the overall result. The computational demands are high, and all the steps involved in processing, like database storage and management, geo-referencing, filtering, interpolation, correction, sensor fusion, map construction and display, or the decision-support system based on artificial intelligence, all stress to the edge the existing software technologies.

Data sets incoming from modern spectral cameras are continuously increasing in complexity. The system must easily handle terabytes of data in a timely fashion, which raises problems for conventional procedures. Lately a deeper focus towards sensor technology is seen in this research field, leaving agriculture-dedicated analysis techniques wanting. The need is more and more felt for specific solutions, developed organically and directionally for precision agriculture. However, [12] remarks growing efforts in the last years and summarizes the current state of this domain.

Machine learning has been assessed and proposed as one of the most potent solutions [13]. The study shows good results in applying this artificial intelligence technique for detecting pathogen effects on plants.

Another useful feat is discrimination between different species, soil and materials etc., also called segmentation. Padua et al. [14] enumerate different methods with which it can be accomplished, like thresholding, via vegetation indices, Hough transform, skeletization, machine learning, K-means vector quantization, FFT analysis, Gabor filters or object-oriented image analysis, and applies them to successfully differentiate individual trees.

Thomas et al. [4] mention also SVM (Support Machine Vector) as a common method of hyperspectral investigation, mainly for its extraordinary capacity of producing good thresholds for feature distinction, even with a reduced training set. A study shows reliable identification of foliar disease before it is visible to the human eye.

Preprocessing for noise elimination is another widely used function of software algorithms, like running-average, Savitzky-Golay smoothing, mean and median filtering [15]. Also derivative analysis is useful for suppressing external influence.

Shi et al. [8] present a summary of used or forecasted technologies for smart agriculture, grouped on a Gartner hype cycle (see Fig. 2). Although in some cases technologies are somewhat mature on their own, their integration into a joint solution is in most cases a domain of discussions.

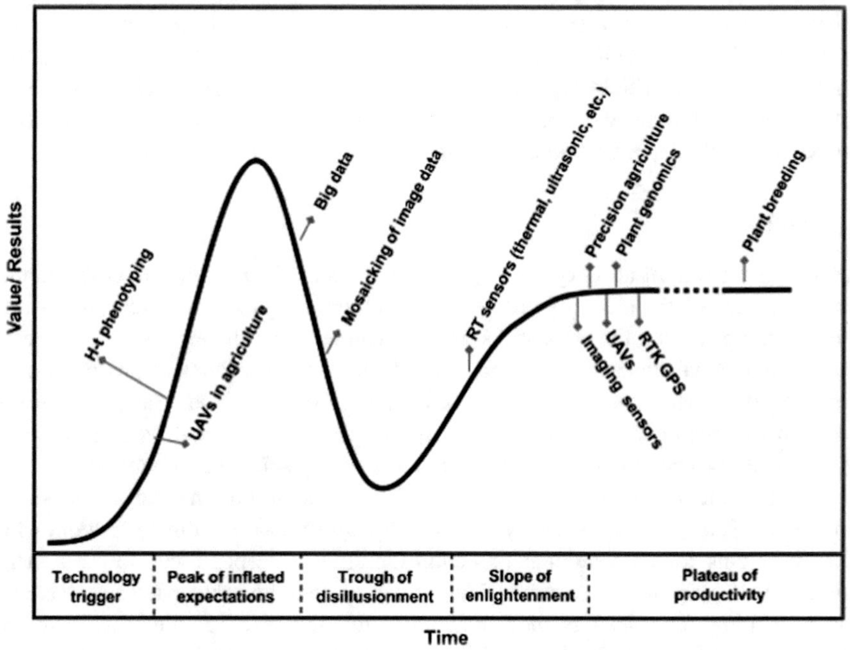

Fig. 2. Smart agriculture technologies [8]

3 The Biggest Challenges Faced

Following this centralization of efforts done in the field of precision agriculture, some common trends and obstacles emerge. Clarifying them will shed new light on the directions to take and encourage further a clearer development and collaboration.

Surveying the literature, one can see that the most mature systems are based on hyperspectral imaging sensors mounted on UAVs. Following, we treat mostly the obstacles found in this case, as it is the most advanced and promising of precision agriculture. However, UAV systems come with some intrinsic disadvantages, like noise issues due to the long atmospheric transmission path, weight and power restrains, geometrical distortions, vibration, spectral ambiguity caused by a larger field of view, low autonomy, high initial cost, robustness, etc. [2, 9].

Globally an overarching impediment in precision agriculture, is the lack of confluence of disciplines. The complexity and costs of the equipment and technologies make this kind of research projects impractical for individual efforts. Shi et al. [8] see until now a lack of collaboration of such proportion, and accentuate the necessity of a sustained effort for organizing an inter-disciplinary project.

3.1 Standardization

As mentioned, one of the great challenges for remote sensing in agriculture is the universalization/standardization of solutions. Many solutions are studied and implemented for specific soils, species, cultures, sensors, mobile platform etc., and when tried out in slightly different conditions the outcomes are disappointing.

Zhang and Kovacs [2] also see this need, especially regarding large-scale image processing procedures (e.g. georeferencing). Ge et al. [15] observed this standardization problem when it comes to sensors. It is technically difficult to respect the mechanical and cost requirements, and also have rich information, although the technology is headed towards powerful sensors that are well portable on an UAV for example.

3.2 Software

One of the major problems in agricultural applications is the sheer volume of data that can be produced with modern sensors, considering the inherently complex processing that has to be done. The amount of data is a factor that slowed the optimization and standardization of processing [16]. By nature, the data volume is terabytes in magnitude and the algorithms are computationally intensive. The issue is felt even more in mobile real-time applications, where the entire processing and reaction must be done in situ, and resources are scarcer (battery life, weight, time, cost etc.).

Kersting et al. [12], by looking at these growing demands, point to the inexistence of software technologies dedicated to precision agriculture. On the contrary, we remark rather a reutilization of technologies developed for other disciplines.

Solving these issues would pave the way for a fluidization of the interpretation process. Palaniswami et al. [17] also argue for finding automatic, generic and mechanized solution for the analysis and interpretation of data. However, this is closely tied to the uniformization of inputs, so the same requirements span to input-capture systems.

3.3 Sensors and Vehicles

Integrating all the steps into one seamless process is a hot topic of precision agriculture. Because the efficiency of the entire agricultural process is given by the quality of the constituent stages, in insuring a proper result we must talk about an equalization of performances. Again, the largest limitation seems to be the data-capture phase. An acquisition system must provide relevant and clean data, feature good spatial, temporal, spectral and radiometric resolution, as defined by [18], be non-invasive, fast, cost-effective, have autonomy etc. It can safely be asserted that sensor technology is mature in most of these individually, but a solution combining all of them is currently missing.

In general, the variations that are present naturally in the environment produce many unknowns and variations in the readings. This is a big obstacle in the face of the uniformity mentioned. The spectral response of a culture is vastly different from that of a single leaf. Shadows, angle of illumination, the background influence (soil), leaf angle, differences between species, atmospheric conditions, vegetation density, sensor calibration, measurement conditions (height, method, sensor etc.), processing methods, phenotype variability and many other variables produce a subject that is hard to model accurately and that causes incongruences between studies [13, 19].

Other issues encountered when utilizing reflectance readings are the calibration of the instrument, vignette effect, band interference, mosaic effect [2]. These variations in readings produce the non-uniformity mentioned before, and inconsistency between researchers is impeding forward progression by building on others' foundation.

Alchanatis et al. [20] stress the issue of natural illumination, which is a source of high variability. It is not always available, it varies in intensity, angle and spectral composition, it's not controllable and it produces shadows and other reading artefacts.

Du et al. [5] mention a pervasive yet untreated problem—the differences in accuracy for identical spectral indices, but derived with different sensors. The study shows that finding the best mathematical formulas for indices is not enough, if not considering the spectral scale effect, i.e. the peculiarities of the spectral response of each sensor.

4 Most Promising Technologies and Future Directions

Precision agriculture stands at the confluence of engineering disciplines, and only large-scale and inter-disciplinary projects can propel the state-of-the-art.

Regarding the lack of standardization, [21] proposes a general, modular and standard SW architecture for mobile acquisition systems. Standardization is closely linked to the collaboration issue described above. Lack of peculiar algorithms is also dependent on collaboration and standardization. They all go together and require large-scale efforts.

The big-data issue is not tackled directly by studies, yet beyond powerful modern computing technology, standardization is the key to handling complexity. We suggest that this problem will find a natural solution once standardization is reached.

Another topic that is highly connected with standardization is fluidization. Shi et al. [8] demonstrate that integration of the sensors, flying crafts and informatics system for data processing, although difficult, is possible. The study proposes as a solution an integrated system that has a cycle of 48 h for obtaining a complete crop mosaic.

The inconsistency of the readings is addressed by [5], which studies the spectral scale effect, and proposes a new vegetation index (VIUPD—Vegetation Index of the Universal Pattern Decomposition) which is not affected by the sensor used. The results validate its independence to the sensor. Regarding consistence in processing, [22] explains that most of the approaches involving Machine Learning have started to build a new model from scratch. The study brings forth a novel Machine Learning approach

—Unsupervised Domain Adaptation—to transfer the model based on one species to another. The results show the capacity to detect hydric stress in incipient phase by reutilizing a model developed for another species.

One of the topics that gets most attention is good-performance sensors, as it is one of the most limited technical sides. Ishida et al. [13] developed a novel way of using a monochrome sensor for gathering a vast quantity of data. The study proposes and tests the use of a LCTF (Liquid Crystal Tunable Filter), of which the center bandwidth is electronically controllable. This also results in a simpler processing of the final data.

Chakravortty and Bhondekar [23] implemented and tested a software solution to mediate the shortcomings of the hyperspectral sensors. The algorithms works on neural networks to correlate readings from a hyperspectral sensor with those from a multi-spectral sensor. The correlation rate of the fusioned images was above 90% for all bandwidths, and the resulting spectral cube had the spectral resolution of the hyper-spectral sensor and the spatial resolution of the multispectral one.

The inherent variability of the environment is a big impediment for standardization. Natural illumination is one of the largest factors. Odabe et al. [9] sees as solution the use of LIDAR and RADAR sensors, but these have limited applicability due to punctual coverage, strict positioning for valid readings, and not providing spectral data.

Moreover, [16] sees as future development the use of active hyperspectral cameras. An active camera is one that contains its own source of electromagnetic energy, precise and controllable. Alchanatis et al. [20] obtain good results with a mobile terrestrial system that blocks any sunlight to obtain data from a much more controlled environment. Symonds et al. [24] also create a static platform that features controlled illumination, to successfully achieve an automatic species discrimination. No application has yet been proposed for UAV active cameras. The obvious reason is the inability to block extern light.

5 Conclusion

In conclusion, the crisis the world is facing will become even more pronounced, if measures are not taken towards a mature, sustainable and technologized agriculture. The largest disruptors identified by the present study are the lack of informatics methods developed specifically for precision agriculture, processing of big data, insufficient collaboration of disciplines for synergic solutions, sensor limitations, low uniformity and standardization leading to low fluidity of the interpretation process, inadequate performance of vehicles (e.g. drones), high noise levels (natural illumination variation, movement, soil etc.), inertia in some research domains (e.g. vegetation indices), resolution.

Given this context, we conclude that any possible solution will necessarily be synergic, combining classical agricultural sciences (horticulture, agronomy, stockbreeding etc.) with other domains, like economy, genetic engineering, mechanical engineering and automatics, bioinformatics, geography, sensor technology, artificial intelligence etc. The agriculture of the future will be without a doubt a discipline that conjoins engineering, management and information technology. Large-scale research endeavors will lead to a maturity of the technological approaches, to development of agriculture-specific methods and equipment, and finally to standardization and uniformity.

References

1. Luculescu, M.: Utilizarea informațiilor multispectrale în managementul culturilor agricole. Universitatea Transilvania Brașov, 01.03.2015 [Online]. Available: http://www.unitbv.ro/Portals/9/Cercetare%20stiintifica/C04/MoniCult/MoniCult_RA_ctr.225_2014_Etapa_1.pdf. Accessed 25 June 2015
2. Zhang, C., Kovacs, J.: The application of small unmanned aerial systems for precision agriculture: a review. Precision Agric. **13**(6), 693–712 (2012)
3. Thorp, K., Tian, L.: A review on remote sensing of weeds in agriculture. Precision Agric. **5** (5), 477–508 (2004)
4. Thomas, S., Kuska, M.T., Bohnenkamp, D., Brugger, A., Alisaac, E., Wahabzada, M., Behmann, J., Mahlein, A.-K.: Benefits of hyperspectral imaging for plant disease detection and plant protection: a technical perspective. J. Plant Dis. Prot. **125**(1), 5–20 (2018)
5. Du, H., Jiang, H., Zhang, L., Mao, D., Wang, Z.: Evaluation of spectral scale effects in estimation of vegetation leaf area index using spectral indices methods. Chin. Geogra. Sci. **26**(6), 731–744 (2016)
6. Lussem, U., Hollberg, J., Menne, J., Schellberg, J., Bareth, G.: Using calibrated RGB imagery form low-cost UAVs for grassland monitoring: case study at the Rengen grassland experiment (RGE), Germany. In: International Conference on Unmanned Aerial Vehicles in Geomatics, Bonn (2017)
7. von Bueren, S.K., Burkart, A., Hueni, A., Rascher, U., Tuohy, M.P., Yule, I.J.: Deploying four optical UAV-based sensors over grassland: challenges and limitations. Biogeosciences **12**, 163–175 (2015)
8. Shi, Y., Thomasson, A., Murray, S., Pugh, A., Rooney, W., Shafian, S., Rajan, N., Rouze, G., Morgan, C., Neely, H., Rana, A., Bagavathiannan, M., Henrickson, Yang, J.C.: Unmanned aerial vehicles for high-throughput phenotyping and agronomic research, 29 July 2016
9. de Paul Odabe, V., Lai, R., Chen, J.: Remote sensing of soil and water quality in agroecosystems. Water, Air Soil Pollut. (Sept 2013)
10. Boon, M.A., Tesfamichael, S.: Wetland vegetation integrity assessment with low altitude multispectral UAV imagery. In: International Conference on Unmanned Aerial Vehicles in Geomatics, Bonn (2017)
11. Nasi, R., Viljanen, N., Kaivosoja, J., Hakala, T., Pandzic, M., Markelin, L., Honkavaara, E.: Assessment of various remote sensing technologies in biomass and nitrogen content estimation using an agricultural test field. In: ISPRS - International Archives of the Photogrammetry, Remote Sensing and Spatial Information Sciences, vols. XLII-3/W3, pp. 137–141 (Oct 2017)
12. Kersting, K., Bauckhage, C., Wahabzada, M., Mahlein, A.-K., Steiner, U., Oerke, E.-C., Roemer, C., Pluemer, L.: Feeding the world with big data: uncovering spectral characteristics and dynamics of stressed plants. In: Computational Sustainability, pp. 99–120. Springer, Cham (2016)
13. Ishida, T., Kurihara, J., Viray, F.A., Namuco, S.B., Paringit, E., Perez, G.J., Takahashi, Y., Marciano, J.J.: A novel approach for vegetation classification using UAV-based hyperspectral imaging. Comput. Electron. Agric. **144**, 80–85 (2018)

14. Padua, L., Adao, T., Hruska, J., Sousa, J., Peres, E., Morais, R., Sousa, A.: Very high resolution aerial data to support multi-temporal precision agriculture information management. Procedia Comput. Sci. **121**, 407–414 (2017)
15. Ge, Y., Thomasson, A., Sui, R.: Remote sensing of soil properties in precision agriculture: a review. Front. Earth Sci. **5**(3), 229–238 (2011)
16. Tan, S.-Y.: Developments in hyperspectral sensing. In: Handbook of Satellite Applications, pp. 1–21. Springer, New York, NY (2016)
17. Palaniswami, C., Gopalasundaram, P., Bhaskaran, A.: Application of GPS and GIS in sugarcane agriculture. Sugar Tech **13**(4), 360–365 (2011)
18. Siama, S., Khorram, S., van der Wiele, C., Koch, F., Nelson, S., Potts, M.: Data aquisition. In: Principles of Applied Remote Sensing, pp. 21–67. Springer International Publishing, Berlin (2016)
19. Kayad, A., Al-Gaadi, K., Tola, E., Madugundu, R., Zeyada, A., Kalaitzidis, C.: Assessing the spatial variability of alfalfa yield using satellite imagery and ground-based data (June 2016)
20. Alchanatis, V., Schmilovitch, Z., Meron, M.: In-field assessment of single leaf nitrogen status by spectral reflectance measurements. Precision Agric. **6**(1), 25–39 (2005)
21. Cosma, A., Preda, C.-I., Luculescu, M.-C., Cristea, L., Zamfira, S.-C.: Data acquisition system used in precision agriculture for vegetation status monitoring—software subsystem. In: 2017 International Conference on Optimization of Electrical and Electronic Equipment (OPTIM), Brasov (2017)
22. Schmitter, P., Steinruecken, J., Roemer, C., Ballvora, A., Leon, J., Rascher, U., Pluemer, L.: Unsupervised domain adaptation for early detection of drought stress in hyperspectral images. ISPRS J. Photogrammetry Remote Sens. pp. 65–76 (Sept 2017)
23. Chakravortty, S., Bhondekar, A.: Spatial and spectral quality assessment of fused hyperspectral and multispectral data. In: Biologically Rationalized Computing Techniques for Image Processing Applications, pp. 133–158. Springer, Cham (2017)
24. Symonds, P., Paap, A., Alameh, K., Rowe, J., Miller, C.: A real-time plant discrimination system utilising discrete reflectance spectroscopy. Comput. Electron. Agric. **117**, 57–69 (2015)

Part VII

Augmented and Mixed Reality Environments for Education and Training

Work-in-Progress: Development of Augmented Reality Application for Learning Pneumatic Control

Brajan Bajči[1(✉)], Vule Reljić[1], Jovan Šulc[1], Slobodan Dudić[1], Ivana Milenković[1], Dragan Šešlija[1], and Hasan Smajić[2]

[1] University of Novi Sad, Novi Sad, Serbia
brajanbajci@uns.ac.rs
[2] University of Applied Sciences, Cologne, Germany
hasan.smajic@th-koeln.de

Abstract. This paper describes the development of an augmented reality application for learning pneumatic control. The goal is to improve the workbook of pneumatic control with additional digital information. Through the implementation of the developed application, the assumption is that it will be easier for students to understand the functionality of complex pneumatic schemes. The augmented reality application was developed for the Android platform. Students are able to download it from the Internet and use it at home, at the university or anywhere else. The application recognizes the solution of the exercise and the pneumatic components in the workbook and covers it with additional information (for example, simulation in form of a video clip). The developed augmented reality application can be easily introduced into teaching activities in two university study courses of pneumatic control. Augmented reality applications, as the one described in this paper, can be a very useful tool for students during their studies, as well as for others who want to learn more about this topic.

Keywords: Augmented reality · Education · Pneumatic control

1 Introduction

Technological achievements from the second half of the twentieth century and their development today have led to the fact that we live in the era of digitization. The use of various type of modern gadgets, smartphones, tablets, computers, Internet and so on, became quite normal and even necessary in our everyday life. Beside entertainment, modernization of processes and digitization become an integral part of industrial environment, medicine, marketing, designing and management, among others. Furthermore, as a condition for the progress of the society in the future, of course, education must follow these trends.

In various areas of scientific and engineering education, understanding of teaching material only from printed literature can be a problem for students. Standard educational methods, such as learning from books, are therefore more and more often enhanced by some form of digital technologies as additional educational materials.

M. E. Auer and K. Ram B. (Eds.): REV2019 2019, LNNS 80, pp. 711–718, 2020.
https://doi.org/10.1007/978-3-030-23162-0_64

Different types of simulations, video clips, dedicated software, online, virtual or remote laboratories, and more and more popular Augmented Reality (AR) are just a few examples of such digital educational materials.

AR represents a technology that combines various digital information with the user's view of the real physical world. In such a system, virtual and real objects create the so-called augmented world/reality in which they can interact in real time [1]. Nowadays, increased emerging of AR applications in educational processes is present [2]. Several AR applications are present in the literature, from different fields of science such as electrical engineering [3, 4], civil engineering [5] biology [6] and medicine [7]. It is a real fact that AR improves students' learning capabilities and leaves a positive impression on them [8]. Furthermore, the authors in [9] proposed several suggestions for future research, that include exploring learning experience and learner characteristics, in order to make AR applications more useful in STEM (Science, Technology, Engineering and Mathematics) education.

The aim of this paper is to present the AR smartphone application for educational purposes. This application improves the pneumatic control workbook with additional digital information intended to facilitate acquiring of theoretical knowledge in this field of engineering education. The application was developed for the Android platform and its beta version has already been tested with a group of students.

2 The Pneumatic Control Workbook

At the University of Novi Sad, at the Faculty of Technical Sciences, pneumatic control is an obligatory course in two study programs, Mechatronics and Industrial Engineering. In the first case, it is studied within the course Components of technological systems, while in the other case, within the course Automation of work processes 1. For the purposes of these courses a workbook is formed under the name "A workbook of solved exercises with theoretical basis of pneumatic control" [10]. The workbook and the courses cover the field of pneumatic control that starts with the basic pneumatic components, describing their working principles, explanation of drawing principles of pneumatic schemes up to the working principles of complex pneumatic systems.

The first part of the workbook is dedicated to the theoretical basis and explanations of pneumatic systems and pneumatic control in order to familiarize the reader with the material. Here, some of the most important pneumatic components are presented as well, from which pneumatic systems are composed of. These components are compressed air preparation units, pneumatic valves and actuators such as cylinders, grippers, motors etc.

The second part of the workbook contains several exercises related to pneumatic control. Each exercise processes a particular, realistic problem and introduces a new component. At the beginning of the exercise, a textual explanation is given together with the presentation in the form of a picture, shown in Fig. 1. A pneumatic press, from the exercise one is presented in Fig. 1a, while in Fig. 1b is shown a pneumatic lift from the exercise two.

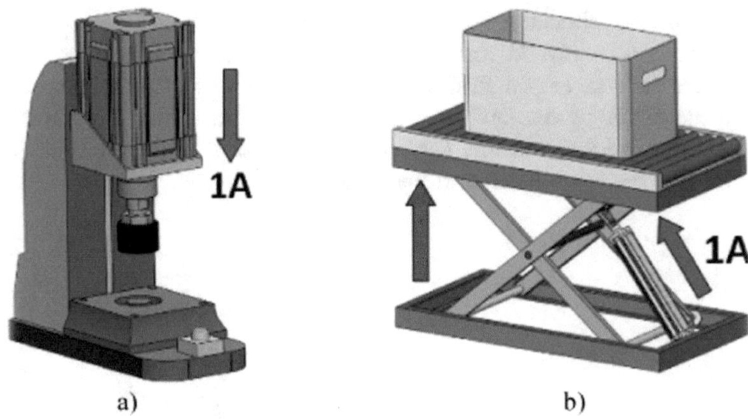

Fig. 1. Presentation of the exercise: **a** pneumatic press; **b** pneumatic lift

After that, all of the necessary components, needed for the solution of the exercise are stated. The new components, that are not previously mentioned, are described in details and their schematic drawings are given. In Fig. 2a, a schematic drawing of a single acting pneumatic cylinder is presented while a 5/2 pneumatic directional valve is presented in Fig. 2b.

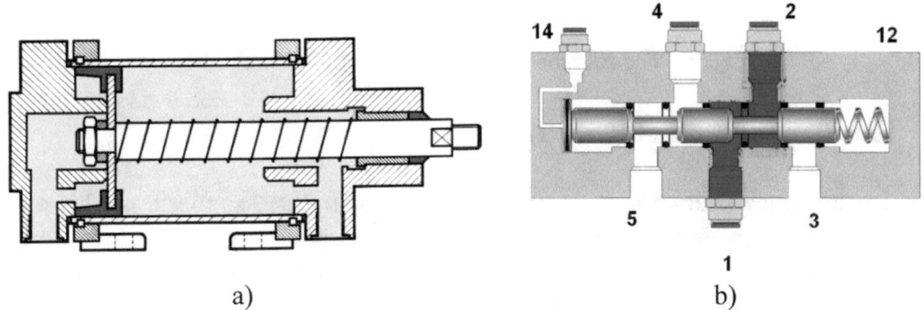

Fig. 2. Component's schematic drawings: **a** single acting pneumatic cylinder; **b** 5/2 pneumatic directional valve

The solution of each task is given after the theoretical and schematic explanations of the components. The solution represents a drawn pneumatic control circuit. In the scheme, the pneumatic components are drawn according to the precisely defined rules, in their initial positions. Such a scheme is shown in Fig. 3.

With such an exercises, it is completely defined what kind of pneumatic elements should be used for the practical realization of these and similar pneumatic systems and how to connect them in practice. However, several difficulties were encountered among students during studying this field of technique. The working principles of pneumatic components often represents a problem. It is not always enough to show a schematic drawing because it is difficult for some students to imagine which parts are moving and which not in an assembly. In some cases, it can also be a problem for students to imagine where the compressed air flows through the component, how the components work in a pneumatic system and how the compressed air flows through the whole system. For that purpose, an AR application is developed and shown in the next section, to make easier to understand these principles.

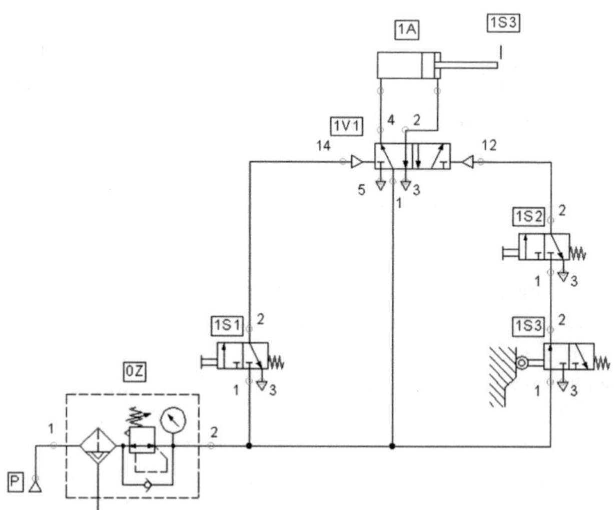

Fig. 3. The solution of an exercise from the workbook

3 The Augmented Reality Application

The developed application is an image based AR smartphone application. This means that the application displays additional digital information to the user, based on the scanned image. The user just needs to open the application, find the images in the workbook that are in the application's database and scan them with their smartphone. As soon as the application recognizes the image, the additional information is displayed automatically. It remains displayed on the display as long as the image is in the focus. Three types of images can be scanned with the application and three different type of additional information are shown to the user, as can be seen in Table 1.

Table 1. The form of additional information in the AR application based on the scanned images in the workbook

Image to scan	Form of additional information
Presentation of the exercise	3D CAD model
Component's cross section drawings	Animation of a component
Solution of an exercise	Simulation of the exercise

Firstly, the user is able to scan the presentation of each exercise. In that case, a 3D Computer Aided Design (CAD) model appears on the display of the smartphone. Rotating the workbook or the smartphone around it, the model rotates as well, and by getting closer to it, the model become bigger. In such a way, the user has a better view of the exercise presentation compared to a just printed figure. In Fig. 4a is shown an example of a 3D CAD model of a pneumatic press that appears on the display using the developed AR application, while a pneumatic lift is shown in Fig. 4b.

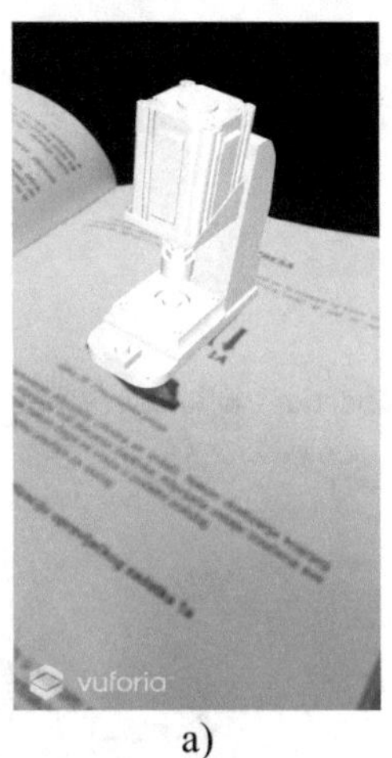

a)

b)

Fig. 4. 3D models in the AR application: **a** pneumatic press; **b** pneumatic lift

On the other side, by scanning the component's schematic drawing, an animation of it appears on the display. In Fig. 5a, the animation of a single acting pneumatic cylinder is shown. The printed schematic drawings are black and white, while the animations are in color that already makes easier the understanding of its working principles.

A cylinder's piston and piston rod can be seen in Fig. 5a extracting and retracting and the chamber that is filled with compressed air become darker every time. The animation of a 5/2 pneumatic command valve is shown in Fig. 5b. The movement of the internal parts of the valve is shown in this animation as well as the flow of the compressed air through it.

In Fig. 6, a simulation of the solution of an exercise from the workbook is shown. After recognizing the solution, the AR application shows a simulation of the solution in form of a video clip. The pneumatic elements on the scheme starts moving according to the exercises' description. Furthermore, the flow of the compressed air through the whole system is indicated with dark blue lines at every moment.

Each of the exercises are augmented with additional information in the same way as it is described above. The application is developed with Unity and Vuforia SDK (Software Development Kit). The mentioned software are chosen due to their user friendly interface that is easy to use for creating such an application.

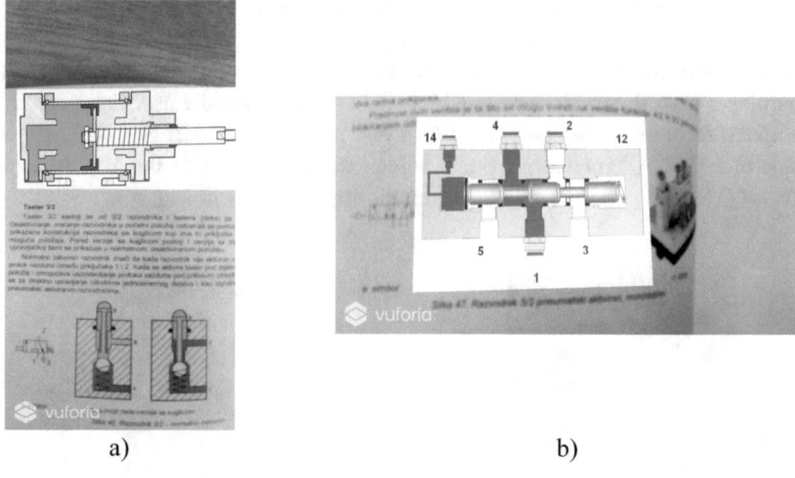

a) b)

Fig. 5. Pneumatic component animations in the AR application: **a** single acting pneumatic cylinder; **b** 5/2 pneumatic command valve

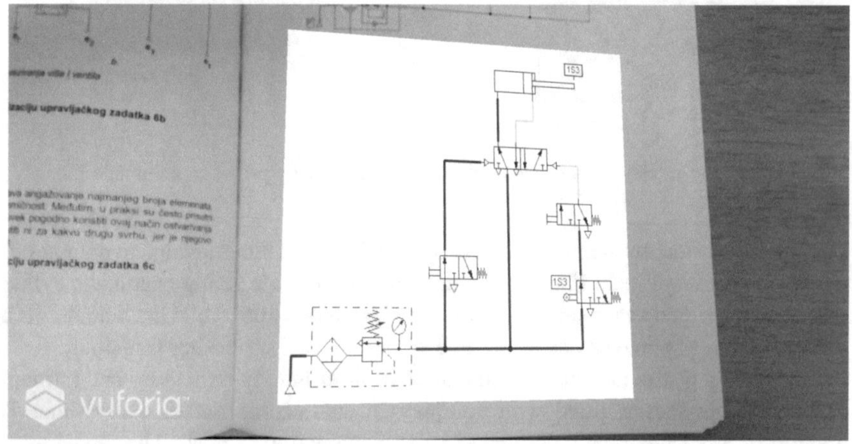

Fig. 6. Simulation of the solution of an exercise from the workbook in the AR application

4 Discussion

The developed application has shown very satisfactory characteristics during the test period. The application is fast and recognizes the images easily in the first two cases. However, one lack of the concept is observed. A problem occurs in case of the recognition process of the solutions. Namely, the solutions of some exercises are very similar and consists of a large number of thin lines and text. Further, the automatic focus does not work at the same way at every smartphone. This makes it difficult for the application to identify some of the images. In the next issue of the workbook, it is planned to print Quick Response (QR) codes along with the solutions of the exercises in order to avoid this lack. Recognition of QR codes provides much greater reliability of the application.

The beta version of the application is already tested with a group of undergraduate students at the Faculty of Technical Sciences in Novi Sad. The application proved to be a good additional material for the workbook and it led to the positive impressions on the test group. The students from the group were asked to give their personal opinions about the application. In addition to their positive impressions, a few comments are stated related to further improvement of the application such as: "*Show the 3D models in color; Create short animations of the 3D models; Further develop the application for the whole workbook, not just for the exercises; Give some information in form of the textual explanations as well*". These comments will be considered during the future development of this application and for applications for other courses as well.

5 Conclusion

The digital era in which we are living today, the modernization of our everyday life and technological developments led to the implementation of new methods into educational processes. One of them is AR, which intends to change completely the concept of learning, as we know it today. In this paper, an AR application developed for the learning of basic principles of pneumatic control is presented. The application is developed for smartphones, for the Android platform. The already formed workbook is augmented with digital information in order to make easier understanding of this field of technique. After opening the application, it is necessary just to scan certain images from the workbook and the information is shown to the user automatically. Three form of the digital information are used for that purpose, 3D models, animations and simulations. The application is primarily developed for the students use but everyone who wants to learn about pneumatic control can use it.

References

1. Azuma, R.T.: A survey of augmented reality. Presence Teleoperators Virtual Environ. **6**(4), 355–385 (1997)
2. Bower, M., Howe, C., McCredie, N., Robinson, A., Grover, D.: Augmented reality in education–cases, places and potentials. Educ. Media Int. **51**(1), 1–15 (2014)
3. Martín-Gutiérrez, J., Fabiani, P., Benesova, W., Meneses, M.D., Mora, C.E.: Augmented reality to promote collaborative and autonomous learning in higher education. Comput. Hum. Behav. **51**, 752–761 (2015)
4. Andujar, J.M., Mejías, A., Márquez, M.A.: Augmented reality for the improvement of remote laboratories: an augmented remote laboratory. IEEE Trans. Educ. **54**(3), 492–500 (2011)
5. Dong, S., Behzadan, A.H., Chen, F., Kamat, V.R.: Collaborative visualization of engineering processes using tabletop augmented reality. Adv. Eng. Softw. **55**, 45–55 (2013)
6. Weng, N.G., Bee, O.Y., Yew, L.H., Hsia, T.E.: An augmented reality system for biology science education in Malaysia. Int. J. Innovative Comput. **6**(2) (2016)

7. Kamphuis, C., Barsom, E., Schijven, M., Christoph, N.: Augmented reality in medical education? Perspect. Med. Educ. **3**(4), 300–311 (2014)
8. Akçayır, M., Akçayır, G., Pektaş, H.M., Ocak, M.A.: Augmented reality in science laboratories: the effects of augmented reality on university students' laboratory skills and attitudes toward science laboratories. Comput. Hum. Behav. **57**, 334–342 (2016)
9. Cheng, K.H., Tsai, C.C.: Affordances of augmented reality in science learning: suggestions for future research. J. Sci. Educ. Technol. **22**(4), 449–462 (2013)
10. Dudić, S., Šešlija, D., Milenković, I., Šulc, J., Reljić, V., Bajči, B.: Zbirka rešenih zadataka sa teorijskim osnovama iz pneumatskog upravljanja. Faculty of Technical Sciences. ISBN: 978-86-7892-926-7 (2017)

Collaborative Augmented Reality in Engineering Education

Nina Schiffeler(✉), Valerie Stehling, Max Haberstroh, and Ingrid Isenhardt

IMA – RWTH Aachen University, Aachen, Germany
{nina.schiffeler, valerie.stehling, max.haberstroh, isenhardt.office}@ima-ifu.rwth-aachen.de

Abstract. A continuous trend in education is the use of new technologies like Augmented Reality (AR). These technologies are assumed to make teaching and learning processes more hands-on and more tangible, particularly in terms of abstract learning contents. Due to this practical nature, AR is assumed to foster the motivation of concerning oneself with a specific learning content and supports, thus, the comprehension of it and its purpose for e.g. the future work life. In academic learning contexts such as engineering education, however, AR can also be used for collaborative purposes despite its currently most common purpose of demonstration and instruction. The present paper investigates the effects of using AR in collaborative team processes with special respect to motivation and emotional activation. A mixed methods approach is chosen in order to examine qualitative and quantitative data to gather both a subjective and an objective perspective on the subject of research. As the investigation shows, the motivation to use this technology is high which derives, amongst other factors, from the better comprehension of the learning content when compared to the results of the control group.

Keywords: Collaboration · Augmented reality · Motivation · AR app

1 Context

Modern technological trends such as Augmented Reality (AR) increasingly find their way into higher education. There already is a large number of AR applications and scenarios for various disciplines and use cases in higher education [1–3], e.g. when presenting an automotive motor and its components in a lecture hall. For such use cases, AR technology provides a major advantage for a large audience: each student—equipped with an AR capable device like a smartphone or tablet—can access the demonstration by scanning the AR marker. Then, an interactive virtual object (e.g. a car motor) is displayed on their device screen.

AR technology influences or enhances the users' perception of reality by supplementing the real world with virtual objects [4]. With AR, thus, reality is augmented, i.e. enhanced, by means of virtual objects and information mapped onto the real environment. The main functions of AR are (1) embedding virtual objects into reality, (2) interaction in real-time, and (3) correct alignment of the virtual objects in the real

M. E. Auer and K. Ram B. (Eds.): REV2019 2019, LNNS 80, pp. 719–732, 2020.
https://doi.org/10.1007/978-3-030-23162-0_65

3D world. AR is often realised with the help of mobile devices (e.g. smartphones, tablets or AR-specific glasses) that have a rear camera at their backs (e.g. all current smartphones and tablets) and a specific AR application. When scanning the environment (i.e. object-based or markerless tracking[1]) or a particular AR marker (i.e. marker tracking[2]), virtual objects can be placed into the environment by appearing on the device screen at the respective point in the environment.

Most of the current AR applications, however, mainly serve demonstrative purposes or are merely interactive in terms of one person being able to interact with the technology. Due to the increased number of collaborative elements in higher education, suitable and particularly effective didactical tools and methods need to be taken into account. Due to the possibility of creating network connections between different devices, for instance, collaborative settings can be realised by means of AR. This way, students can conduct digitally supported group work, e.g. by role-plays, that is even fostered by the collaborative nature of a specifically developed AR application or scenario [8, 9].

2 Purpose and Goal of the Survey

The presented work aims at investigating the extent to which a collaborative AR-app affects the student's general assessment of a role-play used for conceiving the principles of agile management and scrum teamwork. In particular, the motivation as well as the emotional activation via the use of AR during team or group learning processes are put under investigation in the survey available. In the role-play, students form a fictitious scrum team that needs to meet client-specific requirements in a simulated product development process, in order to experience the work of a scrum team and agile principles in project management. An AR-app has also been developed to fit the specific characteristics of an exemplary lecture in the field of engineering education (see Sect. 3.1). While using this app, the research questions on its influence on the collaboration and motivation of concerning oneself with the exercise of a role-play are tackled. The lecture, which the AR-app is used in, focuses on agile management in general and the collaborative work in scrum teams in a specific session as an example of agile working processes. Although this is a topic derived from the field of Software Engineering, its principles and procedures for project management are increasingly

[1] In order to display virtual objects and/or additional information, the object-based tracking uses real-world objects as triggers for displaying the virtual ones by predefining them in the AR-application [5]. Markerless tracking, in contrast, scans the real-world environment and detects e.g. "locations of walls or points of intersection, allowing users to place virtual objects without needing to read an image" [5]. Markerless tracking is used e.g. by the Microsoft HoloLens.

[2] Marker tracking is a common approach that does not require many performance resources in terms of computing power in smartphones and tablets, for instance. It is realised by scanning an AR marker with the back camera of the respective device. An AR marker is a visual trigger that cause the display of virtual objects or additional information [6], e.g. a picture or QR-code. With the help of markers, it is defined where to locate and place virtual objects in the right place in the scene. Recognising predefined markers in the scene is achievable on a vast spectrum of devices [7], since even devices with a lower computing power can detect markers and show virtual objects respectively.

becoming more and more valuable for engineering contents and teamwork processes as well as project management tasks in the field of engineering sciences. With the collaborative AR-app, students are supposed to experience how a scrum team works rather than merely conceiving it theoretically. The app is integrated into a role-play in which the students are asked to collaborate as a scrum team. The scenario chosen for the role-play is the interior furnishing for a shared apartment for three students.

In this role-play, it is necessary to realise specific requirements and also build a concept and model of the shared apartment in order to show the fictitious clients the extent to which the requirements have been met. For both realising a hands-on, realistic model of the shared apartment and handling the requirements and tasks an AR-app is developed. Previous experiences of the authors and literature reviews suggest that AR supports the learning and acquisition of knowledge particularly for abstract or complex learning contents [10, 11]. In order to examine the effects of using the respective AR-app in the collaboration, an empirical survey surrounding the conduct of the role-play is set up and realised.

3 Method

3.1 Role-Play and Collaborative AR-App

For fostering the conception of the learning content "scrum" from the field of agile management, a role-play has been developed that serves the purpose of experiencing the way processes in scrum teams work. Moreover, the role-play aims at showing the principles of agile project management by making it more concrete for the students by setting them into a fictitious, but realistic simulation of a scrum team's working processes. The students, thus, receive different roles of a scrum team that they have already learnt prior to the role-play in the context of the corresponding lecture "Agile Management in Technology and Organisation" (AMITO) at the RWTH Aachen University. These roles are "Product Owner" (PO), "Scrum Master" (SM), "Scrum Team Member (Developer)" (Dev), and "PR Team Member" (PR). While the roles PO and SM are assigned to one student each, the roles Dev and PR can be assigned to up to three students each. Thus, a group of four to eight students can form a scrum team in the role-play.

The aim of the fictitious scrum team is the development of a product that is in alignment with the requirements set up by their clients. The product to be developed is a shared apartment for three students and its interior furnishing in particular. In order to meet those requirements and create an apartment that is visible as a concrete model to both the clients and the scrum team during their collaboration, an AR-app has been developed. The AR-app aims at supporting the collaboration of the scrum team on the one hand in terms of visualising the procedural steps, while furnishing the shared apartment and with respect to the organisation of the collaboration (e.g. status of tasks and competences) on the other hand. It runs on both smartphones and tablets. However, the use of tablets is recommended during the role-play since the display size is large enough to see every detail of the virtual objects and the scene. Respective devices are equipped with the AR-app and then handed to the students for conducting the role-play.

The prototype of the collaborative AR-app allows every participant of the role-play to choose virtual objects, i.e. furniture and walls in this scenario, from a catalogue area displayed on the screens of their mobile devices. Once in the scene, i.e. playing area, the students can manipulate the virtual objects e.g. by turning or moving them. This manipulation is simultaneously visible for all scrum team members on their respective individual mobile devices. With this functionality, every team member can help form the outcome of the product development process by shaping the product itself collaboratively. The AR-app, thus, also allows a parallel collaboration on the same product since every team member can see the virtual objects and changes made to them during the whole process in the second they are made. By means of entering a user name and choosing a colour for each team member, the interaction and comprehensibility of each team member's actions are fostered since the team member's colour is displayed under the virtual object he or she manipulates (see Fig. 1: blue and read markings under the virtual objects).

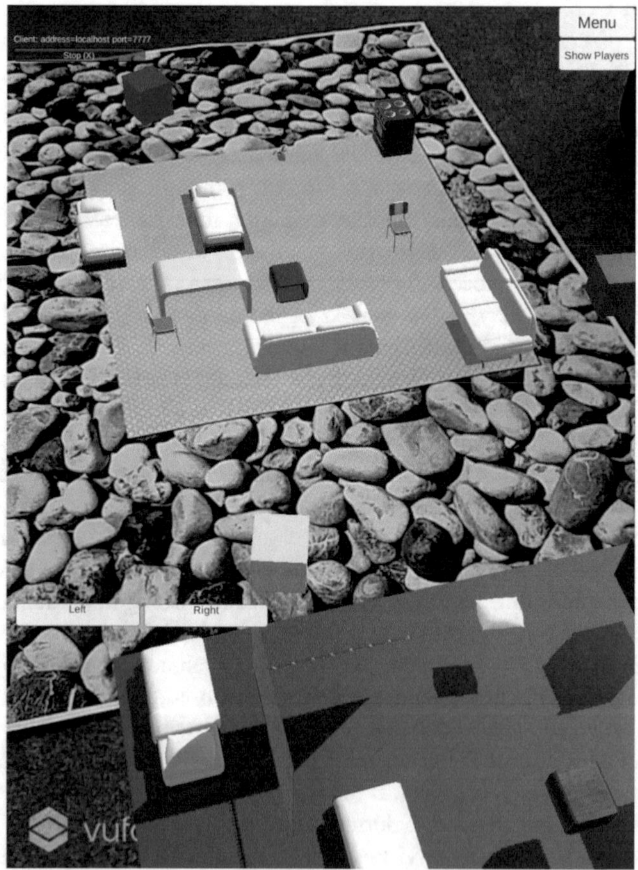

Fig. 1. Screenshot of the collaborative AR-app while in simultaneous use

The collaboration within the AR-app is realised by means of an Internet connection. When starting the app, one device is the host and creates a virtual room the other devices can enter. For including the AR-app into a role-play or a lecture, a stable Internet connection, mobile devices such as tablets, the AR-app and the respective AR-marker (for this scenario: poster with grey, dotted playing area and surrounding pebbles, see Fig. 1) are essential.

When conducting the role-play, the students are assigned their roles first and respective role descriptions are handed to them. They also receive a fact sheet with the task and the requirements (only PO). They then have 30 min to meet the requirements and build a shared apartment by using the virtual objects from the catalogue area of the AR-app. After this 30-min conduct, the PR team is supposed to give a short presentation on the concept for the shared apartment and its interior furnishing in particular. This presentation needs to be prepared also during the 30-min conduct, when the Devs, SM and PO work on the virtual objects.

3.2 Empirical Survey

In order to investigate the effect(s) of using AR in collaborative educational settings a pre- and post-test survey has been designed in combination with an experimental setting and will be iteratively conducted with student groups (see Fig. 2). The results of the first iteration are presented in this paper. The tests are conducted by means of standardised questionnaires. They include items from empirical models like the Task-Technology Fit [12] and Technology Acceptance Model (TAM 3) [13]. The purpose of these models is the investigation of the acceptance of a specific technology and the use of it in a chosen context. With the Technology Acceptance Model, the factors perceived ease of use, the perceived usefulness, and use behaviour are put under investigation. Moreover, the TAM 3 also includes factors external or prior to the test use of the technology under investigation (see Fig. 3). From these, the factors "job relevance", "experience", "computer anxiety", "output quality", "computer playfulness", and "computer self-efficacy" are included in the quantitative questionnaire.

The external factor "voluntariness" was not included in the questionnaire since the participants elected the lecture AMITO on a voluntary basis and, additionally, were asked about their voluntary willingness to take part before the start of the study and before handing the pre-test questionnaire to them. As far as the Task-Technology Fit model is concerned, it is used for investigating to which extent a technology is suitable for efficiently conducting a given task or solving a specific problem. Its factors "utilization", "performance impacts", and "task characteristics" are included in the study. While the task characteristics are part of the post-test questionnaire, the other factors mentioned are gathered by means of a qualitative survey, i.e. a focus group interview collecting feedback on the AR-app in particular and the role-play in general. The standardised questionnaire also comprises scenario-specific items in order to holistically investigate the use of the collaborative AR-app within the role-play. These items deal with the effectiveness of the AR-app for fostering learning processes, the integration of the app into the role-play, and the matching of the app functionalities to the tasks and requirements of the role-play, for instance.

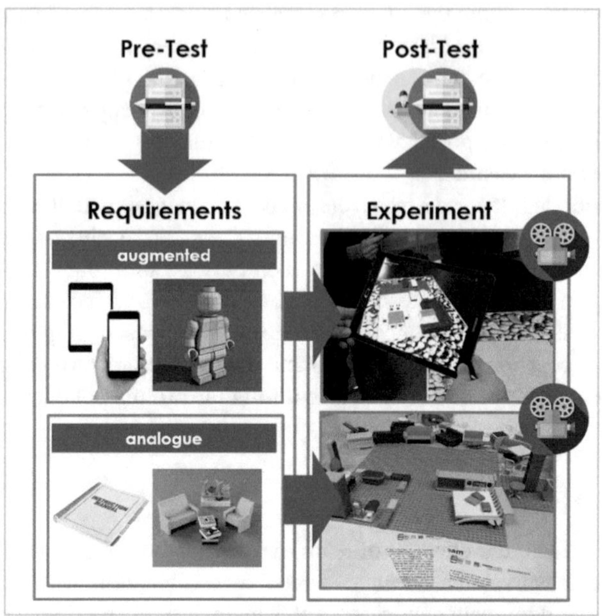

Fig. 2. Empirical method including a pre-post-test and an experimental setting

First, a pre-test has been taken to gather information on the participants' general assessment of role-plays and their prior knowledge to agile management as well as the process of working in scrum teams. After conducting the AR application and scenario, the post-test aims at gathering data on its outcome, conception, and perceived effectiveness. The paper deals with the results of this pre-post-test questionnaire survey.

The pre-test questionnaire deals with the participants' demographic details (i.e. age, gender, subject of studies and number of university semesters) and the prior knowledge and assessment mentioned above. The post-test questionnaire gathers information on the evaluation of the scenario as well as the process and outcome of the role-play. Except for the demographic details, all items are gathered by means of a six-step Likert scale ranging from 1 = "strongly disagree" to 6 = "strongly agree". Also, qualitative feedback by the participants is integrated into the assessment on the scenario, role-play as a method, and the AR-app. Common to both parts of the questionnaire is the Affect Grid [14]. It aims at investigating the emotional activation before and after the use of the collaborative AR-app. The Affect Grid is a tool for self-evaluating a person's emotional status. It comprises four sets of complementary factors that are

- "high arousal" and "sleepiness"
- "stress" and "excitement"
- "depression" and "relaxation", and
- "unpleasant feeling" and "pleasant feeling".

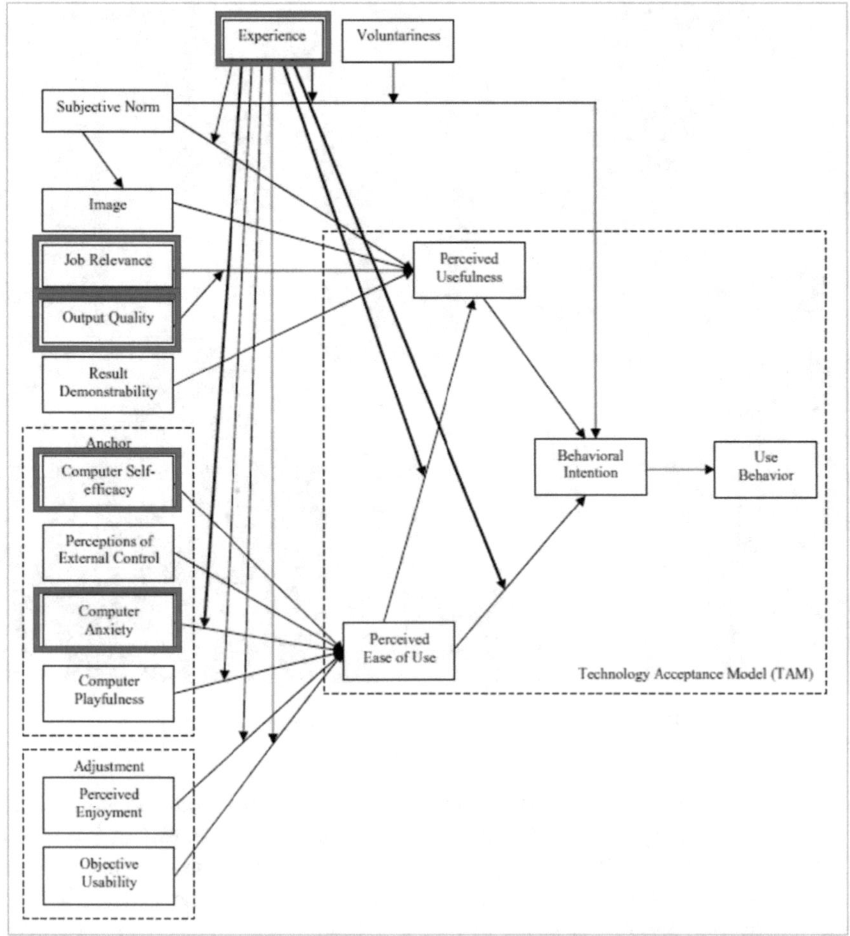

Fig. 3. Selected factors from the Technology Acceptance Model 3 [13] for the pre-post-test questionnaires

Moreover, the study was conducted with a test and a control group. While the test group conducted the role-play with the collaborative AR-app (see Fig. 4), the control group used paper-pencil materials together with Lego and Playmobil stones.

4 Outcomes

4.1 Sample

In the summer term 2018, 13 students have participated in the lecture "Agile Management in Technology and Organisation" (AMITO). All of those students have also voluntarily taken part in the conduct of the role-play and, thus, the empirical survey attached to it. From this sample (n = 13), the test group includes six participants, while

Fig. 4. Student group using the collaborative AR-app

the control group comprises seven participants. These groups were formed randomly by assigning numbers to the students that resemble either the test or the control group. The average age of the sample is 24.08 years. The lecture is open to Master degree students resulting in twelve participants indicating to pursue a Master's degree while merely one participant states to pursue a Bachelor's degree. The participants' subjects of study all derive from engineering sciences but differ in terms of their focus: production technology (PT), automation engineering (AE), general mechanical engineering (GME), computational engineering science (CES), and technology communication (TC). The distribution of the sample to these specialisations is displayed in Fig. 5. With 84.6% male to 15.6% female participants, the distribution of the sample is similar to the gender distribution in the field of engineering sciences of the RWTH Aachen University.

When investigating the composition of the sub-samples for the test group and the control group, the distributions of gender and age are similar. The rest of the demographic details, however, differs only slightly as Table 1 shows.

Consecutive to the general demographic details of the participants, the pre-test questionnaire comprises items on the technical affinity, the locus of control, and the ease of use in terms of digital devices like laptop, smartphone, and tablet [15, 16]. For both sub-samples the technical affinity is high since all items from the KUT model [15] are assessed with 5 ("agree") in average. Also, the ease of use of (private) digital devices is ranked high, with averages of around 5.8 ("very easy") as well. It is assumed,

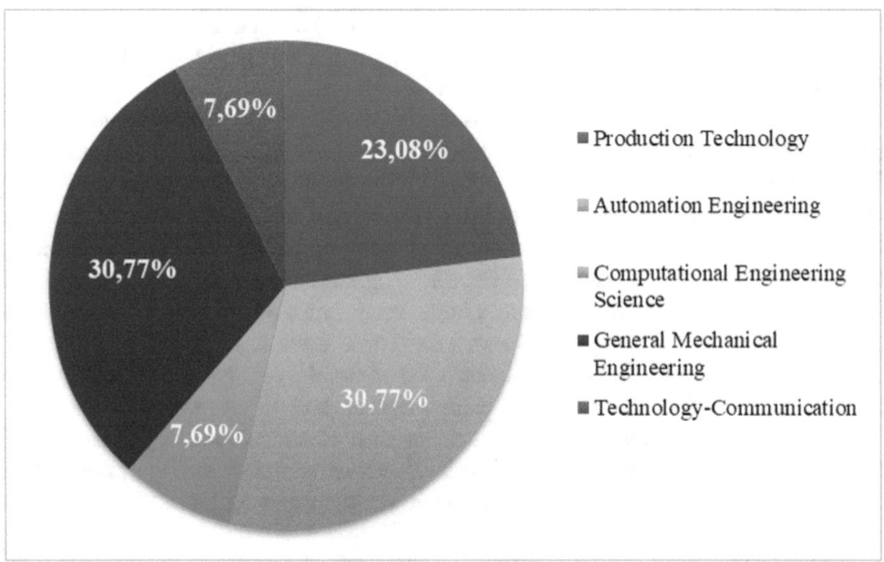

Fig. 5. Sample distribution to specialization of engineering sciences

Table 1. Comparison of the composition of the sub-samples

Item	Test group	Control group
Age	23.7 years	24.4 years
Gender	5 male, 1 female	6 male, 1 female
Degree	1 B.Sc., 5 M.Sc.	7 M.Sc.
Semester	1–2	3–4

thus, that the participants will not be having trouble initially handling the AR-app. Due to a high technical affinity and a high locus of control for digital devices and applications, the transfer of these competencies to new devices and software is expected to be intuitive.

4.2 General Assessment of the AR-App Use and Role-Play

As far as the general conception of the role-play and collaborative AR-app is concerned, the students participating in the study have assessed them very positively in general. Differences in the conception, however, also show when comparing the test and control groups.

The participants of the analogue, control group state to have taken a longer ramp-up time of finding into the role-play than the test group. As the assessment of the respective questionnaire items in pre- and post-test show, even these participants (in both groups) needed time to feel comfortable in the role-play situation according to the post-test who stated in the pre-test that they do not need a ramp-up time in such exercises. In terms of the learning content “scrum” and “agile management”, the post-test of the control group results in the participants' conception of having a deeper understanding of how scrum teams work and agile management works both in comparison to the pre-test and the results of the test group.

With respect to the digital, test group, the results of the post-test questionnaire present a higher consciousness on the processes in scrum teams than before conducting the role-play. The app, thus, helped students visualise the (abstract) processes during the agile collaboration better than for the control group. This conception might derive from the assessment of having less difficulty with comparing tasks and overviewing their status. Due to the visual nature of the AR-app, the students participating in the test group were able to directly see changes in the scene since the app shows the colour of each participant (as described in Sect. 3.1). These colours highlighting the virtual objects during their manipulation increase the comprehensibility of the collaborative process, i.e. students know which team member has which role and moves or manipulates which virtual object simultaneous to the manipulation. By means of this functionality, the participants also state to receive a better understanding of the barriers, obstacles, and chances of using scrum for project management and product development processes.

As far as the usability of the AR-app is concerned, the prototype status used in the role-play becomes also apparent in the participants' evaluation of it. The test group states the handling and usage of the app to not have been intuitive and, thus, not easy. With respect to the technical affinity and the technical locus of control of the test group, this estimation was not expected from the pre-test results. It can, however, be deduced from the current status of the user interface design and the user experience of the AR-app prototype. Since it has come to an unexpected breakdown of the app during the conduct of the role-play, for instance, the students in the test group had to start over with furnishing the apartment from the scratch again, as the app had to be re-started on each mobile device. Moreover, the test group needed some advice and help on the functionalities and handling of the AR-app before and during the role-play. These insecurities and problems might also be the reason for the rather negative assessment of the items "With the app, I was able to finish my tasks quickly" and "With the app, we quickly came to a result". Also, the participants of the test group stated that the AR-app did not include all functionalities they expected, which is another possible reason for the rather negative assessment of the AR-app. It is assumed that the AR-app and its helpfulness in collaborative (learning) team processes, however, will be increased when the user interface is improved and extra functionalities like a saving mode are implemented.

4.3 Motivation and Emotional Activation

In terms of the effects on the learning process, the AR-app is evaluated rather positive. Even though the participants of the test group state the prototype of the AR-app to be unnecessarily complex, it is still assumed to be useful for the purpose of interactive, collaborative learning in the context of project management and product development. It helps the students visualise the collaboration processes and actions of each team member resulting in a positive conception of the idea of using an AR-app in this context. In general, the item "Using the application was fun" is assessed with $\bar{x} = 3.83$

was rated rather positive since it equals the answer option "I rather agree". In comparison to the qualitative statements on the use of the AR-app, this fun and positive assessment are also reflected.

As far as the factor "fun" is concerned, the control group also perceives fun in the context of the role-play, but assigns it to the conception of experiencing the effects and processes of scrum and agile management. It is in alignment with this group's perception of the role-play in general. The results also confirm the estimation of the combination of role-play and collaborative AR-app being a useful exercise to conceive the topics of agile management and scrum even in this software-external field of studies. Moreover, the results of the test group post-test show a satisfaction with the role-play in terms of an equal participation of all team members. This means that all members of the fictitious scrum team have participated equally in the product development and collaboration process. In the control group, however, this item was evaluated less important. They, in turn, have a higher satisfaction with the teamwork and the result of it. This difference in the sub-samples might derive from the aforementioned problems of the test group with the AR-app in terms of the usability. Particularly, events such as the unexpected breakdown of the app or the unintuitive user interface at some points of the use of the AR-app are often causes for frustration or lesser satisfaction with technical devices and applications. Despite these problems, however, the post-test results of the test group point at the immersive capabilities of the AR-app. The item "During the role-play, I have completely forgotten about the world around me" is evaluated with rather agree, which tends to the perception of immersion while using the AR-app, i.e. a strong feeling of diving into a virtual scenario. Thus, the AR-app tends to help the participants (of the test group) concern themselves with the role-play and identify themselves with the task of performing as a scrum team for the time given.

With regards to the Affect Grid, a difference between the test and control group shows: while the control group was rather homogeneously highly activated and had a predominantly pleasant feeling in the post-test, the post-test of the test group shows a heterogeneous and more negative result. The participants in the test group felt less pleasant and tend to be exhausted after using the collaborative AR app (see Figs. 6 and 7).

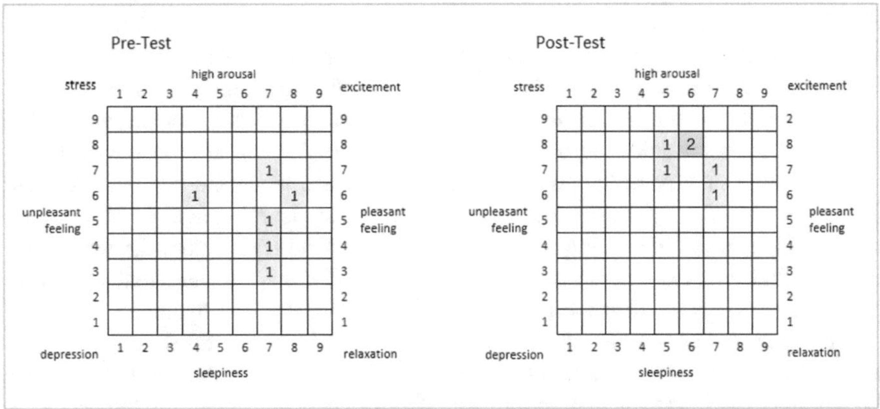

Fig. 6. Affect grid (control group): comparison of the emotional activation in pre- and post-test

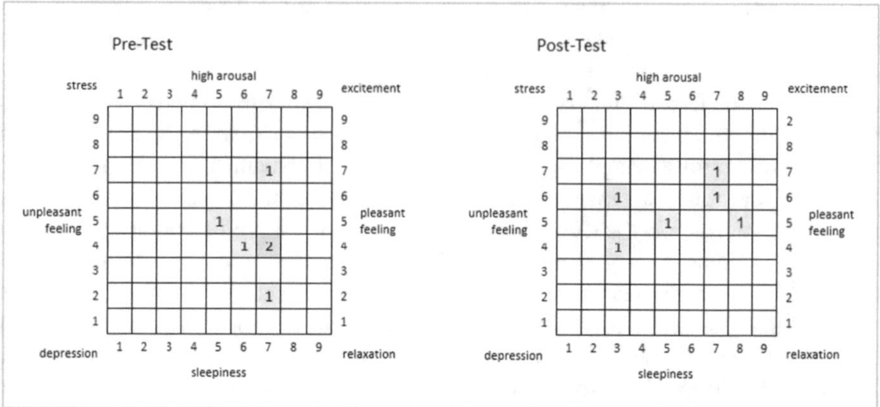

Fig. 7. Affect grid (test group): comparison of the emotional activation in pre- and post-test

In terms of a qualitative gathering of feedback on the AR app, the students stated the app to be hard to handle without a tutorial and not as visually aesthetic as they had wished for it to have a high usability for them. However, the post-test also showed that this technology displays a high motivating factor for students to concern themselves with a specific task, which they actually were not positively disposed at.

5 Discussion and Conclusion

The insights into the effects of using AR in collaborative settings in higher education presented derive from a first iteration of an empirical survey. In this iteration, a random group of students participated resulting in a hardly representative sample. While the age and gender distribution were similar to the composition of the engineering sciences at the RWTH Aachen University in total, the distribution in terms of the subject of studies is not representative. For the role-play, however, the inclusion of different subjects (e.g. production technology, automation engineering, technology-communication—see Fig. 5) was essential in order to create a simulation of a collaborative (scrum) team as authentic as possible, since many modern teams comprise people from different fields of expertise. As the sample available is both small and hardly representative, the results cannot directly be transferred to engineering education in total, but present starting points and hints for adjusting both the survey and the AR-app used in the study.

Since the usability of the AR-app was assessed as a hindering factor for successfully learning and fully concentrating on the task during the role-play, the usability and user interface design needs improvement on the technical level. From the results, it is assumed that the AR-app can actually foster the motivation for participating in exercises like role-plays, which are actually not popular for the participants.

However, the use of the AR-app was not intuitive for most participants and might have led to more negative results than without this confounder. As the problems with the interface and lacking stability of the AR-app reflect a stressing factor for the participants, it might explain the diverse answers on the Affect Grid for the test group. While the control group had a rather pleasant and aroused-excited feeling after the role-play, the results of the test group were dispersed. To some participants of the test group the use of the AR-app was stressful and frustrating while the others were still concentrated on the task and were so immersed with the technology that the problems were hardly conceived.

As the study shows, the collaborative AR-app was evaluated as very motivating and innovative but also hard to use without prior knowledge or tutorial. On this basis, the usability of the AR-app is considered low and will be put under revision for the next stage of the prototype. Although the AR-app has caused interruptions and slight problems during the conduct of the role-play, e.g. in terms of handling the user interfaces or the breakdown of the app, this technology still has a high immersive potential. This potential is considered useful particularly for collaborative learning processes like the one described in the work available, as it can foster the identification with the exercise format role-play even for those participants that do not like to act or that take some ramp-up time finding into the role-play according to the pre-test.

However, the use of AR for learning purposes can be assumed helpful in order to the students' evaluation. As the results of this pre-test-study show, the technology has a high potential of motivating students in a specific, particularly abstract learning process —e.g. the lecture on scrum and agile management in this case available—by making the collaboration more tangible and, thus, the learning contents better conceivable. It can, however, also be assessed e.g. in future studies, to which extent a collaborative AR-app influences the communication and interaction within the team or group work in learning contexts. Moreover, future work and research is also needed in terms of the usability of the collaborative AR-app in order to make its use and handling more intuitive and, thus, investigate the effects on the collaboration in more detail. Also, the composition of the samples for the next study iterations will be chosen more carefully in terms of representativeness in order to be able to deduce recommendations for action that aim at describing the use of AR in engineering education beyond this single use case presented in this work.

Acknowledgements. This work is part of the project "Excellent Teaching and Learning in Engineering Sciences" (ELLI 2) and was funded by the Federal Ministry of Education and Research (BMBF), Germany.

References

1. Kaufmann, H., Steinbügl, K., Dünser, A., Glück, J.: General training of spatial abilities by geometry education in augmented reality. Ann. Rev. CyberTherapy Telemedicine **3**, 65–76 (2005)
2. Schnier, C., Pitsch, K., Dierker, A., Hermann, T.: Collaboration in augmented reality: how to establish coordination and joint attention? In: Proceedings of the 12th European Conference on Computer-Supported Cooperative Work (ECSCW2011), pp. 405–416. Springer, London (2012)

3. Lee, J.-H., Mraz, R., Zakzanis, K.K., Black, S.E., Snyder, P.J., Kim, S.I., Graham, S.J.: Spatial ability and navigation learning in a virtual city. Ann. Rev. CyberTherapy Telemedicine **3**, 151–158 (2005)
4. Schmalstieg, D., Hollerer, T.: Augmented Reality: Principles and Practice. Addison-Wesley Professional, Boston (2016)
5. https://www.marxentlabs.com/what-is-markerless-augmented-reality-dead-reckoning/
6. https://anymotion.com/wissensgrundlagen/augmented-reality-marker
7. Koch, C., Neges, M., König, M., Abramovici, M.: Natural markers for augmented reality-based indoor navigation and facility management. Autom. Constr. **48**, 18–30 (2014)
8. FitzGerald, E., Adams, A., Ferguson, R., Gaved, M., Mor, Y., Thomas, R.: Augmented reality and mobile learning: the state of the art. In: 11th World Conference on Mobile and Contextual Learning (mLearn 2012), 16–18 Oct 2012, Helsinki, Finland (2012)
9. Radu, J.: Augmented reality in education: a meta-review and cross-media analysis. Pers. Ubiquituous Comput. **18**(6), 1533–1543 (2014)
10. Dunleavy, M., Dede, C.: Augmented reality teaching and learning. In: Spector, J., Merrill, M., Elen, J., Bishop, M. (eds.) Handbook of Research on Educational Communications and Technology, pp. 735–745. Springer, New York (2014)
11. Wu, H.-K., Lee, S., Chang, H.-Y., Liang, J.-C.: Current status, opportunities and challenges of augmented reality in education. Comput. Educ. **62**, 41–49 (2013)
12. Goodhue, D.L., Thompson, R.L.: Task-technology fit and individual performance. MIS Q. **19**(2), 213–236 (1995)
13. Venkatesh, V., Bala, H.: Technology acceptance model 3 and a research agenda on interventions. Decis. Sci. **39**(2), 273–315 (2008)
14. Russel, J.A., Weiss, A., Mendelsohn, G.A.: Affect grid: a single-item scale of pleasure and arousal. J. Pers. Soc. Psychol. **57**(3), 493–502 (1998)
15. Beier, G.: Kontrollüberzeugungen im Umgang mit Technik: ein Persönlichkeitsmerkmal mit Relevanz für die Gestaltung technischer Systeme (2003)
16. Lefcourt, H.M.: Locus of control. In: Robinson, J.P., Shaver, P.R., Wrightsman, L.S. (eds.) Measures of Social Psychological Attitudes, Vol. 1. Measures of Personality and Social Psychological Attitudes, pp. 413–499. Academic Press, San Diego, CA, US (1991)

Developing Virtual Labs in Fluid Mechanics with UG Students' Involvement

C. Sivapragasam[1(✉)], B. Archana[1], G. C. Rithuchristy[1], A. Aswitha[2], S. Vanitha[1], and P. Saravanan[1]

[1] Center for Water Technology, Department of Civil Engineering, Kalasalingam Academy of Research and Education (Deemed to be University), Krishnankoil, Virudhunagar (District), Tamilnadu, India
sivapragasam25@gmail.com

[2] Department of Chemical Engineering, Kalasalingam Academy of Research and Education (Deemed to be University), Krishnankoil, Virudhunagar (District), Tamilnadu, India

Abstract. Although there are mixed opinion upon the advantages and disadvantages associated with the use of virtual labs, they are still increasingly being adopted by institutions for their curricular teaching. In this study an attempt is made wherein the undergraduate students are made to directly involve in creating virtual lab using the LabVIEW platform. To the authors' knowledge this is one of the first attempts in the country. The virtual labs are created for tracking the profile of the jet trajectory from an orifice fitted in a tank and for drawing of flownet for a given velocity potential stream function as a part of experiments in fluid mechanics of civil engineering curriculum at undergraduate level. The response of the students indicate that the experience gained by involving in creation of virtual lab gives them a better understanding of the advantages and disadvantages associated and builds confidence in using them.

Keywords: Virtual lab · Jet trajectory · Flownet · Fluid mechanics

1 Introduction

While virtual labs are increasingly being adopted worldwide in higher education sectors to impart deeper understanding and 'experiences' of the subject or topics being taught, the initiative to embrace such application of information technology in education sectors in developing countries like India are still in its inception. Today, the most important and challenging question that all educators have to answer is how to give real time experience to the students of 'digital generation' who are more accustomed to use of technology than their predecessors have ever imagined. This is particularly true in developing countries. It is not practically and economically feasible for many institutions to create all the necessary hardware to teach practical experience to students. Hence, incorporation of technological innovations into teaching with an aim to impart practical experience has almost become a necessity.

Daineko et al. [1] summarizes that while on the one hand the use of virtual labs offer many important advantages viz., purchase of expensive equipments/reagents,

M. E. Auer and K. Ram B. (Eds.): REV2019 2019, LNNS 80, pp. 733–741, 2020.
https://doi.org/10.1007/978-3-030-23162-0_66

modeling processes/phenomena which is outside our experience, simulating scenarios with a wide range of combination of parameters etc. to name a few, on the other hand there is also a lacuna of imparting full practical skills and develop competence for real research. One alternative for this issue can be to allow the students themselves to involve in the development of virtual labs. When the students have some practical experience in the lab in conducting the basic experiments, this experience can be properly nurtured and directed towards developing experiments on virtual platform for which the physical experimental set-ups are not available. This approach also offers another very important advantage. As observed by Ayas and Altas [2], commercial software packages usually are highly user friendly wherein the users can enter data and get results, but students are never aware of what is going on behind the screen, what assumptions are used or which design equations are implemented. When students involve themselves in developing the necessary package/tool at some simple level, they get a much deeper insight on how a given problem can be best implemented under a virtual environment. This study demonstrates such an attempt by two undergraduate students at their second year of Bachelors degree in developing virtual labs for two of the experiments in the fluid mechanics laboratory, which is an undergraduate course at second year of engineering degree in Civil Engineering in India.

The "Virtual Labs" scheme as practiced in India is an initiative of Ministry of Human Resources Development (MHRD), Government of India under the National Mission on Education through ICT (NMEICT). Under this scheme there are 12 participating higher technical institutions including the 7 older Indian Institute of Technology (IITs) who have been actively involved in creating virtual labs catering to various fields of engineering. However, the experiments in fluid mechanics have received very minimal attention under this scheme. There are other reported works in the field of hydraulics, in general, like Gao et al. [3] described two types of hydraulic experiment supporting animations namely schematic diagram based 2D virtual hydraulic circuits using Flash and VRML—based 3D virtual hydraulic equipments and reported that the students' response about the courseware are very encouraging. Similarly, Ribando et al. [4] developed a PC based GUI for solving the forced convection over a heated flat plate wherein the students can take plate temperature and surface temperature gradient measurements using a slider mechanism from which local and overall virtual convection correlations can be developed. To the authors' knowledge engaging the students themselves in the development of virtual experiments is not reported so far. Further, virtual experiments in basic fluid mechanics experiments have also not received sufficient attention.

In this study, creation of virtual experiments by second year undergraduate students in Bachelor's degree programme using LabVIEW is discussed for two experiments viz., jet profile from an orifice fitted to a rectangular tank and flownet construction for a given set of stream function and velocity potential function. While the curriculum prescribes only the estimation of coefficient of discharge (C_d) for the orifice [5], the students were asked to create virtual experiments for jet profile which as a part of extended experiment. Similarly, the construction of flownet is also outside the scope of their curriculum. Currently, this virtual lab has been done for intra university access and soon it will be enabled to inter university also as remote access.

The rest of the manuscript is organized as follows. The next section discusses briefly about the LabVIEW platform. Then a summarized discussion is provided on the basic concepts behind formation of jet profile and the flownet. The implementation of these two experiments in the LabVIEW is described and conclusions are drawn in the subsequent sections with an analysis on the experience and response of the students

2 LabVIEW

LabVIEW (**Lab**oratory **V**irtual **E**ngineering **W**orkbench) is a graphical programming environment which has been adopted in many research labs, academia and industry for analyzing the system for measurement and automation. Its graphical programming language called G *programming* is performed using a graphical block diagram. LabVIEW is embedded with data collections and analysis in a flexible programming environment. The GUI has drag—and—drop user interface on its control palette from which required applications can be created using the built-in functions. It also facilitates easy integration with new technologies in the future. The provision of visualization tools like charts and graphs are particularly suited for creating virtual labs to simulate physical experiments.

There are two main associated windows namely the front panel window and the block diagram window. The front panel is built using controls and indicators and consists of knobs, push buttons, graphs etc. The block diagrams accompany the front panel. The block diagram consists of nodes, terminals, wires and also includes components such as built-in functions, constants and control structures which are used for designing the implementation of the two case studies considered in this study.

3 Jet Profile and Flownet Concept

An orifice is a small opening provided in a vessel through which the liquid flows. The orifice may be provided in the sidewall or bottom of the vessel which is used for discharging fluids into the atmosphere from tanks. As the jet of water projects out of an orifice, the shape of the trajectory will vary according to the variation in head. In many applications, it is of interest to know this change in profile. The equations used in determining the profile of trajectory are

$$Q_a = C_d\sqrt{2gh} \tag{1}$$

$$V_a = \frac{Q_a}{a} \tag{2}$$

$$y = \frac{1}{2}\frac{x^2}{V_a^2} \tag{3}$$

where Q_a is the actual discharge, V_a is the actual velocity and C_d is the coefficient of discharge. 'h' is the head of water under which the orifice functions. 'a' is a the cross sectional area of the orifice. 'x' and 'y' represent the ordinates to plot the profile.

The estimation of seepage and gravitational flow through soil structures or ground very much depends on flownet. A flownet is a grid obtained by drawing a series of streamlines (Ψ) and equipotential lines (ϕ) which gives a simple graphical technique for studying two dimensional, irrotational flows, when the mathematical calculation is difficult and cumbersome. Analytical method, Graphical method, Electrical analogy method are some of the methods to construct the flownets. While the equipotential lines join the points the equal velocity potential ϕ, the streamlines in flow net show the direction of flow. By definition, the equipotential lines and streamlines are defined for a two-dimensional flow as follows:

$$u = -\frac{\partial \phi}{\partial x} \quad v = -\frac{\partial \phi}{\partial y} \tag{4}$$

$$u = \frac{\partial \psi}{\partial y} \quad v = -\frac{\partial \psi}{\partial x} \tag{5}$$

More details about the concepts related to jet trajectory and flownet can be referred to any standard textbook in fluid mechanics [6, 7].

4 Implementation of Jet Profile Construction in LabVIEW

Figure 1 shows the front end designed to assign inputs for the construction of jet profile. The block diagram for this design is shown in Fig. 2. In this experiment head of water (h), C_d and diameter of orifice is given as inputs. The head and diameter are given in centimeters. The given inputs can be controlled by the side icon and the C_d ranges from (0.62–0.67). When the tool is run, area of orifice (m^2), actual discharge (m^3/s), actual velocity (m/s) will be displayed. The water tank level and the resulting virtual trajectory will be plotted in the range of 'x' between 0 and 1.5 m with an equal interval of 0.25 m. The profile for a head of 4 cm is shown in Fig. 3. This tool is user friendly and any student can easily use it to study the variation in the jet profile shape with the variation in the tank head. The students can also easily study whether change in the diameter of the orifice affect the profile shape as well as how the change in C_d effect the profile shape.

5 Implementation of Flownet Construction in LabVIEW

The flownet is constructed for a set of ψ and ϕ as represented below:

$$\phi = 2xy \tag{6}$$

$$\Psi = x^2 + y \tag{7}$$

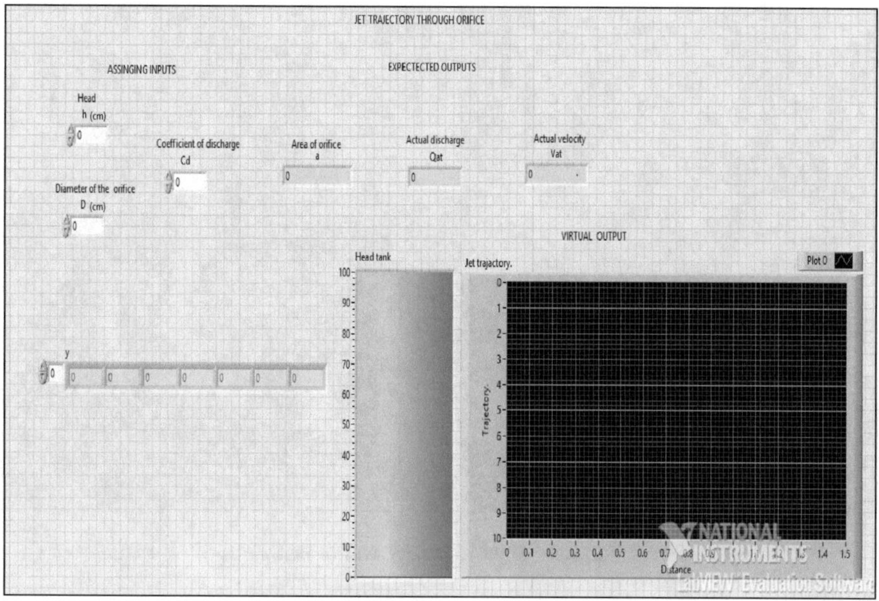

Fig. 1. Front panel for jet trajectory

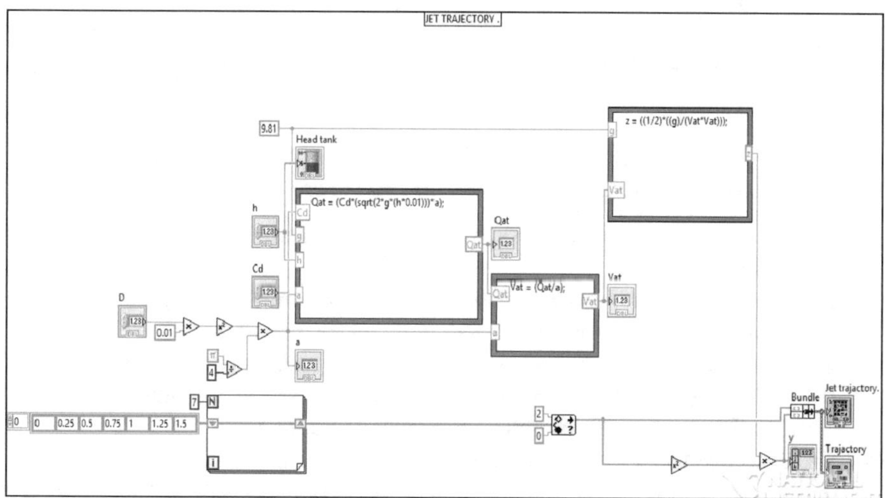

Fig. 2. Block Diagram for jet trajectory

For a given value of ψ and ϕ say 1 and 1, Eqs. (6) and (7) can be plotted for x and y. The range of x used is between 1 and 5 m with an interval of 0.5 m. The front panel is shown in Fig. 4 whereas the block diagram used to design this experiment is shown in Fig. 5. The screenshots obtained when the virtual experiment is run for $\phi = 1$ and $\psi = 1$ is shown in Fig. 6 (the first output window panel). By changing the ϕ and ψ, a

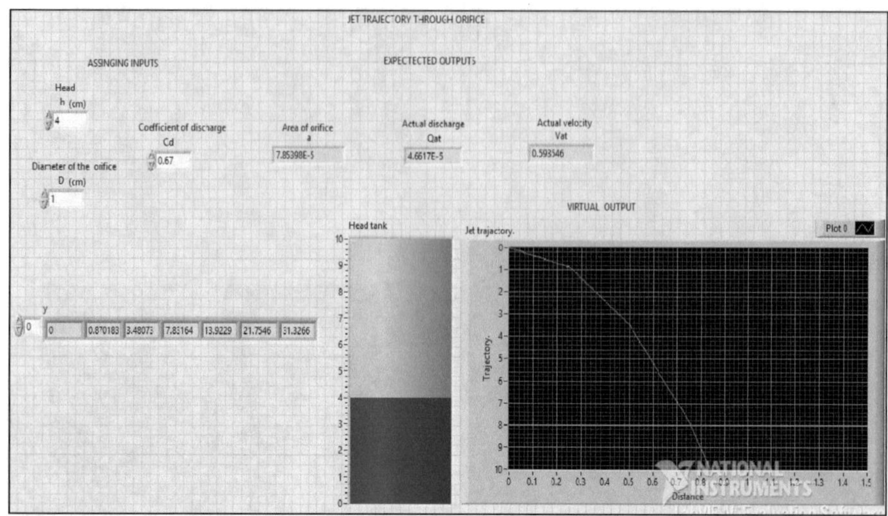

Fig. 3. Simulation run for jet trajectory with a tank head 'h' = 4 cm

set of streamlines and equipotential lines will be generated which when plotted together gives the flownet (as shown in the second output window panel in Fig. 6). This tool is helpful to the students in dynamically visualizing the formation of flownet as the values of ϕ and ψ are gradually changed. However, it has the limitation as it cannot be used for plotting any other set of ϕ and ψ than those given by Eqs. (6) and (7).

Students' response

A survey was carried out to obtain the feedback of the students involved in the design of the virtual experiments. The specific inputs are summarized as below:

(a) The students were very excited to be a part of virtual laboratory development assignment.
(b) Most importantly, the students developed a deeper understanding of the step-by-step process that goes into the development of a software tool.
(c) The efficiency of the software depends on the algorithm designed to implement the problem statement. Students could better understand through the trail-and-error the importance of designing an efficient algorithm.
(d) The students also gave demonstration of the developed virtual lab to other students in their class as well the junior students. This has generated interest in other students as well to use virtual labs for their studies.
(e) The work done by the students is in the process of refinement to make it more user friendly and then offer as online course to students from other institutions. The students are feeling encouraged to be a part of social service to others.
(f) The students realized that use of software without having prior direct practical exposure to the experiments will not be effective in learning process. Software can support and enhance the learning process but cannot fully replace the experience gained by practically doing an experiment.

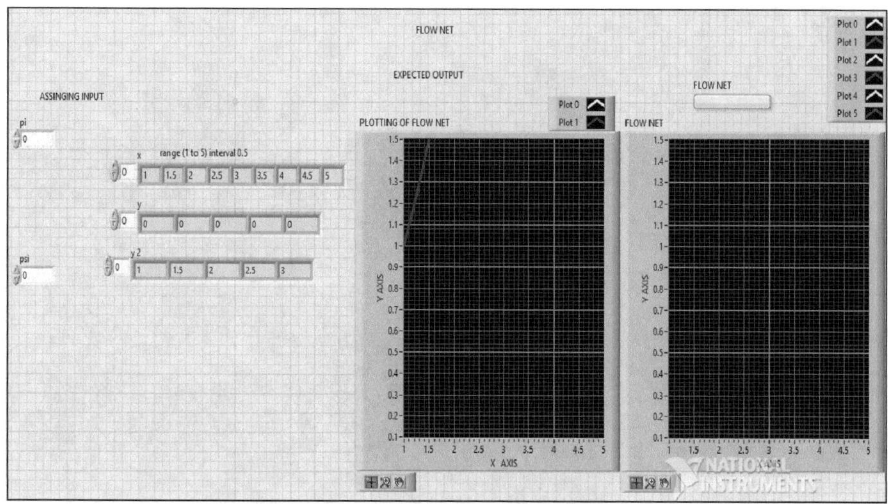

Fig. 4. Front panel for flownet

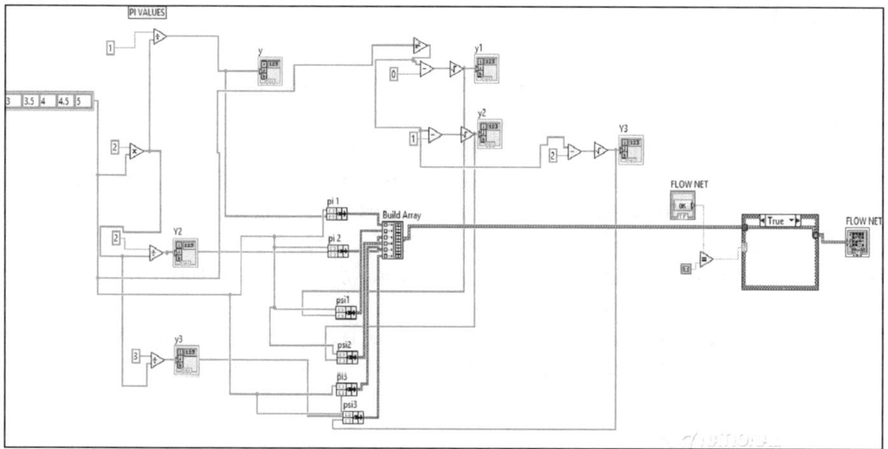

Fig. 5. Block diagram for flownet

(g) The students gained more knowledge and were keen about learning the ideas used in the virtual experiment.

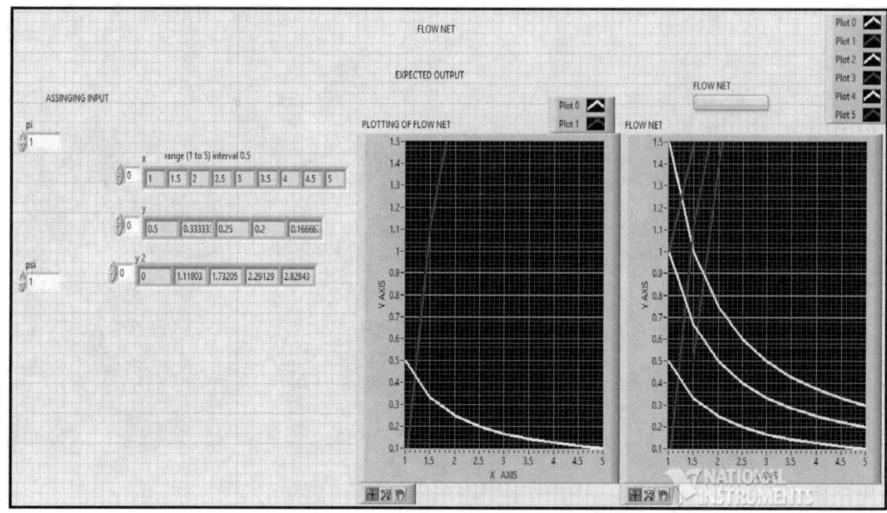

Fig. 6. Simulation results for flownet

6 Conclusions

Based on this study, it is concluded that the involvement of students in the development of virtual labs is an innovative approach to invoke deeper understanding of a given topic and for better understanding of the uses and limitations of software. Such an experience at the undergraduate level of an engineering course will help the students to take up similar assignments in the future in developing solutions for complex engineering problems. LabVIEW seems to be potential platform for the design of virtual experiments and can be made use of more thoroughly by the educational community.

Acknowledgements. The authors wish to thank the coordinators of National Instruments (NI) lab in the institute for their support in the successful implementation of this work.

References

1. Daineko, Y., Dmitriyev, V., Ipalakova, M.: Using virtual laboratories in teaching natural sciences: an example of physics courses in university, pp. 39–47 (2016)
2. Ayas, M.S., Altas, I.H.: A virtual laboratory for system simulation and control with undergraduate curriculum, pp. 122–130 (2015)
3. Gao, Z., Liu, S., Ji, M., Liang, L.: Virtual hydraulic experiments in courseware: 2D virtual circuits and 3D virtual equipments, pp. 315–326 (2008)
4. Ribando, R.J., Coyne, K.A., O'Leary, G.W.: Teaching module for laminar and turbulent forced convection on a flat plate, pp. 115–125 (1998)

5. Sivapragasam, C., Deepak, M., Vanitha, S.: Experiments in Fluid Mechanics and Hydraulic Machinery. Lambert Academic Publishing (2016)
6. Streeter, V.L., Whlie, E.B.: Fluid Mechanics. McGraw Hill, New York (1983)
7. Rama, D.D.: Fluid Mechanics and Machinery, 1st edn, reprint. New Age International Publishers, New Delhi (2006)

Open Education Resources and New Age Teachers

T. Anushalalitha(✉)

Department of Telecommunication Engineering, BMS College of Engineering, Bangalore, India
anusha.tce@bmsce.ac.in

Abstract. The advance of the internet related technologies have created a paradigm shift in the entire world, not to forget the education sector. Today the emphasis is not on the data but how we create, imbibe and make use of them. There is abundance of resources available, but it is important to enhance our contribution to this scenario and to streamline the available resources to the best use of students.

Keywords: OER · Education ecosystem

1 Purpose or Goal

1.1 Open Education Resources

Open Education Resources are one facility available to create a repository of education resources and made accessible to all as an open access platform. This gives the one who creates it a sense of satisfaction in sharing the knowledge that they have and also the one who receives them in the form of getting to understand the topics covered.

Open educational resources help in sharing of information for free and encourage others also to follow suit. Much of valuable resources are lost without being shared in many areas such as medicine, etc.

The Community College Consortium for Open Educational Resources—CCCOER, OpenStax, Free Online Course Materials, MIT OpenCourseWare (OCW), The Open Course Library, OER Commons, The Open Education Resource (OER) Foundation, WikiEducator, The World Digital Library (WDL) etc. are part of OERs working towards these objectives.

1.2 OERs from India

Digital Library of India (http://www.dli.ernet.in/), National Digital Library (NDL) (http://www.ndl.iitkgp.ac.in/), National Knowledge Network (NKN) (http://nkn.gov.in/), Shodhganga (http://shodhganga.inflibnet.ac.in/), Vidyanidhi (http://eprints.uni-mysore.ac.in/), National Programme on Technology Enhanced Learning (NPTEL) (http://nptel.ac.in/), Consortium for Educational Communication (CEC) (http://cec.nic.in/Pages/Home.aspx), Project Ekalavya (http://ekalavya.it.iitb.ac.in/), Project OSCAR (Open Source Courseware Animations Repository) (http://oscar.

M. E. Auer and K. Ram B. (Eds.): REV2019 2019, LNNS 80, pp. 742–747, 2020.
https://doi.org/10.1007/978-3-030-23162-0_67

iitb.ac.in/), National Mission on Education using Information and Communication (NMEICT) (http://www.sakshat.ac.in/), National Repository of Open Educational Resources (NROER) (http://nroer.gov.in/home/repository), National Council of Educational Research and Training (NCERT) (http://www.ncert.nic.in/), SWAYAM (Study Webs of Active-Learning for Young Aspiring Minds) are some of the developments towards achieving these objectives.

2 Approach

Open educational resources(OERs) were created as part of Two week AICTE approved FDP on, "Pedagogy for online and Blend learning Process" conducted by IITBombayX and uploaded on youtube.

2.1 OERs for Microwave and RADAR

OERs were created for selected topics as modules for the course Microwave and RADAR [4]. This was submitted as a group activity. Each module consist of Learning Dialogue (Led), Learning by doing (LBD), Learning Experience Interaction (LxI) and Learning extension resources (Lxt).

2.2 OER Format

For each topic the objectives were defined, some questions raised on the topic to create interest, some concepts were explained and some questions in the form of quiz or numerical problems included for testing the understanding of concepts. Discussion on topics in the forum can give increased scope for learning. A flipped class room activity was conducted based on OER developed and students assessed on the topic. https://wordpress.com/anushalalithablog.wordpress.com.

Some of the screenshots of the OERs are included here as a reference (Figs. 1, 2, 3, 4 and 5).

2.3 OERs for Optical Fiber and Satellite Communication

Efforts on creating OERs for the courses being handled are on and will be uploaded. A quiz has been conducted on these resources for Optical Fiber communication and satellite communication course and a seminar conducted on related topics. This will create a stable mechanism for sharing educational resources and reduce students' time in getting the relevant content (Fig. 6).

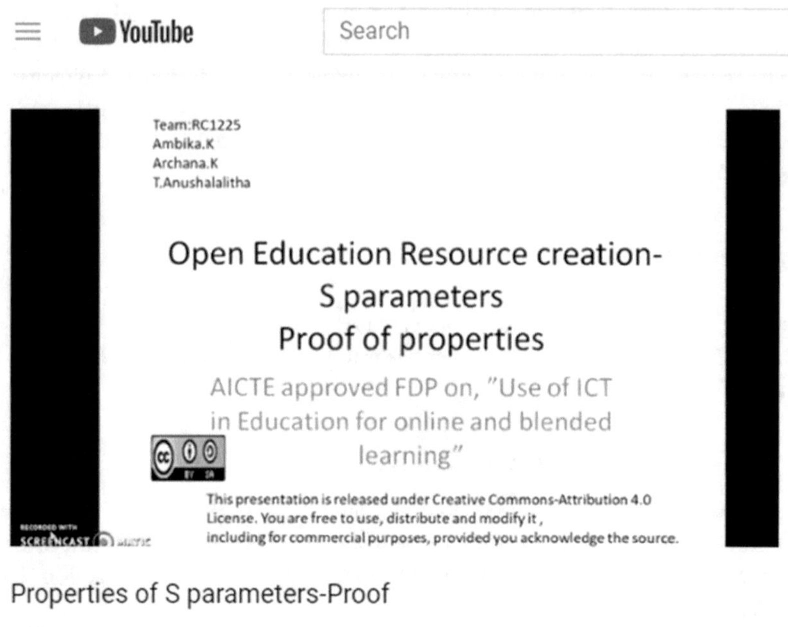

Fig. 1. Video on properties of S parameters

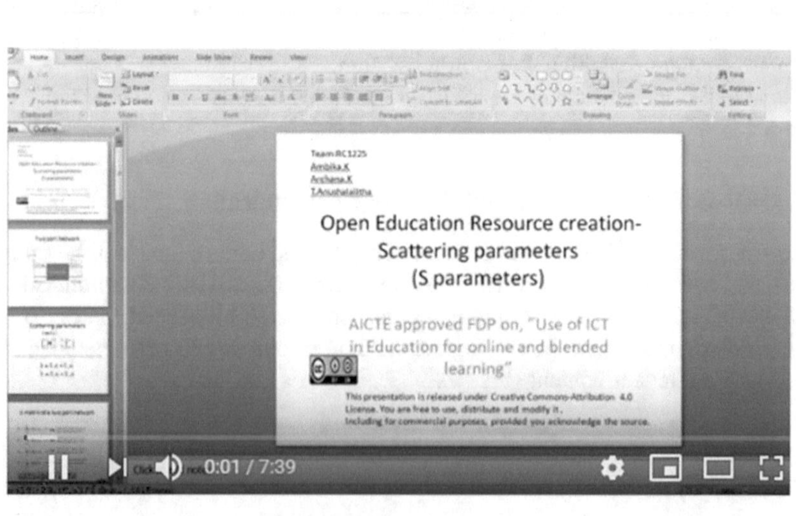

Fig. 2. Video on S-parameters of a two port network

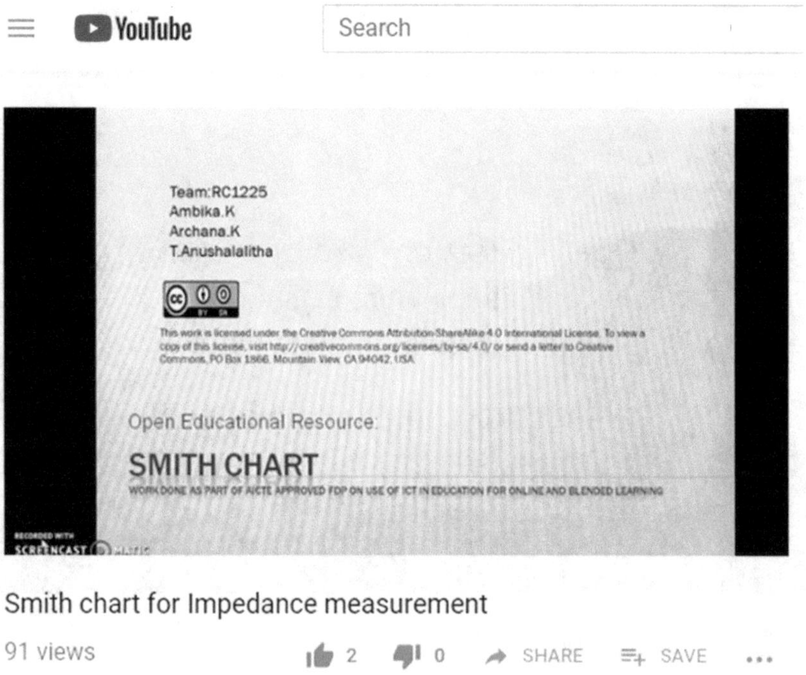

YouTube
Search

Team:RC1225
Ambika.K
Archana.K
T.Anushalalitha

This work is licensed under the Creative Commons Attribution-ShareAlike 4.0 International License. To view a copy of this license, visit http://creativecommons.org/licenses/by-sa/4.0/ or send a letter to Creative Commons, PO Box 1866, Mountain View, CA 94042, USA.

Open Educational Resource:
SMITH CHART
WORK DONE AS PART OF AICTE APPROVED FDP ON USE OF ICT IN EDUCATION FOR ONLINE AND BLENDED LEARNING
RECORDED WITH SCREENCAST-O-MATIC

Smith chart for Impedance measurement
91 views 2 0 SHARE SAVE

Fig. 3. Video on impedance measurement using Smith chart

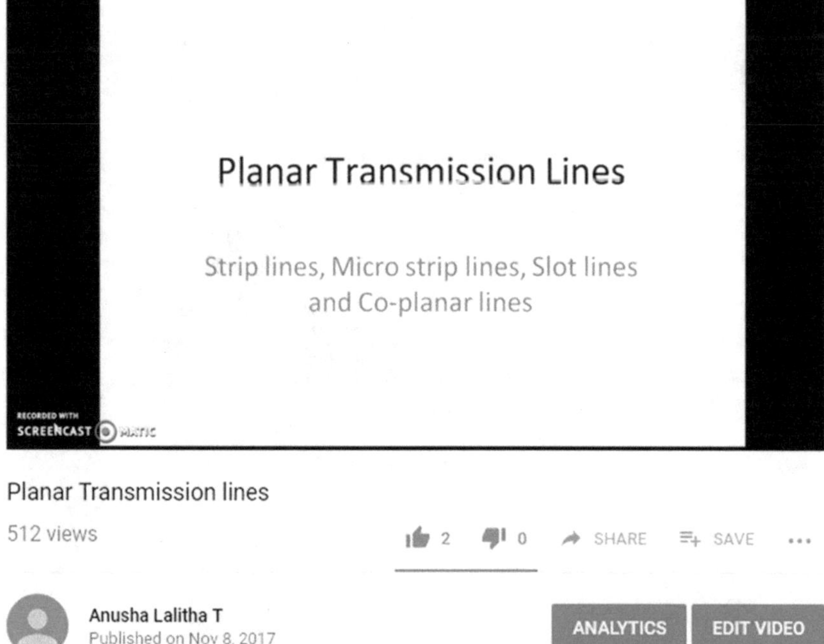

Fig. 4. Video on planar transmission lines

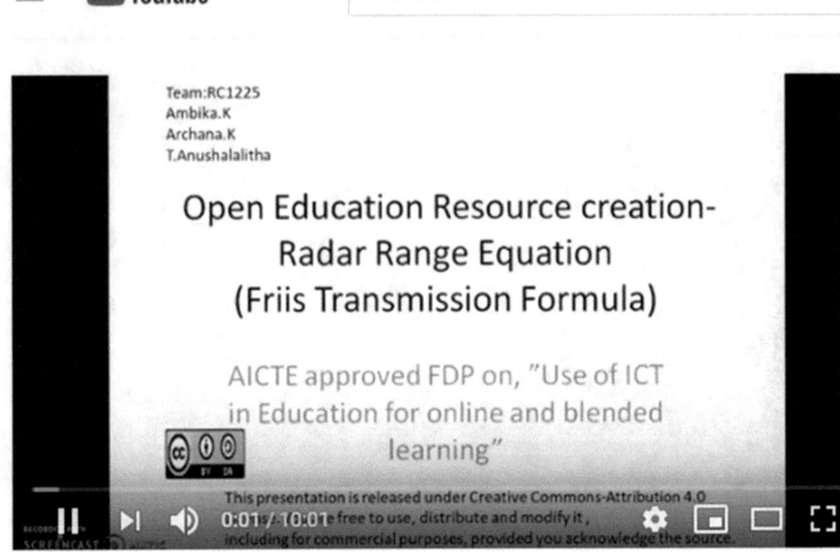

Fig. 5. Video on radar range equation

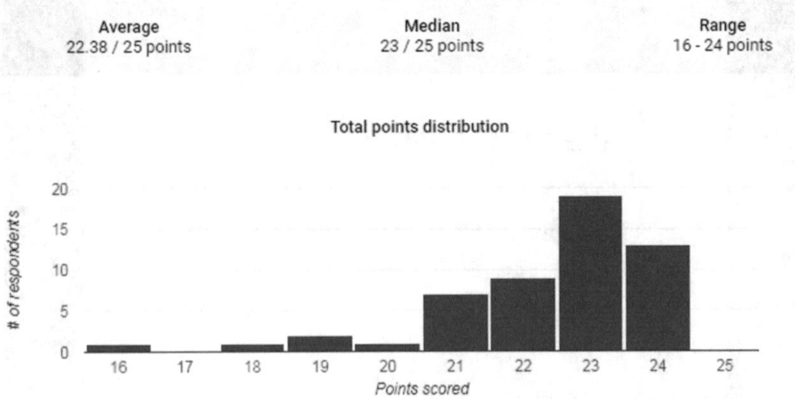

Fig. 6. Results of quiz

3 Actual or Anticipated Outcomes

If scaled up this can be part of a regular process and students can directly access them and focus on higher order skills including open ended experiments on each course and attain better learning outcomes. Exercises like these help in demonstrating the ability of the Teachers against their peers in other institutions. The challenges lie in the creation of original materials and the scope lies in including the outgoing students working in related domains contribute further in improving the ecosystem.

4 Conclusions/Recommendations/Summary

Regular updating of resources from research papers and project work can help institutions develop valuable content and help in developing the faculty also which in turn help in the development of students.

Acknowledgements. This work is submitted as part of TEQIP-III program of BMS college of Engineering, Bangalore.

References

1. Premawardhena, N.C.: Introducing computer aided language learning to Sri Lankan schools: challenges and perspectives. In: 15th International Conference on Interactive Collaborative Learning and 41st International Conference on Engineering Pedagogy (ICL & IGIP), Villach, Austria (2012)
2. https://library.educause.edu/topics/teaching-and-learning/open-educational-resources-oer
3. https://www.iaeme.com/MasterAdmin/uploadfolder/IJLIS_06_05_003/IJLIS_06_05_003.pdf
4. Course details available at www.bmsce.ac.in

Remote Monitoring and Control of Electrical Systems with Augmented Reality and Digital Twins

Vaishnavi Nagesh Kalyavi[1], Vidhya Thandayuthapani[1(✉)], Meena Parathodiyil[1], and Desh DeepakSharma[2]

[1] B.M.S. College of Engineering, Bangalore, India
{vaishnavi2908,vidhyat98,meenabms}@gmail.com
[2] MJP Rohilkhand University, Bareilly, India
deshdeepak101@gmail.com

Abstract. In this paper an attempt is made to bring in an interactive laboratory experience in the working of an electric circuit. The circuit operation can be controlled remotely. The augmented reality system is implemented through an app developed on a smart mobile phone. The whole system involves a physical electric circuit, its digital counterpart and a smart mobile phone app. The mobile app developed is capable of identifying the individual circuit elements and overlaying it on the circuit. It is capable of implementing changes in circuit parameters in real time through the app and the display of the waveforms from the operation of the circuit in real time. The paper also highlights the successful working prototype of an electrical load such as an electric bulb whose on-off and brightness control operation is accomplished remotely through the mobile phone and state of the lamp is visualized, its electrical parameters such as voltage, overlaid and controlled in real time through the app on the mobile phone. The results indicate the reliable and stable operation of such systems in an uninterrupted internet facilitated environment.

Keywords: Augmented reality · LABVIEW · Google firebase · Unity · Vuforia · NodeMCU · Remote control of electric bulb · Digital twin · Education · Cloud data transfer · QR code

1 Introduction

Inventions of new technologies with a wider spectrum of applications have enabled breakthroughs in several fields of engineering. Virtual, augmented and mixed reality have revolutionized the world of experience for humanity at large. Their footprint has been remarkable in overcoming several challenges that existed in various domains. The capability to blend software and hardware to create spatial effects and control through the internet has resulted in several innovations to improve the operating efficiency of systems. The future of these technologies entails professionals with a thorough understanding of concepts and ability to innovate.

This has created a transformation in the world of education and necessitates the development of interesting material to encourage the involvement of individuals in the

M. E. Auer and K. Ram B. (Eds.): REV2019 2019, LNNS 80, pp. 748–760, 2020.
https://doi.org/10.1007/978-3-030-23162-0_68

learning process. The electrical circuits in laboratories are generally systems that operate in standalone mode and one is clueless about the component details through physical observations. The inability of humans to ascertain the details and the lack of ease of repeated interactions with such systems imposes challenges in understanding the underlying concepts and principles of operation of these systems. Incorporation of augmented reality enhances the quality of learning experience through interactions. A decade ago, controlling an electric circuit from distant places was a far fetched idea. With the advent of internet this idea has become a reality now and is extensively used. The access to such experiences from anywhere at any time also encourages participation. The project presented integrates augmented reality and Google Firebase, a cloud data transfer platform to analyze the circuits and provides the capability to the user to control and monitor them from anywhere at any time.

The block diagram shown in Fig. 1 shows the overall features of the augmented reality system. It includes overlay of text over the physical real environment as well as User interfaces such as buttons and sliders for the switching and control operations of the circuit.

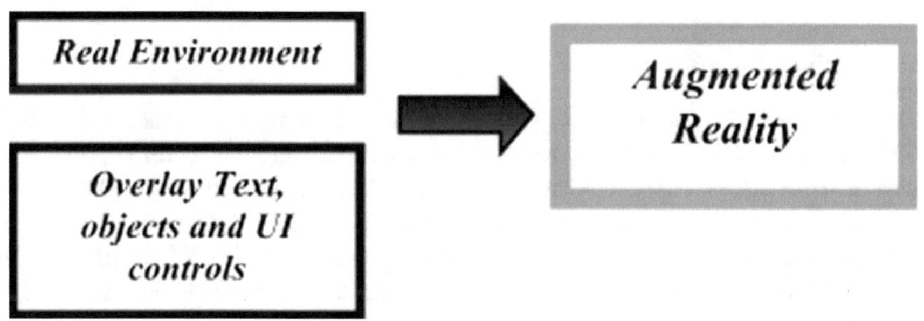

Fig. 1. The fundamental features of the augmented reality system

The physical electrical circuit is controlled through a mobile app based on augmented reality and involves the real time transfer of the circuit control parameters through cloud. Unity and Vuforia platforms are used to build a mobile app that is used to control the electric circuit and visualize a digital twin that has the same function as that of the physical electric circuit. The digital twin is configured to assert the performance of the circuit. There are two physical circuits developed, one being a simple diode rectifier circuit with small signal components and the other being a large signal power circuit with a light bulb.

To distinguish the selection of the two circuits in the app, Quick Response (QR) codes are used.

The two encoded QR codes utilized in the app to choose the two different circuits are as shown in Fig. 2.

Fig. 2. Encoded QR codes for power circuit and half wave rectifier

2 Description of Circuits

There are two circuits rigged up. One being the rectifier and the other is that of powering an electric appliance, an incandescent bulb being considered here.

2.1 Rectifier Circuit

A half wave rectifier circuit is rigged up on a breadboard by connecting a diode 1N4007 in series with a load. Figure 3a shows the block diagram representation of the experiment set-up. A resistive load is connected. Two capacitors 1 and 0.01 μF are connected in parallel with the load through a relay contact such that, when relay is on 1 μF capacitor gets connected in the circuit and when the relay is off 0.01 μF capacitor gets connected. The relay operation selects the capacitor to be included. The relay is activated through the app developed for the purpose.

Figure 3b shows the overlay of the specifications of the circuit components through augmented reality.

2.2 Power Circuit

An incandescent light bulb is used to demonstrate the working of power circuit. The block diagram representation of the set-up is as shown in Fig. 4a. The bidirectional switch used is a TRIAC which is controllable through digital signals.

The digital signals are isolated through an Optocoupler. The control terminal of the TRIAC is connected to NodeMCU which retrieves data sent through a mobile app through Firebase cloud for ON-OFF as well as control of voltage applied to the lamp.

A digital equivalent of the bulb is represented on the interface to the user so that he/she can visualise the state of the apparatus he is controlling from a remote location. In this case, the brightness of the bulb for different voltage levels are replicated.

(a)

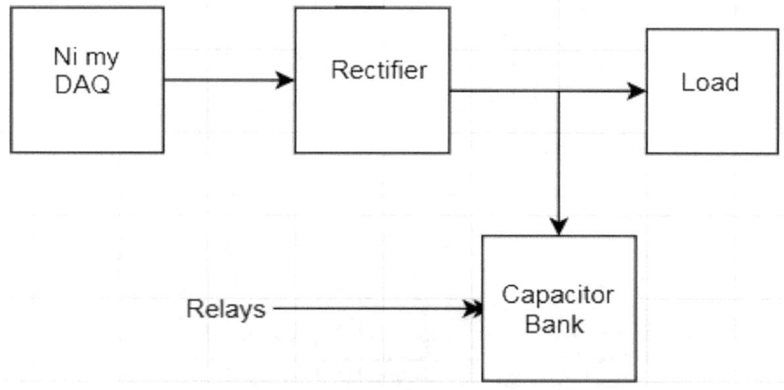

(b)

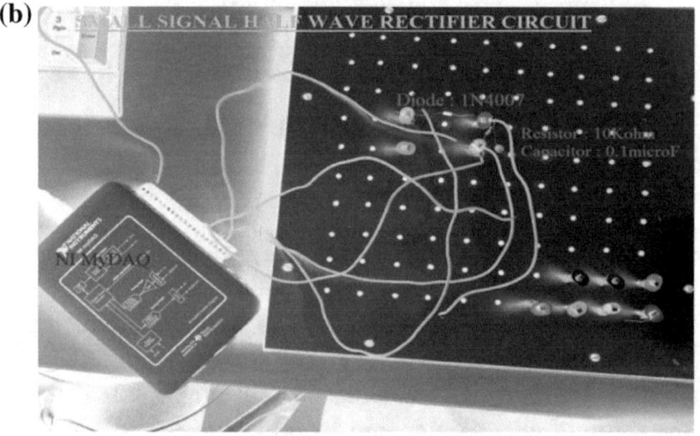

Fig. 3. **a** Block diagram of the rectifier set-up. **b** Specifications of the components of the half wave rectifier circuit overlaid through augmented reality

3 Development of Application

QR codes are coded for two use cases. The first case being, the control of a small signal rectifier circuit and the second case is control of voltage in a power circuit. The QR codes are encoded. They are later decoded to activate the control parameters of the respective circuits namely, Resistors, capacitors and bulb, that are used as image targets. Image targets are used to recognize the circuit through Augmented Reality. Image targets are downloaded as unity packages as the Unity software uses only unity packages. Unity software is used to design the user interface with sliders and buttons to perform required operations. C# coding language acts as the backend coding software for unity. The developed interface is downloaded on the mobile as a mobile app. Figure 5 shows the sequence of process involved in the development of the mobile application.

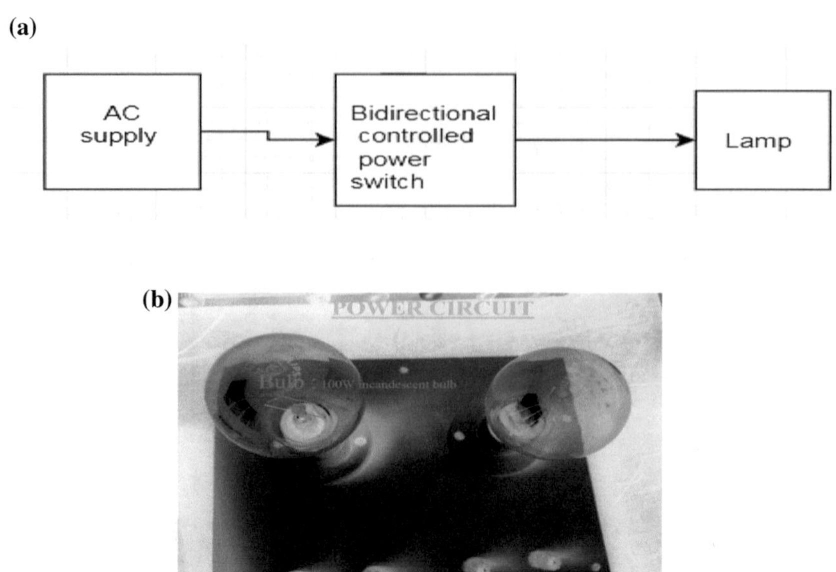

Fig. 4. **a** Block diagram of the power circuit set-up. **b** Information of power circuit overlaid through augmented reality

QR codes are generated and sent to the users only. Nobody can use the app without the QR codes. These QR codes must be scanned by the users for activating the controls of the circuits. QR codes act as additional security to forbid data leak and unauthorized use of the app.

The User Interface window also includes several plug-ins which aid the user in learning about the setup of experiments and its operation.

The instruction manuals are stored as PDF documents. The app is configured such that, on pressing a button the PDF documents are opened and the user can read it before conducting the experiment.

Circuit diagram of the experimental setup is very important to understand the working of the experiment. A button has been configured to open a 2D image of the circuit diagram.

Videos describing the experiments can also be accessed through the app for additional information.

Online video conferencing tool SKYPE is also integrated with the app and the user can get remote assistance or clarify doubts from the professor by pressing the concerned buttons.

While using the app, if the user is unable to understand certain steps or if he gets stuck, then a screenshot button is provided to help the user. The user can take a screenshot of the step he is unable to understand and send it to the concerned professor and get the support from the professor.

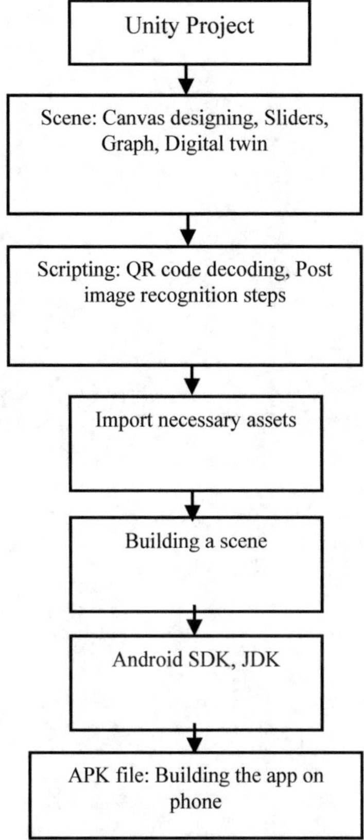

Fig. 5. Flow chart for building the mobile app

A digital twin of the power circuit is also interfaced with the application so that the user can visualize the working of the device he/she is controlling. A digital twin is a digital replica of the process (Fig. 6).

4 Tools Used

4.1 Unity

Unity is a multipurpose game engine that supports 2D and 3D graphics, drag-and-drop functionality and scripting using C# created by Unity Technologies.

This software is used to build an user interface and it provides a platform to implement augmented reality. Augmented reality is used to showcase a digital twin (of the electric bulb) that works the same way as in the physical circuit (Fig. 7).

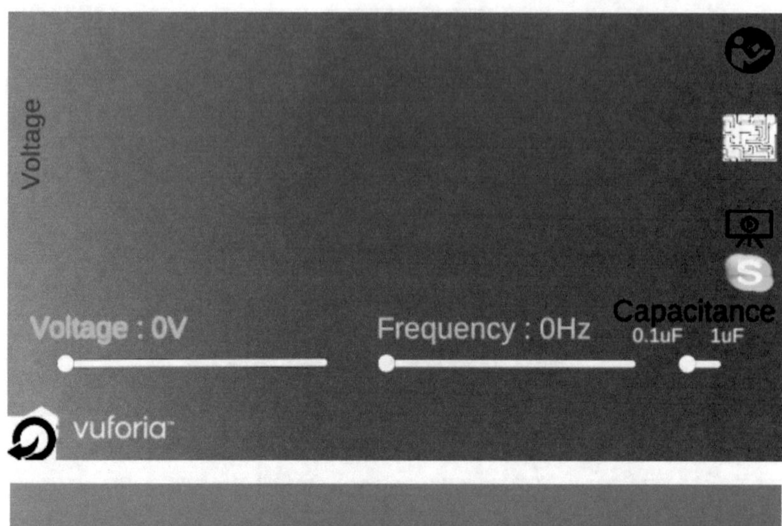

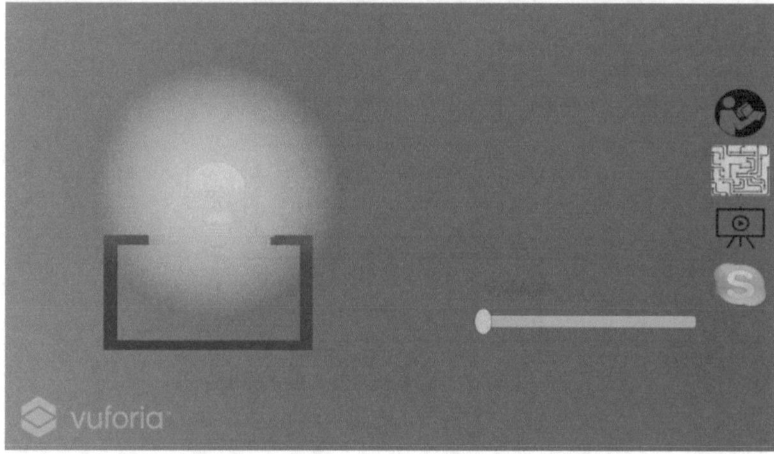

Fig. 6. User interface buttons configured in the mobile app

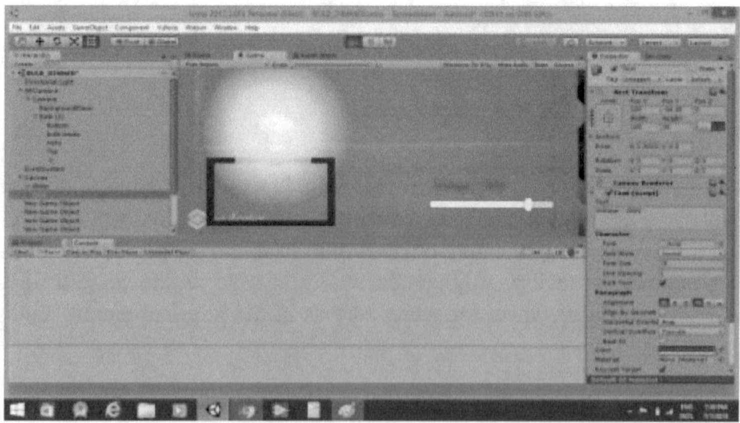

Fig. 7. Unity software

4.2 Vuforia

Vuforia is an Augmented Reality Software Development Kit initially owned by Qualcomm and now taken over by PTC (Parametric Technology Corporation). It can be used with Unity game engine and Android Studio to create Augmented Reality applications.

In the development of this application, the augmented reality (AR) camera feature of Vuforia Unity is used which facilitates overlaying of required data on the camera screen.

4.3 Google Firebase

It is a cloud platform provided by Google. It is one of the fastest real time data transfer service. It provides Application Program Interfaces used to retrieve and feed data to the cloud. It requires a good internet connection for real time transfer of data.

In the developed application, google firebase API package for Unity was used to integrate the cloud services to the application, in order to facilitate exchange of data.

4.4 Labview

LABView is an acronym for Laboratory View. It is a graphical programming software developed by National Instruments. This software has been used as an interface between computer and NI MyDAQ (Fig. 8).

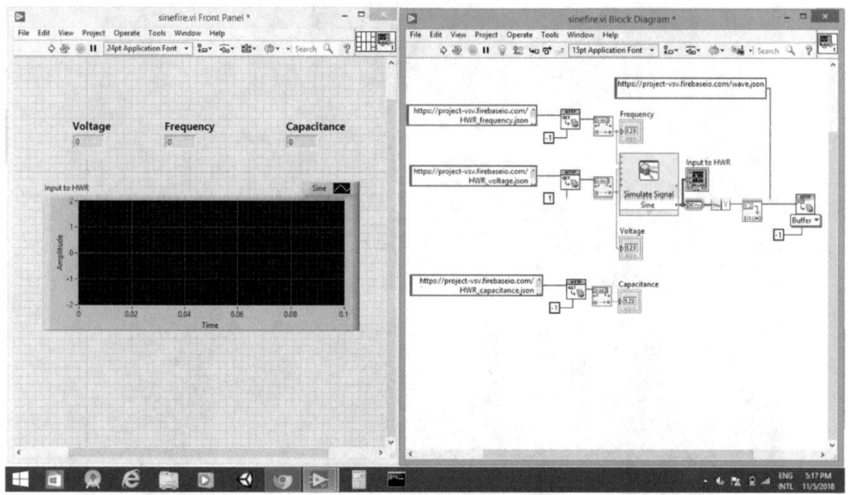

Fig. 8. LABView code for interfacing firebase data with NI MyDAQ

5 Operation

The mobile app is used to activate the controls of either the power circuit or the electric circuit by scanning the QR code. Moving the sliders shown in Fig. 7 changes the input parameters for the rectifier such as voltage, frequency and capacitance variation achieved using the button on the screen that activates the relay. The designed digital twin (the brightness of the bulb) parameters also change accordingly (Figs. 9 and 10).

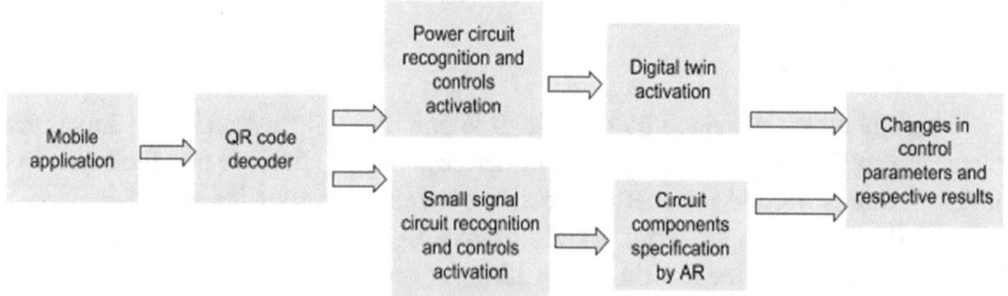

Fig. 9. Block diagram of the mobile application

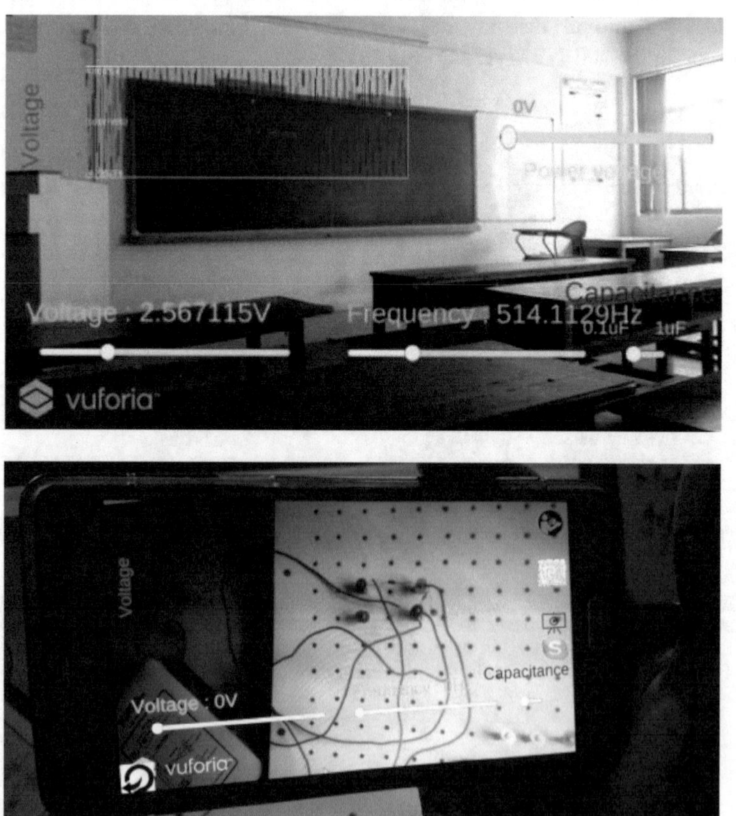

Fig. 10. Controls of half wave rectifier circuit displayed after circuit recognition

The output of the mobile app will be the changes in voltage and frequency of the small signal rectifier circuits. This signal will be stored to the Google Firebase cloud as integer values. Changes in these parameters through the app are transferred to Google firebase in real-time and the data is retrieved by the LabVIEW software. This software controls the DAQ unit and the input parameters are changed to the circuits. Thus, a remote transfer of input parameters is achieved. The rectifier circuit is powered using a computer based lab instrument NI MyDAQ which is a student data acquisition device whose output can be controlled through a LAB VIEW program. It is also capable of acquiring signals from the external rectifier circuit (Fig. 11).

Fig. 11. Block diagram of the operation

6 Components Used

6.1 Ni MyDaQ

It is a simple and intuitive data acquisition device from National Instruments.

- Analog Inputs (AI), Analog Outputs (AO)
- Digital Inputs (DI) and Digital Outputs (DO).

Specifications:

- Two Differential Analog Input and Analog Output Channels (200 ks/s, 16 bits, +/- 10 V)
- Eight Digital Input and Digital Output Lines (3.3 V TTL-Compatible) +5, +15, and-15 V Power Supply Outputs (up to 500 m Watts of Power)
- 60 V Digital Multimeter (DMM) for measuring voltage, current and resistance
- Built-in digital multimeter.
- Used as a power supply.
- Used as an oscilloscope and function generator (Fig. 12).

6.2 Node MCU

NodeMCU is an open source IoT platform. It includes firmware which runs on the ESP8266 Wi-Fi SoC from Espressif Systems, and hardware which is based on the ESP-12 module. The term “NodeMCU” by default refers to the firmware rather than the development kits.

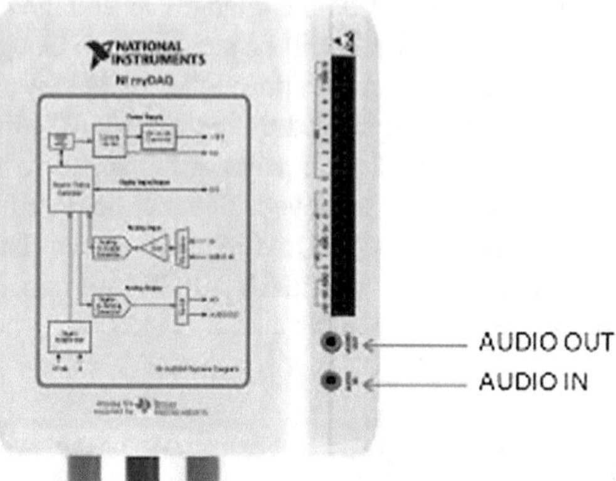

Fig. 12. NI MyDaQ

NodeMCU provides access to the GPIO (General Purpose Input/Output) and for developing purposes. Pin mapping table from the API documentation should be referenced (Fig. 13).

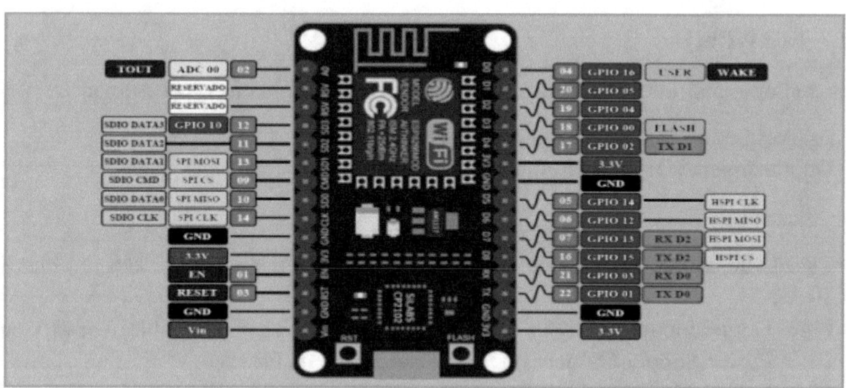

Fig. 13. Pin diagram of NodeMCU

LABVIEW software is used to control and monitor rectifier circuit. It is interfaced with NI myDAQ unit which can send the voltage to circuit and receive the output as well. The output from Firebase is analyzed in LABVIEW and the DAQ is instructed to send voltage based on the Firebase output. Hence, voltage and frequency can be controlled. If relay is off the default capacitor is connected in the circuit and when relay is on, the other capacitor gets connected.

Arduino programming language is used to code NodeMCU. NodeMCU uses an ESP wifi module to receive the data from cloud and it has an embedded arduino to perform control operations. Based on the Firebase data it received, Arduino sends the data required as control signal to vary the input voltage fed into the power circuit.

Arduino is a small signal circuit and bulb is a power device. These two cannot be connected directly. An Optocoupler is used to isolate the control signal and power circuits.

7 Results Obtained

Observation data during the lab test session leads to following conclusions:

1. The functioning of the app was tested from a location far away from the setup. The control of the small signal and power signal circuit was conducted successfully.
2. Real-time control as well as monitor of the circuit was achieved without any time delay and glitches.
3. The performance of the digital representation of the power circuit was also observed and recorded successfully (Fig. 14).

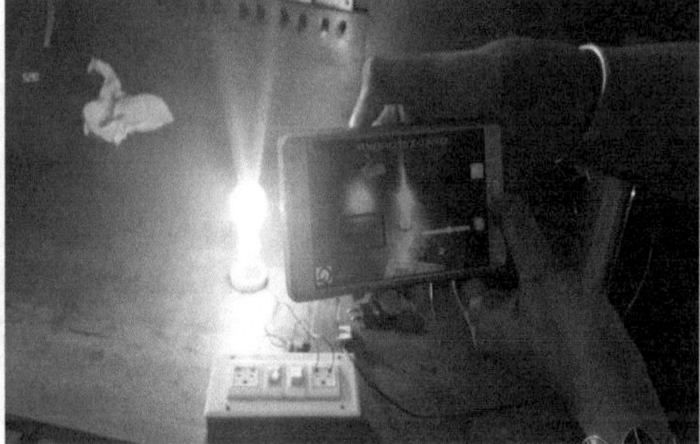

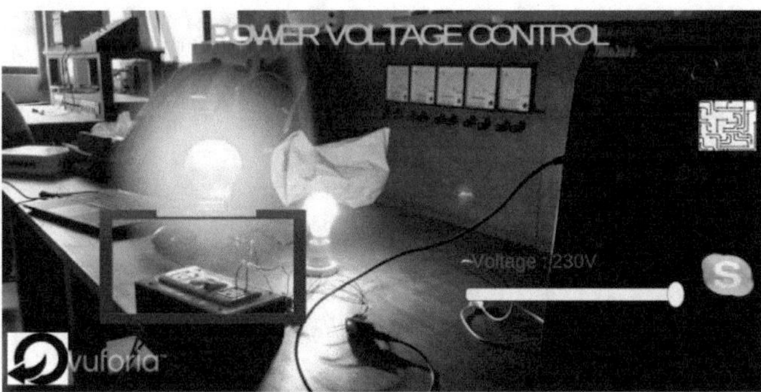

Fig. 14. Power circuit controls and digital equivalent with the bulb in the background

8 Conclusion

Augmented reality is a new and growing platform in the current era. The use of AR in the project emphasises the use of new technology. Augmented reality is an enhanced version of reality created by the use of technology to overlay digital information on an image of something being viewed through a device. The conventional camera displays only the existing world. Through Augmented Reality, the world can be viewed on a whole new platform. Incorporation of Augmented reality to visualize details of the circuit which is otherwise hidden from the user, aid in better understanding of subject concepts. This work focuses on interfacing augmented reality with electrical circuits and simplifies the use of them. It is a tool that can be applied using the methodology discussed in several fields for remote monitoring and controlling of devices. The cloud transfer is implemented through Wi-Fi. Good internet bandwidth requirements is one of the main conditions for data transfer.

Acknowledgements. We thank the Department of Electrical and Electronics Engineering, BMS College of Engineering, Bangalore, India for providing the platform and necessary equipments to carry out our work.

References

1. Aviles′, F.R., Cruz, C.A.: Mobile augmented reality on electric circuits. In: IEEE Computing Conference 2017, 18–20 July 2017. London, UK
2. Google Firebase: Google analytics for firebase for unity. [Online]. https://firebase.google.com/docs/analytics/unity/start
3. Opriş, I., Costinaş, S., Senior Member IEEE, Ionescu, C.S., Nistoran, D.E.G.: IEEE paper on Towards Augmented Reality in Power Engineering, University POLITEHNICA of Bucharest
4. Restivo, M.T., de Fátima Chouzal, M., Rodrigues, J., Menezes, P., Patrão, B., Lopes, J.B.: Augmented reality in electric fundamentals. Int. J. Online Eng. (2014)
5. National Instruments: Learn LABView. [Online]. http://www.ni.com/academic/students/learn-labview/
6. National Instruments: Getting started with LABView. [Online]. http://www.ni.com/pdf/manuals/373427j.pdf
7. National Instruments: LABView web development documents. [Online]. https://forums.ni.com/t5/LabVIEW-Web-Development/NIWeek-2016-Develop-Distributed-Systems-of-the-Future-Using-the/ta-p/3538624
8. Unity: Scripting API reference. [Online]. https://docs.unity3d.com/ScriptReference/
9. Vuforia: Unity API reference. [Online]. https://library.vuforia.com/content/vuforia-library/en/reference/unity/
10. Wikipedia: Augmented reality. [Online]. https://en.wikipedia.org/wiki/Augmented_reality

Public Health Surveillance to Promote Clean and Healthy Life Behaviours Using Big Data Approach (An Indonesian Case Study)

Vitri Tundjungsari[1(✉)], Kholis Ernawati[2], and Nabilah Mutia[1]

[1] Faculty of Information Technology, YARSI University, Jakarta, Indonesia
vibarall@gmail.com
[2] Faculty of Medicine, YARSI University, Jakarta, Indonesia

Abstract. Clean and Healthy Life Behaviour is a social engineering program stated by Ministry of Health Republic of Indonesia that aims to make as many people as agents of change to improve the quality of public health. This program is named as Perilaku Hidup Bersih Sehat (PHBS) in Indonesian language. Clean and healthy life behaviour is very important to prevent infectious disease. In Indonesia, the prevalence of infectious diseases due to high risk daily behaviour (unclean and unhealthy) is high, thus we need a solution to solve this problem. This research discusses the potential use of big data concept (including analytics and data visualization) as an approach for implementing web-based health surveillance related to clean and healthy life behaviour. The main objective of the application is to promote clean and healthy life behaviour as well as to improve the quality of public health through awareness-raising processes which is delivered by data visualization. The application provides areas/districts where required more attention to promote clean and healthy life behaviour. As a result of this, government also has more information and knowledge for producing better policy toward better public health. Another main benefit of this application is the creation of a health conscious society and has the provision of knowledge and awareness to live life behaviours that maintain cleanliness and meet health standards.

Keywords: Web-based · Health surveillance · Public health · Big data

1 Introduction

Clean and Healthy Life Behavior is defined as all behaviors related to health awareness so then family members or him/herself can play an active and healthy role in daily activities [1]. Clean and healthy life behavior is very important for preventing infectious disease. In Indonesia, infectious diseases are categorized as high risk diseases compared to US [2].

There are some major of infectious diseases in Indonesia related to unclean and unhealthy life behavior, for example: hepatitis A, typhoid fever, malaria, dengue fever, chikungunya, leptospirosis, etc. Ministry of Health of Republic Indonesia determines Ten Clean and Healthy Life Behavior indicators [1], as follows:

M. E. Auer and K. Ram B. (Eds.): REV2019 2019, LNNS 80, pp. 761–775, 2020.
https://doi.org/10.1007/978-3-030-23162-0_69

1. Birth delivery by health personnel
2. Infants are given exclusive mother's breast milk
3. Weighing toddlers monthly in health facilities
4. Availability of clean water in every household
5. Washing hands with clean water and soap
6. Availability of health latrines in every household
7. Eradicate mosquito larvae in every household
8. Eat fruits and vegetables in sufficient amount
9. Do physical activities in daily basis
10. Not smoking at home.

The motivation of study is to develop a web-based health surveillance application in order to promote Indonesia public health, specifically for preventing infectious disease. Infectious diseases in Indonesia are the cause of death, such as: dengue fever, tuberculosis, malaria, and poor nutrition. Therefore the application we develop is very important in order to reduce the prevalence of infectious disease by promoting better public health than traditional way (without using the application).

The purpose of this research is to develop a web based health surveillance application, namely PHBS online. The application is developed using big data approach and able to visualize the data. The application offers benefit to public and government, by providing useful information of which places have infectious disease risk. The information then can be used to educate public in order to improve public health in terms of clean and healthy life behavior.

This paper consists of five main sections, as follows: Sect. 1 introduce the background and the purpose of this research; Sect. 2 discusses the big data concept and how big data acts as a potential tool for promoting better public healthcare; Sect. 3 explain the method used in this research; Sect. 4 presents our research findings and discussions; and Sect. 5 concludes our research and defines our future research.

2 Big Data for Healthcare

Big data refers to large volume of data, which may be formed as structured or unstructured data [3]. Gartner [4] defines Big Data as high-volume, high-velocity and/or high-variety information assets that demand cost-effective, innovative forms of information processing that enable enhanced insight, decision making, and process automation.

Big data concept involves three important dimensions, i.e.:

1. Volume of data: refers to size and amount of data, from datasets with sizes of terabytes to zettabyte.
2. Velocity: refers to rate at which data is created, from batch processing to real time streaming.
3. Variety: refers to various sources, types, and formats of data; from unstructured, semi-structured, to structured data.

Figure 1 shows the three important dimensions (three Vs) of big data concept.

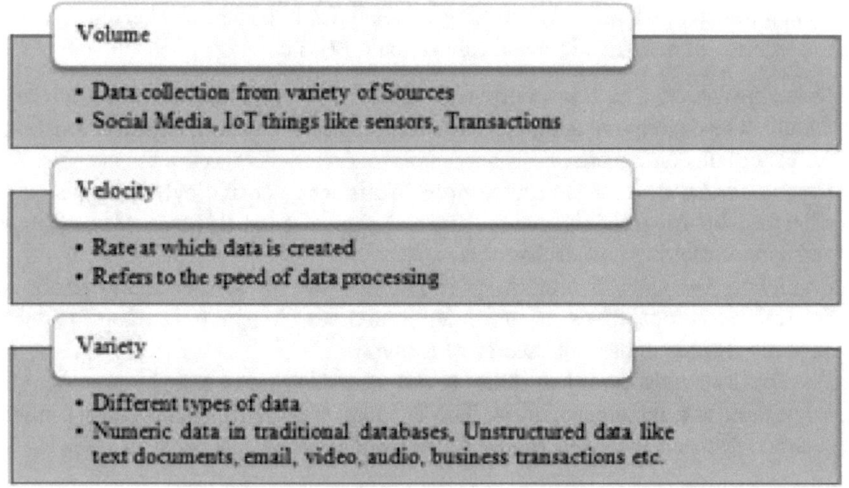

Fig. 1. Three Vs of big data [3]

While authors in [5] mention that Big Data has four aspects, i.e.:

1. Volume: refers to the quantity of data gathered by a company.
2. Velocity: refers to the time in which Big Data can be processed.
3. Variety: refers to the type of data that Big Data can comprise.
4. Veracity: refers to the degree in which a leader trusts the used information in order to take decision.

2.1 Big Data Analytics and Visualization in Healthcare Industry

Patel and Patel [3] mention that Big Data Analytics terminology comes from two distinct concepts, i.e.: Big Data and Analytics. Watson et al. [6] explain how Analytics evolved from Decision Support Systems in 1970s to Business Intelligence in 1990s and finally Analytics in 2010s. Figure 2 shows the development of Analytics since 1970s to present.

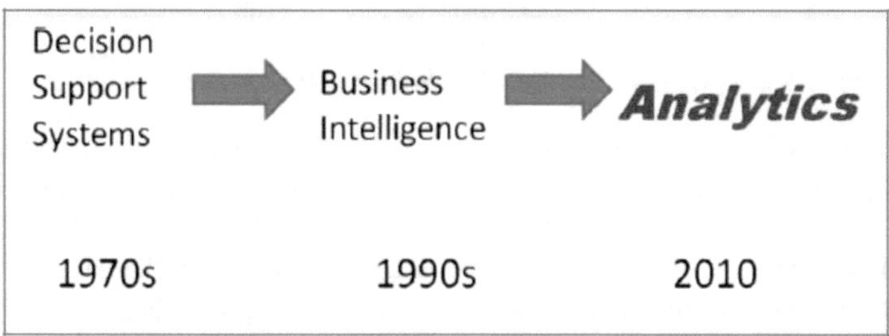

Fig. 2. Evolution of big data [6]

In this research, we are only focusing on analytics in healthcare domain. There are four categories of analytics in healthcare domain [7], i.e.:

1. *Descriptive Analytics*. It has purpose to describe recent situations by having report feature. The descriptive analytics tools can display the data as statistics tools, such as: histograms and charts.
2. *Diagnostic Analysis.* It is used to give details why specific events happened and what variable triggered the events. The technique used in this type of analytics, for example: clustering and decision trees.
3. *Predictive Analytics*. It aims to predict future events by identifying trends and determining probabilities of uncertain outcomes. Machine learning is one of example technique used in predictive analytics.
4. *Prescriptive Analytics*. It has ability to optimize decision-making by providing best suggestion and recommendation. The techniques used for prescriptive analytics, such as: decision trees and Monte Carlo simulation.

Figure 3 illustrates analytics phases for healthcare domain [7].

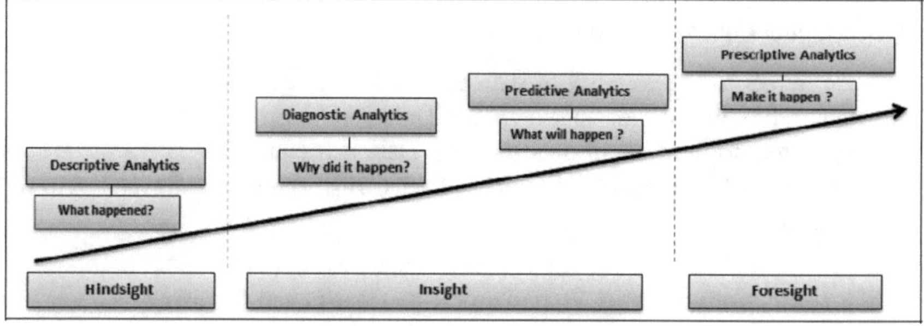

Fig. 3. Analytics type for healthcare domain [7]

Figure 4 illustrates the conceptual architecture of big data analytics, from its sources to its applications. There are several tools can be used to implement big data for healthcare domain, such as: Hadoop, Mapreduce, Jaspersoft, Tableau, Qlik, etc. Those tools offer various big data analytics applications, such as: queries, reports, OLAP, and data mining.

2.2 Big Data Visualization for Public Health Information

Big data analytics offer wide opportunities for having better decision making than before [9]. Nowadays, most of industries' decision making are depend on quantities and qualities of data which belong to them. However, analytics itself is not enough because the complexity, size, speed, and heterogeneity of data make data more difficult to be explored and comprehend [10]. Data visualization supports user to understand the meaning of data in easier and faster way by performing a series of analysis tasks [11].

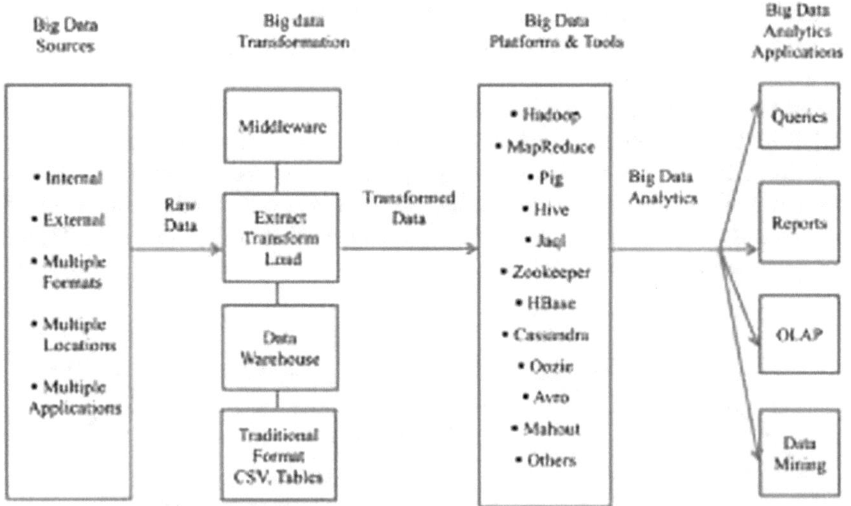

Fig. 4. An applied conceptual architecture of Big Data analytics [8]

Data visualization tools in healthcare domain enable users to recognise and classify patterns, trends, and deviations. Thus, data visualization is very potential tool for improving clinical healthcare delivery and public health policy by providing visual analytics to find out knowledge from data, and as a result it lead to improvements in service delivery, healthcare cost reduction, and healthcare service quality to patients [12, 13].

Figure 5 illustrates the layer architecture of big data [7] which consists of: (1) visualization layer (provide dashboard, prediction, and emergency alerts), (2) big data processing and analytics layer (performs batch processing and stream processing),

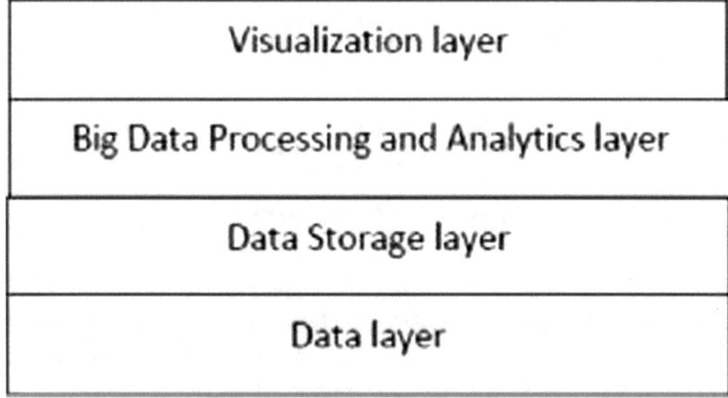

Fig. 5. Layer architectures of Big Data [7]

(3) data storage layer (keeps relational database, JSON documents, etc.), (4) data layer (connects with other devices or systems, such as: Electronic Health Record, body sensors, mobile devices).

Hesse et al. [14] propose a framework to analyse health informatics technologies, which called as "Health 2.0", as shown as in Fig. 6. They recommend health strategies which integrate web-based, participatory approach, and mobile. Health 2.0 utilizes tools and devices, such as: social media, personal sensors, mobile devices, big data analytics and visual analytics integrated with advanced statistical methods. There are three types of domains in Health 2.0, i.e.:

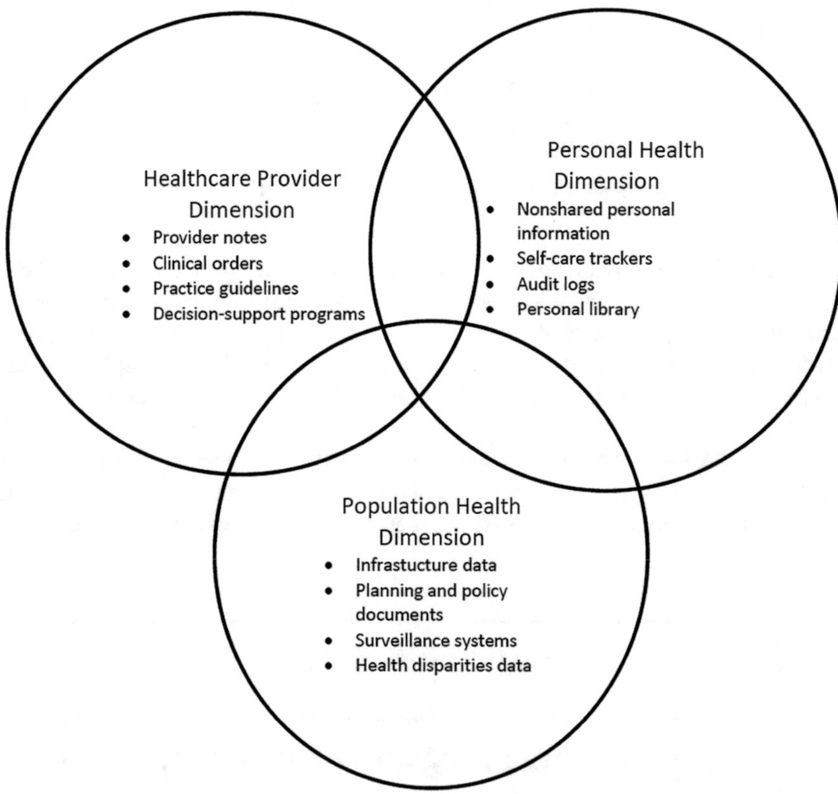

Fig. 6. Health 2.0 framework [14]

1. Personal Health Information. This type of domain keeps information, such as: non-shared personal information, self-care trackers, audit logs, and personal library. Personal patient data is retained in this type of domain and can be combined with other devices, such as: body sensors, mobile devices, etc.

2. Clinical Health Information. This type of domain maintains information, such as: provider notes, clinical orders, practises guidelines, decision support programs. Other information related to Electronic Health Record (EHR) systems also stored in this type of domain.
3. Public Health Information. This type of domain maintains information, such as: infrastructure data, planning and policy documents, surveillance system, health disparities data. The purpose of this type of domain is to enable government collect large volumes of public health data in order to produce better and more reliable decisions.

3 Method: Visual Analytic and Visualization Platform (VAVP)

To develop the web based health surveillance application, we employ Virtual Analytic and Visualisation Platform (VAVP) which proposed by Martinez et al. [15]. VAVP consists of five main components, as follows: (1) Data sources, (2) Data preparation and integration, (3) Data storage and management, (4) Data exploration and visual analytics, (5) Web-based application as services for data dissemination. Figure 7 shows how VAVP works.

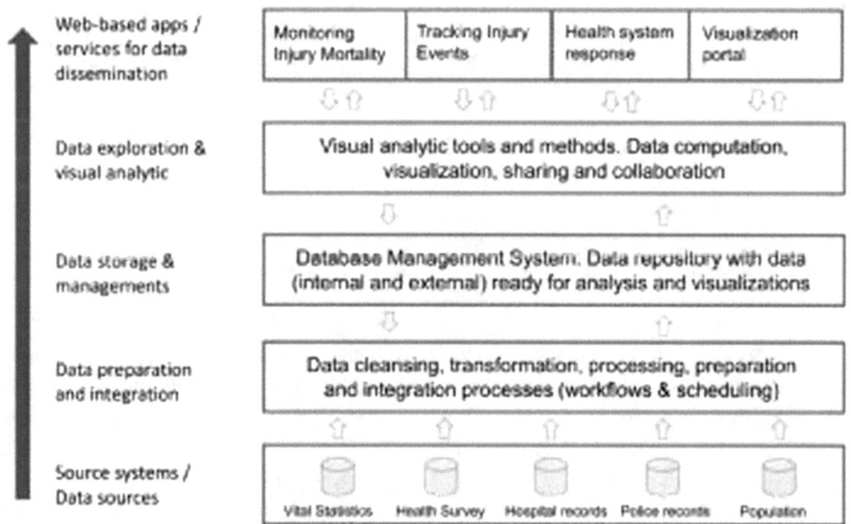

Fig. 7. Visual analytics and visualization platform (VAVP) [15]

In this research, data sources are gathered from external source, by having public health survey. We asked public to fill in the health surveillance related to clean and health life behaviour, provided by web-based application that we developed. Data preparation and integration work as a relational database which connected to data sources and calculating the behaviour score. For every correct answer in the survey, the individual as a respondent will get an additional score. The score then calculates to find out the clean and healthy life behaviour of an individual. The survey's score, identity, and location of the individual will be stored in data storage and management. Data storage and management involves a database management system (DBMS) with methods, procedures and tools for data storage, data querying and database managements. It also stored a data repository or data warehouse, where data are organised, stored and made available for direct access by analysts and analytic applications. After that, data will be produced into data visualisation using data exploration and visual analytic component. Finally, web-based application will display the output of data visualisation to users.

4 PHBS Online: An Indonesian Web Based Health Surveillance for Promoting Clean and Healthy Life Behavior

The web based application named as "PHBS Online". This web application stored big volume of information collected as public health surveillance application. This web application currently still under examining process of Indonesian health experts. However, the application provides valuable information to help government for producing decisions related to public health and infectious diseases.

The application supports five main features, as shown in Fig. 8, which illustrates the Use Case Diagram of PHBS Online. The main features of PHBS Online are as follows:

1. Fill in respondents identity and health surveillance
2. View survey result and suggestion
3. View map of health surveillance result
4. View detail graphic of health surveillance result for each location
5. View related health articles for public education purposes.

The interface of PHBS online is presented in Figs. 9, 10, 11, 12, 13, 14, 15 and 16. Figure 9 shows the index interface of the application. Figure 10 displays the interface where user fills in his/her personal identity before fills in the health surveillance. Figure 11 illustrates the interface of health surveillance, where user has to answer fifteen questions provided by the application. The question is related to clean and health behavior guidelines defined by Ministry of Health Republic of Indonesia. User can check their score and result of health surveillance immediately after they submit his/her answers.

The application's interface of survey result is presented in Fig. 12.

After all the data sources are gathered and compiled, data are stored and can be visualized through the application, as shown in Fig. 13. In Fig. 13, the location/district (or also called as "kecamatan" in Indonesia language) is represented by circle symbol.

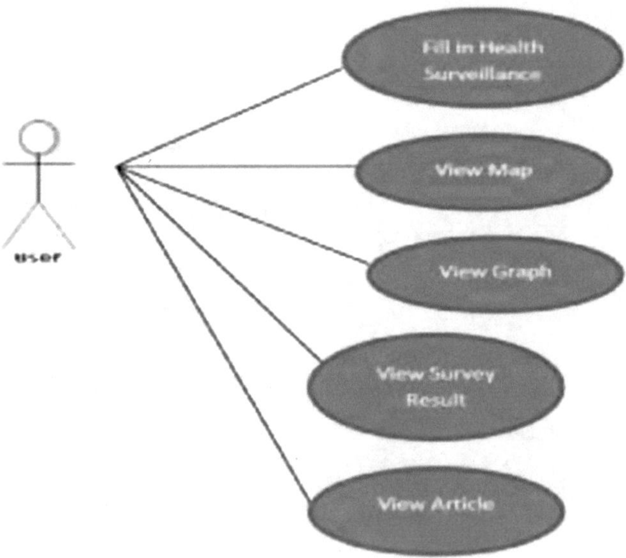

Fig. 8. Use case diagram of PHBS online.

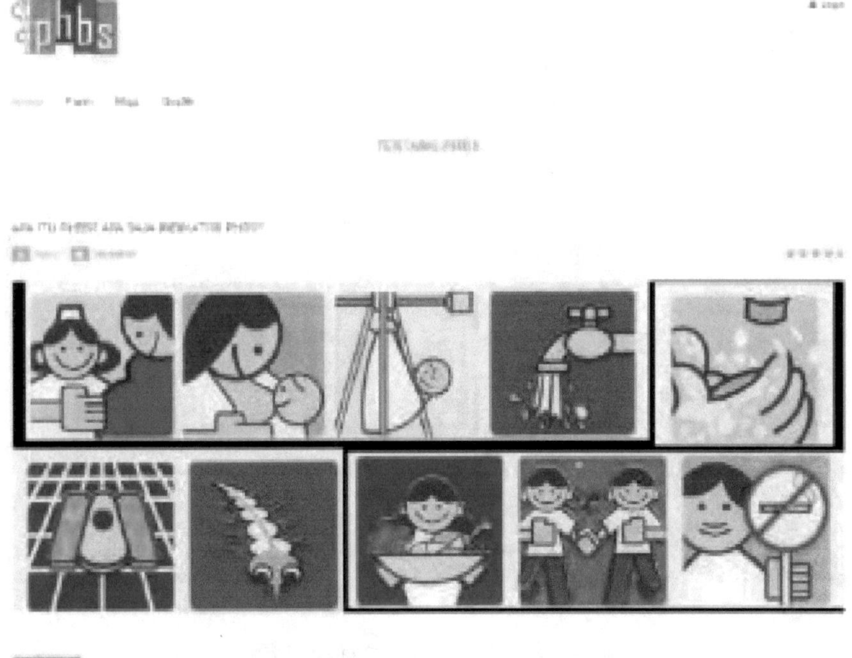

Fig. 9. Index page of PHBS online.

Fig. 10. Identity form page of PHBS online

Fig. 11. Health surveillance page of PHBS

Fig. 12. Score result page of PHBS online

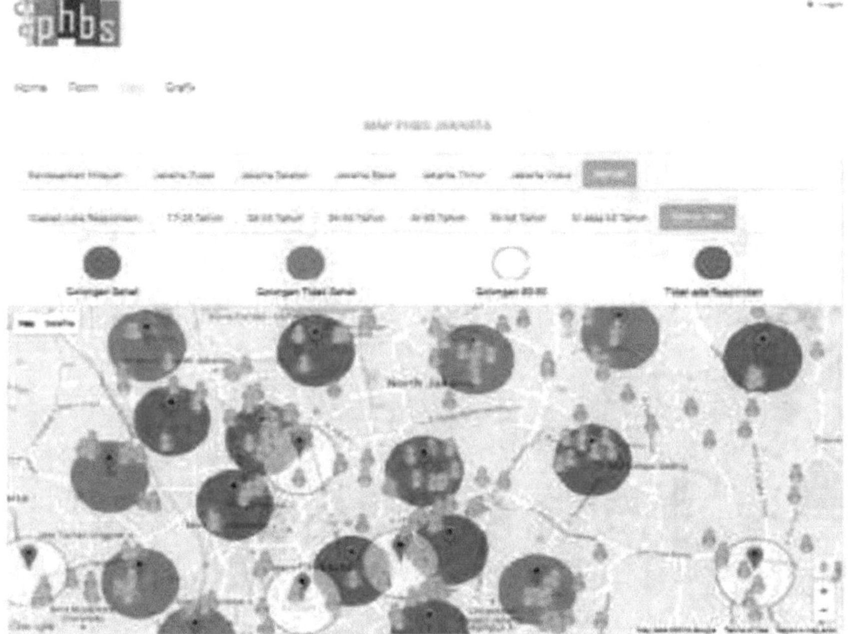

Fig. 13. Map view result page of PHBS online

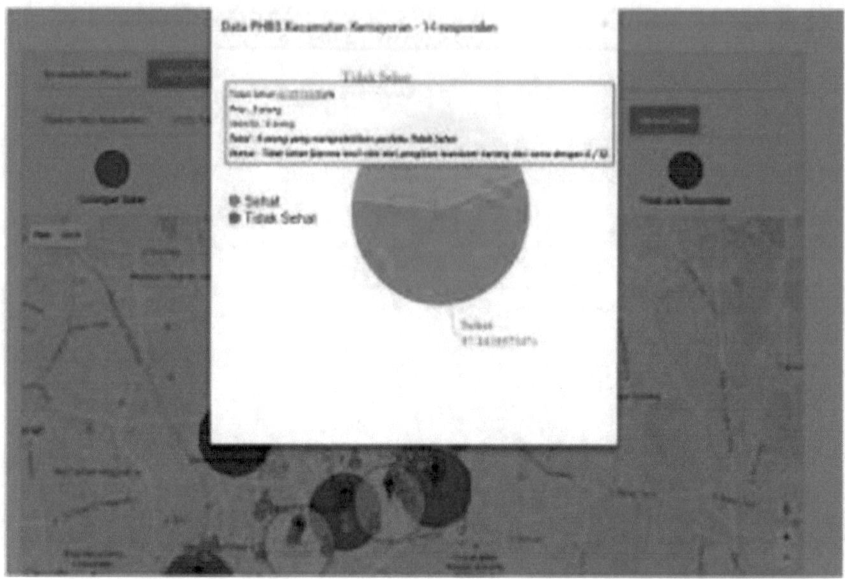

Fig. 14. Graph view result page for each individual of PHBS online.

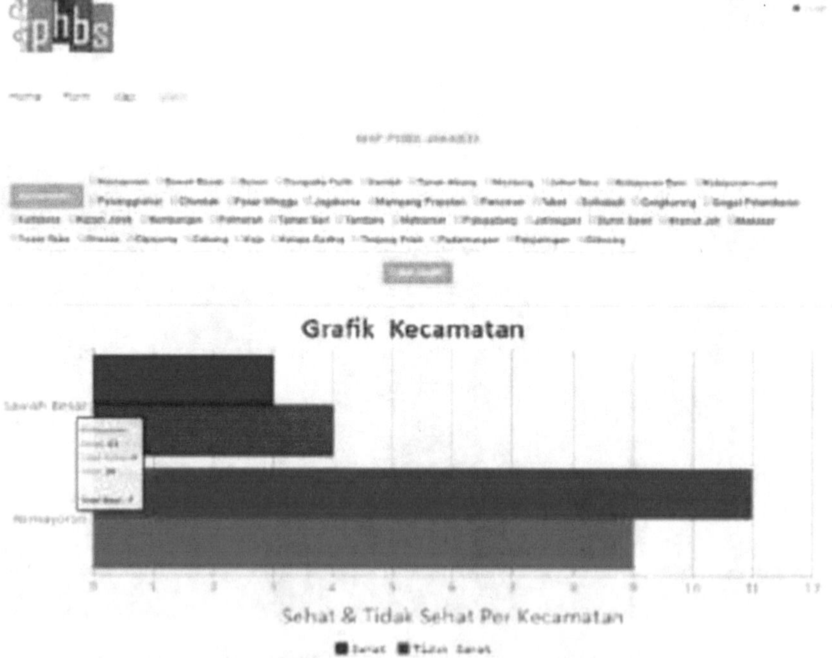

Fig. 15. Graph view result page for each district of PHBS online

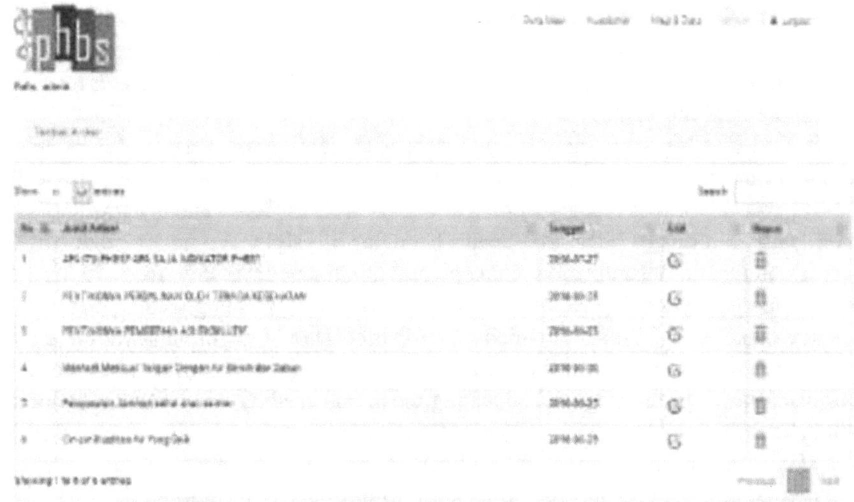

Fig. 16. List of health articles of PHBS online

Red circle symbol represents more than 50% individual who live and fill in the health surveillance are practicing unclean and unhealthy life behavior. This indicates high risk of infectious diseases is occurred in that location, marked by red circle symbol. On the other hand, blue circle symbol indicates that most of individuals who live in that location/district are practicing clean and healthy life behavior. White circle symbol represent that half of individuals who live in that area practicing clean and healthy life behavior but half of them are not. Therefore, area with red circle and white circle symbol should get more attention of government to promote and educate clean and healthy life behavior.

The application also can explore further details of each individual score who live in each district, as shown in Fig. 14. This feature is important to give personal attention to each individual related to clean and health life behavior education and promotion. The application also provide comparison purpose of each location, high risk (indicates with red barchart) versus low risk (indicates with blue barchart), as shown in Fig. 15. Figure 16 is the supplementary features, to give online information and education to public related clean and health behavior, as well as how to prevent infectious disease.

5 Conclusion

In this paper we present how big data analytics and visualization concept is used to promote public health via web-based health surveillance application. The web-based health surveillance application, named as PHBS Online, is developed with VAVP method which employs big data approach. The application has five main features, i.e.: health surveillance; graphic visualization for health surveillance result based on data map (high risk for red circle, low risk for blue circle, and neutral for white circle); graphic visualization for health surveillance result for each district; graphic visualization for each individual in each district; and graphic visualization as comparison among the districts (high risk vs low risk). The application's interfaces for each feature are presented in this paper to show how data visualization enable user to explore and comprehend the data.

We find that big data analytics and data visualization are very useful in area of healthcare and public health, as demonstrated with our application. The application employs web technology and mapping tools, as well as data visualization for health surveillance in terms of clean and health behavior life style. We believe that big data analytics and data visualization concept and platform have key roles to improve decision and policy making related to public health. Using this application, government as policy maker can understand which areas/districts should get more attention in order to decrease unclean and unhealthy life behavior. The government can set the target areas/districts for promoting and educating clean and healthy life behavior to specific communities and in turn will improve the quality of public health. In spite of the promising result of this research, we still have to validate the impact of our application by experimenting and testing the application to more users in Indonesia.

References

1. Promosi Kesehatan Kementerian Kesehatan Republik Indonesia: http://promkes.kemkes.go.id/phbs (2016). Accessed on 1 Sept 2018
2. Indonesia Demographic Profile Report—Index Mundi: https://www.indexmundi.com/indonesia/major_infectious_diseases.html (2018). Accessed on 1 Sept 2018
3. Patel, S., Patel, A.: A big data revolution in health care sector: opportunities, challenges, and technological advancements. Int. J. Inf. Sci. Tech. (IJIST) **6**(1/2), 155–162 (2016)
4. Beyer, M.A., Laney, D.: The Importance of Big Data: A Definition. Gartner Report (2012)
5. Zikipoulos, P., Deutsch, T., Deroos, D.: Harness the Power of Big Data. http://www.ibmbigdatahub.com/blog/harness-power-big-data-book-excerpt (2012)
6. Watson, H.J.: Tutorial: big data analytics: concepts, technologies, and applications. Commun. Assoc. Inf. Syst. **34**(65) (2014)
7. El aboudi, N., Benhlima, L.: Big data management for healthcare systems: architecture, requirements, and implementation. Hindawi Adv. Bioinform. **2018**(4059018) (2018)
8. Raghupathi, W., Raghupathi, V.: Big data analytics in healthcare: promise and potential. Health Inf. Sci. Syst. **2**(3) (2014)
9. Gandomi, A., Haider, M.: Beyond the hype: big data concepts, methods, and analytics. Int. J. Inf. Manage. **35**(2) (2015)
10. Shneiderman, B., Plaisant, C., Hesse, B.W.: Improving health and healthcare with interactive visualization methods. IEEE Comp. (2013)
11. Keim, D.A., Kohlhammer, J., Ellis, G.P., Mansmann, F.: Mastering the Information Age—Solving Problems with Visual Analytics. Eurographics Association (2010)
12. Palanisamy, V., Thirunavukarasu, R.: Implications of big data analytics in developing healthcare frameworks—a review. J. King Saud Univ. Comput. Inf. Sci.

13. Wang, L., Alexander, C.A.: Big data in medical applications and health care. Curr. Res. Med. **6**(1) (2015)
14. Hesse, B.W., Hansen, D., Finholt, T., Munson, S., Kellogg, W., Thomas, J.C.: Social participation in health 2.0. IEEE Comp. **43**(11) (2010)
15. Martinez, R., Ordunez, P., Soliz, P.N., Ballesteros, M.F.: Data visualisation in surveillance for injury prevention and control: conceptual bases and case studies. Inj Prev **22**(1), i27–i33 (2016)

Augmented Reality for Troubleshooting Measurement Devices

N. K. Vaishnavi[1], T. Vidhya[1(✉)], Vinay Kariwala[2], and T. T. Mini[2]

[1] B.M.S. College of Engineering, Bengaluru, India
{vaishnavi2908, vidhyat98}@gmail.com
[2] ABB Ability Innovation Center, Bengaluru, India
{vinay.kariwala, mini.tt}@in.abb.com

Abstract. Augmented Reality (AR) is a rapidly emerging technology used to visualize virtual data superimposed on the real world. It improves the way humans visualize data. Measurement devices form the base of automation pyramid. It is essential that these devices operate accurately and reliably for safe operation of the plant. Troubleshooting these devices can be an arduous task. To this end, finding relevant information in a timely fashion from service manuals can be difficult. Thus, there is a strong need to find alternate approaches to facilitate troubleshooting these devices. This paper presents an approach involving use of AR technology to improve the accuracy and efficiency of troubleshooting measurement device.

Keywords: Augmented reality · Automation · Digitalization · Dynamic QR code · Human-machine interface · Measurement devices · Service · Troubleshooting

1 Introduction

Innovative thinking has fueled several extraordinary technological advancements. What was deemed impossible has been created through the use of novel technologies. Augmented Reality (AR) is one such technology which has seen a rapid increase in number of users for varied applications. AR can be applied to most of the scenarios that the human eye can visualise. AR is continuously expanding its scope in gaming applications, training, shopping, home decor, marketing, medical industry and field service. AR is overlaying of data in digital form onto physical world, in order to enhance the amount of information human beings can retrieve from their vision. An advancement in human-machine interface (HMI), AR seeks to improve the way a user can interact with devices.

We live in an era, where smartphones have become an integral part of lifestyle and many of the tasks are done with this handheld device. Ranging from travel, hotel reservations, shopping, banking, study materials and news, everything is being facilitated by our smartphones. However, the servicing of our devices is still done with the help of offline manuals. Often, the service engineers carry a set of bulky manuals with them for assistance. Within these pages lie the procedural information on troubleshooting errors that are encountered with the malfunctioning device. Therefore, if

M. E. Auer and K. Ram B. (Eds.): REV2019 2019, LNNS 80, pp. 776–786, 2020.
https://doi.org/10.1007/978-3-030-23162-0_70

there is any modification in the troubleshooting procedure or if new error solutions are to be added, the manuals are modified and re-circulated to all the service engineers. This is a cumbersome process.

Another issue with printed manuals is that, while troubleshooting, the engineer has to continuously shift his vision to and fro the manual and device. This reduces the efficiency of the engineer performing the task. For the given textual procedures, their interpretation at certain steps can be unclear or subject to interpretation. In such situations, the engineer needs to either troubleshoot with his/her experience of handling similar cases or refer to an expert for guidance. The service engineer may have reduced level of expertise to debug complex problems.

The earlier discussion mentions only some of the problems faced with troubleshooting of devices using available service manuals. One of the solutions to these issues is the use of AR for troubleshooting the devices. Use of AR in the industrial sector for troubleshooting devices will transform the rudimentary servicing procedure and pave way to Industry 4.0 revolution.

According to an article published in Harvard Business Review, AR is one of the extensively used technology by industrial manufacturers. AR is incorporated by many companies to perform different operations. Alstom Transport uses AR to design trains, BAE systems use AR to design warships and ThyssenKrupp uses AR as a tool for servicing elevators [1].

The AR apps for servicing, help to visualise the tasks to be performed to rectify the errors in the device and get it functional. They form a communication bridge between the engineer and the device; thereby facilitating a better HMI. AR applications not only aid the service engineer in performing his/her tasks efficiently and in a hassle-free way, but also go a step forward in digitalisation. The use of these apps reduces the necessity of printed service manuals for each device. The manufacturer of the device has to just develop an application that holds the digital procedures for servicing of different devices. Any modifications to the procedures can be implemented by changing the corresponding code section of the developed app.

Discussed in this paper is an AR servicing application for measurement devices. Measurement devices are instruments that measure parameters of a specific quantity like gas, level, fluids so on. For example, flow meters are devices that measure the flow of fluids and gas analysers are devices that analyse the concentration of various constituent substances (like SO_2, NO_2, SO_3 and so on) in ppm.

2 Description of Methodology

The AR application aids the service engineer in troubleshooting measurement devices. The application involves identification of device facilitated by Quick Response (QR) codes provided in device or image recognition, identification of presence of error facilitated by decoding QR code and presentation of error solutions through AR. Figure 1 represents a block diagram of various fragments of the application, each of which is discussed in the below sub-sections. Note that the use for QR codes facilitates the troubleshooting, but is not mandatory for application of AR technology.

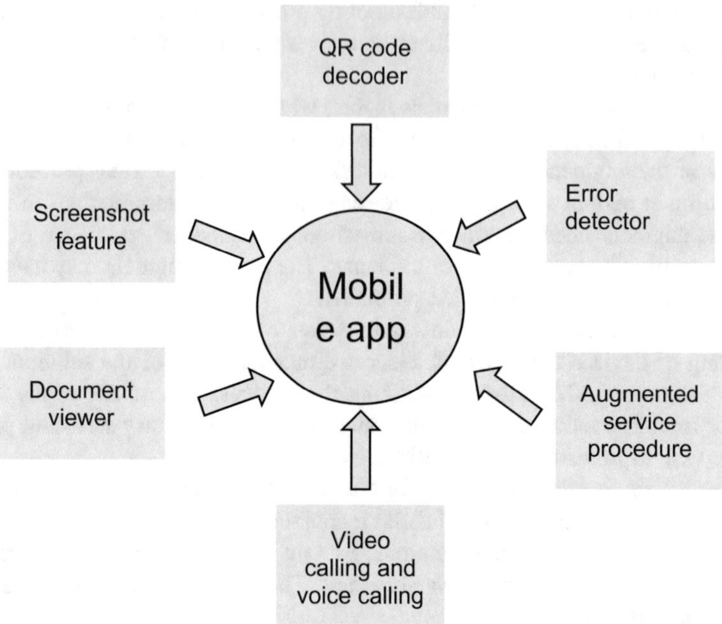

Fig. 1. Features of AR application

2.1 Guide to QR Code

The initial phase in troubleshooting a device is the identification of the device. This has been facilitated through QR codes or image recognition in the developed application.

QR codes serve two purposes, identification of device and identification of error. This code is dynamic and is presented by the measurement device. However, it is not a part of the welcome screen of the device; hence has to be located inside sub menus. This location might not be well known to all users of device and the application introduces a feature of "Guide to QR Code" which navigates the user to the code display.

QR code is a matrix type barcode that usually stores data. If the QR code is scanned through a QR code scanner, the data stored in that specific QR code is displayed. QR codes are extensively used because of their faster readability and greater storage capacity (Fig. 2).

Under study is a gas analyser, namely Easy Line 3020. This device has several features for calibration of device, device status display, gas concentration display and dynamic QR codes containing various device information. The mobile app should include a detailed process to find the QR code through navigation and selection of different features in the display system of the device. Hence, the guide to QR code feature in the mobile app helps the user to navigate to the code and scan it to find the status of the device. This navigation is provided as an augmented feature, wherein the keys of the device that the user has to press in order to reach the QR code menu are highlighted through a digital overlay.

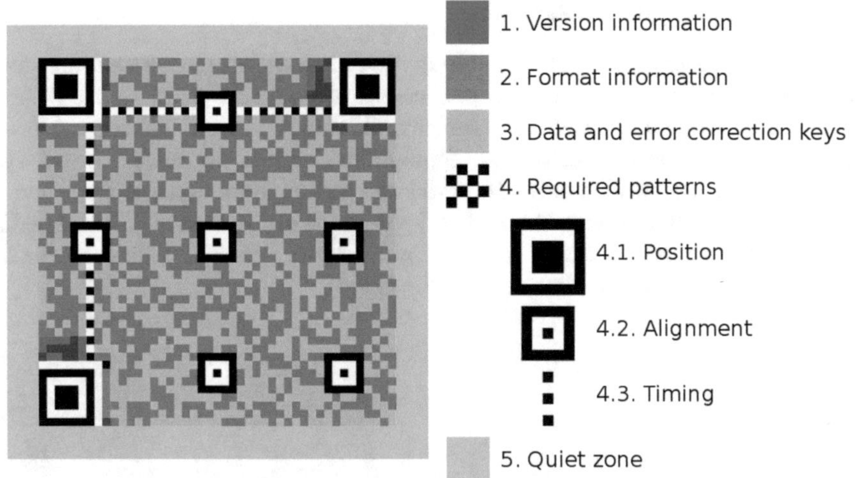

Fig. 2. Structure of QR code, highlighting functional elements [9]

Figure 3 represents a screenshot of the application navigating the user to the QR code menu. The navigation is presented through graphics and text instruction panel. The shown screenshot is the representation of a step in the "Guide to QR Code" feature of the device, Easy Line 3020, a gas analyser manufactured by ABB. In the particular step, the user is at a menu which routes to display of two different QR Codes, System and Uras. The arrow buttons to be pressed for each code are augmented in black. It is required that the user presses the 'OK' button in order to see the code. Hence this button is augmented in blue. The complete instruction to be followed is also provided in the instruction panel at the bottom of the screen, this includes the information on what menu the user should be looking at and what is the further action s/he has to take.

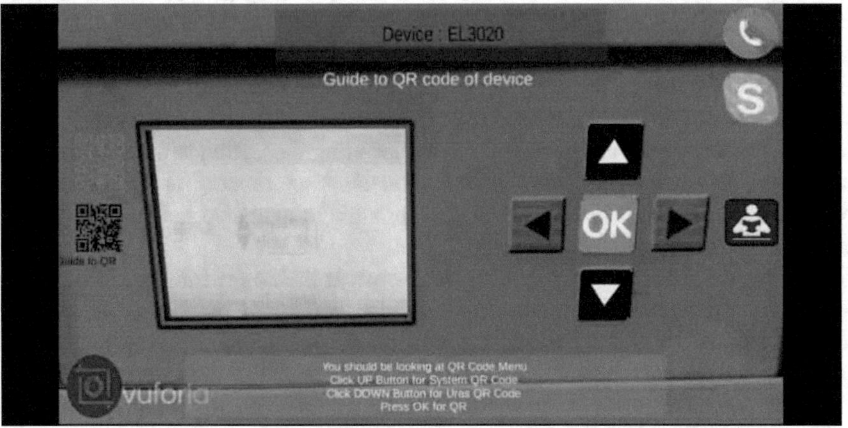

Fig. 3. The procedure for navigation to read QR code

2.2 QR Code Decoding Mechanism

A QR code decoder is used in the mobile application to decode the QR code and detect the error in the device. This decoder extracts text from the 2D code. As the dynamic code is presented by the device, it contains different classes of information distinguished with special characters. The application identifies the section of the code containing error information about the device and further performs necessary actions. If the device is malfunctioning, the mobile application displays an augmented service procedure for the user to rectify the error in the device. The QR code decoding and associated processes are as shown in figure (Fig. 4).

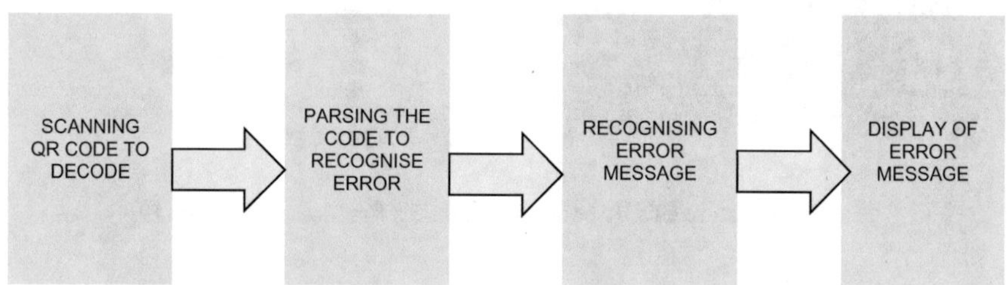

Fig. 4. Block diagram of QR code decoding mechanism

2.3 Augmented Service Procedure

Error identification and rectification is a very tedious process. Service manuals are very descriptive and may not provide specific solutions. Hence, identifying the fallacy in the machine and repairing it becomes very difficult but it has to be done at the earliest for increased efficiency.

Recognising the error in the machine and providing required and timely assistance is very important. The error is encoded in the dynamic QR codes. The QR code decoding mechanism fetches the error code details and recognises the stored data in the machine. The decoded message contains variety of information about the device, the desired error data is extracted by parsing the decoded message.

On identification of the error message, respective servicing assistance has to be provided through the mobile app. The servicing procedure is designed in the AR platform. When the error code is fetched by decoding the QR code through the mobile application, the respective error and the rectification process is displayed in the application. Then, the technician is guided through an animated detailed procedure of servicing the machine for the identified error.

The service procedure is very comprehensive. It is designed in the Unity platform and guides the technician through a descriptive animated procedure. It shows the technician the steps to be performed using an animated procedure so as to aid the him/her in a more methodical and systematic way for servicing the product.

Figure 5 is an example of the augmented service procedure for a faulty display card in Advanced Optima (AO) 2020 gas analyser. A 3D model of the device is augmented beside the real device with the top panel removed, so that the engineer can visualise the orientation of the parts inside the device if required for troubleshooting. The yellow card with a text saying 'Functional Card' is animated to show the engineer how and where the card has to be inserted in the device. In this case, it has to be inserted in the space to the left of the black button array. The two arrows towards the right of the screen provide navigation to next or previous service procedure steps. At the bottom of the screen, an instruction panel is provided that reinforces the augmented procedure clearly via text. The gray 'back' button in the bottom left corner leads the user back to the welcome page of the application, where he can scan and look for service procedures of devices again.

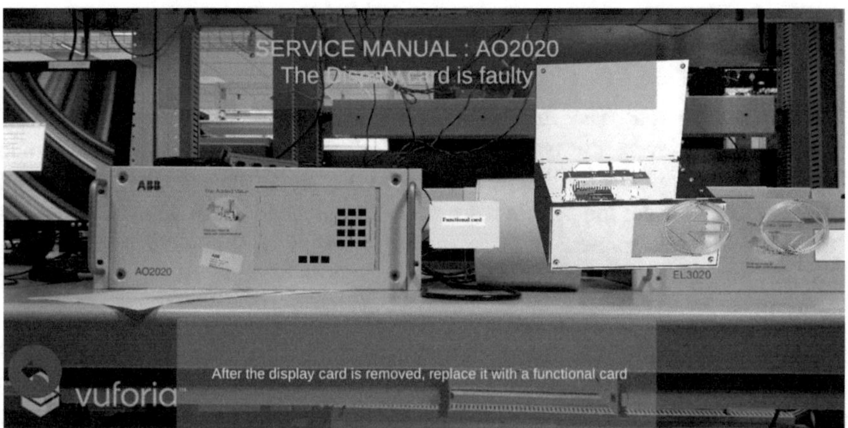

Fig. 5. Augmented servicing procedure

2.4 Assistance

Many UI options have been provided with the application to aid the user at any stage, which are as follows:

- A pdf document option in the application to open a service manual document.
- Voice call facility to contact an expert and seek help.
- Video calling via access to Skype application, so that the user or engineer can seek remote assistance from an expert by showing him the current scenario on field.
- A facility to capture screenshot of what the user is looking at through the app is provided so that if the user wants to share the device view with the augmented instructions through email, WhatsApp or any other media, he/she can do so through the click of a button in the application itself instead of using his gadget's options for the same.

3 App Development

The main stages of application development are represented by block diagram in Fig. 6 and are briefly described next:

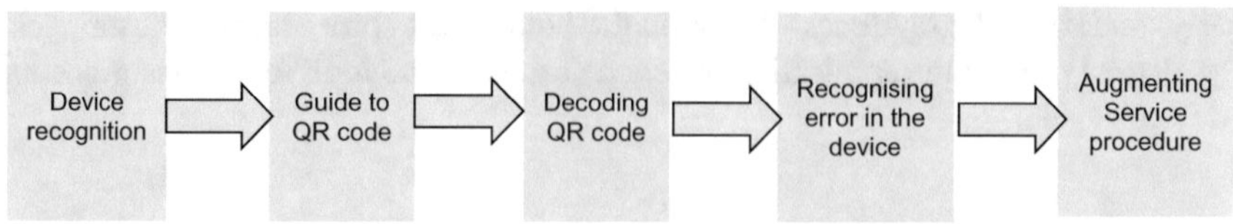

Fig. 6. Block diagram of the mobile application procedure

- Device recognition: The first step dealing with application development is to identify what device the user wants to service, so that the correct troubleshoot procedures and guides for that device can be displayed. This is achieved by image recognition, wherein the pictures of the device uploaded to the app as reference are matched with the camera Vuforia's "Image target" feature, wherein images of devices are uploaded to a personalised database in Vuforia and downloaded as a unity package.

Each navigation menu is also uploaded as image target, so that when the mobile camera is focused on the display unit of the device, the app augments the buttons to be pressed to go to the QR code menu based on the image targets present in the application's database. A feasible alternative to this is the use of optical character recognition algorithms.

- QR code decoder: After QR code detection, the code has to be decoded. On focusing the mobile camera on the code, the decoder embedded in the application reads the data stored in the code. Later, it detects the error message in the device which is encoded in the QR code and navigates to the service procedure.
- Servicing: The 3D models of the device are designed and animated in the Vuforia Unity engine analogous to the error and are encompassed in the servicing procedure. When the mobile application detects the error, the respective service procedure is augmented on the device.

4 Tools Used

4.1 Unity

Unity is a 3D game development platform developed by Unity Technologies. This platform enables the developers to create 2D and 3D environments. The primary scripting background is C#. Microsoft Visual Studio and Monodevelop are common platforms on which C# scripting is done for Unity. The engine is claimed to support over 27 platforms, some of which are Universal Windows Platform, Mac, Linux, Android, iOS, Google Cardboard, Playstation and Windows Mixed Reality.

Unity contains plethora of features for developers some of which include artificial intelligence, animation, texture compression, 3D modelling, design, graphics and text to speech. It also provides cloud based services, some of which are Unity Cloud Build, Unity Ads and Unity Analytics (Fig. 7).

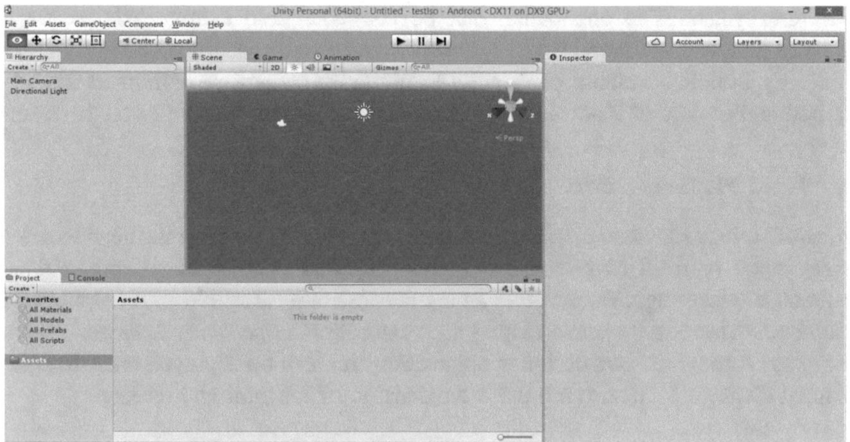

Fig. 7. Unity software window

4.2 Vuforia

Vuforia is an AR Software Development Kit initially owned by Qualcomm and now taken over by PTC (Parametric Technology Corporation). It can be used with Unity game engine and Android Studio to create AR applications. In the development of this application, the AR camera feature of Vuforia Unity is used which facilitates overlaying of required data on the camera screen.

Vuforia facilitates object and image recognition in its framework. In order to identify the device to be serviced, images of the device, called 'image targets' are uploaded to Vuforia, after which a .unitypackage is returned which when interfaced in Unity facilitates image matching. This data can further be used to place the augmentable data in reference to the device coordinates.

4.3 CAD Models

The mobile applications decodes the dynamic QR code and extracts the error information of the device and navigates to the service manual. 3D models of the device is required to overlay the service procedures on the device. In this work, CAD models of the device is used to augment the servicing procedure. The format of the 3D models used is .obj and .dae. The models exactly resemble the corresponding parts of the machine in color, shape, design and size, thereby camouflaging the digital data and physical object and hence facilitating a good user experience.

The designed application can be implemented on various platforms. it can be a mobile application or it can be deployed onto a head mounted display system.

4.4 Smartphones

The apk package is a Android Package is a package file format used in Android Operating system. The application can be converted into a apk package file and deployed onto android mobile. The application can also be used in an iOS or Windows mobile by building suitable packages. A minimum camera resolution of 13 MP is required and in case of Android phones a operating system version of 5+ is required.

4.5 Head Mounted Devices

The HMD (Head mounted display) device is a display unit worn on the head which has display optic in front of one eye or both. This unit is used in several fields like medicine, engineering, aviation and gaming because of its ease of use. An Optical Head Mounted Device is a wearable display that can reflect projected images and allows a user to see through it. The designed application can also be deployed onto these head mounted displays by modifying the file extension of the built app package.

5 Operation

The developed application starts with a 'Scan the device' screen, wherein the user is expected to focus his camera on the device to be troubleshooted. In case of image recognition, complete device has to be present in the camera view so that it can be detected. Using image recognition and extended tracking mechanism, the device is identified and a user interface unique to its repair is set up. This user interface includes options to scan QR codes, guide to QR code menu, AR service procedures, error and diagnostic information display, skype connectivity and manuals.

To use all the functions in the app, ideally the user can start with a guide to the QR code, where the user is navigated through various via augmented instructions on which button to press and a textual instruction panel for the same. On successfully following the augmented instructions, the device will display its dynamic QR code. Now the user can use the scan QR code option to scan the code. Upon scanning, any errors if present are identified and the troubleshoot procedure is displayed to the user either as text or AR procedure depending on the development done in the application.

The AR procedure overlays the steps to be carried out by the user in order to repair the device and make it functional. An example of this is augmentation of rotating screws and screwdriver on the exact location of the screws in the real device to show that he has to remove or fix those screws, an animation of opening and closing panels etc.

In case the user is unable to solve the problem with the device, he can opt for a remote assistance from an expert via a skype call or look into the device's service manual for a written procedure of the same. An option for screenshot is also provided so that the user can send an augmented procedure view to the expert (Fig. 8).

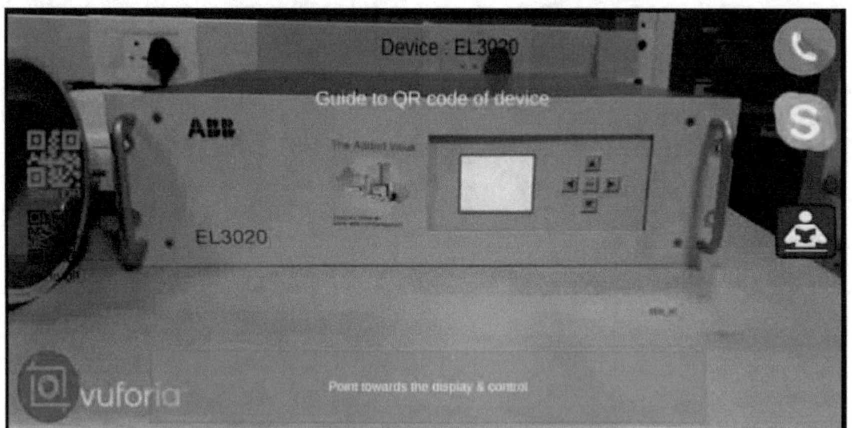

Fig. 8. The user interface as seen through the mobile application

6 Results Obtained

The developed application provides a complete package of operation guides for the measurement devices. Starting with device recognition, the app is tailored to display only the services suitable to that device. Available services include AR navigation to QR code menu, QR code scanner, error detection and solution display, AR troubleshoot procedure, service manual pdf and remote assistance via Skype and voice call. The application can be deployed on a wide range of devices, some of which are Android/iOS devices, smart glasses (e.g., Microsoft Hololens) and Windows machines.

Figure 9 is a screenshot of augmented troubleshoot view. It is a section of the augmented service manual of Advance Optima 3020, a gas analyser manufactured by ABB. After recognising the device, scanning its dynamic QR code and retrieving error data from the code, it has been brought to notice that the device is having a problem of faulty display card. In the application's interface it is seen that the device name and error information are augmented as text. The instruction panel at the bottom explains via text, the steps to be carried out. The same instruction is augmented as a graphic on the device. In the screenshot shown in Fig. 9, a 3D model of the front panel has been exactly placed over the actual device's front panel and an animation indication the way the panel has to be removed is shown. This is one of the steps to be carried out to troubleshoot a device having faulty display card. The two arrows in the right hand side are navigation tools to the previous and next step. The entire service procedure is present in the application to guide the user from unscrewing to open parts of the device to screwing back after placing the parts in the right place.

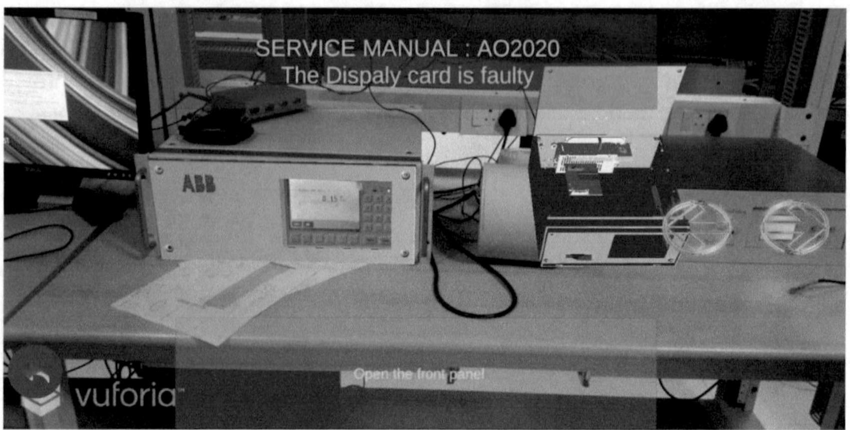

Fig. 9. Augmented service procedure

7 Conclusion

Augmenting graphical information allows the service engineer to understand and complete the service steps in a more effective and efficient way as compared to using service manuals. It not only helps in speeding up the repair time, but also reduces the user errors. The approach provides an immersive experience and is a positive step towards digitalization. Changes in service procedures can be incorporated by updating the app instead of printing new manuals. Field trials of the developed application are pending, but it is expected that this approach will have significant influence on troubleshooting of measurement devices.

References

1. How Augmented reality is changing Industrial Asset Management: https://www.manufacturing.net/article/2018/04/how-augmented-reality-changing-industrial-asset-management
2. Unity: Scripting API reference. [Online]. https://docs.unity3d.com/ScriptReference/
3. Augmented reality and Industry 4.0: https://www.manufacturing.net/article/2018/04/how-augmented-reality-changing-industrial-asset-management
4. Vuforia: https://library.vuforia.com/articles/Training/Image-Target-Guide
5. Fraga-Lamas, P., Fernández-Caramés, T.M., Blanco-Novoa, O., Vilar-Montesinos, M.A.: A review on industrial augmented reality systems for the industry 4.0 shipyard. IEEE Access **6**, 13358–13375 (2018)
6. Nee, A.Y.C., Ong, S.K.: Virtual and augmented reality applications in manufacturing (2004)
7. Friedrich, W.: ARVIKA-augmented reality for development, production and service. IEEE xplore, Digital library
8. Friedrich, W.: ARVIKA-augmented reality for development, production and service. IEEE xplore, Digital library
9. QR code: Wikipedia. https://en.wikipedia.org/wiki/QR_code

Cloud Software Central—A Building Block of Remote Engineering

Kalyan Ram B.(✉), Mahesh Bhaskar,
Arun Kumar Sampangi, and S. Prathap

Electrono Solutions Pvt Ltd, Bengaluru, India
{kalyan, mahesh, arun, prathap}@electronosolutions.com

Abstract. In academic institutions the lab sessions for students are usually limited to 1–2 slots per week and at specified time only. This duration of time will not be enough if one wants to try anything beyond listed experiments. The need of the hour here is round-the-clock lab access to students. The primary goal is to empower academic institutions and industries to utilize their resources (hardware and software tools), round-the-clock to enable their users to access their resources from the comfort of their homes.

Keywords: Cloud software central · Remote engineering · Remote labs · Cloud computing · Lab-as-a-Service

1 Introduction and Architecture

Cloud Software Central is a Lab as a Service Platform which provides access to students and professors from various universities access to a wide variety of software through the cloud platform.

*Cloud Software Central (**CSC**)* comprises of various components:

1. **Slot/Session Booking**: The intention here is to present a 24-h calendar to a user, to be able to choose a suitable slot for lab access. Once the slot gets confirmed by the system (based on various factors like resource availability, scheduled maintenance etc.), then all the resources required are reserved for the user at the specified time.
2. **Database**: All slot/session booking information needs to be saved for later access. This is where database comes into scope, for secure and persistent storage & access.
3. **Lab Access**: During the selected slot duration, the reserved resources are exclusively available for the specified user. The access could be provided based on multiple factors such as type of resource, operating system type, with hardware access or only software access etc.
4. **Statistics**: Over a period, the user as well as the institution would want to look back on the usage of statistics of the resources and measure/gauge the performance of the user. The historical usage data will be handy here.
5. **User Profile Management**: A continuous profile building activity will be useful in cumulatively building a student's profile.

M. E. Auer and K. Ram B. (Eds.): REV2019 2019, LNNS 80, pp. 787–792, 2020.
https://doi.org/10.1007/978-3-030-23162-0_71

6. **Interaction**: User to User interaction, User to Faculty interaction, Faculty to Faculty interaction etc. These interactions are facilitated via 1-to-1 chat mechanism. Audio and video calling functionality with screen sharing will greatly enhance the user experience with all help available at his figure tips.

2 Design and Build

The technology stack used to build the application is as below (well, this is just our choice),

- Java
- Node.JS
- HTML, CSS, Javascript, jQuery
- MongoDB.

a. **Backend**: Java is one of the most powerful programming languages known today. Java's advantage as a cross-platform (a.k.a multi-platform) development tool is the reason behind its choice.
   The entire business-logic is controlled by the backend, this is the reason it is considered the backbone of the architecture. Backend will be involved in almost all interactions fro UI to Database, emails, SMS etc…
b. **Frontend**: Also known as User Interface (UI) provides a beautiful interactive interface for the user. HTML5, CSS3, Vanilla Javascript, jQuery, JSP etc… are the building blocks for frontend.
c. **Database**: All the received data needs to be stored in a database for processing and future computation and metric analysis. In our application, there isn't much of relational data, so we had to use a NoSQL database and MongoDB was a clear winner.
d. **Session Controller**: Session Controller is a desktop-based application, used to access the software. This application takes a single input—Session ID—which is generated at the time of booking the session. The Session ID is a unique key, used to fetch session information such as session start time, software access information detailing machine information, ports etc…, user details, duration for which the session was accessed etc…
e. **Application Manager**: We have here another important component: Application Manager, which runs on the machine where software is hosted (or) machine to which Lab hardware is connected. Node.Js is used as programming language here for faster development and execution.
   Figure 1, gives a clear understanding of the architecture, application manager runs in lab setups, which facilitates the interaction with the backend.

The image below shows a graphical representation of the sessions booked for multiple software (Fig. 2).

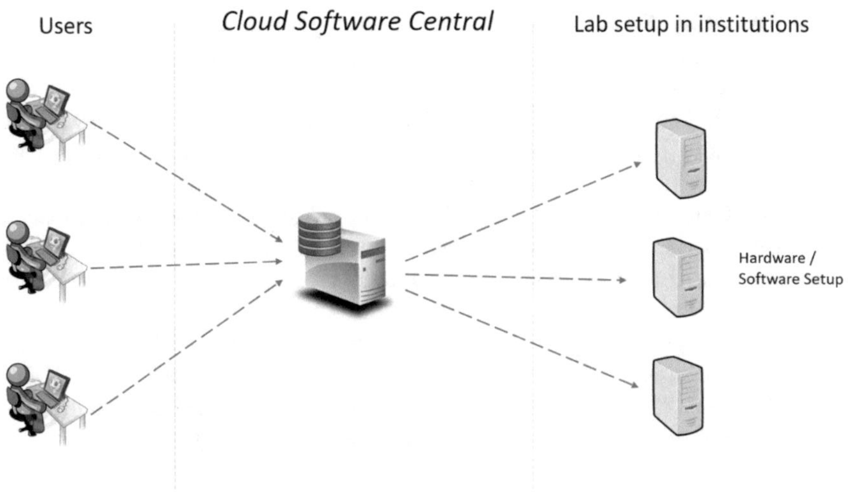

Fig. 1. Architecture of cloud software central

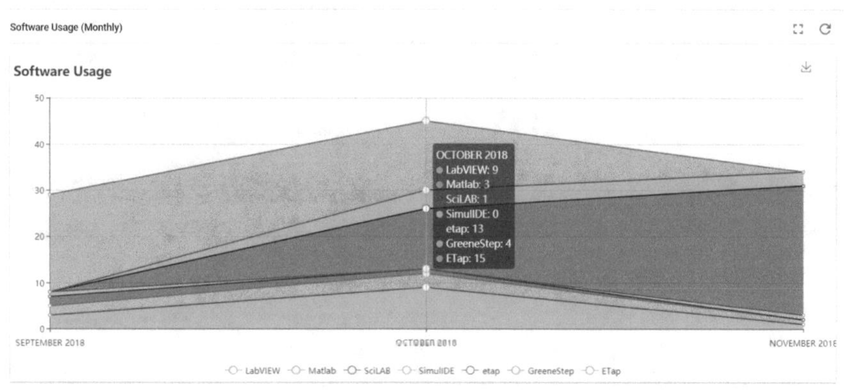

Fig. 2. Usage of different softwares over 3 month period

3 Features

Here are a set of features planned/implemented in *Cloud Software Central*.

- **User Profile**: Basically, any profile consists of user information such as name, contact details, photo etc…, apart from these, *CSC* provides a video profile of the user, where a user can record a video of himself, explaining his strengths and skills. We also record the software usage details (hours of usage) and display that on the user's profile, which is a fool-proof way to gauge a user's strength. The user can later use the information in the profile to generate a CV.

- **Bug Reporting**: No software is bug free, knowing this fact, we have enabled a bug reporting feature where users and report any bugs found. This helps in building an error free application.
- **Metrics**: Stats and metrics is a way of building a healthy competitive environment. A comparative data will create an urge in the students to utilize the application better.
- **System Health Check**: The application manager running in individual machines doubles as a system health checker, which constantly monitor's CPU usage, RAM consumption, Bandwidth usage etc… and report it to the backend which takes necessary action of intimating the authorities.
- **Modes of Access**: The systems used for the application are mostly windows-based systems, hence, RDP (Remote Desktop Protocol) is the primary mode of access here. We provide 3 variations of RDP access, i.e. Direct RDP file, RDP through Session Controller, RDP on Browser directly (Figs. 3 and 4).

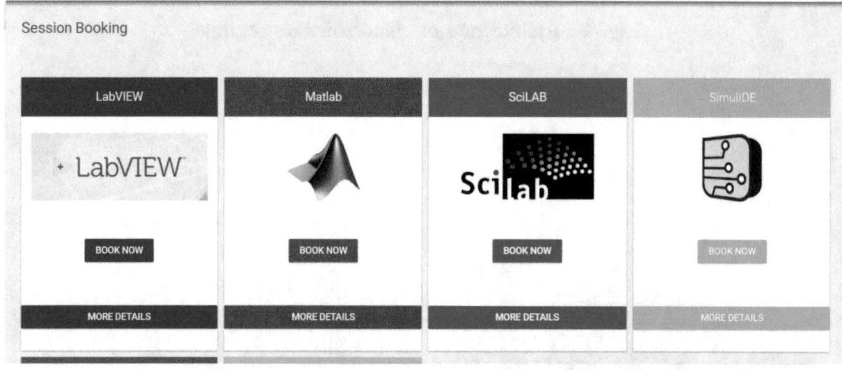

Fig. 3. Tiles showing different available software

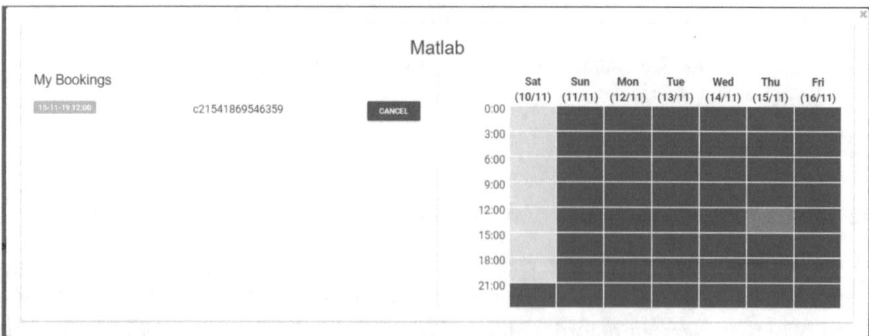

Fig. 4. Slot booking screen

All this while we were discussing features available for a user, now let us see a few related to Admin access.

- **RBAC**: Role Based Access Control is a mechanism in which the functionalities and UI differ based on the role of the logged-in user. Roles in *Cloud Software Central* are broadly classified as DEPARTMENT_ADMIN, SESSION_ADMIN and USER.
  DEPARTMENT_ADMIN is a role provided to the head of the department, who has access to view all the information related to all the students as well faculty, their session information, also has access to alter the roles of users.
  SESSION_ADMIN is the role given to a faculty who can control the session access of the students. He can extend the session of students, control the duration for which a student can access a software or lab experiment etc…
  USER is the lowest role in the hierarchy, which is provided to the students, who can book session, view reports, alter profiles only for themselves.
- **User and Session Management**: As discussed above, the admin will be able to manage the student's access, he will be able to enable/disable student access to the application, control number of sessions one can access per week, extend sessions if necessary, send announcements and notifications to students related to any event.
- **Software Management**: *Cloud Software Central* has an automated process to add/edit and delete software/lab access. This control resides with the admin. He can add multiple machine, multiple software, link the software to the machines and provide access to the necessary software to the users.
- **License Management**: Every organization needs to procure and active license for usage of *Cloud Software Central,* this is facilitated in license management, an admin can request for license, renew or terminate a license.
- **Admin Dashboard**: Dashboard for admin consists of consolidated reports and graphs of all users. A sample graph is as shown in Fig. 5.

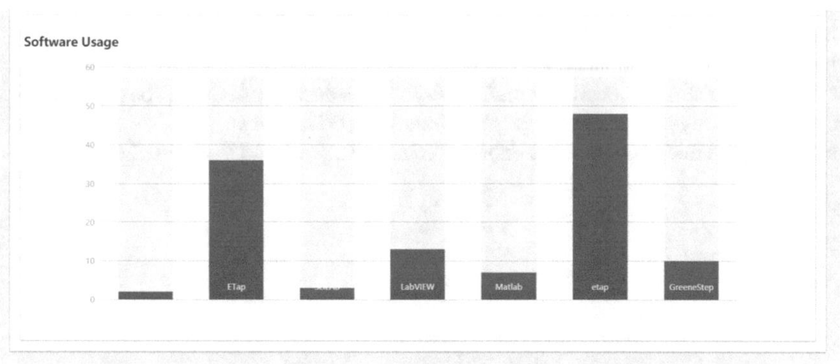

Fig. 5. Total comparative usage by sessions

3.1 Actual Outcomes

Cloud Software Central is deployed and running successfully for last 3 years with 2.X version currently in BETA.

The resource access is broadly classified into two variants i.e. Remote Desktop and Remote App. Remote Desktop is usually used for resources requiring any hardware access. For software only access, Remote App is used.

All historical session booked along with the amount of time a resource is utilized is available with detailed analytics dashboard. The session booked reports are available on daily, weekly, monthly etc. reports.

3.2 Conclusions/Recommendations/Summary

Based on our interactions and understanding so far, *Cloud Software Central* is clearly the need of the hour for Remote Engineering. As more and more applications are moving towards cloud computing, cloud platform is a promising way forward for Remote Engineering.

With *Cloud Software Central*, we bring to you Engineering Lab-as-a-Service (LaaS).

References

1. Bhimavaram, K.R., et al.: A distinctive approach to enhance the utility of laboratories in Indian academia. In: 2015 12th International Conference on Remote Engineering and Virtual Instrumentation (REV). IEEE (2015)
2. http://docs.spring.io/spring/docs/current/spring-frameworkreference/html/overview.html
3. “jar file”, [online] Available: https:lldocs.oracle.com/javase/tutorialldeployment/jar/
4. “REST”, [online] Available: https://www.ics.uci.edu/~fielding/pubs/dissertation/rest_arch_style.htm
5. https://en.wikipedia.org/wiki/Single_sign-on
6. https://en.wikipedia.org/wiki/MongoDB
7. http://economictimes.indiatimes.com/industry/servic/education/AICte-to-soon-recognise-distance-learning-engineeringcourses/articleshow/49937735.cms
8. https://technet.microsoft.com/enus/library/dn383589.aspx#BKMK_RA2012R2

Blended Learning Practices for Improving the Affective Domain of K-12 Learners of Madurai District, TN

P. Karthikeyan(✉), A. M. Abirami, and M. Thangavel

Department of Information Technology, Thiagarajar College of Engineering, Madurai, Tamil Nadu, India
{karthikit,abiramiam,mtit}@tce.edu

Abstract. Learners of countryside background, face different types of problems and challenges when they do their higher education in urban areas. Their culture, custom, and the way they had been taught in their schools are entirely different. These differences are the major concern for some students and they usually perform low in their higher education. It results in de-motivation of individual and impacts their overall performance. Only very few students cope-up with the new environment (College) and adapt to it. This paper details the study that has been carried out to examine whether the ICT-based pedagogical practices motivate the learners and inculcate learning thirst among them for their life-long learning.

Keywords: Blended · Learning · Affective domain · K12 learners · Learning management system

1 Introduction

Blended learning is a mix of traditional classroom presentation and ICT based content delivery which may have higher learners' engagement. Blended learning helps to achieve active learning participation of all students when it is delivered as eContent during the face-to-face interaction in the classrooms. K-12 learners prefer game-based learning which is increasingly becoming an effective training tool within the education and training community. It has its own advantages like simplicity, cost-effectiveness, and involvement of learners through physical movement. eLearning or mLearning enables the learners to participate in the learning by remote access. This liberty of time and space engages all types of learners and creates interest in the topic of learning with enthusiastic participation.

The theoretical framework for learning proposed by Benjamin Bloom is as follows:

- **Cognitive**: Knowledge and critical thinking of learners on a particular subject. Traditional classroom teaching follows this and it improves the Lower Order Thinking Skills (LOTS) of learners.
- **Affective**: Behavioral skills of the learners while learning happens. It improves soft skills like attitude of learners and Higher Order Thinking Skills (HOTS) of learners.

M. E. Auer and K. Ram B. (Eds.): REV2019 2019, LNNS 80, pp. 793–804, 2020.
https://doi.org/10.1007/978-3-030-23162-0_72

– **Psychomotor**: Practice oriented learning. It enhances the skill set of learners by improving his behavior and HOTS.

This paper focuses on improving the affective domain of learners by following blended learning practices. Figure 1 shows the characteristics of affective domain. Bloom's taxonomy divides this domain into five categories like:

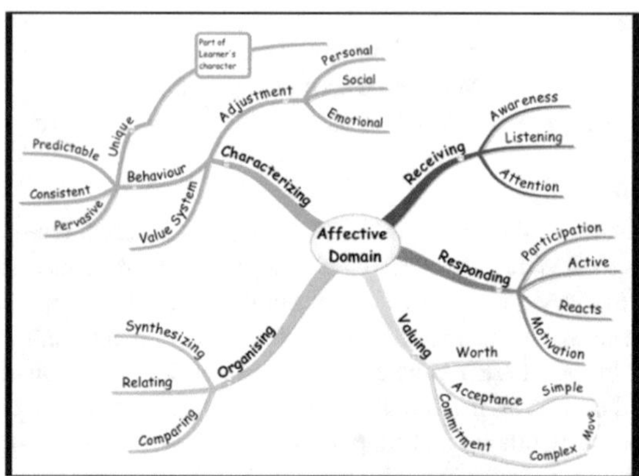

Fig. 1. Characteristics of affective domain

Receive—the learner shows willingness in participation
Respond—the learner shows interest in activity and completes it with pleasure
Value—the learner appreciates the objectives of activity
Organize—the learner is able to resolve discrepancies, if any
Characterize—the learner is motivated for life-long learning.

Ekanayake et al. [2] proposed improved learning architecture for enhancing affective learning. Pandey [3] proposed collaborative and interactive learning platform for Uttarakhand schools for learning mathematics (geometry) and science using open source tools. Abirami [7, 9] proposed collaborative learning for improving higher order thinking skills. Karthikeyan [5, 6] developed mobile-based application to measure the effectiveness of mobile learning and programme outcomes of graduates. Nail and Ammar [4] discussed the advantages of mobile technology and gadgets inside the classroom for the improvement of learning. Bano et al. [1] had done a detailed study on the use of mobile apps for teaching mathematics and science and categorized them into different domain, type and context of use.

It is a common practice that the pedagogy practices with quality and set of well-defined measurable learning outcomes have to be clearly identified in prior, before the implementation of blended learning approach. The assessment and evaluation mechanism have to be developed in order to foster the improvements in blended instructions. The blended learning practices provide focus for teachers while designing learning experiences with the use of technology to cater to the needs of all types of learners.

Having done the preliminary literature survey on this domain, the research questions have been framed as follows:

RQ1: *Does ICT tools usage improve the learning potential of K12 learners?*
RQ2: *Does the blended learning motivate the K12 learners to accommodate themselves in Higher Education?*

With these RQs, a study has been carried out with school students to examine whether the ICT-based active learning strategies motivate them for their life-long learning.

2 Methodology

Teachers of higher education can bring-in the transformations to K-12 learners, who are just moving from pedagogy state to andragogy state. Figure 2 shows the framework of implementing blended learning practices for K-12 learners.

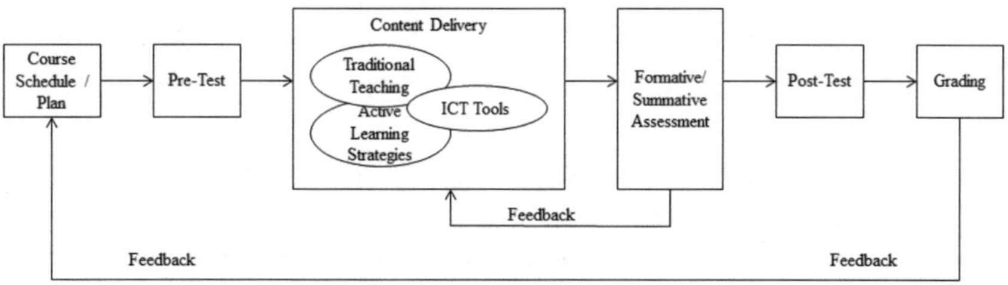

Fig. 2. Proposed framework for implementing blending learning practices

The Higher Education Institute may organize summer or winter training courses during the vacation period of schools. Initially, learners are assessed on their prior knowledge in the domain using pre test. The course contents are delivered in three different modes. They are (i) traditional teaching (ii) active learning strategies to make the learners to understand the concepts, and (iii) the use of ICT based tools for the complete participation of all learners. The active learning strategies include role play, instant quiz, summary, discussions etc. The online Learning Management System (LMS) 'Canvas', is used for discussion, feedback, formative/summative assessment like Quiz, and assignment submissions. Google forms are used for the pre-test and post-test. The effectiveness of training courses is assessed by the post-test and the course exit feedback.

3 Implementation and Results

The Summer Training Course (STC) was organized for school students during April 2018 for the computer domain: Course 1—C & Python Programming and Course 2—Web & Mobile App Design. 16 students from different boards of education, origin, and medium of study participated in Course 1 and 28 in Course 2. The types of learners are shown in Table 1.

Table 1. Demography of learners

Standard	X and below X	27%
	XI	23%
	XII	50%
Specialization	SSLC and below	19%
	HSC—Maths and Biology	15%
	HSC—Maths and Computer Science	62%
	HSC—Vocational/Pure Science	4%
Board of Education	Matriculation/StateBoard	92%
	CBSE	8%

Pre-test had been conducted for the participants before the course was started. The use of blended learning practices for the content delivery improved the performance of learners in the post-test or course exit test. Google Forms had been used for getting the responses.

The concepts were explained using Active Learning Strategies (ALS) and ICT tools. Think Pair Share, Team Pair Solo, Role Play, and Group discussion had been adopted to give better understanding on the concepts. For example, the array concept in C programming was demonstrated and visualized using ***pythontutor.com***, as shown in Fig. 3a. At the end of each session, the feedback from the students was obtained using ***paddlet***, as shown in Fig. 3a, b.

At the end of Course, post-test was conducted to assess the knowledge of learners. For the question 5, the number of responses for the correct answer was improved in post-test, as shown in Fig. 4a, b.

The number of correct responses for the question 3 was much better in post-test when compared with pre-test, as shown in Fig. 5a, b. Similarly, Fig. 6a, b shows the improvement for the question 4. The average scores of pre-test and post-test are shown in Fig. 7a, b. These figures show that the cognitive domain of each learner on the subject has been improved to a greater extent.

The post-test also includes 30 questions from the topics what they have learned to test their knowledge. Nearly 15/28 students scored >50% in this test. Figure 8a shows the analysis of course feedback, shared by the participants in the textual format. It shows 72% words used in the feedback forms are related to positive context and 8.8% words on negative context. Figure 8b also shows the use of words in the feedback forms. All these words prove that the learners had active participation in the assigned

(a)

(b)

Fig. 3. **a** Canvas tool for sharing eContent and Feedback for Course 1. **b** Canvas tool for sharing eContent and Feedback for Course 2

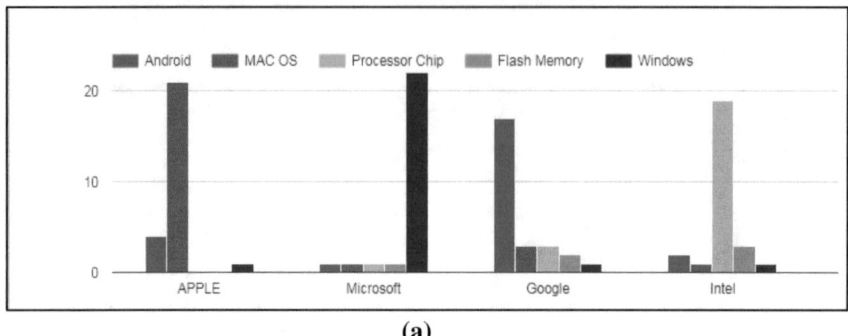

(a)

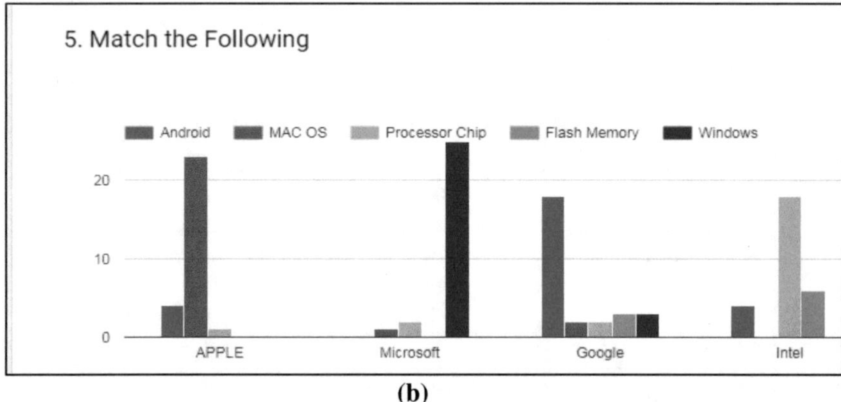

(b)

Fig. 4. **a** Responses of pre-test for the question 5. **b** Responses of post-test for the question 5

activities. Hence, the affective domain of learners is moved to the second level "Respond".

Table 2 shows the Canvas usage of each learner for different purposes like accessing learning materials, participating in quiz, assignment submission, etc. The attribute 'Page Views' shows how many times the learner accessed the content, the attribute 'Participation' shows how many times quiz, feedback links had been used, the attribute 'Submissions' shows how much Online Assignments had been submitted and the attribute 'OnTime' shows how many students completed the submission before deadline. All these attributes show the behavioral aspects of learners.

Figures 9 and 10 show the learners' satisfaction level on the course. Figure 10 shows how this course motivates their learning thirst for their higher education.

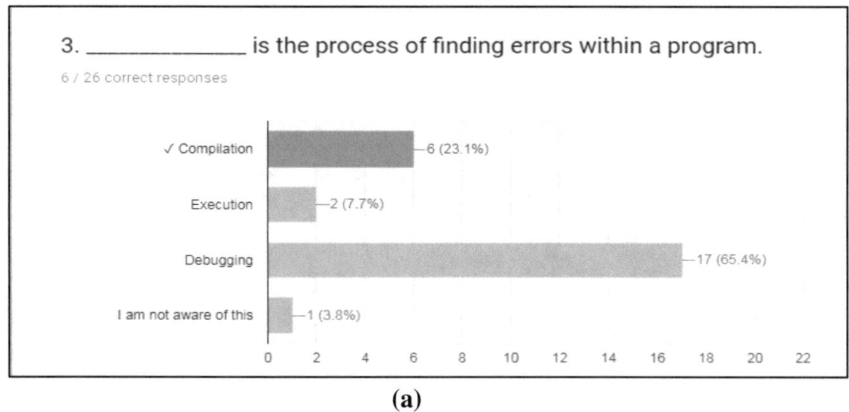

(a)

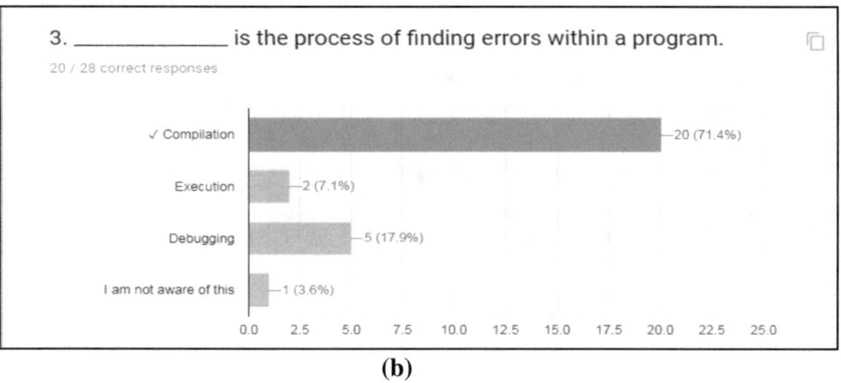

(b)

Fig. 5. **a** Responses of pre-test for the question 3. **b** Responses of post-test for the question 3

4 Discussion

H_0: Null hypothesis states that there is no improvement in the affective domain of K12 learners when blended learning practices are adopted.
H_a: Alternate hypothesis states that there is improvement in the affective domain of K12 learners when blended learning practices are adopted.

The results of post-test showed that there is an impact in the learning among K12 learners when ICT tools are used for content delivery and assessment (RQ1), as shown by Figs. 4, 5, 6 and 7. The motivation of K12 learners is improved when blended learning practices like ALS along with ICT tools had been used for content delivery (RQ2). It is evident from Figs. 8, 9 and 10. Table 2 shows the active participation of learners in the LMS tool 'Canvas' during the course period. With these results, the null hypothesis H_0 can be rejected and the alternate hypothesis H_a can be accepted.

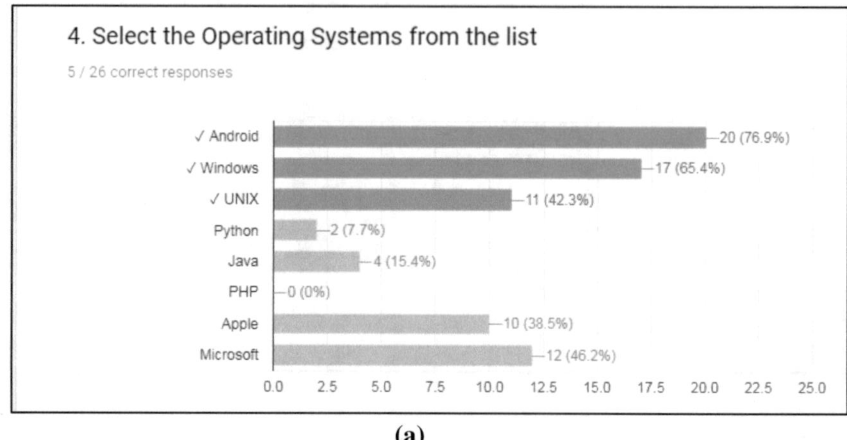

(a)

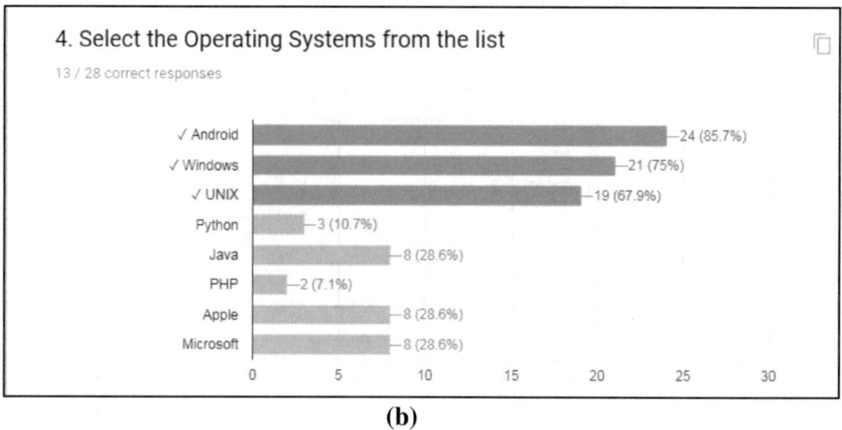

(b)

Fig. 6. **a** Responses of pre-test for the question 4. **b** Responses of post-test for the question 4

5 Conclusion

Blended learning practices enable the learners to access quality content from home or school or college, communicate with a large community of learners and teachers, and work online. It helps all the learners to communicate, collaborate and enhance their learning with new ideas. The study that had been carried out shows that the blended learning practices changed the K12 learners with enthusiastic participation in all the activities. The evidences show that this blended learning improved their affective domain above the respond level. Blended learning motivates the learners to participate in online discussions and it is more significant than in the classroom discussions. This type of blended learning improves the affective domain of learners and thus improves the professional education of learners.

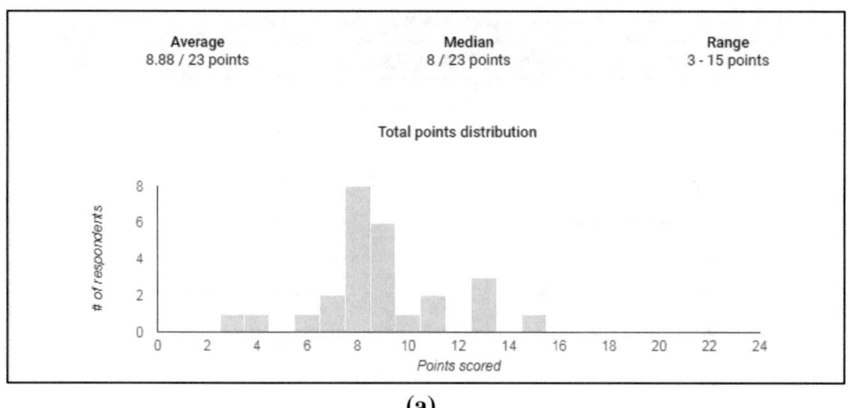

(a)

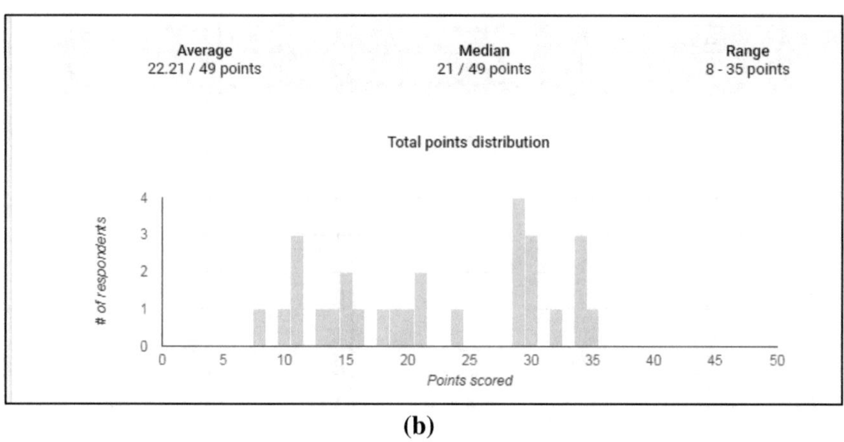

(b)

Fig. 7. **a** Scores of pre-test. **b** Scores of post-test

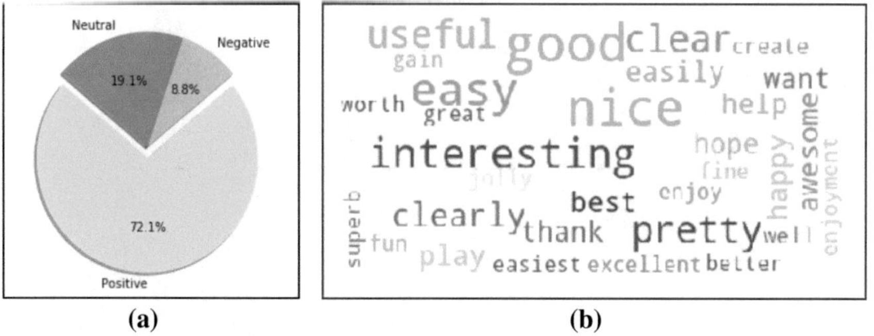

(a) (b)

Fig. 8. **a** Feedback analysis. **b** List and frequency of words

Table 2. Learner's behavior in canvas

Student ID	Page views	Participation	Submissions	OnTime
S1	286	3	2	2
S2	105	2	2	2
S3	91	2	2	2
S4	110	4	2	2
S5	101	3	2	2
S6	10	1	0	0
S7	170	2	2	2
S8	170	1	2	2
S9	97	3	2	2
S10	127	2	2	2
S11	100	2	2	2
S12	92	2	1	1
S13	13	0	0	0
S14	213	2	1	1
S15	192	3	2	2
S16	252	3	2	2
S17	104	3	2	2
S18	25	1	1	1
S19	211	4	1	1
S20	128	3	2	2
S21	109	3	1	1
S22	166	3	2	2
S23	180	2	1	1
S24	197	4	2	2
S25	183	4	2	2
S26	225	2	1	1
S27	131	2	1	1
S28	150	2	1	1
S29	139	3	2	2
S30	155	3	2	2
S31	134	4	2	2
S32	11	0	0	0

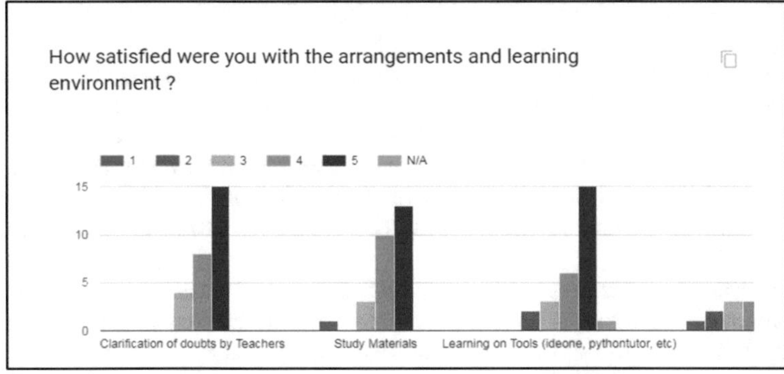

Fig. 9. Learner's satisfaction level on the course

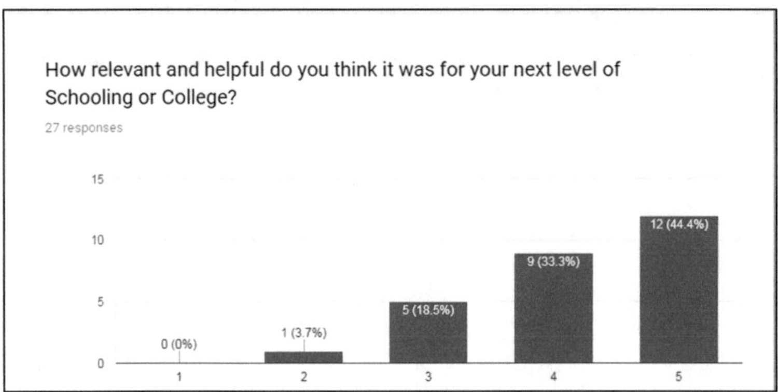

Fig. 10. Learner's feedback on the course

Acknowledgements. We acknowledge Department of Information Technology, Thiagarajar College of Engineering for sharing their data for this study.

References

1. Bano, M., Zowghi, D., Kearney, M., Schuck, S., Aubusson, P.: Mobile learning for science and mathematics school education: a systematic review of empirical evidence. Comput. Educ. **121**, 30–58 (2018)
2. Ekanayake, H., Hewagamage, K.P., Karunaratna, D.: Cognitive architecture for affective learning. Int. J. Comp. Internet Manage. **41**, 32–40 (2006)
3. Pandey, J.: Use of e-learning in Uttarakhand school education system: case study of open source e-learning tools for fundamental mathematics and sciences. Int. J. Comput. Sci. Technol. **3**, 80–82 (2012)
4. Nail, B., Ammar, W.A.: Mobile learning education has become more accessible. Am. J. Comput. Sci. Inf. Technol. **5**, 2 (2017). https://doi.org/10.21767/2349-3917.100005

5. Karthikeyan, P.: Effectiveness of mobile learning at TCE, India: a learner perspective. In: IEEE International Conference on T4E (2015)
6. Karthikeyan, P.: EIAT: Enhanced indirect assessment tool for assessing the graduate degree programme outcomes. In: Fourth IEEE International Conference on MOOCs, Innovations, Technology in Education (MITE) (2016). https://doi.org/10.1109/mite.2016.027
7. Abirami, A.M., Divya, A.: Active learning environment for achieving higher-order thinking skills in engineering education. In: Fourth IEEE International Conference on MOOCs, Innovations, Technology in Education (MITE), pp. 47–53 (2016). https://doi.org/10.1109/mite.2016.020
8. Karthikeyan, P., Abirami, A.M., Thangavel, M.: Efficient assessment methods for improving the programme outcomes of undergraduate-information technology programme in India. In: Fourth IEEE International Conference on MOOCs, Innovations, Technology in Education (MITE), pp. 351–356 (2016)
9. Abirami, A.M.: Collaborative learning tools for data structures. J. Eng. Educ. Transformations **31**(3), 79–83 (2018)
10. https://www.schoology.com/blog/blended-learning-higher-ed-best-practices-getting-started. Accessed on 27 Sept 2018
11. https://smartprimaryed.com/2015/11/15/theory-dap-developmentally-appropriate-practice-part-3-the-affective-domain/. Accessed on 25 Sept 2018

Work-in-Progress: Silent Speech Recognition Interface for the Differently Abled

Josh Elias Joy(✉), H. Ajay Yadukrishnan, V. Poojith, and J. Prathap

The Oxford College of Engineering, Bengaluru, Karnataka, India
{joshej197,ajayyadukrishnan,poojithvp63,Prathu7284}
@gmail.com

Abstract. Silent speech or unvoiced speech can be interpreted by lip reading, which is difficult, or by using EMG (Electromyography) electrodes to convert the facial muscle movements into distinct signals. These signals are processed in MATLAB and matched to a predefined word by using Dynamic Time Warping algorithm. The identified word is then converted to speech and can be used to control a nearby device such as a motorized wheelchair. Thus, a silent speech interface has the potential to enable a differently-abled person to communicate and interact with objects in their surroundings to ease their lives.

Keywords: Silent speech · Signal processing · Human computer interface · Electromyography

1 Introduction

Speech is the most basic and preferred means of communication among humans. Unfortunately, around 18.5 million individuals have a speech, voice, or language disorder [1]. Some of these individuals have great potential but are unable to communicate their thoughts. There are several types of speech disorders such as stuttering, lisps, Dysarthria (weakness or paralysis of speech muscles due to ALS or Parkinson's diseases.). We do not always need our voice to convey our thoughts and ideas to others. There are many problems we face when communicating with people who are not familiar with our language. When we speak, each word has a different type of movement of the facial muscles.

In this paper, we are designing a Human Computer Interface to detect and analyze facial movements and differentiate facial muscle movements into words, by using different signal processing methods to process the input signal retrieved from the hardware component.

2 Methodology

When we speak, there will be movement of facial muscles and each word will have a different type of facial muscle movement. We can recognize words by looking at other peoples facial muscle movements. This method is lip reading. We are applying the same principle to differentiate between words by analyzing these facial muscle

M. E. Auer and K. Ram B. (Eds.): REV2019 2019, LNNS 80, pp. 805–813, 2020.
https://doi.org/10.1007/978-3-030-23162-0_73

movements. Electromyography is the recording of the electrical activity of muscle tissues, or its representation as a visual display, using electrodes attached to the skin. This device uses electrodes attached to facial muscles to acquire the facial muscle movement signals (Fig. 1).

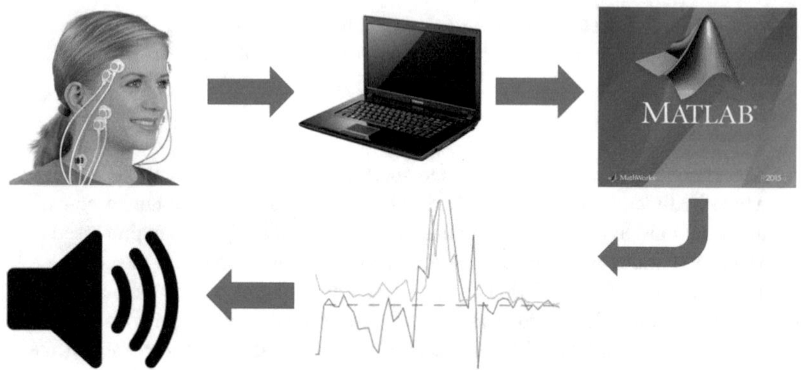

Fig. 1. Block diagram of the silent speech interface

3 Hardware

The hardware components used are Arduino UNO (ATmega328p), AD8232 (Heart Rate monitor), electrodes and a personal computer. Although the AD8232 is a Heart rate monitor, it has proved to be a low-cost alternative to a regular EMG sensor. We are also in the process of testing other integrated circuits for EMG measurement. The sensor has three electrodes attached to it and these electrodes are placed on the facial muscle of a person to read the muscle movement pattern.

The AD8232 (Fig. 2) is designed to extract, amplify, and filter small biopotential signals in the presence of noisy conditions, such as those created by motion or remote electrode placement. The Arduino acquires these EMG signals and sends it serially to the personal computer for signal processing.

Fig. 2. AD8232 Heart rate monitor module

Arduino (Fig. 3) is an open-source hardware and software company, project and user community that designs and manufactures microcontrollers and microcontroller kits for building digital devices and interactive objects that can sense and control objects in the physical and digital world.

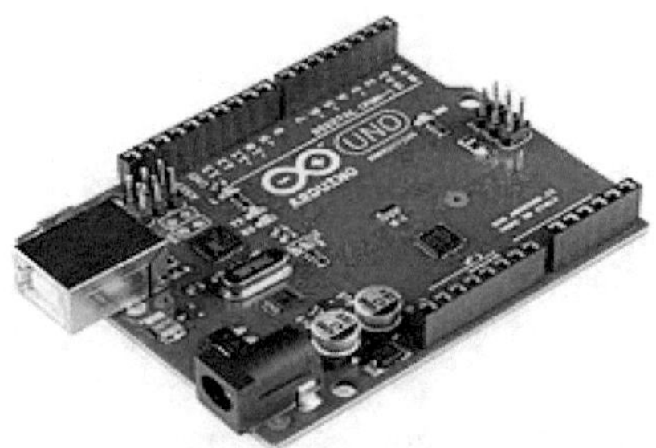

Fig. 3. Arduino uno

A specially designed face mask is used to place the electrodes on the user's face. It has holes at specific locations through which electrodes can be attached. This mask will help in keeping the electrodes properly attached to the skin and also receive a stable signal from the electrodes. The mask is made out of cloth and is based on the pollution masks available in the market. Figure 4 shows the 3D representations of the face mask designed using Autodesk Fusion 360 software.

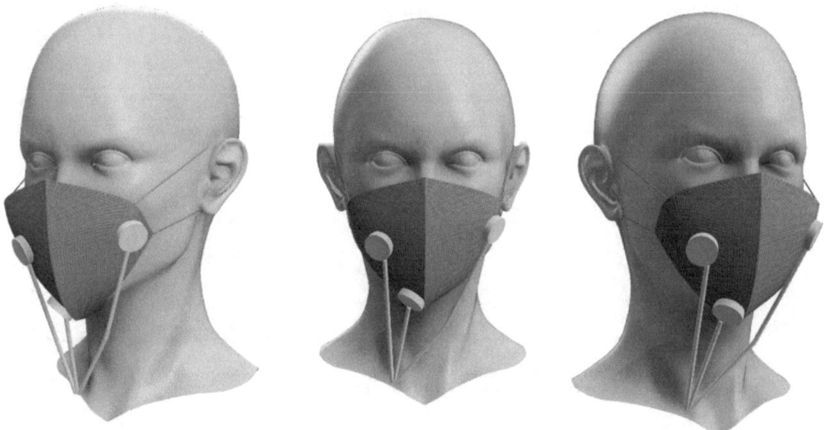

Fig. 4. Right, front and left views of the specially designed face mask

These are the 3D representations of the face mask designed using Autodesk Fusion 360 software.

4 Software

MATLAB R2018a software is used to process the signals acquired by the AD8232 module. A graphical user interface has been developed to ease the functionality, a separate interface for training and identifying signals have been developed using the App Designer tool in MATLAB.

4.1 Graphical User Interface

MATLAB has a very useful and powerful tool called App Designer. App Designer is a rich development environment that provides layout and code views, a fully integrated version of the MATLAB editor, and a large set of interactive components.

The training app (Fig. 5) has an input field to type the word and another number input field to type the number of trials for each word. After the 'START' button has been clicked, the user will say the word and the app will plot the signal obtained from the electrodes and the sensor. This process is repeated until the specified number of trials is complete. These signal values are then stored in an Excel file.

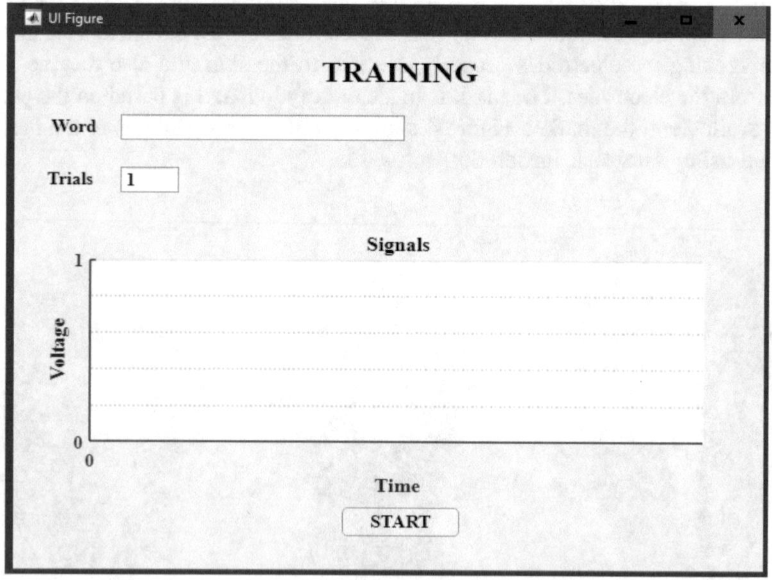

Fig. 5. Training application

The Testing application (Fig. 6) has a button named 'Test'. When this button is clicked, the system will record the signal and compare that with the signals already stored in the database, using a method that will be discussed in the next section.

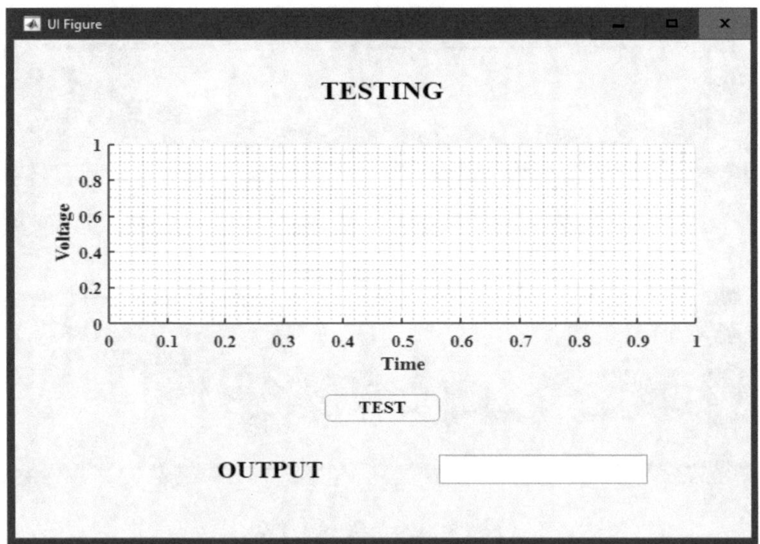

Fig. 6. Testing application

4.2 Dynamic Time Warping

Euclidean distance is the straight-line distance between two points in space. This distance can be used to find similarities between signals if they are perfectly aligned. But, the signals we acquire will not be aligned always because the way we say a word won't be the same every time. The speed may vary each time we say the word. To overcome this problem, we use an algorithm called Dynamic Time Warping.

Dynamic Time Warping (DTW) is one of the algorithms for measuring similarity between two temporal sequences, which may vary in speed. For instance, similarities in walking could be detected using DTW, even if one person was walking faster than the other, or if there were accelerations and decelerations during the course of an observation. It aims at aligning two sequences of feature vectors by warping the time axis iteratively until an optimal match (according to a suitable metrics) between the two sequences is found.

Figure 7 shows the Euclidian distance and DTW distance of two similar signals.

Figure 8 shows the two different utterances of the words 'Hello' and 'Baseball'. These two signals can be visually interpreted as different but we need to programmatically infer that these two signals are indeed different and the words uttered are different.

Dynamic Time Warping function will find the best optimal path between two signals. This is shown by a diagonal line. The lesser the deviation from the diagonal, the higher is the similarity.

By finding this distance, we can calculate the similarity between signals and we can set a threshold for the deviation from the diagonal. If the deviation is less than the threshold, the two signals are said to be similar.

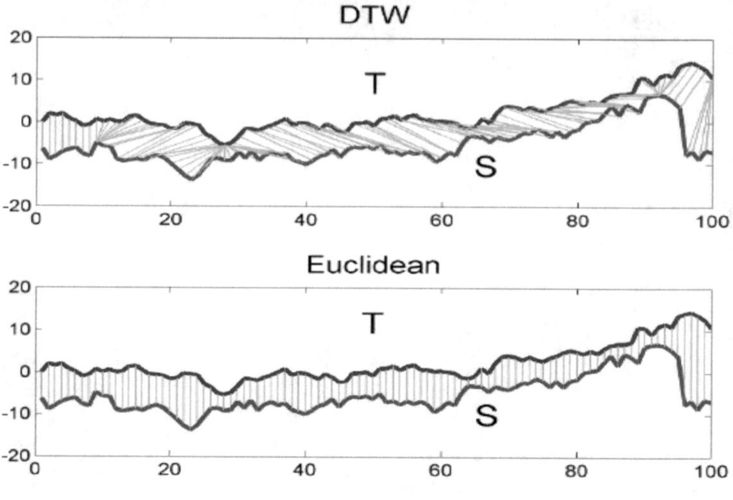

Fig. 7. Example of DTW

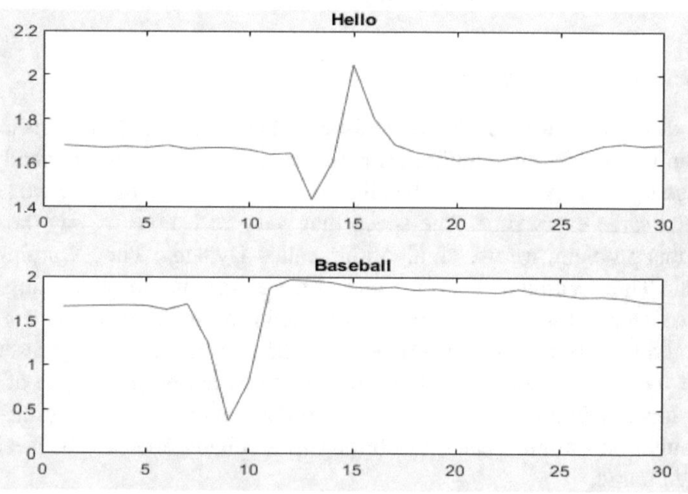

Fig. 8. Utterances of the words 'Hello' and 'Baseball'

Figure 9 shows the warping path of words 'Hello' and 'Baseball'. The warping path exceeds the threshold and thus the output would be that the words are not a match.

Figure 10 shows the two different utterances of the word 'Hello'. These two signals are similar visually, but we need to find the similarity between these signals programmatically.

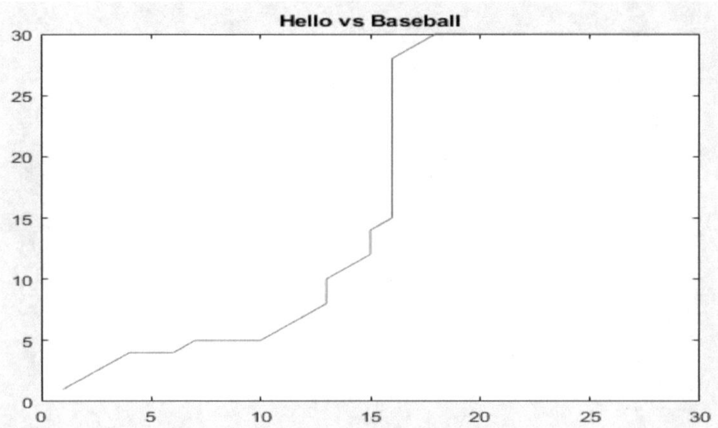

Fig. 9. The warping path for the words 'Hello' and 'Baseball'

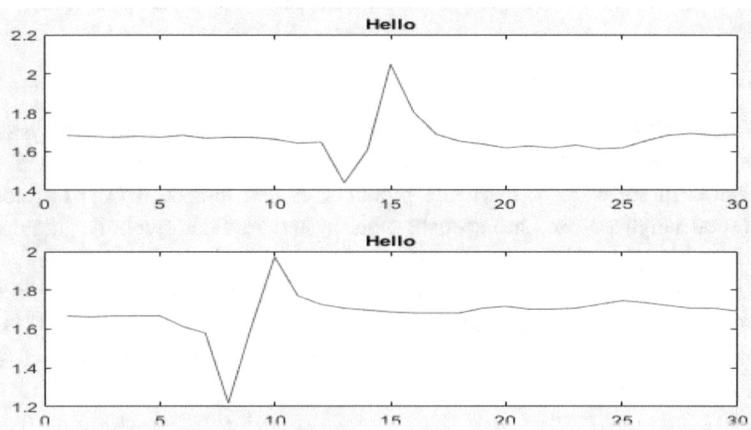

Fig. 10. Two different utterances of the word 'Hello'

Figure 11 shows the warping path of two different utterances of the word 'Hello'. Although the diagonal is not perfectly a straight line, it is still below the threshold deviation. Hence, it is found to be similar to each other.

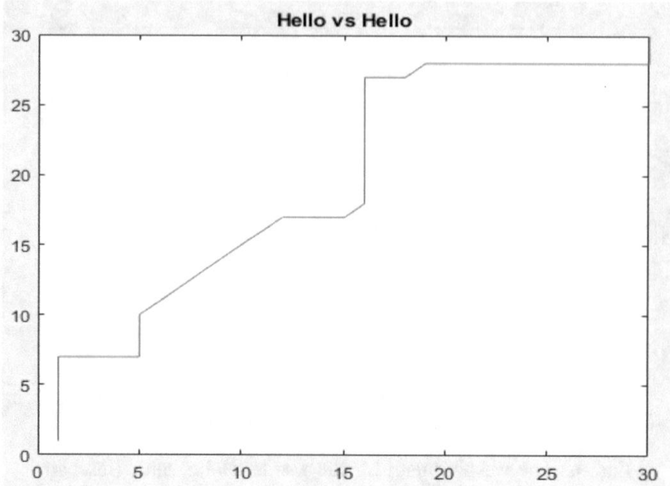

Fig. 11. The warping path of two different utterances of the word 'Hello'

4.3 Text to Speech Conversion

Speech synthesis is the artificial production of human speech. A computer system used for this purpose is called a speech computer or speech synthesizer, and can be implemented in software or hardware products. A text-to-speech (TTS) system converts normal language text into speech; other systems render symbolic linguistic representations like phonetic transcriptions into speech.

We use the built-in text-to-speech function in Windows to receive the audio form of the uttered word. The text obtained from the signal processing in MATLAB is passed to the text to speech converter program which gives the audio output through a speaker.

5 Conclusion

Our work, currently at its initial stage, proposes a method for people with a speech impediment to communicate with others and to control IOT devices in their surroundings. A special abled person with a speech impediment who is wheelchair bound could use the silent speech interface to control his wheelchair and also appliances in his surroundings to ease his life and thus making him more independent. After the completion of the first fully functional prototype, the usability of the prototype will be assessed by a spectacled abled person, whose feedback will be essential for further development. A language translator function can also be added so that the recognized silent speech can be converted to another language.

References

1. National Institute on Deafness and Other Communication Disorders: Retrieved from https://www.nidcd.nih.gov/health/statistics/quick-statistics-voice-speech-language (2010)
2. Juang, B.H.: On the hidden Markov model and dynamic time warping for speech recognition—a unified view. AT&T Bell Laboratories Tech. J. **63**(7), 1213–1243 (1984)
3. Wand, M., Schultz, T.: Session-Independent EMG-Based Speech Recognition (2011)
4. Kapur, A., Kapur, S., Maes, P.: AlterEgo: a personalized wearable silent speech interface. In: 23rd International Conference on Intelligent User Interfaces (IUI 2018), 5 Mar 2018
5. Gaikwad, S.K., Gawali, B.W., Yannawar, P.: A review on speech recognition technique. Int. J. Comput. Appl. (0975–8887) (2010)

Part VIII

Data Science

Deep Learning Frameworks for Convolutional Neural Networks—A Benchmark Test

Andreas Pester(✉), Christian Madritsch, Thomas Klinger, and Xabier Lopez de Guereña

Carinthia University of Applied Sciences, Europastraße 4, Villach, Austria
pester@fh-kaernten.at

Abstract. Deep neural networks are state of the art for many machine-learning problems. The architecture of deep neural networks is inspired by the hierarchical structure of the brain. Deep neural networks feature a hierarchical, layer-wise arrangement of nonlinear activation functions (neurons) fed by inputs scaled by linear weights (synapses). Deep Learning frameworks simplify model development and training by providing high-level primitives for complex and error-prone mathematical transformations, like gradient descent, back-propagation, and inference. The main goal of this study is to compare the performance of two trending frameworks, Caffe and TensorFlow for Deep Machine Learning. As benchmark for the comparison with other approaches in deep learning was a well-analyzed database for handwritten cipher classification chosen. These two frameworks (Caffe and TensorFlow) were chosen out of nine different frameworks, after checking which ones better fits the proposed selection criteria. Caffe and TensorFlow frameworks were selected after nine different frameworks were analyzed using the following selection criteria: programming language, functionality level, algorithms and network architecture, pretrained models and community activity. As the performance is heavily affected by the number of parameters in the networks, four well-known convolutional neural network (CNN) architectures (LeNet, AlexNet, VGG19 and GoogLeNet) were trained and tested using both frameworks. Using architectures with different deepness will allow investigating how the number of hidden layers improves the accuracy of the CNN. As a CNN requires high computational effort, two computers equipped with different NVIDIA GPU were used to ease the effort and they were used to investigate how the hardware improves the performance of the CNN and if it is worthy to invest on it. As the CNN are widely used for image classification, it was defined that the used architecture was going to be used for classification of handwritten numbers, because this example of classification is very well analyzed and can serve as a benchmark for comparison. Due to this and considering that the training of these networks requires a huge amount of data, MNIST database was set for training and testing them. All the architectures were adapted to the3 MNIST database and they were developed for Caffe and TensorFlow frameworks and analyzed on the named architectures. The biggest differences on our hardware we have got in training the VVG19 and the GoogLeNet architectures in TensorFlow and Caffe.

Keywords: Deep learning · Convolutional neural network · TensorFlow · Caffe · LeNet · AlexNet · VGG19 · GoogLeNet

M. E. Auer and K. Ram B. (Eds.): REV2019 2019, LNNS 80, pp. 817–831, 2020.
https://doi.org/10.1007/978-3-030-23162-0_74

1 Introduction

Machine learning (ML) is already around; written in the software on phones, in cars and homes and in the business software that is used at work, helping to access information and make better and more informed decisions, more quickly. In recent years, Machine Learning has significantly changed the world around by drastically reducing computing time and improving the effectiveness of predictions and classifications on big data resources. Although we are still quite far from reliable machine intelligence that would be comparable to human intelligence, many tasks can be delivered to machines.

Machine learning is a branch of data science and artificial intelligence (AI) where systems are provided with the ability to automatically learn and improve from experience without being explicitly programmed for it. Machine learning focuses on the development and implementation of algorithms that are trained with data and are minimizing cost functions or problem dimensions.

A convolutional neural network is a type of an Artificial Neural Networks (ANN), which is a feed-forward network, wherein connections between the nodes do not form a cycle. They are formed by an input, output and hidden layers that uses learnable filters. These hidden layers consist on convolution layers, pooling layers, normalization layers and fully connected layers.

Deep learning is a subfield of a broader family of machine learning methods based on learning data representations, where a set of techniques allows a system to automatically discover the representations needed for feature detection or classification from raw data. One type of these deep learning networks is the CNN, which is the one that will be tested in this thesis. They are mainly used to classify images and to perform object recognition.

Deep learning involves the use of large neural network; they are called deep due to the number of layers that compound it, usually between 3 and 27 layers, which are compound from hundreds of thousand parameters up to millions. Due to this, huge amount of data, from tens thousands of images (like in MNIST database) up to a million (ImageNet database), and a lot of computational effort is required to train them. Nowadays this is feasible due to the development of hardware, like GPUs which allows parallelizing the training process and due to the available amount of data, from databases like MNIST and ImageNet.

The main goal of this analysis is to compare the performance of two trending frameworks (Caffe and TensorFlow) for handwritten numbers classification. As the performance is heavily affected by the number of parameters in the networks, four different architectures from five to twenty-seven hidden layers will be trained and tested using both frameworks. Using architectures with different deepness will allow investigating how the number of hidden layers improves the accuracy of the CNN.

As a CNN requires high computational effort, two computers equipped with different NVIDIA GPU were used to ease the effort and to investigate how the hardware improves the performance of the CNN and if it is worthy to invest on it.

2 Choice of the Frameworks

Deep learning frameworks offer building blocks for designing, training and validation of deep neural networks through a high-level programming interface. These deep learning frameworks rely on GPU-accelerated libraries such as cuDNN [1] and NCCL to deliver high-performance GPU accelerated training. Developers, researchers and data scientists can get easy access to GPU optimized deep learning framework containers, which eliminates the need to manage packages and dependencies or build deep learning frameworks from scratch.

Given the high number of frameworks that are available, selecting one among them can be a tedious task. Each framework is built in a different manner for different purposes. In addition, depending on the goals and resources, different frameworks might be relevant.

Two frameworks were selected out of nine. These frameworks are the ones that were used to create four different convolutional neural networks, the detailed explanation of how were used is given in the fifth chapter. To select the two frameworks five different selection criteria were set: programming language, functionality level, algorithms and network architecture, pretrained models and community activity.

The analyzed frameworks are TensorFlow [2], Torch/Porch [3], Microsoft Cognitive Toolkit/CNTK [4], Caffe [5], MXNet [6], Chainer [7], Keras [8], DeepLearning4j [9], Theano [10].

To evaluate which of the frameworks is the best alternative to create convolutional neural networks for image classification they should meet the following requirements:

Support Python programming language (SP)

Data scientists prefer python [12] and due to the libraries available, it is easier to build highly performing algorithms, it excels in terms of its easy syntactical character as compared to other languages. Many languages like Java, C/C++ and Scala can be equivalently used for scalable machine learning; but when starting from scratch, Python is the easiest language, which can be learnt in a short period and quickly used for implementation. Python also provides a huge set of libraries, which can be easily used for machine learning (e.g. NumPy, SciPy, ScikitLearn and PyBrain).

- Promote: TensorFlow, PyTorch, Microsoft Cognitive Toolkit, Caffe, MXNet, Chainer, Keras and Theano
- Pass: Deeplearning4j.

The functionality level of the framework (LF)

It affects the overall control over the architectures; low functionality level is preferred as it allows more control. Deep learning frameworks have different levels of functionality. In the case of low-level functionality, the framework allows the researchers to define a neural network of arbitrary complexity from the most basic building blocks; this kind of frameworks might be called languages. On the other hand, frameworks with a higher level of functionality behave like drivers whose objective is to boost the developing productivity, but they are limited due to the higher level of abstraction.

When selecting the deep learning framework to develop a neural network, a low-level framework will be chosen. A high-level wrapper is a useful addition but not necessarily required. Even though due to the ease of use of high-level frameworks, more and more low-level frameworks are developing high-level companions.

- Promote: TensorFlow, PyTorch, Microsoft Cognitive Toolkit, Caffe, MXNet, Chainer and Theano
- Pass: Keras.

Neural networks architectures (AN)

For image classification deep convolutional neural networks architectures will be used, due to this the selected framework must support this architecture (AN). In recent years, among all the available architectures, CNNs have become the leading architecture for most image recognition, classification, and detection tasks. The selected framework must support convolutional neural network architecture.

The CNNs have several different filters consisting of trainable parameters (weights) depending on the depth and filters at each layer of a network, which can convolve on a given input spatially to create some activation maps at each layer. During this process, they learn by adjusting those initial values to capture the correct magnitude of a spatial feature on which they are convolving. These filters essentially learn to capture spatial features from the input volumes based on the learned magnitude. Hence, they can successfully boil down a given image into a highly abstracted representation, which is easy for predicting. All selected frameworks support CNNs:

- Promote: TensorFlow, PyTorch, Microsoft Cognitive Toolkit, Caffe, MXNet, Chainer and Theano
- Pass: None.

Availability of pretrained models (PM)

In case of not having a large dataset to train a model, it is crucial to have available pretrained models. Pretrained models can be found in Model Zoos where there are available example scripts for state-of-the-art models and models whose filters already have pretrained weights. Using them will save computation time and help to achieve better results on a new problem by transferring the intelligence acquired on a different data set.

- Promote: TensorFlow, PyTorch, Microsoft Cognitive Toolkit, Caffe, MXNet and Chainer
- Pass: Theano.

Community activity (CA)

Community activity is an essential part of a deep learning framework; an active community will contribute providing documentation, scripts, models and tutorials, as well as help to solve errors that may appear during the development of a neural network.

Each framework was rated based on community activity and member ratings in GitHub and Stack Overflow. After comparing all results of this selection process TensorFlow and Caffe were chosen (see Table 1, details see in [11]).

Table 1. Solution for framework selection

Frame-works/ Criteria	Tensor-Flow	Torch/ PyTorch	CNTK	Caffe	MX Net	Chai-ner	Keras	Deep-learn-ing4j	Theano
SP	✓	✓	✓	✓	✓	✓	✓	×	✓
FL	✓	✓	✓	✓	✓	✓	×	NP	✓
AN	✓	✓	✓	✓	✓	✓	NP	NP	✓
PM	✓	✓	✓	✓	✓	✓	NP	NP	×
CA	✓	×	×	✓	×	×	NP	NP	NP

3 Used Data and Architectures

3.1 Dataset

The dataset is a fundamental part while designing CNN architectures and must be set before choosing or designing them. The dataset will define the number of classes in the last fully connected layer and while designing the architecture for a convolutional neural network is needed to take care of the size of the data, either convolutional layers or pooling layers may reduce the size of it and it could run out.

The CNN, we used for benchmarking, are focused on number classification; their purpose is to detect which handwritten cipher from 0 to 9 appears in an image. To train a deep network a huge amount of data is required; this is the main reason why an already created dataset with thousands of images was chosen, MNIST database [13].

MNIST is a large database of handwritten digits widely used for training and testing image processing systems in the field of machine learning. MNIST is a subset of a larger set available, NIST. It takes half of its samples from NIST's Special Database 1 and the other half from Special Database 3, both contains binary (black and white) images of handwritten digits normalized to 20×20 pixels preserving their original ratio. To create MNIST database NIST images were converted to greyscale levels (from 0 to 250) because of the anti-aliasing technique used by the normalization algorithm and later, they were centered in 28×28 pixels by computing the center of mass of the pixels.

The MNIST data comes in two parts. The first part contains 60,000 images to be used for training. The second part of the MNIST data set is 10,000 images to be used for testing. The images are greyscale and 28 by 28 pixels in size. Both were used.

3.2 CNN Architectures

The below described architectures that are used for this analysis have deepness in the range of five to twenty-seven layers and numbers of learning parameters from hundreds of thousand up to millions.

Most of the architectures here presented were designed for ImageNet database, whose pictures have a size of 224 × 224 × 3 and have 1000 different classes. Due to this, some of them were modified to fit MNIST database.

LeNet

LeNet [14] developed at 1989, was one of the first convolutional neural networks. LeNet architecture was developed mainly for character recognition tasks such as reading zip codes and digits. This architecture helped boosting the field of Deep Learning.

Lot of the new CNN architectures are improvements of this architecture, using the same concept but adding more layers like for example the VGG19.

AlexNet

AlexNet [15] is one of the most influential CNN architectures. It was created by Alex Krizhevsky, Ilya Sutskever, and Geoffrey Hinton and it won the ILSVRC (ImageNet Large-Scale Visual Recognition Challenge [16]) competition in 2012. The outstanding performance (the network achieved a top-5 error of 15.3%, more than 10.8 percentage points ahead of the second) of this architecture using such a difficult dataset (1000 possible outputs) was the beginning of the use of CNN for the computer vision community.

VGG19

This is one of the state-of-the-art models [17], created in 2014 by Karen Simonyan and Andrew Zisserman. This is a very simple model, is called VGG19 because it uses 16 convolution layers and three fully connected layers, it is considered a very deep architecture. VGG19 contains 23 hidden layers and 38.946.762 parameters.

GoogLeNet

This is also a state-of-the-art architecture; GoogLeNet [18] unlike VGG19 does not use a stack of convolution and pooling layers. In this case, Google created a new model, which improves the memory and power usage and should reduce the use of computational resources compared with other architectures, which were using a huge number of filters wand consequently require a high computational effort.

This model uses stacks of inception modules in a sequential structure. An inception module uses convolution layers of three different sizes allowing the network to perform the functions of these different operations [19]. As this will need a huge amount of power, a 1 × 1-convolution layer is used to decrease the input size and remaining computationally considerate. There are nine stacks of this inception module in GoogLeNet architecture, all with a deepness of two hidden layers. Even though all modules share the same structure and size parameters (stride and padding), their convolutional layers differ in the number of filters.

4 Implementation of the Architecture in TensorFlow and Caffe

The approaches for the implementation of CNN Architectures using the TensorFlow and Caffe Frameworks are different. Even though the general steps are on a conceptual level identical, the practical solution is not. Therefore, this section describes both approaches separately.

4.1 CNN Implementation Using TensorFlow

TensorFlow uses configuration scripts, which are being executed with Python. The following steps need to be performed to achieve consistent results.

Data loading

The first step is to load the dataset for training and testing. Once the dataset is loaded, individual entries are labeled are saved as a NumPy array:

```
mnist = tf.contrib.learn.datasets.load_dataset("mnist")
train_data = mnist.train.images
train_labels = np.asarray(mnist.train.labels, dtype = np.
int32)
eval_data = mnist.test.images
eval_labels = np.asarray(mnist.test.labels, dtype = np.
int32)
```

Model definition

The second step is to define the model that is going to be used; this model needs data, labels and the mode (train or test) as inputs and will include the architecture, the predictions, the loss function and the evaluation metrics:

```
def cnn_model(features, labels, mode)
```

The dataset is loaded as an input feature map (features) of one dimension (greyscale) and it has to be reshaped to 28 × 28 pixel size:

```
input_layer = tf.reshape(features["x"], [-1, 28, ;28, 1])
```

Once the data is reshaped, the next step is to build the architecture of the model by creating the appropriate layers. The following API-calls are examples of how this can be done:

```
tf.layers.conv2d(inputs, filters, kernel_size, strides,
padding, activation)
tf.layers.max_pooling2d(inputs, pool_size, strides,
padding)
```

```
tf.layers.dense(inputs, units, activation
tf.layers.dropout(inputs, rate, training)
```

Once all the layers have been set, the next step is to generate predictions. Predictions consist of two variables: classes and probabilities. Classes contain the predicted number by the network and are calculated by picking the number with the higher value in the output layer. Probabilities will contain the probability of each number to be the number in the input image.

For both training and evaluation, it is necessary to define a loss function that measures how closely the model's predictions match the target classes. For multiclass classification problems, cross entropy is typically used as the loss metric:

```
loss = tf.losses.sparse_softmax_cross_entropy(labels =
labels, logits = logits)
```

Once the loss is calculated, the model has to minimize it and it must be done only during the training of the network. To do so the stochastic gradient descent algorithm is implemented:

```
optimizer = tf.train.GradientDescentOptimizer(learn-
ing_rate = 0.01)
train_op = optimizer.minimize(loss = loss, global_step =
tf.train.get_global_step())
tf.estimator.EstimatorSpec(mode = mode, loss = loss, trai-
n_op = train_op)
```

The last step when defining a model is to set the evaluation metrics. TensorFlow has a module call tf.metrics, which includes numerous preconfigured functions to evaluate a network.

Training

To train the model the estimator (a class of TensorFlow that allows the user to manage training, evaluation, and inference in a simple manner) is used, using a numpy_input_fn. This returns a function outputting features and targets based on the dictionary of NumPy arrays.

```
train_input_fn = tf.estimator.inputs.numpy_input_fn(
 x = {"x": train_data},
 y=train_labels,
 batch_size=100,
 num_epochs=None,
 shuffle=True)
```

Where x and y are NumPy array objects of NumPy which are used to set the training feature data and labels. `num_epochs` = `None` means that the model will train until the specified number of steps is reached. Shuffle is used to shuffle the data queue of training data.

Evaluation

The last step is to evaluate the model to determine its accuracy on the MNIST test set. As it was done for the training the evaluation method of the estimator will be used, and to do so a numpy_input_fn function is used:

```
eval_input_fn = tf.estimator.inputs.numpy_input_fn(
 x = {"x": eval_data},
 y=eval_labels,
 num_epochs=1,
 shuffle=False)
eval_results  =  mnist_classifier.evaluate(input_fn=eval_
input_fn)
```

4.2 CNN Implementation Using Caffe

Caffe uses files to configure the parameters and initial settings. Prototxt files are used to define both the structure of the model and the solver of the network and they will be used to serialize and deserialize data.

Data preparation

The first step is to save the input database in the Lightning Memory-mapped Database format. LMDB is a software library that provides a high-performance embedded transactional database in the form of key-value store, where the labels will be the keys and the data of the image will be the values.

Model definition

The second step is to create a prototxt file that will contain the description of the structure of the network. The first layers have to be two input layers, one for the training phase and one for testing. After creating the input layers, the rest of the architecture has to be defined:

```
layer {
 name: "conv1"
 type: "Convolution"
   bottom: "data"
   top: "conv1"
   param {lr_mult: 1
      decay_mult: 1}
 param {lr_mult: 2
```

```
    decay_mult: 0}
  convolution_param {
  num_output: 96
  kernel_size: 3
  stride: 2
  weight_filler {type: " xavier "}
  bias_filler {type: "constant"}}}
```

Model solver

After the architecture is defined, the next step is to create a solver.prototxt, this is a configuration file used to tell Caffe how to train a model.

5 Performance Results of the Benchmarking Tests

The performance of the two deep learning frameworks was measured by the time needed to minimize the loss function (learning process) and by the accuracy of the network after being trained. The performance comparison was done in each of the architectures to see how the deepness affected them.

As a CNN requires a high computational effort the four architectures created with two frameworks were trained and tested on two different hardware platforms. This way the relevance of the hardware and how profitable is it to invest in it could be investigated.

- The first system is a laptop computer MSI GP62 2QE Leopard Pro, Intel Core i7-5700HQ@2.7 GHz and NVIDIA GeForce GTX950 M GPU (640 CUDA cores).
- The second system is a desktop computer, Intel Core i5-2400 CPU@3.1 GHz and NVIDIA GeForce GTX 1050 Ti GPU (768 CUDA cores).

The used Base learning rate was 0.01. For training the networks, the epoch was set to 10, this means that during the training the network was feed with 10 times the training set (60.000 images).

The Batch size was set to 100 except for GoogLeNet as the GPUs memory could not handle it, so it was set to the maximum possible value: 15 for NVIDIA GeForce GTX950 M and 50 for NVIDIA GeForce GTX 1050 Ti. For a Batch size of 100 the maximum number of iterations for 10 epochs was 6.000, for a Batch size of 15 it was 40.000 and for 50 it was 12.000.

Tables 2, 3, 4 and 5 show performance comparisons run on both system using both frameworks with all evaluated architectures.

As expected, the more powerful desktop system leads to proportionally shorter training times on both frameworks and all architectures. TensorFlow as well as Caffe in combination with the LeNet architecture result in short training times.

Table 2. Performance comparison for LeNet

<table>
<tr><th colspan="2">GTX 950M</th></tr>
<tr><th>TensorFlow</th><th>Caffe</th></tr>
<tr><td>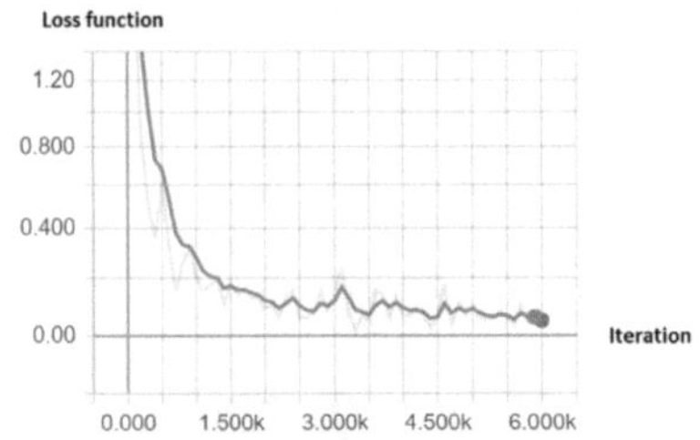
</td><td>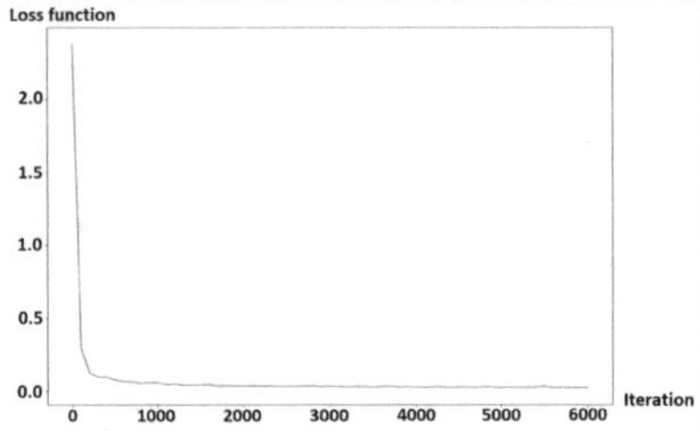
</td></tr>
<tr><td>Trained in 2 min 37 sec</td><td>Trained in 1 min 34 sec</td></tr>
<tr><th colspan="2">GTX 1050 Ti</th></tr>
<tr><th>TensorFlow</th><th>Caffe</th></tr>
<tr><td>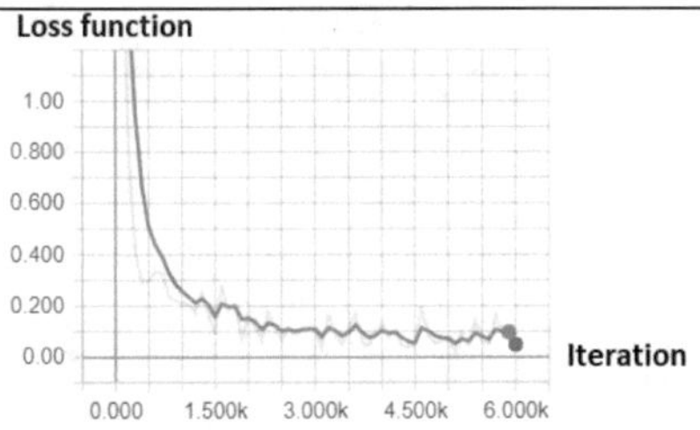
</td><td>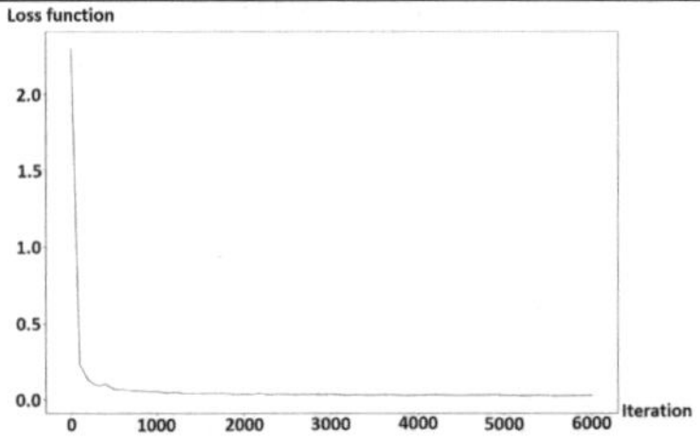
</td></tr>
<tr><td>Trained in 1 min 2 sec</td><td>Trained in 48 sec</td></tr>
</table>

6 Conclusions

As performance is concern, both frameworks achieved similar results in accuracy (around 99%) for every architecture except for VGG19. For this architecture TensorFlow could not perform the learning process, most probably due to the vanishing gradient problem. In terms of learning process Caffe outperform by far TensorFlow. For all the architectures, being more accentuated the more layers there were, Caffe was always much faster and needed less samples to minimize the loss function.

The result could not lead to a clear conclusion about the relation between the deepness of the architectures and the improvement of the accuracy. It should be also taken in consideration, that we tested architectures, which were developed for huge computer vision applications (VGG19, GoogLeNet) on a quite small data set.

Regarding the use of more powerful GPU, it was clear that while training a very deep architecture which performs large numbers of computations on a vast amount of

Table 3. Performance comparison for AlexNet

GTX 950M	
TensorFlow	**Caffe**
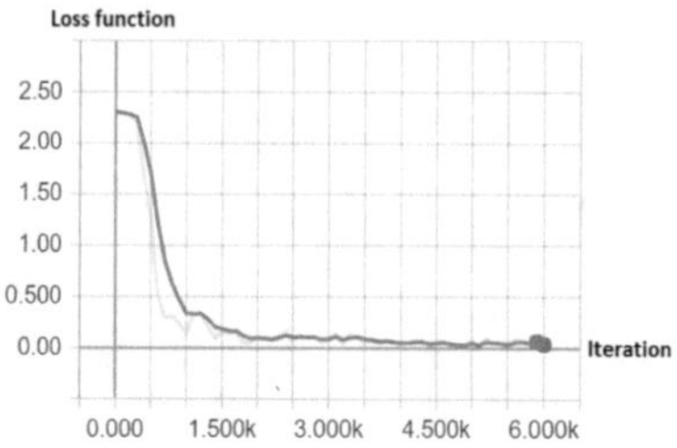 Loss function 2.50 2.00 1.50 1.00 0.500 0.00 0.000 1.500k 3.000k 4.500k 6.000k Iteration	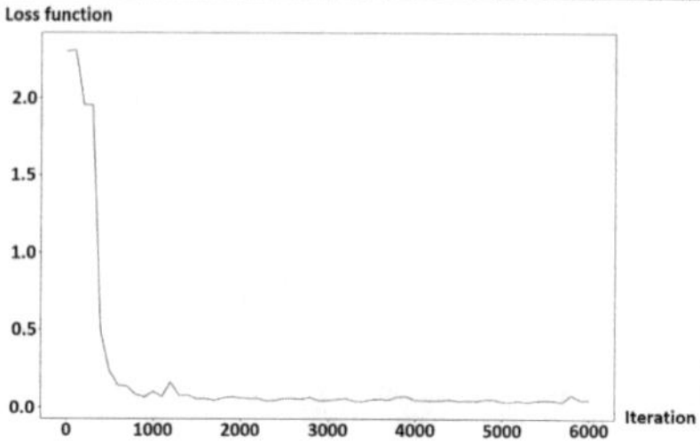 Loss function 2.0 1.5 1.0 0.5 0.0 0 1000 2000 3000 4000 5000 6000 Iteration
Trained in 27 min 29 sec	Trained in 13 min 44 sec
GTX 1050 Ti	
TensorFlow	**Caffe**
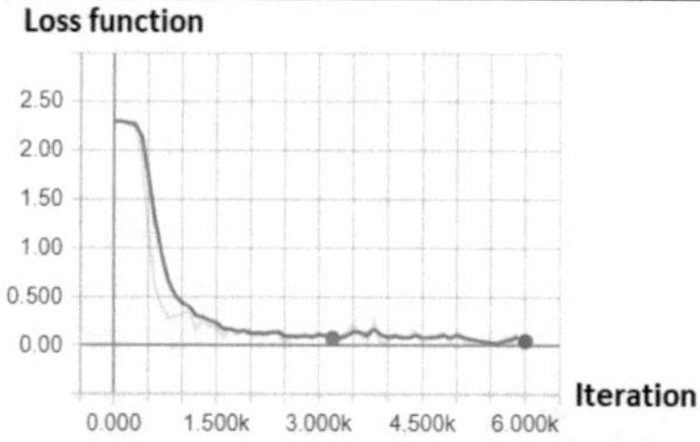 Loss function 2.50 2.00 1.50 1.00 0.500 0.00 0.000 1.500k 3.000k 4.500k 6.000k Iteration	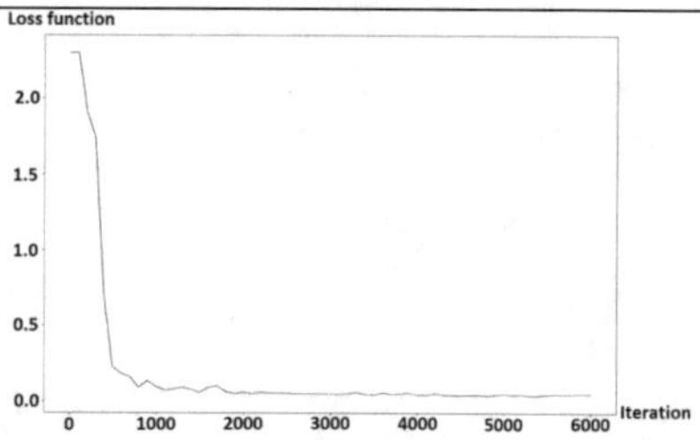 Loss function 2.0 1.5 1.0 0.5 0.0 0 1000 2000 3000 4000 5000 6000 Iteration
Trained in 16 min 56 sec	Trained in 4 min 33 sec

Table 4. Performance comparison for VGG19

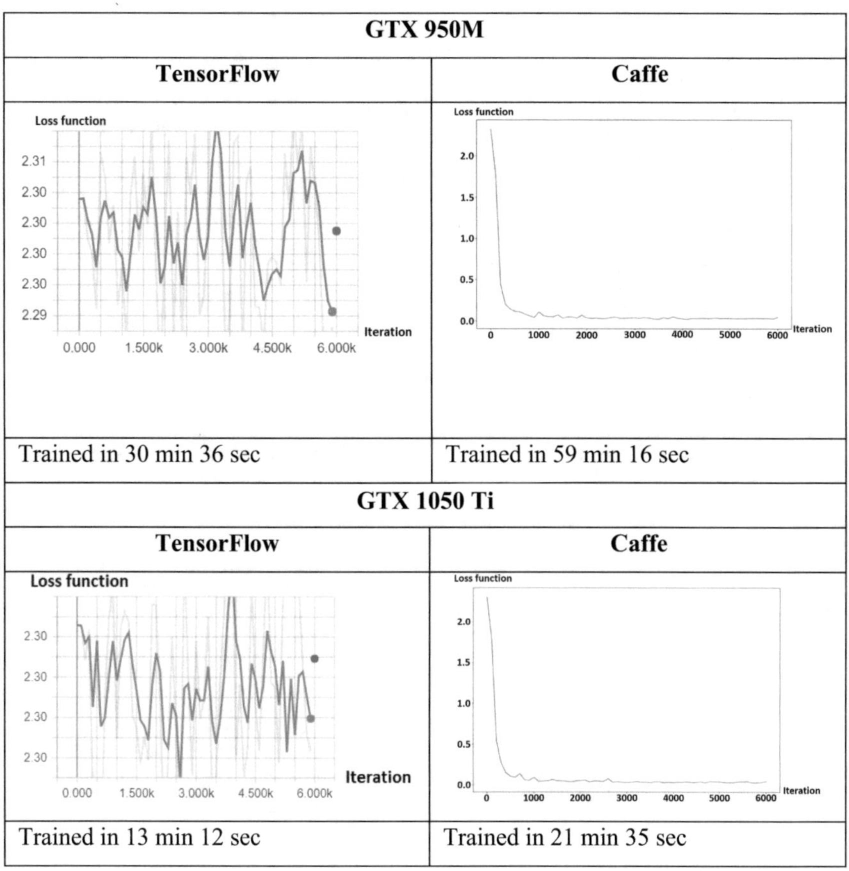

GTX 950M	
TensorFlow	**Caffe**
Trained in 30 min 36 sec	Trained in 59 min 16 sec
GTX 1050 Ti	
TensorFlow	**Caffe**
Trained in 13 min 12 sec	Trained in 21 min 35 sec

Table 5. Performance comparison for GoogLeNet

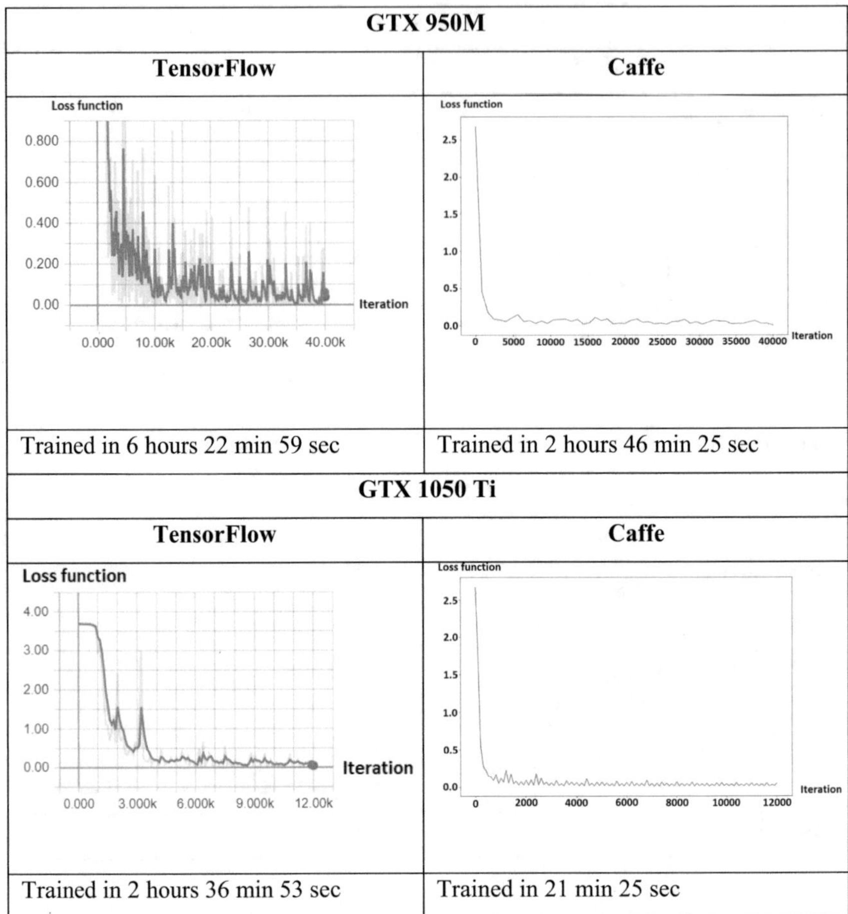

GTX 950M	
TensorFlow	**Caffe**
Trained in 6 hours 22 min 59 sec	Trained in 2 hours 46 min 25 sec
GTX 1050 Ti	
TensorFlow	**Caffe**
Trained in 2 hours 36 min 53 sec	Trained in 21 min 25 sec

data, considerably performance improvements could be seen when running on a more powerful GPU. A conclusion is also to include special machine learning hardware systems in the test.

Next steps are questions which concern the minimization of the systems, the distribution in of the computation in parts, performed directly in enhanced integrated sensor systems, and the computation and storage of parts in the cloud.

References

1. Chetlur, S., et al.: cuDNN: Efficient Primitives for Deep Learning. CoRR abs/1410.0759 (2014)
2. Abadi, M., et al.: TensorFlow: Large-Scale Machine Learning on Heterogeneous Distributed Systems. https://www.tensorflow.org/about/bib, 9 Nov 2015 (14 June 2018)
3. Collobert, R., Kavukcuoglu, K., Farabet, C.: Torch7: A Matlab-Like Environment for Machine Learning (2011)
4. Agarwal, A., Akchurin, E., Basoglu, C.: An Introduction to Computational Networks and the Computational Network Toolkit. https://www.microsoft.com/en-us/research/publication/an-introduction-to-computational-networks-and-the-computational-network-toolkit/, 1 Oct 2014 (20 June 2018)
5. Jia, Y., et al.: Caffe: Convolutional Architecture for Fast Feature Embedding (2014). CoRR. [On-line] abs/1408.5093. Available: https://arxiv.org/abs/1408.5093, 20 June 2018
6. Chen, T., et al.: MXNet: A Flexible and Efficient Machine Learning Library for Heterogeneous Distributed Systems (2015). CoRR. [On-line] abs/1512.01274. Available: https://arxiv.org/abs/1512.01274, 21 June 2018
7. Tokui, S., Oono, K., Hido, S.: Chainer: A Next-Generation Open Source Framework for Deep Learning (2015)
8. Chollet, F.: Keras (2015). https://github.com/fchollet/keras, 21 June 2018
9. Gibson, A., Nicholson, C., Patterson J.: Deeplearning4j (2017). https://deeplearning4j.org/, 22 June 2018
10. Al-Rfou, R., Alain, G., Almahairi, A.: Theano: A Python Framework For Fast Computation Of Mathematical Expressions (2016). https://arxiv.org/abs/1605.02688, 22 June 2018
11. Lopez de Guereña, X.: Deep Learning Frameworks for Convolutional Neural Networks. Master Thesis at CUAS. Unpublished (2018)
12. van Rossum, G.: Python tutorial (1995). https://www.python.org/, 23 June 2018
13. Lecun, Y., et al.: The MNIST Dataset of Handwritten Digits (Images) (1999). http://yann.lecun.com/exdb/mnist/, 23 June 2018
14. Lecun, Y., et al.: Gradient-based learning applied to document recognition. In: Proceedings of the IEEE, vol. 86, no. 11, pp. 2278–2324, Nov 1998. https://doi.org/10.1109/5.726791
15. Krizhevsky, A., Sutskever, I., Hinton, G.: ImageNet classification with deep convolutional neural networks. In: NIPS'12 Proceedings of the 25th International Conference on Neural Information Processing Systems, vol. 1, pp 1097–1105, Lake Tahoe, Nevada, 3 Dec 2012
16. Russakovsky, O., et al.: ImageNet large scale visual recognition challenge. IJCV (2015)
17. Simonyan, K., Zisserman, A.: Very Deep Convolutional Networks for Large-scale Image Recognition (2014). CoRR. [On-line] abs/1409.1556. https://arxiv.org/abs/1409.1556, 24 June 2018
18. Szegedy, C., et al.: Going deeper with convolutions. In: IEEE Conference on Computer Vision and Pattern Recognition (CVPR), Boston, MA, pp. 1–9 (2015). https://doi.org/10.1109/cvpr.2015.7298594
19. Goodfellow, I., Bengio, Y., Courville, A.: Deep Learning. Frechen, MITP (2018)

Foreground Detection Scheme Using Patch Based Techniques for Heavily Cluttered Video

L. R. Karl Marx[1(✉)] and S. Veluchamy[2]

[1] Department of Mechatronics, Thiagarajar College of Engineering, Thirupparankundram, Madurai 625015, India
lrkarlmarx@tce.edu
[2] National Skill Training Institute, Chennai 600032, India
Pvs1834@gmail.com

Abstract. Foreground objects can be extracted effectively by subtracting the background in the image frames, provided that an updated model of the background is available at any time. This is achieved by initialization (also called bootstrapping) of the background followed by its maintenance. In this paper, a patch-based technique for robust background initialization has been proposed, that overcomes the sleeping person problem. The proposed technique is able to cope with heavy clutter, i.e, foreground objects that stand still for a considerable portion of time. The method rests on sound principles in all its stages and only few, intelligible parameters are needed. Experimental results shows that the proposed algorithm is effective.

Keywords: Background initialization · Background modeling · Bootstrapping · Content-based representation

1 Introduction

Background is defined as the static part in a motion-compensated sequence, where foreground objects are parts that have non-zero residual motion, either due to relative motion with respect to the camera or to parallax with respect to the stabilization plane. Background initialization is the extraction of a stationary scene model from a short training sequence where foreground objects may be present.

Background initialization using video inpainting techniques are based on:

(i) identifying an initial background region and then
(ii) filling-in the remaining unknown background incrementally by choosing values from the same time-line. At each step, the patch that maximizes a likelihood measure with respect to the surrounding zone, already identified as background, is selected. This entails that the background should be self-similar (like a building facade) and that the starting region should be large enough to provide sufficient structure information. Background initialization could be cast as video inpainting if the foreground masks were known beforehand, which does not make sense since foreground clutter cannot be known beforehand.

M. E. Auer and K. Ram B. (Eds.): REV2019 2019, LNNS 80, pp. 832–837, 2020.
https://doi.org/10.1007/978-3-030-23162-0_75

Background initialization can be done using the median brightness value for each pixel. Median depends on an assumption that the background for each pixel will have the same brightness value fifty percent more than the time spent during the training sequence. The output will contain large errors if the assumption is not met and hence has not been efficient.

A standard form of adaptive background is a time-averaged background image (TABI), in which a background approximation is obtained by averaging a long time image sequences. Although this method is effective in situations where objects move continuously, it is not robust to scenes with many moving objects especially if they move slowly, and the foreground objects always can be blending into the background image (sleeping person problem).

The proposed approach based on patches for background initialization is able to cope with sequences where foreground objects persist in the same position for a considerable portion of time. First the sequence is subdivided in patches that are clustered along the time-line in order to narrow down the number of background candidates. Then the background is grown incrementally by selecting at each step the best continuation of the current background.

The novelties of the proposed algorithm are as follows. Unlike the above mentioned techniques, it does not expect all frames of the sequence to be stored in memory simultaneously. Here instead, it processes the frames sequentially. Background areas are iteratively filled by selecting the most appropriate candidate patches according to the combined responses of extended versions of the candidate patch and its neighborhood. It is assumed that the most appropriate patch results in the smoothest response, indirectly exploiting the spatial correlations within small regions of a scene. The applications of this technique is in areas such as video surveillance [1], perpetual interfaces [3], and content-based video encoding (MPEG4), tracking, action recognition [7–9].

The remainder of this paper is organized as follows. In Sect. 2, continuity-based method is explained step-by-step. Experimental results are shown in Sect. 3. Finally conclusion is drawn in Sect. 4.

2 Proposed Model

The video sequence is modeled as a 3-D array of pixel values vx, y, t. Each entry contains a color value, which is a triplet (R, G, B). A spatio-temporal patch vs is a sub-array of the video sequence, defined in terms of the ordered set of its pixel coordinates: Ix * Iy * It where Ix, Iy and It are set of indices. The window W = Ix * Iy is the spatial footprint of the patch. An image patch vR is a spatio-temporal patch with a singleton temporal index: R = W * {t}. Our method for background initialization is based on the following hypothesis: (i) the background is constant; (ii) in each spatio-temporal patch the background is revealed at least once; (iii) foreground objects introduce a color discontinuity with the background.

Division of input video into frames, clustering of patches, Representative cluster formation and Background Tessellation (Fig. 1).

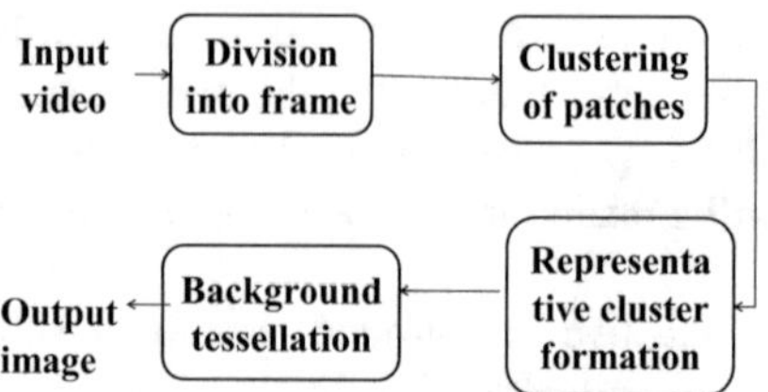

Fig. 1. Basic block diagram

2.1 Clustering

The spatial indices are subdivided into windows W_i of size NxN, overlapping by half of their size in both dimensions as shown in Fig. 2. In each spatio temporal patch, we cluster image patches that depict the same static portion of the scene with single linkage agglomerative clustering algorithm. On each window, the algorithm finds the distance between the image patches using Sum of Squared Distances (SSD). A cutoff distance, i.e., a distance behind which two clusters are not linked is set. The cutoff distance should prevent clustering together image patches that do not have the same content. It is obtained from the chi-square distribution of SSD. The two image patches are clustered if the distance between them is less than the cutoff distance.

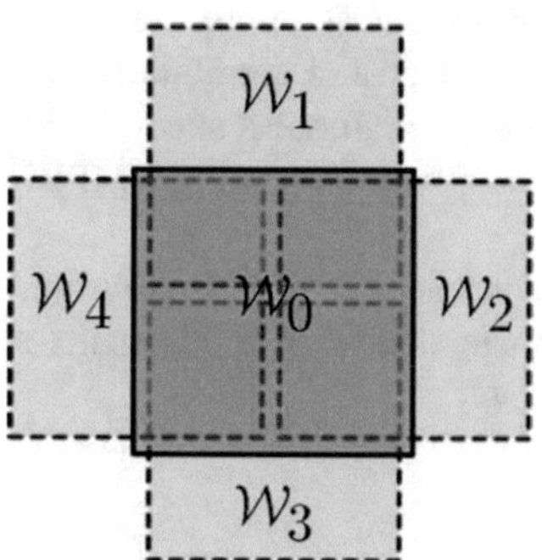

Fig. 2. Each footprint w_0 has four overlapping neighbors: w_1, w_2, w_3 and w_4

2.2 Representative Image Patch

Let $\mathcal{W} \times T_k$ denote cluster k over spatial footprint $\mathcal{W}$, a representative image patch for that cluster is obtained by averaging pixel values along the time-line.

$$u_{x,y,k} = 1/|T_k| \sum v_{x,y,t} \quad \forall\ x,y \in W$$

In each spatial footprint $\mathcal{W}$, we now have a variable number of cluster representatives.

2.3 Background Tessellation

The background is constructed with a sequential approach: starting from seed patches, a tessellation is grown by choosing at each site, the best continuation of the current background. The background seeds are the patches that represent the largest clusters. Since it is assumed that no foreground object is stationary in all frames, if the largest clusters have size L (maximal), they are fully reliable. The growing proceeds as follows. Let W be a spatial footprint where a background patch has not been assigned yet, but it has been assigned to at least one of its neighbors, W_0. This means the W overlaps with some background. The algorithm assigns the background patch to W by choosing one from the cluster representatives with footprint W. The selected patch has to fulfill two requirements:

1. Seamlessness: In the part that overlaps with W_0 it has to depict the same content as the background patch, so that it can be stitched seamlessly to it.
2. Best Continuation: In the non-overlapping part, it has to represent the best continuation of the current background, meaning that, among several candidates, the patch that introduces the least discontinuity is chosen.

This procedure is repeated for all the footprints, until all the background has been assigned.

3 Results and Discussion

The simulations of these sub-blocks were done and the simulated results are given in Fig. 3. All the blocks were simulated using single color channel, double color channels and three color channels. The three color channel mode is preferred due to its clarity.

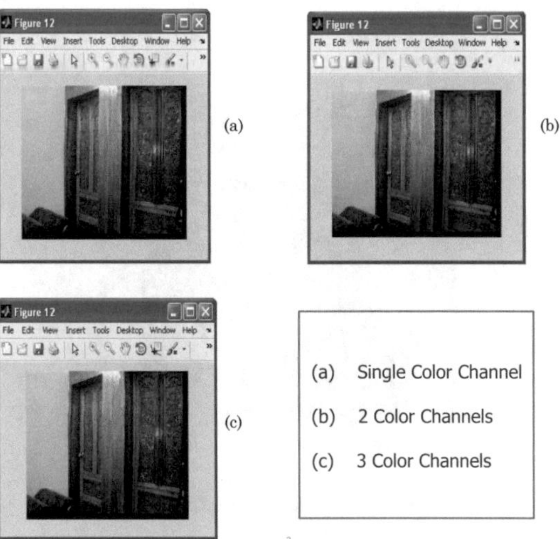

Fig. 3. Output for single and multi-color channels

The background and foreground had been extracted for a video with a heavy clutter which is also shown in Figs. 4 and 5.

The processing time in seconds for various color channels have also been calculated. For single color channel, the time taken was 49.03 s. For two color channels, the time taken was 49.11 s. For three color channels, the time taken was 49.70 s.

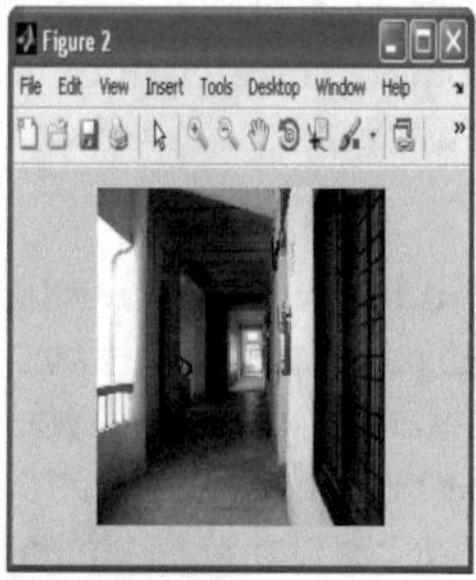

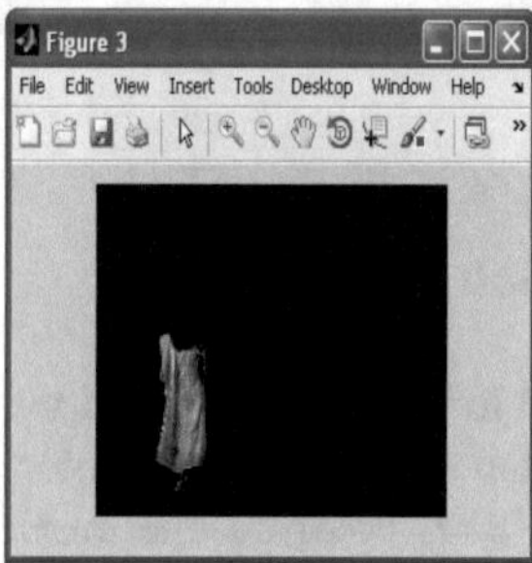

Fig. 4. Output for video with heavy clutter

Fig. 5. Foreground extraction

4 Conclusion

The method is robust, as it can cope with serious occlusions caused by moving objects. It is scalable, as it can deal with any number of frames greater or equal than two. It is effective, as it always recovers the background when the assumptions are satisfied. Moreover, this method rests on sound principles in all its stages, and only few, intelligible parameters are needed, namely the confidence level for the tests and the patch size. The latter can be tuned manually or automatically by a multi resolution approach.

References

1. Colombari, A., Fusiello, A.: Patch-based background initialization in heavily cluttered video. IEEE Trans. Image Process. Intell. (2010)
2. Farin, D., deWith, P.H.N., Effelsberg, W.: Robust background estimation for complex video sequences. In: Proceedings of IEEE International Conference on Image Processing, vol. 1, pp. 145–148 (2003)
3. Shi, J., Malik, J.: Normalized cuts and image segmentation. IEEE Trans. Pattern Anal. Mach. Intell. **22**(8), 888–905 (2000)
4. Jain, A.K., Murty, M.N., Flynn, P.J.: Data clustering: a review. ACM Comput. Surv. **31**(3), 264–323 (1999)
5. Colombari, A., Fusiello, A., Murino, V.: Background initialization in cluttered sequences. Presented at the 5th Workshop on Perceptual Organization in Computer Vision, in conjunction with CVPR 2006 (2006)
6. Nunes, P., Correia, P., Pereira, F.: Coding video objects with the emerging mpeg-4 standard. Presented at the I Conferência Nacional de Telecomunicações, Apr 1997
7. Porikli, F., Tuzel, O.: Human body tracking by adaptive background models and mean-shift analysis. In: IEEE International Workshop on Performance Evaluation of Tracking and Surveillance (2003)
8. Javed, O., Shah, M.: Tracking and object classification for automated surveillance. Lecture Notes in Computer Science, vol. 2353, pp. 439–443 (2006)
9. Lv, F., Nevatia, R.: Single view human action recognition using key pose matching and Viterbi path searching. In: IEEE Conference on Computer Vision and Pattern Recognition, pp. 1–8 (2007)

Remote Access and Automation of SONIC Analysis System

Kalyan Ram B.[1,4(✉)], Raman Ramachandran[2], Aswani Barik[2], Nitin Zanvar[3], Vishwas Apugade[3], Dhawal Patil[3], Gautam G. Bacher[4], Venkata Vivek Gowripeddi[1,5,6], and Nitin Sharma[4]

[1] Electrono Solutions Pvt Ltd, Bengaluru, India
kalyan@electronosolutions.com
[2] Spicer India Pvt Ltd, Satara, India
[3] Ashta Liners Pvt Ltd, Ashta, India
[4] BITS-Pilani KK Birla Goa Campus, Pilani 403726, GOA, India
nitinn@goa.bits-pilani.ac.in
[5] International Institute of Information Technology, Hyderabad, India
[6] Cymbeline Innovation Pvt Ltd, Bangalore, India

Abstract. Non-Destructive Testing (NDT) is an analysis technique to evaluate the properties of a component without causing damage to it. SONIC Analysis is one technique of NDT using sound. This Paper discusses in detail—implementation of Remote Health Monitoring of SONIC Analysis Plant. The system has been developed over the last few years, with each interaction bringing more remote access and automation to the plant. We discuss the plant as it evolved and how different aspects of Remote Engineering and Remote Health Monitoring were implemented with each interaction. The paper also discusses the outcomes of the initial model to the later models. The discussion about the advantages of the Non Destructive method are highlighted. The System of Consists of 3 three parts: (1) The Physical Testbed (2) Cloud Server for Data processing and (3) HMI for user interaction. The Key outcomes include: (1) Ensured Uptime of the system (2) Provide statistical Reports (3) Remote Diagnosis.

Keywords: Remote monitoring · Cloud technology · Cyber physical system · Remote diagnosis · Non-Destructive Testing (NDT)

1 Introduction

Sonic Analysis System is a Non-Destructive Testing Technique to differentiate between the nodularity of Cast iron in a casting environment using the sound. This paper describes the Remote Health monitoring of Sonic Analysis system, different challenges that are presented in the process and finding solutions to them. The paper consists of 5 sections viz.,

- Section 1: **Introduction**: which introduces the reader to the concept of Non Destructive testing

M. E. Auer and K. Ram B. (Eds.): REV2019 2019, LNNS 80, pp. 838–848, 2020.
https://doi.org/10.1007/978-3-030-23162-0_76

- Section 2: **Architecture**: The Description of the physical system
- Section 3: **Implementation** of Remote Engineering to the physical system
- Section 4: **Outcomes**: Different outcomes are discussed
- Section 5: **Conclusions**: This section organization of the whole project.

With the demand for zero rejections becoming the norm in automotive manufacturing sector, several NDT techniques have been emerging to the fore front since over 2 decades.

Though over several years earlier, there have been many papers published about defect identification in metals and non-metals using several techniques, during the PANNDT (Pan-American conference for Non-Destructive Testing) 2003 in Rio, Brazil, Ingolf Hertlin (Germany) and Detlev Schultze (Brazil) presented a paper on "Acoustic Resonance Testing: the upcoming volume-oriented NDT method".

Further to this, there have been several white papers and conference proceedings published by "The Modal Shop Inc." of Cincinnati, Ohio.

2 Architecture

To explain the overall architecture, bottom up approach is followed where the system is explained from the physical level to the top level architecture. We discuss the architecture in three separate subsections as follows:

2.1 Physical System

Physical System consists of an acoustic chamber as shown in Fig. 1 which has 4 loads fitted to the bottom of the mount plate to distribution of weight and check for abnormalities in weight by comparing with standards specified which adjust based on AI algorithms as the system becomes more accurate, a central vibration sensor measures the vibration of the object for analysis when hit by the striker. Four mics are fitted on four corners of the chamber to record audio signals, which is later analyzed for processing.

Depending on inputs from four load cells, vibration sensor and audio mics, a part is determined to be a pass or fail using the Non-Destructive way—SONIC Analysis way of testing.

2.2 SONIC Analysis System v1.1—With Controls and HCI

Figure 2 Shows the architecture of the system with Physical system connected to Panel Box and a monitor and a printer for report viewing and generation respectively.

2.3 SONIC Analysis System v1.2—With Automatic Loading Using Conveyor

The system evolved with addition of conveyor belt and robot for loading and remote access of the system through a computing device placed for Remote Access through internet. Electrical system to power the mechanical systems are as shown in Fig. 3.

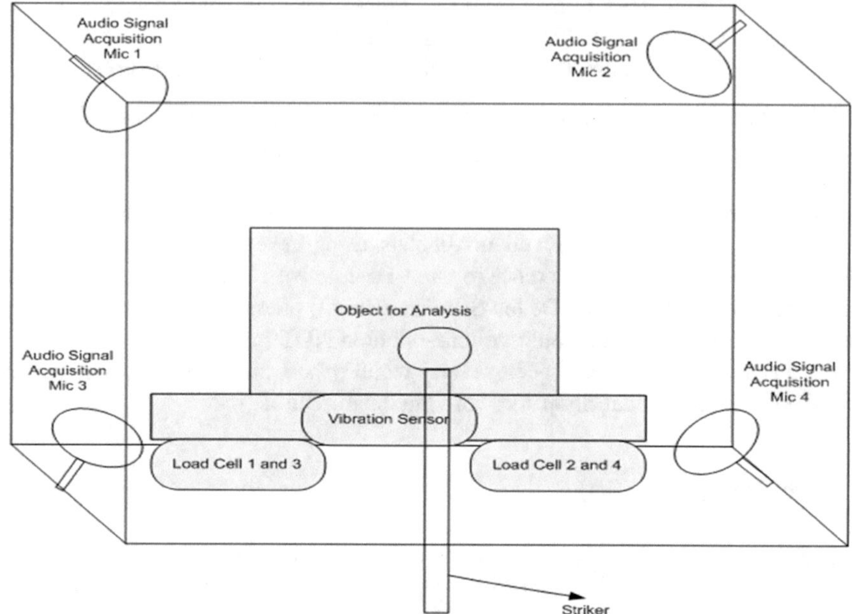

Fig. 1. Physical structure of the acoustic chamber

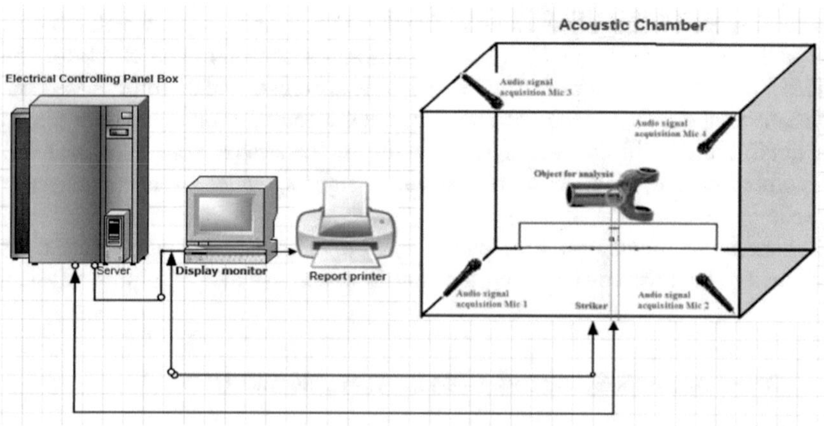

Fig. 2. Architecture of system v1.1

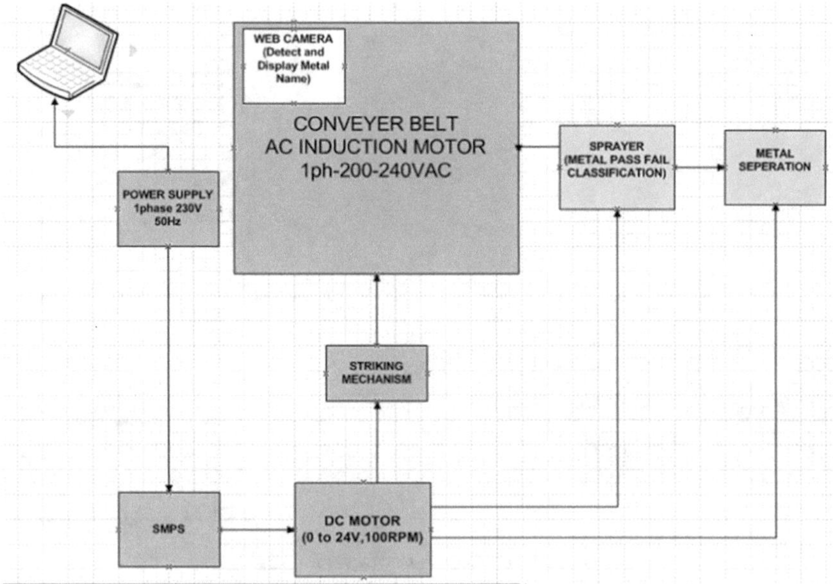

Fig. 3. Architecture of system v1.2 with controls and HCI

2.4 SONIC Analysis System v2.0—v1.2 With Cloud Data Processing and Self Learning AI Algorithms

This shows the current and most advanced architecture of the system, combining elements from previous generations and adding features such as self learning algorithms for fixing thresholds, data storage and processing in cloud and well designed HCI interface for more intuitive communication (Fig. 4).

3 Implementation

The implementation discusses the implementation of different architectures and gives the features of the system.

3.1 Images of the Implemented System

Figures 5 and 6 depict the system that is implemented and different components of the architecture such as sensors, mics and controls can be seen here.

Figure 7 shows the latest version of the system with conveyer belt for automatic loading and cloud data processing for online data analytics.

3.2 Features

- Remote Access to system

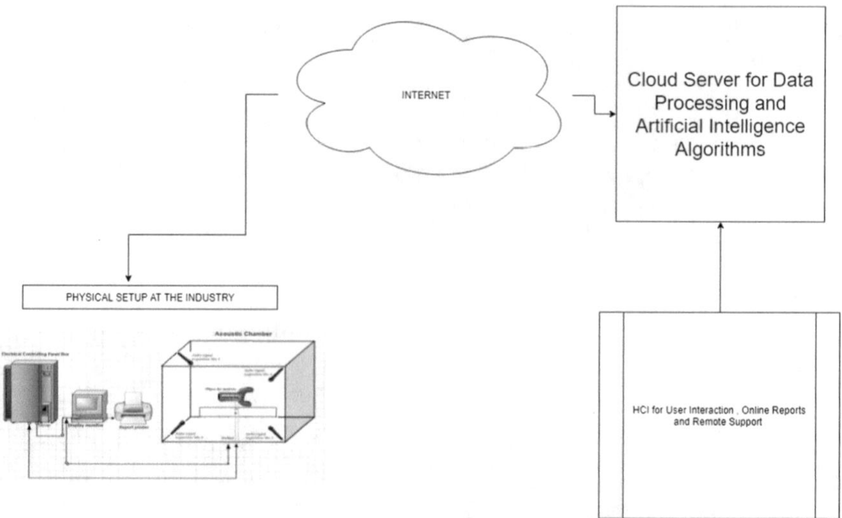

Fig. 4. Architecture of system v2.0

Fig. 5. Sensors and mic inside the chamber

- It can be extended to any other type of metallic objects as well with appropriate calibration
- The end-to-end cycle time is less than 45 s. Could be optimized further on case to case basis
- User friendly interface for differentiating objects as Pass/Fail
- Non-Contact type measurement detection mechanism

Fig. 6. Physical system connected to controls

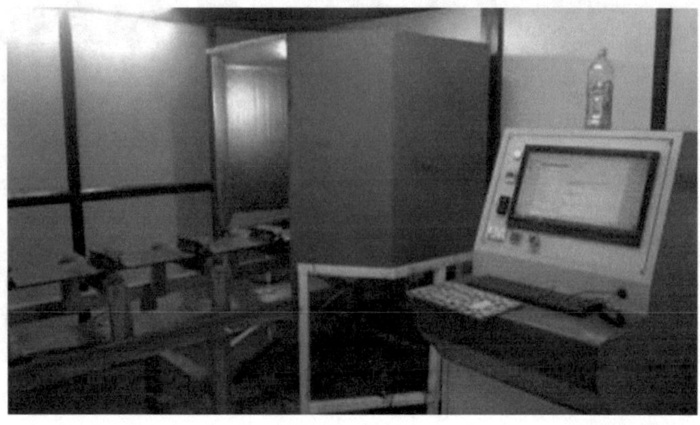

Fig. 7. SONIC analysis system with conveyer belt v2.0 and HCI

- This mechanism doesn't use any consumables
- No replaceable parts due to wear and tear of the measurement system as the measurement mechanism is non-contact type
- Maintain the traceability of parts tested
- Requires very little maintenance.

3.3 HCI Panels

See Figs. 8 and 9.

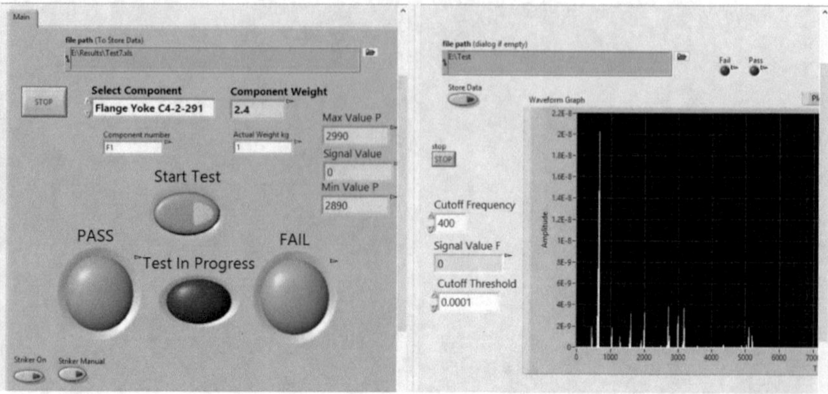

Fig. 8. Manual control user screen with user input thresholds

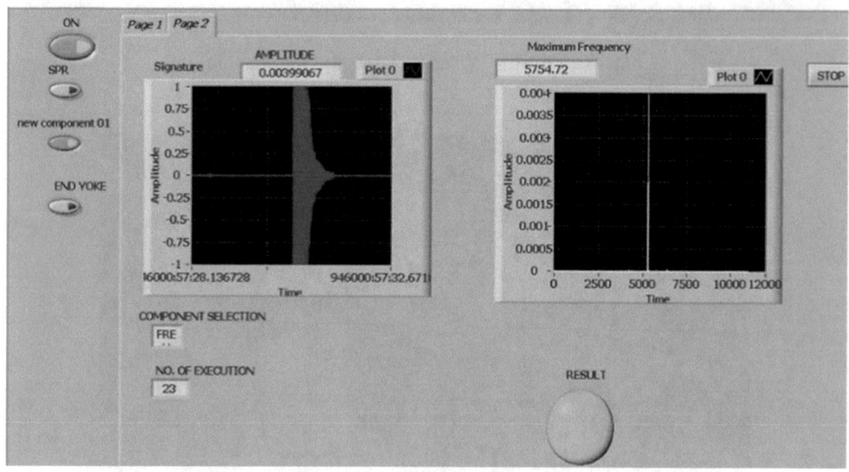

Fig. 9. User screen showing output graphs and results of the test with thresholds calculated through AI algorithms

4 Outcomes

The following are the key outcomes of the system.

4.1 Ensure Appropriate System Uptime and Maintain Part Traceability Through Result Stored in Cloud Database

Traceability is an important outcome of this project as the part can be traced throughout as it moves through different processes and owners. It accelerates the identification of failure by many folds and helps establish responsibility (Fig. 10).

Component 1			
	Sl.NO	Frequency	
		Ductile	Grey
Location 01	1	6942	4807
	2	6943	4816
	3	6938	4810
	4	6938	4808
Location 02	5	5754	3986
	6	5756	3985
	7	5754	3984
	8	5755	3986
Location 03	9	4773	2818
	10	4776	2803
	11	4773	2802
	12	4777	2811

Fig. 10. The above sheet gives the traceability of the component by location and serial number

Date	Time	Component Type	Ladel	Frequency	Result
19-02-16	15:34:03	800	10	2078	Pass
19-02-16	15:34:05	800	10	2078	Pass
19-02-16	15:35:01	800	10	2086	Pass
19-02-16	15:35:06	800	10	2078	Pass
19-02-16	15:35:16	800	10	2075	Pass
19-02-16	15:35:27	800	10	2080	Pass
19-02-16	15:35:38	800	10	2078	Pass
19-02-16	15:35:44	800	10	2088	Pass
19-02-16	15:35:49	800	10	2086	Pass
19-02-16	15:35:54	800	10	2077	Pass
19-02-16	15:36:04	800	10	2075	Pass
19-02-16	15:36:14	800	10	2080	Pass
19-02-16	15:36:26	800	10	2079	Pass
19-02-16	15:36:37	800	10	2088	Pass
19-02-16	15:36:47	800	10	2086	Pass
19-02-16	15:36:52	800	10	2078	Pass
19-02-16	15:37:02	800	10	2075	Pass
19-02-16	15:37:13	800	10	2080	Pass
19-02-16	15:37:23	800	10	2078	Pass
19-02-16	15:37:34	800	10	2088	Pass
19-02-16	15:37:39	800	10	2086	Pass
19-02-16	15:37:50	800	10	2078	Pass
19-02-16	15:38:00	800	10	2074	Pass
19-02-16	15:38:05	800	10	2080	Pass

Fig. 11. Test results of a inline sonic analysis system

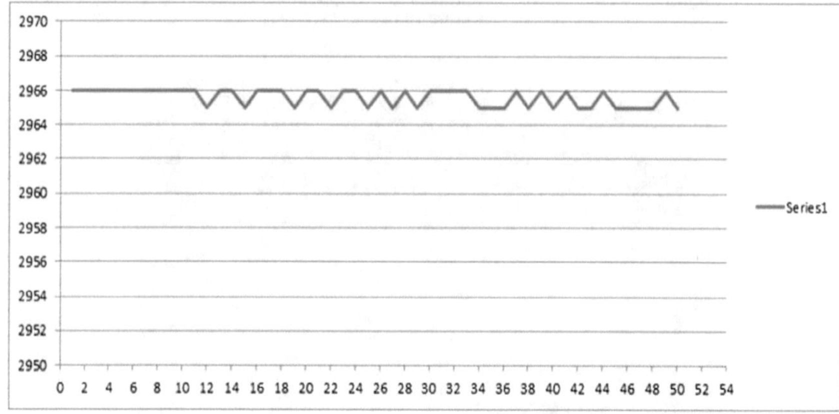

Fig. 12. Data of frequency versus component serial number

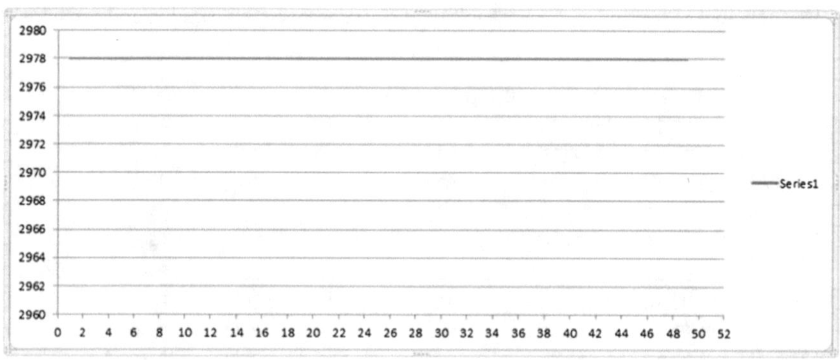

Fig. 13. Data of frequency versus component serial number

4.2 Facility to Provide Statistical Analysis and Reports

Different reports as per the production manager needs can be provided. Figures 11, 12 and 13 provide examples of different kinds of reports.

4.3 Online Diagnostics and Remote Support

Provide reports and access from anywhere through internet thus enabling the system to be 100% Online and Support from Industry experts and technicians can be provided remotely (Fig. 14).

In addition, Sonic Analysis System could be used for

(a) cracks inside of the part,
(b) crack detection during shearing of bars,
(c) nodularity (of cementite or graphite),
(d) inclusions,

	ET	MT/PT	UT	RT	SAS
Defect Type					
Cracks/chips/porosity/voids	Yes	Yes	Yes	Yes/No	Yes
Missed processes/operations	Yes/No	No	Yes/No	Yes/No	Yes
Material property	Yes/No	No	No	No	Yes
Structurally significant	Yes	Yes	Yes	Yes	Yes
Production lot variations	Yes/No	Yes	Yes	Yes	Yes/No
Defect Location					
Surface (external)	Yes	Yes	Yes	No	Yes
Internal	No	No	Yes	Yes	Yes
Brazing/bonding/welding	No	No	Yes/No	Yes/No	Yes
Speed/Training/Cost					
Part throughput	Medium	Low	High	Low	High
Training requirements	High	High	Medium	High	Low
Overall inspection costs	Medium	Medium	High	High	Low
Automation Capacity					
Quantitative results	Yes/No	No	Yes/No	No	Yes
Automation requirements	Medium	N/A	Complex	Complex	Easy
Automation cost	Medium	N/A	High	High	Low/Medium

ET	Eddy current/ Electromagnetic Testing
MT/PT	Magnetic Particle Testing / Dye Penetrant Testing
UT	Ultrasonic Testing
RT	X-Ray / Radiographic Testing
SAS	Sonic Analysis System

Fig. 14. Comparisons of different techniques

(e) density differences in sinter metal products,
(f) hardness differences (heat treated or aged),
(g) bonding (welding, friction welding).

5 Conclusions

Authors feel that Non Destructive testing is the way to go for testing in automotive and manufacturing industry and making that system remote will add tremendous value. The paper highlights the implementation of one such remote system. In the future, different systems and different architectures would be explored and results studied. This work would be a corner stone for the upcoming research and development in this field.

References

1. Hertlin, I., Schultze, D.: Acoustic Resonance Testing: The Upcoming Volume-Oriented NDT Method
2. Baker, N.: Techniques for Optimisation and Analysis of Composite Structures for Damage Tolerance and Buckling Stiffness
3. Stultz, G.R., Bono, R.W., Schiefer, M.I.: Fundamentals of Resonant Acoustic Method NDT
4. Ewins, D.J.: Modal Testing: Theory and Practice
5. Juran, J.M., Godfrey, A.B.: Juran's Quality Handbook, 5th edn. McGraw-Hill, New York
6. Jonsson, K., Westergren, U.H., Holmström, J.: Technologies for value creation: an exploration of remote diagnostics systems in the manufacturing industry. Inf. Syst. J. **18**(3), 227–245 (2008)
7. Smajic, H., Wessel, N.: Remote control of large manufacturing plants using core elements of Industry 4.0. In: Online Engineering & Internet of Things, pp. 546–551. Springer, Cham (2018)
8. Angrisani, L., Cesaro, U., D'Arco, M., Grillo, D., Tocchi, A.: IOT enabling measurement applications in Industry 4.0: platform for remote programming ATES. In: 2018 Workshop on Metrology for Industry 4.0 and IoT, pp. 40–45. IEEE (2018)
9. Steels, L., Brooks, R.: The Artificial Life Route to Artificial Intelligence: Building Embodied, Situated Agents. Routledge (2018)
10. Stergiou, C., Psannis, K.E., Kim, B.G., Gupta, B.: Secure integration of IoT and cloud computing. Future Gener. Comput. Syst. **78**, 964–975 (2018)
11. Fitzgerald, R., Karanassios, V.: The Internet of Things (IoT) for a smartphone-enabled optical spectrometer and their use on-site and (potentially) for Industry 4.0. In: Next-Generation Spectroscopic Technologies XI, vol. 10657, p. 1065705. International Society for Optics and Photonics (2018)
12. Wang, S.: Remote real-time monitoring system based on cloud computing. U.S. Patent No. 9,967,168, 8 May 2018

A Greedy Approach to Hide Sensitive Frequent Itemsets with Reduced Side Effects

B. Suma(✉) and G. Shobha

Department of Computer Science & Engineering, Rashtreeya Vidyalaya College of Engineering, Bengaluru, Karnataka, India
{sumab_rao, shobhag}@rvce.edu.in

Abstract. Frequent itemsets mining discovers associations present among items from a large database. However, due to privacy concerns some sensitive frequent itemsets have to be hidden from the database before delivering it to the data miner. In this paper, we propose a greedy approach which provides an optimal solution for hiding frequent itemsets that are considered sensitive. The hiding process maximizes the utility of the modified database by introducing least possible amount of side effects. The algorithm employs a weighing scheme which computes transaction weight that allows it to select at each stage of iteration candidate transactions, based on side effects measurement. We investigated the effectiveness of proposed algorithm by comparing it with other heuristic algorithm using parameters such as number of sensitive frequent itemsets, length of sensitive frequent itemsets and minimum support on a number of datasets which are publicly available through the Frequent Itemset Mining (FIMI) repository. The experiment results demonstrated that our approach protects more non-sensitive frequent itemsets from being over-hidden than those produced by heuristic approach.

Keywords: PPDM · Frequent itemsets hiding · Candidate transactions · Victim item

1 Introduction

Data mining methods extract hidden patterns, relationships, interdependencies and correlations in large dataset. Sensitive information extracted from data mining methods can disclose critical information about business transactions [1]. Protection of such sensitive knowledge may affect the utility of the data mining result. Privacy Preserving Data Mining (PPDM) has emerged to address this problem that deals with sensitive knowledge without giving up the utility of the data [2–4]. PPDM technologies permit knowledge extraction from data, while preserving privacy. The success of PPDM algorithms is measured in terms of data utility, its performance, and level of resistance to data mining algorithms. Verykios et al. [5] analyzed the state-of the-art, proposed classification hierarchy and clustering of various privacy preserving data mining technique. Bertino et al. [6] proposed a framework which allows evaluating the different features of a privacy preserving algorithm according to a variety of assessment criteria.

M. E. Auer and K. Ram B. (Eds.): REV2019 2019, LNNS 80, pp. 849–858, 2020.
https://doi.org/10.1007/978-3-030-23162-0_77

The foremost approach to PPDM known as "knowledge hiding" ensures privacy by modifying the sensitive data before delivering it to the data miner [7]. The second approach which is based on cryptography technique assumes that the data distributed across different sites and these sites collaborate to learn the global data mining results without disclosing the data at their individual sites [8]. The third approach called query auditing and inference control method modifies the results of a query to preserves the privacy of sensitive data [9–11].

The problem we have taken for solving in this paper, known as frequent itemsets hiding, is to hide certain itemsets in the database that are considered sensitive. The hiding process needs to maximize the data utility of the modified database by introducing least possible amount of side effects, such as (1) hiding of non-sensitive itemsets (2) production of frequent itemsets that were not present in the initial data set. These side effects can be used to evaluate the performance of a sanitization approach.

The remainder of this paper is organized as follows: The next section reviews related work in the field of frequent itemsets. Section 3 gives a formalization of the problem, while in Sect. 4, we describe our proposed greedy algorithm for the sensitive frequent itemsets hiding problem. Section 5 includes the experimental evaluation and followed by some concluding remarks in the final section.

2 Related Work

Many studies in the literature have focused on hiding sensitive data either in the form of frequent itemsets or in the form of sensitive association rules. The first work to address this issue was presented in [12], where the authors proposed a heuristic algorithm for itemsets hiding and proved that the optimal solution to the underlying problem is NP-hard. Dasseni et al. [13, 14] proposed several heuristic algorithms based on two methods in order to conceal sensitive rules. The first method prevents the rules from being generated by hiding frequent itemsets from which they are derived and the second approach hides the rules by setting confidence of sensitive rules below a user-specified threshold. The authors demonstrated that the proposed algorithms exhibit very good behavior both in their complexity and their effectiveness.

Oliveira et al. [15] introduced a sanitization algorithm called Sliding Window Algorithm based on a disclosure which is controlled by a data owner. A strong point of this algorithm is that it does not result in false drops in the database. Amiri [16] proposed three heuristic approaches to hide multiple sensitive rules. The proposed approaches sanitize the database with great data accuracy, thus resulting in insignificant modification of the resulting database. Verikios et al. [17] devised a heuristic hiding technique called WSDA to protect sensitive rules by reducing the confidence of the rule. Wu et al. [18] proposed a template-based method which aims at avoiding the side effects in the rules hiding process instead of hiding all sensitive rules.

Hong et al. [19] devised SIF-IDF algorithm that employs the concept of TF-IDF measure of text mining to compute the similarity between sensitive frequent itemsets and transactions for minimizing the side effects. Sun [20] introduced a border based approach for frequent itemsets hiding which focuses on preserving the quality of the border constructed by non-sensitive frequent itemsets in the itemset lattice.

Menon et al. [21] proposed an integer programming technique to hide sensitive frequent itemsets. The approach considers hiding process as a constraint satisfaction problem that identifies the minimum number of transactions to be modified. Another integer programming approach is presented by Gkoulalas-Divanis et al. [22]. The authors devised an exact methodology, which is guaranteed to identify the least amount of items, whose modifications will cause the minimum impact on the dataset.

Saygin et al. [23] introduced a blocking based technique, which replaces some values of a database with unknown values. Pontikakis et al. [24] and Verikios et al. [25] devised the blocking based method to decrease the risk of data modification from an adversary.

Kagklis et al. [26] proposed a novel methodology to hide sensitive frequent itemsets, by formulating it as an Integer Linear Program (ILP). A heuristic is introduced, which further utilizes the coefficients of the objective function. Stavropoulos et al. [27] presented a methodology relies on a technique that enumerates the minimal transversals of a hyper graph in order to induce the ideal border between frequent and sensitive frequent itemsets.

Menon et al. [28] presented scalable approaches to address privacy issues when database is shared among multiple parties and proposed an optimal procedure leveraging intuition from linear programming based column generation.

3 Problem Formulation

The problem of frequent itemsets mining was introduced by Agrawal et al. [29] and is defined as follows: Let $I = \{I_1, I_2, \ldots, I_k, \ldots, I_m\}$ be a set of all binary attributes called items and $D = \{T_1, T_2, \ldots, T_i, \ldots, T_n\}$ be a set of database transactions. Each transaction T_i characterized by an ordered pair, denoted as $T_i = <Id, Y>$, where Id is a unique transaction identifier and Y represents list of items from I, $Y \subseteq I$, is called an itemset which the transaction contains. A transaction T_i is said to contain an itemset X, if $X \subseteq T_i$. Given an itemset X, support(X) defined as the number of transactions that contain the itemset X and frequency(X) denotes the fraction of transactions that support X. An itemset X is called frequent if its frequency in D is at least equal to a user specific minimum threshold 'minf'. Equivalently, X is frequent in D, if support(X) $\geq \sigma$, where minimum support denoted by σ defined as $\sigma = \text{minf} \times n$. The goal of frequent itemset mining problem is to discover all itemsets F which are frequent in the database D. Frequent itemsets which the data owner want to conceal, are called sensitive frequent itemsets and are denoted as S, where $S \subseteq F$. The set of all the sensitive itemsets, their supersets in F is denoted as P, where $P = \{X \in F \mid \forall Y \in S, X \supseteq Y\}$ and $S \subseteq P \subseteq F$. The set of non-sensitive frequent itemsets, denoted as N, is given by $N = F - P$.

Given a database D, a minimum support σ and a set of sensitive frequent itemsets S, the frequent itemsets hiding problem modifies selected transactions in D in order to decreases the support of sensitive frequent itemsets below σ so that itemsets in S cannot be mined from the sanitized database D′. The hiding process also aims for a transformation of D into D′ that maximizes the number of non-sensitive frequent itemsets in N that can still be mined. The proposed approach reduces the support of sensitive frequent itemsets by applying item deletions and thus no infrequent itemsets of the D can be frequent in D′.

To give a description of the hiding problem, we consider the database D given in Table 1. Let the minimum support σ be 3. Table 2 gives frequent itemsets mined from D. Among these frequent itemsets, let S = {pqs, qrs, qr} is a user specified sensitive itemsets. The objective is to transform D into database D′ in such a way that everything but the sensitive knowledge remains intact. In [13] the authors shown that the problem of decreasing the support of frequent itemsets is an NP-hard problem.

Table 1. Sample database D

TId	Items
1	*pqrst*
2	*prs*
3	*pqsvw*
4	*qrst*
5	*pqs*
6	*qrsvx*
7	*pqrw*
8	*prst*
9	*prsx*

Table 2. Frequent itemsets in D

Itemset : Support
pqs : 3, *prs*: 4, *qrs* : 3, *rst* : 3
pq : 4, *pr* : 5, *ps* : 6, *qr* : 4, *qs* : 5, *rs* : 6, *rt* : 3, *st* : 3
p : 7, *q* : 6, *r* : 7, *s* : 8, *t* : 3

4 Proposed Greedy Approach for Hiding Sensitive Frequent Itemsets

In this section, we propose a greedy approach to hide a given set of sensitive frequent itemsets. The proposed strategy also aims to reduce side effects in terms of the number of non-sensitive frequent itemsets that are affected while hiding sensitive frequent itemsets. The support of sensitive item sets are decreased by applying only item deletions to bring their support values below σ, hence no itemset that was not among the non-sensitive frequent itemsets mined from the original database will be introduced in the sanitized database.

The hiding strategy that we propose employs the concept of border presented in [20] and we present a brief review of border concept here. For a given set of itemsets V the upper border of V, denoted as Bd^+ (V), is a subset of V with the property that (1) $Bd^+(V)$ is an antichain group of sets and (2) $\forall X \in V$, there exists at least one itemset $Y \in Bd^+$ (V) such that $X \subseteq Y$. The lower border of itemsets V, denoted as $Bd^-(V)$, is a subset of V with the property that (1) $Bd^-(V)$, is an antichain group of sets and (2) $\forall X \in V$, there exists at least one itemset $Y \in Bd^-(V)$ such that $X \supseteq Y$. An itemset present in upper border or lower border is termed as a border element.

The fundamental idea of our greedy approach is as follows. From the given set of sensitive frequent itemsets S, non-sensitive itemsets N we compute upper border of non-sensitive itemsets i.e. $Bd^+(N)$ and lower border of sensitive frequent itemsets i.e. $Bd^-(S)$. In the rest of this paper Bd^+ and Bd^- are used to denote $Bd^+(N)$ and $Bd^-(S)$ respectively. The hiding strategy reduces the support of each border element in Bd^- below the minimum support σ with minimal impact on the set Bd^+. During the hiding process we assign weight for each sensitive transaction based on its vulnerability of being affected by an applied item deletion. A transaction T_i is called as sensitive transaction if and only if $\exists si_j | si_j \in S \wedge si_j \subseteq T_i$. A transaction with the lowest weight is selected at each stage for item deletion.

4.1 Algorithm

We now present the greedy algorithm and also describe the strategy to hide user specific sensitive frequent itemsets by selecting victim items such that side effects are minimized. An outline of this algorithm is shown in Fig. 1.

In the running example, we have sensitive and non-sensitive set of itemsets S = {qr, pqs, qrs} and N = {p, q, r, s, t, pq, pr, ps, qs, rs, rt, st, prs, rst} respectively. The lower border comprises of the most general itemsets in S and hence Bd^- = (pqs, qr}. Because of the Apriori property, when all the lower border elements in Bd^- are hidden then all their super set of items also will be hidden. Therefore, instead of hiding sensitive itemsets S , we hide only the lower border elements in Bd^-. The upper border comprises of the most specific itemsets in N and hence Bd^+ = (pq, qs, prs, rst}. During the sanitization process if an element from the set Bd^+ gets concealed then all its subsets will also get concealed. Focusing on the upper border Bd^+ during the hiding process is avoids over-hiding non-sensitive frequent itemsets than concentrating on set of non-sensitive itemsets N.

In order to find candidate transactions for sanitization, each sensitive transaction is assigned a weight, where weight of a sensitive transaction T_i is defined using the Formula (1)

$$\text{weight}(T_i) = \sum_{j=1}^{|Bd^+|} \left| nsi_j \cap T_i \right| \tag{1}$$

The weight of each sensitive transaction is measured based on number of items of every non-sensitive itemsets present in the transaction. The larger the weight of a sensitive transaction T_i is the lower priority of transaction getting selected for an item deletion.

```
Input: A transaction dataset  D = {T_1, T_2, …, T_i, …, T_n}
with a set of m items I = {I_1, I_2, …, I_k, …,,I_m}, a user-
specific minimum support  σ, the set F of frequent item-
sets in D and the set S of user-specific sensitive itemsets

Output: The database D transformed into D′ so that sensi-
tive itemsets S cannot be mined from D′ under the same or
a higher minimum support threshold.

Method:
Compute Bd- from S
Compute Bd+ from F and S
Sort Bd- in the descending order of size and support
T_H = { T_z| Z ∊ Bd- and ∀t ∊ D : Z⊆t}
for each Z in Bd-
   for each transaction T_i in T_z
      weight(T_i)= 0
      for each nsi_j in Bd+
         weight(T_i)= weight(T_i) +|nsi_j ∩ T_i|
      end for
   end for
   sort T_z in ascending  order of  their weight
   N_deletions = |T_z|-σ × |D|
   for i = 1 to N_deletions
      T_v=remove the first transaction from
                    the T_z for item deletion
      I_v=item from Z with maximum support count
      for each si_j  in  Bd-
         if si_j ⊆ T_v and I_v ∊ si_j
             support(si_j)= support(si_j)-1
         if support(si_j)< σ
             Bd- = Bd- - {si_j}
      end for
      delete victim item I_v from transaction T_v
   end for
end for
Output D′ = D
```

Fig. 1. Greedy algorithm

4.2 Hiding Sensitive Frequent Itemsets

After defining weight of each sensitive transaction, we next discuss the hiding all lower border elements in Bd^-.

The sensitive frequent itemsets in Bd^- are initially sorted in the descending order of their length and support. Then, these sensitive frequent itemsets are concealed starting from the largest sensitive frequent itemset. If there are multiple sensitive frequent itemsets with same length, then sensitive frequent itemset with the maximum support is chosen for hiding. During each iteration of hiding a sensitive itemset Z in Bd^-, we compute the weight of each transactions T_i containing itemset Z to estimate the possible effect of deleting an item and a transactions with lowest weight is chosen for sanitization. If there are many sensitive transactions with the lowest weight we sort these transactions in the increasing order of their length, number of sensitive items to break the tie and then we select the first 'N_deletions' transactions for sanitization where 'N_deletions' defined using the Formula (2)

$$N\_deletions = |T_Z| - \sigma \times |D| \quad (2)$$

In each selected transaction an item from the sensitive itemset Z with maximum support count is chosen as victim item for deletion. Deletion of the victim item will have minimum side effects on the set of non-sensitive itemsets. We update the support of those border elements in Bd^- that are affected by the deletion of victim item and then we delete the victim item from database. Deletion of a sensitive item from first 'N_deletions' transactions will bring support of sensitive frequent itemset Z below the user specific minimum support σ.

5 Experiments and Evaluation

The proposed algorithm is tested using different parameters such as number of sensitive frequent itemsets, length of sensitive frequent itemsets and minimum support, on various datasets which are openly accessible through the FIMI repository. These datasets exhibit different characteristics in terms of the average transaction length, the number of transactions and the number of items.

We investigated the effectiveness of proposed algorithm by comparing it with heuristic algorithm 2.b from [14] and presented the comparison results on datasets Chess and Mushroom. The tests are performed on dataset Chess with the percentage of minimum support set as 0.9% and Mushroom with the percentage of minimum support set as 0.4%. We considered several sets of sensitive frequent itemsets such that the average length of sensitive itemsets for each sensitive itemsets is 4. The side effects evaluation in terms of number of lost non-sensitive itemsets as a result of hiding process are shown in Fig. 2.

Based on the experiment results depicted in figures, we can see that greedy approach protects more non-sensitive frequent itemsets from being over-hidden than the heuristic algorithm 2.b. The proposed approach hides sensitive frequent itemsets efficiently and demonstrates good effectiveness on limiting the side effects on the resultant database.

6 Conclusion

In this paper, we presented a greedy algorithm in order to hide sensitive frequent itemsets. The quality of dataset is well maintained by greedily selecting the transitions for sanitization which results minimal side effects on the set of non-sensitive itemsets.

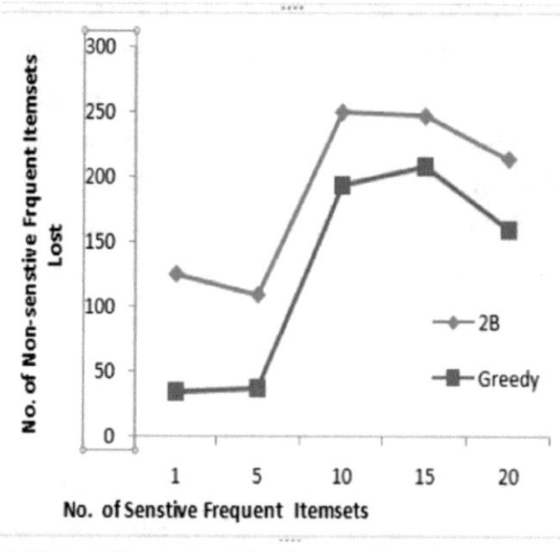

a) Side Effects for Mushroom Database

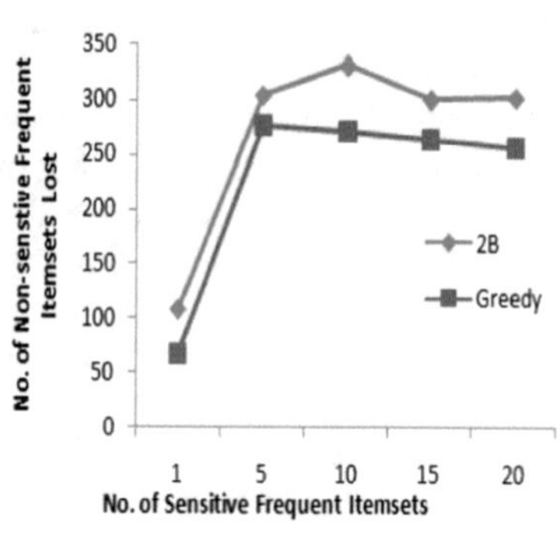

b) Side Effects for Chess Database

Fig. 2. Effectiveness of evaluation

To minimize side effects of hiding approach we efficiently evaluate the impact of any modification to the database transactions and items during the hiding process. A thorough experimental evaluation indicated that solution of greedy algorithm is of higher quality than those produced by the heuristic approach.

References

1. Atallah, M., Bertino, E., Elmagarmid, A., Ibrahim, M., Verykios, V.S.: Disclosure limitation of sensitive rules. In: Scheuermann, P. (ed.) Proceedings of the IEEE Knowledge and Data Exchange Workshop (KDEX'99), pp. 45–52. IEEE Computer Society (1999)
2. Clifton, C., Marks, D.: Security and privacy implications of data mining. In: Proceeding of 1996 ACM SIGMOD Int'l Workshop Data Mining and Knowledge Discovery, pp. 15–19 (1996)
3. Agrawal, R., Srikant, R.: Privacy-preserving data mining In: Proceeding of ACM SIGMOD '00, pp. 439–450 (2000)
4. Lindell, Y., Pinkas, B.: Privacy preserving data mining. In: Advances in Cryptology—CRYPTO 2000, 20th Annual International Cryptology Conference, Santa Barbara, California, USA, pp. 36–54 (2000)
5. Verykios, S., Bertino, E., Fovino, I., Provenza, L., Saygin, Y., Theodoridis, Y.: State-of-the-art in privacy preserving data mining. ACM SIGMOD Rec. **33**(1), 50–57 (2004)
6. Bertino, E., Fovino, I., Provenza, L.: A framework for evaluating privacy preserving data mining algorithms. Data Min. Knowl. Disc. **11**(2), 121–154 (2005)
7. Evfimievski, A.: Randomization in privacy preserving data mining. SIGKDD Explor. **4**(2), 43–48 (2002)
8. Pinkas, B.: Cryptographic techniques for privacy-preserving data. SIGKDD Explor. **4**(2), 12–19 (2002)
9. Aggarwal, C.C., Yu, P.S.: A condensation approach to privacy preserving data mining. In: EDBT Conference, pp. 183–199 (2004)

10. Aggarwal, C.C., Yu, P.S.: On variable constraints in privacy-preserving data mining. In: SIAM Conference, pp. 115–125 (2005)
11. Aggarwal, G., Feder, T., Kenthapadi, K., Motwani, R., Panigrahy, R., Thomas, D., Zhu, A.: Anonymizing tables. In: ICDT Conference, pp. 246–258 (2005)
12. Atallah M., Bertino E., Elmagarmid A., Ibrahim M., Verykios V.: Disclosure limitation of sensitive rules. In: Proceedings of the Workshop on Knowledge and Data Engineering Exchange(KDEX). IEEE, 1999, pp. 45–52
13. Dasseni, E., Verykios, V.S., Elmagarmid, A.K., Bertino,E.: Hiding association rules by using confidence and support. In: Proceedings of the Fourth International Workshop on Information Hiding, vol. 2137, pp. 369–383 (2001)
14. Verykios, V.S., Emagarmid, A.K., Bertino, E., Saygin, Y., Dasseni, E.: Association rule hiding. IEEE Trans. Knowl. Data Eng. **16**(4), 434–447 (2004)
15. Oliveira, S.R.M., Zaiane, O.R.: Protecting sensitive knowledge by data sanitization. In: Proceedings of the 3rd IEEE International Conference on Data Mining (ICDM), pp. 613–616. IEEE
16. Amiri, A.: Dare to share-Protecting sensitive knowledge with data sanitization. Decis. Support Syst. **43**, 181–191 (2007). https://doi.org/10.1016/j.dss.2006.08.007
17. Verykios, V.S., Pontikakis, E.D., Theodoridis, Y., Chang, L.: Efficient algorithms for distortion and blocking techniques in association rule hiding. Distrib. Parallel Databases **22**, 85–104 (2007). https://doi.org/10.1007/s10619-007-7013-0
18. Wu, Y.H., Chiang, C.M., Chen, A.L.: Hiding sensitive association rules with limited side effects. IEEE Trans. Knowl. Data Eng. **19**, 29–42 (2007). https://doi.org/10.1109/TKDE.2007.250583
19. Hong, T.P., Lin, C.W., Yang, K.T., Wang, S.L.: Using TF-IDF to hide sensitive itemsets. Appl. Intell. **38**, 502–510 (2013). https://doi.org/10.1007/s10489-012-0377-5
20. Sun, X., Yu, P.S.: A border-based approach for hiding sensitive frequent itemsets. In: Proceedings of the Fifth IEEE International Conference on Data Mining (ICDM), pp. 426–433 (2005)
21. Menon, S., Sarkar, S., Mukherjee, S.: Maximizing accuracy of shared databases when concealing sensitive patterns. Inf. Syst. Res. **16**(3), 256–270 (2005)
22. Gkoulalas-Divanis, A., Verykios, V.S.: Exact Knowledge Hiding Through Database Extension. Technical report, Department of Computer and Communication Engineering, University of Thessaly, May 2007
23. Saygin, Y., Verykios, V.S., Clifton, C.: Using unknowns to prevent discovery of association rules. ACM SIGMOD Rec. **30**(4), 45–54 (2001)
24. Pontikakis, E.D., Theodoridis, Y., Tsitsonis, A.A., Chang, L., Verykios, V.S.: A quantitative and qualitative analysis of blocking in association rule hiding. In: Proceedings of the ACM Workshop on PRIVACY in the Electronic Society. ACM, pp. 29–30 (2004)
25. Verykios, V.S., Pontikakis, E.D., Theodoridis, Y., Chang, L.: Efficient algorithms for distortion and blocking techniques in association rule hiding. Distrib. Parallel Databases **22**, 85–104 (2007). https://doi.org/10.1007/s10619-007-7013-0
26. Kagklis, V., Verykios, V.S., Tzimas, G., Tsakalidis, A.K.: An integer linear programming scheme to sanitize sensitive frequent itemsets. In: IEEE 26th International Conference on Tools with Artificial Intelligence, pp. 771–775 (2014)
27. Stavropoulos, E.C., Verykios, V.S., Kagklis, V.: A transversal hypergraph approach for the frequent itemset hiding problem. Knowl. Inf. Syst. **47**(3), 625–645 (2016)

28. Menon, S., Sarkar, S.: Privacy and big data-scalable approaches to sanitize large transactional databases for sharing. MIS Q. **40**(4), 963–981 (2016)
29. Agrawal, R., Srikant, R.: Fast algorithms for mining association rules in large databases. In: Proceedings of the 20th International Conference on Very Large Databases (VLDB'94), pp. 487–499 (1994)

Author Index

M. E. Auer and K. Ram B. (Eds.): REV2019 2019, LNNS 80, pp. 859–862, 2020.
https://doi.org/10.1007/978-3-030-23162-0